Metabolic Pathways

Third Edition

VOLUME VII

Metabolism of Sulfur Compounds

Contributors to This Volume

Yasushi Abiko

Romano Humberto De Meio

K. S. Dodgson

Dale E. Edmondson

Max A. Eisenberg

James D. Finkelstein

Martin Flavin

David M. Greenberg

William C. Kenney

Kickiko Koike

Masahiko Koike

Irwin G. Leder

Alton Meister

F. A. Rose

Lewis M. Siegel

Thomas P. Singer

Bo Sörbo

Metabolic Pathways

THIRD EDITION

VOLUME VII
Metabolism
of Sulfur Compounds

EDITED BY

David M. Greenberg

*Department of Biochemistry and
Biophysics and Cancer Research Institute
University of California
San Francisco, California*

ACADEMIC PRESS New York San Francisco London 1975
A Subsidiary of Harcourt Brace Jovanovich, Publishers

ACADEMIC PRESS, INC.
111 Fifth Avenue, New York, New York 10003

United Kingdom Edition published by
ACADEMIC PRESS, INC. (LONDON) LTD.
24/28 Oval Road, London NW1

Library of Congress Cataloging in Publication Data
Main entry under title:

Metabolic pathways.

  First published in 1954 under title: Chemical pathways of metabolism. First-2d editions cataloged under D. M. Greenberg.
  Includes bibliographies.
  CONTENTS: v. 1.   Energetics, tricarboxylic acid cycle, and carbohydrates.—v. 2.   Lipids, steroids, and carotenoids.—v. 3.   Amino acids and tetrapyrroles. [etc.]
  1.   Biological chemistry.   2.   Metabolism.   I.   Greenberg, David Morris, (date)   ed.   II.   Vogel, Henry James, (date)   ed.   III.   Hokin, Lowell E., (date) ed.
QP514.M45   574.1'33   67-23160
ISBN 0–12–299257–1 (v. 7)

Contents

Chapter 1

Metabolism of Coenzyme A

YASUSHI ABIKO

Chapter 2

Biotin

MAX A. EISENBERG

Chapter 3

Thiamine, Biosynthesis and Function

> IRWIN G. LEDER

Chapter 4

Lipoic Acid

> MASAHIKO KOIKE AND KICKIKO KOIKE

Chapter 5

Biochemistry of Glutathione

> ALTON MEISTER

Chapter 6

Covalent Adducts of Cysteine and Riboflavin

WILLIAM C. KENNEY, DALE E. EDMONDSON, AND
THOMAS P. SINGER

Chapter 7

Biochemistry of the Sulfur Cycle

LEWIS M. SIEGEL

Chapter 8

Sulfate Activation and Transfer

ROMANO HUMBERTO DE MEIO

Chapter 9

Sulfohydrolases

K. S. DODGSON AND F. A. ROSE

Chapter 10

Thiosulfate Sulfurtransferase and Mercaptopyruvate
Sulfurtransferase

BO SÖRBO

Chapter 11

Methionine Biosynthesis

MARTIN FLAVIN

Chapter 12

Biosynthesis of Cysteine and Cystine

DAVID M. GREENBERG

Chapter 13

Utilization and Dissimilation of Methionine

DAVID M. GREENBERG

Chapter 14

Oxidative Metabolism of Cysteine and Cystine in Animal Tissues

THOMAS P. SINGER

Chapter 15

Enzyme Defects in Sulfur Amino Acid Metabolism in Man

JAMES D. FINKELSTEIN

List of Contributors

Numbers in parentheses indicate the pages on which the authors' contributions begin.

YASUSHI ABIKO, Research Institute, Daiichi Seiyaku Co., Ltd., Edogawa-ku, Tokyo, Japan (1)

ROMANO HUMBERTO DE MEIO, Department of Physiological Chemistry, Philadelphia College of Osteopathic Medicine, Philadelphia, Pennsylvania (287)

K. S. DODGSON, Department of Biochemistry, University College, Cardiff, Wales (359)

DALE E. EDMONDSON, Department of Biochemistry and Biophysics, University of California, and Molecular Biology Division, Veterans Administration Hospital, San Francisco, California (189)

MAX A. EISENBERG, Department of Biochemistry, College of Physicians and Surgeons of Columbia University, New York, New York (27)

JAMES D. FINKELSTEIN, Veterans Administration Hospital, and Department of Medicine, George Washington University School of Medicine, Washington, D. C. (547)

MARTIN FLAVIN, Laboratory of Biochemistry, National Heart and Lung Institute, National Institutes of Health, Bethesda, Maryland (457)

DAVID M. GREENBERG, Department of Biochemistry and Biophysics and Cancer Research Institute, University of California, San Francisco, California (505, 529)

WILLIAM C. KENNEY, Department of Biochemistry, University of California, San Francisco, California (189)

KICKIKO KOIKE, Department of Pathological Biochemistry, Atomic Disease Institute, Nagasaki University School of Medicine, Sakamoto-cho, Nagasaki-shi, Japan (87)

MASAHIKO KOIKE, Department of Pathological Biochemistry, Atomic Disease Institute, Nagasaki University School of Medicine, Sakamoto-cho, Nagasaki-shi, Japan (87)

IRWIN G. LEDER, National Institute of Arthritis, Metabolism and Digestive Diseases, National Institutes of Health, Bethesda, Maryland (57)

ALTON MEISTER, Department of Biochemistry, Cornell University Medical College, New York, New York (101)

F. A. ROSE, Department of Biochemistry, University College, Cardiff, Wales (359)

LEWIS M. SIEGEL, Department of Biochemistry, Duke University Medical School, and Veterans Administration Hospital, Durham, North Carolina (217)

THOMAS P. SINGER, Department of Biochemistry and Biophysics, University of California, and Molecular Biology Division, Veterans Administration Hospital, San Francisco, California (189, 535)

BO SÖRBO,* Research Institute of National Defence, Department 1, Sundbyberg, Sweden (433)

* Present address: Clinical Chemical Laboratory, St. Erik's Hospital, Stockholm, Sweden.

Preface

With increased knowledge, one has become more aware of the importance of sulfur compounds and their participation in biochemical reactions. It is apparent, but not often stressed, that the sulfur cycle in nature is just as indispensable for the existence of life as are the carbon and nitrogen cycles. The disulfide bonds of proteins contribute significantly to their conformation, and thiol groups, in many instances, are vital for the catalytic functions of enzymes.

Certain sulfur compounds play important roles in the homeostasis of organisms. Other sulfur compounds serve as coenzymes of oxidation−reduction systems. Methionine is the major methyl group donor in most biological transmethylating systems, and biotin is the coenzyme of transcarboxylation. Formation of the many sulfated compounds in nature requires the action of 3′-phosphoadenylsulfate; this compound is also involved in sulfate−sulfite interconversion. It should therefore come as no surprise that an increasing number of the diseases of man are recognized to be the result of genetic deletions of enzymes concerned with reactions of sulfur compounds.

These considerations influenced me to undertake the preparation of this volume. Although previous volumes of this treatise covered many aspects of sulfur biochemistry, data were scattered throughout the volumes and were incomplete. Although much progress has been made in sulfur biochemistry in recent years, many facets of the subject are incomplete and obscure. These gaps of knowledge and understanding have been emphasized by the contributors to this volume. Thus this book can serve as a valuable guide in delineating fruitful areas for future investigation. The literature citations are reasonably complete, so the volume is a good reference source of recent publications on sulfur biochemistry. The historical development and present status of the various areas of sulfur metabolism are also well covered.

DAVID M. GREENBERG

Contents of Other Volumes

METABOLIC PATHWAYS

Third Edition

Editor-in-Chief
DAVID M. GREENBERG

CHAPTER 1

Metabolism of Coenzyme A

Yasushi Abiko

Coenzyme A plays essential roles in the metabolism of carbohydrates and fatty acids as an acyl group activator. This essential coenzyme is synthesized by a series of soluble enzymes and dissimilated by particulate enzymes in the cells. The pathways of biosynthesis and dissimilation of the coenzyme described below were elucidated on the basis of the substrate specificity of the enzymes catalyzing the respective metabolic reactions.

I. BIOSYNTHESIS OF COENZYME A

A. Biosynthetic Route of Coenzyme A from Pantothenic Acid

The biosynthesis of CoA has been studied by many investigators with various living systems: microorganisms such as *Proteus morganii* [1–3], *Lactobacillus arabinosus* [4,5], *L. bulgaricus* [6], *L. helveticus* [6,7], *Acetobacter suboxydans* [1,8–10], and others [2,11–13b], and animals such as pigeon [14–17], pig [18], and rat [3,19–26]. Most extensive studies, however, have been carried out with rat liver and *P. morganii*.

CoA is synthesized from pantothenic acid, L-cysteine, and ATP in the

1

presence of Mg^{2+} in various organisms mainly through the pathway presented in Scheme 1.

Phosphorylation of pantothenic acid at the 4'-OH position is the first step of CoA synthesis [3,19]. The phosphorylation is catalyzed by pantothenate kinase (EC 2.7.1.33) in the presence of ATP and Mg^{2+}. The kinase can also phosphorylate pantetheine indicating utilizability of pantetheine from exogenous sources [16,27] or from the dissimilative pathway of CoA [19,25]. This phosphorylation is the essential step for synthesis of CoA, because the subsequent reactions can proceed only with the 4'-phosphorylated substrates as described below [3,19,20].

The second step is the condensation between 4'-phosphopantothenic acid and L-cysteine in the presence of CTP or ATP and Mg^{2+} to form 4'-phosphopantothenoyl-L-cysteine [3,19,24]. Incorporation of cysteine into the cysteamine moiety of the CoA molecule has been shown by the

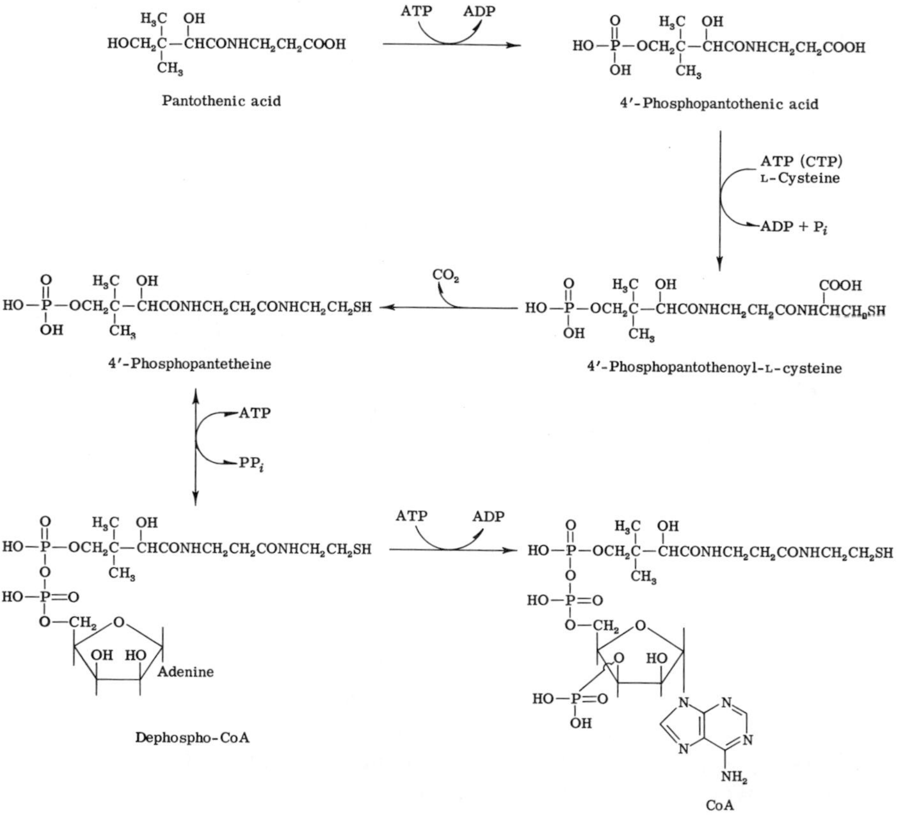

SCHEME 1. Biosynthetic pathway of CoA from pantothenic acid.

experiments with *P. morganii* [1], *L. arabinosus* [4], and rat liver extracts [18]. The enzyme 4′-phosphopantothenoyl-L-cysteine synthetase (EC 6.3.2.5) can also catalyze the condensation between 4′-phosphopantothenic acid and β-mercaptoethylamine, bypassing the decarboxylation step [3]. However, a natural source of β-mercaptoethylamine in the body is not known except for a degradation product of CoA [28]. The synthetase cannot utilize pantothenic acid instead of 4′-phosphopantothenic acid even in the presence of ATP. This was shown by the purified enzyme free of pantothenate kinase activity described later [19].

4′-Phosphopantothenoyl-L-cysteine is then decarboxylated to 4′-phosphopantetheine by the action of phosphopantothenoyl-L-cysteine decarboxylase (EC 4.1.1.36) [3,20]. The enzyme does not seem to require any cofactor for its activity. Pantothenoyl-L-cysteine is not susceptible to decarboxylation at all, although α-carboxy-CoA [29] and dephospho-α-carboxy-CoA (Scheme 2) undergo, to a minor extent, decarboxylation by this enzyme [20].

The final stage of CoA biosynthesis is the pyrophosphate bond formation between 4′-phosphopantetheine and 5′-AMP from ATP to form dephospho CoA and the subsequent phosphorylation of dephospho CoA at the 3′-OH of ribose to form CoA [18,21]. In rat liver, these two reactions are catalyzed by the bifunctional enzyme complex composed of dephospho-CoA pyrophosphorylase (EC 2.7.7.3) and dephospho-CoA kinase (EC 2.7.1.24) activities [22]. The pyrophosphorylase reaction is the only reversible step in the biosynthetic reactions of CoA [18]. The pyrophosphorylase and the kinase are highly specific for 4′-phosphopantetheine and dephospho-CoA, respectively. 4′-Phosphopantothenoyl-L-cysteine cannot replace 4′-phosphopantetheine in this system [21], and dephospho-CoA kinase is different from NAD kinase (EC 2.7.1.23)

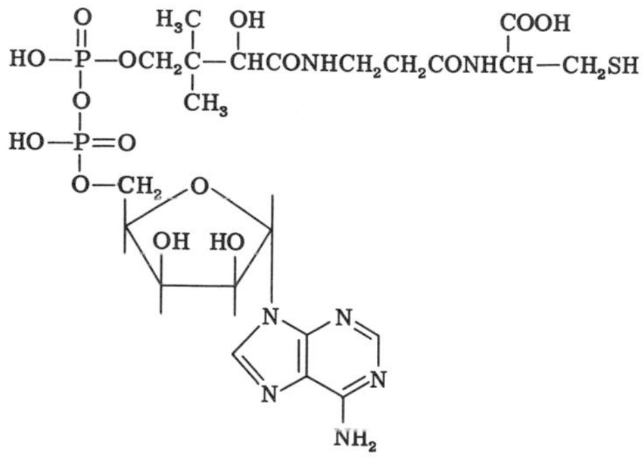

SCHEME 2. Dephospho-α-carboxy-CoA.

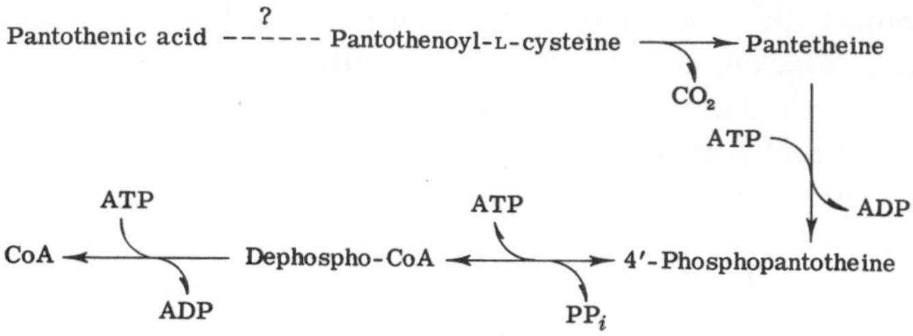

Pantothenic acid $-- \overset{?}{--}--$ Pantothenoyl-L-cysteine $\longrightarrow$ Pantetheine

SCHEME 3. Alternate pathway of biosynthesis of CoA.

[14]. These findings rule out the possibility of a pathway from 4'-phosphopantothenoyl-L-cysteine to CoA via dephospho-α-carboxy-CoA [21].

An alternate pathway to CoA from pantothenic acid (Scheme 3) may be possible in some microorganisms. This partly supported by accumulation of pantetheine in *L. helveticus* cultured on pantothenic acid [7] and by the existence of pantothenoyl-L-cysteine decarboxylase in *A. suboxydans* [8,9], *L. bulgaricus* [6], and a mutant strain from *L. helveticus* [6]. Pantetheine can be phosphorylated by pantothenic acid kinase as described above, but the formation of pantothenoyl-L-cysteine from pantothenic acid remains to be demonstrated as a key step for the alternate pathway.

B. Properties of Individual Enzymes

Enzymological studies on the member enzymes involved in the biosynthesis of CoA have been carried out mainly with rat liver enzymes.

1. PANTOTHENIC ACID KINASE (EC 2.7.1.33) [19,25,30]

Pantothenic acid kinase has been purified about 700-fold over the extracts from rat liver by procedures including fractional precipitation with protamine, pH fractionation, DEAE-Sephadex A-50 chromatography, and Sephadex G-200 gel filtration under successful stabilization by ATP (1 mM) and sucrose (10%). The preparation is free from adenosine triphosphatase and phosphatase activities, although it is not completely homogeneous.

The kinase requires ATP-Mg^{2+} 1 : 1 complex as an active substrate and it has apparent K_m values of 0.011 mM for D-pantothenic acid and 1 mM for ATP-Mg^{2+} at the optimal pH of 6.1. Rat liver pantothenic acid kinase has a relatively broad substrate specificity: it phosphorylates pantothenic acid, pantetheine, pantothenoyl-L-cysteine, and pantothenyl alcohol at a similar rate, but the configurational requirement (D-config-uration of 2'-OH) is strict. Neither L-pantothenic acid nor 2'-keto-pantetheine is a substrate of the kinase, although they are competitive inhibitors (K_i: 0.6 mM for L-pantothenate and 0.83 mM for 2'-keto-pantetheine). Pantothenic acid kinase from rat liver is inhibited, to var-ious extents, by the intermediates of CoA biosynthesis, of which 4'-phosphopantetheine and CoA are the most potent inhibitors (about 50% inhibition at 0.25 mM).

2. 4'-Phosphopantothenoyl-l-cysteine Synthetase (EC 6.3.2.5) [19,24,31]

4'-Phosphopantothenoyl-L-cysteine synthetase catalyzes the conden-sation reaction between 4'-phosphopantothenic acid and L-cysteine in the presence of ATP and Mg^{2+} by a mechanism analogous to the amide bond formation in the synthesis of glutamine or glutathione. A stoichio-metric study of the synthetase reaction has shown the formation of equimolar amounts of ADP and inorganic phosphate accompanied by an equimolar decrease in 4'-phosphopantothenate. The synthetase from *P. morganii* requires CTP in place of ATP [3].

The synthetase has been purified about 120-fold over the extracts from rat liver by procedures including protamine treatment, fractionation with ammonium sulfate, calcium phosphate gel treatment, repeated chro-matography on CM-Sephadex C-50, and Sephadex G-75 gel filtration. Complete separation of the synthetase from pantothenic acid kinase has been achieved by protamine treatment.

The molecular weight of synthetase is around 37,000. The K_m value of the synthetase is 0.071–0.083 mM for 4'-phosphopantothenic acid at the optimal pH of 7.5. Pantothenate cannot be the substrate of this reaction and it does not affect the rate of the condensation reaction. β-Mercap-toethylamine, its disulfide, and α-methylcysteine all can be substrates of the synthetase from *P. morganii* [3]. Phosphopantothenoyl-L-cysteine synthetase requires Mg^{2+} for action and is inhibited about 50% by KCl and NaCl (0.05 M) and completely by chelators such as EDTA (1 mM) and phosphate (0.08 M).

3. 4'-Phosphopantothenoyl-l-cysteine Decarboxylase (EC 4.1.1.36) [20,32]

4'-Phosphopantothenoyl-L-cysteine decarboxylase has been purified about 110-fold over the extracts from rat liver. The purification proce-dures include ammonium sulfate fractionation of the crude extracts, cal-cium phosphate gel treatment, and DEAE-cellulose chromatography after aging in the cold for 3 weeks. The purified preparation is free of phosphatase activity.

The optimal pH of the decarboxylation reaction is about 8. The K_m for 4'-phosphopantothenoyl-L-cysteine is 0.133–0.15 mM at pH 8.0. Pantothenoyl-L-cysteine cannot be the substrate of this reaction and does not affect the rate of the decarboxylation of 4'-phosphopantothenoyl-L-cysteine by this enzyme. Dephospho-α-carboxy-CoA [29] and α-carboxy-CoA [29] are susceptible to decarboxylation by this enzyme, although only to a minor extent. The purified decarboxylase is activated by cysteine or reduced glutathione. Mg^{2+} has no effect. Pyridoxal and pyridoxal phosphate inhibit the reaction. The reaction product, 4'-phosphopantetheine, inhibits the decarboxylase reaction strongly and competitively with a K_i value of 0.43 mM. ATP is also inhibitory (about 40% inhibition at 1 mM), but ADP and AMP have no effect.

4. DEPHOSPHO-CoA PYROPHOSPHORYLASE (EC 2.7.7.3) AND DEPHOSPHO-CoA KINASE (EC 2.7.1.24) [18,22,23]

Dephospho-CoA pyrophosphorylase and dephospho-CoA kinase catalyze the pyrophosphate bond formation between 4'-phosphopantetheine and 5'-AMP and the subsequent phosphorylation at the 3'-OH of the ribose moiety, respectively. The formation of dephospho-CoA from 4'-phosphopantetheine is the only reaction that is freely reversible. This was shown by the dismutative formation of CoA from dephospho-CoA and inorganic pyrophosphate with a coupled enzyme system of the pyrophosphorylase and the kinase.

$$2 \text{ Dephospho-CoA} + PP_i \rightarrow \text{CoA} + \text{phosphopantetheine} + \text{ADP}$$

There has been no report on the separation of these two enzymes from each other. Studies on the isolation and the properties of these enzymes from rat liver suggest the existence of a bifunctional enzyme complex composed of these two enzymes.

Dephospho-CoA pyrophosphorylase and dephospho-CoA kinase have been purified in a complexed form about 250-fold over the extracts from rat liver by procedures including protamine treatment, ammonium sulfate fractionation, calcium phosphate gel treatment, column chromatographies on CM-cellulose and DEAE-cellulose, and Sephadex G-200 gel filtration [22,23].

The purified enzyme preparation has an optimal pH of 8–10 for the pyrophosphorylase reaction and 10 for the kinase reaction. The apparent K_m values of the rat liver pyrophosphorylase are 0.14 mM for 4'-phosphopantetheine and 1 mM for ATP. Those of the kinase are 0.12 mM for dephospho-CoA and 0.36 mM for ATP. The pyrophosphorylase and the kinase are highly specific for 4'-phosphopantetheine and de-phospho-CoA, respectively, as the substrates. The pyrophosphorylase cannot catalyze the formation of NAD from nicotinamide mononucleotide and ATP, and is therefore distinct from NAD pyrophosphorylase. The pyrophosphorylase reaction is freely reversible, but the enzyme does not catalyze the reaction between CoA and pyrophosphate. As described above, 4'-phosphopantothenoyl-L-cysteine and 3'-dephospho-α-carboxy-CoA are not substrates of the pyrophosphorylase and kinase, respectively. The kinase cannot phosphorylate NAD to NADP [14]. These two enzymes are activated by Mg^{2+} and cysteine. The optimal concentration of Mg^{2+} is 0.5 mM for both enzymes.

These two enzymes were also purified about 100-fold from hog liver extracts [18]. The hog liver phosphorylase and the kinase have pH optima of 7.5 and 9.0, respectively, and they require Mg^{2+} for action as well as the respective reduced forms of the substrates [18].

C. Possible Regulation in the Biosynthesis of Coenzyme A

Biosynthesis of CoA from pantothenic acid seems to be regulated at several steps (Scheme 4). The inhibitory action of 4'-phosphopantetheine and CoA on pantothenic acid kinase of rat liver is very meaningful. 4'-Phosphopantetheine is the sole intermediate of CoA biosynthesis, having a considerable pool in rat liver (219 nmoles/gm [34] corresponding to more than 0.22 mM). The CoA level is 363 nmoles/gm [34], corresponding to more than 0.36 mM. These values are close to the effective concentrations of these substances for inhibiting pan-

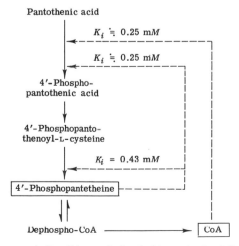

SCHEME 4. Possible regulation in biosynthesis of CoA.

tothenic acid kinase [25]. This suggests that the kinase is not fully active in the liver cells and that its inhibition by 4'-phosphopantetheine and CoA may be involved in regulating the CoA level in the cells. Similar phenomena were also observed with a bacterial pantothenate kinase that was purified from *Brevibacterium ammoniagenes* [34a]. Another possible site for regulation may be 4'-phosphopantothenoyl-L-cysteine decarboxylase which is effectively inhibited again by 4'-phosphopantetheine ($K_i = 0.43$ mM) [20]. Partial deficiency of pantothenate in the diet results in decreased levels of 4'-phosphopantetheine and dephospho-CoA in rat liver while the hepatic CoA level of the rat remains normal, but it causes no significant change in the 4'-phosphopantothenoyl-L-cysteine decarboxylase activity in the liver [35].

The level of CoA and its intermediate precursors are increased during the initial growth after birth [23,36]. In the rat, hepatic CoA reaches a constant level 30 days after birth, and renal and brain CoA 10 and 20 days after birth, respectively. Free pantothenic acid levels of the liver and the kidney gradually decrease with age [36]. The liver CoA level is also affected by dietary protein content [37] and hormonal disturbance [38]. A low protein diet results in a marked reduction of the hepatic CoA level of the rat, and methionine content is most influential on the CoA level [37]. Adrenalectomy causes a significant suppression of the liver CoA level of the rat. The decreased CoA level caused by adrenalectomy can be restored to normal by administration of prednisolone, but the long-term application of the glucocorticoid (more than 10 days, 8 mg/kg/day) causes reduction of the CoA level [38].

II. COENZYME A AS THE PRECURSOR OF THE PROSTHETIC GROUP OF ACYL CARRIER PROTEIN

Coenzyme A, the final product of the biosynthesis from pantothenic acid, serves, in turn, as the obligatory precursor of the prosthetic group (phosphopantetheine) of acyl carrier protein. This protein is a functional unit in the fatty acid synthetase system [39], in addition to functioning as an acyl group activator in many metabolic reactions listed in Table III.

Acyl carrier protein (ACP) or an ACP-like unit is widely distributed in the fatty acid synthetase system of living organisms, among which *Escherichia coli* has been most extensively studied (Scheme 5).

CoA acts as the phosphopantetheine donor in the synthesis of holo-ACP which is catalyzed by holo-ACP synthetase [41]. The synthetase was purified 780-fold over the extract from *E. coli* under stabilization by CoA [41]. The holo-ACP synthetase requires Mg^{2+} or Mn^{2+} for its

```
          1                                   10
NH₂-Ser -Thr- Ile -Glu -Glu -Arg-Val -Lys- Lys - Ile - Ile -Gly -Glu -

                        20
Gln- Leu- Gly - Val - Lys- Gln - Glu - Glu - Val - Thr- Asp- Asn- Ala - Ser -

        30                                          40
Phe- Val- Glu - Asp- Leu- Gly - Ala - Asp- Ser- Leu- Asp- Thr- Val - Glu -
                                        |
                                        O
                                        |
                          HO — P — O- Pantetheine- SH
                                        ||
                                        O

                                50
Leu- Val - Met- Ala - Leu- Glu - Glu - Glu - Phe- Asp- Thr- Glu - Ile - Pro-

                60
Asp- Glu - Glu - Ala - Glu - Lys - Ile - Thr- Thr- Val - Gln - Ala - Ala - Ile -

  70                          77
Asp- Tyr - Ile - Asn- Gly - His - Gln - Ala - COOH
```

SCHEME 5. The primary structure of acyl carrier protein from *Escherichia coli* [40].

activity. The optimum pH of the reaction is 8.5. The synthetase has K_m values of 0.4 μM for apo-ACP and 0.15 mM for CoA. The synthetase is highly specific for CoA: dephospho-CoA and the oxidized form of CoA cannot be the phosphopantetheine donor in this system. In the case of *E. coli,* the intracellular activity of the synthetase has been postulated to be sufficient to account for all synthesis of ACP in exponentially growing cells [41,41a].

On the other hand, holo-ACP undergoes, in its turnover, the action of ACP hydrolase which catalyzes the removal of phosphopantetheine from ACP [42] (Scheme 6). The hydrolase was purified from the plasmatic extract of *E. coli* [42]. The enzyme requires a divalent cation for its activity: Mg^{2+} is most effective (0.025 mM). The hydrolase has its pH optimum at 8.5 and is stimulated by sulfhydryl compounds such as dithiothreitol and β-mercaptoethanol. The ACP hydrolase has a strict specificity for the intact ACP, and any other phosphodiester linkages, including those in CoA, glycerophosphorylserine, and even a trypsin-modified ACP ([19]Gln–[61]Lys), are not susceptible to hydrolysis by this enzyme.

The pulse and chase experiments using a pantothenate auxotroph of *E. coli* showed that the rate of turnover of the phosphopantetheine prosthetic group of ACP was four times the rate of growth of the ACP pool [43]. A similar result was obtained with the rat liver fatty acid synthetase complex: the rate of protein turnover of the complex was one order of magnitude less than that of turnover of phosphopantetheine in the synthetase complex [44]. These findings suggest that the faster turn-

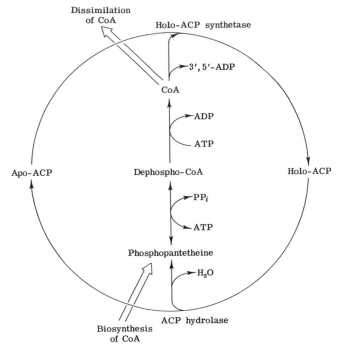

SCHEME 6. Turnover of the prosthetic group (4'-phosphopantetheine) in acyl carrier protein [43].

over of the phosphopantetheine residue may be involved in regulation of the function of ACP or the fatty acid synthetase system.

III. DISSIMILATION OF COENZYME A

A. Route of Degradation of CoA in the Cell

Enzymatic degradation of coenzyme A was first studied by Lipmann and his co-workers [15,45] in their studies on the chemical structure of CoA. However, the investigation on the dissimilative pathway of CoA in living systems has been started only recently. The studies on this subject have been performed mainly with rat liver [46–48], rat kidney [48], and horse kidney [28]. The sequence of the intracellular degradation reactions of CoA seems to be nearly a reverse one of that of biosynthetic reactions (Scheme 7).

In rat liver, CoA is first degraded to dephospho-CoA by lysosomal acid phosphatase (EC 3.1.3.2) [46]. Then, dephospho-CoA is subjected to pyrophosphate bond cleavage to 4'-phosphopantetheine and 5'-AMP by the action of a dephospho-CoA pyrophosphatase (EC 3.6.1.15) [47]. The pyrophosphatase is located on the plasma membrane of microsomal and nuclear fractions of rat liver. CoA can be also a substrate of the pyrophosphatase, but its susceptibility to this reaction is one order of magnitude less than that of dephospho-CoA. The enzyme systems for degrading CoA to 4'-phosphopantetheine seem to be common to nucleotide pyrophosphates and not specific for CoA, because dephospho-CoA pyrophosphatase shares many properties with nucleotide pyrophosphatase [49] as described below. Inorganic pyrophosphatase is not involved in this reaction. The nucleotide pyrophosphatase activity present in plasma membrane fraction may account for all dephospho-CoA degradation in the cell [47].

4'-Phosphopantetheine may undergo dephosphorylation prior to degradation of the pantetheine moiety, because a pantetheine-splitting enzyme acts more effectively on pantetheine than its phosphate [28,48]. Degradation of pantetheine is carried out by amidase [48] or pantethinase (EC 3.5.1.15) [28]. The products of this reaction are pantothenic acid and cysteamine (β-mercaptoethylamine). The amidase or pantethinase is highly specific for the degradation of pantetheine which cannot be hydrolyzed by any proteases [28]. The rat liver amidase is located in the lysosomal-microsomal fraction and hydrolyzes only pantetheine, whereas the rat kidney enzyme distributes to both the soluble and lysosomal-microsomal fractions and can hydrolyze both pantetheine

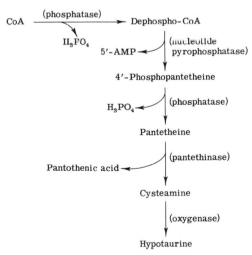

SCHEME 7. Dissimilative pathway of CoA.

and its 4'-phosphate, although the former is more susceptible to the action of the kidney enzyme [48]. Horse kidney enzyme, the intracellular localization of which has not been studied yet, is also much more effective for hydrolyzing pantetheine than its phosphate [28,50]. Cysteamine cannot be split off from CoA or dephospho-CoA molecule, because these nucleotides are not the substrate of the amidase [48].

Pantothenic acid is a final product of CoA degradation in animals. Cysteamine, another product of the amidase reaction, is further oxidized to hypotaurine by a specific oxygenase [51,52] which will be described in detail in Chapter 14.

B. Properties of Individual Enzymes

1. ACID PHOSPHATASE AND DEPHOSPHO-CoA PYROPHOSPHATASE

Enzymes catalyzing degradation of CoA to 4'-phosphopantetheine, acid phosphatase, and dephospho-CoA pyrophosphatase are not specific for CoA but common to all nucleotide pyrophosphates.

Phosphatase activity catalyzing the conversion of CoA to dephospho-CoA was localized in lysosomes of rat liver and was identified with lysosomal acid phosphatase by the inhibition test with three known inhibitors of lysosomal acid phosphatase, inorganic phosphate, L-tartrate, and fluoride, which strongly inhibit the dephosphorylation of CoA [46].

Dephospho-CoA pyrophosphatase, which catalyzes pyrophosphate bond cleavage of dephospho-CoA to 4'-phosphopantetheine and 5'-AMP, has been studied in detail by Skrede [47]. Dephospho-CoA pyrophosphatase is located in nuclear and microsomal fractions of rat liver, and its activity is concentrated on the plasma membranes of these subcellular fractions. The pyrophosphatase has a pH optimum of 7.8. Its K_m value is 0.03 mM for dephospho-CoA and 0.3 mM for CoA. The activity toward CoA is one order of magnitude less than that toward dephospho-CoA. The pyrophosphatase is inactivated by dialysis against ethylenediaminetetraacetate and reactivated by Mn^{2+} and other divalent cations. The enzyme is inhibited strongly by NADH ($K_i = 4 \mu M$), ATP and ADP ($K_i = 0.02$ mM), moderately by 5'-CTP, 5'-AMP, and FAD ($K_i = 0.04$ mM), and weakly by 5'-CMP, UDPG, and inorganic pyrophosphate ($K_i = 0.1$ mM). The inhibition by nucleotides is competitive, whereas the inhibitory action of inorganic pyrophosphate is almost completely blocked in the presence of divalent cations such as Mn^{2+}, indicating chelation of a cation as the mechanism of the inhibition by inorganic pyrophosphate. These properties indicate that dephospho-

CoA pyrophosphatase is probably the known nucleotide pyrophosphatase reported by Decker and Bischoff [49].

2. AMIDASE OR PANTETHINASE

Pantetheine-splitting enzyme has been extensively studied with horse kidney by Duprè and his co-workers [28,50,53,54] and with rat liver and kidney by Kameda and Abiko [48]. Studies of the former group have aimed at the purification and characterization of the enzyme as a cysteamine-producing system in the body, while those of the latter have focused on intracellular localization and substrate specificity of the enzyme.

Pantetheine-splitting enzyme, pantethinase, was highly purified from horse kidney cortex tissue. The purification procedures included homogenization of the frozen-thawed tissue, heating it at 60° for 10 minutes, centrifugation, fractionation of the supernatant with ammonium sulfate (45–65% saturation), CM-cellulose chromatography, DEAE-cellulose chromatography, and zone electrophoresis on a cellulose column [28]. Recently, further purification has been achieved to obtain a preparation homogeneous on ultracentrifugal and disc electrophoretic analyses [50]. Its molecular weight is 55,000. It has an apparent K_m value of 5 mM which was calculated at low concentrations of pantetheine because of substrate inhibition observed at concentrations above 5 mM. Pantethinase also undergoes product inhibition by pantothenate. The optimum pH of the enzyme is around 5. Pantethinase is highly specific for pantetheine and does not hydrolyze any other carboamide bonds such as those in β-alanylcysteamine and pantothenic acid, but it can hydrolyze 4′-phosphopantetheine, although to a minor extent [28]. Pantethinase contains at least one thiol group at its active site and is activated by thiol compounds, most effectively by mercaptoethanol and dithiothreitol, and inactivated by iodoacetamide. The enzyme is most stable in a pH range from 5 to 7.5 [28].

Intracellular distribution of the pantetheine splitting enzyme activity is somewhat different between the liver and the kidney in the rat. In rat liver the activity is exclusively located in the microsomal-lysosomal fraction, whereas it distributes to both the soluble and microsomal-lysosomal fractions in the kidney (Table I) [48].

The apparent K_m value for pantetheine is 0.095 mM with the rat liver enzyme, 0.036 mM with the microsomal enzyme of rat kidney, and 0.031 mM with the rat kidney supernatant enzyme (Fig. 1). Both liver and kidney enzymes of the rat are activated by cysteine; the maximum rate is observed at concentrations of cysteine above 10–20 mM. The

TABLE I

INTRACELLULAR DISTRIBUTION OF THE PANTETHEINE SPLITTING
ENZYME ACTIVITY IN RAT LIVER AND KIDNEY [48]

Fraction[a]	Liver		Kidney	
	Distribution (% of total activity)	Specific activity[b]	Distribution (% of total activity)	Specific activity[b]
Homogenate	100	2.28	100	7.40
Nuclei	1.1	0.11	1.1	0.68
Mitochondria	3.0	0.6	3.4	3.94
Lysosomes	27.5	6.8	19.4	21.4
Microsomes	35.7	8.6	44.8	16.1
Supernatant	5.3	0.53	25.6	7.95

[a] Rat liver and kidney cells were fractionated according to De Duve et al. [55] and Ali and Lack [56], respectively.

[b] The specific activity is represented as nanomoles of pantothenic acid liberated from pantetheine by milligrams of protein of each fraction at pH 7.0 and 37° for 30 minutes.

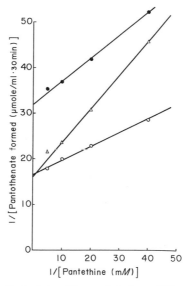

FIG. 1. The Lineweaver-Burk plots of the pantetheine splitting reaction catalyzed by the rat liver and the rat kidney enzymes [48]. The reaction mixture contained cysteine 20 μmoles, pantetheine, and the enzyme fraction in 1 ml of 0.08 M Tris-maleate (pH 7.0). The amounts of protein of each enzyme fraction used were 8, 4, and 4 mg for liver microsome-lysosomes, kidney microsomes, and the ammonium sulfate fraction (precipitates at 25–50% saturation) of the kidney supernatant fractions, respectively. After incubation at 37° for 30 minutes, the reaction was stopped by heating. Pantothenate liberated was separated from the reaction mixture by the use of a Dowex-1 (Cl) column and assayed microbiologically with *Lactobacillus arabinosus*. $\triangle$, Liver microsome-lysosomes; $\bigcirc$, kidney microsomes; and $\bullet$, kidney supernatant fraction.

14

optimum pH is around 6 for the liver enzyme, 7.0 for the microsomal enzyme of the kidney, and around 7.5 for the kidney supernatant enzyme. The pantetheine-splitting enzyme of rat liver microsomal-lysosomal fraction is highly specific for pantetheine; it cannot hydrolyze pantothenoyl-L-cysteine at all. 4'-Phosphopantetheine appears to be a poor substrate of the liver enzyme, but it cannot be a substrate at all if sodium fluoride (10^{-3} M) is present in the reaction mixture. The rate of hydrolysis of pantetheine by the liver microsome-lysosomes in the presence of NaF (1 mM) is not affected by the addition of 4'-phosphopantetheine. These facts indicate that the liver enzyme by itself cannot hydrolyze 4'-phosphopantetheine [48]. On the other hand, the kidney enzymes, both microsomal and supernatant, degrade pantetheine and, to lesser extent, 4'-phosphopantetheine. However, they can hydrolyze neither pantothenoyl-L-cysteine nor CoA (Table II). The enzyme preparation has been obtained in a soluble form, which contains no nucleotide pyrophosphatase activity and only negligible amounts of phosphatase activity, by treatment of rat kidney microsomes with 0.5% Triton X-100 in the presence of 0.1 mM β-mercaptoethylamine, followed by repeated gel filtration on a Sephadex G-200 gel column [48].

TABLE II

SUBSTRATE SPECIFICITY OF PANTETHEINE SPLITTING ENZYMES
FROM RAT LIVER AND KIDNEY [48]

	Pantothenate liberated[a] (nmoles/ml) by			
		Kidney		
		Microsome		
Substrate	Liver microsome-lysosome	Intact	Solubilized	Supernatant
Pantetheine	46.9	89.2	47.6	49.2
Pantothenoyl-L-cysteine	0	4.3	0	1.7
4'-Phosphopantetheine	25.9	90	29.6	48.1
4'-Phosphopantetheine (+0.1 mM NaF)	2.4	–	–	–
CoA	0	–	0	0

[a] The reaction system: substrate 100 nmoles, L-cysteine 20 μmoles, and enzyme fraction in 1 ml of 0.08 M Tris-maleate, pH 7.0. Incubation: 37° for 30 minutes. Protein of each fraction used: 7.5, 7.4, 0.085, and 3.5 mg for the liver microsome-lysosome, the kidney microsome, the solubilized preparation of kidney microsome, and the ammonium sulfate fraction (precipitates at 25–50% saturation) of the kidney supernatant fractions, respectively.

IV. METABOLIC FATE OF EXOGENOUS
COENZYME A IN ANIMALS

Intravenous administration of nucleotides or coenzymes, such as ATP, UDP-glucose, NAD, thiamine pyrophosphate, and CoA, has been introduced for the treatment of liver diseases including hepatic coma [57–60]. However, the metabolic fates of these exogenous coenzymes in animals are still open to investigation. At present, a few reports on the fate of exogenous CoA are available [61–64].

Intravenous injection of CoA (2500 Lipmann units/kg) to dogs resulted in a rapid rise and decline of the plasma level of degradation products of CoA which were active in the Kaplan–Lipmann assay [65] (pantetheine and higher precursors of CoA), but intact CoA was not detected in plasma and urine [61]. This was supported by the experiment reported by Bigler and Thölen [62]. They administered CoA to human subjects by intravenous infusion (240 ml of 250 μM solution of CoA at a rate of 1 ml/minute) and found that a large part of the infused CoA was rapidly converted to degraded forms including pantothenate and that 20–30% of the dose infused was excreted as pantothenate in the urine. The fate of the remainder was speculated to be adsorbed on the cell surface or taken up by the cells either as intact CoA or its breakdown products, such as dephospho-CoA, phosphopantetheine, pantetheine, and pantothenic acid [62].

Rapid breakdown of CoA in the blood was shown by Domschke et al. [63] using an isolated rat liver perfusion system. They observed a rapid degradation of CoA to pantetheine containing fragments, dephospho-CoA, phosphopantetheine, and pantetheine in the erythrocyte-free liver perfusion system, in blood (breakdown rate: 250 nmoles/minute/ml blood), and even in an erythrocyte suspension. Cleavage of the CoA molecule may occur on erythrocyte and liver cell membranes. Pantetheine containing fragments were stable in this liver perfusion system. Domschke et al. [63] also demonstrated the lack of permeation of CoA through liver cell membranes by the fact that the CoA content of the liver remained unaltered during the perfusion with CoA for 60 minutes while the pantetheine containing fragments were increased in the liver cells.

The incorporation of exogenous CoA or pantetheine into the intracellular CoA pool was studied by several workers [27,64] with experimental CoA deficient rats induced by deprivation of pantothenic acid in the diet, homopantothenate feeding, a low protein diet, or prednisolone treatment. Subcutaneous application of CoA, pantetheine, and pantothenate were equally effective in restoring the hepatic CoA level to normal in the pantothenic acid deficient rats [64]. With a consecutive liver biopsy technique, however, a faster normalization of the hepatic CoA level was observed with intravenous injection of pantetheine than with pantothenate injection [27]. On the other hand, in the CoA deficient rats induced by homopantothenate or low protein diet, exogenous CoA and pantetheine were markedly effective in normalizing the liver CoA level, while pantothenate was a relatively poor precursor of intracellular CoA in these animals [64].

These findings show that exogenous CoA administered parenterally is first degraded to pantetheine containing fragments, mainly pantetheine, which then permeate into the cells, where they are resynthesized to CoA or undergo, to some extent, further fragmentation to pantothenic acid, and join the turnover cycle of endogenous CoA.

V. METABOLIC FUNCTIONS OF COENZYME A

Metabolic reactions that involve CoA or ACP as a cofactor are listed in Table III [66–176].

TABLE III

REACTIONS INVOLVING CoA AND ACYL CARRIER PROTEIN

Reaction	Enzyme	Reference
Oxidoreductase		
1. Aldehyde + CoA + NAD = acetyl-CoA + NADH$_2$	Aldehyde dehydrogenase	[66]
2. Glyoxylate + CoA + NADP = oxalyl-CoA + NADPH$_2$	Glyoxylate dehydrogenase	[67]
3. Malonate semialdehyde + CoA + NAD(P) = acetyl-CoA + CO$_2$ + NAD(P)H$_2$	Malonate semialdehyde dehydrogenase	[68,69]
4. RCO—COOH + CoA + NAD = RCO—CoA + CO$_2$ + NADH$_2$	Branched chain α-keto acid dehydrogenase	[70]
5. Butyryl-CoA + oxidized cyt c = crotonyl-CoA + reduced cyt c	Butyryl-CoA dehydrogenase	[71,72]
6. Acyl-CoA + oxidized cyt c = 2,3-dehydroacyl-CoA + reduced cyt c	Acyl-CoA dehydrogenase	[72,73]
7. Acetoacetyl-ACP + NADPH$_2$ = β-hydroxybutyryl-ACP + NADP	β-Ketoacyl-ACP reductase	[74]
8. RCH=CHCO—ACP + NADPH$_2$ − RCH$_2$CH$_2$CO—ACP + NADP	Enoyl-ACP reductase	[75]

TABLE III (*Continued*)

Reaction	Enzyme	Reference
Transferase		
9. Methylmalonyl-CoA + pyruvate = propionyl-CoA + oxalacetate	Methylmalomyl-CoA carboxyltransferase	[76,77]
10. Acetyl-CoA + L-glutamate = CoA + N-acetyl-L-glutamate	Aminoacid acetyltransferase	[78]
11. Acetyl-CoA + imidazole = CoA + N-acetylimidazole	Imidazole acetyltransferase	[79]
12. Acetyl-CoA + 2-amino-2-deoxy-D-glucose = CoA + 2-acetamido-2-deoxy-D-glucose	Glucosamine acetyltransferase	[80]
13. Acetyl-CoA + 2-amino-2-deoxy-D-glucose 6-phosphate = CoA + 2-acetamido-2-deoxy-D-glucose 6-phosphate	Glucosamine phosphate acetyltransferase	[81,82]
14. Acetyl-CoA + arylamine = CoA + N-acetylarylamine	Arylamine acetyltransferase	[83–86]
15. Acetyl-CoA + choline = CoA + O-acetylcholine	Choline acetyltransferase	[87,88]
16. Acetyl-CoA + carnitine = CoA + O-acetylcarnitine	Carnitine acetyltransferase	[89]
17. Palmityl-CoA + carnitine = palmitylcarnitine + CoA	Carnitine palmityltransferase	[90]
18. Acyl-CoA ($<C_{10}$) + carnitine = CoA + acylcarnitine	Carnitine acyltransferase	[91]
19. Acetyl-CoA + orthophosphate = CoA + acetylphosphate	Phosphate acetyltransferase	[92,93]
20. Acetyl-CoA + acetyl-CoA = CoA + acetoacetyl-CoA	Acetyl-CoA acetyltransferase	[94,95]
21. Acetyl-CoA + H_2S = CoA + thioacetate	Hydrogen sulfide acetyltransferase	[96]
22. Acetyl-CoA + thioethanolamine = CoA + S-acetylthioethanolamine	Thioethanolamine acetyltransferase	[96]
23. Acetyl-CoA + dihydrolipoate = CoA + 6-S-acetylhydrolipoate	Lipoate acetyltransferase	[96,97]
24. Acetyl-CoA + glycine = CoA + N-acetylglycine	Glycine acetyltransferase	[98]
25. Phenylacetyl-CoA + L-glutamine = CoA + α-N-phenylacetyl-L-glutamine	Glutamine phenylacetyltransferase	[99]
26. Acyl-CoA + L-glycerol 3-phosphate = CoA + monoglyceride phosphate	Glycerophosphate acyltransferase	[100]
27. Acyl-CoA + acetyl-CoA = CoA + 3-oxoacyl-CoA	Acetyl-CoA acyltransferase	[101,102]

TABLE III (*Continued*)

Reaction	Enzyme	Reference
28. Acetyl-CoA + L-aspartate = CoA + N-acetyl-L-aspartate	Aspartate acetyltransferase	[103]
29. Acetyl-CoA + β-D-galactoside = CoA + 6-acetyl-β-D-galactoside	Galactoside acetyltransferase	[104]
30. Butyryl-CoA + orthophosphate = CoA + butyrylphosphate	Phosphate butyryltransferase	[105]
31. Acyl-CoA + 1,2-diglyceride = CoA + triglyceride	Diglyceride acyltransferase	[106,107]
32. Acetyl-CoA + propionate = acetate + propionyl-CoA	Propionate CoA-transferase	[108]
33. Succinyl-CoA + oxalate = succinate + oxalyl-CoA	Oxalate CoA-transferase	[109]
34. Acetyl-CoA + malonate = acetate + malonyl-CoA	Malonate CoA-transferase	[110]
35. Succinyl-CoA + 3-oxoacid = succinate + 3-oxoacyl-CoA	3-Ketoacid CoA-transferase	[94,111]
36. Succinyl-CoA + 3-oxoadipate = succinate + 3-oxoadipyl-CoA	3-Oxoadipate CoA-transferase	[112]
37. Acetyl-CoA + ACP = acetyl-ACP + CoA	Acetyl-CoA ACP-transferase	[74]
38. Malonyl-CoA + ACP = malonyl-ACP + CoA	Malonyl-CoA ACP-transferase	[74,113]
39. β-Ketoacyl-CoA + ACP = β-ketoacyl-ACP + CoA	β-Ketoacyl-CoA ACP-transferase	[114,115]
Hydrolase		
40. Acetyl-CoA + H_2O = CoA + acetate	Acetyl-CoA hydrolase	[116]
41. Palmityl-CoA + H_2O = CoA + palmitate	Palmityl-CoA hydrolase	[117,118]
42. Succinyl-CoA + H_2O = CoA + succinate	Succinyl-CoA hydrolase	[116]
43. 3-Hydroxyisobutyryl-CoA + H_2O = CoA + 3-hydroxyisobutyrate	3-Hydroxyisobutyryl-CoA hydrolase	[119]
44. 3-Hydroxy-3-methylglutaryl-CoA + H_2O = CoA + 3-hydroxy-3-methylglutarate	3-Hydroxy-3-methylglutaryl-CoA hydrolase	[120]
45. Palmityl-ACP + H_2O = ACP + palmitate	Palmityl-ACP hydrolase	[121]
Lyase		
46. Oxalyl-CoA = formyl-CoA + CO_2	Oxalyl-CoA decarboxylase	[109,122]
47. Malonyl-CoA = acetyl-CoA + CO_2	Malonyl-CoA decarboxylase	[110]
48. L-Malate + CoA = acetyl-CoA + H_2O + glyoxylate	Malate synthase	[123,124]

TABLE III (*Continued*)

Reaction	Enzyme	Reference
49. 3-Hydroxy-3-methylglutaryl-CoA = acetyl-CoA + acetoacetate	Hydroxymethylglutaryl-CoA lyase	[125,126]
50. 3-Hydroxy-3-methylglutaryl-CoA + CoA = acetyl-CoA + H_2O + acetoacetyl-CoA	Hydroxymethylglutaryl-CoA synthase	[126–128]
51. Citrate + CoA = acetyl-CoA + H_2O + oxaloacetate	Citrate synthase	[129,130]
52. ATP + citrate + CoA = ADP + P_i + acetyl-CoA + oxalacetate	ATP citrate lyase (citrate cleavage enzyme)	[131,132]
53. L-3-Hydroxyacyl-CoA = 2,3 (or 3,4)-*trans*-enoyl-CoA + H_2O	Enoyl-CoA hydratase	[133,134]
54. 3-Hydroxy-3-methylglutaryl-CoA = *trans*-3-methylglutaryl-CoA + H_2O	Methylglutaconyl-CoA hydratase	[135]
55. β-Alanyl-CoA = acrylyl-CoA + NH_3	β-Alanyl-CoA ammonia-lyase	[136]
56. D-(−)-β-Hydroxybutyryl-ACP = crotonyl-ACP + H_2O	Enoyl-ACP hydratase	[137,138]
Isomerase		
57. L-3-Hydroxybutyryl-CoA = D-3-hydroxybutyryl-CoA	3-Hydroxybutyryl-CoA epimerase	[139,140]
58. D-Methylmalonyl-CoA = L-methylmalonyl-CoA	Methylmalonyl-CoA racemase	[141,142]
59. Vinylacetyl-CoA = crotonyl-CoA	Vinylacetyl-CoA isomerase	[143,144]
60. Methylmalonyl-CoA = succinyl-CoA	Methylmalonyl-CoA mutase	[141,145,146]
Lygase		
61. ATP + acetate + CoA = AMP + PP_i + acetyl-CoA	Acetyl-CoA synthetase	[83,147–150]
62. ATP + acid $(C_4–C_{11})$ + CoA = AMP + PP_i + acyl-CoA	Acyl-CoA $(C_4–C_{11})$ synthetase	[151–154]
63. ATP + acid $(C_6–C_{20})$ + CoA = AMP + PP_i + acyl-CoA	Acyl-CoA $(C_6–C_{20})$ synthetase	[155,156]
64. GTP + acid + CoA = GDP + P_i + acyl-CoA	Acyl-CoA synthetase (GTP-dependent)	[157,158]
65. GTP + succinate + CoA = GDP + P_i + succinyl-CoA	Succinyl-CoA synthetase (GDP)	[159–161]
66. ATP + succinate + CoA = ADP + P_i + succinyl-CoA	Succinyl-CoA synthetase (ADP)	[162,163]
67. ATP + glutarate + CoA = ADP + P_i + glutaryl-CoA	Glutaryl-CoA synthetase	[164]
68. ATP + cholate + CoA = AMP + PP_i + choloyl-CoA	Choloyl-CoA synthetase	[165,166]

TABLE III (*Continued*)

Reaction	Enzyme	Reference
69. ATP + acetyl-CoA + CO_2 + H_2O = ADP + P_i + methylmalonyl-CoA	Acetyl-CoA carboxylase	[167–169]
70. ATP + propionyl-CoA + CO_2 + H_2O = ADP + P_i + methylmalonyl-CoA	Propionyl-CoA carboxylase	[170,171]
71. ATP + 3-methylcrotonoyl-CoA + CO_2 + H_2O = ADP + P_i + 3-methylglutaconyl-CoA	Methylcrotonoyl-CoA carboxylase	[144,172,173]
72. Acetyl-ACP + malonyl-ACP = acetoacetyl-ACP + ACP + CO_2	β-Ketoacyl-ACP synthetase	[174–176]

REFERENCES

1. G. M. Brown and E. E. Snell, *J. Amer. Chem. Soc.* **75**, 2782 (1953).
2. G. B. Ward, G. M. Brown, and E. E. Snell, *J. Biol. Chem.* **213**, 869 (1955).
3. G. M. Brown, *J. Biol. Chem.* **234**, 370 (1959).
4. W. S. Pierpoint and D. E. Hughes, *Biochem. J.* **56**, 130 (1954).
5. W. S. Pierpoint, D. E. Hughes, J. Baddiley, and A. P. Mathias, *Biochem. J.* **61**, 368 (1955).
6. G. M. Brown, *J. Biol. Chem.* **226**, 651 (1957).
7. G. M. Brown, *J. Biol. Chem.* **234**, 379 (1959).
8. T. E. King and V. H. Cheldelin, *Proc. Soc. Exp. Biol. Med.* **84**, 591 (1953).
9. G. M. Brown and E. E. Snell, *J. Bacteriol.* **67**, 465 (1954).
10. G. M. Brown, M. Ikawa, and E. E. Snell, *J. Biol. Chem.* **213**, 855 (1955).
11. G. D. Novelli and F. Lipmann, *J. Biol. Chem.* **171**, 833 (1947).
12. G. D. Novelli and F. Lipmann, *J. Biol. Chem.* **182**, 213 (1950).
13. T. Wieland, W. Maul, and E. F. Möller, *Biochem. Z.* **327**, 85 (1955).
13a. S. Shimizu, S. Satsuma, K. Kubo, Y. Tani, and K. Ogata, *Agr. Biol. Chem.* **37**, 857 (1973).
13b. S. Shimizu, K. Kubo, H. Morioka, Y. Tani, and K. Ogata, *Agr. Biol. Chem.* **38**, 1015 (1974).
14. T. P. Wang and N. O. Kaplan, *J. Biol. Chem.* **206**, 311 (1954).
15. G. D. Novelli, F. Schmetz, and N. O. Kaplan, *J. Biol. Chem.* **206**, 533 (1954).
16. L. Levintow and G. D. Novelli, *J. Biol. Chem.* **207**, 761 (1954).
17. D. Cavallini, B. Mondovi, C. De Marco, and G. Ferro-Luzzi, *Enzymologia* **20**, 359 (1959).
18. M. B. Hoagland and G. D. Novelli, *J. Biol. Chem.* **207**, 767 (1954).
19. Y. Abiko, *J. Biochem.* (*Tokyo*) **61**, 290 (1967).
20. Y. Abiko, *J. Biochem.* (*Tokyo*) **61**, 300 (1967).
21. Y. Abiko, T. Suzuki, and M. Shimizu, *J. Biochem.* (*Tokyo*) **61**, 309 (1967).
22. T. Suzuki, Y. Abiko, and M. Shimizu, *J. Biochem.* (*Tokyo*) **62**, 642 (1967).
23. T. Nakamura, T. Kusunoki, and K. Soyama, *J. Vitaminol.* (*Kyoto*) **13**, 289 (1967).
24. Y. Abiko, M. Tomikawa, and M. Shimizu, *J. Biochem.* (*Tokyo*) **64**, 115 (1968).
25. Y. Abiko, S. Ashida, and M. Shimizu, *Biochim. Biophys. Acta* **268**, 364 (1972).

26. T. Nakamura, T. Kusunoki, K. Soyama, and M. Kuwagata, *J. Vitaminol. (Kyoto)* **18**, 34 (1972).
27. M. Shimizu and Y. Abiko, *Chem. Pharm. Bull.* **13**, 189 (1965).
28. S. Duprè, M. T. Graziani, M. A. Rosei, A. Fabi, and E. D. Grosso, *Eur. J. Biochem.* **16**, 571 (1970).
29. M. Shimizu, O. Nagase, Y. Hosokawa, and H. Tagawa, *Tetrahedron* **24**, 5241 (1968).
30. Y. Abiko, M. Tomikawa, Y. Hosokawa, and M. Shimizu, *Chem. Pharm. Bull.* **17**, 200 (1969).
31. Y. Abiko, *in* "Methods in Enzymology" (D. B. McCormick and L. D. Wright, eds.), Vol. 18, Part A, p. 350. Academic Press, New York, 1970.
32. Y. Abiko, *in* "Methods in Enzymology" (D. B. McCormick and L. D. Wright, eds.), Vol. 18, Part A, p. 354. Academic Press, New York, 1970.
33. Y. Abiko, *in* "Methods in Enzymology" (D. B. McCormick and L. D. Wright, eds.), Vol. 18, Part A, p. 358. Academic Press, New York, 1970.
34. T. Nakamura, T. Kusunoki, K. Soyama, and M. Kuwagata, *Vitamins* **40**, 412 (1969).
34a. S. Shimizu, K. Kubo, Y. Tani, and K. Ogata, *Agr. Biol. Chem.* **37**, 2863 (1973).
35. T. Nakamura, T. Kusunoki, K. Soyama, K. Tsujita, and K. Tanaka, *Vitamins* **40**, 354 (1969).
36. T. Nakamura, T. Kusunoki, S. Kataoka, K. Soyama, I. Masugi, and K. Tanaka, *J. Vitaminol. (Kyoto)* **13**, 274 (1967).
37. T. Nakamura, T. Kusunoki, and S. Kataoka, *J. Vitaminol. (Kyoto)* **13**, 283 (1967).
38. T. Nakamura, N. Nomoto, R. Yagi, and N. Oya, *J. Vitaminol. (Kyoto)* **16**, 89 (1970).
39. P. R. Vagelos, *in* "The Enzymes" (P. D. Boyer, ed.), 3rd ed., Vol. 8, p. 155. Academic Press, New York, 1973.
40. T. C. Vanaman, S. J. Wakil, and R. L. Hill, *J. Biol. Chem.* **243**, 6420 (1968).
41. J. Elovson and P. R. Vagelos, *J. Biol. Chem.* **243**, 3603 (1968).
41a. D. J. Prescott, J. Elovson, and P. R. Vagelos, *J. Biol. Chem.* **244**, 4517 (1969).
42. P. R. Vagelos and A. R. Larrabee, *J. Biol. Chem.* **242**, 1776 (1967).
43. G. L. Powell, J. Elovson, and P. R. Vagelos, *J. Biol. Chem.* **244**, 5616 (1969).
44. J. Tweto, M. Liberti, and A. R. Larrabee, *J. Biol. Chem.* **246**, 2468 (1971).
45. G. D. Novelli, N. O. Kaplan, and F. Lipmann, *J. Biol. Chem.* **177**, 97 (1949).
46. J. Bremer, A. Wojtczak, and S. Skrede, *Eur. J. Biochem.* **25**, 190 (1972).
47. S. Skrede, *Eur. J. Biochem.* **38**, 401 (1973).
48. K. Kameda and Y. Abiko, unpublished.
49. K. Decker and E. Bischoff, *FEBS (Fed. Eur. Biochem. Soc.) Lett.* **21**, 95 (1972).
50. S. Duprè, M. A. Rosei, L. Bellussi, E. Barboni, and R. Scandurra, *Abstr. Int. Congr. Biochem., 9th, 1973* p. 98 (1973).
51. S. Duprè and C. De Marco, *Ital. J. Biochem.* **13**, 386 (1964).
52. D. Cavallini, C. De Marco, R. Scandurra, S. Duprè, and M. T. Graziani, *J. Biol. Chem.* **241**, 3189 (1966).
53. D. Cavallini, S. Duprè, M. T. Graziani, and M. G. Tinti, *FEBS (Fed. Eur. Biochem. Soc.) Lett.* **1**, 119 (1968).
54. S. Duprè, M. T. Graziani, and M. A. Rosei, *Ital. J. Biochem.* **19**, 132 (1970).
55. C. De Duve, B. C. Pressman, R. Gianetto, R. Wattiaux, and F. Appelmans, *Biochem. J.* **60**, 604 (1955).
56. S. Y. Ali and C. H. Lack, *Biochem. J.* **96**, 63 (1965).
57. K. Boettge, K. H. Jaeger, and H. Mittenzwei, *Arzneim.-Forsch.* **7**, 24 (1957).
58. L. Mascaretti, *Med. Int.* **8**, 42 (1967).
59. A. Pasquale, L. Cerutti, and L. Soranzo, *Minerva Med.* **60**, 1 (1969).

60. H. Thölen, A. Colombi, F. Duckert, F. Huber, H. R. Muller, and F. Bigler, *Helv. Med. Acta* **33**, 492 (1966).
61. W. M. Govier and A. J. Gibbons, *Arch. Biochem. Biophys.* **32**, 349 (1951).
62. F. Bigler and H. Thölen, *Experientia* **22**, 321 (1966).
63. W. Domschke, M. Liersch, and K. Decker, *Hoppe-Seyler's Z. Physiol. Chem.* **352**, 85 (1971).
64. K. Tsujita, *Vitamins* **46**, 255 (1972).
65. N. O. Kaplan and F. Lipmann, *J. Biol. Chem.* **174**, 37 (1948).
66. R. M. Burton and E. R. Stadtman, *J. Biol. Chem.* **202**, 873 (1953).
67. J. R. Quayle and G. A. Taylor, *Biochem. J.* **78**, 611 (1961).
68. E. W. Yamada and W. B. Jakoby, *J. Biol. Chem.* **235**, 589 (1960).
69. O. Hayaishi, Y. Nishizuka, M. Tatibana, and S. Kuno, *J. Biol. Chem.* **236**, 781 (1961).
70. Y. Namba, K. Yoshizawa, A. Ejima, T. Hayashi, and T. Kaneda, *J. Biol. Chem.* **244**, 4437 (1969).
71. D. E. Green, S. Mii, H. R. Mahler, and R. M. Bock, *J. Biol. Chem.* **206**, 1 (1954).
72. J. G. Hauge, F. L. Crane, and H. Beinert, *J. Biol. Chem.* **219**, 727 (1956).
73. F. L. Crane, S. Mii, J. G. Hauge, D. E. Green, and H. Beinert, *J. Biol. Chem.* **218**, 701 (1956).
74. A. W. Alberts, P. W. Majerus, B. Talamo, and P. R. Vagelos, *Biochemistry* **3**, 1563 (1964).
75. G. Weeks and S. J. Wakil, *J. Biol. Chem.* **243**, 1180 (1968).
76. R. W. Swick and H. G. Wood, *Proc. Nat. Acad. Sci. U.S.* **46**, 28 (1960).
77. H. G. Wood and R. Stjernholm, *Proc. Nat. Acad. Sci. U.S.* **47**, 289 (1961).
78. W. K. Maas, G. D. Novelli, and F. Lipmann, *Proc. Nat. Acad. Sci. U.S.* **39**, 1004 (1953).
79. S. C. Kinsky, *J. Biol. Chem.* **235**, 94 (1960).
80. T. C. Chou and M. Soodak, *J. Biol. Chem.* **196**, 105 (1952).
81. D. H. Brown, *Biochim. Biophys. Acta* **16**, 429 (1955).
82. E. A. Davidson, H. J. Blumenthal, and F. Roseman, *J. Biol. Chem.* **226**, 125 (1957).
83. T. C. Chou and F. Lipmann, *J. Biol. Chem.* **196**, 89 (1952).
84. S. P. Bessman and F. Lipmann, *Arch. Biochem. Biophys.* **46**, 252 (1953).
85. H. Tabor, A. H. Mehler, and E. R. Stadtman, *J. Biol. Chem.* **204**, 127 (1953).
86. H. Weissbach, B. G. Redfield, and J. Axelrod, *Biochim. Biophys. Acta* **54**, 190 (1961).
87. R. Berman, I. B. Wilson, and D. Nachmansohn, *Biochim. Biophys. Acta* **12**, 315 (1953).
88. J. F. Berry and V. P. Whittaker, *Biochem. J.* **73**, 447 (1959).
89. S. Friedman and G. Fraenkel, *Arch. Biochem. Biophys.* **59**, 491 (1955).
90. K. R. Norum, *Biochim. Biophys. Acta* **89**, 95 (1964).
91. I. B. Fritz, S. K. Schultz, and P. A. Srere, *J. Biol. Chem.* **238**, 2509 (1963).
92. E. R. Stadtman, *J. Biol. Chem.* **196**, 527 (1952).
93. M. Shimizu, T. Suzuki, K. Kameda, and Y. Abiko, *Biochim. Biophys. Acta* **191**, 550 (1969).
94. F. Lynen and S. Ochoa, *Biochim. Biophys. Acta* **12**, 299 (1953).
95. J. R. Stern, G. I. Drummond, M. J. Coon, and A. del Campillo, *J. Biol. Chem.* **235**, 313 (1960).
96. R. O. Brady and E. R. Stadtman, *J. Biol. Chem.* **211**, 621 (1954).
97. I. C. Gunsalus, L. S. Barton, and W. Gruber, *J. Amer. Chem. Soc.* **78**, 1763 (1956).
98. D. Schachter and J. V. Taggart, *J. Biol. Chem.* **208**, 263 (1954).

 99. K. Moldave and A. Meister, *J. Biol. Chem.* **229**, 463 (1957).
100. A. Kornberg and W. E. Pricer, *J. Biol. Chem.* **204**, 345 (1953).
101. D. S. Goldman, *J. Biol. Chem.* **208**, 345 (1954).
102. J. R. Stern and S. Ochoa, *J. Biol. Chem.* **191**, 161 (1951).
103. F. B. Goldstein, *J. Biol. Chem.* **234**, 2702 (1959).
104. I. Zabin, A. Kepes, and J. Monod, *J. Biol. Chem.* **237**, 253 (1962).
105. R. C. Valentine and R. S. Wolfe, *J. Biol. Chem.* **235**, 1948 (1960).
106. P. Goldman and P. R. Vagelos, *J. Biol. Chem.* **236**, 2620 (1961).
107. S. B. Weiss, E. P. Kennedy, and J. Y. Kiyasu, *J. Biol. Chem.* **235**, 40 (1960).
108. E. R. Stadtman, *Fed. Proc., Fed. Amer. Soc. Exp. Biol.* **11**, 291 (1952).
109. J. R. Quayle, D. B. Keech, and G. A. Taylor, *Biochem. J.* **78**, 225 (1961).
110. O. Hayaishi, *J. Biol. Chem.* **215**, 125 (1955).
111. J. R. Stern, M. J. Coon, A. del Campillo, and M. C. Schneider, *J. Biol. Chem.* **221**, 15 (1956).
112. M. Katagiri and O. Hayaishi, *J. Biol. Chem.* **226**, 439 (1957).
113. P. W. Majerus, A. W. Alberts, and P. R. Vagelos, *Proc. Nat. Acad. Sci. U.S.* **51**, 1231 (1964).
114. D. J. Prescott and P. R. Vagelos, *J. Biol. Chem.* **245**, 5484 (1970).
115. A. W. Alberts, R. M. Bell, and P. R. Vagelos, *J. Biol. Chem.* **247**, 3190 (1972).
116. J. Gergely, P. Hele, and C. V. Ramakrishnan, *J. Biol. Chem.* **198**, 323 (1952).
117. J. W. Porter and R. W. Long, *J. Biol. Chem.* **233**, 20 (1958).
118. P. A. Srere, W. Seubert, and F. Lynen, *Biochim. Biophys. Acta* **33**, 313 (1959).
119. G. Rendina and M. J. Coon, *J. Biol. Chem.* **225**, 523 (1957).
120. E. E. Dekker, M. J. Schlesinger, and M. J. Coon, *J. Biol. Chem.* **233**, 434 (1958).
121. E. M. Barnes and S. J. Wakil, *J. Biol. Chem.* **243**, 2955 (1968).
122. W. B. Jakoby, E. Ohmura, and O. Hayaishi, *J. Biol. Chem.* **222**, 435 (1956).
123. D. T. O. Wong and S. J. Ajl, *J. Amer. Chem. Soc.* **78**, 3230 (1956).
124. Y. Yamamoto and H. Beevers, *Biochim. Biophys. Acta* **48**, 20 (1961).
125. B. K. Bachhawat, W. G. Robinson, and M. J. Coon, *J. Biol. Chem.* **216**, 727 (1955).
126. N. L. R. Bucher, P. Overath, and F. Lynen, *Biochim. Biophys. Acta* **40**, 491 (1960).
127. J. J. Ferguson and H. Rudney, *J. Biol. Chem.* **234**, 1072 (1959).
128. H. Rudney, *J. Biol. Chem.* **227**, 363 (1957).
129. S. Ochoa, J. R. Stern, and M. C. Schneider, *J. Biol. Chem.* **193**, 691 (1951).
130. P. A. Srere and G. W. Kosicki, *J. Biol. Chem.* **236**, 2557 (1961).
131. P. A. Srere, *J. Biol. Chem.* **234**, 2544 (1959).
132. P. A. Srere, *J. Biol. Chem.* **236**, 50 (1961).
133. J. R. Stern, A. del Campillo, and I. Raw, *J. Biol. Chem.* **218**, 971 and 985 (1956).
134. S. J. Wakil, *Biochim. Biophys. Acta* **19**, 497 (1956).
135. H. Hilz, J. Knappe, E. Ringelmann, and F. Lynen, *Biochem. Z.* **329**, 476 (1958).
136. P. R. Vagelos, J. M. Earl, and E. R. Stadtman, *J. Biol. Chem.* **234**, 490 (1959).
137. P. W. Majerus, A. W. Alberts, and P. R. Vagelos, *J. Biol. Chem.* **240**, 618 (1965).
138. M. Mizugaki, G. Weeks, R. E. Toomey, and S. J. Wakil, *J. Biol. Chem.* **243**, 3661 (1968).
139. J. R. Stern and A. del Campillo, *J. Amer. Chem. Soc.* **77**, 1073 (1955).
140. S. J. Wakil, *Biochim. Biophys. Acta* **18**, 314 (1955).
141. P. Overath, G. M. Kellerman, and F. Lynen, *Biochem. Z.* **335**, 500 (1962).
142. R. Mazumder, T. Sasakawa, Y. Kaziro, and S. Ochoa, *J. Biol. Chem.* **237**, 3065 (1962).
143. F. Lynen, J. Knappe, E. Lorch, G. Jütting, and E. Ringelman, *Angew. Chem.* **71**, 481 (1959).

144. H. C. Rilling and M. J. Coon, *J. Biol. Chem.* **235**, 3087 (1960).
145. R. Stjernholm and H. G. Wood, *Proc. Nat. Acad. Sci. U.S.* **47**, 289 (1961).
146. W. S. Bock and S. Ochoa, *J. Biol. Chem.* **232**, 931 (1958).
147. M. A. Eisenberg, *Biochim. Biophys. Acta* **16**, 58 (1955).
148. P. Hele, *J. Biol. Chem.* **206**, 671 (1954).
149. M. E. Jones, F. Lipmann, H. Hilz, and F. Lynen, *J. Amer. Chem. Soc.* **75**, 3285 (1953).
150. A. Millerd and J. Bonner, *Arch. Biochem. Biophys.* **49**, 343 (1954).
151. G. M. Kellerman, *J. Biol. Chem.* **231**, 427 (1958).
152. A. L. Lehninger and G. D. Greville, *Biochim. Biophys. Acta* **12**, 188 (1953).
153. H. R. Mahler, S. J. Wakil, and R. M. Bock, *J. Biol. Chem.* **204**, 453 (1953).
154. C. H. L. Peng, *Biochim. Biophys. Acta* **22**, 42 (1956).
155. B. Borgstrøm and L. W. Wheeldon, *Biochim. Biophys. Acta* **50**, 171 (1961).
156. A. Kornberg and W. E. Pricer, *J. Biol. Chem.* **204**, 329 (1953).
157. C. R. Rossi and D. M. Gibson, *J. Biol. Chem.* **239**, 1694 (1964).
158. L. Galzigna, C. R. Rossi, L. Sartorelli, and D. M. Gibson, *J. Biol. Chem.* **242**, 2111 (1967).
159. S. Kaufman, C. Gilvarg, O. Cori, and S. Ochoa, *J. Biol. Chem.* **203**, 869 (1953).
160. D. R. Sanadi, D. M. Gibson, and P. Ayengar, *Biochim. Biophys. Acta* **14**, 434 (1954).
161. R. Mazumder, D. R. Sanadi, and W. V. Rodwell, *J. Biol. Chem.* **235**, 2546 (1960).
162. S. Kaufman and S. G. A. Alivasatos, *J. Biol. Chem.* **216**, 141 (1955).
163. S. Kaufman, *J. Biol. Chem.* **216**, 153 (1955).
164. C. K. K. Menon, D. L. Friedman, and J. R. Stern, *Biochim. Biophys. Acta* **44**, 375 (1960).
165. W. H. Elliott, *Biochem. J.* **62**, 427 (1956).
166. W. H. Elliott, *Biochem. J.* **65**, 315 (1957).
167. S. J. Wakil, *J. Amer. Chem. Soc.* **80**, 6465 (1958).
168. J. V. Formica and R. O. Brady, *J. Amer. Chem. Soc.* **81**, 752 (1959).
169. M. D. Hatch and P. K. Stumpf, *J. Biol. Chem.* **236**, 2879 (1961).
170. M. D. Lane, D. R. Halenz, D. P. Kosow, and C. S. Hegre, *J. Biol. Chem.* **235**, 3082 (1960).
171. Y. Kaziro, S. Ochoa, R. C. Warner, and J. Chen, *J. Biol. Chem.* **236**, 1917 (1961).
172. J. Knappe, H. G. Schlegel, and F. Lynen, *Biochem. Z.* **335**, 101 (1961).
173. F. Lynen, J. Knappe, E. Lorch, G. Jütting, E. Ringelman, and J. P. Lachance, *Biochem. Z.* **335**, 123 (1961).
174. A. W. Alberts, P. W. Majerus, and P. R. Vagelos, *Biochemistry* **4**, 2265 (1965).
175. R. E. Toomey and S. J. Wakil, *J. Biol. Chem.* **241**, 1159 (1966).
176. M. D. Greenspan, A. W. Alberts, and P. R. Vagelos, *J. Biol. Chem.* **244**, 6477 (1969).

Note Added in Proof

Recently 4'-phosphopantothenoyl-L-cysteine decarboxylase was highly purified from horse liver [R. Scandurra, E. Barboni, F. Granata, B. Pensa, and M. Costa, *Eur. J. Biochem.* **49**, 1 (1974)]. The properties of the horse enzyme were reported to be very similar to those of the rat enzyme except for a higher K_m value (1.43 mM) and lack of the product inhibition in the former.

CHAPTER 2

Biotin

Max A. Eisenberg

I. INTRODUCTION

Biotin, as with other essential metabolites, was discovered as the result of the effects of a nutritional deficiency. Three independent lines of research were brought together when biotin was found to be physiologically active for bacteria, yeast, and higher animals. Its requirement as a growth factor for yeast was discovered at the turn of the present century and eventually led to the isolation and crystallization of biotin by Kögel and Tönnis [1]. It was possible to prove the identity of crystalline biotin with "coenzyme R," the growth factor for *Rhizobium*, and with "vitamin H," the protective factor against "egg-white injury" [2]. Its chemical structure was determined in a relatively short period of time, but knowledge of its physiological role in the cell was slow in coming. It was with the discovery by Wakil *et al.* [3] that biotin was an integral part of the acetyl-CoA carboxylase enzyme, that the definitive role of biotin in carboxylation reactions was established. This generated an intensive search for other biotin enzymes resulting in an extensive literature on the role of biotin in a variety of metabolic processes. It also brought with it a renewed interest in the biosynthesis and metabolism of biotin, with significant advances in these areas over the past ten years. It is only possible to present some of the highlights in each of these areas and a more extensive coverage is available in recent review articles [4–8].

II. CHEMISTRY

The structure of biotin was elucidated by du Vigneaud and his collaborators in 1942 [2,9]. The bicyclic structure was shown to be a fusion of an imidazolidone ring with a tetrahydrothiophene ring bearing a valeric acid side chain. The structure of biotin shown in Fig. 1 indicates the presence of three asymmetric carbon atoms and hence the possibility of 8 stereoisomers. All isomers have been synthesized by the Merck group [10] but only *d*-biotin is biologically active [11]. The absolute stereochemistry of biotin was established by X-ray crystallography [12]. The analysis revealed that the imidazolidone and the tetrahydrothiophene rings are fused in the cis configuration producing a boatlike structure. The ureido ring projects upward at an angle of 62° with respect to the plane formed by the four carbon atoms of the tetrahydrothiophene ring and a second plane, made by the sulfur atom and carbon atoms 2 and 5, projects upward at an angle of 37.6°. The valeric side chain, attached to C-2 of the sulfur ring, is cis with respect to the ureido ring. The apposition of C-6 and 3′-N may result in steric hindrance which could interfere

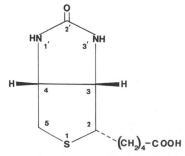

FIG. 1. The absolute stereochemistry of *d*-biotin [12].

with 3'-N carboxylation. The X-ray analysis of the bis-*p*-bromoanalide derivative of the active form of biotin, carboxybiotin, confirmed the location of the carboxyl group on the 1'-N position [13].

The isolation from yeast of biocytin suggested to Wright and Skeggs [14] the possibility that biotin may exist functionally in amide linkage to an amino group of a lysyl residue (Fig. 2). This was verified by Kasow and Lane [15] when they found that the biotin was indeed so linked in the propionyl-CoA carboxylase enzyme. This linkage has been subsequently shown to be true for all biotin-dependent enzymes. A number of biotin analogues were synthesized during the period of structural analysis, and certain of these have since been observed in the medium filtrate of a variety of organisms. Some of these analogues are able to

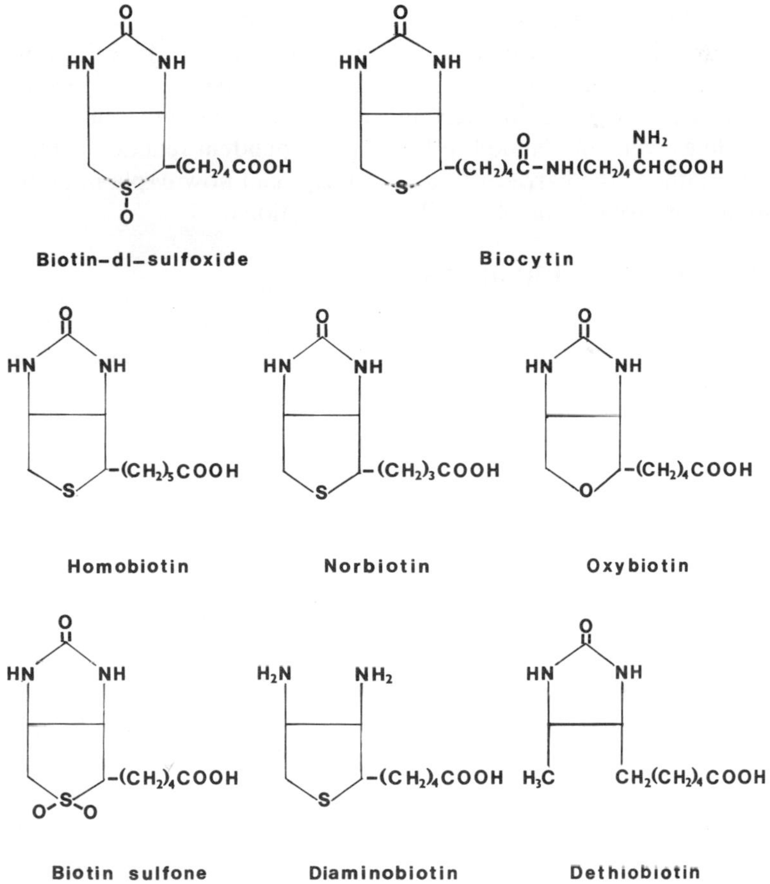

FIG. 2. Biotin analogues.

replace the biotin requirement of certain microorganisms while others have been found to be biologically inactive and may act as biotin antagonists. The structures of some of the common biotin analogues are provided in Fig. 2 and the remainder will be discussed in the sections dealing with the biosynthesis and degradation of biotin. The term "total" biotin has been used to indicate all biotin analogues capable of supporting the growth of yeast cells while "true" biotin refers to vitamers supporting the growth of *Lactobacillus arabinosus* and consists primarily of biotin and biotin *d*-sulfoxide.

III. BIOLOGICAL ROLE OF BIOTIN

As the result of nutritional and biochemical studies over the last fifty years, biotin has been implicated in the metabolism of the four major classes of cellular constituents. The decarboxylation of oxalacetate and succinate, the carboxylation of pyruvate, the biosynthesis of unsaturated fatty acids, aspartic acid, certain non-biotin-containing proteins, and pyrimidines were all claimed to be biotin-dependent reactions. The function of biotin as the carrier of "active CO_2" can now explain, directly or indirectly, its role in the above diverse reactions.

A. Classification of Biotin Enzymes

The biotin-dependent enzymes fall into three functional categories: CO_2 fixation (carboxylase reaction), CO_2 transfer (transcarboxylase reaction), and CO_2 loss (decarboxylase reaction). A classification of the known biotin enzymes and the reactions they catalyze are presented in Table I. All reactions can be subdivided into two defined steps which are coupled through a carboxybiotin–enzyme complex intermediate as shown in a generalized manner below:

$$CO_2 \text{ donor} + \text{biotin-E} \rightleftarrows \text{carboxybiotin-E} + \text{products} \tag{1}$$

$$\text{Carboxybiotin-E} + \text{acceptor} \rightleftarrows \text{carboxylated acceptor} + \text{biotin-E} \tag{2}$$

In all the carboxylase reactions, HCO_3^- is the carboxyl donor in Eq. (1) and is activated by ATP and Mg^{2+}. The observation that the ATP-$^{32}P_i$ exchange requires HCO_3^- and ADP and the ATP-$[^{14}C]$ADP exchange depends on HCO_3^- and P_i is consistent with the above reaction sequence. In the transcarboxylase and decarboxylase reactions ATP is not required and the active CO_2 is transferred either from an acyl-CoA derivative or from an α- or β-keto acid to the biotin enzyme. The exchange reaction between the acyl-CoA donor and acyl-CoA products

TABLE I

CLASSIFICATION OF BIOTIN ENZYMES

I. Carboxylases
   A. Acyl-CoA carboxylase
      ATP + HCO_3^- + acyl-CoA $\rightleftarrows$ carboxylated acyl-CoA + ADP + P_i
      1. Acetyl-CoA $\rightleftarrows$ malonyl-CoA
         [acetyl-CoA : CoA : CO_2 ligase (ADP) (EC 6.4.1.2)]
      2. Propionyl-CoA $\rightleftarrows$ S-methylmalonyl-CoA
         [propionyl-CoA : CO_2 ligase (ADP) (EC 6.4.1.3)]
      3. β-Methylcrotonyl-CoA $\rightleftarrows$ β-glutaconyl CoA
         [3-methylcrotonyl-CoA : CO_2 (ADP) (EC 6.4.1.4)]
      4. Geranyl-CoA $\rightleftarrows$ carboxygeranyl-CoA
         [geranyl-CoA : CO_2 (ADP) (EC 6.4.1.5)]
   B. α-Ketocarboxylase
      ATP + HCO_3^- + pyruvate $\rightleftarrows$ oxalacetate + ADP + P_i
      [pyruvate : CO_2 ligase (ADP) (EC 6.4.1.1)]
   C. Amidocarboxylase
      ATP + HCO_3^- + urea $\rightleftarrows$ allophanate + ADP + P_i
      [urea : CO_2 ligase (ADP) (EC 6.4.1.6)]
II. Transcarboxylase
    S-Methylmalonyl-CoA + pyruvate $\rightleftarrows$ oxalacetate + propionyl-CoA
    [methylmalonyl-CoA : pyruvate carboxyltransferase (EC 2.1.3.1)]
III. Decarboxylase
    1. S-Methylmalonyl-CoA $\rightarrow$ propionyl-CoA + CO_2
       [methylmalonyl-CoA decarboxylase (EC 4.1.1.__)]
    2. Oxalacetate $\rightarrow$ pyruvate + CO_2
       [oxalacetate decarboxylase (EC 4.1.1.3)]

in the transcarboxylase reaction is independent of the keto acids, which is also in accord with the two step sequence. The most definitive evidence for the first reaction [Eq. (1)] was attained with the isolation of the carboxybiotin–enzyme complex by Kaziro and Ochoa [16] after treating propionyl-CoA carboxylase with ATP, Mg^{2+}, and HCO_3^-. This complex was subsequently isolated with other carboxylases and the transcarboxylase, but not with the decarboxylase enzyme because of the rapidity with which CO_2 is lost. With the availability of the carboxybiotin–enzyme complex in substrate quantities, it was possible to demonstrate the formation of ATP on the addition of ADP, Mg^{2+}, and P_i [reverse of Eq. (1)].

The nature of the carboxybiotin–enzyme complex was elucidated by Lynen and his collaborators [17] in a model system where free biotin was carboxylated by β-methylcrotonyl-CoA carboxylase in the presence of ATP and Mg^{2+}. The unstable product when treated with diazomethane yielded methylcarboxybiotin, which on further analysis showed

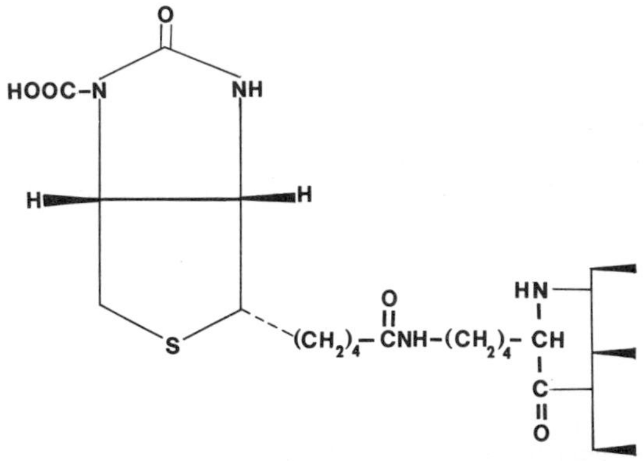

FIG. 3. The active form of biotin, $1'$-N-carboxybiotin, attached to the biotin carrier protein.

the carboxylation to have occurred on the $1'$-N position. As indicated previously, this was verified by X-ray analysis. When the diazomethane treatment was applied to the isolated carboxybiotin–enzyme complex labeled with radioactive bicarbonate, it stabilized the carboxybiotin and permitted digestion of the complex with the proteolytic enzyme pronase. A radioactive compound was isolated and identified as the methylcarboxy derivative of biocytin. On further degradation with biotinadase, the $1'$-N-carboxybiotin was isolated [18]. The structure of active biotin is shown in Fig. 3.

The second step in the sequence [Eq. (2)] is also supported by the exchange reactions between the acceptor and the carboxylated acceptor. In the carboxylase reactions, this exchange is independent of ATP, ADP, P_i, and HCO_3^-, while in the transcarboxylase reaction, the exchange reactions between the keto acids is independent of the acyl-CoA derivatives. In addition, the carboxyl group from the carboxybiotin–enzyme complex is quantitatively transferred to a number of acceptors to form the carboxylated acceptor [Eq. (2)]. Both steps are reversible and sensitive to avidin, indicating the participation of the biotin prosthetic group in both steps.

B. Biotin–Holoenzyme Formation

The biotin prosthetic group, as indicated previously, is covalently linked to an ϵ amino group of a lysyl residue of the biotin-dependent enzymes. Kasow and Lane showed the presence of an enzyme in bacterial cell-free extracts which catalyzed the attachment of biotin to the apoenzyme [15]. The observation that in biotin deficient cells, both animal and bacterial, the apoenzymes accumulated in large quantities enabled the investigators to purify them and study the holoenzyme synthetase reaction in greater detail. The reaction required ATP and Mg^{2+}, with AMP and PP_i as the end products and the exchange reaction between ATP-$^{32}PP_i$ was biotin dependent. This was reminiscent of the activation mechanism of amino acids and fatty acids, indicating a mixed anhydride intermediate involving the carboxyl group of the valeric acid side chain. In support of this mechanism was the finding that biotinyl adenylate could replace the biotin and ATP requirement in formation of propionyl-CoA carboxylase and transcarboxylase holoenzymes [19,20]. The holoenzyme synthetase reaction was postulated to occur in two steps as indicated below:

$$\text{Biotin—COOH + ATP + Mg}^{2+} \rightleftarrows \text{biotin—COO—AMP + PP}_i \qquad (3)$$
$$\text{Biotin—COO—AMP + NH}_2\text{—apoenzyme} \rightleftarrows$$
$$\text{biotin—COO—NH—apoenzyme + AMP} \quad (4)$$
$$\overline{\text{Biotin + ATP + Mg}^{2+} + \text{apoenzyme} \rightleftarrows \text{holoenzyme + AMP + PP}_i \qquad (5)}$$

The specificity of the transcarboxylase holoenzyme synthetase is almost absolute [21]. Although certain biotin analogues were found by competition experiments to bind to the synthetase, it was possible with the aid of ^{14}C to show that only O-heterobiotin (oxybiotin) was activated by the holoenzyme synthetase, but an inactive transcarboxylase was formed. On the other hand, this biotin analogue had been previously shown to support the growth of yeast and to be incorporated into a bound form without exchanging the oxygen atom for sulfur [22]. The nature of the biotin enzymes involved was not determined, but the substitution of oxygen for sulfur apparently did not lead to inactive biotin enzymes.

While the holoenzyme synthetase shows a high degree of specificity for biotin, the specificity for the apoenzyme is rather broad. The synthetases from yeast, liver, and *Propionobacter* are able to utilize the apoenzymes from liver, yeast, and bacteria, but with marked variability in activity. This broad specificity for the apoenzyme would suggest that the environment of the lysine residue to which the biotin is attached may be similar in the various apoenzymes.

C. Structural Aspects

All biotin-dependent enzymes appear to be of high molecular weight $(0.1–6 \times 10^6)$ and of complex quaternary structure. In addition to the covalently bound biotin, those enzymes which have a keto acid as an acceptor also contain tightly bound metals, manganese, zinc, and cobalt, in a 1:1 ratio with biotin. The three biotin enzymes that have been studied in greatest detail have been acetyl-CoA carboxylase, pyruvate carboxylase, and transcarboxylase.

1. ACETYL-CoA CARBOXYLASE

The acetyl-CoA carboxylase as isolated from a variety of animal sources consists of an enzymatically active filamentous form with a molecular weight of 4–6×10^6. This information is based on the hydrodynamic properties of the enzyme and its appearance in electron micrographs [23,24]. In the avian liver enzyme and bovine adipose tissue enzyme, these linear polymers are made up of protomeric units averaging 4–5×10^5 daltons. In the presence of sodium dodecyl sulfate the protomer dissociates into four subunits of 100,000 daltons. These are not identical, as there is only one mole of biotin per 400,000 daltons and only one binding site each for substrate and activator. The protomer–polymer equilibrium is readily shifted in the direction of the polymer on the addition of the activator, citric or isocitric acids. In the absence of the activator, the formation of the carboxylated enzyme shifts the equilibrium toward the protomer. The catalytic activity of the enzyme increases on polymerization and decreases on disaggregation, indicating that the polymeric state of the enzyme controls the catalytic activity. The acetyl-CoA carboxylation reaction is the rate controlling step in fatty acid biosynthesis. The protomer–polymer equilibrium, controlled by the citrate and isocitrate levels of the cell, modulates the level of the active enzyme and hence is an important aspect in the regulation of fatty acid biosynthesis in animals. On the other hand, the carboxylases from plants, bacteria, and yeast are insensitive to the tricarboxylic acids.

The mechanism of action of the activator appears to involve a conformational change in the vicinity of the biotin prosthetic group. Not only are both partial reactions [Eqs. (1) and (2)] affected by the activator, indicating a common center, but the inhibition of both reactions by avidin is eliminated in the presence of citrate, suggesting that the biotin is not readily available to the inhibitor protein. It has been proposed that a conformational change, caused by the addition of the activator, positions the biotin into a recessed substrate site so that it is properly oriented for carboxylation, but unavailable for binding with avidin.

In contrast to the animal enzyme, the *Escherichia coli* acetyl-CoA carboxylase is readily dissociated into its subunit structures which have been studied in great detail in the laboratories of both Vagelos and Lane [25–28]. Three distinct functional subunits have been isolated: (a) the biotin carboxylase (BC), (b) the biotin carboxyl carrier protein (BCC), and (c) the carboxyl transfer protein (CT). The biotin carboxylase, which is a dimer of 100,000 daltons, catalyzes the carboxylation of the biotin prosthetic group on the biotin carboxyl carrier protein of molecular weight 22,000 daltons [Eq. (1)]. This subunit can also catalyze the carboxylation of free biotin in the absence of the BCC and CT proteins. The carboxyl transfer protein, which is also a dimer of 90,000 daltons, effects the carboxyl transfer from the carboxylated BCC protein to acetyl-CoA, forming malonyl-CoA [Eq. (2)]. It can also catalyze the carboxylation of free biotin in the absence of BCC and BC proteins with malonyl-CoA as the carboxyl donor. The BC and CT proteins therefore both possess distinct binding sites for biotin.

2. PYRUVATE CARBOXYLASE

Although a large body of evidence had accumulated to indicate the addition of carbon dioxide to pyruvic acid by both bacterial and animal cells, it was not until 1960 that Utter and Keech isolated an enzyme from avian liver extracts that catalyzed the formation of oxalacetic acid from pyruvic acid CO_2, ATP, and Mg^{2+} [29]. The enzyme is inhibited by avidin, indicating the presence of a biotin prosthetic group and also has an absolute requirement for acetyl-CoA which does not enter into the reaction. The short chain acyl-CoA derivatives are activators for the animal enzymes, with acetyl-CoA being the most effective. The long chain fatty acyl-CoA derivatives are the more effective activators for the yeast enzyme, but an absolute requirement is not observed. Manganese is an integral part of all animal enzymes, whereas the yeast enzyme requires zinc [30]. The acetyl-CoA appears to function only in the carboxylation reaction [Eq. (1)] while the metal functions primarily in the carboxyl transfer reaction [Eq. (2)].

The avian liver enzyme with a molecular weight of 660,000 contains 4 biotin prosthetic groups and 4 atoms of manganese indicating a tetrameric structure. This is compatible with the fact that the enzyme dissociates into protomers of 165,000 daltons and electron micrographs of the purified enzyme preparation show the protomers to be arranged at the corners of a square. On treatment with sodium dodecyl sulfate, the protomers are broken down into subunits of 45,000 daltons. No information is presently available concerning the nature of the subunits, but it is surmised that they are different and possess the functional characteristics observed with the E. coli acetyl-CoA carboxylase, which would be in accord with the proposed reaction mechanism.

The chicken liver enzyme has the interesting property of dissociating into the protomeric form in the cold and reaggregating again when the temperature is raised. The loss and gain of activity parallels the decrease and increase of the temperature. Acetyl-CoA, the positive effector, protects the enzyme against the cold lability. The activation of pyruvate carboxylase by acetyl-CoA is instantaneous and does not involve an aggregation process as was observed in the activation of mammalian acetyl-CoA carboxylase. However, the enzyme becomes more susceptible to avidin inhibition in the presence of acetyl-CoA, suggesting a possible conformational change which exposes the biotin prosthetic group. This is in contrast with the effect observed with citrate in the mammalian acetyl-CoA carboxylase activation.

The pyruvate carboxylase functions as an essential enzyme in gluconeogenesis by funneling the carbon skeletons of alanine, serine, and cysteine, as well as lactic acid, into the formation of phosphoenolpyruvic acid via oxalacetic acid. The enzyme levels fluctuate with the gluconeogenic rate which, in turn, is markedly influenced by the physiological state of the animal. Any change due to an altered nutritional state or to a disease process which results in the mobilization of lipids for cellular oxidation will generate high levels of acetyl-CoA and thereby enhance gluconeogenesis. The administration of glucocoid hormones and glucagon also enhances gluconeogenesis, while adrenalectomy and insulin administration have the opposite effect as they depress acetyl-CoA accumulation.

3. METHYLMALONYL-CoA TRANSCARBOXYLASE

Wood and Werkman made their historic discovery of CO_2 fixation by a nonphotosynthetic organism, *Propionobacter*, during their study of the fermentation of glycerol to propionic acid [31]. In an extended series of studies on the reaction sequence involved in the fermentation process, the CO_2 fixation step was shown to involve the formation of methylmalonyl-CoA from succinyl-CoA. The transcarboxylase enzyme catalyzed the terminal step of the reaction sequence, namely, the transfer of the carboxyl group from methylmalonyl-CoA to pyruvic acid to form propionyl-CoA and oxalacetic acid [32,33].

The enzyme has been purified to homogeneity and contains a mixture of cobalt and zinc. The proportions of each of the metals apparently is determined by the trace metal composition of the growth medium. The enzyme has a molecular weight of 790,000 and dissociates into subunits in a complex manner according to the scheme shown in Fig. 4. The size of the subunits has been determined by sedimentation analysis and disc-gel electrophoresis, and in conjunction with electron microscopy it has been possible to gain some insight as to the organization of the subunits in the holoenzyme [34,35].

The intact enzyme appears in the electron micrographs as a large central core, the "head," surrounded on the periphery by three units called

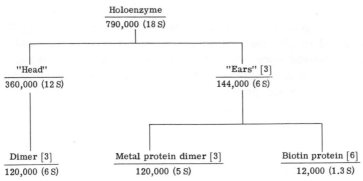

FIG. 4. Schematic representation of the subunit structure of methylmalonyl-CoA trans-carboxylase [35].

the "ears." Occasionally, the central core is observed free of its units or associated with only one or two. The latter was initially referred to as the "Mickey Mouse" structure. The central core can be dissociated under mild conditions into three subunits, each of which is a dimer of probably identical polypeptide chains. The ears are made up of two nonidentical subunits, one containing the biotin prosthetic group and the other containing the firmly bound cobalt or zinc. The metal protein subunit is a dimer of identical polypeptide chains, and in the ear structure is associated with two of the biotin containing proteins. Thus a composite picture of the holoenzyme is that of a central core trimer with each subunit associated with one of the ear structures. The activities of the component parts have not as yet been determined. It is surmised that the biotin protein may have the same function as the BCC subunit of *E. coli* acetyl-CoA carboxylase. The metals have been shown to be involved in the second step of the reaction and therefore this protein may be equivalent to the carboxyl transferase protein, leaving the head protein involved in the carboxylation of biotin.

D. Enzyme Mechanisms

The two-step reaction sequence postulated for all biotin-dependent reactions was formulated on the basis of a carboxybiotin–enzyme complex which was isolated, the isotope exchange data from the partial reactions, and the kinetic studies with the purified enzymes. A "ping-pong" type of mechanism has been shown to apply to the acetyl-CoA carboxylase, the transcarboxylase, and the pyruvate carboxylase reactions, and may be applicable to all biotin-dependent enzymes. In a classical ping-pong mechanism, the products are released from the first step in the reaction sequence prior to binding of the substrate in the second step. It also implies that there are separate binding sites for the substrates of each partial reaction, as well as for the biotin prosthetic group. With the separation of the *E. coli* acetyl-CoA carboxylase into three subunits, it has been possible to demonstrate that the substrate binding sites are on subunits other than the biotin carrier protein. It has been suggested that the biotin prosthetic group functions by oscillating back and forth between the two different subunits. Although attached in covalent linkage, the valeric acid side chain of the biotin molecule is 14 Å long, enabling it to flip from the active site of the biotin carboxylase subunit where carboxylation occurs to the active site on the carboxyl transferase subunit where the carboxyl group is transferred to the acceptor substrate. This is depicted in Fig. 5.

The mechanism of the partial reaction at both active sites has been investigated in a number of laboratories. In the propionyl-CoA carboxylase reaction Kaziro *et al.* [36] found that the addition of $HC^{18}O_3^-$ resulted in the cleavage of ATP with the release of orthophosphate containing one atom of ^{18}O. The other two atoms of ^{18}O were incorporated into the carboxyl group of methylmalonyl-CoA. This distribution of ^{18}O established HCO_3^- rather than CO_2 as the substrate for this reaction. This was confirmed with other biotin enzymes. The data from the ^{18}O experiment in conjunction with the results obtained from the isotope exchange experiments [Eq. (1)] supported a concerted mechanism for the carboxylation reaction as shown in Fig. 6. However, as pointed out by these investigators, the data do not rule out a two-step mechanism which produces in the first step carbonyl phosphate as an intermediate which, in turn, carboxylates the biotin prosthetic group in the second step.

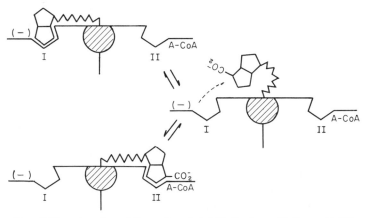

FIG. 5. Translocation of the carboxylated biotinyl prosthetic group [4].

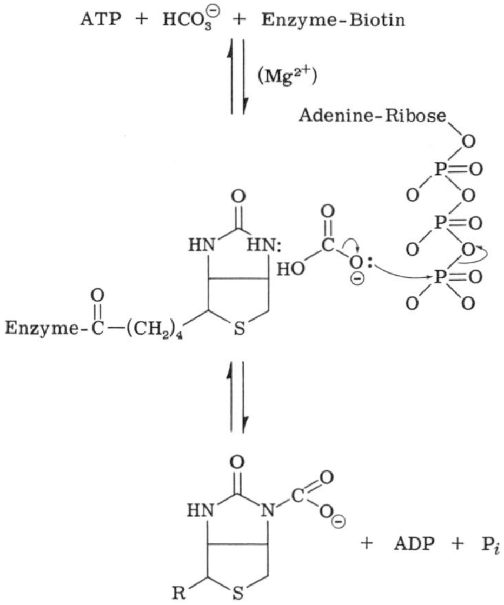

$$\text{ATP} + \text{HCO}_3^{\ominus} + \text{Enzyme-Biotin}$$

FIG. 6. A concerted mechanism for the carboxylation of the biotin prosthetic group [36].

In the carboxyl transferase reaction there are two types of carboxyl acceptors, the acyl-CoA compounds and the keto acids. All biotin-dependent enzymes with keto acids as substrates have tightly bound metal ions for substrate activation. Nuclear magnetic resonance studies with pyruvic carboxylase indicate that the bound manganese functions in activating the methyl hydrogens of pyruvic acid. An enzyme–manganese–pyruvate bridge complex has been established for this enzyme and a concerted mechanism has been proposed for this reaction as shown in Fig. 7(a). The binding of pyruvate as one of the metal ligands activates the α hydrogens. With the abstraction of the activated hydrogen, there is a simultaneous carboxyl transfer from the $1'$-N-carboxy-biotin to the methyl carbon of pyruvic acid [37]. A similar mechanism has also been proposed for the transcarboxylase reaction which requires zinc or cobalt for activity.

When the acyl-CoA derivative is the acceptor, there is also an activation of a hydrogen atom. Investigations of certain carboxylase reactions show the release of an α hydrogen as measured by tritium exchange, either with tritiated water or substrate. The greater lability of the α hydrogen of β-ketothioesters as compared to β-ketoesters is well known. In those carboxylation reactions where the γ position is carboxylated,

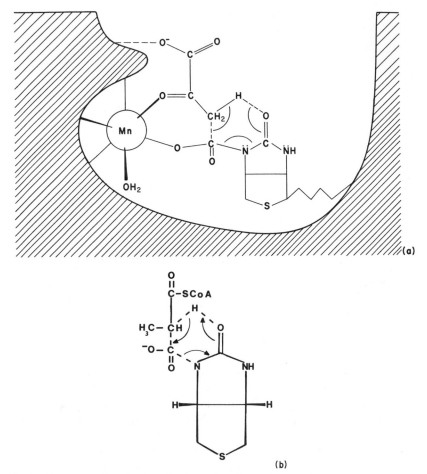

FIG. 7. A concerted mechanism for the carboxyl transfer reaction with (a) pyruvate carboxylase, (b) propionyl-CoA carboxylase [4].

the γ hydrogen is probably activated by the thioester carbonyl group through conjugation with the double bond. In the tritium exchange experiments the presence of the carboxylated biotin prosthetic group is an absolute requirement. The hydrogen abstraction and the carboxylation occur on the same side as evidenced by the fact that the propionyl-CoA carboxylase reaction proceeds with retention of configuration. The evidence supports a concerted mechanism for this reaction as indicated in Fig. 7(b).

IV.   BIOTIN BIOSYNTHESIS

Both pimelic acid and dethiobiotin, degradative products in the structural analysis of biotin, were implicated early in the biogenesis of biotin.
The pimelic acid requirement for the growth of the diphtheria bacillus
could be replaced by biotin. In addition, when pimelic acid was added to
the growth medium of a variety of organisms, it enhanced the excretion
of biotin vitamers, compounds having biotin-like activity. Dethiobiotin
could replace the biotin requirement of yeast, which converted it into
bound biotin, and it also accumulated in the growth medium of biotin
auxotrophs [38,39]. From these early observations the following simplified pathway could be inferred:

$$\text{Pimelic acid} \rightarrow \text{dethiobiotin} \rightarrow \text{biotin} \tag{6}$$

A. Role of Pelargonic Acid Derivatives

An intermediary role for pimelic acid was established simultaneously
by Eisenberg [40] and Elford and Wright [41] with the aid of [1,7-
^{14}C]pimelic acid. The isotope was incorporated into the major excretion
products: biotin and dethiobiotin with *Phycomyces blakesleeanus* and
biotin *l*-sulfoxide with *Aspergillus niger*. The culture filtrate of *P. blakesleeanus* also contained an unknown vitamer which was avidin-uncom-
binable and therefore did not possess the ureido structure which is
essential for avidin binding. Electrophoretic studies established the presence of both a carboxyl and an amino group which appeared to be
widely separated on the basis of their apparent pK_a values. The failure
of the vitamer to incorporate [^{35}S]sulfate eliminated the presence of a
tetrahydrothiophene ring and therefore signified an open chain structure
for the vitamer. That this vitamer may be an intermediate in biotin
biosynthesis was suggested by the fact that it became labeled along with
dethiobiotin and biotin when [1,7-^{14}C]pimelic acid was added to the
growth medium. This would be in accord with the requirement that all
intermediates between [^{14}C]pimelic acid and biotin should be labeled. In
support of this suggestion was the finding that the purified ^{14}C-labeled
vitamer, when added to a growing yeast culture, was directly incorporated into the bound biotin [42]. An early hypothesis to account for the
chemical properties of the vitamer was that it may have arisen as the
result of a condensation of pimelic acid and a three carbon unit, possibly serine or alanine, in a manner similar to the formation of δ-
aminolevulinic acid from succinyl-CoA and glycine. Alanine appeared to
be the candidate of choice when Iwahara *et al.* [43] found that the

dethiobiotin production in *Bacillus sphaericus* was further increased on the addition of alanine with pimelic acid. They subsequently identified the vitamer as 7-keto-8-aminopelargonic acid (7-KAP) by paper chromatography. Eisenberg and Maseda [44] isolated the vitamer in crystalline form from filtrates of *Penicillium chrysogenum* and characterized it by its physical and biological properties.

7-KAP was not the first pelargonic acid derivative to be implicated in biotin biosynthesis. It was very early observed that the growth of yeast and *Aspergillus nidulans* could be supported by 7,8-diaminopelargonic acid (DAPA), and a number of pelargonic acid derivatives, including 7-KAP, could replace biotin to various degrees for the growth of *Brevibacterium*, the biotin auxotroph used for monosodium glutamate fermentation [45]. However, all evidence was still indirect. A more direct approach to the elucidation of the biotin pathway was provided by the mutant technique. The isolation of biotin auxotrophs by chemical mutagenesis enabled two groups of investigators to establish four nutritional groups based on cross-feeding experiments [46,47]. The unequivocal identification of the excretion products in the culture filtrates of each group in conjunction with the evidence obtained from the cross-feeding experiments permitted Rolfe and Eisenberg to propose the biosynthetic pathway for biotin shown in Fig. 8, where the Roman numerals designate the nutritional groups. A mutation in any step of the sequence

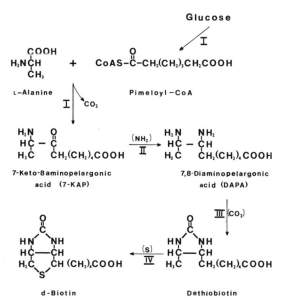

FIG. 8. Proposed mechanism for biotin biosynthesis based on mutant technique [47].

results in the excretion of products prior to the block. Therefore the group I mutants excrete no biotin vitamers; group II only 7-KAP; group III both 7-KAP and DAPA; and group IV 7-KAP, DAPA, and dethiobiotin. Evidence in support of this scheme was obtained with the study of a number of the purified enzymes.

B. Enzymatic Reactions

1. PIMELIC ACID FORMATION

This step in biotin biosynthesis has not been very extensively explored, but evidence from a number of sources would suggest three possible routes: (1) *de novo* synthesis from a common intermediate, probably acetyl-CoA; (2) C_2 degradation of a higher homologue; and (3) C_2 addition to a lower homologue. Lezius *et al.* [48] proposed a *de novo* synthesis involving the condensation of three malonyl-CoA units. This suggestion was based only on the labeling pattern of the biotin molecule obtained from *Achromobacter* cells grown in the presence of $^{14}CO_2$ with isovaleric acid as a carbon source. The formation of pimelic and azelaic acids from the oxidation of long chain hydrocarbons was found to arise by the β-oxidation of dicarboxylic acid intermediates [49]. Similarly, the ability of azelaic acid to replace pimelic acid in stimulating biotin biosynthesis was also attributed to β-oxidation. Ogata *et al.* [50] found that of several odd chain dicarboxylic acids, only glutaric acid appeared to be essential for biotin biosynthesis by strains of *Agrobacterium*, while pimelic acid was inactive. Although these investigators claimed that this may be indicative of a new pathway for biotin biosynthesis, the possibility was not ruled out that this organism can activate glutaric but not pimelic acid to the CoA derivative and subsequently add a C_2 unit. The activation of pimelic acid has only been demonstrated by coupling the activation reaction with the formation of 7-KAP [6].

2. 7-KAP SYNTHETASE

The synthesis of 7-KAP from pimelyl-CoA and alanine was first demonstrated by Eisenberg and Star [51] in a cell-free system. The reaction required pyridoxal phosphate as a cofactor and resembled the condensation reaction of succinyl-CoA and glycine in the formation of δ-aminolevulinic acid. The enzyme has been purified about 300-fold and is extremely unstable. A molecular weight of 45,000 has been estimated for the enzyme by gel filtration. Specificity studies show alanine as the only substrate capable of yielding a biologically active product. The enzyme is present in all mutant groups except group I and therefore *bioF* cistron must code for the protein catalyzing this condensation reaction. (See Section IV,C.)

3. DAPA AMINOTRANSFERASE

The conversion of 7-KAP to DAPA was first demonstrated in *E. coli* by Pai [52] with a coupled cell-free system. Unable to determine the product of the reaction, DAPA, by microbiological assay, Pai coupled the formation of DAPA with the synthesis of dethiobiotin from DAPA. Methionine proved to be the most effective amino donor and pyridoxal phosphate was the required cofactor. The substrate 7-KAP showed inhibition at concentrations greater than 1.3×10^{-2} mM.

In a similar study on DAPA synthesis with resting cells of *E. coli*, Eisenberg and Stoner also found methionine to be the most effective amino donor [53]. In this study, DAPA was directly measured with a strain of *E. coli* which responds to low concentrations of DAPA. However, when methionine and 7-KAP were added to cell-free extracts, DAPA could not be detected. Since the resting cells responded to glucose, a requirement for an energy source was indicated. The addition of ATP and Mg^{2+} to the cell-free system resulted in synthesis of DAPA. The inability of methionine to function in the absence of ATP and Mg^{2+} suggested the possible formation of S-adenosylmethionine (SAM) as the amino donor. When SAM was added in place of optimal concentrations of ATP and L-methionine, it proved to be 10 times more effective in DAPA synthesis.

The enzyme has been purified to near homogeneity and has a molecular weight of 94,000 as determined by gel filtration. The specificity for S-adenosylmethionine is absolute while the 7-KAP analogues, 8-keto-7-aminopelargonic acid and 7,8-diketopelargonic acid, can also serve as substrates for this reaction, but are less active. The kinetics show a "ping-pong" type of mechanism. The enzyme is present in all mutant groups except group II and therefore *bioA* cistron is the gene which codes for the aminotransferase enzyme.

4. DETHIOBIOTIN SYNTHETASE

Studies with resting cells of a mutant unable to grow on dethiobiotin provided the first evidence for the conversion of DAPA into dethiobiotin. Compounds which were either an energy source or a source of CO_2 stimulated this reaction. In cell-free extracts Krell and Eisenberg [54] were able to demonstrate the requirement for CO_2, ATP, and Mg^{2+}, and the direct incorporation of $^{14}CO_2$ into the ureidocarbonyl group [6].

The enzyme, purified over 200-fold from *E. coli,* proved to be cold labile, losing over 90% of its activity when stored at $-20°$ overnight. Gel filtration and SDS gel electrophoresis showed the enzyme to be a dimer of about 42,000 daltons with a subunit of identical polypeptide chains of about 23,000 daltons. The stoichiometry of the reaction as determined with the aid of $H^{14}CO_3^-$ and $[8\text{-}^{14}C]ATP$ indicated the formation of one mole of dethiobiotin for each mole of ATP and HCO_3^- utilized. ADP, the end product of the reaction, was a competitive inhibitor of ATP. CO_2, rather than HCO_3^-, was unequivocally shown to be the substrate for this reaction. Carbamyl phosphate, which was active in the crude preparation in place of bicarbonate and ATP, was completely inert in the purified system. Diaminobiotin could partially replace DAPA with the formation of biotin which would account for its ability to support the growth of the lactobacilli whereas dethiobiotin does not. A mechanism was proposed for dethiobiotin synthesis which envisages a nonenzymatic reaction of DAPA with CO_2 to form a DAPA monocarbamate. Activation of the monocarbamate by ATP would give rise to a substituted carbamyl phosphate. This could cyclize as the result of a nucleophilic attack by the amino group with phosphate as the leaving group. This mechanism would explain the formation of two amide bonds at the expense of one ATP.

5. BIOTIN SYNTHETASE

Information on the mechanism of sulfur incorporation into the tetrahydrothiophene ring is extremely meager as the activity observed with resting cells is completely lost with the disruption of the cell by whatever means. Information from resting or growing cell suspensions has confirmed dethiobiotin as the substrate for this reaction. Niimura [55] made a study of potential sulfur donors in resting yeast suspensions and found the following to be the most effective sulfur donors: methionine sulfoxide, methionine, sodium sulfide, sodium bisulfite, and sodium sulfate. Cysteine, cystine, glutathione, and methionine sulfone, however, were poor donors. Ethionine was an effective inhibitor of the reaction, but this could be overcome with methionine. The sulfur of $[^{35}S]$methionine was incorporated into the covalently bound biotin but the specific activities of the isolated hydrolyzed products were not determined so that no conclusion can be drawn about the direct or indirect incorporation of the sulfur atom from methionine.

The antibiotic acidomycin, also called actithiazic acid, was shown to be a biotin antagonist in *Mycobacterium tuberculosis* [56]. When added to growing cultures of a variety of organisms, it increased the total biotin excretion. Ogata *et al.* [57] in an extensive study with yeast, molds, and bacteria showed that the true biotin excretion was markedly inhibited, and that the increase in total biotin was due primarily to the increase in dethiobiotin. They suggested that the site of action was between dethiobiotin and biotin. Kinetic studies by Eisenberg [6] using a biotin repressor mutant (R^-) of E. coli showed actithiazic acid to be a competitive inhibitor of dethiobiotin in its conversion to biotin in resting cells. In addition, wild-type cells grown on low concentrations of actithiazic acid showed elevated levels of the aminotransferase and dethiobiotin synthetase enzymes. This could be accounted for by the low cellular concentration of biotin produced under these conditions which are insufficient to repress the biotin operon.

C. Genetics and Control

Definitive mapping by complementation analysis [58,59] permitted the following ordering of the genes in the bioA locus:

$$\lambda \, Att \, A \, B \, E \, F \, G \, C \, D \tag{7}$$

This cluster of genes is located next to the λ attachment site at minute 18 on the E. coli map [60]. The enzymes coded for by the above cistrons are bioA–DAPA synthetase, bioB–biotin synthetase, bioF–7-KAP synthetase, bioC–unknown but prior to 7-KAP synthetase, bioD–dethiobiotin synthetase. The question whether E and G constitute separate cistrons is still unresolved. An unlinked gene, bioH, mapping in the maltose region at minute 66, is similar to bioC in its excretion pattern and may also code for an early function enzyme.

The biotin operon was shown to be under biotin control when it was observed by Pai and Lichstein that the total biotin excretion of E. coli decreased with increasing concentrations of biotin in the growth medium [61]. The conversion of dethiobiotin to biotin was inhibited over 90% in resting cells grown in the presence of 5 ng/ml of biotin. Since an active protein synthesizing system was essential for restoring the enzymatic activity, the control exerted by biotin appeared to be due to enzyme repression. The coordinate repression of the biotin enzymes was deduced from the simultaneous decrease in the excretion of all biotin vitamers. This was subsequently verified by studies with the individual enzymes in cell-free systems [62]. The negative control pattern observed was in accord with the Jacob–Monod model which predicts operator constitutive (O^c) and repressor (R^-) mutants. Both were isolated with the aid of the biotin analogue, α-dehydrobiotin [6]. The O^c mutants that mapped in the biotin operon were only half as sensitive to biotin as the parental strain. The R^- mutants showed no enzyme repression with concentrations of biotin up to 500 μM. High vitamer excretion was observed with the R^-

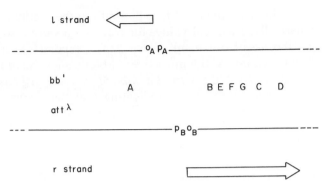

FIG. 9. Transcription map of the *bioA* locus. The genes are denoted by capital letters and arranged by complementation analysis [58,59] while the symbols p and o denote promotors and operators, respectively. The arrows indicate the orientation of transcription from the l and r strands of DNA [65].

mutant reaching levels 50–100 times the parental strain. A mutant with properties similar to the R^- was isolated by Pai and preliminary mapping placed it at minute 79 on the *E. coli* map [63]. Also mapping at minute 79 is the *bir* mutant isolated by Campbell *et al.* [64] which has properties similar to the R^- mutant. However, it is incapable of taking up biotin and requires high biotin concentration for optimal growth, although the biotin enzymes seem to be derepressed. The possibility that this may be a transport mutant has to be considered since a biotin transport system has been reported for *E. coli*.

The position of the operator and the promoter site could not be established by the early mapping studies. This information was recently provided by the hybridization studies of Guha *et al.* [65] which determined the DNA strand from which the biotin mRNA was transcribed. The evidence indicated that cistron *A* was transcribed to the left on the l strand and the remaining cistrons were transcribed to the right on the r strand as shown in Fig. 9. This was the first time that divergent transcription was observed within an operon. Whether there are two functional operators in the *bioA* operon is still an open question.

V.   BIOTIN DEGRADATION

A. Historical

The first study directed toward the metabolism of biotin was carried out by Krueger and Peterson with *Lactobacillus pentosus* grown in media of varying biotin concentrations [66]. As the biotin in the medium increased, the recovery of true biotin from the medium and cells fell off very markedly. Identical values for true biotin were also obtained with yeast as the assay organism. Since yeast responds to a broader spectrum of biotin vitamers, the unrecovered biotin must have been metabolized and not inactivated for the *Lactobacillus* assay organism. Oxybiotin had a fate similar to biotin, indicating that the enzyme system involved in the degradative process was insensitive to the oxygen substitution for sulfur.

The availability of [2-^{14}C]biotin permitted Fraenkel-Conrat and Fraenkel-Conrat [67] to study biotin degradation in mammals. Parenterally administered biotin produced only traces of radioactive carbon dioxide. About 85% of the radioactivity was excreted in the urine within 24 hours and of this only 40% would support the growth of yeast. Thus, while the ureido ring remained intact, the remainder of the molecule was so altered as to render the compound biologically inactive. On the other hand, the aerobic incubation of tissue slices from guinea pig kidney cortex was shown by Baxter and Quastel [68] to degrade the [^{14}C]carboxyl biotin with the loss of both radioactive CO_2 and biological activity. All attempts to obtain a cell-free system were unsuccessful, but it was possible to show that a functional TCA cycle was required. The addition of sodium malonate inhibited the formation of radioactive carbon dioxide while the addition of fumarate enhanced the loss of biological activity. A similar effect of fumarate had been previously noted by investigators in the area of fatty acid oxidation. However, the possible involvement of the fatty acid oxidizing enzymes in biotin degradation was ruled out when it was found that the biotin oxidation could be selectively inhibited. Although the two enzyme systems appeared not to be identical, they did possess certain similarities and Baxter and Quastel suggested a fatty acid oxidation mechanism for biotin degradation including the preliminary activation with CoA. Two steps in addition to the activation process are presently recognized in biotin degradation: β-oxidation of the valeric acid side chain and cleavage of the bicyclic ring structure.

B. Biotinyl-CoA

Experimental evidence to support the activation hypothesis was provided by Christner *et al.* [69] with an organism isolated by selective culture technique to grow on biotin as the major carbon source. The whole-cell and broken-cell preparations were capable of oxidizing carboxyl-labeled biotin to yield radioactive carbon dioxide. In addition, the broken-cell preparations formed a small amount of radioactive acetoacetic acid. Of a number of mammalian organ slices tried, the rat liver was the most effective, but still only 1/100 as active as the bacterial broken-cell preparation. In the latter system, the addition of ATP, biotin, and hydroxylamine resulted in the formation of biotin hydroxamic acid which was enhanced by the addition of coenzyme A. Both the purified bacterial and rat liver enzymes were shown to catalyze the ATP-^{32}PP$_i$ exchange which was dependent on biotin. The addition of synthetic biotinyl adenylate to the enzymes in the presence of coenzyme A yielded biotinyl-CoA and, in the presence of pyrophosphate, produced ATP. All evidence supported the reaction sequence for biotin activation shown below which is identical with carboxyl activation observed for fatty acids:

$$\text{Biotin} + \text{ATP} \rightleftarrows \text{biotinyl-AMP} + \text{PP}_i \qquad (8)$$

$$\text{Biotinyl-AMP} + \text{CoA} \rightleftarrows \text{biotinyl-CoA} + \text{AMP} \qquad (9)$$

The specificity for biotin was demonstrated by the separation of the biotin activating enzyme from the fatty acid activating system specific for octanoic acid. Neither the bacterial nor the liver enzymes possess biotin holoenzyme synthetase activity. Thus while there is a similarity in mechanism between the two biotin activation processes in that biotinyl-AMP is a common intermediate, in the holoenzyme synthetase the biotin is transferred to the apoenzyme, while in the degradative system it is passed onto coenzyme A.

C. β-Oxidation

Although the requirement for a functional TCA cycle, the activation mechanism, and the formation of acetoacetic acid pointed to the β-oxidation of the valeric acid side chain, no intermediates other than biotinyl-CoA were initially isolated and identified. The most definitive series of experiments for the isolation and identification of biotin metabolites were carried out in the laboratory of Wright [5,70–72]. A pseudomonad, isolated on a medium containing biotin as the sole carbon, nitrogen, and sulfur source, degraded biotin during growth to CO_2, ammonia, and hydrogen sulfide. The addition of [^{14}C]carbonyl and [^{14}C]carboxyl biotin to both broken- and whole-cell preparations resulted in a rapid and extensive loss in radioactivity. The broken-cell preparation required the addition of ATP, Mg^{2+}, NAD, and CoA for maximal activity. A number of biotin analogues proved to be moderately inhibitory for the biotin oxidase system. They also provided considerable information as to the substrate requirements when added to the preparation with either the carboxyl group or the carbonyl group labeled. The results shown in Table II indicate that the biotin d-sulfoxide is more active than

TABLE II

SPECIFICITY OF THE BIOTIN OXIDASE SYSTEM[a]

Substrate	CO$_2$ released relative to [^{14}C]carbonyl biotin	
	Carbonyl-labeled	Carboxy-labeled
Biotin	100	76
Biotin d-sulfoxide	114	80
Biotin l-sulfoxide	87	58
Biotin sulfone	5	1
Dethiobiotin	0	57
Oxybiotin	0	–
Norbiotin	0	–
Homobiotin	0	36
Biotinol	0	2
Diaminobiotin	–	1

[a] Data compiled from Brady et al. [71,72].

the l-sulfoxide. The d-sulfoxide is usually the more biologically active isomer and is found less frequently in culture filtrates than the l-sulfoxide. The degradation of the ureido ring is markedly reduced or completely eliminated if the sulfur is removed from the molecule, exchanged for an oxygen atom, or oxidized to the sulfone. An increase or decrease of the side chain by one methylene group also prevents the release of carbon dioxide. The above evidence indicated that the cleavage of the ureido ring is dependent upon the chain length of the side chain and the integrity of the sulfur ring. However, the oxidation of the valeric acid side chain is not as stringent in its requirements since carboxyl labeled homobiotin and dethiobiotin are oxidized by the particulate enzyme system.

The difference in specificity for ring cleavage and side chain oxidation afforded these investigators an opportunity to search for intermediates using large scale preparations with [^{14}C]carbonyl-labeled compounds. With homobiotin as the substrate, it was possible to isolate and identify both bisnorbiotin and trisnorbiotin, derivatives two and four carbons shorter than homobiotin. This provided the first supporting evidence for the hypothesis of β-oxidation of the valeric side chain. A similar degradation was observed with dethiobiotin in growing cultures of Asperigillus. Iwahara et al. [73] isolated bisnordethiobiotin while Wright's group isolated the bisnor- and the tetranordethiobiotin (4 carbons shorter). The ^{14}CO$_2$ formed in the latter experiment was exceedingly low, again supporting the requirement for an intact sulfur ring for ureido cleavage.

The first biotin catabolites isolated from the filtrates of a number of molds grown on high biotin concentrations were identified by Yang *et al.* [74] as bisnorbiotin and its sulfoxide. This was verified by Wright with carbonyl-labeled biotin using growing cultures of the pseudomonad and in addition they were also able to isolate and identify α-dehydrobisnorbiotin and tetranorbiotin. Two unsaturated intermediates would be expected from the β-oxidation of the valeric acid side chain. The first, α-dehydrobiotin, was previously isolated by Hanka *et al.* [75] from the culture filtrates of an *Actinomyces*.

The picture of β-oxidation was completed with the isolation of β-hydroxy derivatives and two keto compounds. One hydroxy derivative was obtained in a culture grown on [^{14}C]carbonyl biotin *l*-sulfoxide with a small amount of added unlabeled biotin. Normally the pseudomonad does not grow on the *l*-sulfoxide but in the presence of biotin it is readily oxidized. As soon as the biotin was utilized, the release of radioactivity ceased and β-hydroxybiotin *l*-sulfoxide accumulated. Biotin *d*-sulfoxide which can support the growth of the organism, however, produced only the metabolites observed with biotin. Therefore the *d*-sulfoxide was not directly oxidized as previously thought, but must first be reduced to biotin. The presence of a biotin *d*-sulfoxide reductase has been reported in a number of organisms and a recent genetic analysis in *E. coli* has indicated that the products of four genes are required for reduction [76]. Two additional hydroxy derivatives, β-hydroxybiotin and β-hydroxybiotin sulfone, were isolated along with two ketones, methyl bisnorbiotinyl ketone and methyl tetranorbiotinyl ketone. These ketones probably arose by nonenzymatic decarboxylation of the corresponding keto acids, β-ketobiotin and β-ketonorbisbiotin. Thus all possible intermediates in the β-oxidation of the side chain could be accounted for and presumably arose within the cell as the CoA derivative as shown in Fig. 10(a).

D. Ring Cleavage

In the study of biotin degradation with the growing bacterium, two radioactive products arising from the ring degradation were obtained for the first time. They were identified as urea with the same specific activity as the carbonyl-labeled biotin and uracil with only one seventh the specific activity. Since no other intermediates were found with an open chain structure, no conclusions could be drawn as to the mechanism of the bicyclic ring cleavage. However, the single label in the carbonyl group of biotin did not allow for the detection of cleavage products with the

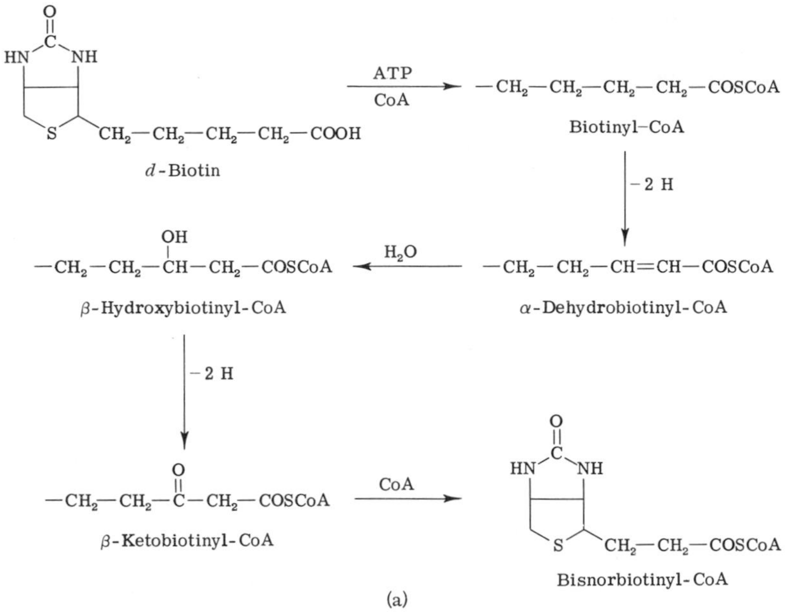

FIG. 10. Scheme for biotin degradation; (a) β-oxidation of the valeric acid side chain, (b) ureido-ring cleavage [77].

concomitant loss of the carbonyl label. Im *et al.* [77] recently used a mixture of carbonyl-labeled biotin and tritiated biotin and found, in addition to the intermediates previously described, a new compound, *d*-allobisnorbiotin. This is an isomer of *d*-bisnorbiotin in which the bridge hydrogens are in the trans configuration rather than the cis. The specific activity for tritium was identical with that for *d*-bisnorbiotin but the specific activity for ^{14}C was only about one seventh, equal to the uracil previously isolated. The latter indicated that the ureido ring structure had opened and reclosed, in the process losing the carbonyl label. The resynthesis of the ureido group from a diluted carbon dioxide pool could account for the dilution observed. In their earlier studies tetranorbiotin was considered the possible precursor for ring degradation, but the present findings supported bisnorbiotin as the precursor. To account for these results the scheme shown in Fig. 10(b) was proposed in which the bridge head carbon is oxidized at the level of *d*-bisnorbiotin with the subsequent decarbamylation to an α-aminoketone derivative which in turn can be recarbamylated or further degraded. Two oxidations on the two bridge head carbons would result in urea with the same specific activity of the carbonyl-labeled biotin.

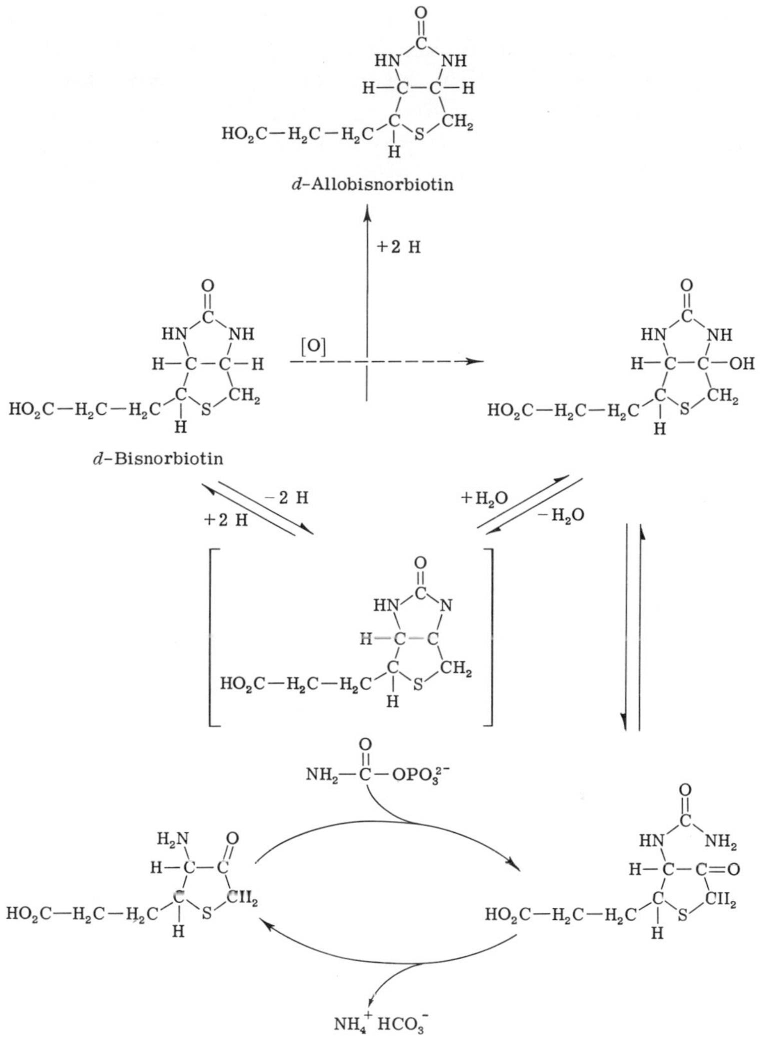

(b)

FIG. 10(b)

VI. CONCLUDING REMARKS

An attempt has been made to survey the extensive progress that has been made in our knowledge of biotin. Most of these developments have been made in the last twenty years and much still remains to be accomplished. In the biosynthesis of biotin, the incorporation of sulfur, the synthesis of pimeloyl-CoA, and the control mechanisms of the biosynthetic pathway require elucidation. In the degradative system, the study of the mechanism of ring cleavage has only begun. No information is available on any of the degradative enzymes so that their isolation, characterization, and control should prove fruitful areas of research. Studies of certain indirect effects of biotin deficiency have revealed new biochemical interrelationships and there are many others that remain to be explored. Finally, the possible existence of other biotin enzymes in living systems still remains a challenging question.

REFERENCES

1. F. Kögel and B. Z. Tönnis, *Hoppe-Seyler's Z. Physiol. Chem.* **242,** 43 (1936).
2. D. B. Melville, *Vitam. Horm.* (*New York*) **2,** 29 (1944).
3. S. J. Wakil, E. B. Titchener, and D. M. Gibson, *Biochim. Biophys. Acta* **29,** 225 (1958).
4. J. Moss and M. D. Lane, *Advan. Enzymol.* **35,** 321 (1971).
5. D. B. McCormick and L. D. Wright, *Compr. Biochem.* **21,** 81 (1971).
6. M. A. Eisenberg, *Advan. Enzymol.* **36,** 317 (1973).
7. J. Knappe, *Annu. Rev. Biochem.* **39,** 757 (1970).
8. J. J. Volpe and P. R. Vagelos, *Annu. Rev. Biochem.* **42,** 21 (1973).
9. D. B. Mellville, A. W. Moyer, K. Hofmann, and V. du Vigneaud, *J. Biol. Chem.* **146,** 487 (1942).
10. W. A. Harris, D. E. Wolf, R. Mozingo, R. C. Anderson, G. E. Arth, N. R. Easton, D. Heyl, A. N. Wilson, and K. Folkers, *J. Amer. Chem. Soc.* **66,** 1756 (1944).
11. S. A. Harris, R. Mozingo, D. E. Wolf, A. N. Wilson, and K. Folkers, *J. Amer. Chem. Soc.* **67,** 2096 (1945).
12. J. Trotter and J. A. Hamilton, *Biochemistry* **5,** 713 (1966).
13. C. Bonnemere, J. A. Hamilton, L. K. Steinrauf, and J. Knappe, *Biochemistry* **4,** 240 (1965).
14. L. D. Wright and H. R. Skeggs, *Proc. Soc. Exp. Biol. Med.* **56,** 95 (1944).
15. D. P. Kasow and M. D. Lane, *Biochem. Biophys. Res. Commun.* **7,** 439 (1962).
16. Y. Kaziro and S. Ochoa, *J. Biol. Chem.* **236,** 3131 (1961).
17. F. Lynen, J. Knappe, E. Lorch, G. Jütting, and E. Ringelman, *Angew. Chem.* **71,** 481 (1959).
18. J. Knappe, E. Ringelman, and F. Lynen, *Biochem. Z.* **335,** 168 (1961).
19. D. P. Kasow and M. D. Lane, *Biochem. Biophys. Res. Commun.* **5,** 191 (1961).
20. L. Siegel, J. L. Foote, J. E. Christner, and M. J. Coon, *Biochem. Biophys. Res. Commun.* **13,** 307 (1963).
21. M. D. Lane, K. L. Rominger, D. L. Young, and F. Lynen, *J. Biol. Chem.* **239,** 2865 (1964).
22. K. Hofmann and T. Winnick, *J. Biol. Chem.* **160,** 449 (1945).
23. P. R. Vagelos, A. W. Alberts, and D. B. Martin, *J. Biol. Chem.* **238,** 533 (1963).
24. A. W. Alberts and P. R. Vagelos, *Proc. Nat. Acad. Sci. U.S.* **59,** 561 (1968).
25. A. K. Kleinschmidt, J. Moss, and M. D. Lane, *Science* **166,** 1276 (1969).
26. A. W. Alberts, A. M. Nervi, and P. R. Vagelos, *Proc. Nat. Acad. Sci. U.S.* **63,** 1319 (1969).

27. P. Dimroth, R. B. Guchhait, E. Stoll, and M. D. Lane, *Proc. Nat. Acad. Sci. U.S.* **68,** 654 (1971).
28. P. Dimroth, R. B. Guchhait, and M. D. Lane, *Hoppe-Seyler's Z. Physiol. Chem.* **352,** 351 (1971).
29. M. F. Utter and D. B. Keech, *J. Biol. Chem.* **235,** PC17 (1960).
30. M. C. Scrutton, M. F. Utter, and A. S. Mildvan, *J. Biol. Chem.* **241,** 3480 (1966).
31. H. G. Wood and C. H. Werkman, *J. Bacteriol.* **30,** 332 (1935).
32. R. W. Swick and H. G. Wood, *Proc. Nat. Acad. Sci. U.S.* **46,** 28 (1960).
33. H. G. Wood, H. Lochmüller, C. Riepertinger, and F. Lynen, *Biochem. Z.* **337,** 247 (1963).
34. H. G. Wood, S. H. G. Allen, R. Stjernholm, and B. Jacobson, *J. Biol. Chem.* **238,** 547 (1963).
35. N. M. Green, R. C. Valentine, N. G. Wrigley, F. Ahmad, B. Jacobson, and H. G. Wood, *J. Biol. Chem.* **247,** 6284 (1972).
36. Y. Kaziro, L. F. Hass, P. D. Boyer, and S. Ochoa, *J. Biol. Chem.* **237,** 1460 (1962).
37. A. S. Mildvan, M. C. Scrutton, and M. F. Utter, *J. Biol. Chem.* **241,** 3488 (1969).
38. V. du Vigneaud, K. Dittmer, E. Hague, and B. Long, *Science* **96,** 186 (1942).
39. E. L. Tatum, *J. Biol. Chem.* **160,** 455 (1945).
40. M. A. Eisenberg, *Biochem. Biophys. Res. Commun.* **8,** 437 (1962).
41. H. L. Elford and L. D. Wright, *Biochem. Biophys. Res. Commun.* **10,** 373 (1963).
42. M. A. Eisenberg and R. Maseda, *Biochem. J.* **101,** 598 (1966).
43. S. Iwahara, M. Kikuchi, T. Tochikura, and K. Ogata, *Agr. Biol. Chem.* **30,** 304 (1966).
44. M. A. Eisenberg and R. Maseda, *Biochemistry* **9,** 108 (1969).
45. S. Okumura, R. Tsugawa, T. Tsunoda, and S. Matozaki, *J. Agr. Chem. Soc. Jap.* **36,** 605 (1962).
46. A. del Campillo-Campbell, G. Kayajanian, A. Campbell, and S. J. Adhya, *J. Bacteriol.* **94,** 2065 (1967).
47. B. Rolfe and M. A. Eisenberg, *J. Bacteriol.* **96,** 515 (1968).
48. A. Lezius, E. Ringelman, and F. Lynen, *Biochem. Z.* **336,** 510 (1963).
49. A. S. Kesten and J. W. Foster, *J. Bacteriol.* **85,** 859 (1963).
50. K. Ogata, Y. Izumi, and K. Tani, *Agr. Biol. Chem.* **37,** 1087 (1973).
51. M. A. Eisenberg and C. Star, *J. Bacteriol.* **96,** 1291 (1968).
52. C. H. Pai, *J. Bacteriol.* **105,** 793 (1971).
53. M. A. Eisenberg and G. Stoner, *J. Bacteriol.* **108,** 1135 (1971).
54. K. Krell and M. A. Eisenberg, *J. Biol. Chem.* **245,** 6558 (1970).
55. T. Niimura, T. Suzuki, and Y. Sahashi, *J. Vitaminol.* (*Kyoto*) **10,** 231 (1964).
56. W. M. McLamore, W. E. Celmer, V. V. Bogert, F. C. Pennington, and I. A. Solamons, *J. Amer. Chem. Soc.* **74,** 2946 (1952).
57. K. Ogata, Y. Itzumi, and K. Tani, *Agr. Biol. Chem.* **37,** 1079 (1973).
58. B. Rolfe, *Virology* **42,** 643 (1970).
59. P. Cleary and A. Campbell, *J. Bacteriol.* **112,** 830 (1972).
60. A. L. Taylor and C. D. Trotter, *Bacteriol. Rev.* **38,** 504 (1972).
61. C. H. Pai and H. C. Lichstein, *Biochim. Biophys. Acta* **65,** 163 (1962).
62. M. A. Eisenberg and K. Krell, *J. Biol. Chem.* **244,** 5503 (1969).
63. C. H. Pai, *J. Bacteriol.* **112,** 1280 (1972).
64. A. Campbell, A. del Campillo-Campbell, and R. Chang, *Proc. Nat. Acad. Sci. U.S.* **69,** 676 (1972).
65. A. Guha, Y. Saturen, and W. Szybalski, *J. Mol. Biol.* **56,** 53 (1971).
66. K. K. Krueger and W. H. Peterson, *J. Bacteriol.* **55,** 693 (1948).
67. J. Fraenkel-Conrat and H. Fraenkel-Conrat, *Biochim. Biophys. Acta* **8,** 66 (1952).

68. R. M. Baxter and J. H. Quastel, *J. Biol. Chem.* **201,** 751 (1953).
69. J. E. Christner, M. J. Schlesinger, and M. J. Coon, *J. Biol. Chem.* **239,** 3997 (1964).
70. R. N. Brady, L. F. Li, D. B. McCormick, and L. D. Wright, *Biochem. Biophys. Res. Commun.* **19,** 777 (1965).
71. R. N. Brady, H. Ruis, D. B. McCormick, and L. D. Wright, *J. Biol. Chem.* **241,** 4717 (1966).
72. H. Ruis, R. N. Brady, D. B. McCormick, and L. D. Wright, *J. Biol. Chem.* **243,** 547 (1968).
73. S. Iwahara, S. Takasawa, T. Tochikura, and K. Ogata, *Agr. Biol. Chem.* **30,** 1069 (1966).
74. H. C. Yang, T. Kusumoto, T. Tochikura, and K. Ogata, *Agr. Biol. Chem.* **34,** 370 (1970).
75. L. J. Hanka, M. E. Bergy, and R. B. Kelly, *Science* **154,** 1667 (1966).
76. D. Dykhuizen, *J. Bacteriol.* **115,** 662 (1973).
77. W. A. Im, D. B. McCormick, and L. D. Wright, *J. Biol. Chem.* **248,** 7798 (1973).

CHAPTER 3

Thiamine, Biosynthesis and Function

Irwin G. Leder

I. INTRODUCTION*

Unlike nicotinic and ascorbic acids, which were well-known organic compounds before there was any interest in them as nutritional factors, thiamine was discovered in the course of a search for an agent that would cure beriberi. Takaki [1] eliminated the disease from the Japanese navy in 1884 by introducing meat and legumes into the diet, but it was the study of experimental polyneuritis in fowl by Eijkman [2] that led to the realization that the curative agent was active in extremely minute quantities. Following the isolation of 3–4 gm of semicrystalline "antineuritic" material from 100 kilos of rice bran by Jansen and Donath [3] in 1926, improved isolation procedures culminated in the synthesis of thiamine a decade later by Williams and Cline [4] and Andersag and Westphal [5].

---

* Abbreviations used in text: AIR, 4-aminoimidazole ribonucleotide; AIR $_s$, 4-aminoimidazole ribonucleoside; hydroxyethylthiazole, 4-methyl-5-(β-hydroxyethyl)thiazole; hydroxymethylpyrimidine, 2-Methyl-4-amino-5-hydroxymethyl-pyrimidine; TMP, thiamine monophosphate; TDP, thiamine diphosphate; TTP, thiamine triphosphate.

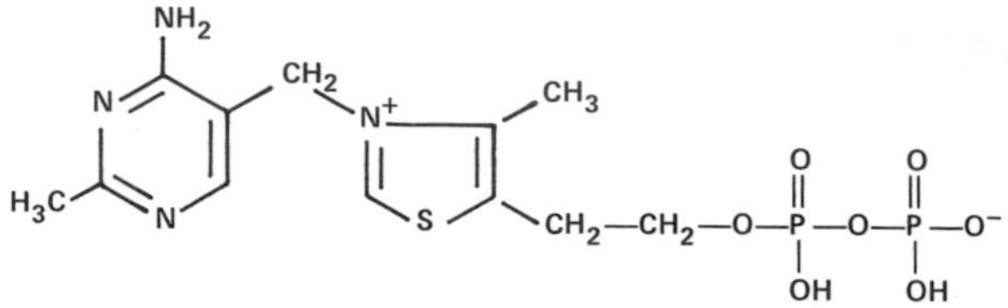

FIG. 1. Thiamine pyrophosphate.

Within a year, Lohmann and Schuster [6] isolated thiamine pyrophosphate (Fig. 1) from yeast and showed that it was the cofactor for the decarboxylation of pyruvic acid [7]. The triphosphate form, discovered in rat liver in 1952 [8] and subsequently in yeast [9] and plant tissues [10], is devoid of cocarboxylase [11] and has no established function in metabolism. In addition to TDP, which represents approximately 80% of total thiamine, free thiamine (4%), TMP (9%), and TTP (7%) are also found in animal tissues. The highest concentration of TTP is found in the liver [12] (Table I).

II. BIOSYNTHESIS OF THIAMINE

A. Formation of Thiamine from Hydroxymethylpyrimidine and Hydroxyethylthiazole

Soon after the elucidation of the structure [13] and chemical synthesis [4] of thiamine, nutritional studies with the fungus *Phycomyces blakesleeanus* [14,15] and excised pea roots [16] revealed that the thiamine requirements of these organisms could be satisfied by a mixture of the hydroxymethylpyrimidine and hydroxyethylthiazole moieties. Using a test organism, *Phytopthora cinnamomi,* that requires intact thiamine, Bonner and Buchman [17] showed that the pyrimidine and

TABLE I

DISTRIBUTION OF THIAMINE AND THIAMINE PHOSPHATES IN RAT TISSUES[a]

Substance	Percent of total thiamine in tissue			
	Brain	Heart	Kidney	Liver
Thiamine	4	2	6	4
TMP	11	6	10	9
TDP	80	86	79	78
TTP	5	6	5	9

[a] From Rindi and de Giuseppe [12].

thiazole components were converted to thiamine by excised pea roots. These early experiments correctly indicated that the biosynthesis of thiamine would involve the independent formation of the two ring structures and their subsequent condensation (Fig. 2).

From studies of the mechanism of action of thiaminase [18] and, in particular, of the synthetic reactions catalyzed by one form of this enzyme, Woolley [19] suggested that thiamine by virtue of its quaternary "onium" structure was a relatively high energy compound whose synthesis would require coupling to other energy yielding reactions. In the chemical synthesis of thiamine, the activated form of the pyrimidine which spontaneously quaternizes the thiazole to form the vitamin is the

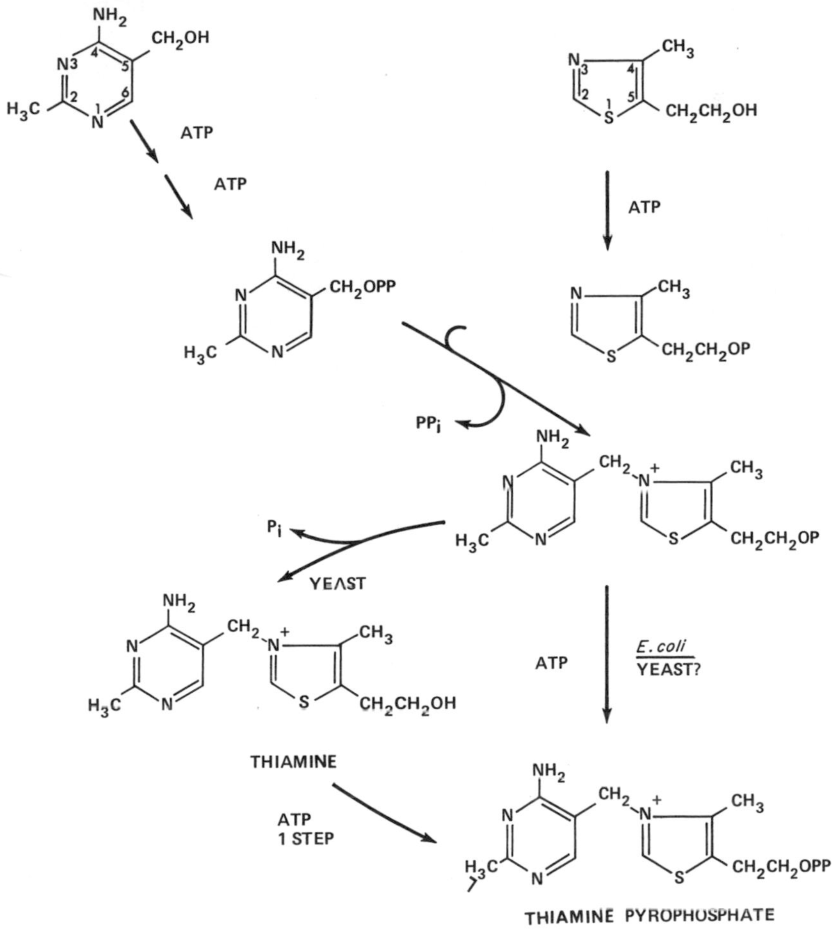

FIG. 2. The biosynthesis of thiamine pyrophosphate.

2-methyl-4-amino-5-bromomethylpyrimidine. The first clue to the nature of the corresponding biochemical activation was the report by Harris and Yavit [20] that the synthesis of the vitamin by cell-free extracts of bakers' yeast proceeded more rapidly and without the lag period previously observed, when the hydroxymethylpyrimidine was replaced by the corresponding pyrimidyl monophosphate. These observations were confirmed by Leder [21], who found that the initial product of the coupling reaction was not free thiamine, but a phosphorylated form subsequently identified [22,23] as thiamine monophosphate. Working with synthetically prepared derivatives, Leder [24] and Nose and co-workers [25] showed that the reactants in the ATP-independent quaternization of the hydroxyethylthiazole were hydroxyethylthiazole monophosphate and hydroxymethylpyrimidine pyrophosphate. Camiener and Brown [26] found that these phosphorylated substrates, as well as free thiamine, thiamine pyrophosphate, and hydroxymethylpyrimidine monophosphate were formed in partially purified extracts of bakers' yeast. Using ^{35}S-labeled thiazole, Suzuoki and Kobata [27] showed that thiazole monophosphate was the sole product of the enzymatic activation of the thiazole moiety by bakers' yeast.

In principle, hydroxymethylpyrimidine monophosphate might be formed from the pyrophosphate by the action of a phosphatase. However, the fact that the rate of thiamine synthesis was increased when hydroxymethylpyrimidine monophosphate replaced hydroxymethylpyrimidine [28,29] implied that the monophosphate is an intermediate in the synthesis of the pyrophosphate. Lewin and Brown [30] showed that hydroxymethylpyrimidine pyrophosphate is formed by two successive phosphorylations, each catalyzed by separate enzymes having significantly different properties. The first phosphorylation is catalyzed by hydroxymethylpyrimidine kinase (EC 2.7.1.49) which is stable at 55° and able to use CTP, GTP, or UTP, as well as ATP as the phosphorylating agent. The enzyme catalyzing the second phosphorylation, phosphohydroxymethylpyrimidine kinase (EC 2.7.4.7), is heat labile and highly specific for ATP.

It is noteworthy that in contrast to the acid stability of the phosphate groups of thiazole phosphate and the α-phosphate of cocarboxylase, both phosphate groups of hydroxymethylpyrimidine pyrophosphate are acid labile [24]. This acid lability reflects the allylic nature of the pyrophosphorylated pyrimidine carbon atom and accounts for the reactivity of this compound in the biosynthetic coupling reaction, as well as the reactivity of the corresponding pyrimidine methyl bromide in the chemical synthesis of the vitamin. Similar allylic pyrophosphate displacements are probably involved in the biosynthesis of folic acid

[31,32] and in the synthesis of squalene, rubber, and higher terpenes [33,34].

The four reactions leading to the synthesis of thiamine monophosphate are not restricted to yeast, but have been demonstrated in *Escherichia coli* [35], as well as in several varieties of plant leaves [36,37]. The coupling enzyme, thiaminephosphate pyrophosphorylase (EC 2.5.1.3), has been purified from *E. coli* and has been crystallized from bakers' yeast [38]. The yeast enzyme is heat stable at 50° and has a molecular weight of approximately 340,000. In contrast, the bacterial enzyme is completely inactivated at 45° and has a molecular weight of approximately 17,000. The enzyme from yeast is inhibited noncompetitively by pyrophosphate [24] and ATP [39]. The kinetics of inhibition by ADP, acetyl phosphate, phosphoenolpyruvate, and phosphocreatine have not been explored [39]. The similar inhibition of the *E. coli* enzyme by ATP and acetyl phosphate appears to be uncompetitive [40]. It has been suggested that this inhibition by high energy compounds may be a form of feedback inhibition indirectly controlling the level of ATP by decreasing cellular concentration of cocarboxylase. A similar explanation has been suggested [41] for the role of ATP in the activation of thiaminepyrophosphate phosphohydrolase in rat liver.

The product of the coupling reaction, thiamine monophosphate, does not function as a coenzyme in carbohydrate metabolism and therefore must be converted to thiamine pyrophosphate. In his studies of thiamine pyrophosphate synthesis in yeast extracts, Weil-Malherbe [42] concluded that thiamine pyrophosphate was formed by a process of transpyrophosphorylation since free thiamine was a far better substrate than thiamine monophosphate. This was convincingly shown by Shimazono and co-workers [43] using ATP labeled with ^{32}P in the γ position. The enzyme, thiamine pyrophosphokinase (EC 2.7.6.2), was purified approximately 100-fold [44] and is completely inactive toward thiamine monophosphate. The existence of this enzyme in yeast, as well as in mammalian tissues [45,46], and the presence in yeast of phosphatase activity which readily converts thiamine monophosphate to free thiamine, has led to the view [47] that free thiamine is an obligate intermediate between thiamine monophosphate and thiamine pyrophosphate. Nevertheless, Steyn-Parvé [48] and Kiessling [49] found evidence for a separate thiaminephosphate kinase in yeast and the partial purification of such an enzyme from bakers' yeast has been reported by Tokuda [50]. Thus, despite the evidence adduced in cell-free systems for the one-step synthesis of cocarboxylase from free thiamine in yeast, the product of the coupling reaction, thiamine monophosphate, may to some extent be directly converted to the coenzyme in the intact yeast cell.

In *E. coli* there is strong biochemical and genetic evidence that free thiamine is not an obligate intermediate in the synthesis of cocarboxylase. Separate enzymes catalyzing the synthesis of TMP from thiamine [51] and TDP from TMP [52] have been demonstrated in extracts of *E. coli* and the latter enzyme, thiamine monophosphate kinase, has been purified [53] approximately 100-fold. The enzyme requires Mg, although Co, Mn, and Zn are partially effective, and is stimulated by a number of monovalent ions.

The excellent experiments of Nakayama and Hayashi [54,55] show that cocarboxylase is synthesized by two successive transphosphorylations. These workers isolated *E. coli* mutants which require phosphorylated forms of thiamine and cannot utilize the free vitamin. One mutant can meet this requirement only with thiamine pyrophosphate. Intact cells of this mutant accumulate thiamine monophosphate when incubated with thiamine, and the presence of thiamine kinase, but not thiaminephosphate kinase, is demonstrable in cell-free extracts. The second mutant grows on either thiamine monophosphate or thiamine pyrophosphate. In this case, cells accumulate free thiamine and cell-free extracts can phosphorylate thiaminephosphate, but not free thiamine. These studies indicate that the synthesis of thiamine pyrophosphate *de novo* or from exogenously supplied thiamine proceeds in *E. coli* with thiamine monophosphate as an intermediate. Evidence for the existence of thiamine pyrophosphokinase in bacterial systems is not compelling. Although the conversion of thiamine to thiamine pyrophosphate by extracts of *Staphylococcus aureus* [56] and by a membrane bound system in *E. coli* [57] seems to suggest the presence of the pyrophosphokinase, a role for thiamine monophosphate in these transformations has not been excluded.

B. Biosynthesis of Hydroxymethylpyrimidine

Despite the close structural relationship between the pyrimidine of thiamine and the pyrimidines of nucleic acid, they do not share a common biosynthetic pathway. Nucleic acid pyrimidines are formed from aspartic acid and carbamyl phosphate through the intermediate orotic acid [58] and none of the carbon atoms are derived from formate. However, [14C]orotic acid is not incorporated into the thiamine pyrimidine moiety [59,60] and formate is the most effective precursor [61–63].

Two phenomena appeared to link the biosynthesis of thiamine to that of the purines: the existence of a group of single site mutations in *E. coli* exhibiting a double requirement for thiamine and purines [64], and the ability of high concentrations of adenine or adenosine to inhibit the synthesis of hydroxymethylpyrimidine [65]. Newell and Tucker [66] studied the effect of adenosine in *Salmonella typhimurium* and found that a secondary consequence of its effect on hydroxymethylpyrimidine synthesis was the derepression of the biosynthetic pathway leading to thiamine. As a result, washed cell suspensions, preincubated with adenosine, synthesize large amounts of thiamine and excrete excess hydroxymethylpyrimidine into the medium. Using amino acid auxotrophs Newell and Tucker found that both methionine and glycine were essential for the synthesis of hydroxymethylpyrimidine [67]. Consistent with earlier observations [59], they found that despite the requirement for methionine, none of the carbon atoms was converted to hydroxymethylpyrimidine. However, both [2-^{14}C]glycine and [1-^{14}C]glycine were incorporated with insignificant radioactive dilution. Since both glycine carbon atoms are used in the biosynthesis of purines, this result clearly indicated that nucleic acid purines and hydroxymethylpyrimidine might share a common origin.

To examine this relationship, Newell and Tucker [68] isolated a large number of adenine mutants and found that those blocked before the synthesis of AIR had a double requirement for adenine and thiamine, and all mutants blocked after this step did not require thiamine. They then demonstrated that AIR$_s$ would overcome the double requirement for adenine and thiamine and that ^{14}C-labeled AIR$_s$ was converted to hydroxymethylpyrimidine in a glycine auxotroph with no significant change in specific activity. The results of these excellent studies clearly establish that in *S. typhimurium,* AIR marks the branch point for the two pathways leading to purines and hydroxymethylpyrimidine as shown in Fig. 3.

The conversion of AIR to hydroxymethylpyrimidine requires the removal of the ribose phosphate, the addition of two single carbon substituents, and the opening of the imidazole ring to add an extra carbon atom. Knowing which specific carbon atoms of the imidazole and pyrimidine rings are derived from common precursors serves to indicate how this process may be carried out. David *et al.* [69] had shown that in yeast formate is incorporated into C-4 of the hydroxymethylpyrimidine. On the basis of this observation Newell and Tucker [68] proposed an ingenious transformation of the imidazole to the pyrimidine which met this constraint. However, Kuomoka and Brown [70] subsequently reported that in *E. coli,* formate is the precursor of C-2 of the pyrimidine ring. This result has been confirmed in the same strain of *S. typhimurium* used by Newell and Tucker and it has also been established [71,72] that the carboxyl and α-carbon atoms of glycine that

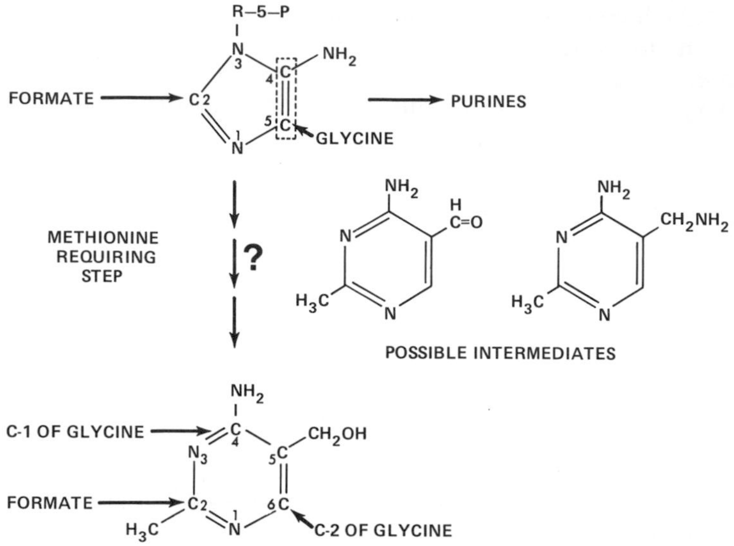

FIG. 3. The transformation of 5-aminoimidazole ribonucleotide to 2-methyl-4-amino-5-hydroxymethylpyrimidine.

are the precursors, respectively, of C-4 and C-5 of the aminoimidazole, become C-4 and C-6 of hydroxymethylpyrimidine. This establishes that the expansion of the aminoimidazole ring is accomplished by cleaving the ring and inserting the extra carbon atom between C-4 and C-5; no complex rearrangement of the carbon atoms of the aminoimidazole ring [68] is required.

Tomlinson et al. [73,74] have confirmed the observation of Nakamura [75] that aspartate is incorporated into hydroxymethylpyrimidine and may be the precursor of C-5 and its substitutent hydroxymethyl group. Although a direct role for aspartic acid is uncertain, the excretion of 5-aminomethyl and 5-formyl derivatives of the pyrimidine moiety by thiamine-requiring mutants of Neurospora and the transformation of these derivatives into hydroxymethylpyrimidine by an extract of yeast [76] suggest that an amino acid may be involved in the transformation of the aminoimidazole ring into the 6-membered pyrimidine ring.

The methyl group of acetate, but not the carboxyl group, appears to be a precursor of hydroxymethylpyrimidine [60]. In addition, pulse labeling experiments with [2-^{14}C]acetate and [2-^{14}C, Me-^{3}H]acetate [73,74] indicate that the methyl carbon and some of the attached hydrogen atoms are incorporated into the C-2 methyl group of hydroxymethylpyrimidine. Kuomoka and Brown [70] found that the incorporation of [2-^{14}C]acetate into hydroxymethylpyrimidine was markedly reduced in the presence of nonradioactive formate, thus raising a question as to whether acetate was indeed a direct precursor. However, the conditions used in this study, involving prolonged incubation with labeled substrate, would facilitate indirect incorporation of acetate and may not be comparable to the pulse labeling experiments of Tomlinson [73,74] in which the exposure of the cells to radioactive acetate was 15–20 minutes.

Newell and Tucker [68] observed that methionine was required for the synthesis of hydroxymethylpyrimidine from AIR_s. Since none of the carbon atoms of methionine are introduced in this process, they suggested that methionine in the form of S-adenosylmethionine might be a cofactor in this transformation. S-Adenosylmethionine may be involved in the generation of the C-2 methyl group from acetate. Methyl cobalamine derivatives have been implicated in the *de novo* synthesis of acetate [77] and a similar intermediate might be formed in the generation of methane from acetate [78]. If a cobalamine enzyme is involved in the biosynthesis of hydroxymethylpyrimidine, S-adenosylmethionine may act catalytically to activate the cobalamine prosthetic group as it does in the B_{12} dependent *de novo* synthesis of methionine from N^5-methyl-H_4-folate and homocysteine in *E. coli* [79].

C. Biosynthesis of Hydroxyethylthiazole

In spite of the numerous studies with ^{14}C- and ^{35}S-labeled potential precursors, neither the nature of the immediate precursors nor even the bare outline of the overall biosynthetic pathway leading to the hydroxyethylthiazole moiety can be stated with any confidence. The sulfur atom and some of the carbon atoms are more directly derived from methionine than cysteine. In yeast, [^{35}S]methionine is nearly 50 times more efficiently incorporated into hydroxyethylthiazole than is cysteine [60], and methionine is far more effective in decreasing the incorporation of ^{35}S from inorganic sulfate [80]. Johnson *et al.* [60] obtained evidence that [Me-^{14}C,^{35}S]methionine was incorporated into hydroxyethylthiazole with retention of the ^{14}C/^{35}S ratio, suggesting that the sulfur and carbon of the ring might be derived from methionine as a unit. This has been supported by the experiments of Tomlinson *et al.* [74] in *Bacillus subtilis,* who found that alanine, as well as methionine, was incorporated into the carbon skeleton of hydroxyethylthiazole, and by the report of Torrence and Tieckelmann [81] that [Me-^{14}C]methionine is incorporated specifically into the C-2 position. There is general agreement from several laboratories [60–62] that formate is not a precursor of the hydroxyethylthiazole ring. On the basis of these and similar studies, it has been proposed that hydroxyethythiazole is formed from methionine and alanine [74] by a pathway related to that of Harrington and Moggridge [82] in which, except for the carboxyl group, all of the carbon skeleton of methionine is directly incorporated into hydroxyethylthiazole. However, the level of incorporation of carbon-labeled methionine achieved in these experiments [74,81] is extremely low, and despite the attractiveness of a biosynthetic pathway based on methionine, there is evidence that glycine and tyrosine may be more directly involved.

Linnett and Walker [83] found that in yeast the most effective precursor of hydroxyethylthiazole was [2-^{14}C]glycine and that the introduction of label from [1-^{14}C]glycine, [3-^{14}C]serine, [1-^{14}C]formate, [Me-^{14}C]-, or [3,4-^{14}C]methionine was insignificant. Using an improved procedure for purifying the thiazole, they showed [84] that 100% of the label introduced from [2-^{14}C]glycine was located at the 2-position of the ring. Linnett and Walker [83] also found a significant incorporation of [^{15}N]glycine into hydroxyethylthiazole, but surprisingly, not into hydroxymethylpyrimidine.

The experiments of Linnett and Walker have been confined to yeast, but there is evidence that glycine and tyrosine are also involved in the biosynthesis of hydroxyethylthiazole in bacteria. Temperature sensitive mutants have been isolated from *S. typhimurium* [85] and *E. coli* [86] which require thiazole when grown on glucose at 37°, but not at 30°, and are prototrophic at either temperature when grown on glycerol. Iwashima and Nose [86] showed that this requirement for thiazole could be partially met by glycine and that thiamine synthesis, by washed cell suspensions provided with hydroxymethylpyrimidine, was stimulated by glycine as well as thiazole. They concluded that glycine may serve as a precursor for hydroxyethylthiazole.

Studies of the derepression of thiamine synthesis by phenylalanine [87] indicate that tyrosine may also be involved in the formation of hydroxyethylthiazole. Phenylalanine appears to interfere with the synthesis of the thiazole moiety, since the derepression of the thiamine pathway by phenylalanine is overcome by hydroxyethylthiazole and not by hydroxymethylpyrimidine. In a thiamine regulatory mutant [88] that has a three- to fourfold higher level of intracellular thiamine than the parent strain, Iwashima and Nose [87] demonstrated the inhibition of thiamine synthesis by phenylalanine and showed that this inhibition could be overcome by hydroxyethylthiazole or tyrosine, but not by hydroxymethylpyrimidine. The aromatic amino acids share a common pathway of biosynthesis leading to chorismic acid which is controlled by a complex interaction of end-product inhibition and repression [89]. Iwashima and Nose [87] suggest that phenylalanine, added at a very high level in these experiments, may be limiting the formation of tyrosine or may be acting on a site of cross-pathway regulation in hydroxyethylthiazole synthesis.

Evidence in support of a precursor role for tyrosine has been developed by Estramareix and Therisod [90] who have shown that the thiamine content of an *E. coli* tyrosine auxotroph is increased tenfold when incubated with hydroxymethylpyrimidine plus hydroxyethylthiazole, tyrosine, or 3,4-dihydroxyphenylalanine, but not with phenylalanine, tryptophan, aminobenzoic acid, or *p*-hydroxybenzoic acid. In experiments with [U-^{14}C]tyrosine, they made the surprising discovery that only the asymmetric carbon of the tyrosine molecule was incorporated into the thiazole, and 85% of the incorporated counts were found in the C-2 position of the thiazole ring. The synthesized hydroxyethylthiazole had half the specific activity of the [2-^{14}C]tyrosine added to the incubation mixture, and although the specific activity of the tyrosine was not determined after reisolation, it appears likely that tyrosine is not the sole precursor of the C-2 carbon. A reversible cleavage of L-phenylserine to form benzaldehyde and glycine has been described in mammalian liver and kidney [91]. Although it has not been reported, a similar cleavage of the corresponding tyrosine derivative would provide a convergence of the tyrosine and glycine pathways leading to the synthesis of the thiazole moiety of thiamine.

D. Control of Thiamine Biosynthesis

One of the difficulties that has plagued investigators of thiamine biosynthesis is that thiamine is not formed in excess and the *de novo* formation of thiamine cannot be demonstrated in nongrowing cell suspensions under normal circumstances. The control mechanism that severely limits thiamine production must regulate two converging pathways in which each of the constituent cyclic moieties is independently synthesized. How this finely tuned regulation is achieved was first demonstrated in *S. typhimurium* by Newell and Tucker [66,92] who observed that when cells were preincubated with adenosine, washed, and reincubated in minimal medium, a burst of thiamine synthesis ensued, increasing the cellular level of thiamine four- to fivefold. The effect of adenosine is to derepress the enzymes of thiamine biosynthesis by inhibiting the synthesis of hydroxymethylpyrimidine and thus ultimately lowering the level of cellular thiamine. As Newell and Tucker have shown, AIR is a common intermediate in the synthesis of hydroxymethylpyrimidine and the purines. Adenine and other purines, in the form of their nucleotides

[58], exert feedback control over the synthesis of AIR by inhibiting the formation of 5-phosphoribosylamine, the first step exclusive to the synthesis of purines and hydroxymethylpyrimidine. The derepression of thiamine synthesis by adenine or adenosine is unaffected by thiazole, but does not occur if thiamine or hydroxymethylpyrimidine is present, or if protein synthesis is prevented during preincubation. Derepression by adenosine in a hydroxyethylthiazole auxotroph is prevented by thiamine, but not by hydroxymethylpyrimidine. This shows that hydroxymethylpyrimidine influences the state of repression only by being converted to thiamine. Derepression in the absence of adenosine also occurs when a hydroxyethylthiazole auxotroph is grown on a limited supply of thiamine. Under these conditions the cells excrete excess hydroxymethylpyrimidine. However, in the converse experiment with a strain auxotrophic for hydroxymethylpyrimidine, an excess of hydroxyethylthiazole is not produced. This indicates that although hydroxyethylthiazole, like hydroxymethylpyrimidine, cannot cause repression, it is able to control its own synthesis, presumably by a process of feedback inhibition. A similar conclusion was reached earlier by Moyed [65], who observed that *Aerobacter aerogenes* did not excrete thiazole when the synthesis of hydroxymethylpyrimidine was inhibited by adenosine.

The cellular thiamine level could, theoretically, be controlled by making the synthesis of either hydroxyethylthiazole or hydroxymethylpyrimidine rate limiting, although this would be wasteful of the component produced in excess. Newell and Tucker [66] showed that this is not the case: the production of both hydroxymethylpyrimidine and hydroxyethylthiazole are controlled at precisely the same concentration. Control is exercised so that the repressed and derepressed states are between 40 and 30 ng of thiamine/mg dry weight and the level at which repression is released is uninfluenced by exogenously supplied hydroxymethylpyrimidine or hydroxyethylthiazole.

Kawasaki, Iwashima, and Nose [93] have extended these studies to *E. coli* where they have studied the regulation of the four enzymes that catalyze the synthesis of thiamine monophosphate from hydroxymethylpyrimidine and hydroxyethylthiazole: hydroxymethylpyrimidine kinase, phosphohydroxymethylpyrimidine kinase, hydroxyethylthiazole kinase (EC 2.7.1.50), and thiaminephosphate pyrophosphorylase. They found that these enzymes were derepressed by adenosine and repressed by thiamine, corresponding to the effects seen in *S. typhimurium* where, however, the assay of specific enzymes has not been studied. They also observed that derepression was brought about by preincubating the growing culture with phenylalanine. The effect of phenylalanine is analogous to that of adenosine except that in this case the derepression is reversed by thiamine or thiazole, but not by hydroxymethylpyrimidine. The inhibition of thiamine synthesis by phenylalanine in a thiamine regulatory mutant and the reversal of this inhibition by thiazole have been demonstrated [87]. Although this inhibition by phenylalanine accounts for the derepression in the wild type, Kawasaki *et al.* [93] made the surprising observation that the derepression caused by phenylalanine is accompanied by a twofold increase in cellular thiamine levels. The paradox of derepression, presumably caused by inhibition, coexisting with increased levels of thiamine, remains unexplained.

It should be noted that essentially all of the thiamine in bacterial cells is in the form of thiamine pyrophosphate. Although the free vitamin was used in all of these studies, thiamine pyrophosphate is equally effective in suppressing the biosynthetic pathway and is undoubtedly the controlling agent within the cell [66].

Two thiamine regulatory mutants have been isolated by Kawasaki and Nose [88] in which the thiamine content is threefold higher than it is in the parent strain. In one of these mutants, *PT-R1*, thiazole kinase and thiaminephosphate pyrophosphorylase are not subject to repression by thiamine, but the enzymes concerned with the activation of the hydroxymethylpyrimidine, hydroxymethylpyrimidine kinase, and phosphohydroxylmethylpyrmidine kinase are. Iwashima *et al.* [94] found that hydroxymethylpyrimidine synthesis was also not controlled in mutant *PT-R1* and was excreted into the medium.

Kawasaki and Nose [88] suggest that *PT-R1* is a constitutive mutant of an operator gene controlling the synthesis of hydroxyethylthiazole kinase and thiaminephosphate pyrophosphorylase and that a separate and intact operator gene governs the synthesis of hydroxymethylpyrimidine kinase and phosphohydroxymethylpyrimidine kinase.

No similar studies of the control of thiamine synthesis have been carried out in *Neurospora* or yeast. It has been observed [38] that the level of thiaminephosphate pyrophosphorylase present in commercial yeast cultivated with added thiamine is less than 15% of that found in the same yeast produced without exogenous thiamine.

III. THE ROLE OF THIAMINE

A. Function as Cocarboxylase

In 1943, Ugai and co-workers [95,96] demonstrated the catalysis of benzoin formation from two moles of benzaldehyde by a series of compounds including thiamine. Mizuhara *et al.* [97,98], recognizing the similarity to the enzymatic synthesis of acetoin, showed that under mildly alkaline conditions, thiamine would catalyze the decarboxylation of pyruvate and the formation of acetoin from pyruvate and acetaldehyde. Proceeding from these observations, Breslow [99] made the striking discovery that the hydrogen at the C-2 position of the thiazole ring readily exchanged with D_2O and proposed that it was by reaction at the C-2 position of a thiazolium zwitterion that active intermediates of thiamine catalyzed reactions were formed. Krampitz and associates [100] synthesized 2α-hydroxyethylthiamine, the active acetaldehyde derivative postulated by Breslow [101,102], and showed that in the presence of thiamine pyrophosphokinase it was converted to acetaldehyde by a soluble pyruvate decarboxylase prepared from yeast. Subsequently, several enzymatically prepared 2α-hydroxyalkylthiamine pyrophosphate compounds have been shown to function as intermediates when added to appropriate apoenzyme preparations [103–105].

Thiamine pyrophosphate serves as coenzyme for enzymatic systems that have in common the cleavage of C—C bonds. Four general types are illustrated in Fig. 4. The 2α-hydroxyethyl and $2\alpha,\beta$-dihydroxyethyl intermediates indicated for reactions A, B, and C have been prepared enzymatically, but it is likely that carbanion forms of these derivatives participate more directly as activated intermediates in cocarboxylase catalyzed reactions [105].

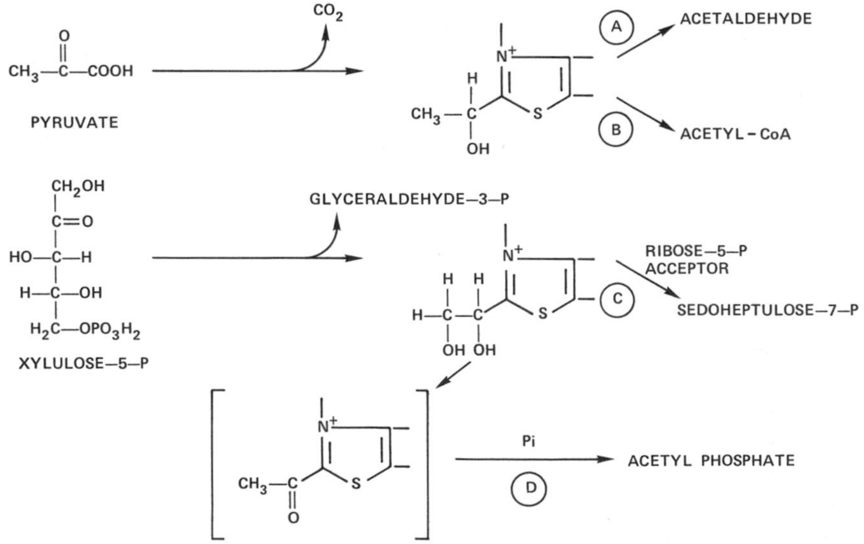

FIG. 4. Examples of reactions for which thiamine pyrophosphate functions as coenzyme. A, Pyruvate decarboxylase; B, pyruvate dehydrogenase; C, transketolase; D, phosphoketolase.

In mammalian systems the intermediate formed from pyruvate is oxidized to acetyl-CoA by a highly organized ternary complex of enzymes constituting pyruvate dehydrogenase (B) in which the oxidation of the active acetaldehyde intermediates involves the reductive acylation of protein bound lipoic acid [106]. Analogous reactions are involved in the oxidative decarboxylation of α-ketoglutarate.

Transketolase (C) and phosphoketolase (D) catalyze the cleavage of 2-keto sugars and they both probably involve the formation of a $2\alpha,\beta$-dihydroxyethyl thiamine pyrophosphate derivative which has been prepared chemically by Krampitz and Votow [107] and isolated by Holtzer and his group [108,109] from incubation mixtures of hydroxypyruvate and pig heart pyruvate oxidase. The transketolase and phosphoketolase reactions differ in the fate of the activated intermediate. In the case of transketolase, the C_2 fragment is transferred to a suitable aldehyde acceptor to form a new ketol and the overall reaction is readily reversible. The phosphoketolase catalyzed synthesis of acetyl phosphate is irreversible and involves a rearrangement of the dihydroxyethyl intermediate; the transient synthesis of an acetyl thiamine pyrophosphate, shown in brackets, has been proposed by Breslow [102].

B. Thiamine Function and Thiamine Deficiency

In his initial studies of acute polyneuritis in fowl, Eijkman referred to the curative factor in rice polishings as the antineuritic principle. Thiamine deficiency in man and in experimental animals is associated with an array of signs of neuropathy which may involve primarily disorders of the peripheral nervous system, as in beriberi, or may chiefly affect the central nervous system, as in Wernicke's encephalopathy. The signs of deficiency, which vary somewhat from species to species, include anorexia, irritability, ataxia, loss of equilibrium, and ophthalmoplegia. Although a prolonged state of deficiency leads to irreversible damage, the early signs are rapidly reversed by the administration of thiamine. A central question in our understanding of the function of thiamine in the nervous system has been whether the observed neuropathy can be linked to the loss of specific enzymes for which thiamine diphosphate serves as cofactor, or whether the evidence supports a role for thiamine in neural activity which is independent of its coenzymatic function.

Our understanding of the role of thiamine in carbohydrate metabolism is derived from the pioneer studies of experimental thiamine deficiency in animals by Peters [110] and his associates. Working with thiamine deficient pigeons in acute opisthotonos, a characteristic retracted position of the head, they discovered [111] that the tissues contained high levels of lactate and pyruvate. Peters guessed that the opisthotonos originated in some defect in the central nervous system and found that brain brei prepared from deficient pigeons had an impaired ability to oxidize lactate [112] or pyruvate [113]. The impairment was corrected by adding thiamine to the diet or directly to the brain brei [112]. This marks the first report of an *in vitro* effect of a vitamin. Since the defect in pyruvate oxidation was manifest when no pathological lesions were demonstrable in the brain by visual microscopy, Peters concluded that the inability of the nerve cell to obtain energy through the oxidation of pyruvate was the "biochemical lesion" [114] which precedes gross physical damage to the cell. Although it might appear that high levels of pyruvate and lactate might be toxic [115,116], studies of individual pigeons in normal and avitaminous states [117] and of different breeds of pigeons [118] indicated that the onset and disappearance of opisthotonos were not directly related to variations in the level of blood pyruvate. Blood pyruvate levels poorly reflect the state of thiamine deficiency, and markedly elevated levels have been noted only after a loading dose of glucose has been administered [119], or in severe beriberi [120].

In experimental animals, thiamine dependent enzymes suffer a much sharper decline in extraneural tissues than they do in brain [121,122]. In pigeons, it is the early fall in pyruvate decarboxylase in heart and skeletal muscle, rather than in brain, that correlates with the high levels of blood pyruvate [123]. Transketolase is extremely sensitive to marginal levels of thiamine deficiency and the assay of erythrocyte transketolase has gained wide acceptance as an indication of functional levels of thiamine in the human as well as in the rat [124,125].

Because of this greater susceptibility of peripheral tissues to thiamine deprivation, it is not unusual for deficient rats to succumb before neurological signs are evident. This resistance of the brain to thiamine deprivation is due, in part, to its ability to retain thiamine under conditions of nutritional deficiency. This has been remarkably visualized in a study of the distribution of [^{14}C]thiamine in normal and deficient rats by whole body radioautography [126]. Although prolonged maintenance of high levels of thiamine in the brain has been reported [127], most studies [128,129] find that the loss of thiamine from the brain begins during the first week, but is appreciably slower than it is in peripheral tissues. Perhaps more significant is the fact that the brain appears to have a substantial reserve of thiamine, since the onset of severe encephalopathy does not occur until the thiamine content has been reduced to approximately 20% of normal and a rapid reversal of neurological signs occurs when the thiamine level has been restored to less than 30% of normal [129,130].

The lateral pontine tegmentum of the brain stem [131] and the cerebellum [132] exhibit the most severe lesions in the deficient rat, and a significant loss of TDP in this area precedes and is proportionate to the seriousness of neurological impairment [133]. Transketolase activity is much more severely affected and falls to 60–70% of normal while the animal is still asymptomatic [134,135]. As signs of neuropathy develop, transketolase activity may be reduced to 40–50% of that observed in pair-fed control rats [130,134]. In contrast, even in symptomatic rats, pyruvate decarboxylase activity does not decrease below 70% of normal [130,136].

McCandless and Schenker [130] attempted to assess the relative importance of these effects by measuring the enzyme levels in paired animals immediately after the rapid reversal of their neurological signs following the intraperitoneal injection of thiamine. Pyruvate decarboxylase activity increased promptly to near normal levels, but the resultant increase in transketolase activity was small. Much more significant increases of approximately 40% have been observed by Dreyfus [134] following the administration of 10 μg of thiamine subcutaneously. A similar, limited response of transketolase to TDP, added *in vitro,* has been reported in separate studies of depressed levels of pyruvate dehydrogenase and transketolase in rat liver [137] and brain [138] homogenates prepared from thiamine deficient rats. As in the *in vivo* studies described above, the addition of TDP restored pyruvate dehydrogenase to normal levels, but limited stimulation of transketolase activity was observed. These results may indicate, as has been suggested [138], that a loss of apotransketolase occurs when thiamine levels are depressed. Alternatively, the reassociation of TDP with the apoenzyme may be rate limiting. It is possible that TTP and ADP may play a role in the reactivation of apotransketolase as reported by Yusa and Maruo [139] for the activation of α-ketoglutarate dehydrogenase prepared from yeast grown on thiamine-poor medium.

Assays of changes in brain metabolites have not proved definitive. The fall in brain stem pyruvate dehydrogenase was accompanied by an increased concentration of lactate in that specific region and lactate levels were reduced to normal in thiamine reversed animals [130]. In Peters' original concept the biochemical lesion was conceived as an inability of the nerve cell to generate energy. However, McCandless and Schenker found that the level of ATP in all areas of the brain of thiamine deficient rats was the same as in asymptomatic controls.

The dysfunction of the central nervous system might also be due to a diminished ability to synthesize acetylcholine. Reports of the effects of thiamine deficiency on acetylcholine levels have been contradictory

[140,141]. Whole brain acetylcholine as well as acetyl-CoA have recently been studied [138] in deficient rats carefully standardized by measuring individual levels of erythrocyte transketolase. Under these conditions, the concentrations of acetyl-CoA and acetylcholine in brain were observed to decrease by 42% and 35%, respectively. Choline acetylase and acetylcholinesterase were unaffected by thiamine deficiency.

In developing thiamine deficiency there are significant neural changes in the levels of pyruvate dehydrogenase and transketolase in the brain stem where the most severe and earliest lesions of thiamine deficiency develop. Although these changes seem insufficient [135] to explain the development of neuropathy, it is possible that slight increments in limiting factors may be of major clinical significance. In the thiamine deficient rat an increase in cerebral thiamine from 20 to 26% of normal is sufficient to restore the deficient animal to the asymptomatic state. Similarly, a remarkable improvement in the physiological state of an homocystinuric individual has been observed to follow the administration of pyridoxine which effects a slight increment in the level of cystathionine synthase in the liver from 1–2 to 3–4% of normal [142].

Although the observed decrements of pyruvate dehydrogenase activity appear to be small, the observed values may not properly reflect the level of activity *in situ*. Pyruvate dehydrogenase in beef kidney [143] and heart [144] has been found to exist in two interconvertable forms — one active, and a second, phosphorylated and inactive. The process of inactivation is inhibited by TDP [145] and reversed by the action of a phosphatase. It is therefore possible that the level of functional pyruvate dehydrogenase in the TDP deficient brain stem is significantly lower than it appears to be when assayed *in vitro*.

A recent report [146] has described the application of extremely sensitive techniques to the study of enzyme and substrate levels in 2-cell and 8-cell stages of the developing mouse embryo. The effects of experimental thiamine deficiency may not be revealed until techniques of similar sensitivity are applied to the study of single cells in affected regions of the central nervous system.

C. Evidence for a Noncoenzymatic Role

Although most studies have focused on the role of thiamine in carbohydrate metabolism, there has coexisted the view that thiamine participates in nerve conduction in a manner independent of its coenzymatic function [147]. The earliest indications of such a dual role was the report of Minz [148], appearing in 1938, at a time of great interest in the

TABLE II

EFFECT OF ELECTRICAL STIMULATION *in Vitro* ON THE PARTITION OF THIAMINE
AND THIAMINE PHOSPHATES IN THE SCIATIC NERVE OF THE RAT[a]

Treatment	Percentage distribution			
	Thiamine	TMP	TDP	TTP
Unstimulated	3.4	29	60.4	7.2
Stimulated	11.3	63.5	23.1	2.1

[a] From Gurtner [149].

role of acetylcholine in the synaptic transmission of the nerve impulse, that electrical stimulation caused the release of thiamine from the cut end of an isolated ox nerve. Using an excised intact peripheral nerve from a rat containing ^{35}S-labeled thiamine, Gurtner [149] showed that electrical stimulation caused an increase in the amount of free and monophosphorylated thiamine and a decrease in the amount of TDP and TTP (Table II).

This same displacement and hydrolysis of thiamine phosphates is induced by several neuroactive drugs. Itokawa and Cooper [150,151] and Itokawa *et al.* [152] showed that such neuroactive drugs as acetylcholine, ouabain, tetrodotoxin, and 5-hydroxytryptamine caused the release of thiamine, largely as free thiamine and TMP, not only from intact perfused nerves, but from membrane fragments prepared from brain, sciatic nerve, and spinal cord. No such release is caused by NaCl or choline.

A third line of evidence which has been interpreted as supporting a special role for thiamine in nerve conduction has been the differential effect of two thiamine antagonists, pyrithiamine, and oxythiamine (Fig. 5). Oxythiamine rarely provokes signs of neuropathy though it accelerates the onset of signs of deficiency in peripheral tissues. Pyrithiamine almost always induces neural dysfunction. This difference was at first interpreted [153] as indicating some special function for thiamine in the

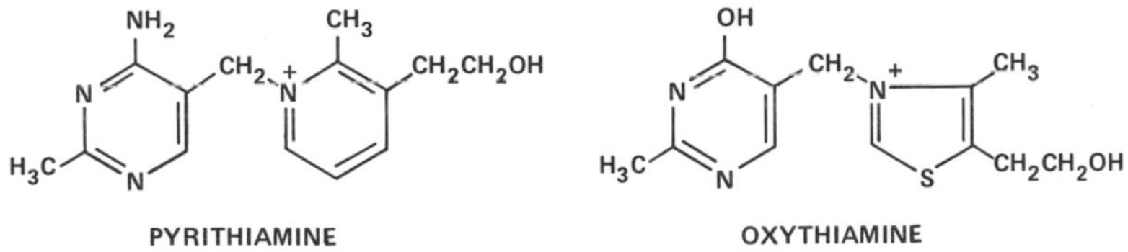

PYRITHIAMINE   OXYTHIAMINE

FIG. 5. Thiamine antagonists.

nervous system which was affected by pyrithiamine and not by oxythiamine. It was subsequently shown that whereas oxythiamine is incapable of penetrating the blood–brain barrier [154], pyrithiamine has a special affinity for the central nervous system [155] and accumulates in the brain after leaving peripheral tissues.

The two agents compete with thiamine in distinctly different ways. Both are converted to the corresponding pyrophosphate derivatives. Oxythiamine pyrophosphate is a powerful inhibitor of cocarboxylase requiring enzymes [156]. Pyrithiamine as the pyrophosphate, is a less effective competitor of TDP [157] but as free pyrithiamine, is a powerful inhibitor of the synthesis of TDP by thiamine pyrophosphorylase [158,159]. This inhibition results in the loss of thiamine from all tissues and a marked reduction of cocarboxylase dependent enzymes [157].

Even before these differential effects of pyrithiamine and oxythiamine were known, experiments with thiamine antagonists had opened up a separate line of evidence which seemed to implicate thiamine in the process of neural excitation. When applied to an isolated node of Ranvier, pyrithiamine [160] and an anti-thiamine factor from fern extracts [147] diminished the amplitude and rate of rise of the action potential of the nerve. These experiments have recently been extended by Armett and Cooper [161] to desheathed vagus nerves of the rabbit. In their preparation, pyrithiamine, but not oxythiamine, increased the action potential and reversed the hyperpolarization after a tetanus had been applied to the nerve. To show that these alterations in electrical response were unrelated to changes in the activity of cocarboxylase dependent enzymes, Cooper [162] exposed the nerves to pyrithiamine as in the experiments of Armett and Cooper [161], but did not include electrical stimulation of the nerve preparation. At appropriate times, the nerves were assayed for thiamine content, for pyruvate and α-ketoglutarate dehydrogenases and for transketolase. There was no change in the activity of any of the three enzymes, but the thiamine content of the nerves was reduced by approximately 44%. Cooper considers that these observations imply that the alterations in electrical activity induced by pyrithiamine were due to the displacement of thiamine esters from the nerve membrane and not by some effect on the coenzymatic function of the vitamin. The experiments of Cooper do not, however, reproduce the conditions of the previous studies in two respects: (a) they introduce a significant dilution of the pyrithiamine in the course of homogenizing and assaying the preparation and (b) the absence of an accompanying electrical stimulation, itself capable of initiating the release and hydrolysis of thiamine esters, alters significantly the potential effect of pyrithiamine as an inhibitor of thiamine pyrophosphokinase. Thus the effects of pyrithiamine on cocarboxylase linked enzymes in the original experiments of Armett and Cooper might be more extensive than is revealed in the corollary experiments performed in the absence of electrical stimulation.

The various forms of thiamine undergo repeated dephosphorylation and rephosphorylation *in vivo* [163,164] and TDP and TTP are rapidly interconverted [165,166] as shown in Eq. (1).

$$TDP + ATP \rightleftharpoons TTP + ADP \qquad\qquad (1)$$

In addition, there are separate phosphatases for TDP [167] in the cell membrane, and for TTP [168,169] in the axoplasm. The specific location of thiamine in the axonal membrane [170,171] rather than in the cytoplasm, as in other cells, is consistent with its participation in neural excitation. Both thiamine phosphates form stable complexes with sodium and calcium ions [172] and it has been suggested that TTP may participate directly in transmembrane ion transport [151,165]. The above described effects of electrical stimulation and neuroactive agents have been interpreted in support of such a role. These stimulants have in common the ability to alter the structure of the nerve membrane. It is far from clear, therefore, that the observed displacement and hydrolysis of thiamine phosphates is not a consequence of alterations in membrane permeability and the resultant exposure of thiamine phosphates to the action of phosphatases—a possibility considered by Itokawa and Cooper [150]—rather than an indication that TDP or TTP are directly coupled to the process of ion transport and electrical excitation in the neural process.

A great deal of interest in TTP has been stimulated by studies of subacute necrotizing encephalomyelopathy (SNE), a familial, probably genetic [173], disease of children first described by Leigh [174]. Although this disease develops while the subjects are receiving normal thiamine intake, they manifest a complex symptomology which generally includes psychomotor retardation, somnolence, visual disturbances, spasticity of limbs, and peripheral neuropathy. Abnormal chemical data include lactic and pyruvic acidemia [175,176] and elevated concentrations of plasma alanine [177]. Nevertheless, patients exhibit normal levels of pyruvate dehydrogenase, α-ketoglutarate dehydrogenase, and transketolase [178]. The necrotizing lesions, generally found in the midbrain and brain stem, resemble those observed in Wernicke's encephalopathy, a classical thiamine deficiency disease [179], although striking differences in the histopathology of the two conditions have been pointed out [180], particularly as regards the regular involvement of the mammilary bodies in Wernicke's encephalopathy, which are almost never affected in SNE.

Interest in a possible connection between SNE and thiamine metabolism was sparked by the discovery by Cooper *et al.* [181] that the urine, blood, and cerebrospinal fluid of SNE patients contained a non-dialyzable factor, possibly a glycoprotein, which inhibits the synthesis of TTP from TDP. The inhibitor did not inhibit TTP synthesis in the liver [182]. On autopsy, the brain, but not other tissues, was found to contain no detectable TTP which normally constitutes from 6 to 12% of the total thiamine of all tissues [181]. In subsequent studies, Cooper and Pincus [183] have found a reasonable, but not consistent, correlation between the level of TTP found in various regions of the brain and the severity of the pathological lesions found on autopsy.

Studies in two laboratories [182,184] have associated the presence of the urinary inhibitory factor with a diagnosis of SNE, proved on autopsy, but the inhibitor is also occasionally found in the urine of patients not suffering from Leigh's disease [183]. High levels of inhibitor have also been found in the urine of asymptomatic parents of four SNE patients [184]. Recently, Murphy *et al.* [185] have been able to extract a factor from normal fibroblasts cultured in thiamine-poor medium that inhibits the synthesis of TTP. Significantly greater amounts of the inhibitor have been obtained from fibroblasts of patients or from their heterozygous parents cultured under the same conditions.

The obvious connection with thiamine metabolism prompted clinical trials of the vitamin and although the usual therapeutic levels of thiamine, 10–50 mg/day were without effect, dramatic clinical improvement followed the use of massive amounts of thiamine and of thiamine propyldisulfide and thiamine furfuryldisulfide ranging up to 4 gm/day for thiamine and 1 gm/day for the disulfide derivatives [186]. The latter compounds are reductively cleaved to thiamine [187], but readily penetrate the central nervous system when administered intravenously. Rapid formation of TTP in rat heart muscle has been reported [188] following the intravenous administration of ^{35}S-labeled thiamine disulfide derivatives.

Although Leigh's disease is subject to variable, spontaneous, and temporary remissions, massive treatment with thiamine raises the level of thiamine in the cerebrospinal fluid, reduces the level of the inhibitor in the urine, and produces marked clinical improvement. However, patients become refractory to this treatment and the observed amelioration cannot be sustained. Although there exists a close association between SNE and the presence of a factor which inhibits the synthesis of TTP in neural tissue, it remains to be established that the production of this factor is a cause rather than a consequence of neural pathology. It is clear that thiamine does not provide a therapeutic solution to Leigh's disease.

In contrast with the postulate that Leigh's disease is due essentially to a deficiency of TTP in the central nervous system, has been the view, chiefly advanced by Hommes and associates [176], that the disease derives from a lack of pyruvate carboxylase (EC 6.4.1.1). A deficiency of this enzyme in the liver of patients manifesting symptoms of SNE has been reported from other laboratories [189–191]. However, in only one of these cases [189] was a specific diagnosis of SNE made.

A deficiency of pyruvate carboxylase would impair the anaplerotic synthesis of glucose, lead to lactic acidosis and hyperpyruvic acidemia and hyperalanemia, and reduce the supply of oxalacetate required for the operation of the Krebs cycle. Operating on the assumption that an ineffectively operating Krebs cycle plays an important role in the pathogenesis of the disease, Tang et al. [189] supplemented the formula of a 4-month-old infant who displayed the signs of SNE and whose sib had died of this disease (proved on autopsy), with regular additions of L-glutamine in an attempt to provide a continuous supply of dicarboxylic 4-carbon acids. Supplements of vitamin B_6 were provided to insure maximal rates of transamination. The result of this therapy was a rapid and marked clinical improvement in the patient, accompanied by significant decreases in blood lactic and pyruvic acids and increases in serum cholesterol and asparagine. Similar results have been reported by De Groot and Hommes [192], in which improvement in a patient followed the administration of thiamine as well as aspartate.

De Groot and Hommes have called attention to a special property of pyruvate dehydrogenase reported by Roche and Reed [145] which they suggest may underlie the beneficial effects of thiamine in treating SNE. Pyruvate dehydrogenase is composed of subunits, one of which is subject to phosphorylation by a specific pyruvate dehydrogenase kinase which inactivates the enzyme. The enzyme is reactivated by dephosphorylation. Roche and Reed have demonstrated that TDP competes for the site of phosphorylation and inhibits the inactivation of the enzyme. In a test of this protective effect of TDP in vivo, Hommes et al. [193] have demonstrated that pyruvate dehydrogenase isolated from rats receiving high doses of thiamine was far less sensitive to inhibition by ATP than the enzyme isolated from control animals. They further suggest that the effect of thiamine therapy in SNE may be, in part, to facilitate the oxidation of pyruvate in the central nervous system.

Brunette et al. [191] recently described a patient in whom a normal component of pyruvate carboxylase, with a low K_m for pyruvate, was missing in the liver. The female infant exhibited many of the symptoms of Leigh's syndrome but no test for urinary inhibitory factor was attempted nor was a diagnosis of SNE made. Although pyruvate dehydrogenase in leukocytes and cultured fibroblasts was normal, thiamine therapy markedly ameliorated the metabolic abnormality. However, pronounced psychomotor and mental retardation have persisted. The authors suggest that thiamine might act by reducing the rate of inactivation of pyruvate dehydrogenase, thus enhancing the basal activity of the enzyme *in vivo*. The protective effect of TDP revealed by Roche and Reed [145] may provide such a mechanism.

Yusa and Maruo [139] have made the interesting observation that the activity of α-ketoglutarate dehydrogenase, prepared from yeast grown on thiamine deficient media and from thiamine deficient rat liver mitochondria, is doubled when incubated with TTP and ADP and not by TDP and ATP or ADP. Using ^{32}P-labeled TTP, they showed that in the process of activation of ketoglutarate dehydrogenase by TTP and ADP, TDP is fixed to the enzyme in nondialyzable form. The significance and generality of these observations, particularly with respect to the role of TTP in neural tissue, is yet to be explored.

Some years ago, von Muralt proposed a specific role for thiamine in the process of neural conduction independent of its coenzymatic function in general metabolism. It may be useful to restate some of the evidence in support of this concept:

1. Thiamine and its esters are specifically located in axonal membranes.
2. Electrical stimulation and neuroactive drugs provoke the release and hydrolysis of TDP and TTP.
3. Thiamine antagonists alter the electrical response of excised nerves.
4. TDP and TTP form strong complexes with sodium and calcium ions.
5. Diminished levels of TTP are associated with severely affected regions of the brain in Leigh's encephalomyelopathy.
6. No coenzymatic function has been established for TTP.

These observations suggest that thiamine may indeed play a direct role in neural excitation, but the evidence remains circumstantial and equivocal. In no case has it been explicitly demonstrated that the transitions of thiamine derivatives within the nerve are involved in the evocation rather than the consequences of neural activity. While the accumulated evidence is provocative, a specific biophysical role for thiamine in the nervous system remains unclarified and uncertain.

REFERENCES

1. K. Takaki, *Lancet* **1,** 1369, 1451, and 1520 (1906).
2. C. Eijkman, *Arch. Pathol. Anat. Physiol. Klin. Med.* **148,** 523 (1897).

3. B. C. P. Jansen and W. F. Donath, *Chem. Weekbl.* **23**, 201 (1926).
4. R. R. Williams and J. K. Cline, *J. Amer. Chem. Soc.* **58**, 1504 (1936).
5. E. Andersag and K. Westphal, *Ber. Deut. Chem. Ges. B* **70**, 2035 (1937).
6. K. Lohmann and P. Schuster, *Naturwissenschaften* **25**, 26 (1937).
7. K. Lohmann and P. Schuster, *Biochem. Z.* **294**, 188 (1937).
8. A. Rossi-Fanelli, N. Siliprandi, and P. Fasella, *Science* **116**, 711 (1952).
9. K. H. Kiessling, *Nature (London)* **172**, 1187 (1953).
10. N. Kochibe, T. Yusa, and K. Hayashi, *Plant Cell Physiol.* **4**, 239 (1963).
11. G. de la Fuente and R. Diaz-Cadavieco, *Nature (London* **174**, 1014 (1954).
12. G. Rindi and L. de Giuseppe, *Biochem. J.* **78**, 602 (1961).
13. R. R. Williams, *J. Amer. Chem. Soc.* **58**, 1063 (1936).
14. W. H. Schopfer and A. Jung, *C. R. Acad. Sci.* **204**, 1500 (1937).
15. W. J. Robbins and F. Kavanagh, *Proc. Nat. Acad. Sci. U.S.* **23**, 499 (1937).
16. J. Bonner and E. R. Buchman, *Proc. Nat. Acad. Sci. U.S.* **24**, 431 (1938).
17. J. Bonner and E. R. Buchman, *Proc. Nat. Acad. Sci. U.S.* **25**, 164 (1939).
18. A. Fujita, *Advan. Enzymol.* **15**, 389 (1954).
19. D. W. Woolley, *Nature (London)* **171**, 323 (1953).
20. S. L. Harris and J. Yavit, *Fed. Proc., Fed. Amer. Soc. Exp. Biol.* **16**, 192 (1957).
21. I. G. Leder, *Fed. Proc., Fed. Amer. Soc. Exp. Biol.* **18**, 270 (1959).
22. I. G. Leder, *Biochem. Biophys. Res. Commun.* **1**, 63 (1959).
23. G. W. Camiener and G. M. Brown, *J. Amer. Chem. Soc.* **81**, 3800 (1959).
24. I. G. Leder, *J. Biol. Chem.* **236**, 3066 (1961).
25. Y. Nose, K. Ueda, T. Kawasaki, A. Iwashima, and T. Fujita, *J. Vitaminol. (Kyoto)* **7**, 98 (1961).
26. G. W. Camiener and G. M. Brown, *J. Biol. Chem.* **235**, 2404 (1960).
27. J. Suzuoki and A. Kobata, *J. Biochem. (Tokyo)* **47**, 262 (1960).
28. Y. Nose, K. Ueda, and T. Kawasaki, *Biochim. Biophys. Acta* **34**, 277 (1959).
29. Y. Nose, K. Ueda, and T. Kawasaki, *J. Vitaminol. (Kyoto)* **7**, 92 (1961).
30. L. M. Lewin and G. M. Brown, *J. Biol. Chem.* **236**, 2768 (1961).
31. L. Jaenicke and P. C. Chan, *Angew. Chem.* **72**, 752 (1960).
32. T. Shiota, M. N. Disraely, and M. P. McCann, *Biochem. Biophys. Res. Commun.* **7**, 194 (1962).
33. B. W. Agranoff, H. Eggerer, V. Henning, and F. Lynen, *J. Amer. Chem. Soc.* **81**, 1254 (1959).
34. F. Lynen, B. W. Agranoff, H. Eggerer, V. Henning, and E. Moslein, *Angew. Chem.* **71**, 657 (1959).
35. Y. Nose, Y. Tokuda, M. Hirabayashi, and A. Iwashima, *J. Vitaminol. (Kyoto)* **10**, 105 (1964).
36. H. Mitsuda, T. Tanaka, and F. Kawai, *J. Vitaminol. (Kyoto)* **16**, 263 (1970).
37. H. Mitsuda, T. Tanaka, Y. Takii, and F. Kawai, *J. Vitaminol. (Kyoto)* **17**, 89 (1971).
38. I. G. Leder, *in* "Methods in Enzymology" (D. B. McCormick and L. D. Wright, eds.), Vol. 18A, p. 207. Academic Press, New York, 1970.
39. T. Kawasaki and K. Esaki, *Biochem. Biophys. Res. Commun.* **40**, 1468 (1970).
40. Y. Kayama and T. Kawasaki, *Arch. Biochem. Biophys.* **158**, 242 (1973).
41. M. Yamazaki, *Biochem. Biophys. Res. Commun.* **16**, 416 (1964).
42. H. Weil-Malherbe, *Biochem. J.* **33**, 1997 (1939).
43. N. Shimazono, Y. Mano, R. Tanaka, and Y. Kaziro, *J. Biochem. (Tokyo)* **46**, 959 (1959).
44. Y. Kaziro, *J. Biochem. (Tokyo)* **46**, 1523 (1959).
45. F. Leuthardt and H. Nielsen, *Helv. Chim. Acta* **35**, 1196 (1952).

46. Y. Mano, *J. Biochem.* (*Tokyo*) **47,** 283 (1967).
47. G. M. Brown, *in* "Metabolic Pathways" (D. M. Greenberg, ed.), 3rd ed., Vol. 4, p. 369. Academic Press, New York, 1970.
48. E. P. Steyn-Parvé, *Biochim. Biophys. Acta* **8,** 310 (1952).
49. K. H. Kiessling, *Ark. Kemi* **10,** 279 (1956).
50. Y. Tokuda, *Bitamin* **29,** 40 (1964).
51. A. Iwashima, H. Nishino, and Y. Nose, *Biochim. Biophys. Acta* **258,** 333 (1972).
52. H. Nishino, A. Iwashima, and Y. Nose, *Biochem. Biophys. Res. Commun.* **45,** 363 (1971).
53. H. Nishino, *J. Biochem.* (*Tokyo*) **72,** 1093 (1972).
54. H. Nakayama and R. Hayashi, *J. Bacteriol.* **109,** 936 (1972).
55. H. Nakayama and R. Hayashi, *J. Bacteriol.* **112,** 1118 (1972).
56. A. K. Sinha and G. C. Chatterjee, *Biochem. J.* **104,** 731 (1967).
57. I. Miyata, T. Kawasaki, and Y. Nose, *Biochem. Biophys. Res. Commun.* **27,** 601 (1967).
58. S. C. Hartman, *in* "Metabolic Pathways" (D. M. Greenberg, ed.), 3rd ed., Vol. 4, p. 1. Academic Press, New York, 1970.
59. G. A. Goldstein and G. M. Brown, *Arch. Biochem. Biophys.* **103,** 449 (1963).
60. D. B. Johnson, D. T. Howells, and T. W. Goodwin, *Biochem. J.* **98,** 30 (1966).
61. M. J. Pine and R. Guthrie, *J. Bacteriol.* **78,** 545 (1959).
62. S. David and B. Estramareix, *Biochim. Biophys. Acta* **42,** 562 (1960).
63. S. David and B. Estramareix, *Biochim. Biophys. Acta* **49,** 411 (1961).
64. T. Yura, *Carnegie Inst. Wash. Publ.* **612,** 62 (1956).
65. H. S. Moyed, *J. Bacteriol.* **88,** 1024 (1964).
66. P. C. Newell and R. G. Tucker, *Biochem. J.* **100,** 517 (1966).
67. P. C. Newell and R. G. Tucker, *Biochem. J.* **106,** 271 (1968).
68. P. C. Newell and R. G. Tucker, *Biochem. J.* **106,** 279 (1968).
69. S. David, B. Estramareix, and H. Hirshfeld, *Biochim. Biophys. Acta* **148,** 11 (1967).
70. H. Kuomoka and G. M. Brown, *Arch. Biochem. Biophys.* **122,** 378 (1967).
71. B. Estramareix and M. Lesieur, *Biochim. Biophys. Acta* **192,** 375 (1969).
72. B. Estramareix, *Biochim. Biophys. Acta* **208,** 170 (1970).
73. R. V. Tomlinson, *Biochim. Biophys. Acta* **115,** 526 (1966).
74. R. V. Tomlinson, D. P. Kuhlman, P. F. Torrence, and H. Tieckelmann, *Biochim. Biophys. Acta* **148,** 1 (1967).
75. M. Nakamura, *Bitamin* **32,** 383 (1965).
76. A. F. Diorio and L. M. Lewin, *J. Biol. Chem.* **243,** 3999 and 4006 (1968).
77. L. G. Ljungdahl and H. G. Wood, *Annu. Rev. Microbiol.* **23,** 515 (1969).
78. T. C. Stadtman, *Annu. Rev. Microbiol.* **21,** 121 (1967).
79. H. A. Barker, *Biochem. J.* **105,** 1 (1967).
80. C. H. S. Hitchcock and J. Walker, *Biochem. J.* **80,** 137 (1961).
81. P. F. Torrence and H. Tieckelmann, *Biochim. Biophys. Acta* **158,** 183 (1968).
82. C. R. Harrington and R. C. E. Moggridge, *Biochem. J.* **34,** 685 (1940).
83. P. E. Linnett and J. Walker, *Biochem. J.* **109,** 161 (1968).
84. P. E. Linnett and J. Walker, *Biochim. Biophys. Acta* **184,** 381 (1969).
85. J. L. Parada and M. V. Ortega, *J. Bacteriol.* **94,** 707 (1967).
86. A. Iwashima and Y. Nose, *J. Bacteriol.* **101,** 1076 (1970).
87. A. Iwashima and Y. Nose, *J. Bacteriol.* **104,** 1014 (1970).
88. T. Kawasaki and Y. Nose, *J. Biochem.* (*Tokyo*) **65,** 417 (1969).
89. F. Gibson and J. Pittard, *Bacteriol. Rev.* **32,** 465 (1968).
90. B. Estramareix and M. Therisod, *Biochim. Biophys. Acta* **273,** 275 (1972).

91. F. H. Bruns and L. Fiedler, *Biochem. Z.* **330,** 324 (1958).
92. P. C. Newell and R. G. Tucker, *Biochem. J.* **100,** 512 (1966).
93. T. Kawasaki, A. Iwashima, and Y. Nose, *J. Biochem.* (*Tokyo*) **65,** 407 (1969).
94. A. Iwashima, K. Takahashi, and Y. Nose, *J. Vitaminol.* (*Kyoto*) **17,** 43 (1971).
95. T. Ugai, S. Tanaka, and S. Dokawa, *J. Pharm. Soc. Jap.* **63,** 269 (1943).
96. T. Ugai, T. Dokawa, and S. Tsubokawa, *J. Pharm. Soc. Jap.* **64,** 3 (1944).
97. S. Mizuhara, *J. Jap. Biochem. Soc.* **22,** 201 (1950).
98. S. Mizuhara, T. Rzohei, and H. Arata, *Proc. Jap. Acad.* **27,** 302 (1951).
99. R. Breslow, *J. Amer. Chem. Soc.* **80,** 3719 (1958).
100. L. O. Krampitz, G. Greull, C. S. Miller, J. B. Bicking, H. R. Skeggs, and J. M. Sprague, *J. Amer. Chem. Soc.* **80,** 5893 (1958).
101. R. Breslow, *Chem. Ind.* (*London*) p. 893 (1957).
102. R. Breslow, *Ann. N. Y. Acad. Sci.* **98,** 445 (1962).
103. H. Holtzer and K. Beaucamp, *Biochim. Biophys. Acta* **46,** 225 (1961).
104. G. L. Carlson and G. M. Brown, *J. Biol. Chem.* **235,** PC3 (1960); **236,** 2099 (1961).
105. B. Deus, J. Ullrich, and H. Holtzer, *in* "Methods in Enzymology" (D. B. McCormick and L. D. Wright, eds.), Vol. 18A, p. 259. Academic Press, New York, 1970.
106. L. J. Reed, *Compr. Biochem.* **14,** 99 (1966).
107. L. O. Krampitz and R. Votow, *in* "Methods in Enzymology" (W. A. Wood, ed.), Vol. 9, p. 65. Academic Press, New York, 1966.
108. F. Da Fonseca-Wollheim, K. W. Bock, and H. Holtzer, *Biochem. Biophys. Res. Commun.* **9,** 466 (1962).
109. F. Pohlandt, G. Kohlhaw, and H. Holtzer, *Z. Naturforsch. B.* **22,** 407 (1967).
110. R. A. Peters, *in* "Thiamin Deficiency" (G. E. W. Wolstenholme and M. O'Connor, eds.), p. 1. Little, Brown, Boston, Massachusetts, 1967.
111. H. W. Kinnersley and R. A. Peters, *Biochem. J.* **23,** 1126 (1929).
112. N. Gavrilescu, A. P. Meiklejohn, R. Passmore, and R. A. Peters, *Proc. Roy. Soc. Ser. B* **110,** 431 (1932).
113. R. H. S. Thompson and R. E. Johnson, *Biochem. J.* **29,** 694 (1935).
114. R. A. Peters, *Lancet* **1,** 1161 (1936).
115. H. G. K. Westenbrink, *Arch. Neer. Physiol.* **19,** 94 (1934).
116. I. Nitzescu and C. Angelescu, *Z. Vitaminforsch.* **12,** 82 (1942).
117. S. de Jong, *Arch. Neer. Physiol.* **21,** 465 (1936).
118. H. B. Lofland, H. D. Goodman, T. B. Clarkson, and R. W. Pritchard, *J. Nutr.* **79,** 188 (1963).
119. E. K. O. Elsom, F. D. W. Lukens, E. H. Montgomery, and L. Jonas, *J. Clin. Invest.* **19,** 153 (1940).
120. B. S. Platt and G. D. Lu, *Biochem. J.* **33,** 1523 (1939).
121. P. M. Dreyfus and G. Hauser, *Biochim. Biophys. Acta* **104,** 78 (1965).
122. C. J. Gubler, *J. Biol. Chem.* **236,** 3112 (1961).
123. C. H. Manfoort, *Biochim. Biophys. Acta* **16,** 219 (1955).
124. H. E. Sauberlich, *Amer. J. Clin. Nutr.* **20,** 528 (1967).
125. M. Brin, *in* "Newer Methods of Nutritional Biochemistry" (A. A. Albanese, ed.), Vol. 3, p. 407. Academic Press, New York, 1967.
126. C. P. Heinrich, D. Hornig, and O. Wiss, *Int. J. Vitam. Nutr. Res.* **43,** 174 (1973).
127. J. Salcedo, Jr., V. A. Najjar, L. M. Holt, Jr., and E. W. Hutzler, *J. Nutr.* **36,** 307 (1948).
128. M. Balaghi and W. U. Pearson, *J. Nutr.* **89,** 127 (1966).
129. P. M. Dreyfus, *J. Neurochem.* **8,** 139 (1959).
130. D. W. McCandless and S. Schenker, *J. Clin. Invest.* **47,** 2268 (1968).

131. P. M. Dreyfus and M. Victor, *Amer. J. Clin. Nutr.* **9**, 414 (1961).
132. C. O. Prickett, *Amer. J. Physiol.* **107**, 459 (1934).
133. J. H. Pincus and I. Grove, *Exp. Neurol.* **28**, 477 (1970).
134. P. M. Dreyfus, *J. Neuropathol. Exp. Neurol.* **24**, 119 (1965).
135. J. H. Pincus and K. Wells, *Exp. Neurol.* **37**, 495 (1972).
136. P. M. Dreyfus and G. Hauser, *Biochim. Biophys. Acta* **104**, 78 (1965).
137. K. Takahashi, A. Nakamura, and Y. Nose, *J. Vitaminol. (Kyoto)* **17**, 207 (1971).
138. C. P. Heinrich, H. Stadler, and H. Weiser, *J. Neurochem.* **21**, 1273 (1973).
139. T. Yusa and B. Maruo, *J. Biochem. (Tokyo)* **60**, 735 (1966).
140. K. V. Speeg, D. Chen, W. W. McCandless, and S. Schenker, *Proc. Soc. Exp. Biol. Med.* **134**, 1005 (1970).
141. D. L. Cheney, C. J. Gubler, and A. Jaussi, *J. Neurochem.* **16**, 1283 (1969).
142. S. H. Mudd, W. A. Edwards, P. M. Loeb, M. S. Brown, and L. Laster, *J. Clin. Invest.* **49**, 1762 (1970).
143. T. C. Linn, F. H. Pettit, and L. J. Reed, *Proc. Nat. Acad. Sci. U.S.* **62**, 234 (1969).
144. O. Wieland and B. Jagow-Westermann, *FEBS (Fed. Eur. Biochem. Soc.) Lett.* **3**, 271 (1969).
145. T. E. Roche and L. J. Reed, *Biochim. Biophys. Res. Commun.* **48**, 840 (1972).
146. E. K. Barbehenn, R. G. Wales, and O. H. Lowry, *Proc. Nat. Acad. Sci. U.S.* **71**, 1056 (1974).
147. A. von Muralt, *Ann. N. Y. Acad. Sci.* **98**, 499 (1962).
148. B. Minz, *C. R. Soc. Biol.* **127**, 1251 (1938).
149. H. P. Gurtner, *Helv. Physiol. Pharmacol. Acta, Suppl.* **11**, 1 (1961).
150. Y. Itokawa and J. R. Cooper, *Biochem. Pharmacol.* **18**, 545 (1969); **19**, 985 (1970).
151. Y. Itokawa and J. R. Cooper, *Biochim. Biophys. Acta* **196**, 274 (1970).
152. Y. Itokawa, R. A. Schulz, and J. R. Cooper, *Biochim. Biophys. Acta* **266**, 293 (1972).
153. D. W. Woolley and R. B. Merrifield, *Fed. Proc., Fed. Amer. Soc. Exp. Biol.* **11**, 458 (1952).
154. G. Rindi, L. de Giuseppe, and U. Ventura, *J. Nutr.* **81**, 147 (1963).
155. G. Rindi and V. Perri, *Biochem. J.* **80**, 214 (1961).
156. A. G. Datta and E. Racker, *J. Biol. Chem.* **236**, 617 (1961).
157. E. P. Steyn-Parvé, in "Thiamin Deficiency" (G. E. W. Wolstenholme and M. O'Connor, eds.), p. 26. Little, Brown, Boston, Massachusetts, 1967.
158. S. Eich and L. R. Cerecedo, *J. Biol. Chem.* **207**, 295 (1954).
159. S. K. Sharma and J. H. Quastel, *Biochem. J.* **94**, 790 (1965).
160. H. A. Kunz, *Helv. Physiol. Acta* **14**, 411 (1956).
161. C. J. Armett and J. R. Cooper, *J. Pharmacol. Exp. Ther.* **148**, 137 (1965).
162. J. R. Cooper, *Biochim. Biophys. Acta* **156**, 368 (1968).
163. J. E. Vincent, *Rec. Trav. Chim. Pays-Bas* **76**, 779 (1957).
164. K. H. Kiessling, *Ark. Kemi* **11**, 451 (1957).
165. T. Eckert and W. Möbus, *Hoppe-Seyler's Z. Physiol. Chem.* **338**, 286 (1964).
166. Y. Itokawa and J. R. Cooper, in "Methods in Enzymology" (D. B. McCormick and L. D. Wright, eds.), Vol. 18A, p. 226. Academic Press, New York, 1970.
167. J. R. Cooper and M. M. Kini, *J. Neurochem.* **19**, 1809 (1972).
168. R. L. Barchi and P. E. Braun, *Biochim. Biophys. Acta* **255**, 402 (1972).
169. Y. Hashitani and J. R. Cooper, *J. Biol. Chem.* **247**, 2117 (1972).
170. A. von Muralt, *Pfluegers Arch. Gesamte Physiol. Menschen Tiere* **247**, 1 (1943).
171. C. Tanaka and J. R. Cooper, *J. Histochem. Cytochem.* **16**, 362 (1968).
172. H. Hoffmann, T. Eckert, and W. Möbus, *Hoppe-Seyler's Z. Physiol. Chem.* **335**, 156 (1964).

173. F. Robinson, G. B. Solitare, J. B. LaMarche, and L. L. Levy, *Neurology* **17**, 472 (1967).
174. D. Leigh, *J. Neurol., Neurosurg. Psychiat.* [N.S.] **14**, 216 (1951).
175. H. E. Worsley, R. W. Brookfield, J. S. Elwood, R. R. Noble, and W. H. Taylor, *Arch. Dis. Childhood* **40**, 492 (1965).
176. F. A. Hommes, H. A. Polman, and J. D. Reerink, *Arch. Dis. Childhood* **43**, 423 (1968).
177. S. M. Podos, *Arch. Ophthalmol.* [N.S.] **81**, 504 (1970).
178. J. H. Pincus, Y. Itokawa, and J. R. Cooper, *Neurology* **19**, 841 (1969).
179. M. Victor, *Postgrad. Med.* **50**, 75 (1971).
180. R. B. Richter, *Neurology* **18**, 1125 (1968).
181. J. R. Cooper, Y. Itowaka, and J. H. Pincus, *Science* **164**, 74 (1969).
182. J. H. Pincus, *Develop. Med. Child Neurol.* **14**, 87 (1972).
183. J. R. Cooper and J. H. Pincus, *in* "Inborn Errors of Metabolism" (F. A. Hommes and C. J. Van den Berg, eds.), p. 119. Academic Press, New York, 1973.
184. J. V. Murphy, *Pediatrics* **51**, 710 (1973).
185. J. V. Murphy, F. Diven, and L. Craig, *Pediat. Res.* **8**, 393 (1974).
186. J. H. Pincus, J. R. Cooper, J. V. Murphy, E. F. Rabe, D. Lonsdale, and H. G. Dunn, *Pediatrics* **51**, 716 (1973).
187. C. Kawasaki, *Vitam. Horm.* (*New York*) **21**, 69 (1963).
188. S. Iida, *Biochem. Pharmacol.* **15**, 1139 (1966).
189. T. T. Tang, T. A. Good, P. R. Dyken, S. D. Johnson, S. R. McCreadie, S. T. Sly, H. A. Lardy, and F. B. Rudolph, *J. Pediat.* **81**, 189 (1972).
190. T. Yoshida, K. Tada, T. Konno, and T. Arakawa, *Tohoku J. Exp. Med.* **99**, 121 (1969).
191. M. G. Brunette, E. Delvin, B. Hazel, and C. R. Scriver, *Pediatrics* **50**, 702 (1972).
192. C. J. De Groot and F. A. Hommes, *J. Pediat.* **82**, 541 (1973).
193. F. A. Hommes, R. Berger, and G. Luit-de-Haan, *Pediat. Res.* **7**, 616 (1973).

CHAPTER 4

Lipoic Acid

Masahiko Koike and Kickiko Koike

I. INTRODUCTION

Lipoic acid was discovered independently in several laboratories in the late 1940's as a growth factor and a requirement for pyruvate oxidation for certain microorganisms. The biologically active substances, prior to their isolation and identification, were designated as Factor II for the ciliated protozoan *Tetrahymena* [1], the acetate replacing factor for *Lactobacillus casei* [2], the pyruvate oxidation factor for *Streptococcus faecalis* 10C1 [3], the protogen-Λ for *Tetrahymena geleii* [4,5], and the B.R. factor for *Butyribacterium rettgeri* [6]. In 1951 the factor was isolated in crystalline form from the water insoluble residue of beef liver by Reed *et al.* [7] who proposed the name "α-lipoic acid." The Lederle group [8] also crystallized active material later the same year (protogen-B, the monosulfoxide form of protogen-A). The following year Bullock *et al.* [9] established the structure of protogen-A as 6,8-dithioctanoic acid and proposed the name "6-thioctic acid" for the parent substance (α-lipoic acid). The American Society of Biological Chemists has recognized priority of the name "lipoic acid" and has adopted it as the trivial designation of 1,2-dithiolane-3-valeric acid. Lipoic acid is widely distributed in living organisms. Most nutritional experiments with higher animals have failed to demonstrate a growth response to lipoic acid. The only well-defined metabolic role of lipoic acid is that of a coenzyme in the α-ketoacid dehydrogenase complexes which catalyze CoA- and NAD-linked oxidative decarboxylation of α-keto acids. The chemistry and function of lipoic acid have been reviewed by Reed [10]. This article is not, therefore, intended to be a comprehensive review of all facets of lipoic acid. Information on biosynthesis and dissimilation of lipoic acid and a summary of its metabolic function will be presented.

II. ISOLATION AND SOME PROPERTIES

A. Brief Sketch of Isolation and Structure

Lipoic acid is widely distributed among microorganisms, plants, and animals in a protein-bound form. In the isolation of crystalline lipoic acid, the water insoluble residue of beef liver was processed as follows [11]: (1) hydrolysis with 6 N sulfuric acid, (2) extraction into benzene, (3) extraction into aqueous bicarbonate, (4) acidification and extraction into benzene, (5) esterification with diazomethane, (6) chromatography on alumina, and then on florisil, (7) saponification, and (8) crystallization of the acidic material from n-hexane to give pale yellow platelets, m.p. 47.5°C, $[\alpha]_D^{25} + 96.7°$ (c 1.88, benzene). A total of approximately 30 mg of the crystalline substance was obtained from 10 tons of liver residue. Patterson et $al.$ [8] also released lipoic acid from liver residue by alkaline hydrolysis. During the course of the isolation, lipoic acid was oxidized and the product was isolated as yellow oil, $[\alpha]_D^{25} + 105°$ (c 0.94, benzene); crystalline S-benzylthiuronium salt, m.p. 132–134°C. This substance was identified as a sulfoxide (β-lipoic acid). Elemental analyses and molecular weight determinations of lipoic acid established the formula $C_8H_{14}O_2S_2$ [11,12]. The structure 1,2-dithiolane-3-valeric acid was established by synthesis [9,11,13]. Absolute configuration of the natural (+)-form has been established [14]. Total synthesis by Bullock et $al.$ [9,13] gave DL-lipoic acid in an overall yield of approximately 8%. Modification of this method improved the yield of lipoic acid to from 30 to 39% [15]. Reed and Niu [16] reported a synthesis which

1,2-Dithiolane-3-valeric acid

gave DL-lipoic acid in a 36% overall yield. By suitable modification of this method, DL-[$^{35}S_2$]lipoic acid of high specific activity (68 μCi/mg) has been made available for biological use [17,18]. Walton *et al.* [19] reported the preparation of synthetic (+)- and (−)-lipoic acids using ephedrine. Some *N*-lipoylamino acids and peptides were also prepared from DL-lipoic-isobutyl carbonic anhydride [20].

B. Properties

Mild oxidation of lipoic acid with *tert*-butyl hydroperoxide or air converts it to a sulfoxide (β-lipoic acid). Under more drastic conditions, such as performic acid oxidation, lipoic acid is converted to 6,8-disulfooctanoic acid [21]. Lipoic acid is reduced easily to dihydrolipoic acid (6,8-dithioloctanoic acid) by sodium borohydride. Dihydrolipoic acid is oxidized readily to lipoic acid by reagents such as iodine, oxygen in the presence of metal cations, ferricyanide, and *o*-iodosobenzoate. Polarographic measurements at pH 7.0 gave the oxidation–reduction potential of −0.325 V for the pair, lipoic acid ⇌ dihydrolipoic acid [22].

Lipoic acid is relatively stable in the solid state. When it is heated above its melting point or when its solutions are exposed to light, it shows a tendency to polymerize. The lipoic acid polymers appear to be linear disulfides [23]. These polymers undergo depolymerization in basic solution, apparently by intramolecular thiol–disulfide interchange. Lipoic acid exhibits an absorption maximum at 330 nm ($\epsilon = 150$), whereas linear disulfides show an absorption maximum at 250 nm. This spectral shift is attributed to an increase in ring strain [24,25]. The unique reactivity and instability of 1,2-dithiolane point up this possibility. Equilibrium measurements of the reaction between 1,2-dithiolane and *n*-butane thiol and thioglycolate gave, respectively, the values of strain energy of 5.3 and 4.2 kcal/mole [25,26]. Values of 4 kcal/mole for the strain energy in 1,2-dithiolane and 3.5 kcal/mole for lipoic acid were obtained by Sunner [27]. It appears from these results that the five-membered ring as found in lipoic acid does not exhibit undue strain.

C. Protein-Bound Form of Lipoic Acid

1. FUNCTIONAL FORM

That lipoic acid in living organisms is bound covalently to protein is indicated by the observation that lipoic acid is not released by extraction of the α-ketoacid dehydrogenase multienzyme complexes with hot alcohol–ether or hot trichloroacetic acid [28]. The pyruvate and α-ketoglutarate dehydrogenase complexes from *Escherichia coli* [28,29] and

mammals [30,31] contain lipoic acid as a prosthetic group that is covalently bound to the transacylase component of these complexes, lipoate
acetyltransferase [32,33] and lipoate succinyltransferase [34,35]. The
lipoyl moiety in *E. coli* pyruvate and α-ketoglutarate dehydrogenase
complexes is bound in amide linkage to the ϵ-amino group of a lysine
residue. Further analyses [36] of the amino acid sequence around the
ϵ-N-lipoyllysine residue indicated that the pyruvate dehydrogenase complex contains the sequence Gly·Asp·ϵ-Lipoyl-Lys·Ala, and that the α-
ketoglutarate dehydrogenase complex contains the sequence Thr·Asp·ϵ-
Lipoyl-Lys·Val·(Val·Leu)·Glu.

2. Lipoic Acid-Incorporating Enzyme System

Reed *et al.* [37,38] observed that incubation of cell-free extracts of
lipoic acid deficient *S. faecalis* with lipoic acid resulted in activation of
the apopyruvate dehydrogenase complex and conversion of lipoic acid
to a protein-bound form. The crude extract was separated by protamine
sulfate fractionation into two fractions, both of which had to be incubated simultaneously with lipoic acid to obtain a holopyruvate dehydrogenase complex. The protamine sulfate supernatant fraction was further separated into PS-1, which contains lipoamidase, and PS-2, which
contains the lipoic acid-incorporating system. The latter fraction was
fractionated by alkaline ammonium sulfate solution into two protein
fractions, PS-2A and PS-2B, both of which are required, together with
ATP, for incorporation of lipoic acid into the apopyruvate dehydrogenase complex. One of the essential protein fractions produces
lipohydroxamate and pyrophosphate when incubated with lipoic acid,
ATP, and hydroxylamine. Lipoic acid and ATP, but neither of the protein fractions, are replaceable by synthetic lipoyl adenylate. Protein fractions PS-2A and PS-2B apparently catalyze Eqs. (1–3). Equation (1) is

$$E_I + ATP + \text{lipoic acid} \rightarrow E_I\text{-lipoyl-AMP} + PP_i \qquad (1)$$

$$E_I\text{-lipoyl-AMP} + E_{II} \rightarrow \text{lipoyl-}E_{II} + AMP + E_I \qquad (2)$$

$$\text{Lipoyl-}E_{II} + APDC \rightarrow \text{lipoyl-APDC} + E_{II} \qquad (3)$$

visualized as an activation of the carboxyl group of lipoic acid through
formation of lipoyl adenylate, which appears to be bound to a lipoic
acid-activating enzyme (E_I = PS-2A). Equations (2) and (3) are visualized as a transfer of the lipoyl moiety to the apopyruvate dehydrogenase
complex (APDC) by E_{II} (PS-2B). A lipoic acid-incorporating system has
also been isolated from *E. coli* extract [37] and has been detected in dog
and beef liver [39].

3. LIPOIC ACID-RELEASING ENZYME (LIPOAMIDASE)

The lipoyl moieties in the *E. coli* pyruvate and α-ketoglutarate dehydrogenase complexes are bound in amide linkage with the ϵ-amino group of lysine residues. A hydrolytic enzyme, lipoamidase, which releases lipoic acid from the *E. coli* complexes, has been purified approximately 100-fold [29,40,41] from *S. faecalis* extracts. Incubation of the *E. coli* complexes with lipoamidase releases about 96% of the protein-bound lipoic acid and results in a loss of the CoA- and NAD-linked oxidation of α-keto acids [29]. When the inactive complexes are incubated with DL-[$^{35}S_2$]lipoic acid, ATP, and the lipoic acid-incorporating system, about as much radioactive lipoic acid is incorporated into the complexes as is found in the native complexes, and the enzymatic activities of the reactivated and native complexes are essentially the same. These results indicate that lipoamidase cleaves the amide bond between lipoic acid and a protein lysine residue. Reactivation of the inactive complexes presumably involves reforming this bond and requires incubation with lipoic acid, ATP, and a lipoic acid-incorporating system. *Streptococcus faecalis* lipoamidase hydrolyzes methyllipoate, lipoamide, and several *N*-lipoylamino acids and peptides including ϵ-*N*-lipoyl-L-lysine. The presence of an apparently similar enzyme in bakers' yeast [42], *Corynebacterium bovis* [43], and various animal tissues [44] has been reported.

III. BIOSYNTHESIS AND DISSIMILATION

A. Biosynthesis

Although the enzymology and chemistry of lipoic acid has been studied in detail, relatively little is known about the biosynthesis and metabolic fate of the compound. Isotopic experiments with [1-^{14}C]octanoic acid and *E. coli* [45] indicate that octanoic acid may be a precursor of lipoic acid. The extent of incorporation of ^{14}C into lipoic acid was proportional to the concentration of [1-^{14}C]octanoic acid. [1-^{14}C]Octanoic acid appeared to be incorporated into lipoic acid as a unit, since C-1 of the biosynthesized lipoic acid corresponded to C-1 of the octanoic acid. No ^{14}C was incorporated into lipoic acid when [1-^{14}C]hexanoic acid, [1,6-^{14}C]adipic acid, or [1-^{14}C]-8-hydroxyoctanoic acid were included in the growth medium. In *E. coli* the source of the sulfur moiety of biosynthesized lipoic acid appears to be inorganic sulfate ion and methionine. However, the pathway of incorporation of sulfur into lipoic acid is not yet known (Y. Nose, personal communication).

B. Dissimilation

The nature of the metabolic need of lipoic acid remains largely unresolved due to the failure to produce its nutritional deficiency in animals. Wada et al. [46] observed that in healthy humans serum levels of lipoic acid were elevated sixfold 30 minutes after oral administration of 10 mg of lipoic acid, suggesting facilitated absorption of lipoic acid from the gut. Experiments with isolated rat intestine [47] and ligated dog intestine [48] showed that lipoic acid was rapidly absorbed into the mesenteric venous blood and was found in a bound form with plasma protein. It was reported earlier [49] that only small amounts of injected DL-lipoic acid were recoverable in some conjugated form from the urine of humans and several species of animals, and that no increase in the lipoic acid content of tissues followed its administration. These studies did not account for, but implied, a rapid distribution of lipoic acid in the body. In vitro, thiamine pyrophosphate, FAD, and lipoic acid are required for oxidative decarboxylation of pyruvate and α-ketoglutarate. However, simultaneous administration of thiamine did not affect the rate and amount of excretion of lipoic acid in the urine of normal animals [49]. Gal and Razevska [50] reported the in vivo distribution, utilization, and rate of elimination of DL-[^{35}S]lipoic acid both in normal and thiamine deficient rats. About 26–40% of the radioactivity of the injected dose (2 mg/100 gm body weight) was found in rat urine 24 hours after injection. Of the injected DL-lipoic acid, 9% was excreted as such, 34% of which was in the (+)-form as determined by microbiological assay. Blood, spleen, and liver of the thiamine deficient animals showed larger amounts of radioactivity. Intraperitoneal injection of DL-lipoic acid was toxic to thiamine deficient animals, but had no effect on thiamine excretion. Wada et al. [46] found that after intravenous injection of rats with 2.5 mg of [^{35}S]lipoic acid about 45% of the radioactivity was excreted in urine in the first 24 hours. Nakata [51] reported that a total of six unidentified biologically active lipoic acid-like substances, in addition to lipoic acid and its sulfoxide, were detected in human urine after oral administration of lipoic acid or lipoamide. Two of these metabolites were partially purified [52]. After acid hydrolysis, lipoic acid and ninhydrin-positive substances were detected. These lipoic acid conjugates were also detected in rat urine after the oral administration of [^{35}S]lipoic acid. Harrison and McCormick [53] have studied the metabolism of DL-[1,6-^{14}C]lipoic acid in rats and isolated rat liver. When labeled lipoic acid at a level of 0.5 mg/100 gm body weight was administered by intraperitoneal injection or stomach tube, the urinary excretion of radioactivity was maximal at 4 hours with 45% and 57% of radioactivity, respectively, recovered within 24 hours. Respiratory $^{14}CO_2$ from the same animals was maximal at 3 hours, after which it fell off markedly. About 30% of radioactivity was recovered as $^{14}CO_2$ within 24 hours and only a maximum of approximately 10% of the injected compound or its metabolite was retained in the body after 24 hours. This apparent discrepancy of retention of [^{35}S]lipoic acid and [$1,6-^{14}C$]lipoic acid has not yet been explained. Radioactivity (^{14}C) in the rat was greatest in liver, intestinal contents, and muscle in all cases. Ion-exchange and paper chromatography of 24 hour pooled urine revealed several water soluble radioactive metabolites. Incubation of [^{14}C]lipoic acid with liver homogenates or mitochondrial preparations resulted in the production of $^{14}CO_2$, which was decreased by incubation with unlabeled fatty acids and unaffected by the addition of carnitine or (+)-decanoylcarnitine. The rat apparently metabolizes lipoic acid via β-oxidation of the valeric acid side chain and by other metabolic reactions on the dithiolane ring, which render the molecule more water soluble.

A strain of the bacterium *Pseudomonas putida,* which grows on lipoic acid as the sole organic substrate that supplies carbon and sulfur, was isolated [54,55]. The bacterium catabolizes lipoic acid, largely via β-oxidation of the valeric acid side chain, to yield the two carbon shorter analogue, bisnorlipoic acid, or 4,6-dithiohexanoic acid as major catabolites present in the medium. As with rat liver, β-oxidation of the side chain appears to be the pathway employed by the pseudomonad to degrade lipoic acid.

IV. SUMMARY OF METABOLIC ROLE

It is well documented that lipoic acid is one of the essential coenzymes in the oxidative decarboxylation of α-keto acids, which is catalyzed by several enzymes through a coordinated sequence of reactions [10,56]. The α-keto acids such as pyruvate and α-ketoglutarate and branched α-keto acids are very important intermediate substances in the citric acid cycle and amino acid metabolism. Two species of the enzyme systems which catalyze the overall reaction [Eq. (9)], namely, the pyruvate and α-ketoglutarate dehydrogenase complexes, have been isolated as soluble functional units of high molecular weight (i.e.) multienzyme complexes [57] from pigeon breast muscle [58,59], bacteria [28,37,60–63], mammals [30,31,34,64–69], and *Neurospora* [70]. The complexes contain protein-bound thiamine pyrophosphate and FAD, besides lipoic acid, as essential coenzymes. Both types of complexes show characteristic shapes in the electron microscope and catalyze a coordinated sequence of reactions shown in Eqs. (4)–(8) (the brackets indicate enzyme-bound coenzyme and their derivatives):

$$R—CO—COOH + [TPP]—E_1 \rightarrow [R—CHO—TPP]—E_1 + CO_2 \quad (4)$$

$$[R—CHO—TPP]—E_1 + [L{<}^S_S]—E_2 \rightarrow [L{<}^{S—CO—R}_{SH}]—E_2 + [TPP]—E_1 \ (5)$$

$$[L{<}^{S—CO—R}_{SH}]—E_2 + CoASH \rightarrow [L{<}^{SH}_{SH}]—E_2 + R—CO—S—CoA \ (6)$$

$$[L{<}^{SH}_{SH}]—E_2 + [FAD]—E_3{<}^S_S] \rightarrow [H\dot{F}AD]—E_3{<}^{\dot{S}}_{SH} + [L{<}^S_S]—E_2 \quad (7)$$

$$[H\dot{F}AD]—E_3{<}^{\dot{S}}_{SH} + NAD^+ \rightarrow NADH + H^+ + [FAD]—E_3{<}^S_S] \quad (8)$$

Sum: $R—CO—COOH + CoA—SH + NAD^+ \rightarrow$

$$R—CO—S—CoA + CO_2 + NADH + H^+ \ (9)$$

where R is $CH_3—$ or $HOOC—CH_2—CH_2—$, E_1 is pyruvate or α-ketoglutarate dehydrogenase, E_2 is lipoate acetyltransferase or lipoate succinyltransferase, E_3 is lipoamide dehydrogenase, TPP is thiamine pyrophosphate, and

$$—E_3{<}^S_S]$$

indicates intramolecular disulfide linkage of lipoamide dehydrogenase.

Confirmation of this sequence has been achieved by dissociation of the complexes into their component enzymes and reconstitution of the complexes. As reported previously [32], the E. coli pyruvate dehydrogenase complex was separated into three enzymes, pyruvate dehydrogenase, lipoate acetyltransferase, and lipoamide dehydrogenase, and reconstituted from these enzymes. Recently, in this laboratory [33,34], the pig heart pyruvate and α-ketoglutarate dehydrogenase complexes were separated into their three component enzymes and reconstituted from them. The data [32,71] indicate that the E. coli pyruvate dehydrogenase complex (MW 4.6×10^6) contains a basic structural core consisting of one molecule of lipoate acetyltransferase (MW 1.7×10^6), to which 12 molecules of pyruvate dehydrogenase (MW 192,000) and 6 molecules of lipoamide dehydrogenase (MW 112,000) are noncovalently bound. The data in this laboratory indicate that the pig heart pyruvate dehydrogenase complex (MW 7.4×10^6) consists of one molecule of lipoate acetyltransferase (MW 1.98×10^6), 30 molecules of pyruvate dehydrogenase (MW 150,000), and 6 molecules of lipoamide dehydrogenase (MW 110,000) [33]. The *E. coli* and pig heart lipoate acetyltransferases consist of 24 apparently identical polypeptide chains with molecular weights of $\sim 65,000$–$70,000$ and 74,000, respectively, and each containing one molecule of covalently bound lipoic acid. The α-ketoglutarate dehydrogenase complex (MW 2.7×10^6) consists of one molecule of lipoate succinyltransferase (MW 1×10^6), 6 molecules of α-ketoglutarate dehydrogenase (MW 216,000) [71a], and 6 molecules of lipoamide dehydrogenase (MW 108,000) [34]. The subunit composition of *E. coli* α-ketoglutarate dehydrogenase complex [72] is apparently similar to that of the pig heart complex.

It is apparent that the lipoyl moiety, which is bound to lipoate acetyltransferase or lipoate succinyltransferase, undergoes a cycle of transformation, acyl-generation, acyl-transfer, and electron-transfer. The lipoyl moiety interacts with "aldehyde-TPP" which is bound to α-ketoacid dehydrogenase, with FAD, which is bound to lipoamide dehydrogenase, and with CoA, which is presumably bound to lipoate acyltransferase. The acyl-generation reaction [Eq. (5)] may be regarded as being a reductive acylation of enzyme-bound lipoic acid. In particular, "active aldehyde" is believed to attack the dithiolane ring of enzyme-bound lipoic acid in a nucleophilic displacement reaction, followed by a reverse condensation to form 6-acyldihydrolipoyl-lipoate acyltransferase (E_2), as illustrated in Eq. (10). Recently, Roche and Reed [73] reported that the bovine kidney pyruvate dehydrogenase catalyzes the reductive acylation of the lipoyl moieties of lipoate acetyltransferase.

$$\tag{10}$$

The acyl-transfer reaction [Eq. (6)] is a nucleophilic displacement of dihydrolipoyl-lipoate acyltransferase by CoA to produce acyl-S-CoA. Evidence for Eq. (6) is based largely on a model reaction carried out with substrate amounts of lipoic acid and *E. coli* lipoate acetyltransferase in the presence of phosphotransacetylase with acetyl phosphate and a catalytic amount of CoA. The thioester produced in this reaction has been isolated and characterized as (+)-6-S-acetyldihydrolipoic acid [74]. Reversal of this model reaction appears to be a process represented by Eq. (6). Recently, Barrera *et al.* [75] reported that [^{14}C]acetyl groups, which were incorporated into the bovine kidney pyruvate dehydrogenase complex in the presence of [2-^{14}C]pyruvate, TPP, and Mg^{2+}, were quantitatively released by exposure of the protein to performic acid vapor [76], and that most of the protein-bound radioactivity was also released when the ^{14}C-labeled complex was incubated with CoA and arsenite. These data suggest that the incorporated radioactivity was due to [^{14}C]acetyl groups bound to the protein in thioester linkage. Mono-S-succinyldihydrolipoic acid has been produced enzymatically by coupling succinic thiokinase with *E. coli* lipoate succinyltransferase in the presence of dihydrolipoic acid, succinate, ATP, and Mg^{2+} [77]. Recently, Pettit *et al.* [72] reported evidence, obtained by isotopic experiments with α-[5-^{14}C]ketoglutarate and the *E. coli* α-ketoglutarate dehydrogenase complex, that the S-succinyldihydrolipoyl moiety covalently bound to the lipoate succinyltransferase is an intermediate in the formation of succinyl-CoA.

In the overall oxidation of pyruvate [Eq. (9)] the interactions among coenzymes occur within a multienzyme complex in which the movement of the component enzymes is limited, and from which the intermediates do not appear to dissociate during the reaction. Therefore, it is assumed that these three component enzymes must have a special position in the complex permitting efficient interactions of the three coenzymes to achieve multistage sequence reaction. A possible solution to this enigma is provided by the discovery by Nawa *et al.* [21] that the lipoyl moieties are bound to the ϵ-amino groups of lysine residues. The efficient coupling among the enzyme-bound coenzymes within the complex may be achieved by rotation of the flexible arm of the lipoyllysine moiety (14 Å) (Fig. 1). The net charge on the lipoyl moiety during the reaction may be 0, minus 1, or minus 2. This change in the net charge may provide the driving force for displacement of the lipoyl moiety from one site to the

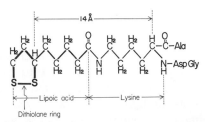

FIG. 1. The lipoyllysine moiety in the *E. coli* pyruvate dehydrogenase complex.

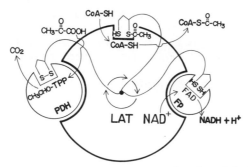

FIG. 2 Hypothetical scheme for intercoenzyme translocation of the flexible lipoyllysine moiety of lipoate acetyltransferase (LAT) in the pyruvate dehydrogenase multienzyme complex. PDH, Pyruvate dehydrogenase; Fp, lipoamide dehydrogenase.

next within the complex as shown in Fig. 2. A similar suggestion has been made independently by Green and Oda [78].

REFERENCES

1. G. W. Kidder and V. C. Dewey, *Arch. Biochem.* **8,** 293 (1945).
2. B. M. Guirard, E. E. Snell, and R. J. Williams, *Arch. Biochem.* **9,** 361 (1946).
3. D. J. O'Kane and I. C. Gunsalus, *J. Bacteriol.* **56,** 499 (1948).
4. E. L. R. Stokstad, C. E. Hoffmann, M. A. Regan, D. Fordham, and T. H. Jukes, *Arch. Biochem.* **20,** 75 (1949).
5. G. W. Kidder and V. C. Dewey, *Arch. Biochem.* **20,** 433 (1949).
6. L. Kline and H. A. Barker, *J. Bacteriol.* **60,** 349 (1950).
7. L. J. Reed, B. G. DeBusk, I. C. Gunsalus, and C. S. Hornberger, *Science* **114,** 93 (1951).
8. E. L. Patterson, J. A. Brockman, F. P. Day, J. V. Pierce, M. E. Macchi, C. E. Hoffmann, C. T. O. Fong, E. L. R. Stokstad, and T. H. Jukes, *J. Amer. Chem. Soc.* **73,** 5919 (1951).
9. M. W. Bullock, J. A. Brockman, E. L. Patterson, J. V. Pierce, and E. L. R. Stokstad, *J. Amer. Chem. Soc.* **74,** 3455 (1952).
10. L. J. Reed, *Compr. Biochem.* **14,** 99 (1966).
11. L. J. Reed, I. C. Gunsalus, G. H. F. Schnakenberg, O. F. Soper, H. E. Boaz, S. F. Kern, and T. V. Parke, *J. Amer. Chem. Soc.* **75,** 1267 (1953).
12. J. A. Brockman, Jr., E. L. R. Stokstad, E. L. Patterson, J. V. Pierce, and M. E. Macchi, *J. Amer. Chem. Soc.* **76,** 1827 (1954).
13. M. W. Bullock, J. A. Brockman, Jr., E. L. Patterson, J. V. Pierce, M. H. von Saltza, F. Sanders, and E. L. R. Stokstad, *J. Amer. Chem. Soc.* **76,** 1828 (1954).
14. K. Mislow and W. C. Meluch, *J. Amer. Chem. Soc.* **78,** 5920 (1956).
15. M. W. Bullock, J. J. Hand, and E. L. R. Stokstad, *J. Amer. Chem. Soc.* **79,** 1975 (1957).
16. L. J. Reed and C.-I. Niu, *J. Amer. Chem. Soc.* **77,** 416 (1955).
17. P. T. Adams, *J. Amer. Chem. Soc.* **77,** 5357 (1955).
18. R. C. Thomas and L. J. Reed, *J. Amer. Chem. Soc.* **77,** 5446 (1955).

19. E. Walton, A. F. Wagner, F. W. Bachelor, L. H. Patterson, F. W. Holly, and K. Folkers, *J. Amer. Chem. Soc.* **77**, 5144 (1955).
20. K. Daigo, W. T. Brady, and L. J. Reed, *J. Amer. Chem. Soc.* **84**, 662 (1962).
21. H. Nawa, W. T. Brady, M. Koike, and L. J. Reed, *J. Amer. Chem. Soc.* **82**, 896 (1960).
22. B. Ke, *Biochim. Biophys. Acta* **25**, 650 (1957).
23. R. C. Thomas and L. J. Reed, *J. Amer. Chem. Soc.* **78**, 6148 (1956).
24. M. Calvin, *Fed. Proc., Fed. Amer. Soc. Exp. Biol.* **13**, 697 (1954).
25. J. A. Barltrop, P. M. Hayes, and M. Calvin, *J. Amer. Chem. Soc.* **76**, 4348 (1954).
26. A. Fava, A. Iliceto, and E. Camera, *J. Amer. Chem. Soc.* **79**, 833 (1957).
27. S. Sunner, *Nature (London)* **176**, 217 (1955).
28. M. Koike, L. J. Reed, and W. R. Carroll, *J. Biol. Chem.* **235**, 1924 (1960).
29. M. Koike and L. J. Reed, *J. Biol. Chem.* **235**, 1931 (1960).
30. T. Hayakawa, M. Hirashima, S. Ide, M. Hamada, K. Okabe, and M. Koike, *J. Biol. Chem.* **241**, 4694 (1966).
31. M. Hirashima, T. Hayakawa, and M. Koike, *J. Biol. Chem.* **242**, 902 (1967).
32. M. Koike, L. J. Reed, and W. R. Carroll, *J. Biol. Chem.* **238**, 30 (1963).
33. T. Hayakawa, T. Kanzaki, T. Kitamura, Y. Fukuyoshi, Y. Sakurai, K. Koike, T. Suematsu, and M. Koike, *J. Biol. Chem.* **244**, 3660 (1969).
34. N. Tanaka, K. Koike, M. Hamada, K-I. Otsuka, T. Suematsu, and M. Koike, *J. Biol. Chem.* **247**, 4043 (1972).
35. N. Tanaka, K. Koike, K-I. Otsuka, M. Hamada, K. Ogasahara, and M. Koike, *J. Biol. Chem.* **249**, 191 (1974).
36. K. Daigo and L. J. Reed, *J. Amer. Chem. Soc.* **84**, 666 (1962).
37. L. J. Reed, F. R. Leach, and M. Koike, *J. Biol. Chem.* **232**, 123 (1958).
38. F. R. Leach, *in* "Methods in Enzymology" (D. B. McCormick and L. D. Wright, eds.), Vol. 18A, p. 282. Academic Press, New York, 1970.
39. J. N. Tsunoda and K. T. Yasunobu, *Arch. Biochem. Biophys.* **118**, 395 (1967).
40. L. J. Reed, M. Koike, M. L. Levitch, and F. R. Leach, *J. Biol. Chem.* **232**, 143 (1958).
41. K. Suzuki and L. J. Reed, *J. Biol. Chem.* **238**, 4021 (1963).
42. G. R. Seaman, *J. Biol. Chem.* **234**, 161 (1959).
43. M. Nakamura, *Vitamins* **26**, 217 (1962).
44. J.-I. Saito, *Vitamins* **39**, 317 (1969).
45. L. J. Reed, T. Okaichi, and I. Nakanishi, *Abstr., Int. Symp. Chem. Natur. Prod., 1964,* p. 218 (1964).
46. M. Wada, Y. Shigeta, and K. Inamori, *J. Vitaminol.(Kyoto)* **7**, 237 (1960).
47. T. Masuda, *Vitamins* **38**, 62 (1968).
48. T. Masuda, *Vitamins* **38**, 67 (1968).
49. E. L. Patterson, H. P. Broquist, M. H. von Saltza, A. Albrecht, E. L. R. Stokstad, and T. H. Jukes, *Amer. J. Clin. Nutr.* **4**, 269 (1956).
50. E. M. Gal and D. E. Razevska, *Arch. Biochem. Biophys.* **89**, 253 (1960).
51. T. Nakata, *Vitamins* **40**, 179 (1969).
52. T. Nakata, *Vitamins* **40**, 185 (1969).
53. E. H. Harrison and D. B. McCormick, *Arch. Biochem. Biophys.* **160**, 514 (1974).
54. J. C. H. Shih, L. D. Wright, and D. B. McCormick, *J. Bacteriol.* **112**, 1043 (1972).
55. J. C. H. Shih, M. L. Rozo, P. B. Williams, L. D. Wright, and D. B. McCormick, *Biochemistry* (submitted for publication).
56. I. C. Gunsalus, *in* "The Mechanism of Enzyme Action" (W. B. McElroy and H. B. Glass, eds.), p. 545. Johns Hopkins Press, Baltimore, Maryland, 1954.

57. L. J. Reed and D. J. Cox, *Annu. Rev. Biochem.* **35,** 57 (1966).
58. J. Jagannathan and R. S. Schweet, *J. Biol. Chem.* **196,** 551 (1952).
59. R. S. Schweet, B. Katchman, and R. M. Bock, *J. Biol. Chem.* **196,** 563 (1952).
60. D. S. Goldman, *Biochim. Biophys. Acta* **27,** 506 (1958).
61. G. Dennert and S. Höglund, *Eur. J. Biochem.* **12,** 502 (1970).
62. T. Hirabayashi and T. Harada, *Biochem. Biophys. Res. Commun.* **45,** 1369 (1971).
63. T. Hirabayashi and T. Harada, *J. Biochem. (Tokyo)* **71,** 797 (1972).
64. D. R. Sanadi, J. W. Littlefield, and R. M. Bock, *J. Biol. Chem.* **197,** 851 (1952).
65. V. Massey, *Biochim. Biophys. Acta* **38,** 447 (1960).
66. H. Holzer and K. Beaucamp, *Biochim. Biophys. Acta* **46,** 225 (1961).
67. E. Ishikawa, R. M. Oliver, and L. J. Reed, *Proc. Nat. Acad. Sci. U.S.* **56,** 534 (1966).
68. T. C. Linn, J. W. Pelley, F. H. Pettit, F. Hucho, D. D. Randall, and L. J. Reed, *Arch. Biochem. Biophys.* **148,** 327 (1972).
69. E. Junger and H. Reinauer, *Biochim. Biophys. Acta* **250,** 478 (1972).
70. R. W. Harding, D. F. Caroline, and R. P. Wagner, *Arch. Biochem. Biophys.* **138,** 653 (1970).
71. M. H. Eley, G. Namihira, L. Hamilton, P. Munk, and L. J. Reed, *Arch. Biochem. Biophys.* **152,** 655 (1972).
71a. K. Koike, M. Hamada, N. Tanaka, K-I. Otsuka, K. Ogasahara, and M. Koike, *J. Biol. Chem.* **249,** 3836 (1974).
72. F. H. Pettit, L. Hamilton, P. Munk, G. Namihira, M. H. Eley, C. R. Willms, and L. J. Reed, *J. Biol. Chem.* **248,** 5282 (1973).
73. T. E. Roche and L. J. Reed, *Biochem. Biophys. Res. Commun.* **48,** 840 (1972).
74. I. C. Gunsalus, L. S. Barton, and W. Gruber, *J. Amer. Chem. Soc.* **78,** 1763 (1956).
75. C. R. Barrera, G. Namihira, L. Hamilton, P. Munk, M. H. Eley, T. C. Linn, and L. J. Reed, *Arch. Biochem. Biophys.* **148,** 343 (1972).
76. F. Lynen, *Biochem. J.* **102,** 381 (1967).
77. I. C. Gunsalus and R. A. Smith, *Proc. Int. Symp. Enzyme Chem., 1957,* p. 77 (1958).
78. D. E. Green and T. Oda, *J. Biochem. (Tokyo)* **49,** 742 (1961).

CHAPTER 5

Biochemistry of Glutathione

Alton Meister

I. INTRODUCTION

Glutathione (L-γ-glutamyl-L-cysteinylglycine), a peptide which occurs in almost all living cells in appreciable concentrations, has attracted the attention of many investigators. Hundreds of publications have appeared on various aspects of the cellular disposition and function of glutathione in animals, plants, and microorganisms. Its ubiquitous distribution suggests that glutathione may play a fundamental role in virtually all living cells. There is also evidence that in the course of evolution glutathione has been adapted by various cells for specialized functions. It has been suggested that glutathione may be an important participant in many basic cell processes such as cell division and growth [1-4], regulation of energetic and other catalytic functions [3], and in memory and learning [5]. Though much effort has been expended, we still do not fully understand the biological functions of glutathione and the mechanisms that regulate its biosynthesis and utilization. However, information about the biochemistry of glutathione is accumulating rapidly, and it seems probable that many aspects of the physiological functions of glutathione will soon be understood at the molecular level.

Glutathione has two interesting structural features: a γ-glutamyl linkage and a sulfhydryl group. Much of the literature about glutathione has dealt with its sulfhydryl moiety, and indeed, most of the methods which have been used for its quantitative determination are based on reactions involving the sulfhydryl group. It is generally believed that the sulfhydryl group of glutathione is of importance in the maintenance of the sulfhydryl groups of other molecules including proteins, as a catalyst for disulfide exchange reactions, and in the detoxification of foreign compounds, hydrogen peroxide, and free radicals. There is evidence that the γ-glutamyl moiety of glutathione (and of closely related compounds) functions in the transport of amino acids and possibly also of peptides and amines. Still other possible functions of glutathione are considered below. The γ-glutamyl linkage in glutathione seems to protect it against breakdown catalyzed by the usual peptidases, which are specific for α-peptide bonds; on the other hand, glutathione is a substrate for certain peptidases which can cleave the cysteinyl–glycine bond.

Glutathione was discovered in 1888 by deRey-Pailhade [6,7], who isolated a crude material ("philothion") from yeast which appeared to be involved in the conversion of sulfur to hydrogen sulfide. Later, Hopkins [8] isolated a crystalline compound from yeast which he considered to be identical with deRey-Pailhade's compound, and which he thought to be a dipeptide containing glutamic acid and cysteine; Hopkins proposed the name glutathione. Later work [9,10] showed that glutathione is a tripeptide; the structure was established by synthesis [11]. Many interesting details of the early work on the isolation and characterization of glutathione have been reviewed [12].

The literature on glutathione is now very extensive; fortunately, progress in this field has been summarized in part at three conferences which were held in Ridgefield in 1953 [13], London in 1959 [14], and in Tübingen in 1973 [15]. Knox has written an excellent review covering material up to about 1959 [16]. A number of current ideas about glutathione were foreshadowed by Barrón, whose review in 1951 [3] stimulated many investigators.

An attempt has been made here to review the biochemistry of glutathione, perhaps with some tendency to emphasize relatively recent developments and ideas.

II. NATURAL OCCURRENCE OF GLUTATHIONE AND RELATED COMPOUNDS

Glutathione occurs in plants, bacteria, and animal tissues in concentrations that range from about 0.4 mM to about 12 mM. It is well known that yeast is an excellent source of glutathione [9,17], but it has also been isolated from such diverse sources as erythrocytes [9], rubber latex [18], orange juice [19], thymus [20], and peanuts [21], as well as from other biological materials. Rat liver, brain, muscle, and kidney have been reported to contain concentrations of 5.5–9.9, 1.7–2.3, 0.74–1.4, and 3.8–4.6 mM, respectively [22]. Human erythrocytes contain about 2 mM glutathione [23], and the concentration of glutathione in normal human lens decreases with age from about 3.5 mM at age 20 to about half of this value at age 65 [24]. Glutathione is present in *Escherichia coli* [25] and in several other microorganisms [26]. Values reported for the intracellular concentrations of glutathione vary depending upon the method used, and also with the physiological state [16,26]. Thus, there is evidence that the state of nutrition may influence cellular glutathione concentrations [27–30a]. Glutathione is usually the most abundant sulfhydryl compound present in tissues. Virtually all of the glutathione in animals is intracellular and present in the cytosol. However, there is evidence that some glutathione is present in the form of mixed disulfides with the sulfhydryl groups of proteins [31–37]. Small amounts of glutathione disulfide may occur intracellularly and in body fluids.

Glutathione is often the most abundant intracellular γ-glutamyl compound (excluding glutamine). However, a wide variety of other γ-glutamyl compounds have been isolated from various biological materials [38,39], and in some instances the concentrations of such compounds may exceed that of glutathione. For example, it has been reported that Mung bean seedlings contain large amounts of γ-glutamylcysteinyl-β-alanine (homoglutathione); the concentration of homoglutathione in this plant is much greater than that of glutathione [40]. Other glutathione analogues and derivatives which have been isolated include γ-glutamyl-α-aminobutyrylglycine (ophthalmic acid) and γ-glutamyalanylglycine (norophthalmic acid), which have been obtained from bovine lens [41,42]. In addition, S-sulfoglutathione [43] and S-(α,β-dicarboxyethyl)-glutathione [44] have been isolated from lens, and S-methylglutathione from brain [45]. The mixed disulfide of glutathione and coenzyme A has been found in yeast [46] and liver [47,48]. Many γ-glutamylamino acids have been isolated from plant sources. For example, Virtanen and his collaborators [49–58] obtained the γ-glutamyl derivatives of valine, isoleucine, leucine, methionine, phenylalanine, and several cysteine derivatives from the onion. γ-Glutamyl-β-aminoisobutyric acid and γ-glutamyl-β-alanine have been isolated from the ornamental crucifer [59] and iris bulbs [60,61]. β-N-(γ-L-Glutamyl)-4-hydroxymethylphenyl-hydrazine (agaritine) [62–64] and N-(γ-glutamyl)-4-hydroxyaniline [65] have been obtained from mushrooms. β-(N-γ-L-Glutamyl) aminopropionitrile, a compound that produces skeletal lathyrism, was isolated from *Lathyrus odoratus* seeds [66,67]. γ-Glutamyl-β-(methylenecycloprope) alanine (γ-glutamylhypoglycinyl A) [68–75], γ-glutamyl-β-cyanoalanine [76], γ-glutamyl-β-cyanoalanylglycine [77], and γ-glutamyl-β-(N-pyrazolyl alanine) [78] have also been found in plants. Glutathionyl spermidine (γ-glutamylcysteinylglycyl spermidine) has been found in high concentrations in *E. coli*. After cultures of this organism have attained maximal growth, about half of the glutathione is converted to glutathionyl spermidine [25]. The γ-glutamyl linkage also occurs in other natural compounds, for example, folic acid derivatives and capsular polyglutamate material from *Bacillus anthracis* and related organisms [38].

It is noteworthy that the γ-glutamyl derivatives of a number of the protein amino acids have been found in human urine [79] and also in mammalian brain [45,80–82]. In view of the many technical difficulties involved in the isolation of γ-glutamyl compounds, it seems probable that the occurrence of γ-glutamylamino acids and related compounds in biological materials may be much more extensive than indicated by currently available data. On the basis of what is known about the enzymology of glutathione, it would appear that many γ-glutamyl compounds are formed either by transpeptidation reactions with glutathione or by reactions analogous to that catalyzed by γ-glutamyl-cysteine synthetase.

III. DETERMINATION OF GLUTATHIONE

That many methods have been devised for determining glutathione and glutathione disulfide suggests that at least some of the published procedures are not entirely satisfactory. Indeed, difficulties are still encountered in the accurate and specific determination of glutathione in various biological materials. The procedures used for the preparation of samples for analysis must be such as to prevent the destruction of glutathione by transpeptidase as well as the oxidation of glutathione to glutathione disulfide, which has been reported to form complexes with tissue components [22,83].

The methods that have been used for the determination of glutathione were reviewed several years ago [84,85]; a few comments concerning these and some of the methods developed more recently seem appropriate. Procedures based on chemical reactions of various reagents with sulfhydryl groups, while generally not specific, have often given satisfactory results because glutathione is usually the major thiol present. Enzymatic methods are more specific and therefore often preferable to purely chemical ones. The glyoxylase reaction has often been used [86]. Since glutathione is a highly specific cofactor of glyoxylase, conditions can be obtained in which glutathione concentration is proportional to the rate of conversion of methylglyoxal to lactic acid catalyzed by glyoxylases I and II. Glutathione may also be determined with glyoxylase I; conversion of the glutathione adduct of methylglyoxal to S-lactoylglutathione can be determined from the increase in absorbance at 240 nm [87]. Davidson and Hird [22] developed a procedure based on thiol–disulfide exchange [88] between glutathione and cystamine in the presence of glutathione reductase and TPNH; the rate of the reaction is determined spectrophotometrically at 340 nm. The reactions involved may be represented as follows, where RSSR represents cystamine:

$$\text{GSH} + \text{RSSR} \rightarrow \text{GSSR} + \text{RSH} \qquad (1)$$

$$\text{GSH} + \text{GSSR} \rightarrow \text{GSSG} + \text{RSH} \qquad (2)$$

$$\text{GSSG} + \text{TPNH} + \text{H}^+ \xrightarrow[\text{reductase}]{\text{glutathione}} 2\,\text{GSH} + \text{TPN}^+ \qquad (3)$$

$$\overline{\text{RSSR} + \text{TPNH} + \text{H}^+ \longrightarrow 2\,\text{RSH} + \text{TPN}^+} \qquad (1) + (2) + (3)$$

Under appropriate conditions the rate of change of absorbance at 340 nm is proportional to glutathione concentration. In modifications of this procedure, cystamine is replaced by another disulfide [89–91]. Thus, in one procedure [91] the sample is incubated with glutathione reductase,

TPNH, and 5,5′-dithiobis-(2-nitrobenzoic acid), and the rate of formation of 2-nitro-5-thiobenzoic acid is determined from the rate of increase
in absorbance at 412 nm.

Other procedures which have been used for the determination of glutathione include a fluorometric assay based on reaction at pH 8 of glutathione with o-phthalaldehyde to yield a product whose fluorescence at
420 nm is proportional to the glutathione concentration; this method is
reported to be highly sensitive and specific [92]. Glutathione has been
determined by coulometric titration with silver (Ag^+) in ammoniacal
medium [93]. 5,5′-Dithiobis-(2-nitrobenzoic acid) [94] has been used for
the determination of glutathione by a number of investigators. Although
the reagent reacts with many thiol compounds, its use in specific cases,
for example in the determination of erythrocyte glutathione [95], offers
a rapid and convenient method. Glutathione disulfide can be determined
with 5,5′-dithiobis-(2-nitrobenzoic acid) after reduction with potassium
borohydride [96]. The determination of glutathione disulfide by paper chromatography has been described [97]. "Total" glutathione
(GSH + GSSG) has been determined by automated column chromatography of S-acetamidoglutathione [98]. In this procedure, the sample
is treated with potassium borohydride followed by treatment with iodoacetamide. Glutathione has also been determined by a procedure involving the use of formaldehyde dehydrogenase (purified from beef
liver); glutathione is a specific cofactor for this reaction [16,98a].

IV. BIOSYNTHESIS OF GLUTATHIONE

A. Introduction

In view of the very wide distribution of glutathione, and in the absence of substantial evidence for the transport of glutathione across cell
membranes, it would appear that virtually all cells can synthesize glutathione. There is now direct evidence that the synthesis of glutathione
occurs in liver, kidney, erythrocytes, $E.$ $coli,$ certain plants, and in other
cells. The synthesis of glutathione is catalyzed by the successive actions
of γ-glutamyl-cysteine synthetase and glutathione synthetase which,
respectively, catalyze the following reactions (where $M^{2+} = Mg^{2+}$ or
Mn^{2+}):

$$\text{L-Glutamate} + \text{L-cysteine} + \text{ATP} \underset{}{\overset{M^{2+}}{\rightleftharpoons}} \text{L-}\gamma\text{-glutamyl-L-cysteine} + \text{ADP} + P_i \quad (4)$$

$$\text{L-Glutamyl-L-cysteine} + \text{ATP} + \text{glycine} \underset{}{\overset{M^{2+}}{\rightleftharpoons}} \text{glutathione} + \text{ADP} + P_i \quad (5)$$

In 1941 Waelsch and Rittenberg [99] fed ^{15}N-labeled glutamic acid,
cysteine, or glycine to rats and observed extensive labeling of glu-

tathione. They found that the incorporation of [¹⁵N]glycine into glutathione was more rapid than into proteins. Bloch and Anker [100] subsequently observed incorporation of [¹⁵N]glycine into glutathione in rat liver slices, and Braunstein et al. [101] demonstrated synthesis of glutathione from its component amino acids in experiments on rat liver slices. Bloch and his colleagues [102–112] subsequently established the occurrence of the two enzyme catalyzed reactions given above in experiments on cell-free systems obtained from liver and yeast. Similar enzymatic activities have been found in *E. coli* [113], lens [114–116], plants [117–119], and erythrocytes [120–122]. The incorporation of labeled glutamate, cysteine, or glycine into glutathione has often been observed; for example, in studies on erythrocytes [123–131], brain [132–134], intestine [99,135,136], lens [137], muscle [138], and liver [99,132,135,136,138–140]. Some of the properties of γ-glutamyl-cysteine synthetase and glutathione synthetase are considered below; a more detailed review of the biosynthesis of glutathione is given elsewhere [141].

B. γ-Glutamyl-Cysteine Synthetase

γ-Glutamyl-cysteine synthetase has been purified from hog liver [108,112], bovine lens [115,116], wheat germ [118], human [121] and bovine [120] erythrocytes, rat liver [142], *Xenopus laevis* liver [142], and rat kidney [143]. The most highly purified and most active preparation of the enzyme is that from rat kidney. This enzyme preparation [143–145], which is homogeneous on acrylamide gel electrophoresis and analytical ultracentrifugation, exhibits a molecular weight of about 90,000. On acrylamide gel electrophoresis in the presence of sodium dodecyl sulfate and dithiothreitol, the enzyme dissociates into two components exhibiting molecular weights of about 20,000 and 70,000 [145]; the significance of the apparent subunits of this enzyme is not yet clear. Recent work [145] has shown that the component of molecular weight 70,000 is active in the absence of the other component. It may, perhaps, be relevant that a molecular weight of 78,000 was found for the enzyme from bovine erythrocytes [120]. The enzyme requires sulfhydryl groups for activity and is inhibited by sulfhydryl reagents. γ-Glutamyl-cysteine synthetase is relatively specific for L-glutamate, but some activity was observed with α-methylglutamate and β-methylglutamate. D-Glutamate reacts only very slightly as a substrate for dipeptide synthesis. L-Cysteine and L-α-aminobutyrate are about equally active as acceptor amino acids; substantial activity was also observed with S-methyl-L-cysteine, DL-C-allylglycine, β-chloro-L-alanine, L-norvaline, DL-homocysteine, L-threonine, and L-alanine when the reaction was carried out in the presence of magnesium ions. Substitution for magnesium ions by manganese ions leads to even less specificity for the acceptor amino acid. Those amino acids which react relatively slowly in the presence of magnesium ions are much more active with manganese ions. The ratio of the activity observed with manganese ions to that observed with magnesium ions is greater than unity for many of the amino acids that are poor substrates with magnesium ions, and less than unity for amino acids that are good substrates with magnesium ions [144]. β-Aminoisobutyrate was found to be a substrate for the lens enzyme. One isomer of this amino acid is two to three times more active than the other [146].

γ-Glutamyl-cysteine synthetase catalyzes several partial reactions [144]. Thus, the enzyme exhibits ATPase activity which is inhibited by L-glutamate and activated by L-α-aminobutyrate and L-cysteine. The enzyme can also catalyze the synthesis of pyrophosphate when it is incubated with ATP and inorganic phosphate according to the following reaction:

$$ATP + P_i \rightarrow PP_i + ADP \tag{6}$$

This reaction is inhibited by L-glutamate and activated by L-α-aminobutyrate. The ATPase activity of the enzyme is completely inhibited by 2 mM dithiothreitol, and when L-glutamate is added to the dithiothreitol-inhibited ATPase system, inorganic phosphate is formed. This L-glutamate-dependent cleavage of ATP is associated with cyclization of L-glutamate to form 5-oxoproline. When D-glutamate is substituted for L-glutamate in this system, the formation of 5-oxoproline is about five times greater than with L-glutamate. The cyclization of glutamate catalyzed by γ-glutamyl-cysteine synthetase closely resembles that catalyzed by glutamine synthetase and suggests that the enzymatic synthesis of γ-glutamylcysteine, like that of glutamine [147–149], involves intermediate formation of enzyme-bound γ-glutamyl phosphate. A further similarity between γ-glutamyl-cysteine synthetase and glutamine synthetase is that both enzymes are inhibited by methionine sulfoximine. Studies on glutamine synthetase showed that methionine sulfoximine competitively inhibits with respect to glutamate and also that it inactivates the enzyme in a reaction in which methionine sulfoximine is phosphorylated by ATP to yield methionine sulfoximine phosphate [150–154]. γ-Glutamyl-cysteine synthetase is also inhibited competitively by methionine sulfoximine and inactivation associated with stoichiometric formation of ADP and methionine sulfoximine phosphate occurs; both of these products bind tightly to the enzyme [155]. It is of interest that of the four diastereoisomers of methionine sulfoximine, only one (L-methionine-S-sulfoximine) inhibits both γ-glutamyl-cysteine synthetase [155] and glutamine synthetase [153]. Only this same isomer produces convulsions in mice [156]. Methionine sulfoximine phosphate is very tightly bound (noncovalently) to glutamine synthetase, and can only be released by denaturation of the enzyme; however, methionine sulfoximine phosphate may be removed from γ-glutamyl-cysteine synthetase by relatively mild procedures which lead to restoration of enzymatic activity. As discussed elsewhere [149,157], the findings indicate that methionine sulfoximine phosphate is an analogue of the tetrahedral intermediates (or transition states) formed on the enzymes in the reaction of γ-glutamyl phosphate with ammonia and cysteine, respectively. The formation of methionine sulfoximine phosphate in both reactions thus appears to reflect an aspect of the normal catalytic mechanism. When γ-glutamyl-cysteine synthetase was incubated with glutamate, ATP, and magnesium ions, evidence was obtained by gel filtration for the formation of a glutamate-containing enzyme complex, presumably enzyme-bound γ-glutamyl phosphate [144]. The intermediate formation of γ-glutamyl phosphate is consistent with earlier studies that showed that ^{18}O is transferred from glutamate to inorganic phosphate during enzymatic synthesis of the dipeptide [112,158].

γ-Glutamyl-cysteine synthetase does not catalyze exchange between ATP and inorganic phosphate or between ATP and ADP in the absence of all of the other components of the synthesis system [144]. However, since the reaction is reversible, incorporation of inorganic phosphate and of ADP into ATP can occur. Incorporation of α-aminobutyrate and of glutamate into γ-glutamyl-α-aminobutyrate has also been demonstrated. A scheme which accounts for all of the reactions catalyzed by γ-glutamyl-cysteine synthetase has been proposed, and a mechanism of action based on these and related findings has been presented and discussed [141,159]. According to this proposal, the enzyme can combine with ATP in the presence of divalent metal ion to yield a complex which can interact with (a) water, leading to cleavage of ATP; (b) inorganic phosphate, to yield ADP and inorganic pyrophosphate; (c) L-methionine-S-sulfoximine to yield methionine sulfoximine phosphate and ADP; or (d) glutamate, to form enzyme-bound γ-glutamyl phosphate. The interaction of enzyme-bound γ-glutamyl phosphate with L-cysteine (or L-α-aminobutyrate) leads to formation of the dipeptide. It is of interest that L-cysteine and L-α-aminobutyrate activate the ATPase activity of the enzyme and also its ability to phosphorylate inorganic phosphate to form pyrophosphate, as well as its ability to phosphorylate methionine sulfoximine. These findings suggest that the binding of L-cysteine or L-α-aminobutyrate facilitate in some manner the phosphorylation step of the synthetase reaction.

C. Glutathione Synthetase

This enzyme has been studied in preparations obtained from pigeon liver [104,106], *E. coli* [113], plants [119], yeast [110,160–163], erythrocytes [120,121,164,165], and rat kidney [166]. Highly purified preparations of glutathione synthetase have been obtained from yeast [160,161], bovine [164] and human [121] erythrocytes, and rat kidney [166]. The enzyme from yeast seems to be much more active than those obtained from other sources; however, in many respects the several purified preparations are similar. The yeast enzyme has a molecular weight of about 123,000 [160]; values of about 150,000 and 122,000 were obtained for the human and bovine erythrocyte enzymes, respectively. Evidence that the bovine erythrocyte enzyme contains two subunits of molecular weight about 60,000 has been reported [164].

Glutathione synthetase can catalyze the formation of γ-glutamyl-α-aminobutyrylglycine (ophthalmic acid) and γ-glutamylalanylglycine (norophthalmic acid) from the corresponding γ-glutamylamino acids; the γ-glutamyl derivatives of glycine, leucine, and β-alanine were not active when tested with the yeast enzyme. No activity was observed when glycine was replaced by β-alanine, L-aspartate, or L-alanine [160]; aminomethanesulfonic acid is active in place of glycine with pigeon liver glutathione synthetase [106].

There is good evidence that the reaction catalyzed by glutathione synthetase involves intermediate formation of an enzyme-bound dipeptide phosphate [141,159,160,162,162a,163]. In this respect, the mechanism resembles those involved in the synthesis of γ-glutamylcysteine and glutamine in which an acyl phosphate intermediate is also formed. Incubation of glutathione synthetase with ATP, magnesium ions, and γ-glutamyl-α-aminobutyrate leads to formation of γ-glutamyl-α-aminobutyryl phosphate. It was found that the enzyme can utilize chemically synthesized γ-glutamyl-α-aminobutyryl phosphate for the synthesis of tripeptide and, in the reverse direction, for the formation of ATP [162,163]. It was shown that enzyme-bound γ-glutamyl-α-aminobutyryl phosphate is formed rapidly and that it exchanges rapidly with free dipeptide in the absence of an acceptor. Such exchange occurred at a rate of about the same order of magnitude as that of the overall reaction, a finding consistent with the view that the intermediate acyl phosphate lies on the major catalytic pathway [160]. A scheme summarizing the reactions catalyzed by glutathione synthetase has been presented [141,159] in which it is proposed that an enzyme-ATP–magnesium ion complex is formed which interacts with and phosphorylates the dipeptide to yield enzyme bound dipeptidyl phosphate. This complex can then react with glycine (or hydroxylamine) to yield the tripeptide (or the dipeptidyl hydroxamate). This scheme explains several partial reactions catalyzed by the enzyme; these include the dipeptidyl transfer reaction, the arsenolysis of glutathione, and the incorporation of glycine into glutathione. These reactions require the presence of nucleotide and divalent metal ions and the available data indicate that the nucleotide functions in a catalytic role and can thus be considered to represent a part of the enzyme-bound activated substrate complex. A series of kinetic studies on purified bovine erythrocyte glutathione synthetase has been carried out [165]; these and related studies [120] have also shed light on the reaction mechanism.

D. Regulation of Glutathione Synthesis

Little information seems to be available about the occurrence of γ-glutamylcysteine in tissues; it may be present, but apparently does not occur in very high concentration. A relatively low tissue concentration of this dipeptide may well be expected because there are at least three highly active enzymes that could dispose of it. Thus, γ-glutamylcysteine is not only a substrate of glutathione synthetase, but it is also a substrate of γ-glutamyltranspeptidase and γ-glutamylcyclotransferase (see Section V). The synthesis of glutathione could be regulated by the intracellular concentrations of cysteine and γ-glutamylcysteine, which normally appear to be rather low. The utilization of γ-glutamylcysteine by transpeptidase or cyclotransferase could conceivably play a part in the regulation of glutathione synthesis. It has been suggested [167] that the synthesis of glutathione is protected in some way so as to prevent destruction of the intermediate γ-glutamylcysteine, and that there might be coordination or linkage between the two synthetases. On the other hand, as discussed below (Section V), it is possible that γ-glutamylcysteine participates extensively in transpeptidation and that only a fraction of the γ-glutamylcysteine produced is used for glutathione synthesis.

The levels of activities of γ-glutamylcysteine and glutathione synthetases present in human erythrocyte hemolysates have been reported both to be about the same [122,168] and to be markedly different [169,170]. Thus, Jackson [122] found glutathione synthetase to be slightly higher than γ-glutamyl-cysteine synthetase and Blume et al. [168] found γ-glutamyl-cysteine synthetase to be slightly higher than glutathione synthetase. In another study in which the rate of phosphate released from ATP in the presence of glutamate and cysteine was compared to that found with glutamate, cysteine, and glycine, the results suggested that the rate of γ-glutamylcysteine formation was about four times greater than that of glutathione formation [169]. Minnich et al. [170], using a more reliable procedure, found that γ-glutamyl-cysteine synthetase activity was about twice that of glutathione synthetase. Some patients with myeloproliferative disorders (e.g., myelofibrosis, leukemia) have increased erythrocyte glutathione concentrations [171], and data consistent with moderately increased erythrocyte glutathione synthetase activity have been obtained [172]. Administration of diaminodiphenyl-sulfone or methylene blue to rabbits leads to a rapid increase of erythro-cyte glutathione. These compounds seem to increase the activity of glu-tathione synthetase but not to affect γ-glutamyl-cysteine synthetase; the effect is thought to be related to an increase in the affinity of glutathione synthetase for γ-glutamylcysteine [173].

It has been reported that glutathione, but not glutathione disulfide, inhibits γ-glutamylcysteine synthesis somewhat in human erythrocyte hemolysates [122]. Patients with inherited erythrocyte glutathione synthetase deficiency were stated to exhibit higher than normal γ-glu-tamyl-cysteine synthetase activity [174], but actual determinations of enzyme activity were not given. Davis et al. [142] reported that purified rat liver γ-glutamyl-cysteine synthetase is inhibited by glutathione, glu-tathione disulfide, TPN, and TPNH, and is activated by glycine; these effects were not seen with the enzyme obtained from the X. laevis liver. Recent studies on highly purified rat kidney γ-glutamyl-cysteine synthe-tase have elucidated what seems to be an important mechanism for the regulation of glutathione biosynthesis [175]. It was found that this enzyme is substantially inhibited by glutathione under conditions very similar to those which prevail in vivo, thus strongly suggesting a physio-logically significant feedback control mechanism. The inhibition by glu-tathione is not allosteric in nature. Glutathione binds to the glutamate site of the enzyme and also to another enzyme site; the latter binding seems to require a sulfhydryl group since ophthalmic acid is only a weak inhibitor. The data suggest that glutathione binds both to the glutamate and cysteine sites of the enzyme. The apparent K_i value for glutathione is 2.3 mM, which is about equivalent to the average concentration of glutathione in rat kidney. Although inhibition of γ-glutamyl-cysteine synthetase by glutathione can be overcome by increasing the concentra-tion of L-glutamate, inhibition is not substantially reduced by L-α-aminobutyrate or L-cysteine. Studies with a number of compounds struc-turally related to glutathione indicate that the sulfhydryl group and γ-glutamyl moiety are essential for inhibition. These studies and con-siderations that have come from investigations on patients with 5-oxoprolinuria (see Section V,E) are consistent with the idea that glu-tathione biosynthesis is controlled in large measure by the intracellular concentration of glutathione and perhaps, also by the intracellular con-centration of L-cysteine (Fig. 1).

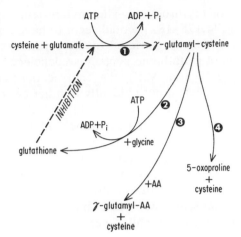

FIG. 1. Reactions leading to the synthesis and utilization of γ-glutamylcysteine [from 175]. 1, γ-Glutamyl-cysteine synthetase; 2, glutathione synthetase; 3, γ-glutamyltranspeptidase; 4, γ-glutamylcyclotransferase.

E. Genetic Deficiencies of γ-Glutamyl-Cysteine and Glutathione Synthetases

Many individuals throughout the world are afflicted with various forms of inherited erythrocyte glutathione deficiency [176]. The vast majority of these have a deficiency of erythrocyte glucose-6-phosphate dehydrogenase [177]; they are thus able to synthesize glutathione from its constitutent amino acids normally, but may fail to maintain glutathione in the reduced form. Only a few patients with deficiencies of γ-glutamyl-cysteine synthetase or glutathione synthetase have thus far been found. In 1961, Oort et al. [178,179] reported studies on a large family in which a number of individuals had markedly reduced erythrocyte glutathione concentrations and well compensated hemolytic disease. Boivin and Galand [174] studied two unrelated patients who had reduced erythrocyte glutathione concentrations as well as mild hemolytic anemia, and found decreased incorporation of glycine into glutathione in erythrocyte hemolysates. However, they also found apparently increased glutamate incorporation into a compound thought to be γ-glutamylcysteine [see also 180,181]. Studies by Mohler et al. [170,182] on a patient with similar symptoms and findings demonstrated a marked deficiency of glutathione synthetase activity in the erythrocytes; erythrocytes from the patient's parents and children exhibited glutathione synthetase activity levels that were about half of those of normal controls. Both the patient and his heterozygous relatives exhibited normal levels of erythrocyte γ-glutamyl-cysteine synthetase activity. The patients [174,178–182] would seem to have a glutathione synthetase deficiency which is restricted to the erythrocyte. In an apparently different form of glutathione synthetase deficiency associated with 5-oxoprolinuria (see Section V,E), the enzyme deficiency is reflected not only in the erythrocytes, but evidently in other cells as well.

Konrad *et al.* [183] described two siblings who exhibited severe deficiency of erythrocyte glutathione associated with a deficiency of erythrocyte γ-glutamyl-cysteine synthetase activity. The erythrocytes of these patients show substantial glutathione synthetase activity. These patients have central nervous system disease (spinocerebellar degeneration, mental retardation). In addition, these patients were found to have markedly reduced muscle and white blood cell glutathione levels [183a]; these patients also exhibit generalized aminoaciduria [167, 183a].

The available data indicate that the patients with glutathione deficiency exhibit erythrocyte glutathione concentrations which are no lower than about 5–10% of normal. It seems possible that complete absence of glutathione from the erythrocyte would be inconsistent with life. Further studies on individuals who have much less than normal amounts of glutathione would of course be of great interest and might shed additional light on the biochemical functions of glutathione. Studies on glutathione deficient animals or microorganisms might also be of importance. Certain sheep have been reported to exhibit decreased erythrocyte concentrations of glutathione [184–187]; while the nature of the defect requires further study, animals with decreased erythrocyte γ-glutamyl-cysteine synthetase activity have been described [187a].

V. UTILIZATION OF GLUTATHIONE AND RELATED γ-GLUTAMYL COMPOUNDS BY THE γ-GLUTAMYLTRANSPEPTIDASE-CYCLOTRANSFERASE-5-OXOPROLINASE PATHWAY

A. Introduction

A major pathway — perhaps the most significant quantitatively — of glutathione breakdown involves transpeptidation with amino acids to yield γ-glutamylamino acids and cysteinylglycine, a reaction catalyzed by γ-glutamyltranspeptidase [Eq. (7)].

$$\text{Glutathione} + \text{L-amino acid} \rightleftharpoons \text{L-}\gamma\text{-glutamyl-L-amino acid} + \text{L-cysteinylglycine} \quad (7)$$

$$\text{L-Cysteinylglycine} + H_2O \rightarrow \text{L-cysteine} + \text{glycine} \quad (8)$$

$$\text{L-}\gamma\text{-Glutamyl-L-amino acid} \rightarrow \text{5-oxo-L-proline} + \text{L-amino acid} \qquad (9)$$

$$\text{5-Oxo-L-proline} + \text{ATP} + 2\ H_2O \rightarrow \text{L-glutamate} + \text{ADP} + P_i \qquad (10)$$

Cysteinylglycine (or the corresponding disulfide) is cleaved by peptidase to yield the corresponding free amino acids [Eq. (8)]. γ-Glutamylcyclotransferase catalyzes the conversion of γ-glutamylamino acids to 5-oxoproline and the corresponding amino acid [Eq. (9)]. 5-Oxoproline is converted to glutamate in an ATP-requiring reaction catalyzed by 5-oxoprolinase [Eq. (10)]. These reactions, which account for the conversion of glutathione to glutamate, cysteine, and glycine, are coupled with those which lead to the synthesis of glutathione; these six reactions constitute the γ-glutamyl cycle (Section V,E). γ-Glutamyltranspeptidase also catalyzes analogous transpeptidation reactions with other γ-glutamyl compounds and reactions in which the glutamyl moiety is transferred to peptides; there is also evidence that this enzyme can, at least to some extent, catalyze the hydrolysis of glutathione to yield glutamate. γ-Glutamylcyclotransferase acts also on γ-glutamylglutathione and other di-γ-glutamyl derivatives (which can be formed by transpeptidation) to yield 5-oxoproline and the corresponding γ-glutamyl compounds.

B. γ-Glutamyltranspeptidase

Transpeptidation phenomena were observed many years ago in studies in which it was found that proteases can catalyze the formation of peptide bonds [188]. Bergmann and Fraenkel-Conrat [189] showed that the papain catalyzed formation of benzoylglycinanilide from benzoylglycinamide and aniline occurred by direct replacement and without prior hydrolysis of benzoylglycinamide to benzoylglycine and ammonia. Fruton and collaborators [190–195] obtained evidence that such transfer reactions involve formation of activated enzyme–substrate complexes which can react either with water, resulting in hydrolysis, or with a replacement agent, resulting in transfer. The ability of enzymes to catalyze both transfer and hydrolysis reactions is fairly general and has been observed not only with peptidases and amidases but also with a large number of other enzymes. According to Fruton's concept, one would expect competition between water and acceptor molecule for reaction with the activated enzyme–substrate complex. In view of the very high concentration of water compared to that of acceptor, it appears that the enzyme can interact preferentially with the acceptor molecule.

In early studies on the enzymatic hydrolysis of glutathione, it was found that pancreatic carboxypolypeptidase could split the cysteinylglycine bond [196]. An activity had been found earlier in pancreas (which may have been carboxypeptidase, another protease, or γ-glu-tamyltranspeptidase), which had an antiglyoxalase effect [197]. A simi-lar antiglyoxalase activity was found in rat kidney by Platt and Schroeder [198], and this activity was shown to destroy glutathione. Thus, Woodward *et al.* [199] and Schroeder *et al.* [200,201] found that glutathione and glutathione disulfide are completely hydrolyzed by en-zyme activities present in the kidneys of several species. Woodward and Reinhart [202] found that rat kidney extracts split both peptide bonds of glutathione to yield glycine, cysteine, glutamate, and 5-oxoproline. In these studies, which seem to represent the first reported evidence for the enzymatic formation of 5-oxoproline, it was found that the relative amounts of glutamic acid and 5-oxoproline formed varied depending upon the pH of the reaction mixture. Thus, more glutamate was found at pH values lower than about 6.6 while more 5-oxoproline was found at higher pH values. Binkley and collaborators [203,204] found that the breakdown of glutathione involved two enzymatic steps in which cys-teinylglycine was formed as an intermediate. The enzyme responsible for cleavage of the γ-glutamyl linkage was thought to be restricted to kidney and intestine, while the one responsible for the hydrolysis of cys-teinylglycine was shown to have a broad tissue distribution. Binkley suggested that cysteinylglycinase is not a protein and presented evidence that this enzyme is a form of ribonucleic acid [205,206], but subsequent studies [207] have not supported this idea.

Hanes *et al.* [208,209] clarified considerably the enzymology of glu-tathione degradation by demonstrating that the first step in glutathione breakdown is a transpeptidation reaction between glutathione and amino acids leading to the formation of γ-glutamylamino acids and cys-teinylglycine according to Eq. (7). Hanes *et al.* found that their enzyme preparations (from kidney and pancreas) also slowly catalyzed the liber-ation of glutamate from glutathione and also from the γ-glutamylamino acid product, that glutathione could be replaced by several γ-glu-tamylamino acids, and that a number of amino acids could serve as ac-ceptors. Subsequent studies have confirmed and extended these findings. Thus, Fodor *et al.* [210,211] showed that the cleavage of the glutamyl moiety of glutathione by preparations obtained from kidney, liver, and brain is greatly accelerated by various amino acids and also by glycylgly-cine and glycyl-L-alanine; glycine, glycylglycylglycine, and several other glycyl peptides were less active. It was also shown that the activation of glutathione hydrolysis by glutamine observed earlier [212] was due to transfer of the glutamyl moiety of glutathione to glutamine, resulting in formation of γ-glutamylglutamine. The reversibility of the transpeptida-tion reaction was demonstrated; thus, glutathione was formed when purified kidney preparations were incubated with γ-glutamyl peptides and cysteinylglycine [211]. Hird and Springell [213,214] purified the transpeptidase and showed that it did not catalyze formation of 5-oxoproline; these workers also obtained evidence that the enzyme was active with most of the protein amino acids. Kinoshita and Ball [215] demonstrated that kidney transpeptidase could form γ-glutamylarginine, and found that glutathione may be replaced by glutathione disulfide or by γ-glutamylglycine, but not by isoglutathione, glutamine, glutamate, or β-aspartyltyrosine. Ball *et al.* [216] obtained evidence, based on the pH-dependence of transpeptidation with various amino acids, that the un-ionized amino group of the acceptor is the reactive species, but studies on a large number of amino acids and peptides failed to reveal a simple relationship between structure and reactivity [217].

Several purified preparations of γ-glutamyltranspeptidase have been obtained from kidney [218–222a]. Highly purified kidney γ-glutamyltranspeptidase preparations (about 1000-fold purified) were found to be glycoproteins containing about 30% carbohydrate [219,220]. Thus, purified γ-glutamyltranspeptidase from hog kidney was found to contain 17.8% neutral sugar (as galactose), 8% amino sugar (as glucosamine), and 9.7% sialic acid. The purified enzymes are active with glutathione and a variety of γ-glutamyl compounds including γ-glutamylnaphthylamide and γ-glutamyl *p*-nitroanilide. In contrast, another preparation of the enzyme from pig kidney was reported not to act on γ-glutamylnaphthylamide [222]. Treatment of purified γ-glutamyltranspeptidase with neuraminidase was shown to yield products which included *N*-acetyl- and *N*-glycolylneuraminic acids; activity was not affected by incubation with neuraminidase [223]. L-Serine in the presence of borate competitively inhibits γ-glutamyltranspeptidase [217]. Serine and borate protects the enzyme against irreversible inactivation by iodoacetamide [224] suggesting that iodoacetamide alkylates a group at or near the active center of the enzyme. Hydrolysis of the alkylated enzyme was shown to yield glycolic acid.

A highly purified preparation of rat kidney γ-glutamyltranspeptidase similar in activity to those previously obtained from hog and beef kidney [219,220] has been studied in detail with respect to its γ-glutamyl donor and acceptor specificities [222a]. Many γ-glutamylamino acids were active as donors. The most active compounds were γ-glutamylglutamine and γ-glutamylmethionine. Glutathione disulfide was about 5% as active as glutathione, but considerable activity was observed with *S*-methyl glutathione and ophthalmic acid. S-Substituted glutathione derivatives were also active, and certain of these were more active than glutathione; the use of these compounds has led to development of convenient spectrophotometric methods for the determination of γ-glutamyltranspeptidase activity. Kinetic studies indicated a "ping-pong" mechanism consistent with the intermediate formation of a γ-glutamyl enzyme. A large number of amino acids were found to be active as acceptors of the γ-glutamyl moiety. In addition, a number of dipeptides were active, e.g., glycylglycine, glycyl-L-alanine, L-methionyl-L-serine, L-glutaminyl-L-glutamine, and L-α-aminobutyrylglycine. Aminoacyl glycine derivatives were, in the instances studied, more active than the corresponding free N-terminal amino acids, while the corresponding aminoacyl L-alanine derivatives were less active than the aminoacyl glycines. γ-Glutamyltranspeptidase activity was found to be inhibited substantially by the γ-glutamylhydrazones of a number of α-keto acids [224a], and such inhibition was found to be competitive with respect to the γ-glutamyl donor [222a] (see also pp. 135, 136).

Several observations have been reported concerning the activation of γ-glutamyltranspeptidase by cations. In one study, moderate activation (about 30%) by magnesium ions was found [220]. Later, much greater activation was reported in the presence of monovalent cations including Na^+ and K^+ [224b]. However, these effects seem to occur only in reaction mixtures containing model substrates such as γ-glutamyl-p-nitroanilide and γ-glutamylnaphthylamide; they were not found in studies in which enzyme activity was assayed with glutathione and an amino acid acceptor [222a,224c].

Glutamine, which is a poor γ-glutamyl donor substrate for γ-glutamyltranspeptidase as compared to glutathione, is slowly hydrolyzed and converted to γ-glutamylglutamine by the enzyme. However, in the presence of maleate, the hydrolysis of glutamine is increased by about tenfold, as is its conversion to γ-glutamylhydroxamate in the presence of hydroxylamine [224d]. The transpeptidase catalyzes γ-glutamylhydroxamate formation from a wide variety of γ-glutamyl compounds and hydroxylamine and this reaction is stimulated four- to fivefold by maleate. Studies on the effect of maleate on γ-glutamyltranspeptidase have led to the conclusion that this reagent decreases reaction of the intermediate γ-glutamyl enzyme with acceptor, thus markedly increasing hydrolysis of the γ-glutamyl donor, probably by affecting the enzyme so as to facilitate reaction of the γ-glutamyl enzyme with water. These [224d] and related observations [224e] suggest that the activity previously described as "maleate-stimulated phosphate-independent glutaminase" [224f,224g] is actually a catalytic function of γ-glutamyltranspeptidase. The latter activity had previously been purified from kidney by a procedure involving the use of proteolytic enzymes [224h,224i]. Similar treatment of the highly purified γ-glutamyltranspeptidase from rat kidney [222a] led to a preparation of the enzyme which exhibited about sixfold higher activity [224j], and which exhibited all of the catalytic properties (including maleate-stimulated glutaminase activity) of less active preparations.

Although mammalian kidney is an excellent source of γ-glutamyltranspeptidase, several investigations had indicated that the enzyme is also present in pancreas [208,209,217], brain [210], intestine [225], and liver [210,217], and subsequent studies show that it is present in virtually all mammalian tissues [226] and in certain insects [227–227b] and plants [228]. It has also been found in bacteria [229–231b]. Early studies on transpeptidase of kidney revealed that this enzyme is associated with tissue particulates [209,213,214,232], and that special procedures are required for its solubilization. The lack of specificity of γ-glutamyltranspeptidase with respect to the γ-glutamyl donor made it possible to develop simple colorimetric methods for the determination of the enzyme analogous to those which had been applied in the study of various peptidases [233,234]. Thus, procedures were designed for the colorimetric determination of γ-glutamyltranspeptidase involving the use of chromogenic substrates such as N-(γ-glutamyl)aniline [235] and L-γ-glutamyl-α-naphthylamide [236]. A very useful substrate is γ-glutamyl-p-nitroanilide [237], which is currently used for the clinical determination of serum γ-glutamyltranspeptidase activity; in this reaction, the formation of free p-nitroaniline is determined from the increase in absorbance at 405 nm. Similar substrates (e.g., various aminoacyl naphthylamides) have been useful for histochemical demonstration of aminopeptidases [238–240], and have also been used effectively in histochemical studies of γ-glutamyltranspeptidase activity [241,242].

In this method, the tissue slice is incubated in a buffered solution containing substrates such as γ-glutamylnaphthylamide and glycylglycine, together with a diazonium salt. The naphthylamine liberated by the action of the transpeptidase reacts with the diazonium salt to yield a colored product. The procedure has been modified for electron microscopic localization of the enzyme [243–245]. Histochemical studies using the light microscope indicated membrane-bound localization of γ-glutamyltranspeptidase, and membraneous localization has been confirmed by electron microscopic studies.

γ-Glutamyltranspeptidase was found to be localized in the brush border of the proximal convoluted tubules of the kidney in several species [241,242,246]. In confirmation and extension of the histochemical studies, investigations on the isolated brush border fraction of rat kidney indicate that γ-glutamyltranspeptidase represents about 1.5% of the membrane protein [247,247a]. Histochemical studies on other tissues indicate similar localization [225,241,242,246,248–250]. In the liver, activity was observed in the bile duct epithelium with slight staining of periportal hepatic cells. The enzyme is localized in both the nucleus and the apical portion of the epithelial cells covering the jejunal villi, and more activity was found in the jejunal epithelium than in the epithelium of other regions of the gastrointestinal tract. Staining activity was noted in granules in the acinar cells of the pancreas; none was seen in the islets. Transpeptidase activity was also found in the epithelium of bronchioles, in the hypophysis, in the salivary glands, and in the epithelia of the uterus, ovaries, fallopian tubes, epididymis, prostate, and seminal vesicles. Histochemical studies have shown localization of the transpeptidase in the apical portions of the epithelial cells of the choroid plexus. In addition, activity was found in the capillaries of the cerebrum, cerebellum, and spinal cord in several mammals [250], but capillary localization was not seen in a recent investigation [251]. The ciliary epithelium was found to exhibit an intense reaction, especially the distal regions of the ciliary processes [252]. The distribution and intensity of transpeptidase activity was similar to that noted in the choroid plexus. Transpeptidase activity was also found in the nonpigmented epithelial cells of the iris at the posterior pupillary margin. Transpeptidase localization has also been reported in certain Purkinje cells and in anterior horn cells of the spinal cord [251]. Histochemical and enzymatic studies have shown that the γ-glutamyltranspeptidase activity of fetal rat kidney is much lower than that of adult kidney. In contrast, the activity of fetal liver, brain, and lung was found to be greater than those of the newborn or adult animals; similar results were obtained in studies on human tissues [248].

In most of the reported studies on the histochemical localization of γ-glutamyltranspeptidase, the tissues were prepared in the usual manner by procedures involving use of fixatives and paraffin infiltration. It is remarkable that the transpeptidase activity is evidently not completely destroyed by such procedures, but the question arises as to whether some fraction of the enzyme initially present may be inactivated. Further development and refinement of histochemical procedures may lead to additional elucidation of the intracellular localization of this enzyme.

Bodnaryk [252a] has recently cited evidence indicating that in house fly larva γ-glutamyltranspeptidase is localized on the brush border of the proximal portion of the Malpighian tubules and the striated border of the epithelial cells of the midgut. The Malpighian tubules are analogous to the mammalian kidney and the striated border of the midgut is believed to be involved in absorption of amino acids in insects.

Although the evidence indicates that most of the γ-glutamyltranspeptidase activity of mammalian tissues is bound to membranes, it appears that a soluble form of the enzyme also exists. Thus, γ-glutamyltranspeptidase activity has been found in human urine and blood serum [235,253–255]. The serum γ-glutamyltranspeptidase activity is elevated in certain liver diseases and in certain other conditions, and therefore has diagnostic value [256–261]. Evidence for the occurrence of a soluble form of γ-glutamyltranspeptidase in human liver and kidney has been reported [262]. Both the soluble and particulate-bound forms of the enzyme appear to contain sialic acid and also exhibit several other properties in common. This suggests that the two forms are related; it is possible that the conversion of the particulate-bound form to the soluble form may be enzyme catalyzed. It has been suggested that the serum enzyme arises largely from the liver, probably from the epithelial cells lining the biliary ductules, but some serum transpeptidase may arise from the pancreas [257]. Most of the clinical interest has been directed to the causes of increased serum transpeptidase activity. However, a deficiency (about 5% of normal) of serum transpeptidase was found in a mentally retarded patient who also exhibits evidence of glutathionemia and glutathionuria [263]. The nature of this interesting disease is not yet known.

The γ-glutamyltranspeptidase of kidney bean fruit has also been reported to be a soluble enzyme [228]. The enzyme was purified to apparent homogeneity and its catalytic properties were found to resemble those of the mammalian kidney enzyme. The existence of this enzyme in plants could account for the formation of γ-glutamylamino acids, many of which have been found in plants (Section II).

C. γ-Glutamylcyclotransferase and the Formation of 5-Oxoproline

Evidence for γ-glutamylcyclotransferase was first indicated by the studies of Woodward and Reinhart [202] who observed the enzymatic formation of 5-oxoproline from glutathione. Connell and Hanes [264] subsequently found an activity in liver that converted γ-glutamylglycine, γ-glutamylglutamic acid, γ-glutamylphenylalanine, and glutathione into 5-oxoproline and the corresponding amino acids or peptide. (The activity observed toward glutathione may probably be ascribed to the presence of some γ-glutamyltranspeptidase activity in the enzyme preparation used.) Subsequent studies [265–270] have confirmed and extended these observations. The enzyme is widely distributed in mammalian tissues [266,270] and has been obtained in highly purified form from human and sheep brain [268] and from pig liver [269]. Studies on the purification of γ-glutamylcyclotransferase from rat liver indicate that the enzyme undergoes modification during isolation, a finding that may probably be explained by limited proteolysis of the enzyme. The purified rat liver enzyme has a molecular weight of about 27,000 [270]. The preparations of the enzyme that have thus far been obtained act rapidly on the γ-glutamyl derivatives of glutamine, cysteine, alanine, glycine, α-aminobutyrate, and methionine, but are much less active toward the γ-glutamyl derivatives of most of the other protein amino acids. The enzyme is very active toward γ-glutamyl-γ-glutamyl amino acids, and it also acts on γ-glutamyl-γ-glutamyl-p-nitroanilide, which is converted to 5-oxoproline and γ-glutamyl-p-nitroanilide [268]; a similar model substrate is γ-glutamyl-γ-glutamyl-α-naphthylamide [269]. Di-γ-glutamyl-amino acids may be formed by the action of γ-glutamyltranspeptidase. Thus, a γ-glutamylamino acid may be converted to 5-oxoproline and the free amino acid by the direct action of γ-glutamylcyclotransferase (γ-GCT), and the same overall result may take place by coupled reactions involving the participation of γ-glutamyltranspeptidase (γ-GTP). Two pathways are possible:

Pathway A:

$$\gamma\text{-Glutamylamino acid} + \gamma\text{-glutamylamino acid} \xrightarrow{\gamma\text{-GTP}}$$
$$\gamma\text{-glutamyl-}\gamma\text{-glutamylamino acid} + \text{amino acid} \tag{11}$$

$$\gamma\text{-Glutamyl-}\gamma\text{-glutamylamino acid} \xrightarrow{\gamma\text{-GCT}} \gamma\text{-glutamylamino acid} + \text{5-oxoproline} \tag{12}$$

Pathway B:

$$\gamma\text{-Glutamylamino acid} + \text{glutamine} \xrightarrow{\gamma\text{-GTP}} \gamma\text{-glutamylglutamine} + \text{amino acid} \tag{13}$$

$$\gamma\text{-Glutamylglutamine} \xrightarrow{\gamma\text{-GCT}} \text{5-oxoproline} + \text{glutamine} \tag{14}$$

The inclusion of glutamine as a substrate in the reactions given under pathway B is somewhat arbitrary because a few other amino acids (e.g., alanine, methionine, cystine, α-aminobutyrate) would be expected to participate equally well. However, the high concentrations of glutamine found in various tissues suggests that this amino acid may be of special importance in the metabolic degradation of γ-glutamylamino acids.

A γ-glutamylcyclotransferase that is relatively specific for L-γ-glutamyl-L-phenylalanine has been obtained from house fly pupae [271]; the molecular weight of this enzyme (about 30,000) is not far from that of the rat liver enzyme [270]. House fly γ-glutamylcyclotransferase is thought to play a role in the breakdown of L-γ-glutamyl-L-phenylalanine which accumulates during growth of the larvae; after transformation into a white pupae, the accumulated phenylalanine disappears coincident with hardening and darkening of the cuticle of the pupa. It is of interest that while the purified enzyme also acts on γ-glutamylmethionine and di-γ-glutamyl-*p*-nitroanilide, a wide variety of other γ-glutamylamino acids are inactive. The enzyme would seem to be highly adapted to the function of liberating phenylalanine at a particular stage in the insect's life cycle.

It seems probable that most of the 5-oxo-L-proline formed in mammalian tissues is produced by the action of γ-glutamylcyclotransferase. However, 5-oxoproline may be formed by several other pathways [272]. For example, L-glutamine cyclotransferase, first found in papaya latex [273–275], converts glutamine and glutaminyl peptides to 5-oxoproline and 5-oxoprolyl peptides, respectively. The enzyme was also found to catalyze the formation of 5-oxoprolyl-tRNA from glutaminyl-tRNA [276]. Such an activity may conceivably also occur in animal tissues. A bacterial L-glutamate cyclotransferase has also been reported [277] which is not active toward L-glutamine or glutathione. The conversion of L-glutamate to 5-oxoproline was reported to be catalyzed by rat liver nuclear preparations [278–280]. The nature of this reaction, which requires energy, has not yet been established, but it is conceivable that it involves synthesis of a γ-glutamyl compound followed by cyclization.

Ratner [281] recovered most of D-glutamic acid administered to rats as urinary 5-oxo-D-proline. Subsequently, an enzyme that catalyzes conversion of D-glutamate to 5-oxo-D-proline was found in kidney and liver of several species including man [282,283]. This enzyme seems to account for the appearance of 5-oxo-D-proline in the urine of animals fed D-glutamate. The 5-oxo-D-proline found in human urine [283] probably arises from dietary or bacterial sources. It appears probable that humans and other mammals are continually exposed to D-glutamate. This D-amino acid is a very poor substrate for D-amino acid oxidase and the available data indicate that it is removed from the organism by cyclization to 5-oxo-D-proline followed by excretion of the latter compound in the urine. Studies in which labeled L- and D-glutamates were administered to rats showed that these amino acids are taken up from the blood much more rapidly by the kidney than by the liver. Small amounts of 5-oxo-L-proline were formed after administration of labeled L-glutamate, and very much larger amounts of 5-oxo-D-proline were found in the kidney after administration of labeled D-glutamate [283a]. Commercial samples of labeled L-glutamate have occasionally been found to contain D-glutamate as an impurity, and this probably explains the reported finding that 8 minutes after administration of labeled L-glutamate to rats, 68% of the nonglutamate radioisotope found in the kidney was 5-oxoproline [283b].

Both glutamine synthetase [149,284–288] and γ-glutamyl-cysteine synthetase [141,144] can catalyze the formation of both isomers of 5-oxoproline from the corresponding isomers of glutamate in the absence of an acceptor (ammonia and cysteine, respectively); however, it does not seem likely that these reactions are normally of quantitative significance in the production of 5-oxoproline. Enzyme activity capable of cleaving 5-oxoprolyl residues from polypeptide chains has been obtained from bacterial [289–292] and mammalian sources [293,294]. Some 5-oxo-L-proline may normally be formed by this pathway. It should also be mentioned that 5-oxo-L-proline may be formed in animals by the bacterial flora present, and may be ingested by animals. Although little 5-oxoproline occurs in fresh tomato juice, about 30% of the total organic acid present in stored tomato juices can be accounted for as 5-oxoproline [295].

The 5-oxo-L-proline content of mammalian body fluids and most tissues is exceedingly low, probably less than 0.05 mM [296]. The determination of 5-oxoproline offers certain experimental difficulties because γ-glutamyl compounds, especially glutamine, that are present in tissues tend to undergo nonenzymatic cyclization during preparation of the body fluid or tissue for analysis. Thus, the finding of as much as 0.22 and 0.33 mM 5-oxoproline in human and guinea pig plasma [297] may probably be ascribed to some cyclization of glutamine. Further work on analytical methods for the determination of 5-oxoproline is needed. In addition to methods that involve nonenzymatic or enzymatic conversion of 5-oxoproline to glutamate [296–299], procedures in which gas–liquid chromatography of esters of 5-oxoproline have been employed [300–304]. It is of interest that the concentration of 5-oxoproline in the skin of various animals including man [298,299,305,306] is much greater than that found in other tissues. The high concentration of 5-oxo-L-proline in skin may be associated with the high activity of γ-glutamylcyclotransferase in skin [272]. Reduced quantities of 5-oxoproline have been found in the scales of psoriatic plaques [307]. Increased urinary excretion of 5-oxoproline has been found after burns, in certain allergic diseases [301], and, as discussed below, in the disease 5-oxoprolinuria [302,303].

D. Utilization of 5-Oxoproline; 5-Oxoprolinase

5-Oxoproline was first described by Haitinger [308] just a few years prior to the initial publication on glutathione [6]. Haitinger prepared pyroglutamic acid in 1882 by heating glutamic acid at 180°–190°; the first enzymatic formation of this compound was described 60 years later in 1942 [202]. Convincing evidence that 5-oxoproline is a major metabolite of glutathione has only recently become available [167,

296,309–311,311a]. However, interest in the metabolism of 5-ox-oproline was evident as early as 1912, when Abderhalden and Hanslian [312] administered 5-oxo-DL-proline to rabbits and found a small fraction of the administered material in the urine; the urinary 5-oxoproline was predominantly of the D configuration suggesting that the L isomer had been metabolized. Later studies on the metabolism of 5-oxoproline gave somewhat conflicting results [313–319]. Bethke and Steenbock [315] fed partly racemized 5-oxoproline to pigs; although evidence for increased urea formation was obtained, substantial amounts of 5-ox-oproline were found in the urine. Butts *et al.* [316] gave racemic 5-ox-oproline to rats and found evidence for glycogen formation, a result similar to that found after feeding glutamate. Pedersen and Lewis [317] fed racemic 5-oxoproline to rabbits and found some extra urea in the urine, but less than that observed after administration of glutamate. Subsequent work suggested the possibility that orally administered 5-ox-oproline might be converted to glutamate in the acid medium of the stomach [318]. Lange and Carey [320] administered [^{14}C]5-oxoproline orally to mice and rabbits; labeled 5-oxoproline and glutamate were found in the blood serum of rabbits, and labeled glutamate and γ-aminobutyrate were found in brain and kidney, respectively, of mice. In studies on the metabolism of glutamate by mouse cells grown in tissue culture, Kitos and Waymouth [321] found that 5-oxoproline was produced, but no utilization of this compound was demonstrated. Several studies have shown that 5-oxoproline can be utilized by various microorganisms [322–328].

The question of whether 5-oxoproline is metabolized by animal tissues arose again in the course of studies in which it was found that kidney contained very high γ-glutamyl-cysteine synthetase activity, and it became apparent that the catalytic potential of the kidney to synthesize γ-glutamylcysteine and glutathione was about equivalent to the very high γ-glutamyltranspeptidase and γ-glutamylcyclotransferase activities of this organ [309]. Since there was no evidence that 5-oxoproline accumulates in the kidney or that it is excreted to an appreciable extent in the urine, the possibility that an enzyme existed that could act on 5-ox-oproline was investigated. Studies in which [^{14}C]5-oxo-L-proline was administered intraperitoneally to mice, showed that the label appeared promptly in the respiratory carbon dioxide [329], and subsequent experiments showed that a major product of 5-oxoproline metabolism in kidney slices is glutamate [330]. Studies on slices of rat kidney, spleen, liver, intestine, heart muscle, and brain showed that these tissues could convert uniformly labeled [^{14}C]5-oxo-L-proline to ^{14}CO$_2$ [310,330]. When rat kidney homogenates were incubated with [^{14}C]5-oxo-L-proline, magnesium ions, ATP, and an ATP-regenerating system, labeled glutamate and glutamine were formed. However, such preparations catalyzed only the formation of glutamate when methionine sulfoximine, an irreversible inhibitor of glutamine synthetase [150], was added. With more highly purified preparations of the enzyme, 5-oxo-L-proline was converted to L-glutamate in the presence or absence of methionine sulfoximine [310]. The reaction was shown to proceed to more than 90% of complete conversion of 5-oxo-L-proline to L-glutamate and to be accompanied by stoichiometric formation of ADP and inorganic phosphate in accordance with the following reaction:

$$\underset{\substack{\text{H}}}{\underset{\text{O}}{\bigodot}}\substack{\text{H}\\\text{COOH}} \;+\; \text{ATP} \;+\; 2\,\text{H}_2\text{O} \;\longrightarrow\; \begin{array}{c}\text{COOH}\\ |\\ (\text{CH}_2)_2\\ |\\ \text{CHNH}_2\\ |\\ \text{COOH}\end{array} \;+\; \text{ADP} \;+\; \text{P}_i \quad (15)$$

The reaction involves hydrolysis of both ATP and an amide bond. Energy is required for the hydrolysis of this amide bond since the equilibrium of the glutamate-5-oxoproline reaction markedly favors cyclization. 5-Oxoprolinase activity has been found in a number of other mammalian tissues including liver, intestine, and brain [310]. The mechanism of this interesting reaction requires further study. This reaction and the one involved in the carbon dioxide-dependent utilization of urea [331,331a] seem to be the only known reactions involving peptide bond cleavage that require ATP.

It is of interest that independent studies by Ramakrishna et al. [332] showed that when [^{14}C]5-oxo-L-proline was injected intraperitoneally in rats, more than half of the radioactivity appeared in the respiratory carbon dioxide within 30 minutes. These investigators also found that slices of liver and kidney converted 5-oxo-L-proline to glutamate. Rush and colleagues [333,334], in the course of an investigation designed to explore the possibility that 5-oxoproline might be an initiator imino acid in mammalian protein biosynthesis, found that 5-oxoproline is not incorporated directly into tRNA, but must first be converted to glutamate. These workers detected the presence of an enzyme that converted 5-oxoproline to glutamate; thus, they found that a soluble rat liver preparation catalyzed the conversion of labeled 5-oxoproline to labeled glutamate and glutamine in the presence of ATP. It is also of interest to recall the much earlier studies of Braunstein et al. [335] and Shamshikova and Ioffe [336], which showed that 5-oxoproline could be effectively utilized by liver slices for the synthesis of glutathione. It is notable that under certain experimental conditions, 5-oxoproline was used as effectively or even better than glutamate. These investigators [336] considered the idea that the cleavage of 5-oxoproline and the synthesis of γ-glutamylcysteine occurred at the same time.

The reaction catalyzed by 5-oxoprolinase [Eq. (15)] seems to account for the utilization of 5-oxoproline in these and the other studies on animals, which have been reported over a period of almost 60 years. However, it is possible that there are other pathways of 5-oxoproline metabolism. Interconversion of 5-oxoproline and proline has often been considered, but there seems to be no evidence for such transformations. It is conceivable that 5-oxoproline might be converted by decarboxylation to pyrrolidone or perhaps to α-ketoglutaramate (via the cyclic form 2-pyrrolidone-5-hydroxy-5-carboxylate) [337–341]. Certain bacteria might utilize 5-oxoproline by such pathways. However, 5-oxoprolinase has recently been isolated from a bacterium and partially purified; the properties of this enzyme are very similar to those of the rat kidney enzyme [342].

Rat kidney 5-oxoprolinase is highly specific for the L isomer [310, 311]; however, this enzyme also interacts with several 5-oxoproline analogues including L-piperidone-2-carboxylic acid, L-2-imidazolidone-4-carboxylate, and L-dihydroorotic acid. It is of interest that when the enzyme is incubated with L-2-imidazolidone-4-carboxylate, this compound does not disappear although ATP is cleaved to ADP and inorganic phosphate [343]. L-2-Imidazolidone-4-carboxylate is an effective competitive inhibitor of the reaction *in vitro* and, as discussed below, *in vivo* [311]. Bacterial 5-oxoprolinase exhibits a much lower affinity for 2-imidazolidone-4-carboxylate [342] and this enzyme is therefore useful for the determination of 5-oxoproline in the presence of the analogue [296].

E. Evidence for the γ Glutamyl Cycle

The recognition that 5-oxoproline, formed by the successive actions of γ-glutamyltranspeptidase and γ-glutamylcyclotransferase, is actively metabolized *in vivo*, and the discovery of 5-oxoprolinase indicate that 5-oxoproline is not an artifact but is actually a quantitatively significant metabolite of glutathione (Section V,D). The finding of 5-oxoprolinase not only provided a link between the reactions that catalyze the breakdown of glutathione with those that catalyze its synthesis, but also made it possible to visualize a cycle [167,309,310] (Fig. 2). Evidence consistent with the view that the γ-glutamyl cycle functions in several mammalian tissues has come from studies in which animals were treated with L-2-imidazolidone-4-carboxylate, an effective competitive inhibitor of 5-oxoprolinase [296,311,343]. Addition of the inhibitor to rat kidney slices led to marked inhibition of the conversion of labeled 5-oxo-L-proline to $^{14}CO_2$, while in similar studies in which the oxidation of labeled L-glu-

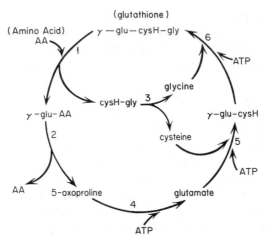

Fɪɢ. 2. The γ-glutamyl cycle. 1, γ-Glutamyltranspeptidase; 2, γ-glutamylcyclotrans-ferase; 3, cysteinyl-glycinase; 4, 5-oxoprolinase; 5, γ-glutamyl-cysteine synthetase; 6, glutathione synthetase.

tamate was examined in the presence of the inhibitor, there was no decrease in carbon dioxide formation. As discussed above, injection of labeled 5-oxo-ʟ-proline into mice is followed by prompt excretion of the administered label as respiratory carbon dioxide; thus, more than half of the administered dose is recovered in less than 2 hours. When mice were treated with the inhibitor, the formation of respiratory carbon dioxide from 5-oxoproline was markedly reduced, and in addition, labeled 5-oxoproline was excreted in the urine in large amounts; normally, only very small amounts of 5-oxoproline appear in the urine. It was also found that when mice were injected with the inhibitor there was marked accumulation of 5-oxoproline in the tissues (kidney, liver, brain, eye), a finding consistent with decreased conversion of 5-oxoproline, normally formed in the tissues, to glutamate. When the animals were given the inhibitor together with one of several ʟ-amino acids, the accumulation of tissue 5-oxoproline and its urinary excretion were much greater than when the inhibitor was given alone. Such augmentation of 5-oxoproline accumulation and excretion in response to administration of amino acid is in accord with the proposed function of the γ-glutamyl cycle. Thus, increased amino acid concentrations in the tissues would be expected to be accompanied by increased transpeptidation, thus leading to increased 5-oxoproline formation, and in the presence of a 5-oxoprolinase inhibitor, 5-oxoproline would accumulate. Additional evidence in support of the function of the γ-glutamyl cycle is the finding of a small, but significant, amount of labeled 5-oxo-ʟ-proline in kidney and liver after the administration of labeled ʟ-glutamate to rats [283a, 311a].

Evidence consistent with the function of the γ-glutamyl cycle has also come from studies on patients who appear to have metabolic lesions in the γ-glutamyl cycle. Thus, Jellum *et al.* [302] described a 19-year-old mentally retarded and partially paralyzed boy who excreted 25–35 gm of 5-oxoproline per day. A second patient with an apparently similar syndrome was reported by Hagenfeldt *et al.* [303]; this patient, about 3 years old, also excretes large amounts of 5-oxoproline. Both patients have high blood and cerebrospinal fluid 5-oxoproline levels (2–4 mM; normal, 0.01–0.1 mM). In studies on the first of these patients, the excretion of 5-oxoproline was increased about two-fold when the blood concentration of amino acids was increased greatly by intravenous administration of amino acids [344]; this finding is consistent with an increase in the turnover of the γ-glutamyl cycle. This patient, who excretes markedly reduced amounts of urea and abnormally large amounts of ammonia even when the urinary pH is close to 7, was at first thought to have a defect in the urea cycle [302]. However, decreased synthesis of urea by this patient may be explained in terms of a metabolic adaptation in which glutamine is hydrolyzed to compensate for the deficit in glutamate [167]. According to this interpretation there would be preferential utilization of ammonia in the liver for glutamate and glutamine synthesis which would be expected to decrease the availability of ammonia for urea formation. This interpretation is consistent with the finding [344] that the patient's formation of urea was increased when given glutamine. Thus a metabolic adaptation of this type would allow sufficient glutamate to be available so as to permit the γ-glutamyl cycle to function.

Some of the observations made on the first patient discovered to have 5-oxoprolinuria were in accord with the view that this patient is blocked in the utilization of 5-oxoproline [344]; thus administration of a mixture of amino acids to this patient led to a marked increase of 5-oxoproline excretion as discussed above. After the patient was given labeled 5-oxoproline, he expired very little labeled carbon dioxide in contrast to extensive labeled carbon dioxide expiration by a normal control. Although this observation is consistent with a block at the 5-oxoprolinase step of the γ-glutamyl cycle, it would be expected that the administered labeled 5-oxoproline would be greatly diluted by the very high blood plasma concentration of 5-oxoproline (50 mg/100 ml [344a]); this value was initially reported erroneously to be one-tenth of this value [302,344,344b]. Furthermore, studies on intact fibroblasts cultured from skin biopsies obtained from the first patient were shown to convert 5-oxoproline to carbon dioxide at an essentially normal rate, and such fibroblasts were also shown to exhibit normal levels of 5-oxoprolinase activity [345]. Similar studies were carried out on the white blood cells obtained from the second patient, and the presence of substantial 5-oxoprolinase activity was demonstrated [303]. It may be calculated from studies in which labeled 5-oxoproline was given to patients, that only about 25% of the 5-oxoproline formed is excreted in the urine. These considerations indicate that the patients have a substantial capacity to utilize 5-oxoproline and that there is an overproduction of this compound. The findings could be explained by a block at the glutathione synthetase step at the cycle, if much more than normal amounts of γ-glutamylcysteine were formed and converted to 5-oxoproline, and if the amounts of 5-oxoproline formed were large enough to exceed the capacity of 5-oxoprolinase [167]. Such an interpretation of the metabolic defect in 5-oxoprolinuria is strongly supported by enzyme studies on placenta, cultured skin fibroblasts, and erythrocytes from the second patient and her recently acquired younger sister [345a]. Both of these patients developed metabolic acidosis early in the neonatal period and have required therapy with bicarbonate since. Both patients exhibit plasma concentrations of 5-oxoproline in the range of 2–5 mM and they excrete substantial amounts of 5-oxoproline in the urine. They also exhibit decreased erythrocyte concentrations of glutathione and an increased rate of hemolysis. While substantial amounts of γ-glutamyl-cysteine synthetase and γ-glutamylcyclotransferase activities were found in the third patient's placenta, the glutathione synthetase activity was only about 2% of that of a control. Extracts of cultured fibroblasts from the second patient exhibited increased activities of γ-glutamyl-cysteine synthetase, γ-glutamylcyclotransferase, and 5-oxoprolinase compared to the fibroblasts from a control subject, while the glutathione synthetase activity was decreased to less than 5% of the control. The erythrocytes from the second and third patients and their parents exhibited γ-glutamyl-cysteine synthetase and γ-glutamylcyclotransferase activity, and the levels of these enzymes were similar to those found in controls. On the other hand, a marked deficiency (5–10% of control values) of glutathione synthetase activity was found in the erythrocytes from the patients. The erythrocyte glutathione synthetase activity of the father was intermediate between that of the patients and that of the controls, while the value for the mother's erythrocytes was slightly lower than the controls. The findings indicate that the patients have a generalized glutathione synthetase deficiency, and that their 5-oxoprolinuria is secondary to this enzyme defect. The patients exhibited anemia early in life, evidently associated with low erythrocyte glutathione concentrations. It seems probable that the first patient discovered to have 5-oxoprolinuria also has glutathione synthetase deficiency; it is notable that this patient was jaundiced during the early neonatal period and was very ill for two weeks after birth [345b].

It is of interest to consider the situation in 5-oxoprolinuria in relation to other patients who have been reported to have erythrocyte glutathione synthetase deficiency associated with decreased erythrocyte glutathione concentrations and compensated hemolytic disease (see Section IV,E). The published information does not indicate whether the latter patients experienced serious neonatal morbidity, nor are there data available relating to the presence of acidosis or 5-oxoprolinuria. It is possible that these patients have a form of glutathione synthetase deficiency which is restricted to the erythrocytes; this suggests the possibility of heterogeneity among patients with glutathione synthetase deficiency.

The metabolic defect in 5-oxoprolinuria can be explained in terms of the γ-glutamyl cycle. Thus, a block in glutathione synthetase would lead to 5-oxoprolinuria if excessive amounts of γ-glutamylcysteine were formed and converted to 5-oxoproline and if such overproduction exceeded the capacity of 5-oxoprolinase. One may then visualize in this disease a γ-glutamyl cycle involving the actions of only four (rather than six) enzymes, i.e., γ-glutamyltranspeptidase, γ-glutamylcyclotransferase, 5-oxoprolinase, and γ-glutamyl-cysteine synthetase (Fig. 3). Since γ-glutamylcysteine is a substrate of γ-glutamylcyclotransferase (while glutathione is not), a marked deficiency of glutathione synthetase would lead to a futile cycle of γ-glutamylcysteine synthesis followed by its conversion to 5-oxoproline and cysteine. Under these circumstances, increased γ-glutamylcysteine synthesis would be necessary to produce sufficient amounts of this peptide for transpeptidation. Studies on the mechanism of regulation of glutathione biosynthesis indicate that glutathione exerts a feedback inhibitory effect on γ-glutamyl-cysteine synthetase [175]. This regulatory system would be inoperative in patients with 5-oxoprolinuria and thus increased rates of γ-glutamylcys-

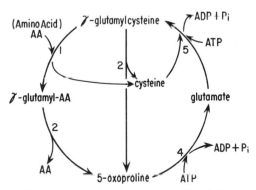

FIG. 3. Proposed modified γ-glutamyl cycle in 5-oxoprolinuria [from 345a]. Enzymes are numbered as in Fig. 2.

teine synthesis would be expected. The biochemical investigations on these patients [345a] and the *in vitro* enzyme studies [175] are consistent with the interpretation given in Fig. 3 (see p. 113).

F. Function of the γ-Glutamyl Cycle

γ-Glutamyltranspeptidase and γ-glutamylcyclotransferase catalyze a major pathway of glutathione utilization. While the reactions catalyzed by these enzymes have long been known, their physiological significance has not been clear. Hanes *et al.* [208,209] suggested that γ-glutamyltranspeptidase might be involved in protein synthesis and this idea was also considered by others [193,346]. That the reaction catalyzed by γ-glutamyltranspeptidase might be involved in amino acid transport was also suggested in the 1950's [205,347–350] and also later [16,188,219,225,236,254,255]. The possibility that γ-glutamyltranspeptidase and γ-glutamylcyclotransferase might function in collagen formation was also mentioned [268]. A direct role of γ-glutamyltranspeptidase in protein synthesis does not seem likely when one considers the detailed information about protein synthesis now available; furthermore, the relatively specific membranous localization of γ-glutamyltranspeptidase would seem to argue against this. It seems unlikely that the transpeptidase functions in folate metabolism because the enzyme does not act on glutamate derivatives in which the α-amino group is substituted. The idea that γ-glutamyltranspeptidase might function in amino acid transport was strengthened considerably by the findings that led to formulation of the γ-glutamyl cycle [167,309,310]. Thus, the recognition that kidney contains high and almost equivalent levels of both glutathione synthesizing enzymes and *amino acid-dependent* degradative enzymes and the discovery of an enzyme activity that links these reactions made it possible to envision a series of chemical events that could explain amino acid transport. The reactions of the cycle may be considered as steps in the overall process of transport, analogous to those which have often been postulated by investigators who have studied transport [see, for example, 351,352]. Membrane-bound γ-glutamyltranspeptidase is postulated to function in translocation by interacting with extracellular amino acid and intracellular glutathione. It has been suggested [167] that the amino acid might bind noncovalently to a site located over a hole in the membrane. A group on the enzyme then reacts with glutathione to form a γ-glutamyl enzyme, and attack of the amino acid nitrogen atom on the γ-carbon atom of the γ-glutamyl enzyme yields a γ-glutamylamino acid, thus removing the amino acid from its binding site and bringing it into the cell. Release of the amino acid from its γ-glutamyl carrier is catalyzed by γ-glutamylcyclotransferase. The cycle includes three energy requiring recovery steps associated with the resynthesis of glutathione from cysteine, glycine, and 5-oxoproline. That transpeptidase is membrane-bound and that it interacts with almost all of the protein amino acids is in accord with its proposed function. The presence of small amounts of γ-glutamylamino acids in normal urine [79] is also consistent with this idea.

Although transport of amino acids by the γ-glutamyl cycle would appear to require more energy than one would think would be minimally necessary, such an expensive transport system may not be altogether unreasonable [167]. The high energy requirement of the cycle may reflect the need for high efficiency; it is notable that other biological processes (e.g., urea synthesis, protein synthesis, glycogen synthesis) also require considerable energy.

It is possible that γ-glutamylcysteine functions as a substrate for the transpeptidase, which is known to be able to interact with many γ-glutamylamino acids. This would lead to a γ-glutamyl cycle involving the actions of only 4 enzymes (transpeptidase, cyclotransferase, 5-oxoprolinase, and γ-glutamyl-cysteine synthetase); glutathione might serve as a reservoir for the γ-glutamyl moiety. The synthesis of glutathione, provided that some glutathione synthetase is membrane-bound, might facilitate the transport of glycine. Other modifications and ramifications of this cycle are possible. For example, there may be pathways involving a number of successive transpeptidation reactions. Many studies on amino acid transport in the kidney and in other systems indicate that there may be separate systems involved in the transport of various groups of amino acids. The function of the γ-glutamyl cycle might involve isozymic forms of the several enzymes which exhibit different amino acid specificities. There may be specific binding proteins for various amino acids and the amino acids may be transferred from these to the transpeptidase. If this is true, the transpeptidase might function primarily in translocation rather than recognition. On the other hand, a single system of enzymes exhibiting different affinities for different amino acids would also explain the findings. The data on the relative rates of the reaction of various amino acids and γ-glutamylamino acids with γ-glutamyltranspeptidase and γ-glutamylcyclotransferase, respectively, suggest that substantial differences might exist between the various amino acids in relation to their affinity for, and rates of utilization by, the γ-glutamyl cycle. It is notable that glutamine, one of the best substrates for transpeptidase and (as the γ-glutamyl derivative) for γ-glutamylcyclotransferase, is present in very high concentrations in the blood and in many mammalian tissues.

The findings cited above (Section V,E) on the patients who exhibit 5-oxoprolinuria seem to support the function of the γ-glutamyl cycle *in vivo.* It is of interest that these patients do not exhibit aminoaciduria. Although one might think that a block in the γ-glutamyl cycle should produce aminoaciduria if the cycle functions in renal amino acid transport, the evidence seems to indicate that these patients maintain an active γ-glutamyl cycle. The loss of glutamate as 5-oxoproline is readily compensated for, as discussed above, by the derivation of glutamate from other sources, probably mainly by hydrolysis of glutamine; this explanation is consistent with the finding of increased urinary ammonia. Although these patients are blocked at the glutathione synthetase step they still can synthesize substantial amounts of γ-glutamylcysteine, which can function in transpeptidation. On the other hand, a deficiency of γ-glutamyl-cysteine synthetase would decrease both the availability of γ-glutamylcysteine and of glutathione and would be expected to produce aminoaciduria if the proposal that the γ-glutamyl cycle functions in amino acid transport is correct. Thus, it is of interest that the patients who appear to have generalized deficiency of γ-glutamyl-cysteine synthetase [183] exhibit aminoaciduria [167,183a,353]. The absence of aminoaciduria in animals treated with 5-oxoprolinase inhibitor may be associated with increased formation of glutamate by other pathways in a manner similar to that discussed above.

As discussed above (Section V,B), purified γ-glutamyltranspeptidase preparations are markedly affected by maleate. Thus maleate induces dissociation of the catalytic functions of γ-glutamyltranspeptidase leading to impairment of the ability of the enzyme to use amino acid acceptors so that the γ-glutamyl moiety is converted to glutamate rather than to a γ-glutamylamino acid. Such an effect, if it were to take place *in vivo,* would be expected to interfere with the proposed function of the γ-glutamyl cycle in amino acid transport. It is thus of some interest that Harrison and Harrison [353a], as well as other workers [353b–353d], have shown that administration of maleate to animals leads to extensive aminoaciduria. A recent study on amino acid transport *in vivo* in which a microinjection procedure was employed led to the suggestion that maleate may produce aminoaciduria by increasing the efflux of amino acid, leading to, or secondary to, a loss of cellular amino acid accumulation [353d]. When these studies are considered in relation to the *in vitro* studies on the effect of maleate on purified transpeptidase, they seem to offer support for the proposed function of the transpeptidase and of the γ-glutamyl cycle in amino acid transport. Although the studies on maleate are clearly consistent with the γ-glutamyl cycle, it must be noted that maleate also produces increased excretion of phosphate and glucose [353a,353e]; in addition there is evidence that maleate inhibits other enzymes and that it reacts with glutathione and other thiols [353f,353g].

Thus additional studies are needed, especially in an effort to develop a highly specific inhibitor of γ-glutamyltranspeptidase.

The hypothesis that the γ-glutamyl cycle functions in amino acid transport does not yet seem to explain the apparent requirement for sodium ions in amino acid transport. In addition, it does not fully explain in detail the various specificity phenomena that have been observed in transport of amino acids. Since proline and various unmetabolizable amino acids such as α-aminoisobutyric acid are not substrates for γ-glutamyltranspeptidase, it does not seem that these amino acids could be transported by the cycle. As emphasized elsewhere [167], it has not been proposed that the γ-glutamyl cycle mediates all amino acid transport; there are many data indicating that a number of different amino acid transport systems exist. It seems probable that the activity of the γ-glutamyl cycle varies considerably in different tissues and also that it varies in a particular tissue at different stages of development. In this respect it is noteworthy that adult kidney exhibits higher γ-glutamyltranspeptidase activity than fetal kidney, while the reverse situation obtains in liver, brain, and lung [248]. The γ-glutamyl-cysteine synthetase and γ-glutamylcyclotransferase activities of fetal kidney are substantially lower than those of the adult kidney [353h]. These observations suggest that the γ-glutamyl cycle may be relatively inactive in adult liver but that it plays a significant role in amino acid transport in fetal liver. The marked increase in γ-glutamyl cycle activity in the kidney after birth may reflect the conversion of an organ which is relatively inactive in amino acid transport in the fetus to one which is highly active in the extrauterine animal.

It is noteworthy that glycylglycine and certain other peptides are good acceptors of the γ-glutamyl group of glutathione in the transpeptidase reaction. The significance of these observations is not yet clear. It is possible that these peptides interact with the enzyme by virtue of their similarity to cysteinylglycine. On the other hand, one cannot exclude the interesting possibility that the transpeptidase may function in the transport, secretion, or degradation of peptides.

It is of interest that glutathione is a much better γ-glutamyl donor than glutathione disulfide [222a]. When tested at concentrations of 20 mM, L-glutamine, L-methionine, O-carbamyl-L-serine, S-methyl-L-cysteine, and L-cystine are among the most active amino acid acceptors of the γ-glutamyl moiety [222a]; under comparable conditions, but at much lower concentrations, L-cystine is more active than L-cysteine, whereas L-tyrosine is much less active [353i]. It is, of course, difficult to interpret the results of such *in vitro* studies in physiological terms. Nevertheless, the relatively high activities found with glutamine and certain dipeptides may reflect a significant physiological role of the transpeptidase in relation to these acceptors. Similarly, the relatively high activity of L-cystine, whose intracellular concentration is usually very low, may indicate that the transpeptidase plays a significant role in the transport or metabolism of this amino acid.* The much higher activity of glutathione as compared to glutathione disulfide suggests that various phenomena that affect the relative intracellular concentrations of these forms may influence the rates of transpeptidation phenomena.

Studies on the enzymes of the γ-glutamyl cycle in the brain have indicated that these activities are substantially higher in the choroid plexus than in other regions [354,354a]. This suggests that the cycle might play a significant role in the transport of amino acids between the blood and the cerebrospinal fluid. The concentrations of almost all the amino acids are substantially lower in the cerebrospinal fluid than in the blood plasma [355–357], indicating that an active transport system functions in the choroid plexus. There is, perhaps, an analogy between the formation of urine by the nephron and the secretion of cerebrospinal fluid by the choroid plexus [358,359]. γ-Glutamyltranspeptidase is localized in the apical portions of the epithelial cells of the choroid plexus in a manner similar to that found in the proximal convoluted tubules of the kidney. Other studies have shown that the ciliary body of the eye also contains high concentrations of the γ-glutamyl cycle enzymes and all of the enzymes except for 5-oxoprolinase are found in the lens [360]. γ-Glutamyltranspeptidase is localized in the basal portions of the epithelial cells of the ciliary body in a manner similar to that found in the choroid plexus. Such cells are thought to be involved in secretory and absorptive activities associated with formation of the aqueous humor. It is thus possible that the cycle functions in the transport of amino acids across the blood–aqueous humor barrier (see Section IX).

Although the enzymes of the γ-glutamyl cycle are very highly localized in the choroid plexus, they are found elsewhere in the brain as well [354], and this suggests that the cycle may have additional functions in the central nervous system. There are data indicating localization of γ-glutamyltranspeptidase in the perikarya of certain neuronal groups within the brain stem [354], in certain Purkinje cells, and in the anterior horn cells of the spinal cord [251]. There is of course much evidence for

---

* It seems likely that the metabolism of glutathione is substantially affected in the condition cystinosis, in which there is extensive intracellular deposition of cystine. Accumulation of cystine may inhibit the activity of enzymes that require free sulfhydryl groups. Increased concentrations of cystine would also be expected to decrease intracellular glutathione concentrations; this might offer an explanation, consistent with the γ-glutamyl cycle, for the aminoaciduria characteristic of cystinosis. Schneider and Seegmiller [353j] have recently reviewed the literature on cystinosis.

the function of a number of amino acids as neurotransmitters in the central nervous system [361–364]. The localization of γ-glutamyl cycle enzymes in various neurons suggests the interesting possibility that membrane bound transpeptidase might function in the intracellular transport of amino acid neurotransmitters [365].

Bodnaryk [227] has examined the specificity of γ-glutamyltranspeptidase of fly larvae toward a series of phenylalanine analogues containing various substituents in the benzene ring. The order of reactivity observed was the same as that found by Hajjar and Curran [366] for the transport of these analogues across the rabbit ileal mucosa. This coincidence of results is consistent with the proposed role of the transpeptidase in amino acid transport; it is certainly of interest that a transpeptidase from fly larvae resembles so closely, in this respect, the affinity of a rabbit ileum transport site. It was suggested [227] that transpeptidase functions in transport of phenylalanine in the fly larva, but that there are other amino acid transport systems as well.

Many years ago Christensen et al. [367] reported that when glutamate was fed to animals there was an increase in the tissue–extracellular fluid distribution ratios of the other amino acids and a decrease in the amino acid concentrations of the plasma. These workers concluded that L-glutamate contributes in some way to the mechanism by which amino acids are concentrated in tissues. It would be of interest to determine whether this remarkable effect of glutamate, which has apparently not been further studied, may be mediated by reactions involving the γ-glutamyl cycle.

VI. INTERCONVERSION OF GLUTATHIONE AND GLUTATHIONE DISULFIDE

A. Introduction

The phenomena considered in this section relate to the participation of glutathione in disulfide exchange reactions, its function as a hydrogen atom donor to free radicals and to hydrogen peroxide and related compounds, and as a compound which can function to maintain the sulfhydryl groups of various molecules including proteins. Some possible physiological functions of glutathione disulfide are also considered.

B. Glutathione Reductase

The intracellular concentrations of glutathione are vastly higher than those of glutathione disulfide, indicating the presence of enzyme activity capable of effectively catalyzing the reduction of glutathione disulfide.

Several investigators obtained evidence for reduction of glutathione disulfide in studies carried out in the 1930's [368–371]. Meldrum and Tarr [371] observed the reduction of glutathione disulfide by erythrocytes and yeast in the presence of zwischenferment (glucose-6-phosphate dehydrogenase), and TPN. Many years later, the enzyme responsible for reduction of glutathione disulfide was demonstrated in pea seeds [372], wheat germ [373], yeast, and liver [374,375]. It is now recognized that the enzyme is widely distributed and probably present in all cells; a number of highly purified preparations of the enzyme have been obtained and there is evidence that in all of these flavin adenine dinucleotide (FAD) is the prosthetic group. The enzyme catalyzes the following essentially irreversible reaction:

$$\text{Glutathione disulfide} + \text{TPNH} + \text{H}^+ \rightarrow 2 \text{ glutathione} + \text{TPN}^+ \qquad (16)$$

The reaction proceeds less readily with DPNH and the available data indicate that the enzyme is highly specific for glutathione disulfide. The glutathione reductase from yeast [376–378] resembles that from germinated peas [379] in having a molecular weight of about 120,000 and in containing 2 moles of FAD per mole of protein. Both FAD moieties can be released from the enzyme by dialysis, indicating a noncovalent linkage between enzyme and cofactor. The enzyme can be dissociated in 5 M guanidine hydrochloride into two apparently identical polypeptide chains. Several studies have been carried out on the kinetic behavior and mechanism of action of the enzyme [see, for example, 377,380–384], and the enzyme has been shown to exhibit β-stereospecificity for the pyridine nucleotide [385]. Glutathione reductases from a variety of sources are reversibly inhibited by various nitrofuran and nitrobenzene derivatives at concentrations as low as 10^{-7} M [386]. Highly purified rat liver glutathione reductase was found to have a molecular weight of 44,000 and to contain one active site per molecule of enzyme [387,388]. The enzyme has also been obtained from *E. coli;* this enzyme has a molecular weight of about 105,000 and contains 2 flavins per enzyme molecule [389,390].

The glutathione reductase of human erythrocytes has also been purified; the molecular weight of this enzyme is not far from that of the yeast enzyme, and the erythrocyte enzyme also contains 2 molecules of FAD per molecule of enzyme [391,392]. A number of studies have been carried out on the activity of erythrocyte glutathione reductase in various anemias. Evidence for an association between glutathione reductase deficiency and a wide variety of hematological disorders has been suggested [393,394]. Staal *et al.* [395] have isolated erythrocyte glutathione reductase from a patient exhibiting decreased activity of this enzyme and found that the affinity of the apoprotein for FAD was diminished as compared to the apoenzyme obtained from normal erythrocytes. Beutler and his colleagues [396–398] have found that the glutathione reductase of normal erythrocytes is often only partially saturated with FAD and that addition of FAD to a hemolysate usually increases glutathione reductase activity. Such an increase is blocked by nucleotides such as ATP, DPN, and TPN. Apparently erythrocytes can synthesize FAD from riboflavin and the feeding of even small amounts of riboflavin results in increased glutathione reductase activity in the erythrocytes. Paniker *et al.* [399] have found that riboflavin deficiency leads to decreased erythrocyte glutathione reductase activity in man and in rats. These workers have concluded that the activity of glutathione reductase does not limit the rate of the hexose monophosphate pathway. The observation that glutathione reductase is frequently unsaturated with FAD under conditions usually employed for enzyme assay raises questions about the significance of some of the earlier findings on erythrocytes of patients with various conditions [398,400]. The low erythrocyte glutathione reductase activities observed in such individuals might reflect a decreased affinity of the apoenzyme for FAD, dietary deficiency of riboflavin, or a deficiency of the enzymes necessary to convert riboflavin to FAD.

C. Glutathione Transhydrogenases

Several enzymes are now known that catalyze reactions of glutathione with disulfides. The first of the glutathione transhydrogenases was described by Racker [401], who obtained an enzyme from acetone-dried preparations of beef liver that catalyzes the reduction of homocystine according to the following reaction:

$$\text{2 Glutathione} + \text{homocystine} \rightarrow \text{glutathione disulfide} + \text{2 homocysteine} \qquad (17)$$

The reaction, which did not occur when homocystine was replaced by cystine, was studied in the presence of glutathione reductase and a reduced pyridine nucleotide coenzyme regenerating system. Although disulfide exchange reactions of this type are known to occur nonenzymatically [31], Racker [401] suggested that the spontaneous reaction would be quite slow under physiological conditions in the presence of low concentrations of reactants. A similar thiol-disulfide transhydrogenase was subsequently obtained from yeast by Nagai and Black [402]. This enzyme, when coupled with glutathione reductase, catalyzes thiol–disulfide exchange between glutathione and several low molecular weight disulfides of which L-cystine is the most active; the enzyme is less active toward L-homocystine.

The ability of tissue preparations to catalyze the reduction of various disulfides in the presence of glutathione can also be explained by nonenzymatic exchange reactions. For example, the reduction of cystamine (RSSR) by intact erythrocytes [31,403] probably takes place as follows:

$$RSSR + GSH \rightarrow GSSR + RSH \text{ (nonenzymatic)} \qquad (18)$$

$$GSSR + GSH \rightarrow GSSG + RSH \text{ (nonenzymatic)} \qquad (19)$$

$$GSSG + TPNH + H^+ \rightarrow 2\ GSH + TPN^+ \text{ (enzymatic)} \qquad (20)$$

That glutathione may function in the metabolism of coenzyme A was first suggested by the isolation of the mixed disulfide of coenzyme A and glutathione from yeast by Stadtman and Kornberg [46]. Subsequently this compound was isolated from liver [47,48]. An enzyme has been obtained from bovine kidney [404] that catalyzes the reversible formation of the mixed disulfide according to the following reaction:

$$GSH + CoASSG \rightleftarrows GSSG + CoA—SH \qquad (21)$$

This activity has been detected also in rat pancreas, brain, liver, lung, muscle, and heart [405,406]. The findings suggest that this enzyme is neither identical to that which catalyzes Eq. (17), nor that which acts on protein sulfhydryl groups (see below). In relatively recent studies [407] it was found that whereas mixed disulfides of glutathione and coenzyme A, glutathione and pantetheine, and glutathione and cysteine are poor substrates for glutathione reductase, they react readily with glutathione to give glutathione disulfide, which of course can be reduced by glutathione reductase in the presence of TPNH. While the reactions between glutathione and the mixed disulfide occur nonenzymatically, there is evidence also for enzymatic catalysis; such enzyme activity was found in rat liver supernatant fractions.

Liver contains an activity, glutathione-insulin transhydrogenase, which promotes the reductive cleavage of the disulfide bonds of insulin by glutathione and closely related thiols [408–414]. A similar enzyme has been found in pancreatic islets [415]. The findings indicate that the action of glutathione-insulin transhydrogenase is coupled with glutathione reductase; Katzen et al. [408,409] found that all three disulfide bonds of insulin are susceptible to reduction. Investigations on the specificity and mechanism of action of the transhydrogenase show that the transhydrogenase undergoes reduction by glutathione and that the reduced enzyme is autoxidizable, suggesting that the reduced form of the enzyme is an active intermediate in the reaction. The enzyme apparently does not act on low molecular weight disulfides, but it does catalyze the reduction of ribonuclease. During reoxidation of the reduced forms of insulin and ribonuclease, the transhydrogenase was found to enhance the rates of regeneration of the native proteins from their reduced precursors. While the transhydrogenase promoted reconstitution of active ribonuclease from a reduced form of ribonuclease, the enzyme was found to have no activity in specifically directing preferential reestablishment of the native configuration of insulin.

It has been suggested that glutathione-insulin transhydrogenase catalyzes a physiologically significant inactivation of insulin; the first step in insulin degradation is the reduction of disulfide bonds, and this is followed by proteolysis of the resultant A and B polypeptide chains. Chandler and Varandani [416] have reported that glutathione-insulin transhydrogenase can also utilize the sulfhydryl groups of various proteins as cosubstrates for the reduction of insulin. It is notable that the transhydrogenase is associated with the plasma membrane of liver cells [417]. There is evidence [417a] that glutathione-insulin transhydrogenase is probably identical with the enzyme that catalyzes sulfhydryl–disulfide interchange (ribonuclease-reactivating enzyme [417b]). The enzyme may therefore function in the assembly of proteins as well as in their degradation.

D. Glutathione Peroxidase

Glutathione peroxidase catalyzes the decomposition of hydrogen peroxide according to the following reaction:

$$2 \text{ GSH} + H_2O_2 \rightarrow \text{GSSG} + 2 \text{ H}_2O \qquad (22)$$

This enzyme was first discovered in erythrocytes by Mills [418,419], who demonstrated that this activity is distinct from catalase, and that it is also present in other tissues [419,420]. Mills proposed that the reduction of hydrogen peroxide in erythrocytes in the presence of glutathione and glutathione peroxidase is coupled to the oxidation of glucose 6-phosphate (G-6-P) [and of 6-phosphogluconate (6-P-G)] (see Scheme 1).

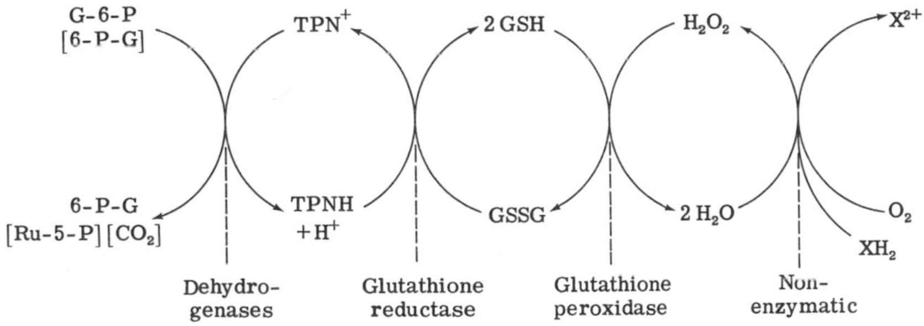

SCHEME 1.

In certain cells TPNH can be formed by the action of other dehydrogenases. Subsequent studies have provided substantial support for Mill's scheme. Thus, there is evidence that the oxidation of glutathione by glutathione peroxidase is the major pathway of hydrogen peroxide metabolism in intact erythrocytes and that catalase plays a less significant role [421]. In studies in which competition between erythrocyte catalase and glutathione peroxidase for hydrogen peroxide was evaluated, oxidation of glutathione accounted for the major fraction of hydrogen peroxide disappearance when extracellular hydrogen peroxide concentrations were maintained at an upper limit of 10^{-6} M. At higher hydrogen peroxide concentrations, decomposition of hydrogen peroxide by catalase was observed. Evidence was obtained that glutathione peroxidase activity protected normal and catalase deficient erythrocytes against methemoglobin formation and osmotic fragility changes. These observations led to the conclusion that glutathione peroxidase, linked to hexose monophosphate shunt activity, catalyzes the major pathway of hydrogen peroxide metabolism under physiological conditions. However, there is evidence that both catalase and glutathione peroxidase function in the destruction of hydrogen peroxide. It is noteworthy that in hereditary acatalasia there is an apparent compensatory increase in the rate of glutathione oxidation and of hexose monophosphate shunt activity [422]. Aebi and Suter [423] have provided evidence which supports the idea that glutathione peroxidase and catalase complement each other so that at low hydrogen peroxide concentrations the main pathway is catalyzed by glutathione peroxidase; catalase acts to remove high concentrations or bursts of hydrogen peroxide formation.

In the mature erythrocyte the hexose monophosphate pathway is of crucial importance in maintaining the concentration of TPNH. There is also evidence that the activity of the hexose monophosphate pathway is influenced by the concentration of TPNH, which is the cofactor for glutathione reductase. Jacob and Jandl [424] have reported evidence that the activity of the hexose monophosphate pathway in erythrocytes is regulated primarily by glutathione. Thus, they found that an increase in the ratio of glutathione disulfide to glutathione led to an increase in the rate of hexose monophosphate pathway metabolism. Sustained low levels of hydrogen peroxide stimulated the hexose monophosphate pathway; such stimulation was potentiated by blocking catalase and prevented by blocking the sulfhydryl group of glutathione by addition of N-ethylmaleimide. Evidence was also obtained that oxidative denaturation of oxyhemoglobin by hydrogen peroxide was promoted by glutathione deficiency. These and related studies support the view that glutathione functions normally to protect hemoglobin and the erythrocyte membrane. Although there seems to be general correlation between the intracellular oxidation of glutathione and erythrocyte hemolysis, marked depletion of erythrocyte glutathione may not invariably be associated with hemolysis. Kosower et al. [425] observed in vitro hemolysis of erythrocytes containing carbon monoxide hemoglobin after treatment with quantities of azoester greater than that needed to oxidize all the glutathione. These workers explain hemolysis as a consequence of membrane damage produced by the free radicals generated by excess azoester. The formation of hydrogen peroxide by erythrocytes is considered further in Section IX.

The function of glutathione peroxidase in other mammalian tissues appears to be similar to that in erythrocyte. Thus, it has been concluded that liver glutathione peroxidase functions to protect sensitive membrane lipids from oxidation [426–428]. The extensive investigations in this area, which include studies on the swelling of mitochondria, have been reviewed in detail by Flohé [429–431]. Lehninger and Gotterer [432] identified a mitochondrial protein which prevents the glutathione induced swelling of mitochondria. This protein was later shown to be glutathione peroxidase [433].

The interrelationships between the metabolism of glutathione, TPNH, and the hexose monophosphate pathway suggested by Mills [418–420] for the erythrocyte seem also applicable to leukocytes [434], Ehrlich ascites tumor cells [435], brain [436–439], and lens [440]. Glutathione peroxidase has been found in the lenses of several species. It has been suggested that hydrogen peroxide, produced nonenzymatically in the aqueous humor by the oxidation of ascorbic acid by molecular oxygen, interacts with glutathione in the lens in a reaction catalyzed by glutathione peroxidase. The studies on lens glutathione are consistent with the pathways previously studied in erythrocytes. Thus, the action of glutathione peroxidase in lens could be coupled with the oxidation of ascorbate (Section IX). It is notable that lens exhibits only weak catalase activity.

Glutathione peroxidase preparations have been obtained from a number of sources including erythrocytes [433,441,442], rat liver [433], and beef lens [443]. While the enzyme is highly specific for glutathione, certain other thiols react at relatively low rates. On the other hand, the enzyme is not specific with respect to the peroxide substrate; thus, linoleic acid hydroperoxide, cumene hydroperoxide, and related compounds are good substrates. Accurate determination of glutathione peroxidase has thus far been difficult and values given in the literature for the activity of this enzyme must therefore be taken with caution. A detailed evaluation of the available methods and an improved assay procedure have been presented by Flohé and Brand [444; see also 431,445]. A number of detailed studies on the preparation and properties of glutathione peroxidase have been carried out by Flohé and his collaborators [446–451]. The molecular weight of glutathione peroxidase of beef erythrocytes has been estimated to be 85,000 [441]; the enzyme from human erythrocytes was reported to have a molecular weight of about 100,000 [452]. Glutathione peroxidase from bovine erythrocytes consists of four apparently identical subunits of molecular weight 21,000. It has recently been reported that glutathione peroxidase contains about 4 gm-atoms of selenium per mole of enzyme [453]. Since there is evidence that the enzyme does not contain a prosthetic group such as heme or flavin, one may consider the possibility that selenium serves a catalytic function. The finding that selenium deficient rats exhibit decreased glutathione peroxidase activity [454] is in accord with the view that selenium is an essential component of the enzyme. Furthermore, there is recent evidence that the glutathione peroxidase activity of several rat tissues depends critically on dietary selenium [454a]. It may be relevant to note that the nonenzymatic oxidation by hydrogen peroxide and oxygen of various thiols (including glutathione) is accelerated by selenocystine [455]. Selenodiglutathione and other "selenotrisulfides" (R—S—Se—S—R) have been described [456,456a], and it has been found that selenite at concentrations of 10^{-5} to 10^{-6} M is a highly effective catalyst for the reduction of cytochrome c by glutathione [456b].

Studies by Necheles and collaborators [457–459] indicate that erythrocyte glutathione peroxidase levels are normally low in newborns. However, they are much lower in certain patients who have mild hemolytic anemia and hyperbilirubinemia. The condition may occur in homozygous or heterozygous forms. Heterozygous individuals may be asymptomatic except during periods of stress. The homozygous form of erythrocyte glutathione peroxidase deficiency is associated with mild compensated hemolytic anemia. A deficiency of leukocyte glutathione peroxidase has been found in one form of chronic granulomatous disease; the enzyme may play a role in destroying bacteria after phagocytosis.

E. Formation and Function of Glutathione Disulfide

Glutathione disulfide may be formed by the action of glutathione transhydrogenases and of glutathione peroxidase as discussed above. Conversion of glutathione to the disulfide may also occur nonenzymatically by disulfide exchange reactions or in reactions involving oxygen:

$$2 \text{ GSH} + \tfrac{1}{2} \text{ O}_2 \rightarrow \text{GSSG} + \text{H}_2\text{O} \tag{24}$$

Such oxidative reactions may be catalyzed by metal ions, metal derivatives such as copper–histidine chelates [460], or by enzymes. Aerobic oxidation of glutathione is commonly observed on homogenization of tissues. There is evidence that the oxidation of glutathione by rat liver homogenates is linked through glutathione peroxidase to the reaction of endogenous substrates with xanthine oxidase, and of uric acid with peroxisomal urate oxidase [461,462].

The enzyme dehydroascorbate reductase, which catalyzes the following reaction,

$$\text{Dehydroascorbate} + 2\ \text{GSH} \rightarrow \text{GSSG} + \text{ascorbate} \qquad (25)$$

has long been known to occur in plants and microorganisms [463–466]. The reduction of dehydroascorbate by erythrocyte hemolysates and preparations of the tissues of several animal species [467,468] has been reported, indicating that animals may also have an enzyme activity that catalyzes Eq. (25).

Evidence that the intracellular concentration of glutathione disulfide functions in the regulation of the hexose monophosphate pathway in the erythrocyte has been cited above [424] (Section VI,D). Thus, glutathione disulfide may be a limiting factor in controlling the utilization of TPNH.

That glutathione disulfide may function in oxidative phosphorylation is suggested by studies which indicate that the formation of ADP from AMP and P_i (or of ATP from AMP and PP_i or from ADP and P_i) can be coupled to the oxidation of glutathione by cytochrome c [469–473]. Glutathione disulfide, possibly in the form of a complex with glutathione (structure as yet undefined), appears to be required as a catalyst in this reaction. The findings suggest the synthesis of one high energy phosphate bond per one electron transferred or possibly a two electron transfer mechanism in which the second electron is transferred to an as yet unknown acceptor.

There is currently no convincing evidence for transport of glutathione across cell membranes, nor is there evidence of detectable glutathione in extracellular fluids. The finding that a wide variety of cell types can catalyze the synthesis of glutathione is thus consistent with the view that glutathione is synthesized within each cell and that there is no transfer of glutathione between cells. Some modification of this idea may be necessary in view of data which indicate that glutathione disulfide can be actively transported across cell membranes [474,475]; thus, it is conceivable that glutathione disulfide can serve as a transport form of glutathione (see, also Section IX). Srivastava and Beutler [474] treated normal and glucose-6-phosphate dehydrogenase deficient human erythrocytes with hydrogen peroxide or methyl phenyl azoformate to oxidize glutathione. A fraction of the glutathione disulfide formed passed out of the erythrocytes and could be recovered from the medium when the normal cells were incubated without glucose or when the glucose 6-phosphate deficient cells were incubated with or without glucose. The transport of glutathione disulfide from the erythrocytes proceeded against a concentration gradient and was shown to be energy-dependent. No evidence was obtained for transport of glutathione disulfide into the cells. Such active transport of glutathione disulfide from erythrocytes may play a role in the pathology of diseases in which glucose-6-phosphate dehydrogenase activity is deficient; for example, it might protect the cells from the effects of high glutathione disulfide concentrations. That glutathione disulfide may be transported out of other cells is indicated by studies on lens glutathione; here too, oxidation of glutathione to the disulfide is followed by loss of the disulfide [476]. Sies *et al.* [477,478] reported that perfusion of rat liver by a solution containing *tert*-butyl hydroperoxide was accompanied by formation of glutathione disulfide and its transport into the perfusate.

Evidence consistent with the possibility that glutathione, synthesized in gut and liver, is transported to the periphery in erythrocytes has come from studies on dogs in which arterial, portal, and hepatic venous blood glutathione and hepatic blood flow were measured [479]. In these studies the changes in concentration of total blood glutathione as the blood passed through gut or liver were found to be as much as twofold. These observations led to the suggestion that glutathione may be a storage form for cysteine as well as for glycine and glutamate; thus, glutathione synthesis might function to conserve cysteine in a relatively unreactive form from which it can be readily recovered. These findings might be explained by direct transport of glutathione between erythrocytes and tissues, possibly as glutathione disulfide. It seems possible, however, that glutathione biosynthesis in the erythrocytes of the portal blood might be increased in response to increased plasma concentrations of glutamate, cysteine, and glycine; a similar phenomenon might also take place in the liver. Such an interpretation would seem to be consistent with the experimental observations of increased output of glutathione from these tissues. However, additional studies on the apparent transport of glutathione from gut and liver via erythrocyte glutathione would be of considerable importance.

Kosower, Kosower, and colleagues [480–490] have studied a new class of thiol-oxidizing agents which are useful for the oxidation of glutathione. These compounds, which have the general formula $RN = NCOX$ or $XCON = NCOX$, produce rapid stoichiometric oxygen-independent oxidation of glutathione to glutathione disulfide.

These agents readily penetrate cell membranes and produce temporary conversion of glutathione to the disulfide. The reaction between glutathione (GS^-) and one such compound ("diamide") may be written as follows:

$$GS^- + H^+ + (CH_3)_2NCON{=}NCON(CH_3)_2 \longrightarrow (CH_3)_2NCO\underset{\underset{H}{|}}{N}{-}NCON(CH_3)_2 \quad (26)$$

(diamide)

$$GS^- + (CH_3)_2NCO\underset{\underset{H}{|}}{N}{-}NCON(CH_3)_2 \xrightarrow{\ H^+\ } GSSG + (CH_3)_2NCONHNHCON(CH_3)_2$$

$$(27)$$

Although diamide reacts with other thiols, including protein sulfhydryl groups and reduced pyridine nucleotides [490–493], glutathione is the major thiol in most cells and is therefore most affected by treatment with diamide. Several interesting studies have been carried out with diamide and related compounds of this series. For example, experiments with diamide have led to the suggestion that glutathione disulfide may play a role in the regulation of protein synthesis [485–488]. In the initial studies it was found that protein synthesis in intact rabbit reticulocytes is inhibited by diamide [485]. Subsequently it was found that 5×10^{-5} M diamide in the presence of much larger concentrations of glutathione produced inhibition of initiation of protein synthesis in rabbit reticulocyte lysates. Recovery from inhibition required reduction of glutathione disulfide and the addition of a postribosomal supernatant factor (factor O) [493]. These and other studies suggest that the intracellular concentration of glutathione disulfide may be an important physiological quantity [493a]. It is generally recognized that addition of high concentrations of thiols to *in vitro* systems is usually accompanied by some disulfide formation, which may, in some cases, lead to inhibitory phenomena. Thus, there is evidence that a number of "sulfhydryl" enzymes can be inhibited by disulfides, including glutathione disulfide; presumably, these effects are related to the formation of mixed disulfides. It is conceivable that glutathione disulfide functions in this way in the regulation of certain enzyme activities. This idea is in accord with that expressed by Racker [494] "that glutathione . . . keep(s) . . . enzymes in a happy state either by preventing their oxidation or by protecting them against toxic heavy metals." It is well known that many enzymes are activated by glutathione as well as by other thiols. Presumably such activation is related to the protection or production of enzyme sulfhydryl groups essential for optimal enzymatic activity or to removal of inhibiting substances, for example, metal ions. Although such effects of glutathione may not be specific when tested *in vitro* with purified enzyme preparations, it is reasonable to suppose that glutathione, the major intracellular thiol, does function this way *in vivo*. The essence of this idea was expressed many years ago by Barrón [3].

It is well known that radiation can produce a decrease in the concentration of glutathione and lead to formation of glutathione disulfide [495]. There is evidence that administration of various thiols can protect animals against the effects of radiation, and that sulfhydryl groups in general are the preferred targets of oxidizing radicals produced by radiation. The roles of glutathione, glutathione reductase, and other related cellular components in protection against radiation damage have been discussed recently [496].

VII. FUNCTION OF GLUTATHIONE IN MERCAPTURIC ACID FORMATION

It has been known for many years that administration of certain halogenated aromatic hydrocarbons to animals leads to urinary excretion of N-acetyl-S-substituted cysteine derivatives of the administered compounds. In 1879, Baumann and Preusse [497] and Jaffe [498] found that administration of bromobenzene or chlorobenzene to dogs was followed by urinary excretion of compounds which Baumann and Preusse called mercapturic acid. Subsequent work established that similar compounds are excreted after administration of a variety of aromatic and aliphatic compounds including, for example, naphthalene, aniline, benzylchloride, and bromoethane. The structures of several mercapturic acids are given in Fig. 4. Boyland and Chasseaud [499] and Wood [500] have recently reviewed the very extensive literature now available on the formation of mercapturic acids.

Early studies showed that administration of naphthalene to animals led to decreased concentrations of glutathione in the liver and ocular lens [501,502]. Later studies in several laboratories [499,503] showed that the concentration of glutathione in the liver decreased when rats were given mercapturic acid precursors, and also that this decrease was about equivalent to the amount of mercapturic acid formed. The subsequent observation that the bile of rats treated with naphthalene contained naphthalene conjugates of glutathione, cysteinylglycine, cysteine, and N-acetyl cysteine [504] suggested that the initial step in mercapturic acid formation involves interaction of the administered compound with glutathione, and that the subsequent steps involve cleavage of the

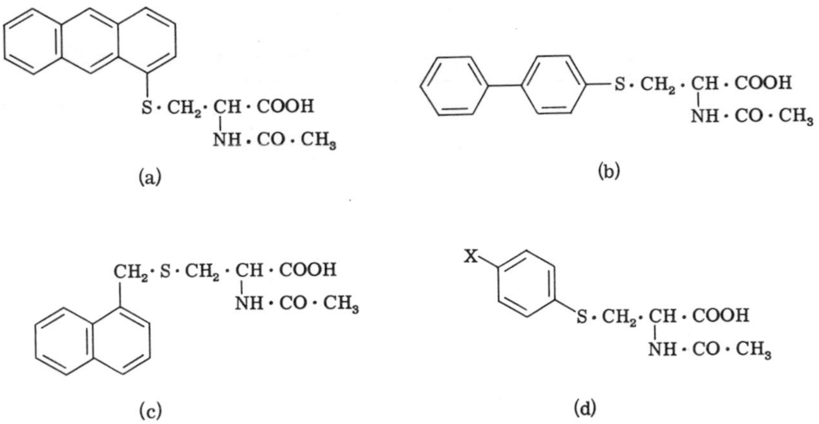

FIG. 4. Structures of several mercapturic acids. (a) 1-Anthrylmercapturic acid (from anthracene); (b) p-biphenylmercapturic acid (from biphenyl); (c) 1-menaphthyl-mercapturic acid (from 1-menaphthyl alcohol); (d) p-X (Br, Cl)-phenylmercapturic acid [from bromo-(or chloro-)benzene].

glutamyl and glycyl moieties followed by N-acetylation of the substituted cysteine.

The reaction of glutathione with a variety of other compounds is catalyzed by a group of enzymes which have been called glutathione S-transferases [505]. There is evidence that there are a number of glutathione S-transferases in liver; the glutathione S-transferases are located mainly in the cytosol. Glutathione S-transferases are widely distributed in vertebrate species and there are many data indicating that they interact with a great variety of chemical compounds [see, for example, 506–512]. Insects can also conjugate foreign compounds with glutathione [513–516]; an interesting reaction involving conversion of O-methylphosphorus compounds to the corresponding O-demethylated organophosphate and S-methyl glutathione has been reported [516].

Chasseaud [517,517a] has recently reviewed the variety of reactions that are catalyzed by the glutathione S-transferases; some of these are listed in Table I. At this time questions relating to the multiplicity and specificity of glutathione S-transferases are difficult to answer because relatively little work has been carried out with purified enzyme preparations. The work that has been done with purified enzymes has been quite interesting and has raised new questions. Thus, Fjellstedt et al. [518] obtained an apparently homogeneous preparation of an epoxide glutathione S-transferase from rat liver that exhibited a molecular weight of 40,000 and is composed of 2 subunits. These workers obtained several different enzymes (as judged by isoelectric focusing and by chroma-

TABLE I

Some Known Glutathione S-Transferases[a]

Enzyme	Second substrate	Reaction products
G—SH S-alkyltransferase	Methyliodide	$H_3C—S—G + H^\oplus + I^\ominus$
G—SH S-aralkyltransferase	Benzyl chloride	$H_5C_6—CH_2—S—G + H^\oplus + Cl^\ominus$
G—SH S-aryltransferase	1,2-Dichloro-4-nitrobenzene	
G—SH S-epoxidetransferase	2,3-Epoxypropylphenyl ether	$H_5C_6—O—CH_2—CH—CH_2—OH$ $\quad\quad\quad\quad\quad\quad\; \|$ $\quad\quad\quad\quad\quad\quad S—G$
G—SH S-alkenetransferase	Diethyl maleate	$H_5C_2OOC—CH—CH_2—COOC_2H_5$ $\quad\quad\quad\quad\; \|$ $\quad\quad\quad\quad S—G$

[a] From Chasseaud [517].

tography) of this type; these were interconvertible by freezing or by treatment with EDTA. Pabst *et al.* [519] purified three glutathione S-transferases from rat liver. Each of these was active with *p*-nitrobenzyl chloride; one was also active with the epoxides and with *p*-nitrophenethyl bromide, the second enzyme was active with methyl iodide, and the third was active with 3,4-dichloronitrobenzene and 4-nitropyridine *N*-oxide. The present indications are that liver contains several glutathione S-transferases of relatively broad and overlapping specificities. Some of the apparent multiplicity of enzymes observed may possibly be explained by limited proteolysis.

While the metabolism of the various S-substituted glutathione derivatives has not been investigated in detail, the available data indicate that these follow the same pathway as that of glutathione breakdown. Bray *et al.* [520] found that liver homogenates liberated S-(*p*-chlorobenzyl) cysteine from S-(*p*-chlorobenzyl) glutathione; glycine was formed simultaneously in the course of this reaction. These workers also found that a partially purified preparation of hog kidney γ-glutamyltranspeptidase liberated S-(*p*-chlorobenzyl) cysteine and glycine from S-(*p*-chlorobenzyl) glutathione; it is of interest that this cleavage reaction took place more rapidly than that observed with glutathione. Both the reaction with glutathione and that with S-(*p*-chlorobenzyl) glutathione were activated by addition of glutamine and were inhibited by phenol red. In studies on the interaction of fluoropyruvate with thiols, Avi-Dor [521] found that the compound formed by the reaction of glutathione with fluoropyruvate is cleaved by an activity apparently identical to γ-glutamyltranspeptidase in a rat kidney particulate preparation. The reaction was followed spectrophotometrically by determining the increase in absorbance at 300 nm due to the formation of S-pyruvoyl cysteinylglycine; β-aminothiols possessing an unsubstituted amino group exhibit high characteristic ultraviolet absorbance, presumably associated with internal Schiff base formation. The rate of reaction was increased greatly by addition of glycylglycine, glutamine, and other amino acids; the pattern of specificity followed closely that of γ-glutamyltranspeptidase. Paper chromatographic evidence was obtained for γ-glutamylamino acid formation, and the appearance of S-pyruvoyl cysteinylglycine was shown by paper electrophoresis. These studies, those of Boyland et al. [504] and of Bray et al. [520] are in accord with the scheme for mercapturic acid formation given in Fig. 5.

Although it has been amply demonstrated that a variety of chemical compounds can be detoxified by the mercapturic acid pathway, it is still uncertain whether there are endogenously formed compounds which normally interact with the sulfhydryl group of glutathione and then follow the mercapturic acid pathway. Recent evidence suggests that estradiol-17β, through the intermediate 2-hydroxy derivative, may be a substrate for this pathway [517], and further studies may uncover additional endogenous substrates. It is also possible, as suggested by Boyland [522], that nature has adapted existing enzymatic mechanisms for glutathione synthesis and utilization for the purposes of detoxification.

It is of interest that glutathione and cysteine conjugates, and also mercapturic acids, have been detected infrequently after administration of known mercapturic acid precursors. Chasseaud [512,517] has emphasized that glutathione conjugates have physicochemical properties that might be expected to favor biliary excretion. Indeed, there is substantial evidence for biliary excretion of such compounds and it seems notable also that histochemical studies indicate localization of liver γ-glutamyltranspeptidase in the bile duct epithelium (Section V,B). In this connection, it seems relevant to note that the test compound sulfobromophthalein, whose rate of disappearance from blood plasma is often used as an index of liver function [523], is excreted in the bile chiefly in the form of its glutathione conjugate [524–530]. It is interesting to note that the rate of conjugation with glutathione does not seem to be the rate-limiting step in hepatic dye clearance [529,530]. Administration of sulfobromophthalein is followed by a decrease in hepatic glutathione. Although sulfobromophthalein is evidently not excreted in the urine in an N-acetylated form, other glutathione conjugates may be metabolized or reabsorbed in the intestine and thus ultimately reappear in the urine as classic mercapturic acid derivatives.

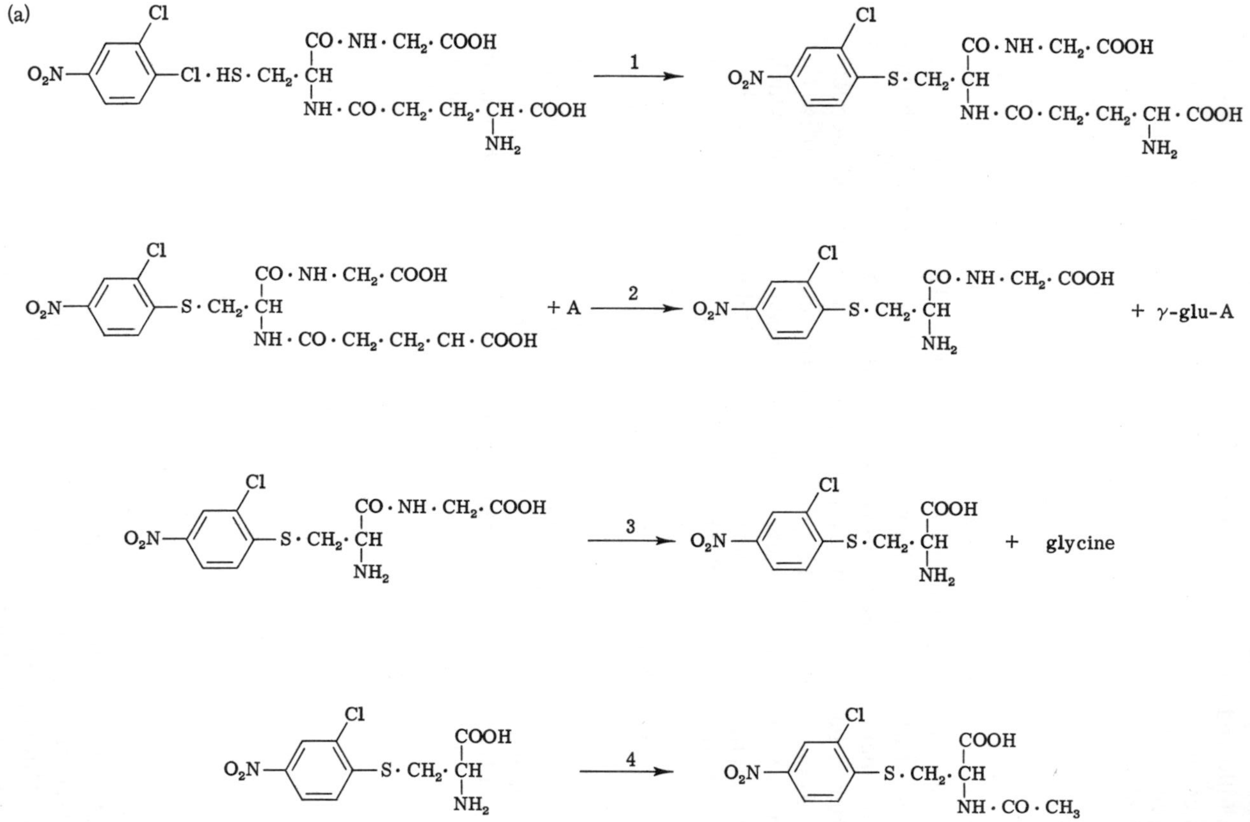

(a)

FIG. 5. Mercapturic acid formation. (a) From 1,2-dichloro-4-nitrobenzene. (b) From naphthalene. The first step involves conversion to the epoxide by a microsomal system in the presence of molecular oxygen and TPNH; in the last step the "premercapturic acid" is converted to the mercapturic acid on acidification [499]. 1, Glutathione S-transferase; 2, γ-glutamyltranspeptidase (A = Acceptor); 3, peptidase; 4, acetylation.

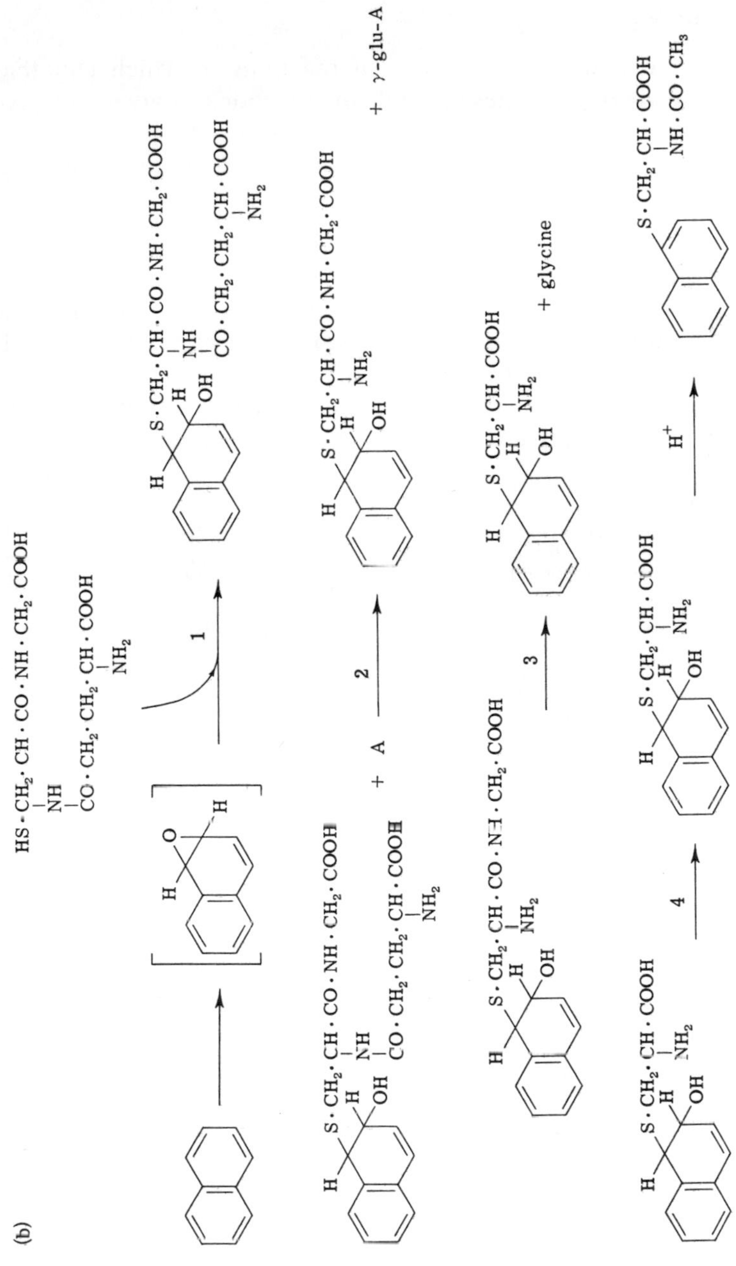

(b)

$HS \cdot CH_2 \cdot CH \cdot CO \cdot NH \cdot CH_2 \cdot COOH$
　　　　　|
　　　　　NH
　　　　　|
　　　　　$CO \cdot CH_2 \cdot CH_2 \cdot CH \cdot COOH$
　　　　　　　　　　　|
　　　　　　　　　　　NH_2

1

2

3

4

+ A

+ glycine

+ γ-glu-A

H^+

Fig. 5(b).

VIII.   FUNCTION OF GLUTATHIONE AS A COENZYME

A. Introduction

There are several enzyme catalyzed reactions in which glutathione functions as a coenzyme; these include the reactions catalyzed by glyoxalase, maleylacetoacetate isomerase, formaldehyde dehydrogenase, and several others. Knox [16] has reviewed the coenzyme functions of glutathione and has summarized the earlier work on glyoxalase.

B. Glyoxalase

The biological significance of the glyoxalase reaction, which is catalyzed by two widely distributed enzymes, is still a mystery; in fact, it is not certain that the substrate usually used in studying this reaction (methylglyoxal) is of physiological importance. The reactions involved in the conversion of methylglyoxal to D-lactic acid may be represented as follows:

$$G{-}SH + H{-}\underset{\underset{O}{\|}}{C}{-}\underset{\underset{O}{\|}}{C}{-}R \xrightarrow{\text{(nonenzymatic)}} G{-}S{-}\underset{\underset{OH}{|}}{C}H{-}\underset{\underset{O}{\|}}{C}{-}R \qquad (29)$$

$$G{-}S{-}\underset{\underset{OH}{|}}{C}H{-}\underset{\underset{O}{\|}}{C}{-}R \xrightarrow{\text{glyoxalase I}} G{-}S{-}\underset{\underset{O}{\|}}{C}{-}\overset{\overset{OH}{|}}{\underset{\underset{H}{|}}{C}}{-}R \qquad (30)$$

$$G{-}S{-}\underset{\underset{O}{\|}}{C}{-}\overset{\overset{OH}{|}}{\underset{\underset{H}{|}}{C}}{-}R \xrightarrow{\text{glyoxalase II}} G{-}SH + HOOC{-}\overset{\overset{OH}{|}}{\underset{\underset{H}{|}}{C}}{-}R \qquad (31)$$

Equation (29) can occur nonenzymatically; thus, the hemimercaptal formed by interaction of methylglyoxal with glutathione can serve as the substrate for glyoxalase I, which catalyzes the oxidoreductive isomerization of the hemimercaptal to form S-lactylglutathione. Glyoxalase II catalyzes the conversion of S-lactylglutathione to glutathione and D-lactic acid. In the reaction catalyzed by glyoxalase I, the transfer of the hydrogen atom on carbon atom 1 to carbon atom 2 of the ketoaldehyde takes place without exchange of hydrogen with the medium [531,532]. It has been demonstrated that S-D-lactylglutathione is formed in Eq. (30). Thus, the enzyme transfers the hydrogen atom stereospecifically [533]; this is in accord with earlier work which showed that glyoxalase II splits S-DL-lactylglutathione to completion [534]. Studies by Vander

Jagt *et al.* [535], in which a number of substituted phenylglyoxals were used as substrates for glyoxalase I, have provided evidence in support of the scheme given in Eqs. (29–31), which had been indicated by earlier work [536,537].

The glyoxalase reaction was independently discovered in 1913 by Neuberg [538] and by Dakin and Dudley [539,540]. Lohmann [541] discovered in 1932 that glutathione is the coenzyme for glyoxalase. Subsequent work established that the reaction is highly specific for glutathione, although asparthione and isoglutathione exhibit activity [542]. Racker [543] showed that two enzymes are involved in the reaction. Later work suggested that the first step is nonenzymatic and that glyoxalase I acts on the hemimercaptal [535–537]. However, kinetic studies by Mannervik *et al.* [544,545] have been interpreted in terms of a random pathway mechanism consisting of a two substrate branch involving glutathione and methylglyoxal as first and second substrates, and a one substrate branch involving the hemimercaptal.

Although glyoxalase has often been isolated from yeast, the enzyme occurs in virtually all cells; purified preparations of glyoxalase I have been obtained from calf liver [537] and also from pig erythrocytes [546]. Like the liver and yeast enzymes [547], the pig erythrocyte enzyme is inhibited by chelating agents, and after such inhibition, is activated by Mg^{2+}, Mn^{2+}, or Ca^{2+}. Analysis of the yeast enzyme showed 1 gm-atom of magnesium per 93,000 gm of protein [547]. Glyoxalase is also known to be active toward hydroxypyruvic aldehyde [548], phosphohydroxypyruvic aldehyde [549], and γ,δ-dioxovalerate [550].

Methylglyoxal was at first considered as a possible intermediate in glycolysis, but Lohmann [541] showed that glyoxalase was not required for glycolysis. Later consideration [16,494,551] suggested that methylglyoxal might be formed nonenzymatically, and that glyoxalase might function to remove methylglyoxal, which is toxic. There is evidence that methylglyoxal may be formed by certain bacteria [552–555], and in certain instances, accumulation of methylglyoxal is associated with increased activity of glyoxalase I. A nonenzymatic formation of methylglyoxal from glycerol and dihydroxyacetone has been observed in Ringer's phosphate suspensions of avian spermatozoa [556], and these findings have again raised the question as to the possible nonenzymatic origin of methylglyoxal.

It has been suggested in a series of provocative papers that ketoaldehydes are physiologically significant in the regulation of cell division [557–563]. For example, it was found that methylglyoxal at a concentration of 1 mM inhibits the division of a number of different cells, and that such inhibition is reversed by cysteine or ethylenediamine

[560,561]. These and other studies (564) are in general accord with a hypothesis according to which methylglyoxal retards cell growth and glyoxalase promotes growth by destruction of methylglyoxal. There is evidence that methylglyoxal inhibits cell growth by interfering with protein synthesis [561,564]. Other studies have shown that methylglyoxal and β-ethoxy-α-ketobutyraldehyde can combine with exposed guanine residues of tRNA [565–567]. Several S-substituted glutathione derivatives have been explored as possible inhibitors of glyoxalase I with the thought that such compounds might be useful as antitumor agents [568–570]. Methylglyoxal has been found to substantially inhibit the growth of a number of experimental tumors [571–573] and such inhibition was enhanced by glutathione, urea, and glyceraldehyde [571,572]. Glyceraldehyde itself also inhibited tumor growth [574,575]. β-Ethoxy-α-ketobutyraldehyde also exhibits an antitumor effect [576].

Methylglyoxal may be formed enzymatically from aminoacetone [577], which in turn may be formed by condensation of acetyl coenzyme A with glycine followed by decarboxylation, or by decarboxylation of β-keto-α-aminobutyric acid which may be formed from threonine. Another reaction that may be relevant to the question of methylglyoxal metabolism is that catalyzed by α-ketoaldehyde dehydrogenase; this enzyme catalyzes the pyridine nucleotide-dependent conversion of methylglyoxal to pyruvate [578].

Recent studies have shown that γ,δ-dioxovalerate is a substrate for glyoxalase I and that the complete glyoxalase reaction occurs with this substrate leading to the formation of D-α-hydroxyglutarate [550]. γ,δ-Dioxovalerate might be formed from δ-aminolevulinate; if it were a substrate for the enzyme described by Monder [578], which converts methylglyoxal to pyruvate, α-ketoglutarate would be formed directly. It was suggested that glyoxalase may offer a pathway for the metabolism of γ,δ-dioxovalerate via the citric acid cycle, either by direct formation of α-ketoglutarate or by oxidation of D-α-hydroxyglutarate. The action of glyoxalase on γ,δ-dioxovalerate might conceivably function in the regulation of porphyrin synthesis.

C. Coenzyme Functions of Glutathione in Other Reactions

Knox and Edwards [579] obtained evidence that maleylacetoacetate is the primary product formed in the oxidation of homogentisate in liver and that a specific isomerase is required for conversion of this 3,5-diketoacid to fumarylacetoacetate. Thus, the characteristic spectrum of maleylacetoacetate remained unchanged in the presence of fumarylacetoacetate hydrolase and glutathione; addition of purified isomerase led to disappearance of the spectrum and formation of acetoacetate and fumarate. The reaction was shown to be dependent on glutathione, although several other thiols exhibit some, but much lower, activity. It is of interest that glutathione catalyzes the isomerization reaction at a very slow rate in the absence of enzyme. The mechanism of the enzymatic isomerization, which requires further study, has been discussed [16]. An analogous cis–trans isomerization also occurs in the degradation of gentisic acid in bacteria [580], and it was reported that glutathione is a coenzyme for the isomerization of maleylpyruvate to fumarylpyruvate, which is the substrate for the hydrolase. It is of interest that these maleyl derivatives are not hydrolyzed by the enzyme, which exhibits a very broad substrate specificity [581–583].

Glutathione is apparently also a coenzyme for formaldehyde dehydrogenase, a pyridine nucleotide-dependent system that converts formaldehyde to formic acid. The enzyme has been obtained from chick [584], beef [584], rat [585], and human [585] liver, and also from the retina of several species [586]. The substrate is believed to be the hemimercaptal formed between formaldehyde and glutathione; this is converted to formate, and glutathione is regenerated. This enzyme might serve in a protective function in the retina [586].

The dehydrochlorination of 1,1,1-trichloro-2,2-bis(p-chlorophenyl) ethane (DDT) to a nontoxic product is catalyzed by another enzyme that requires glutathione as a cofactor [587–590]. The enzyme, which has been obtained from house flies, catalyzes the following reaction:

$$(p\text{-ClC}_6\text{H}_4)_2\,\text{CHCCl}_3 \xrightarrow{\text{GSH}} (p\text{-ClC}_6\text{H}_4)_2\,\text{C} = \text{CCl}_2 + \text{H}^+ + \text{Cl}^- \qquad (32)$$

There is evidence that DDT interacts with glutathione to give an alkylated intermediate which breaks down to yield the product and glutathione.

IX. FUNCTIONS OF GLUTATHIONE IN CERTAIN TISSUES AND CELLS

A. Introduction

The information about glutathione reviewed above has been considered in relation to the various biochemical pathways of glutathione synthesis and utilization. Although there are many similarities between the enzymology of glutathione in various cells, there are some significant quantitative and qualitative differences which seem to reflect specific cellular functions. For example, the glutathione of the erythrocyte appears to function to protect hemoglobin and the cell membrane, and abnormalities of the enzymes involved in glutathione metabolism are associated with hemolytic anemia. The glutathione of the lens appears to be essential for the functional integrity of this organ and decreases in the concentration of glutathione in the lens are associated with the formation of cataracts. Still other functions of glutathione are indicated by studies which have been performed on liver, kidney, and the central nervous system.

B. Lens

The lens contains substantial concentrations of glutathione; thus, values ranging from 2.7 and 3.5 mM in monkeys and man, respectively, to 6.7, 12, and 12.7, in the rat, rabbit, and cat, respectively, have been reported [591]. Much lower but significant concentrations of ophthalmic acid have been found in the lens; values of 0.06, 0.31, 0.08, and 0.98 mM have been reported for the monkey, rabbit, rat, and cat, respectively [591]. The concentration of glutathione disulfide in lens is probably at most no more than a few percent of that of glutathione. About 20% of lens glutathione is found in the free state; the remainder is presumably bound to lens protein, probably by mixed disulfide bonds [592]. It has long been known that lens contains the enzymes that catalyze glutathione biosynthesis [114,265]. While glycine and, to a lesser extent, glutamate are present in substantial concentrations in lens, there is evidence that the concentration of cysteine (or cystine) may be quite low; this may limit the rate of glutathione biosynthesis. The enzymes of the γ-glutamyl cycle, except for 5-oxoprolinase, have been found in rabbit lens [360]. The degradation of lens glutathione probably involves the activities of γ-glutamyltranspeptidase and γ-glutamylcyclotransferase; presumably the 5-oxoproline formed is transported out of the lens and metabolized by other intraocular tissues.

Lens contains glutathione reductase and glutathione peroxidase; these enzymes appear to function essentially by the pathway suggested by Mills [418] (Section VI,D). It is well known that the aqueous humor contains ascorbate, and Pirie [440] found that traces of hydrogen peroxide are also present. Her findings suggested that hydrogen peroxide is formed in the oxidation of ascorbate and that the reaction is catalyzed by light and riboflavin. Although ascorbate, dehydroascorbate, glutathione, and hydrogen peroxide can react nonenzymatically, Pirie's data indicate that glutathione peroxidase (and to a much lesser extent, catalase) functions in the removal of hydrogen peroxide. The interrelationships between ascorbate, dehydroascorbate, hydrogen peroxide, and glutathione may be considered in terms of the following reactions, all of which can occur nonenzymatically; the last reaction is also catalyzed by glutathione peroxidase:

$$\text{Ascorbate} + O_2 \rightarrow \text{dehydroascorbate} + H_2O_2 \tag{33}$$

$$\text{Ascorbate} + H_2O_2 \rightarrow \text{dehydroascorbate} + 2\ H_2O \tag{34}$$

$$2\ \text{GSH} + \text{dehydroascorbate} \rightarrow \text{GSSG} + \text{ascorbate} \tag{35}$$

$$2\ \text{GSH} + H_2O_2 \rightarrow \text{GSSG} + 2\ H_2\dot{O} \tag{36}$$

It is of particular interest that the development of cataracts, produced by a variety of causes, is associated with decreased concentrations of glutathione in the lens. In fact, the glutathione concentration of the lens may decrease before lens opacities become evident [593–597]. In relation to the problem of senile cataract, it seems relevant to note that the concentration of glutathione in human lens decreases substantially with age; thus about 50% of the glutathione present at age 20 is found at age 65 [24]. Much has been written about the development of experimental cataracts [see, for example, 598]; while a detailed discussion of this problem is beyond the scope of this review, a few points may be mentioned. The production of cataracts by naphthalene is associated with the formation of 1,2-dihydroxynaphthalene in the lens and ciliary body from 1,2-dihydro-1,2-dihydroxynaphthalene, which is, in turn, formed from naphthalene in the liver. 1,2-Dihydroxynaphthalene may be oxidized to form hydrogen peroxide and 1,2-naphthoquinone. The quinone may react with lens proteins, glutathione, and with other compounds present in the lens [599,600].

Many studies have been carried out on the production of cataracts by galactose [598]. Administration of galactose leads to a prompt and marked decrease in the concentration of glutathione in the lens. There is evidence that the effects of galactose are at least in part related to alteration of the permeability of the lens membrane associated with an increase in hydration and accumulation of dulcitol [601,602]. The latter is formed in considerable quantities by the reduction of galactose:

$$\text{Galactose} + \text{TPNH} + H^+ \xrightarrow{\text{aldose reductase}} \text{dulcitol} + \text{TPN}^+ \tag{37}$$

Dulcitol is not metabolized and therefore accumulates; the resulting osmotic effect leads to transport of water into the lens [603]. The utilization of TPNH for galactose reduction may also reduce the formation of glutathione by glutathione reductase. It is known that patients with the inborn error galactosemia commonly develop cataracts; such patients frequently exhibit aminoaciduria as well [604]. Indeed it is of interest that a relatively high percentage of patients who have congenital ocular cataracts also have aminoaciduria [605–609]. This association, considered in relation to the fact that high concentrations of the enzymes of the γ-glutamyl cycle are present in the ciliary body and lens, led to the suggestion that decreased amino acid transport across both the renal tubule and the ciliary epithelium might be associated with a similar enzyme defect involving the utilization of glutathione by way of the γ-glutamyl cycle [360].

C. Erythrocyte

The role of glutathione in the erythrocyte has attracted the attention of many investigators, especially after it was shown that the sensitivity of the erythrocytes of certain patients to various drugs (e.g., primaquine) is associated with a decrease in erythrocyte glutathione and a deficiency in ability to maintain glutathione levels when subjected to oxidative stress [176]. The effect is usually due to an inherited deficiency of glucose-6-phosphate dehydrogenase [177]. Presumably the sensitivity to various drugs is associated with the formation of hydrogen peroxide, other peroxides, and free radicals formed in the metabolism of the drugs [610–612]; however, other mechanisms are not excluded. In the absence of sufficient glutathione, whose concentration normally is influenced by activity of the hexose monophosphate pathway via TPNH, glutathione reductase, and glutathione peroxidase (Section VI), damage to erythrocyte membranes and hemoglobin occurs. Thus, hemoglobin may be oxidized to methemoglobin, a reversible process, but under severe oxidative stress, reactions involving the sulfhydryl groups of hemoglobin may occur leading to mixed disulfide formation between glutathione and hemoglobin; such altered hemoglobin may appear as Heinz bodies in the erythrocytes or may be precipitated as hemochromes. Oxidative destruction of erythrocyte membranes may also occur leading to hemolysis. Since the erythrocyte contains about half as many molecules of glutathione as of hemoglobin [611], it would appear that for glutathione to provide efficient protection the rate of interconversion of glutathione and glutathione disulfide must be substantial. Studies by Cohen and Hochstein [612] have provided evidence that hydrogen peroxide is generated in intact erythrocytes when hemolytic agents are added. In this work the hydrogen peroxide formed was detected by demonstrating the formation of catalase–hydrogen peroxide complexes using 3-amino-1,2,4-triazole [613,614]. Thus, the presence of hydrogen peroxide was shown after addition of 8-aminoquinolines such as primaquine, and also phenylhydrazine. The formation of hydrogen peroxide from 8-aminoquinolines required the presence of oxyhemoglobin, but evidence was obtained that phenylhydrazine may react directly with oxygen or with oxyhemoglobin to yield hydrogen peroxide. Although the details of the chemical transformations involved in the formation of hydrogen peroxide from 8-aminoquinolines are not yet entirely clear, it would appear that metabolites of these drugs play a role in the process. Although almost all of the erythrocyte glutathione can be alkylated with N-ethylmaleimide without impairment of erythrocyte survival on reinjection [615], there is evidence that alkylation of glutathione by N-ethylmaleimide is reversible [616]; presumably the regeneration of glutathione occurs at a sufficient rate to afford the cell protection against destruction.

The concentration of glutathione disulfide in erythrocytes of patients with glucose-6-phosphate dehydrogenase deficiency is somewhat higher then normal, and as discussed above (Section VI) there is evidence for the active transport of glutathione disulfide out of erythrocytes [474,475,617–619]. While this pathway could account for some of the observed amino acid turnover of glutathione in erythrocytes, the discovery that erythrocytes possess membrane bound γ-glutamyltranspeptidase as well as intracellular γ-glutamylcyclotransferase [620] leads to an alternative explanation for glutathione turnover. Thus, evidence was obtained for the formation of 5-oxoproline by the γ-glutamyltranspeptidase-cyclotransferase pathway in rabbit erythrocyte and in hemolysates of these cells. The formation of 5-oxoproline by this pathway occurs at a rate which is sufficient to account for the observed turnover of erythrocyte glutathione. This pathway would thus seem to explain the normal turnover of erythrocyte glutathione; the transport of glutathione disulfide out of erythrocytes may be of greater quantitative significance in pathological states in which there is increased tendency to form the disulfide.

It is interesting to note that neither the erythrocyte nor the lens exhibits 5-oxoprolinase activity. Other similarities between these structures, such as their dependency on anaerobic energy metabolism and the apparent requirement for glutathione to protect protein (hemoglobin, lens crystallines) have been noted [476].

D. Liver, Kidney, and Other Tissues

It is probable that glutathione functions in many tissues to maintain the sulfhydryl groups essential for the structure and function of various proteins, including enzymes, to remove hydrogen peroxide, other peroxides, and free radicals, and to serve as a catalyst for disulfide exchange reactions and certain other reactions. The wide distribution of glutathione reductase and glutathione peroxidase, and the occurrence of various glutathione transhydrogenase activities are consistent with such a view. However, the liver appears to be a major site for another function of glutathione, i.e., detoxification of foreign compounds (Section VII). In this pathway, the foreign compound reacts with the sulfhydryl group of glutathione to form a glutathione derivative, which may or may not undergo subsequent metabolism via the mercapturic acid pathway. As discussed above, the presence of γ-glutamyltranspeptidase activity in the epithelium of the bile ducts of the liver (Section V,B) and other findings (Section VII) indicate that compounds of this type can be excreted in the bile. They may ultimately appear in the feces or be reabsorbed, and be excreted as mercapturic acid in the urine. A major unsolved question concerns the extent to which reactions of this type occur normally and a related question is whether there is a quantitatively significant utilization of endogenously formed compounds by this pathway.

In contrast to the protective reactions of glutathione which involve interconversion between glutathione and glutathione disulfide, the role of glutathione in the pathway leading to mercapturic acid formation involves cleavage of both peptide bonds of glutathione; thus, the occurrence of such reactions must be accompanied by resynthesis of glutathione from its constitutent amino acids. A number of studies have been carried out on the turnover of tissue glutathione. Values of less than 4 hours for the biological half-life of glutathione were calculated from studies on rat and rabbit liver [99,135], and similar [132,136], as well as higher [138] and lower [140], values have been obtained. There are of course various technical and theoretical difficulties involved in obtaining such estimates; these have generally been based on observations on the rate of decrease of the specific isotopc content of glutathione after labeling with one of its constitutent amino acids. The result obtained may reflect the overall turnover or the turnover of the least rapidly turning over function of intracellular glutathione. In general, the values for the biological half-life of glutathione that have been obtained for the liver are significantly lower than those which have been obtained in studies on muscle [138], brain [132], lens [137], and erythrocytes [123–125]. The biological half-life for lens glutathione has been estimated to be about 3 days [137], a value which is just slightly lower than that found for the erythrocyte [123–125]; a half-life value in this range was also found for brain [132]. Values of about 18 hours and 6–8 hours were found, respectively, for the glutathione of muscle [138] and fibroblasts [621]. While the liver is evidently more active in the synthesis and degradation of glutathione than a number of other tissues, it is notable that the enzymatic capacity of the kidney to synthesize and to degrade glutathione is substantially higher than that of the liver, and indeed recent data [622] indicate that the biological half-life of kidney glutathione is appreciably less than that of the liver. The latter finding is generally supportive of the proposed role of the γ-glutamyl cycle in the renal handling of amino acids (Section V,E and F). However, the data thus far obtained do not indicate a very close correlation between the overall rates at which amino acids are normally reabsorbed in the renal tubule and at which total glutathione turns over in the kidney. This may reflect the existence of a small pool of glutathione–perhaps localized in the renal tubule itself–whose turnover is very high. On the other hand, as suggested above (Section V,F), alternative explanations of the function of the cycle such as those in which γ-glutamylcysteine plays a major role in transpeptidation or in which there are successive transpeptidation reactions must also be considered. Although a number of investigators have studied the biosynthesis, degradation, and turnover of glutathione in the liver, erythrocyte, and other tissues, investigations involving the kidney have been carried out only relatively recently; these were stimulated by enzymatic studies which emphasized the significance of glutathione in kidney metabolism [167]. It is well known that the kidney contains high concentrations of several enzymes that can cleave peptide bonds; such enzymes may be involved in the tubular reabsorption and utilization of peptides and it is possible that glutathione and γ-glutamyltranspeptidase play a role in the renal metabolism of peptides. Similar considerations may apply also to the role of glutathione in intestinal mucosa, an organ that also contains substantial amounts of γ-glutamyltranspeptidase. The localization of γ-glutamyltranspeptidase in certain epithelial structures (Section V,B) suggests that glutathione may be intimately involved in amino acid and peptide transport and metabolism in these anatomical sites. The relatively high concentration of 5-oxoproline and of γ-glutamylcyclotransferase activity in mammalian skin (Section V,C) suggests that the metabolism of glutathione in this organ may differ substantially from that in others; further work in this area is needed.

E. Nervous System

It has long been known that glutathione is present in the central nervous system and there are ample data indicating that the brain contains enzymatic activity capable of synthesizing and degrading glutathione and catalyzing reactions involved in the interconversion of glutathione and glutathione disulfide [see, for example, 623,624]. The possible function of the γ-glutamyl cycle in the transport of amino acids across the blood–cerebrospinal fluid barrier has been considered above (Section V,F). It should be mentioned, however, that there is substantial evidence for the existence of a number of amino acid transport systems in brain [625,626]. It has also been suggested that glutathione may function in the intracellular transport of amino acids in brain ([365]; Section V,F). This would be consistent with the finding of γ-glutamyl-cysteine synthetase [354,527], glutathione synthetase, γ-glutamyl-transpeptidase, γ-glutamylcyclotransferase, and 5-oxoprolinase [310] as well as a variety of γ-glutamylamino acids [45,80–82] in the brain.

The suggestion that γ-glutamyltranspeptidase might function in the utilization of amino acid neurotransmitters [365] recalls findings which indicate that glutathione is a specific activator for the feeding reaction in hydra [628–630]; this phenomenon, in which a specific chemical compound induces a specific behavioral response, could be a useful model for the study of an elementary form of neurotransmission. In the presence of glutathione, hydra open their mouths and their tentacles writhe and twist toward their mouths. The response is induced by glutathione, but not by γ-glutamylcysteine, cysteinylglycine, glycylcysteine, cysteine, glutathione disulfide, or aspartathione. The inactivity of glutathione disulfide was at first interpreted to indicate that the sulfhydryl group of glutathione is essential for the feeding reaction; later studies, however, showed that ophthalmic acid and norophthalmic acid were about as effective as glutathione in inducing the feeding reaction [631]. This interesting observation demonstrates dramatically that one cannot necessarily conclude that the sulfhydryl group of glutathione (or of another thiol) is essential for a particular reaction simply because the corresponding disulfide is inactive. It might be of interest, as suggested elsewhere [632], to examine some of the effects that are currently ascribed to the sulfhydryl group of glutathione with glutathione analogues that do not have a sulfhydryl group. Recent studies have shown that hydra exhibit substantial γ-glutamyltranspeptidase and γ-glutamylcyclotransferase activities [632a]; it would be of interest to learn whether such enzymatic activities function in relation to the glutathione feeding reaction.

Another role of glutathione (and of glutathione disulfide) in synaptic transmission has been proposed by Kosower [5,633–635], who observed that the thiol oxidizing agent diamide produced rapid cessation of contractile activity in muscle fibers of primary cultures of muscle cells after intracellular oxidation of glutathione to glutathione disulfide had occurred. Diamide dramatically increased the rate of release of miniature end plate potentials and also an increase in size of the neurally evoked end plate potential. It was proposed that the excess glutathione disulfide formed functions as an oxidizing agent converting particular membrane dithiols into disulfides. A hypothesis was proposed according to which the release of acetylcholine is coupled with the effect of glutathione disulfide. This idea, which suggests that there is a new step in transmitter release at the myoneural junction [635], has led to an additional hypothesis concerning the molecular basis for learning and memory [5].

X. OTHER STUDIES ON GLUTATHIONE

Although the facts and ideas reviewed above indicate that glutathione has a number of significant cellular functions, the documentation is incomplete. For example, proof that the γ-glutamyl cycle functions in amino acid or peptide transport is lacking, and the full significance of the function of mercapturic acid formation is still not clear. In addition, there are significant gaps in our knowledge of the apparent function performed by glutathione in maintaining the sulfhydryl groups of proteins, and although it is evident that glutathione is a highly specific coenzyme of glyoxalase, the physiological function of this enzyme is very much in doubt. There are still other questions that remain to be answered about glutathione including some that relate to the chemistry of this molecule.* Recent studies on the dissociation equilibria of glutathione [636] and on the rates of intramolecular proton transfer in glutathione [637] have been reviewed. Studies on the calorimetric determination of the microionization constants of glutathione have suggested that glutathione has no unusual intramolecular interactions [638]. In earlier work the possibility was considered that glutathione may, under certain circumstances, exist in a form possessing a thiazoline ring [639]; subsequent work [640–645] indicates that although the thiazoline derivative of glutathione can be prepared, it is very unstable in neutral aqueous solution [645]. Work has also progressed on the study of metal complexes of glutathione [see, for example, 641,646,647], and structures for such complexes have been proposed and discussed. These and earlier studies indicate pK_a values for glutathione of 2.12, 3.53, 8.66, and 9.62 [648]. Among the interesting derivatives of glutathione which have been prepared are glycolyl glutathione [649] and glutathione trisulfide (G—S—S—S—G) [650,651].

While there has been much interest in glutathione disulfide, other oxidation products of glutathione have received somewhat less attention. However, glutathione sulfinic acid (G—SO$_2$H) and glutathione sulfonic acid (G—SO$_3$H) have been prepared by fission of glutathione disulfide with silver salts, and oxidation of glutathione with performic acid,

---

* A review emphasizing the chemical and physical properties of glutathione will soon appear [635a].

respectively [652]. The thiol sulfonate derivative of glutathione (GSO₂—SG) has also been prepared [652,653]. Glutathione thiosulfate (S-sulfoglutathione) (GSSO₃H) was first found in lens by Waley [43]. It was subsequently found in rat small intestine; after intraperitoneal injection of sodium [³⁵S]sulfate, radioactive glutathione thiolsulfate was isolated [654]. Enzyme activity capable of reducing glutathione thiosulfate to glutathione was found in peas [655] and also in rat liver [656,657]. Although these studies suggested that the conversion of glutathione thiosulfate to glutathione is catalyzed by a TPNH-dependent enzyme different from glutathione reductase, subsequent work [658] indicates that glutathione thiosulfate reacts with glutathione present in the enzyme preparation to yield glutathione disulfide, which is reduced by glutathione reductase. The reaction between glutathione and glutathione thiosulfate may be catalyzed by a transhydrogenase.

The possibility that glutathione can participate in still other types of chemical transformations is suggested by various studies which have been reported in the literature. For example, S-alkyl derivatives of glutathione have been prepared and shown to be substrates for an enzyme activity present in mouse liver [659] (see also [660]). S-Phospho derivatives are also conceivable and protein-bound derivatives of this type have often been considered as possible intermediates in enzyme catalyzed reactions. The α-keto acid analogue of glutathione does not seem to have been prepared, although evidence suggesting that glutathione is a substrate for snake venom L-amino acid oxidase has been published [339]; it is conceivable that glutathione participates in enzymatic transamination. Presumably, the α-keto analogue of glutathione would tend to cyclize to the corresponding lactam [339].

A function of glutathione in amine transport or metabolism is indicated by the finding that spermidine is converted to glutathionyl spermidine in *E. coli* [25,661,662]. Relatively high concentrations of glutathionyl spermidine are formed in this organism under certain conditions; this product may represent as much as 50% of the total sulfur content of the trichloroacetic acid-soluble fraction and about 7–10% of the total sulfur in the bacterium. The data indicate that glutathione is linked to spermidine through the carboxyl group of the glycyl moiety. This appears to be the only glutathione derivative of this type that has thus far been found in nature. Crude extracts of *E. coli* can synthesize glutathionyl spermidine when incubated with high concentrations of glutathione, magnesium ions, ATP, and spermidine at pH 8. Another activity has been found that catalyzes the conversion of glutathionyl spermidine to spermidine.

The propensity of glutathione to form mixed disulfides with proteins and other compounds that have sulfhydryl groups has been mentioned several times above (see Sections V and VI). It is notable that hydrolysis of glutathione can occur slowly and nonenzymatically in aqueous solution to yield products that include the mixed disulfide of glutathione and cysteinylglycine [663]. Mixed disulfides are likely to form readily and nonenzymatically in various *in vitro* systems in which glutathione is a reactant, e.g., in transpeptidation reactions involving glutathione in which cysteinylglycine is a product. The susceptibility of various mixed disulfides to the action of γ-glutamyl-transpeptidase and γ-glutamyl-cyclotransferase remains to be thoroughly explored.

The enzymatic conversion of glutathione to γ-glutamylcysteine and glycine was observed long ago by Grassmann *et al.* [196], and more recently this reaction (on glutathione disulfide) has been effectively used as a preparative procedure for γ-glutamylcysteine disulfide [664]. It is of interest that cathepsin preparations from normal rat liver, spleen, and kidney, as well as from an experimental hepatoma, were found to hydrolyze the cysteinyl–glycine bond of glutathione [665,666]; in these studies the formation of γ-glutamylcysteine was shown. The catheptic activity toward glutathione exhibited a sharp optimum at about pH 4.8; however, some activity was observed at pH values of 7–9, suggesting that some glutathione may be broken down under physiological conditions to γ-glutamylcysteine and glycine [in contrast to breakdown by the γ-glutamyltranspeptidase route which yields cysteinylglycine and a glutamyl derivative (Section V,B)]. However, there seems to be no conclusive evidence that the apparent catheptic activity observed in the pH range 7–9 represents hydrolysis to γ-glutamylcysteine and glycine; thus, the studies in which γ-glutamylcysteine formation was demonstrated were carried out at pH 5.0, and hydrolysis at the other values of pH was determined by relatively unspecific procedures. Nevertheless, the possibility that some glutathione may be broken down to γ-glutamylcysteine under certain circumstances cannot be excluded.

Several glutathione analogues have been described that do not have sulfhydryl groups and a number of studies have been carried out on two of these, i.e., ophthalmic and norophthalmic acids. The available data indicate that these compounds are synthesized by the same enzymes that catalyze glutathione synthesis. Indeed, studies on γ-glutamyl-cysteine synthetase indicate that the amino acid acceptor specificity of this enzyme is quite low, especially in the presence of manganese ions [144]. Further evidence that the amino acid sequence of glutathione is very poorly controlled as compared to that of proteins was obtained in studies in which rats and chicks were fed β-cyanoalanine [77]. Such animals accumulated γ-glutamyl-β-cyanoalanylglycine in their livers, and evidence was obtained for the presence of γ-glutamyl-β-cyanoalanine in the liver of chicks fed β-cyanoalanine. In studies on the metabolism of S-methyl cysteine in yeasts, a major component present in the cells was found to be S-methyl glutathione [667]. These studies suggest the possibility that a number of additional glutathione derivatives may occur in nature and may have thus far escaped detection. Clearly, most of the methods used for the determination of glutathione would not be useful for the detection of glutathione analogues that do not have sulfhydryl groups. The substantial activity of γ-glutamyl-cysteine synthetase toward a number of protein amino acids suggests that glutathione analogues containing these amino acids might be formed. The available information indicates that such compounds would probably be active in transpeptidation reactions.

Several studies have been carried out on the metabolism of glutathione in experimental tumors. Thus, the concentration of glutathione in male rat liver was found to increase following administration of 3'-methyl-4-dimethylaminoazobenzene [668], and a good correlation was found between the relative carcinogenic activities of azo dyes and the ability of these compounds to increase the concentration of liver glutathione [669]. These findings led to the suggestion that the simultaneous occurrence of dye binding and increased glutathione concentration are essential for azo dye hepatocarcinogenesis. Certain transplantable chemically induced rat hepatomas exhibit high levels of γ-glutamyltranspeptidase activity as compared to the relatively low activity of normal adult liver [670]. In studies in which 3'-methyl-4-dimethylaminoazobenzene, 2-acetylaminofluorene, thioacetamide, DL-ethionine, or dimethylnitrosamine were administered to rats, the liver γ-glutamyltranspeptidase activity was also found to increase [671,672]. These observations and the earlier studies [668,669] seem to indicate a significant role of glutathione and γ-glutamyltranspeptidase in carcinogenesis in the liver.

Yoshida ascites sarcoma cells from a strain sensitive to chemotherapy with alkylating agents were found to exhibit higher activities of the glutathione synthesizing enzymes than found in a resistant cell strain [673]. The γ-glutamyltranspeptidase activities of the two strains were similar. After administration of chlorambucil to rats bearing these tumors, the glutathione synthesizing enzymes increased in the sensitive cells and remained unchanged in the resistant. The findings thus demonstrate a relationship between sensitivity to the drug and increased capacity to synthesize glutathione.

The concentrations of glutathione in various tissues may be influenced by a number of factors, which have not yet been thoroughly explored.

Thus, there is evidence that the concentration of glutathione in the liver is decreased during protein deprivation [28]. Fluctuations of liver glutathione levels have been observed over a 24 hour period [30], and also in erythrocytes after meals [479]. A variety of hormonal effects have been observed on tissue glutathione levels, and the effects of various forms of stress have been recorded [674]. Several anesthetics (e.g., ether, barbiturates, nitrous oxide) decrease glutathione levels in the liver and kidney. Epinephrine, chlorpromazine, and hypothermia also have this effect [675].

REFERENCES

1. L. Rapkine, *C. R. Acad. Sci.* **191**, 871–874 (1930).
2. L. Rapkine, *J. Chim. Phys., Physicochim. Biol.* **34**, 416–427 (1937).
3. E. S. G. Barrón, *Advan. Enzymol.* **11**, 201–266 (1951).
4. D. Mazia, *in* "Glutathione" (S. Colowick, A. Lazarow, E. Racker, D. R. Schwarz, E. Stadtman, and H. Waelsch, eds.), pp. 209–222. Academic Press, New York, 1954.
5. E. M. Kosower, *Proc. Nat. Acad. Sci. U.S.* **69**, 3292–3296 (1972).
6. J. deRey-Pailhade, *C. R. Acad. Sci.* **106**, 1683–1684 (1888).
7. J. deRey-Pailhade, *C. R. Acad. Sci.* **107**, 43–44 (1888).
8. F. G. Hopkins, *Biochem. J.* **15**, 286–305 (1921).
9. F. G. Hopkins, *J. Biol. Chem.* **84**, 269–320 (1929).
10. E. C. Kendall, B. F. McKenzie, and H. L. Mason, *J. Biol. Chem.* **84**, 657–674 (1929).
11. C. R. Harington and T. H. Mead, *Biochem. J.* **29**, 1602–1611 (1935).
12. L. J. Harris, *in* "Hopkins and Biochemistry" (J. Needham and E. Baldwin, eds.), pp. 72–79. Cambridge Univ. Press, London and New York, 1949.
13. S. Colowick, A. Lazarow, E. Racker, D. R. Schwarz, E. Stadtman, and H. Waelsch, eds., "Glutathione." Academic Press, New York, 1954.
14. E. M. Crook, ed., "Glutathione," Biochem. Soc. Symp. No. 17. Cambridge Univ. Press, London and New York, 1959.
15. L. Flohé, H. Ch. Benöhr, H. Sies, H. D. Waller, and A. Wendel, eds., "Symposium on Glutathione." Thieme, Stuttgart, 1973.
16. W. E. Knox *in* "The Enzymes" (P. D. Boyer, H. Lardy, and K. Myrbäck, eds.), 2nd ed., vol. 2, Part A, pp. 253–294. Academic Press, New York, 1960.
17. E. C. Kendall, H. L. Mason, and B. F. McKenzie, *J. Biol. Chem.* **87**, 55–79 (1930).
18. A. I. McMullen, *Biochim. Biophys. Acta* **41**, 152–154 (1960).
19. E. F. Jansen and R. Jang, *Arch. Biochem. Biophys.* **40**, 358–363 (1952).
20. N. K. Schaffer and W. M. Ziegler, *Proc. Soc. Exp. Biol. Med.* **42**, 93–94 (1939).
21. W. A. Reeves and J. D. Guthrie, *Arch. Biochem. Biophys.* **26**, 316–318 (1950).
22. B. E. Davidson and F. J. R. Hird, *Biochem. J.* **93**, 232–236 (1964).
23. E. Beutler, O. Duron, and B. M. Kelly, *J. Lab. Clin. Med.* **61**, 882–888 (1963).
24. J. J. Harding, *Biochem. J.* **117**, 957–960 (1970).
25. H. Tabor and C. W. Tabor, *Advan. Enzymol.* **36**, 203–268 (1972).
26. P. C. Jocelyn, "Biochemistry of the SH Group." Academic Press, New York, 1972.
27. H. Barford and E. Eden, *Aus. J. Exp. Biol.* **34**, 269–276 (1956).
28. S. Edwards and W. W. Westerfeld, *Proc. Soc. Exp. Biol. Med.* **79**, 57–59 (1952).

29. P. Batalden, W. R. Swaim, and J. T. Lowman, *J. Lab. Clin. Med.* **71,** 312–318 (1968).
30. L. V. Beck, V. D. Rieck, and B. Duncan, *Proc. Soc. Exp. Biol. Med.* **97,** 229–231 (1958).
30a. R. J. Jaeger, R. B. Conolly, and S. D. Murphy, *Res. Commun. Chem. Pathol. Pharmacol.* **6,** 465–471 (1973).
31. L. Eldjarn and A. Pihl, *J. Biol. Chem.* **225,** 499–510 (1957).
32. L. Revesz and H. Modig, *Nature (London)* **207,** 430–431 (1965).
33. I. Krimsky and E. Racker, *J. Biol. Chem.* **198,** 721–729 (1952).
34. T. H. J. Huisman and A. M. Dozy, *J. Lab. Clin. Med.* **60,** 302–319 (1962).
35. T. H. J. Huisman, A. M. Dozy, B. F. Horton, and C. M. Nechtman, *J. Lab. Clin. Med.* **57,** 355–373 (1966).
36. R. C. Jackson, K. R. Harrap, and C. A. Smith, *Biochem. J.* **110,** 37P (1968).
37. H. Modig, *Biochem. Pharmacol.* **17,** 177–186 (1968).
38. A. Meister, "Biochemistry of the Amino Acids," 2nd ed., Vol. 1, pp. 119–123. Academic Press, New York, 1965.
39. S. G. Waley, *Advan. Protein Chem.* **21,** 2–112 (1966).
40. P. R. Carnegie, *Biochem. J.* **89,** 459–471 (1963).
41. S. G. Waley, *Biochem. J.* **64,** 715–726 (1956).
42. S. G. Waley, *Biochem. J.* **68,** 189–192 (1958).
43. S. G. Waley, *Biochem. J.* **71,** 132–137 (1959).
44. D. H. Calam and S. G. Waley, *Biochem. J.* **86,** 226–231 (1963).
45. A. Kanazawa, Y. Kakimoto, T. Nakajima, and J. Sano, *Biochim. Biophys. Acta* **111,** 90–95 (1965).
46. E. R. Stadtman and A. Kornberg, *J. Biol. Chem.* **203,** 47–54 (1953).
47. S. H. Chang and D. R. Wilken, *J. Biol. Chem.* **240,** 3136–3139 (1965).
48. R. N. Ondarza, *Biochim. Biophys. Acta* **107,** 112–119 (1965).
49. A. I. Virtanen, *Angew. Chem., Int. Ed. Engl.* **1,** 229 (1962).
50. A. I. Virtanen and E. J. Matikkala, *Suom. Kemistilehti B* **34,** 53–54 (1961).
51. A. I. Virtanen and E. J. Matikkala, *Hoppe-Seyler's Z. Physiol. Chem.* **322,** 8–20 (1960).
52. A. I. Virtanen and E. J. Matikkala, *Suom. Kemistilehti B* **34,** 84 and 114 (1961).
53. A. I. Virtanen and E. J. Matikkala, *Suom. Kemistilehti B* **33,** 83–84 (1960).
54. A. I. Virtanen and T. Ettala, *Suom. Kemistilehti B* **31,** 272 (1958).
55. A. I. Virtanen and T. Ettala, *Acta Chem. Scand.* **12,** 787–789 (1958).
56. A. I. Virtanen, M. Hatanaka, and M. Berlin, *Suom. Kemistilehti B* **35,** 52 (1962).
57. A. I. Virtanen and I. Mattila, *Suom. Kemistilehti B* **34,** 73 (1961).
58. A. I. Virtanen and I. Mattila, *Suom. Kemistilehti B* **34,** 44 (1961).
59. P. O. Larsen, *Acta Chem Scand.* **16,** 1511–1518 (1962).
60. C. J. Morris, J. S. Thompson, S. Asen, and F. Irreverre, *J. Biol. Chem.* **236,** 1181–1182 (1961).
61. C. J. Morris, J. F. Thompson, S. Asen, and F. Irreverre, *J. Biol. Chem.* **237,** 2180–2181 (1962).
62. B. Levenberg, *J. Amer. Chem. Soc.* **83,** 503–504 (1961).
63. E. G. Daniels, R. B. Kelly, and J. W. Hinman, *J. Amer. Chem. Soc.* **83,** 3333–3334 (1961).
64. R. B. Kelly, E. G. Daniels, and J. W. Hinman, *J. Org. Chem.* **27,** 3229–3231 (1962).
65. J. Jadot, J. Casimir, and M. Renard, *Biochim. Biophys. Acta* **43,** 322–328 (1960).
66. E. D. Schilling and F. M. Strong, *J. Amer. Chem. Soc.* **77,** 2843–2845 (1955).
67. G. F. McKay, J. J. Lalich, E. D. Schilling, and F. M. Strong, *Arch. Biochem. Biophys.* **52,** 313–322 (1954).

68. C. H. Hassall and K. Reyle, *Biochem. J.* **60**, 334–339 (1955).
69. C. H. Hassall and D. I. John, *J. Chem. Soc., London* pp. 4112–4115 (1960).
70. U. von Renner, A. Jöhl, and W. G. Stoll, *Helv. Chim. Acta* **41**, 588–592 (1958).
71. J. A. Carbon, W. B. Martin, and L. R. Swett, *J. Amer. Chem. Soc.* **80**, 1002 (1958).
72. C. von Holt, W. Leppla, B. Kroner, and L. von Holt, *Naturwissenschaften* **43**, 279 (1956).
73. C. von Holt and W. Leppla, *Angew. Chem.* **70**, 25 (1958).
74. R. S. De Ropp, J. C. Van Meter, E. C. De Renzo, K. W. McKerns, C. Pidacks, P. H. Bell, E. Ullman, S. R. Safir, W. J. Fanshawe, and S. B. Davis, *J. Amer. Chem. Soc.* **80**, 1004–1005 (1958).
75. A. Jöhl and W. G. Stoll, *Helv. Chim. Acta* **42**, 716–718 (1959).
76. C. Ressler, S. N. Nigam, Y.-H. Giza, and J. Nelson, *J. Amer. Chem. Soc.* **85**, 3311–3312 (1963).
77. K. Sasaoka, C. Lauinger, S. N. Nigam, and C. Ressler, *Biochim. Biophys. Acta* **156**, 128–134 (1968).
78. P. M. Dunnill and L. Fowden, *Biochem. J.* **86**, 388–391 (1963).
79. D. L. Buchanan, E. E. Haley, and R. T. Markiw, *Biochemistry* **1**, 612–620 (1962).
80. Y. Kakimoto, T. Nakajima, M. Kanazawa, M. Takesada, and J. Sano, *Biochim. Biophys. Acta* **93**, 333–338 (1964).
81. K. L. Reichelt, *J. Neurochem.* **17**, 19–25 (1970).
82. D. H. G. Versteeg and A. Witter, *J. Neurochem.* **17**, 41–52 (1970).
83. M. D. Sass, *Clin. Chem.* **16**, 466–471 (1970).
84. J. W. Patterson and A. Lazarow, *Methods Biochem. Anal.* **2**, 259–278 (1955).
85. C. G. Thomson and H. Martin, *in* "Glutathione" (E. M. Crook, ed.), Biochem. Soc. Symp. No. 17, pp. 17–27. Cambridge Univ. Press, London and New York, 1959.
86. G. E. Woodward, *J. Biol. Chem.* **109**, 1–10 (1935).
87. E. Racker, *J. Biol. Chem.* **190**, 685–696 (1951).
88. A. Pihl, L. Eldjarn, and J. Bremer, *J. Biol. Chem.* **227**, 339–345 (1957).
89. C. W. I. Owens and R. V. Belcher, *Biochem. J.* **94**, 705–711 (1965).
90. D. R. Grassetti and J. F. Murray, Jr., *Anal. Biochem.* **21**, 427–434 (1967).
91. F. Tietze, *Anal. Biochem.* **27**, 502–522 (1969).
92. V. H. Cohn and J. Lyle, *Anal. Biochem.* **14**, 434–440 (1966).
93. J. H. Ladenson and W. C. Purdy, *Clin. Chem.* **17**, 908–914 (1971).
94. G. L. Ellman, *Arch. Biochem. Biophys.* **82**, 70–77 (1959).
95. S. K. Srivastava and E. Beutler, *Anal. Biochem.* **25**, 70–76 (1968).
96. G. L. Ellman and H. Lysko, *J. Lab. Clin. Med.* **70**, 518–527 (1967).
97. H. Güntherberg and J. Rost, *Anal. Biochem.* **15**, 205–210 (1966).
98. R. Sekura, unpublished work in the author's laboratory (1973).
98a. M. Koivusalo and L. Uotila, *Anal. Biochem.* **59**, 34–45 (1974).
99. H. Waelsch and D. Rittenberg, *J. Biol. Chem.* **139**, 761–774 (1941).
100. K. Bloch and H. S. Anker, *J. Biol. Chem.* **169**, 765–766 (1947).
101. A. E. Braunstein, G. A. Shamshikova, and A. L. Ioffe, *Biokhimiya* **13**, 95–100 (1948).
102. R. B. Johnston and K. Bloch, *J. Biol. Chem.* **179**, 493–494 (1949).
103. K. Bloch, *J. Biol. Chem.* **179**, 1245–1254 (1949).
104. R. B. Johnston and K. Bloch, *J. Biol. Chem.* **188**, 221–240 (1951).
105. J. E. Snoke and K. Bloch, *J. Biol. Chem.* **199**, 407–414 (1952).
106. J. E. Snoke, S. Yanari, and K. Bloch, *J. Biol. Chem.* **201**, 573–586 (1953).
107. J. E. Snoke and K. Bloch, *J. Biol. Chem.* **213**, 825–835 (1955).
108. S. Mandeles and K. Bloch, *J. Biol. Chem.* **214**, 639–646 (1955).
109. S. Yanari, J. E. Snoke, and K. Bloch, *J. Biol. Chem.* **201**, 561–571 (1953).

110. J. E. Snoke, *J. Biol. Chem.* **213**, 813–824 (1955).
111. J. E. Snoke and K. Bloch, *in* "Glutathione" (S. Colowick *et al.*, eds.), pp. 129–141. Academic Press, New York, 1954.
112. D. H. Strumeyer, Doctoral thesis, Harvard University, Cambridge, Massachusetts, (1959).
113. P. J. Samuels, *Biochem. J.* **35**, 441–444 (1953).
114. S. G. Waley, *in* "Glutathione" (E. M. Crook, ed.), Biochem. Soc. Symp. No. 17, pp. 79–92. Cambridge Univ. Press, London and New York, 1959.
115. W. B. Rathbun, *Arch. Biochem. Biophys.* **122**, 62–72 (1967).
116. W. B. Rathbun, *Arch. Biochem. Biophys.* **122**, 73–84 (1967).
117. G. C. Webster, *Arch. Biochem. Biophys.* **47**, 241–250 (1953).
118. G. C. Webster and J. E. Varner, *Arch. Biochem. Biophys.* **52**, 22–32 (1954).
119. G. C. Webster and J. E. Varner, *Arch. Biochem. Biophys.* **55**, 95–103 (1955).
120. A. Wendel, *in* "Symposium on Glutathione" (L. Flohé, ed.), pp. 69–76. Thieme, Stuttgart, 1973.
121. P. W. Majerus, M. J. Brauner, M. B. Smith, and V. J. Minnich, *J. Clin. Invest.* **50**, 1637–1643 (1971).
122. R. C. Jackson, *Biochem. J.* **111**, 309–315 (1969).
123. E. Dimant, E. Landsberg, and I. M. London, *J. Biol. Chem.* **213**, 769–776 (1954).
124. H. A. Elder and R. A. Mortensen, *J. Biol. Chem.* **218**, 261–267 (1955).
125. R. A. Mortensen, M. I. Haley, and H. A. Elder, *J. Biol. Chem.* **218**, 269–273 (1955).
126. D. K. Kasbekar and A. Sreenivasan, *Biochem. J.* **72**, 389–395 (1959).
127. A. Koj, *Acta Biochim. Pol.* **9**, 11–25 (1962).
128. J. Horejsi and L. Mircevova, *Acta Biochim. Pol.* **11**, 107–111 (1964).
129. A. Hochberg, M. Rigbi, and E. Dimant, *Biochim. Biophys. Acta* **90**, 464–471 (1964).
130. A. Hochberg and E. Dimant, *Biochim. Biophys. Acta* **104**, 53–62 (1965).
131. J. Niv, A. Hochberg, and E. Dimant, *Biochim. Biophys. Acta* **127**, 26–34 (1966).
132. G. W. Douglas and R. A. Mortensen, *J. Biol. Chem.* **221**, 581–585 (1956).
133. S. Berl, A. Lajtha, and H. Waelsch, *J. Neurochem.* **7**, 186–197 (1961).
134. Y. Machiyama, R. Balazs, and T. Merei, *J. Neurochem.* **17**, 449–453 (1970).
135. H. Waelsch and D. Rittenberg, *J. Biol. Chem.* **144**, 53–58 (1942).
136. E. I. Anderson and W. A. Mosher, *J. Biol. Chem.* **188**, 717–722 (1951).
137. P. J. McMillan, S. J. Ryerson, and R. A. Mortensen, *Arch. Biochem. Biophys.* **81**, 119–123 (1959).
138. O. B. Henriques, S. B. Henriques, and A. Neuberger, *Biochem. J.* **60**, 409–424 (1955).
139. S. B. Henriques, O. B. Henriques, and F. R. Mandelbaum, *Biochem. J.* **66**, 222–227 (1957).
140. S. C. Kalser and L. V. Beck, *Biochem. J.* **87**, 618–623 (1963).
141. A. Meister, *in* "The Enzymes" (P. D. Boyer, ed.), 3rd ed., Vol. 10, pp. 671–697. Academic Press, New York, 1974.
142. J. S. Davis, J. B. Balinsky, J. S. Harrington, and J. B. Shepherd, *Biochem. J.* **133**, 667–678 (1973).
143. M. Orlowski and A. Meister, *Biochemistry* **10**, 372–380 (1971).
144. M. Orlowski and A. Meister, *J. Biol. Chem.* **246**, 7095–7105 (1971).
145. P. Richman and P. Trotta, unpublished work in the author's laboratory.
146. W. B. Rathbun and H. D. Gilbert, *Anal. Biochem.* **54**, 153–160 (1973).
147. A. Meister, *Advan. Enzymol.* **31**, 183–218 (1968).
148. A. Meister, *Harvey Lect.* **63**, 139–178 (1969).
149. A. Meister, *in* "The Enzymes" (P. D. Boyer, ed.), 3rd ed., Vol. 10, pp. 699–754. Academic Press, New York, 1974.

150. R. A. Ronzio and A. Meister, *Proc. Nat. Acad. Sci. U.S.* **59**, 164–170 (1968).
151. R. Ronzio, W. B. Rowe, and A. Meister, *Biochemistry* **8**, 1066–1075 (1969).
152. W. B. Rowe, R. A. Ronzio, and A. Meister, *Biochemistry* **8**, 2674–2680 (1969).
153. J. M. Manning, S. Moore, W. B. Rowe, and A. Meister, *Biochemistry* **8**, 2681–2685 (1969).
154. W. B. Rowe and A. Meister, *Biochemistry* **15**, 1578–1582 (1973).
155. P. G. Richman, M. Orlowski, and A. Meister, *J. Biol. Chem.* **248**, 6684–6690 (1973).
156. W. B. Rowe and A. Meister, *Proc. Nat. Acad. Sci. U.S.* **66**, 500–506 (1970).
157. J. D. Gass and A. Meister, *Biochemistry* **9**, 1380–1390 (1970).
158. D. H. Strumeyer and K. Bloch, *J. Biol. Chem.* **235**, PC27 (1960).
159. A. Meister, *in* "Symposium on Glutathione" (L. Flohé, ed.), pp. 56–67. Thieme, Stuttgart, 1973.
160. E. D. Mooz and A. Meister, *Biochemistry* **6**, 1722–1734 (1967).
161. E. D. Mooz and A. Meister, *in* "Methods in Enzymology" (H. Tabor and C. W. Tabor, eds.), Vol. 17B, pp. 483–495. Academic Press, New York, 1971.
162. J. S. Nishimura, E. A. Dodd, and A. Meister, *Fed. Proc., Fed. Amer. Soc. Exp. Biol.* **22**, 536 (1963).
162a. J. S. Nishimura, E. A. Dodd, and A. Meister, *J. Biol. Chem.* **238**, PC1179–PC1180 (1963).
163. J. S. Nishimura, E. A. Dodd, and A. Meister, *J. Biol. Chem.* **239**, 2553–2558 (1964).
164. A. Wendel, E. Schaich, U. Weber, and L. Flohe, *Hoppe-Seyler's Z. Physiol. Chem.* **353**, 514–522 (1972).
165. A. Wendel and L. Flohé, *Hoppe-Seyler's Z. Physiol. Chem.* **353**, 523–530 (1972).
166. N. Chernov, V. Wellner, and A. Meister unpublished.
167. A. Meister, *Science* **180**, 33–39 (1973).
168. K.-G. Blume, N. V. Paniker, and E. Beutler, *Clin. Chim. Acta* **45**, 281–285 (1973).
169. M. D. Sass, *Clin. Chim. Acta* **22**, 207–210 (1968).
170. V. Minnich, M. B. Smith, M. J. Brauner, and P. W. Majerus, *J. Clin. Invest.* **50**, 507–513 (1971).
171. F. Goswitz, G. R. Lee, G. E. Cartwright, and M. M. Wintrobe, *J. Lab. Clin. Med.* **67**, 615–623 (1966).
172. K. G. Blume, N. V. Paniker, and E. Beutler, *in* "Symposium on Glutathione" (L. Flohé, ed.), pp. 157–164. Thieme, Stuttgart, 1973.
173. N. V. Paniker and E. Beutler, *J. Lab. Clin. Med.* **80**, 481–487 (1972).
174. P. Boivin and C. Galand, *Nouv. Rev. Fr. Hematol.* **5**, 707–720 (1965).
175. P. Richman and A. Meister, *J. Biol. Chem.* **250**, 1422–1426 (1975).
176. E. Beutler, *in* "The Metabolic Basis of Inherited Disease" (J. B. Stanbury, J. B. Wyngaarden, and D. S. Fredrickson, eds.), 3rd ed., p. 1358. McGraw-Hill, New York, 1972.
177. P. E. Carson, C. Larkin Flanagan, C. E. Ickes, and A. S. Alving, *Science* **124**, 484–485 (1956).
178. M. Oort, J. A. Loos, and H. K. Prins, *Vox Sang.* **6**, 370–373 (1961).
179. H. K. Prins, M. Oort, J. A. Loos, C. Zurcher, and T. Beckers, *Blood* **27**, 145–166 (1966).
180. P. Boivin, C. Galand, R. André, and J. Debray, *Nouv. Rev. Fr. Hematol.* **6**, 859–866 (1966).
181. P. Boivin, C. Galand, and J. F. Bernard, *in* "Symposium on Glutathione" (L. Flohé, ed.), pp. 146–156. Thieme, Stuttgart, 1973.
182. D. N. Mohler, P. W. Majerus, V. Minnich, C. E. Hess, and M. D. Garrick, *N. Engl. J. Med.* **283**, 1253 (1970).

183. P. N. Konrad, F. Richards, W. N. Valentine, and D. E. Paglia, *N. Engl. J. Med.* **286,** 557 (1972).

183a. F. Richards, II, M. R. Cooper, L. A. Pearce, R. J. Cowan, and C. L. Spurr, *Arch. Intern. Med.* **134,** 534–537 (1974).

184. J. E. Smith and B. I. Osburn, *Science* **158,** 374–375 (1967).

185. J. C. Ellory, E. M. Tucker, and E. V. Deverson, *Biochim. Biophys. Acta* **279,** 481–483 (1972).

186. J. E. Smith, *Amer. J. Vet. Res.* **34,** 847–848 (1973).

187. N. S. Agar and J. E. Smith, *Proc. Soc. Exp. Biol. Med.* **142,** 502–505 (1973).

187a. J. E. Smith, M. S. Lee, and A. S. Mia, *J. Lab. Clin. Med.* **82,** 713–718 (1973).

188. A. Meister, "Biochemistry of the Amino Acids," 2nd ed., Vol. 1, pp. 473–482. Academic Press, New York, 1965.

189. M. Bergmann and H. Fraenkel-Conrat, *J. Biol. Chem.* **119,** 707–720 (1937).

190. J. S. Fruton, *Yale J. Biol. Med.* **22,** 263–271 (1950).

191. R. B. Johnston, M. J. Mycek, and J. S. Fruton, *J. Biol. Chem.* **185,** 629–641 (1950).

192. R. B. Johnston, M. J. Mycek, and J. S. Fruton, *J. Biol. Chem.* **187,** 205–211 (1950).

193. J. S. Fruton, R. B. Johnston, and M. Fried, *J. Biol. Chem.* **190,** 39–53 (1951).

194. J. Durell and J. S. Fruton, *J. Biol. Chem.* **207,** 487–500 (1954).

195. M. E. Jones, W. R. Hearn, M. Fried, and J. S. Fruton, *J. Biol. Chem.* **195,** 645–656 (1952).

196. W. Grassmann, H. Dyckerhoff, and H. Eibeler, *Hoppe Seyler's Z. Physiol. Chem.* **189,** 112–120 (1930).

197. H. D. Dakin and H. W. Dudley, *J. Biol. Chem.* **15,** 463–474 (1913).

198. M. E. Platt and E. F. Schroeder, *J. Biol. Chem.* **106,** 179–190 (1934).

199. G. E. Woodward, M. P. Munro, and E. F. Schroeder, *J. Biol. Chem.* **109,** 11–27 (1935).

200. E. F. Schroeder, M. P. Munro, and L. Weil, *J. Biol. Chem.* **110,** 181–200 (1935).

201. E. F. Schroeder and G. E. Woodward, *J. Biol. Chem.* **120,** 209–217 (1937).

202. G. E. Woodward and F. E. Reinhart, *J. Biol. Chem.* **145,** 471–480 (1942).

203. F. Binkley and K. Nakamura, *J. Biol. Chem.* **173,** 411–421 (1948).

204. C. K. Olson and F. Binkley, *J. Biol. Chem.* **186,** 731–735 (1950).

205. F. Binkley, *Nature (London)* **167,** 888–889 (1951).

206. F. Binkley, *Exp. Cell Res., Suppl.* **2,** 145–160 (1952).

207. G. Semenza, *Biochim. Biophys. Acta* **24,** 401–413 (1957).

208. C. S. Hanes, F. J. R. Hird, and F. A. Isherwood, *Nature (London)* **166,** 288–292 (1950).

209. C. S. Hanes, F. J. R. Hird, and F. A. Isherwood, *Biochem. J.* **51,** 25–35 (1952).

210. P. J. Fodor, A. Miller, and H. Waelsch, *J. Biol. Chem.* **202,** 551–565 (1953).

211. P. J. Fodor, A. Miller, A. Neidle, and H. Waelsch, *J. Biol. Chem.* **203,** 991–1002 (1953).

212. F. Binkley and C. K. Olson, *J. Biol. Chem.* **188,** 451–457 (1951).

213. F. J. R. Hird and P. H. Springell, *Biochem. J.* **56,** 417–425 (1954).

214. F. J. R. Hird and P. H. Springell, *Biochim. Biophys. Acta* **15,** 31–37 (1954).

215. J. H. Kinoshita and E. G. Ball, *J. Biol. Chem.* **200,** 609–617 (1953).

216. E. G. Ball, J. P. Revel, and O. Cooper, *J. Biol. Chem.* **221,** 895–908 (1956).

217. J. P. Revel and E. G. Ball, *J. Biol. Chem.* **234,** 577–582 (1959).

218. F. Binkley, *J. Biol. Chem.* **236,** 1075–1082 (1961).

219. A. Szewczuk and T. Baranowski, *Biochem. Z.* **338,** 317–329 (1963).

220. M. Orlowski and A. Meister, *J. Biol. Chem.* **240,** 338–347 (1965).

221. R. Richter, *Arch. Immunol. Ther. Exp.* **17,** 476–495 (1969).

222. F. H. Leibach and F. Binkley, *Arch. Biochem. Biophys.* **127,** 292–301 (1968).

222a. S. S. Tate and A. Meister, *J. Biol. Chem.* **249,** 7593–7602 (1974).

223. A. Szewczuk and G. E. Connell, *Biochim. Biophys. Acta* **83,** 218–223 (1964).

224. A. Szewczuk and G. E. Connell, *Biochim. Biophys. Acta* **105**, 352–367 (1965).

224a. A. J. L. Cooper and A. Meister, *J. Biol. Chem.* **248**, 8489–8498 (1973).

224b. M. Orlowski, P. O. Okonkwo, and J. P. Green, *FEBS (Fed. Eur. Biochem. Soc.) Lett.* **31**, 237–240 (1973).

224c. J. S. Elce, J. Bryson, and L. G. McGirr, *Can. J. Biochem.* **52**, 33–41 (1974).

224d. S. S. Tate and A. Meister, *Proc. Nat. Acad. Sci. U. S.* **71**, 3329–3333 (1974).

224e. N. P. Curthoys, *Int. Conf. Isozymes, 3rd, 1974* (1975) (in press).

224f. N. Katunuma, T. Katsunuma, T. Towatari, and I. Tomino, in "The Enzymes of Glutamine Metabolism" (E. R. Stadtman and S. Prusiner, eds.), pp. 227–258. Academic Press, New York, 1973.

224g. N. P. Curthoys, R. W. Sindel and O. H. Lowry, in "The Enzymes of Glutamine Metabolism" (E. R. Stadtman and S. Prusiner, eds.), pp. 259–275. Academic Press, New York, 1973.

224h. N. Katunuma, I. Tomino, and H. Nishino, *Biochem. Biophys. Res. Commun.* **22**, 321–327 (1966).

224i. N. Katunuma, A. Huzino, and I. Tomino, *Advan. Enzyme Regul.* **5**, 55–69 (1967).

224j. S. S. Tate and A. Meister, *J. Biol. Chem.* **250** (1975) (in press).

225. E. Greenberg, E. E. Wollaeger, G. A. Fleisher, and G. W. Engstrom, *Clin. Chim. Acta* **16**, 79–89 (1967).

226. L. W. DeLap, unpublished studies in this laboratory (1973).

227. R. P. Bodnaryk, *Can. J. Biochem.* **50**, 524–528 (1972).

227a. R. P. Bodnaryk and J. R. Skillings, *Insect Biochem.* **1**, 467–479 (1971).

227b. R. P. Bodnaryk, *J. Insect Physiol.* **16**, 919–929 (1970).

228. M. Y. Goore and J. F. Thompson, *Biochim. Biophys. Acta* **132**, 15–26 (1967).

229. P. S. Talalay, Doctoral dissertation, Cambridge University (1953).

230. P. S. Talalay, *Nature (London)* **174**, 516–517 (1954).

231. R. Milbauer and N. Grossowicz, *J. Gen. Microbiol.* **41**, 185–194 (1965).

231a. A. Szewczuk and M. Mulczyk, *Arch. Immunol. Ther. Exp.* **18**, 515–526 (1970).

231b. R. V. Krishna, unpublished studies in author's laboratory (1973).

232. F. Binkley, J. Davenport, and F. Eastall, *Biochem. Biophys. Res. Commun.* **1**, 206–208 (1959).

233. H. A. Ravin, P. Bernstein, and A. M. Seligman, *J. Biol. Chem.* **208**, 1–15 (1954).

234. C. J. Martin, J. Golubow, and A. E. Axelrod, *Biochim. Biophys. Acta* **27**, 430–431 (1958).

235. J. A. Goldbarg, O. M. Friedman, P. Pineda, E. E. Smith, R. Chatterji, E. H. Stein, and A. M. Rutenberg, *Arch. Biochem. Biophys.* **91**, 61–70 (1960).

236. M. Orlowski and A. Szewczuk, *Acta Biochim. Pol.* **8**, 189–200 (1961).

237. M. Orlowski and A. Meister, *Biochim. Biophys. Acta* **73**, 679–681 (1963).

238. G. Gomori, *Proc. Soc. Exp. Biol. Med.* **87**, 559–561 (1954).

239. M. M. Nachlas, B. Monis, D. Rosenblatt, and A. M. Seligman, *J. Biophys. Biochem. Cytol.* **7**, 261–264 (1960).

240. D. H. Rosenblatt, M. M. Nachlas, and A. M. Seligman, *J. Amer. Chem. Soc.* **80**, 2463–2465 (1958).

241. Z. Albert, M. Orlowski, and A. Szewczuk, *Nature (London)* **191**, 767–768 (1961).

242. G. G. Glenner and J. E. Folk, *Nature (London)* **192**, 338–339 (1961).

243. A. M. Rutenburg, H. Kim, J. W. Fischbein, J. S. Hanker, H. L. Wasserkrug, and A. M. Seligman, *J. Histochem. Cytochem.* **17**, 517–526 (1969).

244. J. S. Hanker, C. Deb, H. L. Wasserkrug, and A. M. Seligman, *Science* **152**, 1631–1634 (1966).

245. A. M. Seligman, H. L. Wasserkrug, R. E. Plapinger, T. Seito, and J. S. Hanker, *J. Histochem. Cytochem.* **18**, 542–551 (1970).

246. G. G. Glenner, J. E. Folk, and P. J. McMillan, *J. Histochem. Cytochem.* **10**, , 481–489 (1962).
247. H. Glossman and D. M. Neville, Jr., *FEBS (Fed. Eur. Biochem. Soc.) Lett.* **19**, 340–344 (1972).
247a. S. G. George and A. J. Kenny, *Biochem. J.* **134**, 43–57 (1973).
248. Z. Albert, Z. Rzucidlo, and H. Starzyk, *Acta Histochem.* **37**, 34–39 and 74–79 (1970).
249. Z. Albert, J. Orlowska, M. Orlowski, and A. Szewczuk, *Acta Histochem.* **18**, 78–89 (1964).
250. Z. Albert, M. Orlowski, Z. Rzucidlo, and J. Orlowska, *Acta Histochem.* **25**, 312–320 (1966).
251. L. L. Ross, unpublished.
252. L. L. Ross, L. Barber, S. S. Tate, and A. Meister, *Proc. Nat. Acad. Sci. U.S.* **70**, 2211–2214 (1973).
252a. R. P. Bodnaryk, J. F. Bronskill, and J. R. Fetterly, *J. Insect. Physiol.* **20**, 167–181 (1974).
253. A. M. Rutenberg, J. A. Goldbarg, and E. P. Pineda, *Gastroenterology* **45**, 43–48 (1963).
254. M. Orlowski, *Arch. Immunol. Ther. Exp.* **11**, 1–61 (1963).
255. F. Kokot, J. Kuska, and H. Grzybek, *Arch. Immunol. Ther. Exp.* **13**, 549–556 (1965).
256. A. Agostoni, G. Ideo, and R. Stabilini, *Brit. Heart J.* **27**, 688–690 (1965).
257. L. Naftalin, V. J. Child, D. A. Morley, and D. A. Smith, *Clin. Chim. Acta* **26**, 297–300 (1969).
258. G. Szasz, *Clin. Chem.* **15**, 124–136 (1969).
259. E. L. Coodley, *J. Amer. Med. Ass.* **220**, 217–219 (1972).
260. W. L. W. Jacobs, *Clin. Chim. Acta* **38**, 419–434 (1972).
261. V. P. Cook and N. K. Carter, *Clin. Chem.* **19**, 774–776 (1973).
262. A. Szewczuk, *Clin. Chim. Acta* **14**, 608–614 (1966).
263. S. I. Goodman, J. W. Mace, and S. Pollak, *Lancet* **1**, 234–235 (1971).
264. G. E. Connell and C. S. Hanes, *Nature (London)* **177**, 377–378 (1956).
265. E. E. Cliffe and S. G. Waley, *Biochem. J.* **79**, 118–128 (1961).
266. G. E. Connell and A. Szewczuk, *Clin. Chim. Acta* **17**, 423–430 (1967).
267. Y. Kakimoto, A. Kanazawa, and I. Sano, *Biochim. Biophys. Acta* **132**, 472–480 (1967).
268. M. Orlowski, P. G. Richman, and A. Meister, *Biochemistry* **8**, 1048–1055 (1969).
269. E. D. Adamson, A. Szewczuk, and G. E. Connell, *Can. J. Biochem.* **49**, 218–226 (1971).
270. M. Orlowski and A. Meister, *J. Biol. Chem.* **248**, 2836–2844 (1973).
271. R. P. Bodnaryk and L. McGirr, *Biochim. Biophys. Acta* **315**, 352–362 (1973).
272. M. Orlowski and A. Meister, in "The Enzymes" (P. D. Boyer, ed.), 3rd ed., Vol. 4, pp. 123–151. Academic Press, New York, 1971.
273. M. Messer, *Nature (London)* **197**, 1299 (1963).
274. M. Messer and M. Ottesen, *Biochim. Biophys. Acta* **92**, 409–411 (1964).
275. M. Messer and M. Ottesen, *C. R. Trav. Lab. Carlsberg* **35**, 1–24 (1965).
276. M. R. Bernfield and L. Nestor, *Biochem. Biophys. Res. Commun.* **33**, 843–849 (1968).
277. S. Akita, K. Tanaka, and S. Kinoshita, *Biochem. Biophys. Res. Commun.* **1**, 179–181 (1959).
278. T. Niwaguchi, N. Motohashi, and H. J. Strecker, *Biochem. Z.* **342**, 469–484 (1965).

279. T. Niwaguchi, N. Motohashi, and H. J. Strecker, *Biochim. Biophys. Acta* **82**, 635–636 (1964).
280. T. Niwaguchi and H. J. Strecker, *Biochem. Biophys. Res. Commun.* **16**, 535–540 (1964).
281. S. Ratner, *J. Biol. Chem.* **152**, 559–564 (1944).
282. A. Meister and M. W. Bukenberger, *Nature (London)* **194**, 557–559 (1962).
283. A. Meister, M. W. Bukenberger, and M. Strassburger, *Biochem. Z.* **338**, 217–229 (1963).
283a. P. Van Der Werf and A. Meister, unpublished observations.
283b. W. L. Nyhan and H. Busch, *Cancer Res.* **18**, 385–393 (1958).
284. P. R. Krishnaswamy, V. Pamiljans, and A. Meister, *J. Biol. Chem.* **235**, PC39 (1960).
285. P. R. Krishnaswamy, V. Paniljans, and A. Meister, *J. Biol. Chem.* **237**, 2932–2940 (1962).
286. A. Meister, P. R. Krishnaswamy, and V. Pamiljans, *Fed. Proc., Fed. Amer. Soc. Exp. Biol.* **21**, 1013–1022 (1962).
287. V. P. Wellner, M. Zoukis, and A. Meister, *Biochemistry* **5**, 3509–3514 (1966).
288. A. Meister, *in* "The Enzymes" (P. D. Boyer, H. Lardy, and K. Myrbäk, eds.), 2nd ed., Vol. 6, pp. 443–468. Academic Press, New York, 1962.
289. R. F. Doolittle and R. W. Armentrout, *Biochemistry* **17**, 516–521 (1968).
290. A. Szewczuk and M. Mulczyk, *Eur. J. Biochem.* **8**, 63–67 (1969).
291. R. W. Armentrout and R. F. Doolittle, *Arch. Biochem. Biophys.* **132**, 80–90 (1969).
292. J. A. Uliana and R. F. Doolittle, *Arch. Biochem. Biophys.* **131**, 561–565 (1969).
293. R. W. Armentrout, *Biochim. Biophys. Acta* **191**, 756–759 (1969).
294. A. Szewczuk and J. Kwaitkowska, *Eur. J. Biochem.* **15**, 92–96 (1970).
295. A. C. Rice and C. S. Pederson, *Food Res.* **19**, 106–114 (1954).
296. P. Van Der Werf, R. A. Stephani, and A. Meister, *Proc. Nat. Acad. Sci. U.S.* **71**, 1026–1029 (1974).
297. M. G. Wolfersberger and J. Tabachnick, *Experientia* **29**, 346–347 (1973).
298. J. Tabachnick and J. H. LaBadie, *J. Invest. Dermatol.* **54**, 24–31 (1970).
299. M. G. Wolfersberger, J. Tabachnick, B. S. Finkelstein, and M. Levin, *J. Invest. Dermatol.* **60**, 278–281 (1973).
300. P. Polgar and A. Meister, *Anal. Biochem.* **12**, 338–343 (1965).
301. R. Tham. L. Nystrom, and B. Holmstedt, *Biochem. Pharmacol.* **17**, 1735–1738 (1968).
302. E. Jellum, T. Kluge, H. C. Borresen, O. Stokke, and L. Eldjarn, *Scand. J. Clin. Lab. Invest.* **26**, 327–335 (1970).
303. L. Hagenfeldt, A. Larsson, and R. Zetterstrom, *Acta Paediat. Scand.* **62**, 1–8 (1973).
304. S. Wilk and M. Orlowski, *FEBS (Fed. Eur. Biochem. Soc.) Lett.* **33**, 157–160 (1973).
305. G. Pascher, *Arch. Klin. Exp. Dermatol.* **203**, 234–238 (1956).
306. K. Laden and R. Spitzer, *J. Soc. Cosmet. Chem.* **18**, 351–360 (1967).
307. S. Marstein, E. Jellum, and L. Eldjarn, *Arch. Dermatol.* **108**, 579 (1973).
308. L. Haitinger, *Monatsh. Chem.* **3**, 228–229 (1882).
309. M. Orlowski and A. Meister, *Proc. Nat. Acad. Sci. U.S.* **67**, 1248–1255 (1970).
310. P. Van Der Werf, M. Orlowski, and A. Meister, *Proc. Nat. Acad. Sci. U.S.* **68**, 2982–2985 (1971).
311. P. Van Der Werf, R. A. Stephani, M. Orlowski, and A. Meister, *Proc. Nat. Acad. Sci. U.S.* **70**, 759–761 (1973).
311a. P. Van Der Werf and A. Meister, *Advan. Enzymol.* **43**, (1975) (in press).
312. E. Abderhalden and R. Hanslian, *Hoppe-Seyler's Z. Physiol. Chem.* **81**, 228–232 (1912).

313. R. Murachi, *Acta Schl. Med. Univ. Kioto* **7**, 445 (1925).
314. L. D. Greenberg and C. L. A. Schmidt, *Univ. Calif., Berkeley, Publ. Physiol.* **8**, 129–143 (1936).
315. R. M. Bethke and H. Steenbock, *J. Biol. Chem.* **58**, 105–115 (1923).
316. J. S. Butts, H. Blunden, and M. S. Dunn, *J. Biol. Chem.* **119**, 247–255 (1937).
317. S. Pedersen and H. B. Lewis, *J. Biol. Chem.* **154**, 705–712 (1944).
318. G. O. Schültz, *Biochem. Z.* **324**, 295–300 (1953).
319. I. Chmielewska, B. Bulhak, and K. Toczko, *Bull. Acad. Pol. Sci.* **15**, 719–721 (1967).
320. W. E. Lange and E. F. Carey, *J. Pharm. Sci.* **55**, 1147–1149 (1966).
321. P. A. Kitos and C. Waymouth, *J. Cell. Physiol.* **67**, 383–398 (1967).
322. M. Forbes and M. G. Sevag, *Arch. Biochem. Biophys.* **31**, 406–415 (1951).
323. P. Simonart and K. Y. Chow, *Antonie van Leeuwenhoek; J. Microbiol. Serol.* **19**, 121–134 (1953).
324. Y. Maruyama and M. Nomura, *J. Biochem. (Tokyo)* **43**, 327–335 (1956).
325. T. Tosa and I. Chibata, *J. Bacteriol.* **89**, 919–920 (1965).
326. Y. Kawai, Y. Kawai, and T. Uemura, *Agr. Biol. Chem.* **29**, 395–402 (1965).
327. Y. Kawai and T. Uemura, *Agr. Biol. Chem.* **30**, 438–446 (1966).
328. Y. Kawai, K. Aida, and T. Uemura, *Agr. Biol. Chem.* **33**, 212–219 (1969).
329. P. Richman, unpublished data (1969), cited in Orlowski and Meister [309].
330. P. Van Der Werf, data cited in Orlowski and Meister [309]; see also P. Van Der Werf, M. Orlowski, and A. Meister, *Fed. Proc., Fed. Amer. Soc. Exp. Biol. Chem.* **30**, 933 (1971).
331. R. J. Roon and B. Levenberg, *J. Biol. Chem.* **247**, 4107–4113 (1972).
331a. R. J. Roon, J. Hampshire, and B. Levenberg, *J. Biol. Chem.* **247**, 7539–7545 (1972).
332. M. Ramakrishna, P. R.Krishnaswamy, and D. R. Rao, *Biochem. J.* **118**, 895–897 (1970).
333. E. A. Rush and J. L. Starr, *Biochim. Biophys. Acta* **199**, 41–55 (1970).
334. E. A. Rush, C. L. McLaughlin, and A. Soloman, *Cancer Res.* **31**, 1134–1139 (1971).
335. A. E. Braunstein, G. A. Shamshikova, and A. L. Ioffe, *Biokhimiya* **13**, 95–100 (1948).
336. G. A. Shamshikova and A. L. Ioffe, *Biokhimiya* **14**, 74–78 (1949).
337. A. Meister, *J. Biol. Chem.* **200**, 571–589 (1953).
338. A. Meister, *J. Biol. Chem.* **210**, 17–35 (1954).
339. T. T. Otani and A. Meister, *J. Biol. Chem.* **224**, 137–148 (1957).
340. C. Monder and A. Meister, *Biochim. Biophys. Acta* **28**, 202–203 (1958).
341. A. J. L. Cooper and A. Meister, *J. Biol. Chem.* **248**, 8499–8505 (1973).
342. P. Van Der Werf and A. Meister, *Biochem. Biophys. Res. Commun.* **56**, 90–95 (1974).
343. P. Van Der Werf, R. A. Stephani, and A. Meister, *Fed. Proc., Fed. Amer. Soc. Exp. Biol.* **32**, 1969 (1973).
344. L. Eldjarn, E. Jellum, and O. Stokke, *Clin. Chim. Acta* **40**, 461–476 (1972).
344a. L. Eldjarn, E. Jellum, and O. Stokke, *in* "Inborn Errors of Metabolism" (F. A. Hommes and C. J. Van Den Berg, eds.), pp. 255–268. Academic Press, New York, 1973.
344b. L. Eldjarn, O. Stokke, and E. Jellum, *in* "Organic Acidurias" (J. Stern and C. Toothill, eds.), pp. 113–120. Williams & Wilkins, Baltimore, Maryland, 1972.
345. J. H. Strömme and L. Eldjarn, *Scand. J. Clin. Lab. Invest.* **29**, 335–342 (1972).
345a. V. P. Wellner, R. Sekura, A. Meister, and A. Larsson, *Proc. Nat. Acad. Sci. U.S.* **71**, 2505–2509 (1974).
345b. T. Kluge, H. C. Borresen, E. Jellum, O. Stokke, L. Eldjarn, and B. Fretheim, *Surgery* **71**, 104–109 (1972).

346. H. Waelsch, *Phosphorus Metab., Symp. 2nd, 1952* Vol. 2, pp. 109–128 (1952).
347. F. J. R. Hird, Doctoral dissertation, Cambridge University, England (1950).
348. F. Binkley, *in* "Glutathione" (S. Colowick *et al.*, eds.), p. 160. Academic Press, New York, 1954.
349. P. H. Springell, Doctoral thesis, Melbourne University, Australia (1953).
350. E. G. Ball, O. Cooper, and E. C. Clarke, *Biol. Bull.* **105**, 369–370 (1953).
351. A. B. Pardee, *Science* **162**, 632–637 (1968).
352. E. Heinz, *in* "Metabolic Pathways" (L. E. Hokin, ed.), 3rd ed., Vol. 6, pp. 455–501. Academic Press, New York, 1972.
353. F. Richards, unpublished data; personal communication (1972).
353a. H. E. Harrison and H. C. Harrison, *Science* **120**, 606–608 (1954).
353b. S. Angielski, R. Niemiro, W. Makarewicz, and J. Rogulski, *Acta Biochim. Pol.* **5**, 431–436 (1958).
353c. L. E. Rosenberg and S. Segal, *Biochem. J.* **92**, 345–352 (1964).
353d. M. Bergeron and M. Vadeboncoeur, *Nephron* **8**, 367–374 (1971).
353e. R. W. Berliner, T. J. Kennedy, and J. G. Hilton, *Proc. Soc. Exp. Biol. Med.* **75**, 791–794 (1950).
353f. J. L. Webb, "Enzyme and Metabolic Inhibitors" Vol. 3, pp. 285–335. Academic Press, New York, 1966.
353g. E. J. Morgan and E. Friedmann, *Biochem. J.* **32**, 733–742 (1938).
353h. S. S. Tate, unpublished work in this laboratory.
353i. G. A. Thompson and A. M. Meister, *Proc. Nat. Acad. Sci. U.S.* (1975) (in press).
353j. J. A. Schneider and J. E. Seegmiller, *in* "The Metabolic Basis of Inherited Disease" (J. B. Stanbury, J. B. Wyngaarden, and D. S. Fredrickson, eds.), 3rd ed., pp. 1581–1604. McGraw-Hill, New York, 1972.
354. S. S. Tate, L. L. Ross, and A. Meister, *Proc. Nat. Acad. Sci. U.S.* **70**, 1447–1449 (1973).
354a. P. O. Okonkwo, M. Orlowski, and J. P. Green, *J. Neurochem.* **22**, 1053–1058 (1974).
355. T. L. Perry and R. T. Jones, *J. Clin. Invest.* **40**, 1363–1373 (1961).
356. J. C. Dickenson and P. B. Hamilton, *J. Neurochem.* **13**, 1179–1187 (1966).
357. M. Van Sande, Y. Mardens, K. Andriaenssens, and A. Lowenthal, *J. Neurochem.* **17**, 125–135 (1970).
358. H. Davson, "Physiology of the Cerebrospinal Fluid." Churchill, London, 1967.
359. T. Z. Csaky, *in* "Handbook of Neurochemistry" (A. Lajtha, ed.), Vol. 2, Chapter 4, pp. 49–69. Plenum, New York, 1969.
360. L. L. Ross, L. Barber, S. S. Tate, and A. Meister, *Proc. Nat. Acad. Sci. U.S.* **70**, 1447–1449 (1973).
361. C. Hebb, *Annu. Rev. Physiol.* **32**, 165–192 (1970).
362. E. De Robertis, *Science* **171**, 963–971 (1971).
363. S. H. Snyder, A. B. Young, J. P. Bennett, and A. H. Mulder, *Fed. Proc., Fed. Amer. Soc. Exp. Biol.* **32**, 2039–2047 (1973).
364. R. J. Baldessarini and M. Karobath, *Annu. Rev. Physiol.* **35**, 273–304 (1973).
365. A. Meister, *Res. Publ., Ass. Res. Nerv. Ment. Dis.* **53**, 273–291 (1974).
366. J. J. Hajjar and P. F. Curran, *J. Gen. Physiol.* **56**, 673–691 (1970).
367. H. N. Christensen, J. A. Streicher, and R. L. Elbinger, *J. Biol. Chem.* **172**, 515–524 (1948).
368. F. G. Hopkins and K. A. C. Elliott, *Proc. Roy. Soc., Ser. B* **109**, 58–90 (1931).
369. P. J. G. Mann, *Biochem. J.* **26**, 785–790 (1932).
370. B. Vennesland and E. E. Conn, *in* "Glutathione" (S. Colowick *et al.*, eds.), pp. 105–127. Academic Press, New York, 1954.

371. N. U. Meldrum and H. L. A. Tarr, *Biochem. J.* **29**, 108–115 (1935).
372. L. W. Mapson and D. R. Goddard, *Biochem. J.* **49**, 592–601 (1951).
373. E. E. Conn and B. Vennesland, *J. Biol. Chem.* **192**, 17–28 (1951).
374. T. W. Rall and A. L. Lehninger, *J. Biol. Chem.* **194**, 119–130 (1952).
375. E. Racker, *J. Biol. Chem.* **217**, 855–865 (1955).
376. R. F. Colman and S. Black, *J. Biol. Chem.* **240**, 1796–1803 (1965).
377. V. Massey and C. H. Williams, Jr., *J. Biol. Chem.* **240**, 4470–4480 (1965).
378. R. D. Mavis and E. Stellwagen, *J. Biol. Chem.* **243**, 809–814 (1968).
379. L. W. Mapson and F. A. Isherwood, *Biochem. J.* **86**, 173–191 (1963).
380. J. E. Bulger and K. G. Brandt, *J. Biol. Chem.* **246**, 5570–5577 (1971).
381. J. E. Bulger and K. G. Brandt, *J. Biol. Chem.* **246**, 5578–5587 (1971).
382. G. Moroff and K. G. Brandt, *Arch. Biochem. Biophys.* **159**, 468–474 (1973).
383. B. Mannervik, *in* "Symposium on Glutathione" (L. Flohé, ed.), pp. 114–120. Thieme, Stuttgart, 1973.
384. A. L. Icen, *FEBS* (*Fed. Eur. Biochem. Soc.*) *Lett.* **16**, 29–32 (1971).
385. B. K. Stern and B. Vennesland, *J. Biol. Chem.* **235**, 209–212 (1960).
386. J. A. Buzard and F. Kopko, *J. Biol. Chem.* **238**, 464–468 (1963).
387. C. E. Mize and R. G. Langdon, *J. Biol. Chem.* **237**, 1589–1595 (1962).
388. C. E. Mize, T. E. Thompson, and R. G. Langdon, *J. Biol. Chem.* **237**, 1596–1600 (1962).
389. C. H. Williams, Jr., G. Zanetti, L. D. Arscott, and J. K. McAllister, *J. Biol. Chem.* **242**, 5226–5231 (1967).
390. C. H. Williams, Jr. and L. D. Arscott, *in* "Methods in Enzymology" (H. Tabor and C. W. Tabor, eds.), Vol. 17B, pp. 503–509. Academic Press, New York, 1971.
391. E. M. Scott, I. W. Duncan, and V. Ekstrand, *J. Biol. Chem.* **238**, 3928–3933 (1963).
392. G. E. J. Stahl, J. Visser, and C. Veeger, *Biochim. Biophys. Acta* **185**, 39–48 (1969).
393. G. E. J. Stahl, and C. Veeger, *Biochim. Biophys. Acta* **185**, 49–62 (1969).
394. H. D. Waller, *City Hope Symp. Ser.* **1**, 185–208 (1968).
395. G. E. J. Stahl, P. W. Hellerman, J. DeWael, and C. Veeger, *Biochim. Biophys. Acta* **185**, 63–69 (1969).
396. E. Beutler, *Science* **165**, 613–615 (1969).
397. E. Beutler, *J. Clin. Invest.* **48**, 1957–1966 (1969).
398. E. Beutler, *in* "Symposium on Glutathione" (L. Flohé, ed.), pp. 109–112. Thieme, Stuttgart, 1973.
399. N. V. Paniker, S. K. Srivastava, and E. Beutler, *Biochim. Biophys. Acta* **215**, 456–460 (1970).
400. G. W. Lohr, K. G. Blume, H. W. Rudiger, and H. Arnold, *in* "Symposium on Glutathione" (L. Flohé, ed.), pp. 165–172. Thieme, Stuttgart, 1973.
401. E. Racker, *J. Biol. Chem.* **217**, 867–874 (1955).
402. S. Nagai and S. Black, *J. Biol. Chem.* **243**, 1942–1947 (1968).
403. L. Eldjarn, J. Bremer, and H. C. Borresen, *Biochem. J.* **82**, 192–197 (1962).
404. S. H. Chang and D. R. Wilken, *J. Biol. Chem.* **241**, 4251–4260 (1966).
405. B. Eriksson, *Acta Chem. Scand.* **20**, 1178–1179 (1966).
406. R. E. Dyar and D. R. Wilken, *Arch. Biochem. Biophys.* **153**, 619–626 (1972).
407. B. Mannervik, *in* "Symposium on Glutathione" (L. Flohé, ed.), pp. 120–131. Thieme, Stuttgart, 1973.
408. H. M. Katzen, F. Tietze, and S. DeWitt, Jr., *J. Biol. Chem.* **238**, 1006–1011 (1963).
409. H. M. Katzen and F. Tietze, *J. Biol. Chem.* **241**, 3561–3570 (1966).
410. H. H. Tomizawa, *J. Biol. Chem.* **237**, 3393–3396 (1962).
411. H. H. Tomizawa and P. T. Varandani, *J. Biol. Chem.* **240**, 3191–3194 (1965).

412. H. T. Narahara and R. H. Williams, *J. Biol. Chem.* **234,** 71–77 (1959).
413. P. T. Varandani and H. Plumley, *Biochim. Biophys. Acta* **151,** 273–275 (1968).
414. P. Varandani, *Biochim. Biophys. Acta* **320,** 249–257 (1973).
415. O. B. Kotoulas, G. R. Morrison, and L. Recant, *Biochim. Biophys. Acta* **97,** 350–352 (1965).
416. M. L. Chandler and P. Varandani, *Biochim. Biophys. Acta* **320,** 258–266 (1973).
417. P. T. Varandani, *Biochem. Biophys. Res. Commun.* **55,** 689–696 (1973).
417a. S. Ansorge, P. Bohley, H. Kirschke, J. Langner, I. Marquardt, B. Wiederanders, and H. Hanson, *FEBS (Fed. Eur. Biochem. Soc.) Lett.* **37,** 238–240 (1973).
417b. R. F. Goldberger, C. J. Epstein, and C. B. Anfinsen, *J. Biol. Chem.* **238,** 628–635 (1963).
418. G. C. Mills, *J. Biol. Chem.* **229,** 189–197 (1957).
419. G. C. Mills, *Arch. Biochem. Biophys.* **86,** 1–5 (1960).
420. G. C. Mills, *J. Biol. Chem.* **234,** 502–506 (1959).
421. G. Cohen and P. Hochstein, *Biochemistry* **2,** 1420–1428 (1963).
422. H. S. Jacob, S. H. Ingbar, and J. H. Jandl, *J. Clin. Invest.* **44,** 1187–1199 (1965).
423. H. Aebi and H. Suter, *in* "Symposium on Glutathione" (L. Flohé, ed.), pp. 192–201. Thieme, Stuttgart, 1973.
424. H. S. Jacob and J. H. Jandl, *J. Biol. Chem.* **241,** 4243–4250 (1966).
425. N. S. Kosower, K-R. Song, and E. M. Kosower, *Biochim. Biophys. Acta* **192,** 23–28 (1969).
426. C. Little and P. J. O'Brien, *Biochem. Biophys. Res. Commun.* **31,** 145–150 (1968).
427. B. O. Christophersen, *Biochim. Biophys. Acta* **164,** 35–46 (1968).
428. B. O. Christophersen, *Biochim. Biophys. Acta* **176,** 463–470 (1969).
429. L. Flohé, *Klin. Wochenschr.* **49,** 669–683 (1971).
430. L. Flohé and R. Zimmerman, *in* "Symposium on Glutathione" (L. Flohé, ed.), pp. 245–259. Thieme, Stuttgart, 1973.
431. L. Flohé and W. A. Gunzler, *in* "Symposium on Glutathione" (L. Flohé, ed.), pp. 132–142. Thieme, Stuttgart, 1973.
432. A. L. Lehninger and G. S. Gotterer, *J. Biol. Chem.* **235,** PC8–PC9 (1960).
433. V. L. Flohé, W. Schlegel, and E. Schlegel, and E. Schach, *Z. Klin. Chem. Klin. Biochem.* **8,** 149–155 (1970).
434. P. W. Reed, *J. Biol. Chem.* **244,** 2459–2464 (1969).
435. S. Hosoda and W. Nakamura, *Biochim. Biophys. Acta* **222,** 53–64 (1970).
436. S. S. Hotta, *Arch. Biochem. Biophys.* **113,** 395–398 (1966).
437. S. S. Hotta, *J. Neurochem.* **9,** 43–51 (1962).
438. S. S. Hotta, *Arch. Biochem. Biophys.* **122,** 524–525 (1967).
439. S. S. Hotta and J. M. Seventko, *Arch. Biochem. Biophys.* **123,** 104–108 (1968).
440. A. Pirie, *Biochem. J.* **96,** 244–253 (1965).
441. V. F. Schneider and L. Flohé, *Hoppe-Seyler's Z. Physiol. Chem.* **348,** 540–552 (1967).
442. C. Little, R. Olinescu, K. G. Reid, and P. J. O'Brien, *J. Biol. Chem.* **245,** 3632–3636 (1970).
443. N. J. Holmberg, *Exp. Eye Res.* **7,** 570–580 (1968).
444. L. Flohé and I. Brand, *Z. Klin. Chem. Klin. Biochem.* **8,** 156–161 (1970).
445. W. A. Gunzler, *in* "Symposium on Glutathione" (L. Flohé, ed.), pp. 180–183. Thieme, Stuttgart, 1973.
446. L. Flohé and I. Brand, *Biochim. Biophys. Acta* **191,** 541–549 (1969).
447. L. Flohé, B. Eisele, and A. Wendel, *Hoppe-Seyler's Z. Physiol. Chem.* **352,** 151–158 (1971).

448. L. Flohé, W. Gunzler, G. Jung, E. Schaich, and F. Schneider, *Hoppe-Seyler's Z. Physiol. Chem.* **352**, 159–169 (1971).
449. L. Flohé, E. Schaich, W. Voelter, and A. Wendel, *Hoppe-Seyler's Z. Physiol. Chem.* **352**, 170–180 (1971).
450. L. Flohé, G. Loshen, W. A. Günzler, and E. Eichele, *Hoppe-Seyler's Z. Physiol. Chem.* **353**, 987–999 (1972).
451. W. A. Günzler, H. Vergin, I. Müller, and L. Flohé, *Hoppe-Seyler's Z. Physiol. Chem.* **353**, 1001–1004 (1972).
452. D. E. Paglia and W. N. Valentine, *J. Lab. Clin. Med.* **70**, 158–169 (1967).
453. L. Flohé, W. A. Günzler, and H. H. Shock, *FEBS (Fed. Eur. Biochem. Soc.) Lett.* **32**, 132–134 (1973).
454. J. T. Rotruck, A. L. Pope, H. E. Ganther, A. B. Swanson, D. G. Hafeman, and W. G. Hoekstra, *Science* **179**, 588–590 (1973).
454a. P. J. Smith, A. L. Tappel, and C. K. Chow, *Nature (London)* **247**, 392–393 (1974).
455. K. A. Caldwell and A. L. Tappel, *Arch. Biochem. Biophys.* **112**, 196–200 (1965).
456. H. E. Ganther, *Biochemistry* **7**, 2898–2905 (1968).
456a. L. N. Vernie, W. S. Bont, and P. Emmelot, *Biochemistry* **13**, 337–341 (1974).
456b. O. A. Levander, V. C. Morris, and D. J. Higgs, *Biochemistry* **12**, 4591–4595 (1973).
457. T. F. Necheles, T. A. Boles, and D. M. Allen, *J. Pediat.* **72**, 319–324 (1968).
458. T. F. Necheles, N. Maldonado, A. Barquet-Chediak, and D. M. Allen, *Blood* **33**, 164–169 (1969).
459. T. F. Necheles, in "Symposium on Glutathione" (L. Flohé, ed.), pp. 173–179. Thieme, Stuttgart, 1973.
460. I. G. Fels, *Exp. Eye Res.* **12**, 227–229 (1971).
461. P. C. Jocelyn, *Biochem. J.* **117**, 947–949 (1970).
462. P. C. Jocelyn, *Biochem. J.* **117**, 951–956 (1970).
463. E. M. Crook and E. J. Morgan, *Biochem. J.* **38**, 10–15 (1944).
464. W. B. Esselen, Jr. and J. E. Fuller, *J. Bacteriol.* **37**, 501–521 (1939).
465. E. Gero, *Biochim. Biophys. Acta* **92**, 160–163 (1964).
466. L. W. Mapson, in "Glutathione" (E. M. Crook, ed.), Biochem. Soc. Symp. No. 17 pp. 28–42. Cambridge Univ. Press, London and New York, 1959.
467. R. E. Hughes, *Nature (London)* **203**, 1068–1069 (1964).
468. R. F. Grimble and R. E. Hughes, *Experientia* **23**, 362 (1967).
469. A. A. Painter and F. E. Hunter, Jr., *Science* **170**, 552–553 (1970).
470. A. A. Painter and F. E. Hunter, Jr., *Biochem. Biophys. Res. Commun.* **40**, 360–368 (1970).
471. A. A. Painter and F. E. Hunter, Jr., *Biochem. Biophys. Res. Commun.* **40**, 369–377 (1970).
472. A. A. Painter and F. E. Hunter, Jr., *Biochem. Biophys. Res. Commun.* **40**, 378–386 (1970).
473. A. A. Painter and F. E. Hunter, Jr., *Biochem. Biophys. Res. Commun.* **40**, 387–395 (1970).
474. S. K. Srivastava and E. Beutler, *J. Biol. Chem.* **244**, 9–16 (1969).
475. S. K. Srivastava and E. Beutler, *Biochem. J.* **114**, 833–837 (1969).
476. E. Beutler and S. K. Srivastava, in "Symposium on Glutathione" (L. Flohé, ed.), pp. 201–205. Thieme, Stuttgart, 1973.
477. H. Sies, C. Gerstenecker, H. Menzel, and L. Flohé, *FEBS (Fed. Eur. Biochem. Soc.) Lett.* **27**, 171–175 (1972).
478. H. Sies, in "Symposium on Glutathione" (L. Flohé, ed.), pp. 261–275. Thieme, Stuttgart, 1973.

479. D. H. Elwyn, H. C. Parikh, and W. C. Shoemaker, *Amer. J. Physiol.* **215**, 1260–1275 (1968).
480. E. M. Kosower and N. S. Kosower, *Nature (London)* **224**, 117–120 (1969).
481. N. S. Kosower, E. M. Kosower, and B. Wertheim, *Biochem. Biophys. Res. Commun.* **37**, 593–596 (1969).
482. N. S. Kosower, K.-R. Song, and E. M. Kosower, *Biochim. Biophys. Acta* **192**, 1–7 (1969).
483. N. S. Kosower, K.-R. Song, E. M. Kosower, and W. Correa, *Biochim. Biophys. Acta* **192**, 8–14 (1969).
484. N. S. Kosower, K.-R. Song, and E. M. Kosower, *Biochim. Biophys. Acta* **192**, 15–22 (1969).
485. T. Zehavi-Willner, N. S. Kosower, T. Hunt, and E. M. Kosower, *Biochem. Biophys. Res. Commun.* **40**, 37–42 (1970).
486. T. Zehavi-Willner, E. M. Kosower, T. Hunt, and N. S. Kosower, *Biochim. Biophys. Acta* **228**, 245–251 (1971).
487. N. S. Kosower, G. A. Vanderhoff, B. Benerofe, and T. Hunt, *Biochem. Biophys. Res. Commun.* **45**, 816–821 (1971).
488. N. S. Kosower, G. A. Vanderhoff, and E. M. Kosower, *Biochim. Biophys. Acta* **272**, 623–627 (1972).
489. N. S. Kosower, Y. Marikovsky, B. Wertheim, and D. Danon, *J. Lab. Clin. Med.* **78**, 533–545 (1971).
490. E. M. Kosower, W. Correa, B. J. Kinon, and N. S. Kosower, *Biochim. Biophys. Acta* **264**, 39–44 (1972).
491. J. W. Harris, N. P. Allen, and S. S. Teng, *Exp. Cell Res.* **68**, 1–10 (1971).
492. J. W. Harris and J. E. Biaglow, *Biochem. Biophys. Res. Commun.* **46**, 1743–1749 (1972).
493. N. S. Kosower and E. M. Kosower, *in* "Symposium on Glutathione" (L. Flohé, ed.), pp. 276–286. Thieme, Stuttgart, 1973.
493a. N. S. Kosower and E. M. Kosower, *Isr. J. Med. Sci.* **9**, 1475–1483 (1973).
494. E. Racker, *in* "Glutathione" (S. Colowick *et al.*, eds.), pp. 165–183. Academic Press, New York, 1954.
495. D. B. Hope, *in* "Glutathione" (E. M. Crook, ed.), Biochem. Soc. Symp. No. 17, pp. 93–114. Cambridge Univ. Press, London and New York, 1959.
496. H. Rink, *in* "Symposium on Glutathione" (L. Flohé, ed.), pp. 206–215. Thieme, Stuttgart, 1973.
497. E. Baumann and C. Preusse, *Ber. Deut. Chem. Ges.* **12**, 806–810 (1879).
498. M. Jaffe, *Ber. Deut. Chem. Ges.* **12**, 1092–1098 (1879).
499. E. Boyland and L. F. Chasseaud, *Advan. Enzymol.* **32**, 173–219 (1969).
500. J. L. Wood, *in* "Metabolic Conjugation and Metabolic Hydrolysis" (W. H. Fishman, ed.), Vol. 2, pp. 261–299. Academic Press, New York, 1970.
501. T. Nakashima, *J. Biochem. (Tokyo)* **19**, 281–314 (1934).
502. K. Yamamoto, *Mitt. Med. Akad. Keijo* **29**, 431 (1940); *Chem. Abstr.* **35**, 4484 (1941).
503. M. M. Barnes, S. P. James, and P. B. Wood, *Biochem. J.* **71**, 680–690 (1959).
504. E. Boyland, G. S. Ramsay, and P. Sims, *Biochem. J.* **78**, 376–384 (1961).
505. J. Booth, E. Boyland, and P. Sims, *Biochem. J.* **79**, 516–524 (1961).
506. P. L. Grover and P. Sims, *Biochem. J.* **90**, 603–606 (1964).
507. M. K. Johnson, *Biochem. Pharmacol.* **14**, 1383–1385 (1965).
508. E. Boyland and L. F. Chasseaud, *Biochem. J.* **104**, 95–102 (1967).
509. E. Boyland and L. F. Chasseaud, *Biochem. J.* **109**, 651–661 (1968).
510. E. Boyland and L. F. Chasseaud, *Biochem. Pharmacol.* **19**, 1526–1528 (1970).
511. D. H. Hutson, B. A. Pickering, and C. Donninger, *Biochem. J.* **127**, 285–293 (1972).

512. L. F. Chasseaud, *Biochem. J.* **131**, 765–769 (1973).
513. A. J. Cohen, J. N. Smith, and H. Turbert, *Biochem. J.* **90**, 457–464 (1960).
514. A. J. Cohen and J. N. Smith, *Biochem. J.* **90**, 449–456 (1964).
515. H. Ohkawa, R. Ohkawa, I. Yamamoto, and J. E. Casida, *Pestic. Biochem. Physiol.* **2**, 95–112 (1972).
516. T. Shishido, K. Usui, M. Sato, and J. Fukami, *Pestic. Biochem. Physiol.* **2**, 51–63 (1972).
517. L. F. Chasseaud, *in* "Symposium on Glutathione" (L. Flohé, ed.), pp. 90–108. Thieme, Stuttgart, 1973.
517a. L. F. Chasseaud, *Drug Metab. Rev.* **2**, 185–219 (1973).
518. T. A. Fjellstedt, R. H. Allen, B. K. Duncan, and W. B. Jakoby, *J. Biol. Chem.* **248**, 3702–3707 (1973).
519. M. J. Pabst, W. H. Habig, and W. B. Jakoby, *Biochem. Biophys. Res. Commun.* **52**, 1123–1128 (1973).
520. H. G. Bray, T. J. Franklin, and S. P. James, *Biochem. J.* **71**, 690–696 (1959).
521. Y. Avi-Dor, *Biochem. J.* **76**, 370–374 (1960).
522. E. Boyland, *Proc. Int. Pharmacol. Meet., 1st, 1961* Vol. 16, pp. 65–76 (1963).
523. S. M. Rosenthal and E. C. White, *J. Pharmacol. Exp. Ther.* **24**, 265–288 (1925).
524. N. B. Javitt, H. O. Wheeler, K. J. Baker, O. L. Ramos, and S. E. Bradley, *J. Clin. Invest.* **39**, 1570–1577 (1960).
525. G. M. Grodsky, J. V. Carbone, and R. Fanska, *J. Clin. Invest.* **38**, 1981–1988 (1959).
526. N. B. Javitt, *Amer. J. Physiol.* **208**, 555–562 (1965).
527. K. D. G. Edwards, N. B. Javitt, H. O. Wheeler, and S. E. Bradley, *Australas. Ann. Med.* **17**, 118–126 (1968).
528. N. B. Javitt, *Proc. Soc. Exp. Biol. Med.* **117**, 254–257 (1964).
529. E. Boyland and P. L. Grover, *Clin. Chim. Acta* **16**, 205–213 (1967).
530. N. B. Javitt, *Progr. Liver Dis.* **3**, 110–117 (1970).
531. V. Franzen, *Chem. Ber.* **89**, 1020–1023 (1956).
532. I. A. Roze, *Biochim. Biophys. Acta* **25**, 214–215 (1957).
533. K. Ekwall and B. Mannervik, *Biochim. Biophys. Acta* **297**, 297–299 (1973).
534. E. Racker, *in* "Glutathione" (S. Colowick *et al.*, eds.), p. 208. Academic Press, New York, 1954.
535. D. L. Vander Jagt, L.-P. B. Han, and C. H. Lehman, *Biochemistry* **11**, 3735–3740 (1972).
536. E. E. Cliffe and S. G. Waley, *Biochem. J.* **79**, 475–482 (1961).
537. K. A. Davis and G. R. Williams, *Can. J. Biochem.* **47**, 553–556 (1969).
538. C. Neuberg, *Biochem. Z.* **49**, 502–506 (1913).
539. H. D. Dakin and H. W. Dudley, *J. Biol. Chem.* **14**, 155–157 (1913).
540. H. D. Dakin and H. W. Dudley, *J. Biol. Chem.* **14**, 423–431 (1913).
541. K. Lohmann, *Biochem. Z.* **254**, 332–354 (1932).
542. O. K. Behrens, *J. Biol. Chem.* **141**, 503–508 (1941).
543. E. Racker, *J. Biol. Chem.* **190**, 685–696 (1951).
544. B. Mannervik, B. Gorna-Hall, and T. Bartfai, *Eur. J. Biochem.* **37**, 270–281 (1973).
545. B. Mannervik, *in* "Symposium on Glutathione" (L. Flohé, ed.), pp. 78–89. Thieme, Stuttgart, 1973.
546. B. Mannervik, L. Lindstrom, and T. Bartfai, *Eur. J. Biochem.* **29**, 276–281 (1972).
547. K. A. Davis and G. R. Williams, *Biochim. Biophys. Acta* **113**, 393–395 (1966).
548. H. C. Reeves and S. J. Ajl, *J. Biol. Chem.* **240**, 569–573 (1965).
549. R. H. Weaver and H. A. Lardy, *J. Biol. Chem.* **236**, 313–317 (1961).
550. T. Jerzykowski, R. Winter, and W. Matuszewski, *Biochem. J.* **135**, 713–719 (1973).

551. H. H. Strain and H. A. Spoehr, *J. Biol. Chem.* **89**, 527–534 (1930).
552. R. A. Cooper and A. Anderson, *FEBS (Fed. Eur. Biochem. Soc.) Lett.* **11**, 273–276 (1970).
553. D. J. Hopper and R. A. Cooper, *FEBS (Fed. Eur. Biochem. Soc.) Lett.* **13**, 213–217 (1971).
554. W. B. Freedberg, W. S. Kistler, and E. C. C. Lin, *J. Bacteriol.* **108**, 137–144 (1971).
555. D. D. Rekarte, N. Zwaig, and T. Isturiz, *J. Bacteriol.* **115**, 727–731 (1973).
556. V. M. Riddle and F. W. Lorenz, *Biochem. Biophys. Res. Commun.* **50**, 27–34 (1973).
557. A. Szent-Györgyi, *Science* **149**, 34–37 (1965).
558. L. G. Egyud, *Proc. Nat. Acad. Sci. U.S.* **54**, 200–202 (1965).
559. A. Szent-Györgyi, L. G. Egyud, and J. A. McLaughlin, *Science* **155**, 539–541 (1967).
560. L. G. Egyud and A. Szent-Györgyi, *Proc. Nat. Acad. Sci. U.S.* **55**, 388–393 (1966).
561. L. G. Egyud and A. Szent-Györgyi, *Proc. Nat. Acad. Sci. U.S.* **56**, 203–207 (1966).
562. G. Fodor, J. P. Sachetto, A. Szent-Györgyi, and L. G. Egyud, *Biochem. J.* **57**, 1644–1650 (1967).
563. A. Szent-Györgyi, *Science* **161**, 988–990 (1968).
564. C. T. Gregg, *Exp. Cell Res.* **50**, 65–72 (1968).
565. M. Staehelin, *Biochim. Biophys. Acta* **31**, 448–454 (1959).
566. M. Litt, *Biochemistry* **8**, 3249–3253 (1969).
567. R. Shapiro and J. Hachmann, *Biochemistry* **5**, 2799–2807 (1966).
568. R. Vince and W. B. Wadd, *Biochem. Biophys. Res. Commun.* **35**, 593–598 (1969).
569. R. Vince, S. Daluge, and W. B. Wadd, *J. Med. Chem.* **14**, 402–404 (1971).
570. R. Vince and S. Daluge, *J. Med. Chem.* **14**, 35–37 (1971).
571. M. A. Apple and D. M. Greenberg, *Cancer Chemother. Rep.* **51**, 455–464 (1967).
572. M. A. Apple and D. M. Greenberg, *Life Sci.* **6**, 2157–2160 (1967).
573. L. G. Egyud and A. Szent-Györgyi, *Science* **160**, 1140 (1968).
574. M. A. Apple and D. M. Greenberg, *Cancer Chemother. Rep.* **52**, 687–696 (1968).
575. M. A. Apple, F. C. Ludwig, and D. M. Greenberg, *Oncology* **24**, 210–222 (1970).
576. F. A. French and B. L. Freedlander, *Cancer Res.* **18**, 172–175 (1958).
577. W. H. Elliott, *Nature (London)* **185**, 467–468 (1960).
578. C. Monder, *J. Biol. Chem.* **242**, 4603–4609 (1967).
579. W. E. Knox and S. W. Edwards, *J. Biol. Chem.* **216**, 489–498 (1955).
580. L. Lack, *J. Biol. Chem.* **236**, 2835–2840 (1961).
581. A. Meister and J. P. Greenstein, *J. Biol. Chem.* **175**, 573–588 (1948).
582. A. Meister, *J. Biol. Chem.* **178**, 577–589 (1949).
583. R. F. Witter and E. Stotz, *J. Biol. Chem.* **176**, 501–510 (1948).
584. P. Strittmatter, and E. G. Ball, *J. Biol. Chem.* **213**, 445–461 (1955).
585. J. I. Goodman and T. R. Tephly, *Biochim. Biophys. Acta* **252**, 489–505 (1971).
586. J. H. Kinoshita and A. B. Masurat, *Amer. J. Ophthalmol.* [3] **46**, 42–46 (1958).
587. H. Lipke and C. W. Kearns, *J. Biol. Chem.* **234**, 2123–2128 (1959).
588. B. Goodchild and J. N. Smith, *Biochem. J.* **117**, 1005–1009 (1970).
589. S. Balabaskaran and J. N. Smith, *Biochem. J.* **117**, 989–996 (1970).
590. M. I. Dinamarca, L. Levenbook, and E. Valdes, *Arch. Biochem. Biophys.* **147**, 374–383 (1971).
591. V. N. Reddy, *Exp. Eye Res.* **11**, 310–328 (1971).
592. J. H. Kinoshita and T. Masurat, *Arch. Ophthalmol.* [N.S.] **57**, 266–274 (1957).
593. J. G. Bellows, *Arch. Ophthalmol.* [N.S.] **16**, 762–769 (1936).
594. J. G. Bellows and L. Rosner, *Amer. J. Ophthalmol.* [3] **20**, 1109–1114 (1937).
595. J. G. Bellows and D. E. Shoch, *Amer. J. Ophthalmol.* [3] **33**, 1555–1564 (1950).
596. V. E. Kinsey and F. C. Merriam, *Arch. Ophthalmol.* [N.S.] **44**, 370–380 (1950).
597. A. Pirie, R. van Heyningen, and J. W. Boag, *Biochem. J.* **54**, 682–688 (1953).

598. J. F. R. Kuck, Jr., in "Biochemistry of the Eye" (C. N. Graymore, ed.), pp. 319–371. Academic Press, New York, 1970.
599. R. van Heyningen and A. Pirie, Biochem. J. 102, 842–852 (1967).
600. J. R. Rees and A. Pirie, Biochem. J. 102, 853–863 (1967).
601. R. van Heyningen, Nature (London) 184, 194–195 (1959).
602. J. H. Kinoshita, L. O. Merola, K. Satoh, and E. Dikmark, Nature (London) 194, 1085–1087 (1962).
603. J. H. Kinoshita, L. O. Merola, and B. Tung, in "Biochemistry of the Eye," pp. 373–382. Karger, Basel, 1968.
604. S. Segal, in "The Metabolic Basis of Inherited Disease" (J. B. Stanbury, J. B. Wyngaarden, and D. S. Frederickson, eds.), 3rd ed., pp. 174–195. McGraw-Hill, New York, 1972.
605. S. Duke-Elder, Syst. Ophthalmol. 11, 192–193 (1969).
606. A. Franceschetti and C. Avanza, Atti Soc. Oftalmol. Ital. 17, 530–536 (1957).
607. V. François, "Les Cataractes Congenitales," pp. 331–344. Paris, 1959.
608. A. Franceschetti, Bull. Schweiz. Akad. Med. Wiss. 17, 414–422 (1961).
609. W. Isola, C. A. Bauza, J. Ferrer, N. Temesio, and M. E. Drets, Arch. Pediat. Uruguay 31, 144–232 (1966).
610. D. W. Allen and J. H. Jandl, J. Clin. Invest. 40, 454–475 (1961).
611. E. R. Jaffe, Exp. Eye Res. 11, 306–309 (1971).
612. G. Cohen and P. Hochstein, Biochemistry 3, 895–900 (1964).
613. E. Margoliash, A. Novogrodsky, and A. Schechter, Biochem. J. 74, 339–350 (1960).
614. E. Margoliash and A. Novogrodsky, Biochem. J. 68, 468–475 (1958).
615. H. S. Jacob and J. H. Jandl, J. Clin. Invest. 41, 1515 (1962).
616. E. Beutler, S. K. Srivastava, and C. West, Biochem. Biophys. Res. Commun. 38, 341–347 (1970).
617. P. C. Jocelyn, Biochem. J. 77, 363–368 (1960).
618. P. C. Jocelyn, Biochem. J. 77, 368–380 (1960).
619. S. K. Srivastava, Exp. Eye Res. 11, 294–305 (1971).
620. A. G. Palekar, S. S. Tate, and A. Meister, Proc. Nat. Acad. Sci. U.S. 71, 293–297 (1974).
621. J. D. Schulman, J. A. Schneider, K. H. Bradley, and J. E. Seegmiller, Clin. Chim. Acta 35, 383–388; 37, 53–58 (1971).
622. R. Sekura and A. Meister, Proc. Nat. Acad. Sci. U.S. 71, 2969–2972 (1974).
623. H. McIlwain, "Biochemistry of the Central Nervous System." Little, Brown, Boston, Massachusetts 1955.
624. H. Martin and H. McIlwain, Biochem. J. 71, 275–280 (1959).
625. W. H. Oldendorf, Amer. J. Physiol. 221, 1628–1639 (1971).
626. S. R. Cohen and A. Lajtha, in "Handbook of Neurochemistry" (A. Lajtha, ed.), Vol. 7, pp. 543–572. Plenum, New York, 1972.
627. P. Richman, unpublished data (1970); see also Doctoral Dissertation, Cornell University Medical College, New York (1974).
628. W. F. Loomis, Ann. N.Y. Acad. Sci. 62, 211–227 (1955).
629. H. M. Lenhoff, Science 161, 434–442 (1968).
630. H. M. Lenhoff, E. M. Kosower, and N. S. Kosower, Nature (London) 224, 717–718 (1969).
631. E. E. Cliffe and S. G. Waley, Nature (London) 182, 804–805 (1958).
632. A. Meister, in "Symposium on Glutathione" (L. Flohé, ed.), p. 298. Thieme, Stuttgart, 1973.
632a. S. S. Tate, unpublished work in this laboratory.

633. E. M. Kosower, *Experientia* **26**, 760 (1970).

634. R. Werman, P. L. Carlen, M. Kushnir, and E. M. Kosower, *Nature (London), New Biol.* **233**, 120–121 (1971).

635. E. M. Kosower and R. Werman, *Nature (London), New Biol.* **233**, 121–122 (1971).

635a. N. S. Kosower and E. M. Kosower, in "Free Radicals in Biology" (W. A. Pryor, ed.). Academic Press, New York, 1976 (in press).

636. G. Jung, E. Breitmaier, W. A. Grinzler, M. Ottnad, W. Voelter, and L. Flohé, in "Symposium on Glutathione" (L. Flohé, ed.), pp. 1–14. Thieme, Stuttgart, 1973.

637. G. Maass, and F. Peters, in "Symposium on Glutathione" (L. Flohé, ed.), pp. 15–19. Thieme, Stuttgart, 1973.

638. D. L. Vander Jagt, L. D. Hansen, E. A. Lewis, and L.-P. B. Han, *Arch. Biochem. Biophys.* **153**, 55–61 (1972).

639. M. Calvin, in "Glutathione" (S. Colowick *et al.*, eds.), pp. 3–30. Academic Press, New York, 1954.

640. D. Garfinkel, *J. Amer. Chem. Soc.* **80**, 4833–4845 (1958).

641. R. B. Martin, S. Lowey, E. L. Elson, and J. T. Edsall, *J. Amer. Chem. Soc.* **81**, 5089–5095 (1959).

642. I. Goodman and L. Salce, *Biochim. Biophys. Acta* **100**, 283–286 (1965).

643. J. S. Reitman, and N. W. Cornell, *Biochim. Biophys. Acta* **208**, 159–162 (1970).

644. P. C. Jocelyn, *Anal. Biochem.* **18**, 493–498 (1967).

645. Y. Hirotsu, T. Shiba, and T. Kaneko, *Biochim. Biophys. Acta* **222**, 540–541 (1970).

646. E. Bayer, H. Giesecke, P. Krauss, and A. Röder, in "Symposium on Glutathione" (L. Flohé, ed.), pp. 34–43. Thieme, Stuttgart, 1973.

647. D. D. Perrin and A. E. Watt, *Biochim. Biophys. Acta* **230**, 96–104 (1971).

648. N. W. Pirie and K. G. Pinkey, *J. Biol. Chem.* **84**, 321–333 (1929).

649. G. W. Jourdian and S. Roseman, *J. Biol. Chem.* **237**, 2442–2446 (1962).

650. J. C. Fletcher and A. Robson, *Biochem. J.* **87**, 553–567 (1963).

651. J. W. Purdie and D. E. Hanafi, *J. Chromatogr.* **59**, 181–184 (1971).

652. D. H. Calam and S. G. Waley, *Biochem. J.* **85**, 417–419 (1962).

653. B. Eriksson and B. Sorbo, *Acta Chem. Scand.* **21**, 958–960 (1967).

654. H. C. Robinson and C. A. Pasternak, *Biochem. J.* **93**, 487–492 (1964).

655. O. Arrigoni and G. Rossi, *Biochim. Biophys. Acta* **46**, 121–125 (1961).

656. B. Eriksson, *Acta Chem. Scand.* **21**, 1119 (1967).

657. B. Eriksson and M. Rundfelt, *Acta Chem. Scand.* **22**, 562–570 (1968).

658. M. Winell and B. Mannervik, *Biochim. Biophys. Acta* **184**, 374–380 (1969).

659. W. Kielley and L. B. Bradley, in "Glutathione" (S. Colowick *et al.*, eds.), pp. 205–207. Academic Press, New York, 1954.

660. H. J. Strecker, in "Glutathione" (S. Colowick *et al.*, eds.), pp. 137–141. Academic Press, New York, 1954.

661. D. T. Dubin, *Biochem. Biophys. Res. Commun.* **1**, 262–265 (1959).

662. C. W. Tabor and H. Tabor, *Biochem. Biophys. Res. Commun.* **41**, 232–238 (1970).

663. E. Wikberg, *Nature (London)* **172**, 398 (1953).

664. D. Strumeyer and K. Bloch, *Biochem. Prep.* **9**, 52–55 (1962).

665. M. E. Maver, J. M. Johnson, and J. W. Thompson, *J. Nat. Cancer Inst.* **1**, 675–686 (1940–1941).

666. M. E. Maver and J. W. Thompson, *J. Nat. Cancer Inst.* **3**, 383–387 (1942–1943).

667. G. A. Maw and C. M. Coyne, *Arch. Biochem. Biophys.* **127**, 241–251 (1968).

668. W. J. P. Neish and A. Rylett, *Biochem. Pharmacol.* **12**, 893–903 (1963).

669. W. J. P. Neish, H. M. Davies, and P. M. Reeve, *Biochem. Pharmacol.* **13**, 1291–1303 (1964).

670. S. Fiala, A. E. Fiala, and B. Dixon, *J. Nat. Cancer Inst.* **48**, 1393–1401 (1972).

671. S. Fiala and E. S. Fiala, *J. Nat. Cancer Inst.* **51**, 151–158 (1973).

672. S. Fiala and E. S. Fiala, *Naturwissenschaften* **4**, 330–331 (1971).

673. K. R. Harrap, R. C. Jackson, and B. T. Hill, *Biochem. J.* **111**, 603–606 (1969).

674. P. C. Jocelyn, *in* "Glutathione" (E. M. Crooke, ed.), Biochem. Soc. Symp. No. 17, pp. 43–65. Cambridge Univ. Press, London and New York, 1959.

675. R. G. Bartlett, Jr. and U. D. Register, *Proc. Soc. Exp. Biol. Med.* **90**, 500–502 (1955).

CHAPTER 6

Covalent Adducts of Cysteine and Riboflavin

William C. Kenney, Dale E. Edmondson,
and Thomas P. Singer

I. INTRODUCTION

The occurrence in plant and animal tissues of riboflavin covalently linked to proteins has been known for nearly two decades, but for a long time succinate dehydrogenase was the only enzyme known to contain this form of flavin. In fact, for many years the terms "covalently bound flavin" and "succinate dehydrogenase flavin" were used synonymously in the biochemical literature. Before the structure of the flavin linkage in succinate dehydrogenase was shown to be 8α-[N(3)-histidyl]-FAD [1], several reports appeared indicating the presence of covalently bound flavin in several other enzymes. Only a few of these turned out to contain histidyl adducts of riboflavin, and not all enzymes containing histidylflavin are at the oxidation level of hydroxyriboflavin, as in succinate dehydrogenase. By 1971 it was established that covalent adducts of cysteine and riboflavin also constitute another important class of covalently bound flavins [2]. These, in turn, may be subdivided into cysteinyl flavin thioethers and cysteinyl flavin thiohemiacetals. Naturally occurring cysteinyl derivatives of riboflavin share several properties in common with histidyl flavins. Substitution, where clearly defined, is, in each case, at the 8α-position of the flavin ring system (Fig. 1); the substituent amino acid is an integral part of the polypeptide chain; the flavin is in all known instances FAD; and in the amino acyl flavin adducts, the properties of both the flavin and of the substituent amino acid are usually modified to a major extent. In several respects, however, the chemical and physical properties of histidyl and cysteinyl flavins are quite different, as are, consequently, procedures for their identification and quantitative determination.

This chapter is intended to call attention to the important new class of cysteine derivatives of flavins and to present current information on the chemistry, biophysics, and metabolism of cysteinyl flavins. As a guide to the properties of cysteinyl flavin adducts, it is useful to summarize first the properties of histidyl flavins and how they were established, since much of the information on the former was deduced from prior knowledge of the latter.

The occurrence of covalently bound flavins in nature was discovered in 1955 by Boukine [3] and by Green *et al.* [4]. Boukine found that a substantial part of the vitamin B_2 content of plant and animal tissues was not acid-extractable but became water-soluble on autolysis or tryptic digestion. Green *et al.* [4] made similar observations on a membrane fragment derived from heart mitochondria. Both workers predicted that this tightly bound form of riboflavin is associated with succinate dehydrogenase. This prediction was experimentally verified the same year by

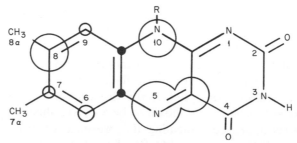

FIG. 1. Map of flavin submolecular structure. The size of the open circles indicates relative spin densities. Full dots indicate sites of unknown spin density. R, Rest of flavin. Taken in part from Salach *et al.* [14].

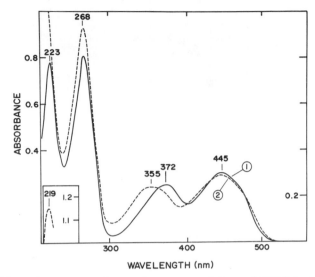

FIG. 2. Absorption spectra of neutral, aqueous solutions of FMN (—) and of acid-hydrolyzed SD-flavin (at the riboflavin level, – – – –) in the oxidized state. From Singer *et al.* [2].

the demonstration [5–7] that purified, soluble preparations of heart muscle succinate dehydrogenase contain FAD in covalent linkage with the protein, and thus the flavin becomes extractable only after digestion with proteolytic enzymes, liberating an FAD peptide [7–9]. The features distinguishing flavin peptides from "free" or "normal" flavins were soon thereafter recognized, primarily as a result of studies in Kearney's laboratory [7,10,11]. Thus, the flavin peptides showed a hypsochromic shift of the second absorption band of flavins from 372 nm to 345–355 nm (depending on the pH and the length of the peptide chain) (Fig. 2); the fluorescence of the flavin peptide at the monophosphate level, and even after dephosphorylation, was strongly quenched at neutral pH, with a pK_a of 4.5 ± 0.1 (Fig. 3); and the dephosphorylated flavin peptide was far more water soluble than riboflavin itself. The first of these anomalies was later recognized [12–14] as a general feature of 8α-substituted flavins and thus provides a good diagnostic test not only for histidyl but also for cysteinyl flavins. The second property is characteristic of histidyl flavin, although the pK_a value seems to vary slightly depending upon the nature of the linkage between the flavin and the imidazole ring of histidine, and thus provides the basis for the determination of histidyl flavin and of enzymes containing this structure in tissues [15,16].

Isolation of the first pure flavin peptide and its analytical composition

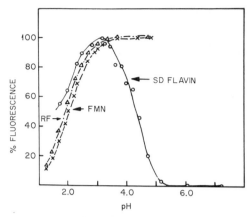

FIG. 3. pH fluorescence curve of succinate dehydrogenase (SD) flavin (FMN level), riboflavin (RF), and FMN in 0.03 *M* buffers. The buffers used were glycine, pH below 3.62; acetate, pH from 3.62 to 5.11; phosphate, pH from 5.11 to 7.30. From Kearney [10].

and properties were reported in 1960 [10]. By that time it was established that the peptide chain must be attached at some point in the isoalloxazine ring system, but not at N(10), since irradiation in alkaline media produced a lumiflavin peptide, rather than free lumiflavin [10,11]. Several years later it was shown [17] that positions 1, 2, and 3 also cannot be the site of peptide substitution, since Ba(OH)$_2$ hydrolysis liberated urea, rather than a ureido peptide.

The fact that the 8α-methyl group is the site of attachment of the peptide was proved many years later as a result of biophysical studies of the flavin. This, in turn, had to await (a) the isolation of pure flavin peptides from succinate dehydrogenase by a far simpler procedure [14] and in greater yield than had been used in earlier studies [10]; (b) the synthesis of several flavin models systematically substituted in the flavin ring; and (c) the assignment of spin densities to the various positions in the ring system on the basis of such models, thus laying the grounds for subsequent EPR studies on flavin peptides derived from succinate dehydrogenase.

Once these developments had occurred, it could be readily shown [12–14] that the hyperfine EPR spectrum of the cationic form of the flavin radical of peptides derived from succinate dehydrogenase differed in characteristic ways from similar EPR spectra of free flavin radical cations, pinpointing the site of attachment to a position in the isoalloxazine ring system endowed with high spin density. Of the positions satisfying this criterion, all but the 8α-carbon could be eliminated either on the basis of previous work or on the basis of the chemical properties of synthetic flavin models substituted in these positions [12,14]. In contrast,

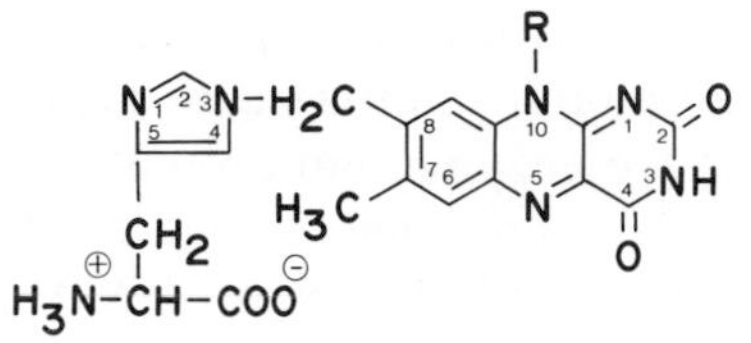

FIG. 4. Structure of covalently bound flavin from succinate dehydrogenase. R, Rest of FAD. From Walker and Singer [1].

8α-substituted flavins showed EPR spectra closely resembling that of the flavin peptide from succinate dehydrogenase. Supporting the conclusion that the 8α-position is the site of attachment were the facts that (a) only flavins substituted at this point showed the hypsochromic shift of the second absorption band (Fig. 2) and (b) that ENDOR and NMR studies confirmed that the 8α-methyl group was, indeed, substituted in the flavin peptide [14,18]. On systematic degradation of the flavin peptide by acid hydrolysis or digestion with aminopeptidase M the immediate substituent was shown to be histidine and the site of substitution the N(3) position of the imidazole ring (Fig. 4). This assignment of the structure was confirmed by chemical synthesis of 8α-[N(3)-histidyl]-riboflavin [19,20].

In the intervening years evidence accumulated that monoamine oxidase from liver and kidney mitochondria [22–25], cytochrome c_{552} from *Chromatium* [26,27], sarcosine dehydrogenase, and, probably, dimethylglycine dehydrogenase from liver mitochondria [28] also contain covalently bound flavin. Subsequently, D-hydroxynicotine oxidase [29], thiamine dehydrogenase [30], and β-cyclopiazonate oxidocyclase [31] were also shown to contain covalently bound flavin. The latter two enzymes contain histidyl flavin but the nature of the linkage appears to be different than in succinate dehydrogenase [32,33]. D-Hydroxynicotine dehydrogenase contains the same flavin structure as succinate dehydrogenase [34], and this is likely to be true also of sarcosine dehydrogenase [35,36]. On the other hand, monoamine oxidase and *Chromatium* cytochrome c_{552} contain cysteinyl adducts of FAD, the structure, properties, and metabolism of which are presented in the sections to follow.

II. 8α-CYSTEINYLFLAVIN THIOETHERS

A. Occurrence

In mitochondria from certain tissues (liver, kidney) a form of covalently linked flavin exists that is associated with monoamine oxidase

[22–25]. The presence of covalently bound flavin was ascertained from the findings that denaturation failed to release the flavin component from purified preparations of the enzyme, but proteolytic digestion resulted in solubilization of the flavin. Upon high voltage electrophoresis this flavin derivative did not migrate with FMN or FAD. A flavopeptide was isolated from a pronase digest of monoamine oxidase which showed a hypsochromic shift (see Section II,C) of the 372 nm maximum (of riboflavin) in the absorption spectrum [25].

A number of enzymes, other than mitochondrial monoamine oxidase from kidney and liver, also oxidize monoamines, but contain pyridoxal phosphate as a cofactor rather than flavin derivatives [37]. An exception, however, is mitochondrial monoamine oxidase from brain, which is known to be a flavoprotein [38,39], but is claimed to contain FAD in noncovalent linkage [39]. This conclusion comes from the findings that after the final purification step (but not earlier) of the enzyme from pig brain, FAD can be dissociated by acid-ammonium sulfate and the resulting apoenzyme partly reactivated by adding FAD. Attempts to extract fluorescent material from less pure preparations of the enzyme using trichloroacetic acid were, however, unsuccessful. These findings are somewhat contradictory for ascertaining whether or not FAD is covalently linked to the protein. These results suggest that at least a part of the enzyme in pig brain may contain covalently bound flavin, as in liver and kidney, and clearly indicate the need to carefully reexamine Tipton's [39] findings. Preliminary investigations in this laboratory* indicate that the turnover number of the enzyme extracted from pig brain mitochondria per mole of proteolytically released cysteinyl flavin is comparable to that obtained from the outer membrane of beef liver mitochondria.

B. Isolation and Sequence of Flavin Peptide

Earlier procedures used for the isolation of monoamine oxidase tended to inactivate, but not necessarily remove, succinate dehydrogenase, which also contains covalently linked flavin. Consequently, the possibility remained that flavin peptides derived from proteolytic digestion of monoamine oxidase preparations were contaminated with flavin peptides originating from succinate dehydrogenase.

In order to overcome this problem, a different purification procedure for monoamine oxidase was undertaken [40,41]. Since monoamine oxidase is exclusively located in the outer mitochondrial membrane and succinate dehydrogenase is an inner membrane constituent, a procedure

* J. I. Salach, unpublished results (1972).

for the large scale preparation of outer membranes was devised and the monoamine oxidase obtained from this source material was nearly completely free of succinate dehydrogenase. This permitted the unambiguous demonstration of the properties of the covalently bound flavin of mitochondrial monoamine oxidase and identification of its chemical nature, as described in subsequent sections.

The isolation of flavin peptides involved heat denaturation of the protein, followed by cold trichloroacetic acid treatment to remove any free flavins, chloroform-methanol extraction of lipids, and acid-acetone treatment. The resulting residue was incubated with trypsin-chymotrypsin, acidified, and subjected to Florisil chromatography. The crude peptide mixture eluted from Florisil was further purified by chromatography on cellulose phosphate and a homogeneous flavin peptide was obtained after descending paper chromatography [41].

The properties of this flavin peptide from monoamine oxidase are given in Table I. The positive chloroplatinate test (for reduced sulfur),

TABLE I

CHARACTERISTICS OF THE CYSTEINYL FLAVIN PEPTIDE FROM MONOAMINE
OXIDASE AND OF SYNTHETIC CYSTEINYL RIBOFLAVIN[a]

Treatment	Cysteinyl flavin peptide from monoamine oxidase	Synthetic cysteinyl riboflavin
Chloroplatinate test[b]	Positive	Positive
Chloroplatinate test[c]	Negative	Negative
Iodine-azide test	Negative	Negative
Mobility[d]	$R_f = 0.6$ relative to FMN	
Mobility[e]	$R_f = 0.65$ relative to FMN	
Absorption ratio 367/448 nm[b]	0.72	0.71
Absorption ratio 354/448 nm[c]	0.84	0.83
Zn reduction		
carboxymethylation, and acid hydrolysis	1 mole carboxymethylcysteine	
performic acid oxidation and acid hydrolysis	1 mole cysteic acid	
	1 mole serine	
	2 mole glycine	
	1 mole tyrosine	

[a] From Walker et al. [42]

[b] Before performic acid oxidation.

[c] After performic acid oxidation.

[d] On paper chromatograms, Whatman No. 1 paper, descending, n-butanol : acetic acid : H_2O (4 : 2 : 2, v/v/v).

[e] Paper electrophoresis, 250 mM pyridinium acetate, pH 5.5.

but negative iodine-azide test (for sulfhydryl and disulfide) are compatible with the presence of a thioether linkage. In addition, on performic acid oxidation the chloroplatinate test becomes negative, as expected for a sulfone, although the peptide remains attached to the flavin, as shown by thin-layer chromatography [42].

On acid hydrolysis (6 N HCl, 110°) one mole each of serine and tyrosine and two moles of glycine were obtained per mole of flavin. Similar results were obtained after digestion with aminopeptidase M [42,43]. As shown in Table I, performic acid oxidation followed by acid hydrolysis gives, in addition to the above amino acids, one mole of cysteic acid. These results are all consistent with a thioether linkage of the sulfur of cysteine to the flavin.

The thioether is readily cleaved upon zinc reduction and the flavin released as FMN, as shown by paper chromatography. The cysteinyl residue is then readily identified by carboxymethylation and amino acid analysis after acid hydrolysis (Table I).

As mentioned in Section I, evidence that the peptide is attached to the 8α position of the isoalloxazine ring system has come from the electron spin resonance and the optical absorption spectra and supporting evidence from its fluorescence characteristics (see Section II,C), using methods previously established for characterizing the flavin structure in succinate dehydrogenase.

Cysteinyl riboflavin has been synthesized [44] (see Section II,D) and shown to be identical with the corresponding derivative from monoamine oxidase [42], as shown in Table I. The structure of the cysteinyl flavin derivative is given in Fig. 5.

Dansylation of the pentapeptide obtained from tryptic-chymotryptic digestion of monoamine oxidase showed serine to be the amino terminal residue. Subsequent sequential analysis by the dansyl–Edman technique indicated the next two residues to be Gly-Gly. After the third Edman step no dansyl derivative was obtained; however, after the fourth Edman step tyrosine was found as the free amino acid. Hence this residue is carboxy terminal in the peptide. The sequence of this pentapep-

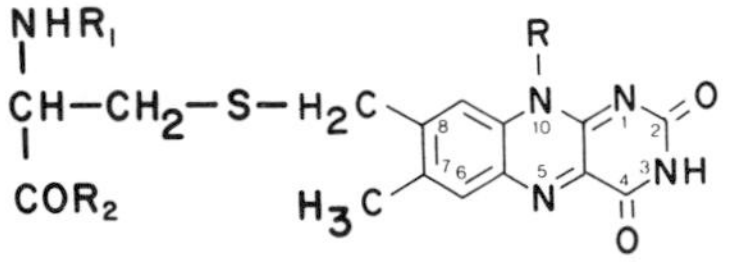

FIG. 5. Structure of the flavin peptide from monoamine oxidase. R, Rest of FAD in native enzyme or rest of FMN in pure peptide; R_1, serylglycylglycyl; R_2, tyrosyl. From Walker et al. [42].

tide from the data above [41,43] is

Ser-Gly-Gly-Cys(FAD)-Tyr

C. Spectral Properties

The visible and ultraviolet absorption spectrum of the pure flavin pentapeptide from monoamine oxidase has maximal absorption at 448, 367, and 270 nm (Fig. 6). The second absorption band is hypsochromically shifted relative to that of riboflavin, as is characteristic of 8α-substitution. Oxidation of the 8α-thioether to the sulfone has little influence on the absorption band at 448 nm but the 367 nm band is shifted to 354 nm (Fig. 5) with an increase in absorption relative to that of the 448 nm band. Very similar spectral properties are also shown by synthetic 8α-cysteinylriboflavin [44].

The fluorescence of the monoamine oxidase flavin peptide, like that of riboflavin, is independent of pH in the range of 3.2–8.5, but the fluorescence quantum yield is only 10% of that of riboflavin [41]. This decreased fluorescence is due to the π-electron donor properties of the divalent 8α-sulfur. Oxidation of the thioether to the sulfone results in an increase in fluorescence to about 70–80% that of riboflavin [41], since the oxidized sulfur no longer has nonbonding electron pairs for donation to the isoalloxazine ring system. As would be expected, the spectral shifts seen in the absorption spectra upon thioether oxidation are also seen in the corrected fluorescence excitation spectra (Fig. 7) [41].

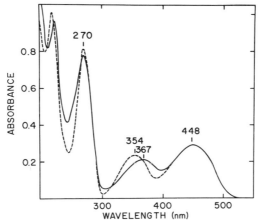

FIG. 6. Absorption spectra of the pure flavin pentapeptide from monoamine oxidase in the neutral flavoquinone form. —, Before performic acid oxidation; – – –, after performic acid oxidation. Spectra are measured in water. From Walker et al. [42].

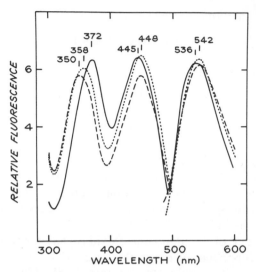

FIG. 7. Corrected fluorescence excitation and emission spectra of flavins and flavin peptides. —, Riboflavin (0.08 nmole/ml); ·····, pure monoamine oxidase pentapeptide before performic acid oxidation (0.8 nmole/ml): – – –, pure monoamine oxidase pentapeptide after performic acid oxidation (0.11 nmole/ml). Solvent: 0.1 M citrate-phosphate buffer pH 3.4. For excitation spectra emission at 523 nm and an excitation slit of 8 nm, for emission spectra excitation at 445 nm and emission slit 8 nm were used. From Kearney et al. [41].

Proof that the observed absorption and fluorescence spectral changes on oxidation of the 8α-thioether are indeed due to the sulfone rather than the sulfoxide comes from infrared spectral data [44]. Oxidation of synthetic 8α-cysteinylriboflavin gives a compound which exhibits infrared stretching vibrations at 1315 and 1225 cm^{-1}, which are characteristic of sulfones [44]. Sulfoxides, on the other hand, exhibit an absorption at 1050 cm^{-1} which is absent in the infrared spectrum of oxidized 8α-cysteinylriboflavin.

D. Chemical Synthesis

8α-Thiol substituted flavins are synthesized using procedures that were developed for the synthesis of 8α-histidylflavin analogues [19,20]. The nucleophilic thiol group of cysteine readily displaces the bromine substituent of 8α-bromotetraacetylriboflavin to form 8α-S-cysteinyltetraacetylriboflavin [44]. The reaction proceeds under mild conditions (room temperature, 48 hours, dimethylformamide) without the addition of base to ionize the sulfhydryl group. The synthesis of 8α-S-glutathionyltetraacetylriboflavin has been reported [45] using the same procedure and, as expected, exhibits properties very similar to those of 8α-cysteinylriboflavin.

The 8α-S-cysteinylsulfoneriboflavin analogue has been synthesized directly in 30–35% yield [46] by incubating 8α-bromotetraacetylriboflavin with cysteinesulfinic acid. The riboflavin analogue is formed by mild acid hydrolysis to remove the ribityl acetyl groups. The spectral and chromatographic properties of this compound are identical with those of the flavin sulfone formed by performic acid oxidation of 8α-cysteinylriboflavin.

E. Chemical Reactivity of the Thioether Bond

The thioether bond of 8α-cysteinylriboflavin gives a positive chloroplatinic acid but a negative iodine-azide test [42], as expected. Upon oxidation to the sulfone the chloroplatinate test becomes negative (Table I).

Cysteine can be liberated from the thioflavin by reductive cleavage with zinc in acid media [42,44]. The products of this cleavage were shown by paper chromatography to be cysteine and unsubstituted flavin (FMN in the case of the monoamine oxidase flavin peptide and riboflavin in the case of 8α-cysteinylriboflavin) [42,44]. In contrast to simple thioethers like methionine, the thioether bond of 8α-cysteinylriboflavin shows an increased acid lability [42]. Under standard conditions of peptide hydrolysis (15 hours, 6 N HCl) extensive cleavage of the cysteinyl flavin bond occurred at 75°, with complete cleavage at 105°C.

The thioether bond of 8α-cysteinylflavins is very susceptible to aerobic oxidation. Although 8α-cysteinylflavin peptides are more stable than 8α-cysteinylriboflavin [42], extensive oxidation can occur during peptide isolation, unless precautions are taken (4°C, anaerobic conditions). Presumably, air oxidation initially converts the thioether to a sulfoxide, followed by oxidation to the sulfone, although this sequence remains to be experimentally verified. The oxidizing ability of the isoalloxazine ring may also play a role in thioether oxidation through intermolecular and/or intramolecular electron transfer. Extensive breakdown of 8α-cysteinylriboflavin to 8-formylriboflavin has been observed on storage in aqueous solution at room temperature or at 0°, under either anaerobic or aerobic conditions.

The sulfoxide form of 8α-cysteinylflavin has not yet been investigated as oxidation methods currently applied yield the sulfone. Reagents effecting this 4 electron oxidation include performic acid [42], peracetic acid [44], and dilute permanganate ion [44]. The sulfone of 8α-cysteinylriboflavin gives cysteic acid and presumably 8-hydroxyriboflavin on acid hydrolysis (6 N HCl, 15 hours, 105°). A reductive method for cleavage of 8α-flavin sulfones has been described [46] and is discussed in the following section.

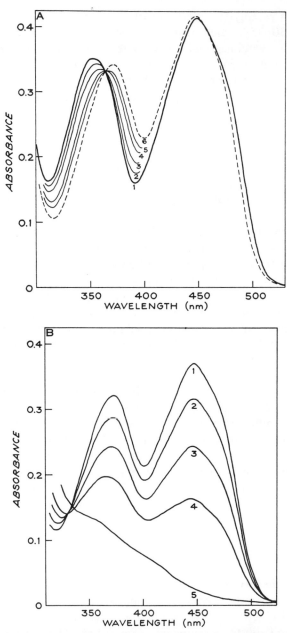

FIG. 8. (A) Spectral changes during the anaerobic addition of the first molar equivalent of dithionite to 8α-*S*-cysteinylsufonetetraacetylriboflavin in 0.1 *M* phosphate, pH 7.0. Upon each addition of dithionite, the spectrum was recorded after no further spectral changes were apparent. The molar equivalents of dithionite for the respective curves are curve 1,

F. Reductive Cleavage of 8α-S-Cysteinylsulfoneriboflavin

Reduction of 8α-S-cysteinylsulfoneriboflavin by 1 mole of dithionite/mole flavin or by photoreduction, using EDTA as the electron donor, results in the formation of riboflavin, as shown by thin-layer chromatography [46]. The spectral changes in the near ultraviolet region during the addition of the first mole of dithionite to 8α-S-cysteinylsulfonetetraacetylriboflavin are illustrated in Fig. 8A. The absorption maximum at 352 nm shifts to 372 nm, with an isosbestic point at 363 nm, showing the presence of only two flavin species in solution. The constant intensity of the 448 nm absorption band shows no flavin reduction at equilibrium during the addition of the first molar equivalent of dithionite. Flavin reduction is observed during the addition of the second molar equivalent of dithionite, however (Fig. 8B). The graphical representation of this anaerobic reductive titration is shown in Fig. 9.

This reductive cleavage occurs via electron transfer from the isoalloxazine ring, since the 450 nm absorption band is rapidly bleached upon the initial anaerobic addition of one molar-equivalent of dithionite. Following this rapid bleaching, the 450 nm band increases slowly under anaerobic conditions by a first-order reaction (Fig. 10). The rate of reductive cleavage is pH-dependent, with maximal rate of pH 6.0 (Fig. 10). A mechanism for this reductive cleavage has been suggested by Edmondson and Singer [46], based on the experimental findings (Fig. 11). It must be pointed out that the reductive cleavage upon flavin reduction is only seen with the sulfone form of 8α-flavin thioethers. Reduction of 8α-cysteinylriboflavin by a stoichiometric amount of dithionite does not cleave the 8α-thioether bond. The reason for this difference may be that the electron deficiency of the 8α-sulfone causes electron migration resulting in oxidation of the flavin and elimination of cysteinesulfinic acid (Fig. 10). The electron density of the sulfur atom in the thioether linkage prevents such electron migration and thus the reduced form of 8α-cysteinylriboflavin is quite stable.

G. Redox Properties of 8α-Cysteinylflavins

To investigate the effect of 8α substitution on the chemical properties of the flavin ring system, a study of the electron affinity of various 8α-

none; curve 2, 0.15; curve 3, 0.3; curve 4, 0.6; curve 5, 0.75; curve 6, 1.0. The spectra are uncorrected for dilution. (B) Spectral changes during the anaerobic addition of a second molar equivalent of dithionite to the flavin solution in 8A. The total molar equivalents of dithionite added for the respective curves are: curve 1, 1.1, curve 2, 1.26; curve 3, 1.43; curve 4, 1.63; curve 5, 2.40. The spectra are uncorrected for dilution. From Edmondson and Singer [46].

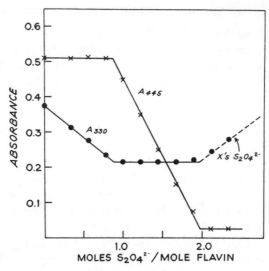

FIG. 9. Dithionite titration of 8α-S-cysteinylsulfonetetraacetylriboflavin in 0.1 M phosphate, pH 7.0. All points are corrected for dilution. The increase in absorption at 330 nm after 2.0 molar equivalents of dithionite is due to excess dithionite. From Edmondson and Singer [46].

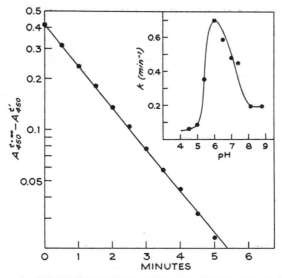

FIG. 10. First-order kinetic plot of the reappearance of absorption at 450 nm after the anaerobic addition of 1 molar equivalent of dithionite to 8α-S-cysteinylsulfonetetraacetylriboflavin in 0.1 M phosphate, pH 6.5 at 25°. The inset shows the dependence of the rate constant on pH. The buffers for the various pH values are 0.1 M acetate, pH 4.5 and 5.0; 0.05 M phosphate-0.05 M acetate, pH 5.5; 0.1 M phosphate, pH 6.0, 6.5, 7.0, and 7.5; 0.1 M pyrophosphate, pH 8.05 and 8.8. From Edmondson and Singer [46].

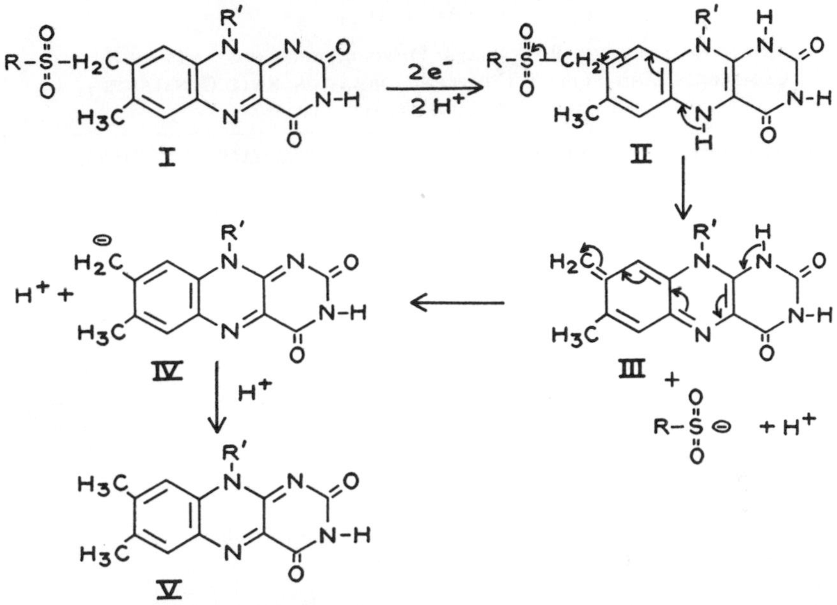

FIG. 11. Scheme depicting the proposed mechanism for elimination of the 8α-cysteinyl-sulfone to form cysteinesulfinate and unsubstituted flavin. From Edmondson and Singer [46].

substituted flavins has been published [46]. Methods for measuring the electron affinities of 8α-flavin analogues include redox potential measurements and determination of sulfite affinities. It has been shown that flavins reversibly form sulfite adducts at the N(5) position with an affinity paralleling their respective redox potentials [47]. The data in Table II show that 8α substitution raises the redox potential of the ribo-flavin ring system by 0.02–0.03 V. Little difference in redox potential is observed among the various 8α-substituted flavins. The small difference in potential (0.007 V) observed between the 8α-cysteinyl and -glutathionyl analogues is likely due to differences in the ribityl side chain and to experimental error.

The increase in electron affinity resulting from 8α substitution of flavins is indicated both by redox potential measurements and by comparison of sulfite affinities with that of riboflavin (Table II). The K_D values of 8α-substituted flavins for sulfite are ten- to twentyfold greater than their corresponding unsubstituted flavin. The fourfold increase in sulfite affinity upon oxidation of the 8α-thioether to the sulfone form is another indication of their differences in electron withdrawing capacity. Due to the reductive cleavage reaction discussed in Section II,F, the redox potential of the flavin sulfone cannot be determined.

TABLE II

OXIDATION-REDUCTION POTENTIALS, DISSOCIATION CONSTANTS OF SULFITE
COMPLEXES, AND SULFITE COMPLEX FORMATION RATE CONSTANTS FOR
VARIOUS 8α-SUBSTITUTED FLAVINS[a]

8α Substituent	Flavin[b]	$E_{m,7}$ (V)	K_D (M)	k (mole^{-1} min^{-1})
H	RF	-0.190^c	1.16^c	1.36^c
Hydroxy	RF	-0.170	0.22	0.98
Carboxy	RF	-0.165	0.081	1.91
N-3-Histidyl	RF	-0.160	0.056	1.94
S-Cysteinyl	RF	-0.169	0.104	1.85
Formyl	RF	-0.159	—	—
H	TARF	-0.195^c	1.21^c	1.12^c
S-Glutathione	TARF	-0.168	0.110	1.74
Sulfonyl	TARF	-0.159	0.100	2.49
S-Cysteinyl sulfone	TARF	—	0.027	3.67

[a] From Edmondson and Singer [46].
[b] RF, riboflavin derivative; TARF, tetraacetylriboflavin derivative.
[c] Taken from Müller and Massey [47].

III. 8α-CYSTEINYLFLAVIN THIOHEMIACETALS

In addition to cysteinyl flavin thioethers, another form of cysteinyl flavin, with different properties, occurs in nature. Although rigorous proof of the structure of this second form is not yet available, as summarized in Section III,C, available evidence is consistent with a 8α-cysteinyl-FAD thiohemiacetal structure.

A. Occurrence

Chromatium cytochrome c_{552} contains a flavin component that is not extracted from the enzyme by acid ammonium sulfate treatment or by trichloroacetic acid [26], but the flavin becomes acid-soluble on digestion with trypsin [27]. Unlike covalently linked flavins in other enzymes, however, the flavin from cytochrome c_{552}, was reported to be released by exposure to pH above 9, by treatment with *p*-chloromercuribenzoate at pH 4.5, or by prolonged incubation with saturated urea solutions in the cold [27]. The flavin obtained by the latter procedure had a modified absorption spectrum relative to riboflavin and was nonfluorescent. This dissociation of the flavin from the enzyme by urea has recently been confirmed [48], but cyanate, present in aged urea solutions, is considered to be the agent that enhances the dissociation, rather than urea itself.* Studies in this laboratory [48a] failed to verify that al-

* J. Cronin, unpublished results (1972).

kaline pH or mercurials dissociate the flavin from the denatured protein.

Cytochrome c_{553} from *Chlorobium thiosulfatophilum* has also been reported to contain covalently bound flavin [27,50]. Although only fragmentary information is available on this flavin, the results reported indicate that this flavin has properties in common with that present in *Chromatium* cytochrome c_{552}. It will be of interest to ascertain whether the nature of the flavin–protein bond in these two enzymes is similar, if not identical.

B. Isolation and Sequence of Flavin Peptides

The flavin released by prolonged incubation in the presence of urea was purified and shown to be at the FAD level [48]. The hypsochromic shift of the 372 nm maximum of the absorption spectrum further suggested that the linkage of the apoenzyme to the FAD is by way of the 8α-carbon, as in other enzymes containing covalently bound flavin.

Conclusive evidence for this assignment and identification of the 8α substituent as cysteine were obtained in a collaborative investigation between J. Cronin's and the author's laboratories [48a,50,51]. Purification and characterization of the flavin peptide released by digestion with proteolytic enzymes proved to be easier than isolation of the product of incubation with urea in pure form. Homogeneous flavin peptides have been isolated from both peptic and tryptic-chymotryptic digests of the enzyme [51,52,52a]. Purification of the peptide involved chromatography on Florisil, chromatography on phosphocellulose columns, and high voltage electrophoresis at pH 1.6. The second fluorescence excitation maximum of the resulting peptides were hypsochromically shifted from 372 nm to 365 nm, as expected for an 8α linkage, and these peptides had a characteristic strongly quenched fluorescence (Fig. 12). This assignment was confirmed by the hyperfine EPR spectrum of the free radical of the flavin obtained by urea treatment of the enzyme [50] as well as by the demonstration that performic acid oxidation of the enzyme liberated 8-carboxyriboflavin in good yield [48a,50].

The findings that oxidation of the tryptic-chymotryptic flavin peptide with performic acid results in a shift of the second fluorescence excitation maximum from 365 nm to 352 nm (Fig. 12) and an increase from 5 to 50% of the fluorescence of an equivalent amount of riboflavin (Table III) are similar to the behavior of peptides containing 8α-S-cysteinylriboflavin thioether. Evidence for the presence of sulfur was further supported by a positive chloroplatinate test (Table III). The presence of cysteine was confirmed by amino acid analysis after acid hydrolysis of the purified flavin peptides (Table IV) [48a].

The amino acid sequence of the peptic and of the tryptic-chymotryptic

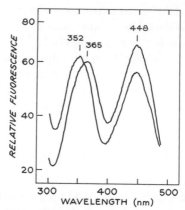

FIG. 12. Fluorescence excitation spectra of *Chromatium* flavin. The product obtained from proteolytic digestion in presence of dithiothreitol without hydrolysis is that with λ_{max} of 365 nm and after performic acid oxidation that with λ_{max} of 352 nm, but using 10% as much flavin as in A. From Hendriks *et al.* [50].

flavin peptides (Fig. 13) are similar, with the former containing an additional tyrosyl residue, N-terminal in this peptide [52a]. The presence of this additional residue is responsible for the significant differences in the properties of these two peptides, as discussed in Section III,D.

C. Evidence for a Thiohemiacetal Bond

The chloroplatinate test is a general test for sulfur in the reduced form, and thus a positive reaction is given by thiols, thioethers, disul-

TABLE III

PROPERTIES OF FLAVIN PEPTIDES ISOLATED FROM *Chromatium* CYTOCHROME c_{552}[a]

	Result	
Test	Tryptic-chymotryptic peptide	Peptic peptide
Molar fluorescence[b]	0.05	0.01
Same[b] after performic acid oxidation at 0°	0.50	0.05
Same[b] after performic acid oxidation at 40°		0.50
Ninhydrin reaction	−	+
Iodine–azide test	−	−
Chloroplatinate test	+	+

[a] From Kenney *et al.* [52].
[b] Relative to equimolar solution of riboflavin.

TABLE IV

AMINO ACID COMPOSITION OF FLAVIN PEPTIDES FROM
Chromatium CYTOCHROME c_{552}[a]

Amino acid[b]	Tryptic-chymotryptic peptide (nmoles)	Peptic peptide (nmoles)
Thr	9.5	9.4
Cys	10.4	8.0
Tyr	11.6	20.6
Flavin[c]	14.2	13.2

[a] From Kenney *et al.* [52].
[b] Determined after hydrolysis at 110° in 6 N HCl for 24 hours.
[c] Based on $\epsilon = 11.3 \times 10^3$.

fides, thioesters, and thiohemiacetals. All but one of these structures are eliminated by the following experiments. The negative iodine-azide test of the peptides (Table III), as well as the inability of reducing agents to release flavin from the enzyme [50], eliminate the possibility of a disulfide linkage. Since hydroxylamine fails to liberate the flavin from the protein, a thioester linkage may also be ruled out [50]. In a number of tests the cysteinyl flavin component of cytochrome c_{552} differs from the thioether present in monoamine oxidase. Thus, 8-carboxyriboflavin is liberated from denatured cytochrome c_{552} but not from monoamine oxidase under the same conditions. Second, the tryptic-chymotryptic peptide from cytochrome c_{552} is considerably more prone to oxidation than is the similar flavin peptide from monoamine oxidase. Third, oxidation of the latter by performic acid yields 80% of the molar fluorescence of riboflavin, while the former gives only 50% fluorescence. Fourth, while the autooxidation product (presumed sulfoxide) of the thioether type of

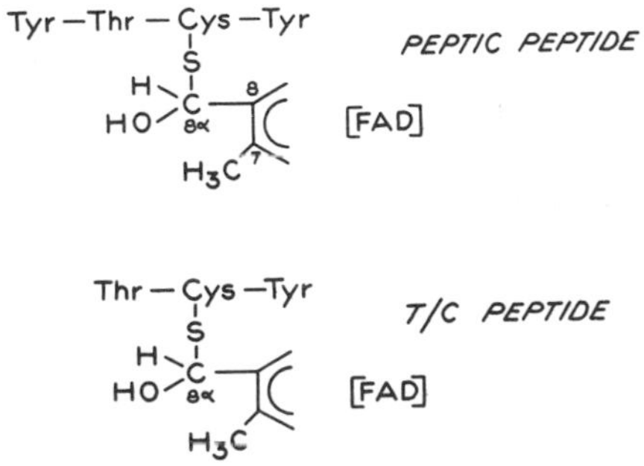

FIG. 13. Proposed structures and amino acid sequences of flavin peptides from *Chromatium* cytochrome c_{552}. T/C, Tryptic-chymotryptic. From Kenney *et al.* [52].

flavin peptide from monoamine oxidase is reduced to the thioether, but not cleaved, by dithionite, the autooxidized tryptic-chymotryptic flavin peptide from *Chromatium* cytochrome c_{552} is cleaved on dithionite treatment, as judged by continued high fluorescence and the presence of new components on high voltage electrophoresis. Although 8α-cysteinyl-riboflavin is stable to reductive titration with dithionite, the tryptic-chymotryptic flavin peptide from cytochrome c_{552} is not, and 8-formylriboflavin has been detected spectrophotometrically. Due to the stabilizing influence of the N-terminal tyrosyl residue (see Section III,D), the peptic peptide is stable to reductive titration with dithionite.

This leaves a cysteinyl thiohemiacetal (8α-hydroxy-8α-S-cysteinyl-riboflavin) as the only remaining possibility (Fig. 13). In accord with this structure, acid hydrolysis releases a flavin with an R_f value in thin-layer chromatography, comparable to 8-formylriboflavin, but not 8α-hydroxyriboflavin or 8-carboxyriboflavin, indicating the 8α-position of the flavin to be at the oxidation level of an aldehyde, as expected for this proposed structure [52a]. Moreover, consistent with the chromatographic data, the acid-hydrolyzed *Chromatium* flavin shows a hypochromic shift of the second fluorescence excitation maximum from 367 to 355 nm. This behavior distinguishes the compound from riboflavin, 8-hydroxy-riboflavin, and 8-carboxyriboflavin (maxima at 372, 363, and 367 nm, respectively). The fluorescence excitation parameters are identical to those obtained when synthetic 8-formylriboflavin is subjected to acid hydrolysis in the presence of a molar ratio of the amino acids within the peptide. Additional evidence came from a highly specific test for 8-for-mylriboflavin. Of all flavins tested, only 8-formylflavins have been found to yield a characteristic blue color with absorption maxima at 565 and 605 nm on reduction with $TiCl_3$ in 6 N HCl. On reduction under similar conditions, the flavin peptide gave little or no blue color, however, preparations in which cleavage of the cysteinyl flavin bond had occurred (urea treatment, long term storage) as ascertained by chromatographic and electrophoretic techniques, the flavin component gave identical spectral properties with those of cationic 8-formylflavin hydroquinones. In considering further evidence for this structure, it is useful to compare the properties of the peptic and tryptic-chymotryptic peptides derived from cytochrome c_{552}.

The flavin peptide linkage in the tryptic-chymotryptic peptide is very labile to aerobic oxidation and yields 8-carboxy-FAD, unless strictly anaerobic conditions are maintained during isolation. Fortunately, the peptic flavin peptide is considerably more resistant to aerobic oxidation and permits obtaining evidence for the proposed structure [52]. An example of this resistance to oxidation is shown in Table III. As mentioned earlier, the fluorescence intensity of the tryptic-chymotryptic peptide increases from 5 to 50% after performic acid oxidation at 0°. In comparable experiments on the peptic peptide, the fluorescence intensity of the reduced form is trivial and increases only to about 5% of that of riboflavin. Under more rigorous conditions of oxidation (40°), however, the same 50% maximum fluorescence relative to riboflavin was obtained as with the tryptic-chymotryptic peptide [52,52a].

After digestion of the flavin peptide (FAD level) with aminopeptidase M and subsequent high voltage electrophoresis at pH 6.5, two flavin components were detected. One had 0.95 of the mobility of FAD, as expected for an amino acyl FAD derivative, but the second component had a much greater anionic mobility than free FAD, suggesting that the α-amino group of cysteine was blocked. This would be expected if cyclization to a thiazolidine derivative (Fig. 14) had occurred during aminopeptidase digestion. Both of these components exhibited the characteristically quenched fluorescence, but like the tryptic-chymotryptic peptide, they were readily oxidized, resulting in destruction of the linkage [48a]. Although 8-formylriboflavin exists in a stable intramolecular hemiacetal form with the 5'-hydroxyl group [52b], thiohemiacetals are considered to be chemically unstable compounds, with a tendency to dissociate. Such dissociation has been observed with the tryptic-chymotryptic flavin peptide and the cysteinyl flavin derivative arising from aminopeptidase digestion of the peptic peptide. The synthetic thiohemiacetal of cysteine and 8-formyllumiflavin also dissociates quite readily. The ability to isolate flavin peptides from *Chromatium* cytochrome c_{552} is probably not the result of mixed acetal formation since the 5'-hydroxyl group is substituted (FAD form) and for steric reasons, the other ribityl hydroxyl groups could not participate in such a linkage. Instead, the stability of the flavin peptides probably arises from interaction with the adenine in FAD as well as tyrosyl residues in the peptides—especially the N-terminal tyrosyl residue of the peptic peptide [52a].

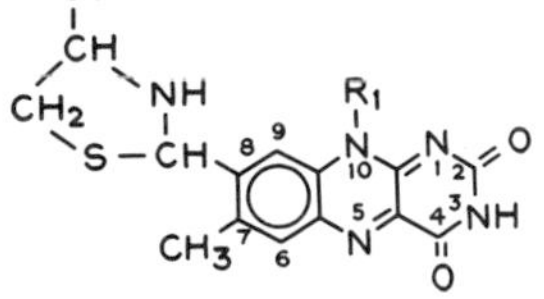

Fig. 14. Proposed structure of cysteinyl flavin from *Chromatium* cytochrome c_{552} in the thiazolidine form.

D. Influence of Tyrosine on the Flavin Properties in the Peptic Peptide

The amino acid sequences of the peptic and tryptic-chymotryptic flavin peptides (Fig. 13) differ by an N-terminal tyrosyl residue. The presence of this residue is responsible for the fact that on oxidation with cold performic acid the fluorescence is not significantly enhanced. As documented in Table V, the low fluorescence is due to a tyrosine–flavin interaction rather than protection of the sulfur from oxidation. This is evident from the fact that on hydrolysis of the peptide with aminopeptidase, following performic acid treatment, the fluorescence increases to nearly the level observed with the tryptic-chymotryptic peptide or on performic acid oxidation of the peptic peptide at 40° [52a].

Direct evidence for the flavin–tyrosine interaction was obtained from circular dichroism (CD) studies [52a]. Figure 15 compares the CD spectra of the two FAD peptides and of FAD. The peptic peptide exhibits a broad, positive Cotton effect with a maximum at 484–490 nm and negative bands at 305 and 375 nm. The latter band has a twofold greater intensity than FAD. The Cotton effects are dramatically altered in the tryptic-chymotryptic peptide, lacking the N-terminal tyrosine, resulting in a positive band at 340 nm as in free FAD; however, only a small negative band at 380 nm is observed [52]. The results are consistent with a direct tyrosine–flavin interaction in the peptic peptide. Removal of the adenine moiety has little effect on the shape and intensity of the CD spectrum. The Cotton effects observed are due to coupling of the individual electronic transition of the flavin and tyrosine by dipole-dipole interactions. The general features of the CD spectrum of the peptic peptide resemble those reported for the flavodoxins [53,54], where a tyrosyl residue in a planar stacking arrangement with the flavin has been demonstrated by X-ray crystallography [55,56].

TABLE V

AMINOPEPTIDASE M DIGESTION OF *Chromatium* CYTOCHROME c_{552} FLAVIN PEPTIDE AFTER PERFORMIC ACID OXIDATION

Time of digestion (hour)	Percent fluorescence[a] after peptidase digestion	
	No prior oxidation	After prior oxidation[b]
0	3	10
4	9	39

[a] Relative to riboflavin, assuming equivalent extinction coefficients at 450 nm.

[b] Peptic peptide was performic acid oxidized at 0°, lyophilized, and then digested with aminopeptidase.

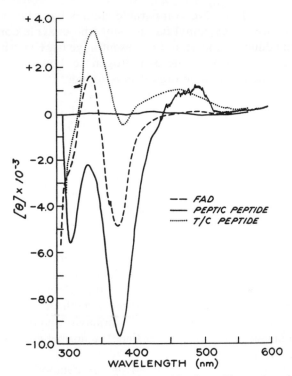

Fig. 15. CD spectra of FAD and of flavin peptides from *Chromatium* cytochrome c_{552}. T/C, Trypsin-chymotrypsin. The spectra were measured at 1.65×10^{-4} *M* concentration at pH 6.8 at 25° in 10 nm cylindrical quartz cells, using a JASCO UV 5 instrument with a Sproul Scientific SS 10 modification. From Kenney *et al.* [52].

IV. METABOLIC SYNTHESIS AND DEGRADATION OF COVALENTLY BOUND FLAVINS

Although covalently bound flavins have been known to be widely distributed in nature since 1955, until recently no information was available on their biosynthesis and catabolism. Recent studies have concentrated on the biosynthesis and degradation of succinate dehydrogenase and of its 8α-histidyl-FAD component [57,58]. The experimental system used for biosynthetic studies has been anaerobic yeast undergoing mitochondrial biogenesis on exposure to O_2, since in this system the cellular level of succinate dehydrogenase and of histidyl flavin rises, within a few hours, from nearly zero to the high level characteristic of aerobic yeast cells [59]. The assembly of protein-bound 8α-histidyl-FAD is at least in part a mitochondrial event, since chloramphenicol blocks the process [58]. No comparable studies have been undertaken on the biosynthesis of cysteinyl flavins, since no enzyme containing this form of flavin adduct is known to be present in yeast. Experiments paralleling investigations on succinate dehydrogenase, however, would likely be considered when a suitable organism is found.

Using riboflavinless mutants, which grow only in the presence of added vitamin B_2, it has been shown that several analogues of riboflavin are incorporated into the covalently bound FAD fraction [57]. Among these 7-ethyl, 8-methylriboflavin and 8-ethyl, 7-methylriboflavin deserve mention. These analogues, as sole sources of "vitamin B_2" in the medium, permit nearly normal growth and the synthesis of catalytically competent succinate dehydrogenase. These analogues also replace riboflavin in weanling rats but the succinate dehydrogenase activity of the tissues of rats grown on these analogues is lower than normal [60,61]. Deazaflavin (riboflavin with N(5) replaced by a C atom) can apparently also serve as a precursor of the flavin component of succinate dehydrogenase, provided that a suboptimal level of riboflavin, along with deazaflavin, is present in the growth medium. Deazaflavin is incorporated into covalent linkage during aerobic growth of the cells and succinate dehydrogenase, containing this riboflavin analogue, appears to have a much higher turnover number than the normal enzyme containing 8α-histidyl-FAD [57].

Studies have been recently initiated on the catabolism of covalently bound flavins, using aerobic yeast cells undergoing catabolite repression [58]. It has been known for some time that the addition of glucose to aerobic yeast cells induces decay of mitochondrial structure, disappearance of mitochondrial enzymes, and a precipitous decline in the covalently bound flavin content of the mitochondria [59]. The process involves derepression of the extramitochondrial synthesis of certain lytic enzymes, since cycloheximide prevents the derepression [58]. It has been recently demonstrated that the addition of glucose to aerobic yeast causes a major rise in phospholipase D activity; this rise is prevented by cycloheximide [58]. It seems possible, therefore, that phospholipase D may be involved in the destruction of the membrane structure, exposing succinate dehydrogenase to attack by proteolytic enzymes. This experimental system appears to provide an opportunity to trace the pathway of metabolic degradation of protein-bound 8α-histidyl-FAD and to delineate the enzymes involved in its catabolism.

It may be seen from the foregoing that studies of the metabolic pathways involving covalently bound flavin have just begun and that, so far, no concrete information is available on the nature of the intermediates. It is a reasonable assumption, however, that the incorporation of the flavin into covalent linkage with proteins would follow the assembly of apoenzyme. It remains to be seen whether prior to incorporation into the protein the flavin is first enzymatically activated in the 8α position, so as to enhance the tendency of this group to react with nucleophiles, such as an imidazole ring nitrogen or the –SH group of cysteine. Noting that the 8α group is in different oxidation states in different covalent structures (at the hydroxyl level in succinate dehydrogenase and in monoamine oxidase but at the carbonyl level in *Chromatium* cytochrome c_{552}, thiamine dehydrogenase, and β-cyclopiazonate oxidocyclase), although in each case the 8α group is involved, it may be supposed that different activated forms of FAD would be required for biosynthesis of the corresponding holoenzymes. Just as in the hypothetical activation step, the incorporation of FAD into the protein is likely to be an enzymatic process.

Studies of the biosynthesis and catabolism of 8α-histidyl-FAD is a logical starting point for investigations of the metabolism of covalently bound flavins. This is because of the availability of a convenient system in which accumulation and decay of the compound can be rapidly induced under controlled conditions. In order to judge the applicability of information to emerge from these studies to the metabolism of cysteinyl flavins, a different system will have to be found. Possibly regenerating liver in conjunction with [^{14}C]riboflavin might provide a test system for the biogenesis of flavin thioethers and growing *Chromatium* cells for that of flavin thiohemiacetals.

ACKNOWLEDGMENTS

The original data reported in this chapter were obtained with the support of Program Project No. 1 PO 1HL 16251-01 of the National Heart Institute and a research grant (No. GB 30078) from the National Science Foundation.

REFERENCES

1. W. H. Walker and T. P. Singer, *J. Biol. Chem.* **245**, 4224 (1970).
2. W. H. Walker, E. B. Kearney, R. Seng, and T. P. Singer, *Biochem. Biophys. Res. Commun.* **44**, 287 (1971).
3. V. N. Boukine, *Congr. Int. Biochim., Resumes Commun., 3rd, 1955* p. 61 (1955); see also V. N. Boukine, *Proc. Int. Congr. Biochem., 3rd, 1955* p. 260 (1956).
4. D. E. Green, S. Mii, and P. M. Kohout, *J. Biol Chem.* **217**, 557 (1955).
5. E. B. Kearney and T. P. Singer, *Congr. Int. Biochim., Resumes Commun. 3rd, 1955* p. 55 (1955).
6. E. B. Kearney and T. P. Singer, *Biochim. Biophys. Acta* **17**, 596 (1955).
7. T. P. Singer, E. B. Kearney, and V. Massey, *in* "Enzymes: Units of Biological Structure and Function" (O. H. Gaebler, ed.), p. 417. Academic Press, New York, 1956.

214   W. C. KENNEY, D. E. EDMONDSON, AND T. P. SINGER

8. T. P. Singer, E. B. Kearney, and V. Massey, *Arch. Biochem. Biophys.* **60**, 255 (1955).
9. E. B. Kearney and T. P. Singer, *Fed. Proc., Fed. Amer. Soc. Exp. Biol.* **15**, 286 (1956).
10. E. B. Kearney, *J. Biol. Chem.* **235**, 865 (1960).
11. T. P. Singer and E. B. Kearney, *in* "Vitamin Metabolism" (W. Umbreit and M. Molitor, eds.), p. 209. Pergamon, Oxford, 1960.
12. P. Hemmerich, A. Ehrenberg, W. H. Walker, L. E. G. Eriksson, J. Salach, P. Bader, and T. P. Singer, *FEBS (Fed. Eur. Biochem. Soc.) Lett.* **3**, 37 (1969).
13. T. P. Singer, J. Salach, W. H. Walker, M. Gutman, P. Hemmerich, and A. Ehrenberg, *in* "Flavins and Flavoproteins" (H. Kamin, ed.), p. 607. Univ. Park Press, Baltimore, Maryland, 1971.
14. J. Salach, W. H. Walker, T. P. Singer, A. Ehrenberg, P. Hemmerich, S. Ghisla, and U. Hartmann, *Eur. J. Biochem.* **26**, 267 (1972).
15. T. P. Singer, J. Hauber, and E. B. Kearney, *Biochem. Biophys. Res. Commun.* **9**, 146 (1962).
16. T. P. Singer, J. Salach, P. Hemmerich, and A. Ehrenberg, *in* "Methods in Enzymology" (D. B. McCormick and L. D. Wright, eds.), Vol. 18B, p. 416. Academic Press, New York, 1971.
17. T. F. Chi, Y. L. Wang, C. L. Tsou, Y. C. Fong, and C. H. Yu, *Scientia Sinica* **14**, 1193 (1965).
18. W. H. Walker, J. Salach, M. Gutman, T. P. Singer, J. S. Hyde, and A. Ehrenberg, *FEBS (Fed. Eur. Biochem. Soc.) Lett.* **5**, 237 (1969).
19. S. Ghisla, U. Hartmann, and P. Hemmerich, *Angew. Chem. Int. Ed.* **9**, 642 (1970).
20. W. H. Walker, T. P. Singer, S. Ghisla, and P. Hemmerich, *Eur. J. Biochem.* **26**, 279 (1972).
21. T. P. Singer, W. H. Walker, W. Kenney, and E. B. Kearncy, *in* "Structure and Function of Oxidation Reduction Enzymes" (Å. Åkeson and Å Ehrenberg, eds.), p. 501. Pergamon, Oxford, 1972.
22. S. Nara, I. Igaue, B. Gomes, and K. T. Yasunobu, *Biochem. Biophys. Res. Commun.* **23**, 324 (1966).
23. I. Igaue, B. Gomes, and K. T. Yasunobu, *Biochem. Biophys. Res. Commun.* **29**, 562 (1967).
24. V. G. Erwin and L. Hellerman, *J. Biol. Chem.* **242**, 4230 (1967).
25. B. Gomes, I. Igaue, H. G. Kloepper, and K. T. Yasunobu, *Arch. Biochem. Biophys.* **132**, 16 (1969).
26. R. G. Bartsch, *Fed. Proc., Fed. Amer. Soc. Exp. Biol.* **20**, 43 (1961).
27. R. G. Bartsch, T. E. Meyer, and A. B. Robinson, *in* "Structure and Function of Cytochromes" (K. Okunuki, M. D. Kamen, and I. Sekuzu, eds.), p. 443. Univ. of Tokyo Press, Tokyo, 1968.
28. W. R. Frisell and C. G. Mackenzie, *J. Biol. Chem.* **237**, 94 (1962).
29. M. Brühmüller, H. Möhler, and K. Decker. *Z. Naturforsch.* **27b**, 1073 (1972).
30. R. A. Neal, *J. Biol. Chem.* **245**, 2599 (1970).
31. J. C. Schabort and D. J. J. Potgieter, *Biochim. Biophys. Acta* **250**, 329 (1971).
32. W. C. Kenney, D. E. Edmondson, and T. P. Singer, *Biochem. Biophys. Res. Commun.* **57**, 106 (1974).
33. W. C. Kenney, D. E. Edmondson, and T. P. Singer, *FEBS (Fed. Eur. Biochem. Soc.) Lett.* **41**, 111 (1974).
34. H. Möhler, M. Brühmüller, and K. Decker, *Eur. J. Biochem.* **29**, 152 (1972).
35. D. R. Patek and W. R. Frisell, *Arch. Biochem. Biophys.* **150**, 347 (1972).

36. D. R. Patek, C. R. Dahl, and W. R. Frisell, *Biochem. Biophys. Res. Commun.* **46**, 885 (1972).
37. E. Costa and M. Sandler, "Monoamine Oxidases—New Vistas." Raven, New York, 1972.
38. M. Harada, K. Mizutani, and T. Nagatsu, *J. Neurochem.* **18**, 559 (1971).
39. K. F. Tipton, *Biochim. Biophys. Acta* **159**, 451 (1968).
40. E. B. Kearney, J. I. Salach, W. H. Walker, R. Seng, and T. P. Singer, *Biochem. Biophys. Res. Commun.* **42**, 490 (1971).
41. E. B. Kearney, J. I. Salach, W. H. Walker, R. L. Seng, W. Kenney, E. Zeszotek, and T. P. Singer, *Eur. J. Biochem.* **24**, 321 (1971).
42. W. H. Walker, E. B. Kearney, R. L. Seng, and T. P. Singer, *Eur. J. Biochem.* **24**, 328 (1971).
43. W. H. Walker, E. B. Kearney, R. Seng, and T. P. Singer. *Biochem. Biophys. Res. Commun.* **44**, 287 (1971).
44. S. Ghisla and P. Hemmerich, *FEBS (Fed. Eur. Biochem. Soc.) Lett.* **16**, 229 (1971).
45. W. C. Kenney and W. H. Walker, *FEBS (Fed. Eur. Biochem. Soc.) Lett.* **20**, 297 (1972).
46. D. E. Edmondson and T. P. Singer. *J. Biol. Chem.* **284**, 8144 (1973).
47. F. Müller and V. Massey, *J. Biol. Chem.* **244**, 4007 (1969).
48. R. Hendriks and J. R. Cronin, *Biochem. Biophys. Res. Commun.* **44**, 313 (1971).
48a. W. H. Walker, W. C. Kenney, D. E. Edmondson, T. P. Singer, J. R. Cronin, and R. Hendriks, *Eur. J. Biochem.* **48**, 439 (1974).
49. T. E. Meyer, R. G. Bartsch, M. A. Cusanovich, and J. H. Mathewson, *Biochim. Biophys. Acta* **153**, 854 (1968).
50. R. Hendriks, J. R. Cronin, W. H. Walker, and T. P. Singer. *Biochem. Biophys. Res. Commun.* **46**, 1262 (1972).
51. W. C. Kenney, W. H. Walker, E. B. Kearney, R. Seng, T. P. Singer, J. R. Cronin, and R. Hendriks, *Z. Naturforsch.* **27b**, 1069 (1972).
52. W. C. Kenney, D. Edmondson, R., Seng, and T. P. Singer, *Biochem. Biophys. Res. Commun.* **52**, 434 (1973).
52a. W. C. Kenney, D. E. Edmondson, and T. P. Singer, *Eur. J. Biochem.* **48**, 449 (1974).
52b. D. E. Edmondson, *Biochemistry* **13**, 2817 (1974).
53. D. E. Edmondson and G. Tollin, *Biochemistry* **10**, 113 (1971).
54. J. A. D'Anna and G. Tollin, *Biochemistry* **11**, 1073 (1972).
55. R. D. Anderson, P. A. Apgar, R. M. Burnett, G. D. Darling, M. E. LeQuesne, S. G. Mayhew, and M. L. Ludwig, *Proc. Nat. Acad. Sci. U.S.* **69**, 3189 (1972).
56. K. D. Watenpaugh, L. C. Sieker, L. H. Jensen, J. LeGall, and M. Dubourdieu, *Proc. Nat. Acad. Sci. U.S.* **69**, 3185 (1972).
57. S. Grossman, J. Goldenberg, E. B. Kearney, G. Oestreicher, and T. P. Singer, *in* "Flavins and Flavoproteins" (T. P. Singer, ed.). ASP Publishers, Amsterdam 1975 (in press).
58. S. Grossman, J. Gobley, P. K. Hogue, E. B. Kearney, and T. P. Singer, *Arch. Biochem. Biophys.* **158**, 744 (1973).
59. T. P. Singer, E. Rocca, and E. B. Kearney, *in* "Flavins and Flavoproteins" (E. C. Slater, ed.), p. 391. Elsevier, Amsterdam, 1966.
60. Y. S. Kim and J. P. Lambooy, *Arch. Biochem. Biophys.* **122**, 644 (1967).
61. J. J. Dombrowski and J. P. Lambooy, *Arch. Biochem. Biophys.* **159**, 378 (1973).

CHAPTER 7

Biochemistry of the Sulfur Cycle

Lewis M. Siegel

I. THE SULFUR CYCLE IN NATURE

Sulfur exists in multiple oxidation states in nature, ranging from 6+ in H_2SO_4 and its derivatives ("hexavalent" or "oxidized" sulfur) to 2− in H_2S and its derivatives ("divalent" or "reduced" sulfur). Reduced sulfur appears in organic compounds essential to all organisms as constituents of protein (cysteine, methionine), coenzymes (e.g., coenzyme A, lipoic acid, biotin, thiamine pyrophosphate), and major cellular metabolites (e.g., glutathione). Oxidized sulfur, in the form of sulfate esters, is also prevalent in a number of organisms. Although in some cases (e.g., the choline O-sulfate formed by filamentous fungi) sulfate esters may serve no physiological role other than sulfur storage, in animals and some algae, at least, sulfate esters can serve important structural functions, as in the polysaccharide sulfates of connective tissue and algal cell walls or the sulfatides of the nervous system. The pools of oxidized and reduced sulfur in the biosphere can be interconverted by metabolic processes which, taken together, constitute the "sulfur cycle" outlined in Fig. 1. No single organism, of course, can catalyze all the reactions shown.

Animals can utilize ingested reduced sulfur compounds to satisfy their requirements for both divalent and hexavalent sulfur, since animals, like plants and many microorganisms, possess the capacity to oxidize organic thiols, such as cysteine, to the level of sulfate. Although animals and some heterotrophic microorganisms can oxidize inorganic sulfide to sulfate, this pathway is probably of minor quantitative importance in these organisms when compared to the oxidation of cysteine itself.

Animals are unable to reduce hexavalent sulfur compounds to the level of thiols. They are also unable, generally, to synthesize the carbon skeletons of methionine and the sulfur containing coenzymes and are thus dependent on a continuous supply of reduced sulfur in organic linkage in the diet both for growth and to replace thiol sulfur which has undergone oxidation. Animals obtain the necessary organic thiols by eating other animals or, ultimately, plants and microorganisms that are capable

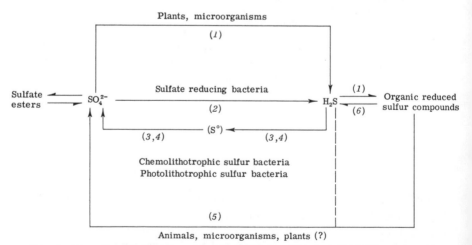

FIG. 1. An outline of the sulfur cycle. (*1*) Sulfate assimilation; (*2*) dissimilatory sulfate reduction; (*3*) chemolithotrophic oxidation of reduced sulfur compounds; (*4*) photolithotrophic oxidation of reduced sulfur compounds; (*5*) oxidation of thiols to sulfate; and (*6*) desulfuration of organic thiols.

of reducing hexavalent sulfur and incorporating the resulting H_2S into organic linkage. Since the major form of sulfur available to plants and many microorganisms is sulfate, the process of organic thiol biosynthesis by these organisms is frequently termed "sulfate assimilation." The rate of sulfate reduction is regulated by the need of these organisms for organic thiols, so that neither H_2S nor the thiols themselves are produced in excess and excreted into the environment.

During the process of microbial decomposition of plant and animal bodies, organic thiols are metabolized and the sulfur largely liberated in the form of H_2S (hence the term "putrefaction"). H_2S is also formed in relatively large amounts through the action of a small group of "sulfate reducing bacteria" of the genera *Desulfovibrio* and *Desulfotomaculum*. These obligately anaerobic bacteria oxidize organic compounds (or molecular hydrogen, which may be produced during the oxidation of organic compounds by other anaerobic bacteria) by using sulfate (or, more precisely, metabolic derivatives of sulfate) as the terminal electron acceptor in much the same way that aerobic organisms utilize O_2. Since sulfate reduction is linked to energy metabolism in the sulfate reducing bacteria, only a small proportion of the H_2S produced can be utilized for synthesis of cysteine and other cellular materials, and most of the sulfide is excreted into the surroundings. This process of respiratory, or "dissimilatory" sulfate reduction is particularly apparent in the mud at the bottom of ponds and streams, in bogs, marine sediments, and waterlogged soils.

Most of the H_2S produced in the biosphere is oxidized to sulfate, although some is sequestered for long periods of time in the form of insoluble sulfides or elemental sulfur. The latter material can readily form by spontaneous oxidation of H_2S in air, or as a result of microbial activity. The biological oxidation of H_2S and elemental sulfur is largely brought about by the action of lithotrophic bacteria. These "sulfur oxidizing bacteria" fall into two major groups, both of which are largely autotrophic, i.e., able to utilize CO_2 as sole carbon source: (a) the colorless chemosynthetic sulfur oxidizers, generally aerobes, which utilize electron flow from reduced sulfur compounds to O_2 (or to NO_3^- for the anaerobic strains) coupled to respiratory chain phosphorylation to generate the ATP and reducing power for CO_2 assimilation; and (b) the photosynthetic purple and green sulfur bacteria, which utilize reduced sulfur compounds as reducing agents for CO_2 assimilation only in the presence of light and the absence of O_2 (the ATP required for CO_2 assimilation is believed to be produced by cyclic photophosphorylation). Taken together, the sulfur oxidizing bacteria occupy a wide variety of aquatic and terrestrial environments. The colorless sulfur oxidizers, which

belong to the families Thiobacteriaceae, Beggiatoaceae, and Achroma-
taceae, may be found, for example, in sewage, sulfur springs, aerated
soils, industrial effluents and mine waste waters, and other generally
aerobic habitats exposed to H_2S or elemental sulfur. The photosynthetic
sulfur bacteria, which belong to the families Chromatiaceae and
Chlorobiaceae, are commonly found in mud and relatively stagnant
waters containing H_2S and exposed to light.

Since the sulfate reducing bacteria can obtain energy from the oxidation of organic compounds using sulfate as electron acceptor, while the
sulfur oxidizing bacteria can utilize H_2S as an electron donor for CO_2
fixation, an environment containing both types of organism together with
conditions suitable for mutual growth can generate a continuous cycle of
sulfur reduction and oxidation within that environment. Such a situation
occurs in nature and is termed a "sulfuretum"; the Dead Sea is a sulfuretum. A more typical example, described by Postgate [1], may be a
pond or polluted stream containing decaying organic matter, e.g., an
abundant fall of leaves in the autumn. The oxidation of this matter by
bacteria renders the lower depths of the pond anaerobic, while the surface remains aerobic. Sulfate reducing bacteria can flourish in the lower
portions of the pond, utilizing the organic matter as electron donor for
production of H_2S from the dissolved sulfate. Chemosynthetic sulfur oxidizing bacteria are active near the surface of the pond, oxidizing the
H_2S produced by the sulfate reducers and by microbial desulfuration of
the organic material. Photosynthetic sulfur bacteria are found in levels of
the anaerobic zone where light can still penetrate, oxidizing H_2S to
sulfur and sulfate.

The sulfur cycle of the sulfuretum, then, represents in principle a situation analogous to the O_2/H_2O cycle utilized by aerobically respiring
organisms and green plants. A great deal of interest in the microbiological sulfur cycle derives from the possibility that sulfate reduction and
photolithotrophic sulfur oxidation represent primitive forms of respiration and photosynthesis which developed on an anaerobic planet prior to
the evolution of the H_2O-splitting reaction of green plant photosynthesis
[2–4]. Evidence on the distribution of stable sulfur isotopes in natural
sulfur deposits dated by independent means suggests that biological sulfate reduction may have been occurring as early as 3.5 billion years ago
[5], which is soon after the probable time of emergence of life on the
earth and almost certainly antedates the appearance of O_2 in the atmosphere (the latter event is thought to have occurred about 2 billion years
ago). Peck [6] has presented data on the sequences of bacterial ferredoxins which suggest that the photosynthetic sulfur oxidizers may
even have preceded the sulfate reducers, both forms being antedated by

anaerobes that obtained energy solely by fermentation and accompanying substrate-level phosphorylations.

Detailed reviews of the biochemistry of the sulfur cycle covering the literature up to 1961 and 1968, respectively, have been presented by Peck [7] and Roy and Trudinger [8,9]. Reviews emphasizing the biology of the cycle [1] and its impact on evolutionary theory [2,4,6] have also appeared. Other recent reviews have treated specialized aspects of the sulfur cycle, including sulfate assimilation [10], dissimilatory sulfate reduction [11], chemolithotrophic sulfur oxidation [12,13], photolithotrophic sulfur oxidation [14], and cysteine oxidation in higher organisms [15].

II. ASSIMILATORY SULFATE REDUCTION

A. Sulfate Activation

The initial step in sulfate metabolism in all organisms is its reaction with ATP to form adenosine 5′-phosphosulfate (APS), catalyzed by the enzyme ATP-sulfurylase (EC 2.7.7.4; ATP : sulfate adenylyltransferase):

$$ATP^{4-} + SO_4^{2-} + H^+ \xrightleftharpoons{Mg^{2+}} APS^{2-} + PP_i^{3-}$$

From the equilibrium data of Robbins and Lipmann [16] and Akagi and Campbell [17], a $\Delta G_0'$ of +9.5 kcal/mole can be calculated for this reaction in the direction of APS formation at 5 mM Mg^{2+}, 25°, and pH 7. (Similar binding of Mg^{2+} by ATP and PP_i has been assumed. Curiously, no systematic study of the equilibrium position of this reaction as a function of [Mg^{2+}] or pH has been reported, in spite of the fact that magnesium complexes of both ATP and PP_i are clearly involved in the enzymatic reaction mechanism [18].) Since the $\Delta G_0'$ for hydrolysis of ATP to AMP and PP_i under these conditions is −10 kcal/mole [19], the $\Delta G_0'$ for hydrolysis of the phosphosulfate anhydride bond must be approximately −19.5 kcal/mole:

$$APS^{2-} + H_2O \rightleftharpoons AMP^{2-} + SO_4^{2-} + 2\,H^+$$

Adenylyl sulfates (a term which we will use to include both APS and its phosphorylated derivative PAPS) thus contain sulfate at a high group potential suitable to drive either (a) formation of sulfate esters, or (b) reduction of the sulfate moiety to the level of sulfite. The importance of sulfate "activation" in the latter process derives from the fact that physiologically important reducing agents, e.g., NADPH ($E_0' = -0.32$ V) are too positive in E_0' to reduce inorganic sulfate itself. Thus, while the E_0'

for the reaction

$$SO_4^{2-} + 2\ H^+ + 2\ e^- \rightleftharpoons SO_3^{2-} + H_2O$$

is -0.48 V[20], the E_0' for the analogous reduction of adenylyl sulfate

$$APS^{2-} + 2\ e^- \rightleftharpoons SO_3^{2-} + AMP^{2-}$$

is calculated to be -0.06 V, the reduction reaction being "pulled" by the free energy released on cleavage of the phosphosulfate anhydride bond. Reduction of an adenylyl sulfate by NADPH, then, would be an exergonic reaction, whereas reduction of sulfate itself by NADPH would be endergonic.

The subject of sulfate activation in animals, where the sole metabolic fate of sulfate is ester formation, is treated elsewhere in this volume, and we shall restrict the present discussion to the process of sulfate activation in organisms capable of reducing sulfate. It should be kept in mind, however, that a number of organisms utilize adenylyl sulfates for both reduction and sulfate ester formation: e.g., in the filamentous fungi and some algae a great deal of sulfate can be stored as choline O-sulfate [21–25], while some algae can produce sulfated polysaccharides [26].

ATP-sulfurylase has been highly purified from a number of sulfate assimilating organisms, including bacteria (*Nitrobacter agilis* [27]), fungi (yeast [16] and *Penicillium chrysogenum* [28]), and higher plants (spinach [29]). The *Penicillium* enzyme, the only one for which there is substantial evidence for homogeneity, has a molecular weight of 425,000, and is composed of eight subunits of molecular weight 56,000 [28]. Steady state kinetic studies with the *Penicillium,* yeast, and spinach enzymes [18,28,30] have led to the conclusion that catalysis by ATP-sulfurylase involves the simultaneous presence of ATP and sulfate on the enzyme prior to phosphosulfate anhydride formation. Thus, the $^{32}PP_i$-ATP exchange catalyzed by ATP-sulfurylase is completely dependent on the presence of SO_4^{2-} (or SeO_4^{2-}) [18,28–31]. The rate of $^{32}PP_i$-ATP exchange is consistent with the rate of reaction of APS with PP_i to give ATP.

ATP-sulfurylase activity is dependent on the presence of Mg^{2+} or other divalent cations [28–32]. In the presence of Mg^{2+}, the actual substrate is $MgATP^{2-}$, with free ATP serving as a competitive inhibitor [18,28,30]. In contrast to the narrow substrate specificity of ATP-sulfurylases for nucleotide phosphoryl donor (only dATP can substitute for ATP [17,29,33,34]), a number of Group VI anions, including SeO_4^{2-}, CrO_4^{2-}, WO_4^{2-}, MoO_4^{2-}, and SO_3^{2-}, can replace SO_4^{2-} in stimulating the conversion of ATP to AMP and PP_i [17,27–31,34]. With SO_4^{2-} and SeO_4^{2-}, the enzyme catalyzes $^{32}PP_i$-ATP exchange, indicating formation of relatively stable adenosine phosphosulfate and phosphoselenate compounds. Wilson and Bandurski [34] detected APSe in small amounts upon incubating Mg^{2+}, ATP, and SeO_4^{2-} with ATP-sulfurylase. They suggested, however, that this compound is very unstable in the presence of the enzyme, and Shaw and Anderson [31] were unable to detect APSe at all, even when using ATP-sulfurylase derived from plants that can accumulate selenium containing amino acids if grown in the presence of SeO_4^{2-}. As might be expected if an unstable APSe were formed, SeO_4^{2-}, in contrast to SO_4^{2-}, catalyzes the virtually complete hydrolysis of ATP to AMP and PP_i in the presence of ATP-sulfurylase. (CrO_4^{2-}, WO_4^{2-}, MoO_4^{2-}, and SO_3^{2-} also catalyze ATP hydrolysis with the enzyme; however, with these anions, there is no $^{32}PP_i$-ATP exchange, and no adenylate anhydride can be detected.) Alternatively, it is possible that anion binding to the enzyme–ATP complex triggers ATP hydrolysis, and the resulting enzyme-bound AMP is attacked by either an anion suitable for anhydride formation or by H_2O. As would be expected by either hypothesis, Group VI anions other than SO_4^{2-} markedly inhibit APS formation from SO_4^{2-} [18,27,34]. It should be noted that $S_2O_3^{2-}$, the thiol analogue of SO_4^{2-}, does not serve as either a substrate or an inhibitor for ATP-sulfurylase.

If adenylyl sulfate is to serve as a substrate for enzymes catalyzing transfer or reduction of the sulfate moiety, it must accumulate within the cell despite the unfavorable equilibrium constant for the ATP-sulfurylase reaction. A general means of pulling APS formation is to hydrolyze the PP_i formed in the ATP-sulfurylase reaction to inorganic phosphate via the action of inorganic pyrophosphatase (pyrophosphate phosphohydrolase, EC 3.6.1.1):

$$PP_i^{3-} + H_2O \xrightleftharpoons{Mg^{2+}} 2\ P_i^{2-} + H^+$$

The $\Delta G_0'$ for this reaction is -4.5 kcal/mole in the presence of excess Mg^{2+} at pH 7 [35]; the $\Delta G_0'$ becomes more negative as the Mg^{2+} concentration is reduced since this cation complexes preferentially to the PP_i rather than the P_i anion. Using the $\Delta G_0'$ values for the ATP-sulfurylase and pyrophosphatase reactions, one can calculate that the equilibrium constant for APS formation from ATP and SO_4^{2-} in the presence of both enzymes and excess Mg^{2+} should be approximately $10^{-4}\ M$ at 25° and pH 7. Thus, if ATP is present at $10^{-4}\ M$, while SO_4^{2-} and P_i are each present at $10^{-3}\ M$, accumulation of APS to the level of approximately $10^{-5}\ M$ could be expected at equilibrium. This equilibrium value might be slow to achieve, however, since ATP-sulfurylases are subject to strong product inhibition by APS [28,31].

A method for further increasing the intracellular concentration of adenylyl sulfate involves the phosphorylation of APS at the 3′ position of its ribose moiety to form 3′-phosphoadenosine 5′-phosphosulfate (PAPS):

$$ATP^{4-} + APS^{2-} \overset{Mg^{2+}}{\rightleftharpoons} PAPS + ADP^{3-} + H^+$$

By analogy with the ATP-dependent phosphorylation of other sugars, this reaction may be expected to proceed with a $\Delta G_0'$ of -3 to -5 kcal/mole; the actual equilibrium constant has not been measured. APS-kinase (EC 2.7.1.2.5; ATP:adenylylsulfate 3′-phosphotransferase) activity has been detected in sulfate assimilating bacteria [36–39], fungi [33,40,41], algae [42–44], and higher plants [45–47]. The enzyme has been partially purified from yeast and *N. agilis* [33,39]. The *Nitrobacter* APS-kinase activity was not separated from PAPS reductase following a 100-fold purification, and complex formation between these enzymes in this organism has been suggested; the molecular weight of the "complex," by gel filtration, is reported to be 280,000 [39].

Yeast APS-kinase requires divalent cations for activity; Mg^{2+}, Mn^{2+}, and Co^{2+} are all effective [32]. APSe may be able to substitute for APS as a substrate for this enzyme, since Wilson and Bandurski [34] found that addition of APS-kinase to a reaction mixture containing ATP, SeO_4^{2-}, Mg^{2+}, ATP-sulfurylase, and pyrophosphatase stimulated release of P_i from ATP. No PAPSe could be detected, however. Since a high affinity of APS-kinase for APS would be important if the enzyme is to serve a physiological role in pulling accumulation of adenylyl sulfate, kinetic studies of this enzyme are of interest. The *Nitrobacter* APS-kinase–PAPS reductase complex has a relatively high K_m for APS, $2 \times 10^{-4}\ M$ [39]. The reaction measured, however, in determining this K_m value was the overall reduction of APS by NADPH to form SO_3^{2-}, and APS-kinase may not have been limiting at all times. Yeast APS-kinase exhibited maximal activity at the lowest APS concentration tested, $5 \times 10^{-6}\ M$; higher concentrations were inhibitory [32]. The true affinity of any APS-kinase for its substrate, therefore, remains to be determined.

In view of the fact that PAPS can accumulate at concentrations approaching millimolar in extracts containing ATP-sulfurylase, pyrophosphatase, and APS-kinase [38,48], it is not surprising that a number of organisms, including sulfate assimilating bacteria and fungi (but not algae or plants) apparently utilize PAPS as the actual substrate for sulfate reduction [36,38,41,49]. Thus, mutants of *Salmonella typhimurium* deficient in APS-kinase activity are unable to grow on SO_4^{2-} as sulfur source, and are also unable to reduce either SO_4^{2-} or APS to H_2S *in vitro* [38,49]. APS-kinase is not present, however, in bacteria which utilize sulfate as a terminal electron acceptor for anaerobic respiration [36], the action of pyrophosphatase alone being apparently sufficient to pull APS formation in quantities suitable for its reduction.

B. Reduction of Adenylyl Sulfate

1. The APS Pathway: Algae

Extracts of *Chlorella* contain two enzymes, termed APS:thiol sulfotransferase [50–52] and thiosulfonate reductase [53], respectively, which acting together can catalyze reduction of the sulfonate moiety of APS to the level of thiol with reduced ferredoxin (Fd_{red}) as electron donor in the presence of a carrier thiol (*Car*SH):

$$APS^{2-} + CarSH \xrightarrow{\text{APS:thiol sulfotransferase}} AMP^{2-} + CarS-SO_3^- + H^+$$

$$6 \ Fd_{red} + 7 \ H^+ + CarS-SO_3^- \xrightarrow{\text{thiosulfonate reductase}} 6 \ Fd_{ox} + 3 \ H_2O + CarS-SH$$

GSH can serve as the carrier thiol for these reactions, and the formation of GS—SO_3^- and GS—SH has been demonstrated with *Chlorella* extracts [50–54]. It is unlikely that GSH is the physiological carrier; Schmidt [53] and Abrams and Schiff [52] have recently identified a heat stable factor, which migrates similarly to bacitracin (MW 1400) upon gel filtration on Sephadex G-50, in *Chlorella* extracts that may serve this role. In the presence of *O*-acetylserine, a reduced ferredoxin generating system, and carrier, *Chlorella* extracts can catalyze incorporation of the sulfur atom of $AP^{35}S$ into [^{35}S]cysteine [53,55]. It is not clear whether free $H_2^{35}S$ is formed during this process (via reduction of the *Car*S—^{35}SH) or whether the carrier-bound reduced ^{35}S is first transferred to the serine carbon skeleton (e.g., with intermediate formation of *Car*S—^{35}S—$CH_2CH(NH_2)COOH$) and the resultant disulfide then reduced. Although *Chlorella* extracts contain an enzyme (sulfite reductase) that catalyzes the Fd_{red}-dependent reduction of inorganic SO_3^{2-} to H_2S [54,55], a *Chlorella* mutant strain that lacks thiosulfonate reductase activity but retains sulfite reductase activity is unable to grow with SO_4^{2-} as sulfur source (although it will grow with thiol containing sulfur compounds) or to incorporate the sulfur atom of $AP^{35}S$ into [^{35}S]cysteine *in vitro* [55]. Similarly, *Chlorella* mutants which specifically lack the APS:thiol sulfotransferase enzyme can neither assimilate SO_4^{2-} *in vivo* nor reduce it *in vitro* [52,56]. When extracts of the two types of mutant are mixed, a complete sulfate reducing system can be reconstituted *in vitro* [55].

The *Chlorella* APS:thiol sulfotransferase has been highly purified [51]. It has a molecular weight, by gel filtration, of approximately 300,000 [51,57]. The enzyme is specific for APS (vs. PAPS), but rather nonspecific for thiol acceptor [51,52,58,59]. With monothiols such as GSH, the reaction product is the S-sulfonate of the thiol, although at high concentrations of thiol, some free SO_3^{2-} is formed [52,59], probably through displacement of the thiol-bound sulfonate moiety by a second molecule of thiol:

$$APS^{2-} + GSH \rightarrow AMP^{2-} + GS—SO_3^- + H^+$$

$$GS—SO_3^- + GSH \rightleftharpoons GSSG + H^+ + SO_3^{2-}$$

The latter reaction, if carrier is a monothiol, might explain the need for sulfite reductase in *Chlorella*. With dithiol compounds capable of internal disulfide bond formation, such as dihydrolipoamide or dithiothreitol, the reaction product is almost exclusively SO_3^{2-} [52,59]:

$$APS^{2-} + R\overset{SH}{\underset{SH}{\big\langle}} \longrightarrow R\overset{SH}{\underset{S—SO_3^-}{\big\langle}} + AMP^{2-} + H^+$$

$$R\overset{SH}{\underset{S—SO_3^-}{\big\langle}} \rightleftharpoons R\overset{S}{\underset{S}{\big\langle}}| + H^+ + SO_3^{2-}$$

Interestingly, when 2,3-dimercaptopropan-1-ol (BAL) serves as thiol acceptor, the reaction product, in addition to AMP, is $S_2O_3^{2-}$ [52,59]. The sulfonate moiety of the thiosulfate is derived exclusively from APS, while the sulfane moiety derives entirely from BAL. Since the internal disulfide of BAL produced by the displacement mechanism envisioned for the S-sulfonates of dithiothreitol or dihydrolipoamide would be quite strained sterically, it has been suggested [59] that the reaction instead involves formation of an episulfide in which one of the thiol groups of BAL is lost together with the sulfonate moiety.

Although the *Chlorella* thiosulfonate reductase has been highly purified by Schmidt [53], little has been reported concerning its properties. The molecular weight of the enzyme, by gel filtration, is reported to be greater than 200,000 [53]. With Fd_{red} and GS—$^{35}SO_3^-$ as substrates, the reaction product was identified by chromatography as GS—^{35}SH rather than $H_2^{35}S$ [53,54]. As expected if a persulfide (but not free H_2S) is formed, incubation of GS—SO_3^- with Fd_{red} and enzyme in the presence of CN^- results in the formation of SCN^-. Although Schmidt and Schwenn [54] have stated that thiosulfonate reductase is inactive with inorganic sulfite as substrate, it is curious that the purified *Chlorella* enzyme can catalyze production of H_2S from $S_2O_4^{2-}$ in the presence of the electron-carrying dye methyl viologen (MV) [53]. This reaction is characteristic of purified sulfite reductase [60], as might be anticipated, since the products of chemical reaction between $S_2O_4^{2-}$ and MV should be SO_3^{2-} and reduced methyl viologen (MVH). The "self reduction" of dithionite by sulfite reductase is in reality a reduction of SO_3^{2-} by electrons derived from dithionite. It is evident that the relationship, if any, between sulfite reductase and thiosulfonate reductase must be more fully explored.

The pathway of sulfate reduction in spinach chloroplasts may be identical to that in *Chlorella,* since Schmidt and Schwenn [54] have shown that the radioactivity of $^{35}SO_4^{2-}$ becomes incorporated into compounds of the structure RS—$^{35}SO_3^-$ and RS—^{35}SH, where RSH represents a heat stable protein fraction of molecular weight 5000 by gel filtration, during incubation of intact chloroplasts with $^{35}SO_4^{2-}$ in the light. The kinetics of appearance and disappearance of radioactivity in these compounds was consistent with a role for them as intermediates in sulfate assimilation. Sulfate reduction *in vitro* by spinach chloroplasts was markedly stimulated by addition of GSH [46,50], and the formation of GS—SO_3^- from APS and GSH in the presence of such extracts has been reported [50].*

Although APS appears to be the immediate substrate for sulfate reduction in *Chlorella* and, possibly, in spinach as well, both of these organisms possess APS-kinase and form PAPS from SO_4^{2-} and ATP. Isotope dilution experiments show that in order for PAPS to be reduced by these organisms it must first be converted to APS [50,51]. This conversion is catalyzed by a Mg^{2+}-dependent 3'-nucleotidase (EC 3.1.3.30; 3'-phosphoadenylylsulfate 3'-phosphohydrolase) activity found in extracts of both *Chlorella* and spinach chloroplasts [47,51,57]. The presence of this enzyme (which has also been reported in bacteria [37]) may explain the relative difficulty in observing PAPS formation from SO_4^{2-} in many higher plants under conditions in which formation of APS is readily detectable [46,47]. It has been suggested [52] that PAPS is used for sulfate ester formation in algae and higher plants, while APS is

---

* The presence of a pathway for sulfate assimilation into cysteine involving "free" SO_3^{2-} and H_2S in spinach and other higher plants is indicated by the fact that the enzymes sulfite reductase [61–66] serine acetyltransferase [67,68], and O-acetylserine sulfhydrylase [67,68] have all been found in plant tissue. The relationship of these enzymes to the bound sulfur intermediates found by Schmidt and Schwenn [54] in spinach chloroplasts is unclear.

used for sulfate reduction. If this is true, some form of regulation and/or compartmentalization of the 3'-nucleotidase is to be anticipated in these organisms.

2. THE PAPS PATHWAY: BACTERIA AND FUNGI

In enterobacteria and fungi, PAPS can be reduced to H_2S with NADPH as electron donor by the successive action of the enzymes PAPS reductase [38,49,69–71] and sulfite reductase (EC 1.8.1.2; $H_2S:NADP^+$ oxidoreductase) [38,49,71–78]:

$$PAPS^{4-} + NADPH \rightarrow PAP^{4-} + NADP^+ + SO_3^{2-} + H^+$$

$$SO_3^{2-} + 5 H^+ + 3 NADPH \rightarrow H_2S + 3 H_2O + 3 NADP^+$$

Both free $^{35}SO_3^{2-}$ and $H_2^{35}S$ have been detected in *Escherichia coli* cells during assimilation of $^{35}SO_4^{2-}$ *in vivo* [79]. Enterobacterial mutants specifically deficient in any one of the enzymes APS-kinase, PAPS-reductase, and sulfite reductase cannot reduce sulfate to the level of thiol either *in vivo* or *in vitro* [38,49]. Mutants deficient in sulfite reductase accumulate large quantities of inorganic SO_3^{2-} in the medium as the product of SO_4^{2-} reduction *in vivo* [49,79]. Since *S. typhimurium* mutant strains lacking APS-kinase are able to reduce PAPS, but not APS, to H_2S, while extracts of mutants lacking ATP-sulfurylase are able to reduce both APS and PAPS, it is clear that PAPS is the obligatory substrate for SO_4^{2-} reduction in this organism. Studies by Peck [80] on the ability of a number of microorganisms to reduce APS or PAPS to acid volatile sulfur compounds *in vitro* (with MVH as electron donor) led to the conclusion that PAPS is the only form of adenylyl sulfate which can be reduced by a wide variety of sulfate assimilating bacteria. The PAPS-reductase of yeast, however, is reportedly capable of using either PAPS or APS as substrate [70], although a higher turnover number was obtained with PAPS.

A number of thiols, including dihydrolipoate and BAL, can substitute for NADPH as reductant in the PAPS reducing systems of enterobacteria and yeast [58,70,81,82]. With dihydrolipoate as reductant, the sulfate moiety of PAPS is converted to SO_3^{2-} [58]. With BAL, the product is $S_2O_3^{2-}$ [82]. These results indicate that the process of NADPH-dependent PAPS reduction involves participation of a PAPS:thiol sulfotransferase enzyme analogous to the APS:thiol sulfotransferase of *Chlorella*. Indeed, Hodson and Schiff [82] have shown that a *S. typhimurium* mutant specifically blocked in NADPH-PAPS-reductase activity is also unable to form $S_2O_3^{2-}$ from BAL and PAPS *in vitro*.

Wilson *et al.* [70,83] have fractionated the NADPH-PAPS-reductase system of yeast into three protein components. One, "Fraction B," is presumably the PAPS:thiol sulfotransferase. The other two components, termed "Fraction C" and "Fraction A," are, respectively, a heat-stable protein, molecular weight approximately 10,000, which contains a reducible disulfide bond, and an enzyme which catalyzes reduction of the Fraction C disulfide with NADPH as reductant. Since the latter reaction is stimulated by added flavin, and Fraction A contains NADPH diaphorase activity, it is presumed that the NADPH–Fraction C(SS) reductase is a flavoprotein. Porque *et al.* [84] have recently found that yeast thioredoxin, a relatively heat stable disulfide-containing protein of molecular weight 12,000, and NADPH thioredoxin reductase (EC 1.6.4.5; NADPH:oxidized-thioredoxin oxidoreductase), a flavoprotein that catalyzes the reduction of thioredoxin disulfide to the dithiol with NADPH, can substitute for Fractions C and A, respectively, in permitting PAPS reduction to SO_3^{2-} by purified yeast Fraction B and NADPH. Thioredoxin, which functions catalytically in the NADPH-dependent reduction of ribonucleoside diphosphates to deoxyribonucleoside diphosphates in many organisms, was shown to function catalytically in the yeast NADPH-PAPS-reductase system [84]. These results suggest the following reaction mechanism for PAPS reduction by NADPH:

$$\text{Thioredoxin}\begin{matrix}S\\|\\S\end{matrix} + NADPH + H^+ \xrightarrow{\text{thioredoxin reductase (A)}} NADP^+ + \text{thioredoxin}\begin{matrix}SH\\\\SH\end{matrix}$$

$$\text{Thioredoxin}\begin{matrix}SH\\\\SH\end{matrix} + PAPS^{4-} \xrightarrow{\text{PAPS:thiol sulfotransferase (B)}} PAP^{4-} + \text{thioredoxin}\begin{matrix}SH\\\\S\;SO_3^-\end{matrix} + H$$

$$\text{Thioredoxin}\begin{matrix}SH\\\\S-SO_3\end{matrix} \rightleftharpoons SO_3^{2-} + H^+ + \text{thioredoxin}\begin{matrix}S\\|\\S\end{matrix}$$

Torii and Bandurski [85] have demonstrated that the sulfonate moiety of PAP^{35}S can be found in linkage with a heat stable protein during the process of NADPH-dependent PAPS reduction in yeast extracts. Since the ^{35}S is recovered as $^{35}SO_3^{2-}$ following incubation of the radioactive protein with nonradioactive sulfite, presumably via exchange, it has been suggested that the protein-bound ^{35}S is in RS—$^{35}SO_3^-$ linkage. It has not been shown that the protein containing this "bound sulfite" is identical to either Fraction C or thioredoxin. In view of the broad specificity of (P)APS:thiol sulfotransferase enzymes for thiol acceptors [59], it is not certain at this time that either thioredoxin or Fraction C (which may or may not be identical to thioredoxin) is the physiological thiol for PAPS reduction. *In vivo* involvement of thioredoxin in sulfate assimilation is amenable to test, however, since such involvement predicts that a single step mutation should lead to a simultaneous growth requirement for both reduced sulfur compound and deoxyribonucleotides.

C. Sulfite Reduction

Enzymes that catalyze the stoichiometric reduction of SO_3^{2-} to H_2S in the presence of an appropriate electron donor have been highly purified from bacteria [60,86], fungi [87,88], algae [61], and higher plants [62,63]. In all cases examined the complete six electron reduction reaction (a) proceeds without release of sulfur compounds intermediate in oxidation state between SO_3^{2-} and H_2S [89] and (b) is catalyzed by a protein which purifies as a single macromolecular species. The apparent physiological electron donor for sulfite reduction in fungi and aerobic bacteria is NADPH [49,72,73,76–78]. In anaerobic bacteria and photosynthetic organisms, the physiological reductant is probably ferredoxin [65,66,90]. In higher plants, sulfite reductase activity is localized in the chloroplasts [66]. Whatever the physiological donor may be, all sulfite reductases are capable of utilizing the reduced form of MV, a dye which transfers single electrons at an E_0' of -0.4 V, as electron donor for stoichiometric reduction of sulfite to sulfide.

Highly purified sulfite reductases from sulfate assimilating organisms exhibit absorption spectra with a maximum in the 582–589 nm region [60–63,86–88]. An absorption band of greater intensity is also generally present in the 385–410 nm region. These maxima show marked shifts in position and intensity when sulfite reductase is treated with typical heme ligands, such as CO and CN^- [60,63,87,91]; these agents are potent inhibitors of sulfite reductase activity [61–64,73,75,86–89,92]. Electron paramagnetic resonance (EPR) spectra of the *E. coli* and *S. typhimurium* sulfite reductases are characteristic of a high spin ferriheme in a rhombically distorted environment [60,92], a result which confirms the presence of a heme component in these enzymes. The heme prosthetic group could be extracted with acetone-HCl [60,93]. The absorption spectra of the extracted heme and its complexes with ligands were unlike those of any heme compound previously described from natural sources [60].

Murphy *et al.* [94] have identified the *E. coli* sulfite reductase heme prosthetic group as an iron tetrahydroporphyrin of the isobacteriochlorin type (i.e., adjacent pyrrole rings are reduced) of molecular formula $FeC_{42}H_{44}N_4O_{16}$. The compound contains eight carboxylic acid groups, and the mass spectrometry cleavage pattern indicates that these groups

are associated with acetyl and propionyl side chains. Recent unpublished data from our laboratory have shown that the *E. coli* sulfite
reductase heme contains two methyl groups which can be derived *in
vivo* from the methyl group of methionine. The stability of the compound
and its derivatives to photooxidation suggests that the "reduced" pyrrole
rings are in fact substituted by some group other than hydrogen, and the
structure shown in Fig. 2, which represents a dimethyl derivative of
uroporphyrin III, has been postulated for the sulfite reductase heme
[94]. This structure can be derived, in principle, from uroporphyrinogen
III, the first tetrapyrrole intermediate in protoheme biosynthesis, by a
process involving only a single enzymatic step unique to the synthesis of
the sulfite reductase heme, i.e., methylation (twice) of the hexahydropor-
phyrin uroporphyrinogen. Subsequent oxidation of the dimethyl uroporphyrinogen to the tetrahydroporphyrin level and insertion of iron might
occur spontaneously or be catalyzed by enzymes operating in the normal
pathway of protoheme biosynthesis. The sulfite reductase heme, which
has been given the name "siroheme" by Murphy and Siegel [95], represents the first protein associated heme or chlorophyll that is not derived
metabolically from protoporphyrin IX. A possible relationship of siroheme to the corrinoids (B_{12}) has been pointed out [94]. Siroheme has
been identified as the prosthetic group of dissimilatory sulfite reductases
[95,96] as well as of enzymes of the assimilatory type. Siroheme is also
found in a second type of "multielectron" reductase, the nitrite reductases of spinach and *Neurospora* [97,98]. Both of these enzymes catalyze the stoichiometric reduction of NO_2^- to NH_3, a six electron transfer
reaction. It should be noted that all sulfite reductases except those

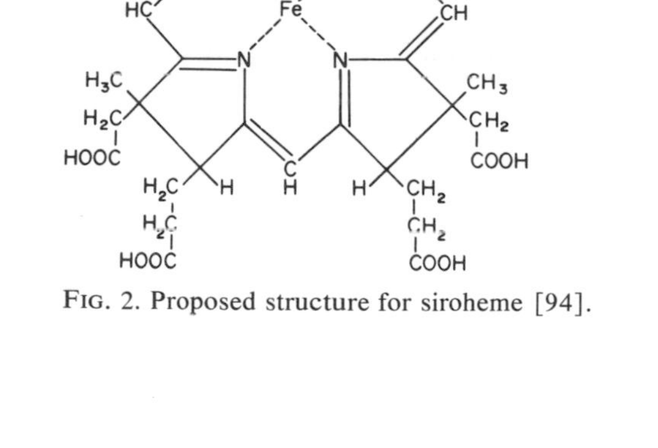

FIG. 2. Proposed structure for siroheme [94].

derived from photosynthetic organisms can also catalyze this reaction, although the physiological distinction between sulfite reductase and nitrite reductase *in vivo* has been amply demonstrated [75,97].

Available evidence indicates that the siroheme prosthetic group is essential for sulfite reduction and suggests, moreover, that siroheme may provide the site of interaction of sulfite reductase with sulfite itself. Thus, with the yeast and enterobacterial enzymes, agents that destroy the siroheme prosthetic group or that complex to it [89,99] cause losses in ability to reduce sulfite which parallel the extent of disappearance of free siroheme. Dissociation of CO from its complex with enzyme-bound siroheme, observed spectrally, is paralleled by reappearance of sulfite reductase activity [89]. Sulfite itself, when added to the *E. coli* or spinach enzymes in the presence of reducing agents, causes marked shifts in the absorption spectrum of the siroheme prosthetic group [63,89], and the EPR spectrum characteristic of the native *E. coli* enzyme disappears [89]. These spectral shifts are accompanied, in the *E. coli* sulfite reductase, by a tight physical binding of $^{35}SO_3^{2-}$ to the enzyme [89]. The bound radioactivity can be released from the enzyme as $H_2^{35}S$ upon addition of excess reductant (NADPH).

Inhibition of the MVH-sulfite reductase activity of the spinach enzyme by PCMB, and the protection afforded by SO_3^{2-} against this inhibition, led Asada [100] to postulate that a thiol group may provide the site of SO_3^{2-} binding to this enzyme. The spinach sulfite reductase, however, requires added thiol compounds (e.g., cysteine) for activity [63], and it is probable that the mercurial inhibition reflects this peculiar requirement, since neither requirement for added thiols nor inhibition by mercurials of MVH-sulfite reductase activity is characteristic of any other sulfite reductase tested, including the highly purified enzyme from garlic leaves [62,87,89].

The molecular size and prosthetic group composition of sulfite reductases are correlated with the type of electron donor which the enzyme is capable of using for sulfite reduction. Thus, sulfite reductases which can utilize MVH, but not NADPH, as reductant are comparatively small, with sedimentation coefficients of 4–6 S [61–63,88,101–103]; these enzymes, when highly purified, generally do not contain flavin. The spinach sulfite reductase, molecular weight 84,000 by gel filtration, is reported to contain no more than 1 gram-atom of Fe per mole of enzyme [63], and the intensity of the absorbance at the Soret maximum is consistent with all of this iron being present as siroheme. A sulfite reductase which can utilize MVH, but not NADPH, as reductant has also been prepared from the NADPH-sulfite reductase of *E. coli* following dissociation of that enzyme in 5 *M* urea [103]. The protein which catalyzes the MVH-sulfite reductase activity exists in solution as a monomeric single polypeptide chain (termed β) of molecular weight 54,000–57,000. Siroheme and nonheme iron/acid labile sulfide (Fe/S) are present as prosthetic groups, but flavin is absent. An Fe/S-containing hemoprotein with similar spectral properties has been purified as an MVH-sulfite reductase from extracts of yeast and *S. typhimurium* mutants which are unable to reduce sulfite with NADPH as electron donor [101,102]. It is of interest that two distinct cistrons (*cys G* and *cys I*) are required for formation of functional MVH-sulfite reductase activity in *S. typhimurium* (a third cistron, *cys J,* is required to form NADPH-sulfite reductase), despite the fact that the MVH-sulfite reductase activity appears to be dependent on only a single polypeptide chain [86,102,103]. Immunological studies have shown that the *cys I* gene codes for the β polypeptide, while the *cys G* gene codes for a component required for its activity. Siegel and Davis [103] have speculated that this component may be an enzyme required for siroheme biosynthesis.

Sulfite reductases which can utilize NADPH, in addition to MVH, as electron donor are soluble proteins of high molecular weight, with sedimentation coefficients of 13–18 S [60,86,87,104]. The yeast, *E. coli,* and *S. typhimurium* enzymes, which have been highly purified, contain FMN and FAD in addition to Fe/S and siroheme [60,86]. The structure of the enterobacterial sulfite reductase has been studied in some detail [86,103]; the enzyme has a molecular weight of 670,000 and contains, per molecule, 4 FMN, 4 FAD, 16 Fe/S, and 3–4 siroheme. This complex system of electron carriers is accommodated on 12 peptide chains, which appear to be of only two types, termed α and β. The 4 β chains bind all of the iron containing prosthetic groups of the enzyme. It is these chains, with their siroheme prosthetic groups, that are responsible for catalysis of sulfite reduction itself. The remaining 8 polypeptides, all of the α type, bind the FAD and FMN groups. The α octamer flavoprotein moiety catalyzes electron transfer between NADPH and the iron prosthetic groups of the enzyme; it can also catalyze NADPH-dependent reduction of cytochrome c, O_2, and a number of diaphorase acceptors. It has been prepared, in partially denatured form, by urea dissociation of the *E. coli* holoenzyme. A more fully functional α_8-flavoprotein, with NADPH diaphorase activity but no NADPH- or MVH-sulfite reductase activity, has been prepared in highly purified form from extracts of *S. typhimurium cys G* and *cys I* mutants [86]. This flavoprotein is immunologically cross-reactive with wild-type *S. typhimurium* and *E. coli* sulfite reductases and can combine with either the hemoprotein moiety of urea dissociated *E. coli* holoenzyme or with the 5 S hemoprotein component which catalyzes MVH-sulfite reductase activity in extracts of *S. typhimurium cys J* mutants to reconstitute a complete NADPH-sulfite reductase activity [86,102]. There is evidence that the NADPH-sulfite reductase of yeast is also composed of separable flavoprotein and hemoprotein components [99,101].

Studies with site-specific inhibitors [87,89] and with enzyme specifically freed of its FMN, but not its FAD prosthetic groups [99,105], have shown that the sequence of electron flow in the enterobacterial and yeast NADPH-sulfite reductases is

$$\text{NADPH} \longrightarrow \text{FAD} \longrightarrow \text{FMN} \cdots\cdots\rightarrow \text{siroheme} \longrightarrow \text{SO}_3{}^{2-}$$

diaphorase   MVH
acceptors

Although the Fe/S prosthetic groups of the *E. coli* enzyme, like the other enzyme prosthetic groups, are demonstrably reducible by NADPH [89], their role in the electron transfer sequence has not been established (hence the dashed arrow between FMN and siroheme in the electron transfer sequence). When the Fe/S centers are fully reduced, they can accommodate one electron per four Fe/S, suggesting the presence of Fe/S prosthetic groups similar to those present in bacterial rather than plant ferredoxins [106]. It should be noted that enterobacterial NADPH-sulfite reductase may be considered a tetramer of the basic electron transferring unit: 1 FAD, 1 FMN, 4 Fe/S, and 1 siroheme. Such a unit would be physically associated with 2 α and 1 β peptide chains, and is capable in principle of carrying at one time the entire six electron "package" needed for reduction of one molecule of $SO_3{}^{2-}$ to H_2S. The physiological utility of such an arrangement is not clear at present since the stoichiometric reduction of $SO_3{}^{2-}$ to H_2S with MVH as electron donor can be carried out by enzymes, such as that of spinach, which contain too few prosthetic groups to permit electron "storage." Sulfite reduction by these enzymes, then, must proceed through a series of discrete enzyme-bound intermediates; the nature of such intermediates is not known.

Siegel *et al.* [86,92,107], on the basis of rapid kinetic and titration studies with enterobacterial sulfite reductase and its flavoprotein moiety, have proposed that the FMN prosthetic group serves a role similar to that of bacterial flavodoxins [108], accepting electrons from the reduced FAD moiety and donating them at an E_0' more negative than that of NADPH to the iron containing groups of the enzyme. Like flavodoxin, the FMN of NADPH-sulfite reductase appears to undergo oxidation and reduction, during catalysis of electron transfer between NADPH and diaphorase acceptors (or O_2), only between the hydroquinone

(FMNH$_2$) and semiquinone (FMNH·) oxidation states. The FAD moiety, on the other hand, is postulated to accept an electron pair from NADPH, the resulting FADH$_2$ being oxidized sequentially by the FMNH· moiety to the semiquinone and fully oxidized states. Siegel et al. [92,106] have found that the Fe/S, siroheme, and at least a portion of the enzyme flavin prosthetic groups require considerable excesses of NADPH for anaerobic reduction, a result consistent with the hypothesis that the iron prosthetic groups, as well as the FMN, function at a redox potential more negative than that of NADPH. It is perhaps pertinent to note, in this regard, that only reductants of E_0' more negative than -0.3 V, i.e., NADPH, viologen dyes, and ferredoxin, can function as electron donors for assimilatory sulfite reduction [88]. Although the average E_0' for the SO_3^{2-}/H_2S redox couple is -0.12 V [20], it is probable that certain steps in the six electron transfer process require electrons at much more negative potential (e.g., addition of a single electron to sulfite converts it to SO_2^-, the dimer of which is the potent reducing agent $S_2O_4^{2-}$), and that the complex electron transport system of the NADPH-sulfite reductases is designed not to store electrons, but rather to deliver them to heme associated SO_3^{2-}, or reduction products thereof, at the appropriate negative E_0'.

D. Sulfate Transport

The active uptake of SO_4^{2-} from the medium against a concentration gradient has been observed in many microorganisms, including *S. typhimurium* [109], *Aspergillus* [110–112], *Penicillium* [111–113], *Neurospora* [114], and *Chlorella* [115,116]. In each case the uptake process is temperature and energy dependent and is inhibited by agents which block respiratory chain electron transport (e.g., CN$^-$, N$_3^-$) or phosphorylation (2,4-dinitrophenol). Interestingly, although dinitrophenol or azide instantaneously block sulfate uptake by *Aspergillus nidulans* or *P. chrysogenum* mycelia, arsenate, which may be expected to inhibit both substrate and respiratory chain phosphorylation, is totally noninhibitory. This result has led Bradfield et al. [111] to suggest that ATP, per se, may not be involved in sulfate transport in the fungi, although some high energy compound or state generated by electron transport is clearly required. Each of the SO_4^{2-} transport systems examined is inhibited by $S_2O_3^{2-}$, SO_3^{2-}, and a number of Group VI anions of general structure XO_4^{2-}, of which CrO_4^{2-} is generally the most potent inhibitor. SO_4^{2-} and $S_2O_3^{2-}$, its thiol analogue, are common substrates of the *Salmonella* and fungal transport systems [109,114]; thus, mutants defective in sulfate transport in these organisms cannot use thiosulfate (unless present at very high concentrations) as a sulfur source for growth. SeO_4^{2-}, MoO_4^{2-}, and CrO_4^{2-} are also taken up via the SO_4^{2-} transport system in fungi [112,117] and *Chlorella* [118].

Only in *S. typhimurium* has sulfate uptake been shown to be reversible, so that internal SO_4^{2-} will exchange with SO_4^{2-} in the medium [109]. The maximum intracellular sulfate concentration which can be achieved in this organism is equivalent to about 1×10^{-4} M. The fungal and *Chlorella* systems are reported not to be reversible, even when ATP-sulfurylase negative mutants are used [111,114,116]. In most organisms, only a single sulfate transport system has been demonstrated. However, in *Neurospora,* two separate sulfate transport systems, encoded for by unlinked genes, predominate at different stages of the life cycle [114,119]. Thus, "permease I," which has a high K_m for SO_4^{2-} (approximately 1×10^{-4} M), is present in conidia, while "permease II," which has a much lower K_m for SO_4^{2-} (8×10^{-6} M), predominates in mycelia. Both systems show similar sensitivity to inhibitors.

The process of sulfate uptake has been separated into two phases in *Salmonella* [120,121] and *Aspergillus* [110]. The initial phase, termed sulfate "binding," is (a) rapid compared to the overall rate of transport at low sulfate concentrations; (b) reversible; (c) energy independent (i.e., not inhibited by dinitrophenol); and (d) saturated at sulfate concentrations considerably below the K_m for transport. Sulfate binding is strongly inhibited by anions structurally similar to SO_4^{2-}, such as CrO_4^{2-}, and is subject to repression and derepression by the same agents that repress and derepress overall sulfate transport.

Since both SO_4^{2-} transport and binding activity are lost in *S. typhimurium* spheroplasts, Dreyfuss and Pardee [120] suggested that a sulfate binding substance was lost from the bacterium when its cell wall was removed. Indeed, a sulfate binding protein has been purified and crystallized from osmotically shocked cells of *S. typhimurium* [122–124]. This protein has a molecular weight of 32,000, and binds one sulfate per molecule with a dissociation constant on the order of 10^{-7} M, depending on the ionic strength. Its primary amino acid sequence has been determined [125]; interestingly, it contains no sulfur amino acids. Pardee and Watanabe [126] have shown that the sulfate binding protein can be localized to the periplasmic space (i.e., between the cell wall and protoplasmic membrane) of *Salmonella.*

The actual involvement of the sulfate binding protein in transport has yet to be demonstrated. The correlation between the levels of binding activity observed in different *Salmonella* strains under varying growth conditions and the overall rates of sulfate transport in those strains is quite poor [127]. However, transport has never been observed under conditions where no binding protein was detectable, and it is possible that the binding protein is normally present in amounts which exceed that required to saturate the transport system [121,127]. In keeping with this idea, Pardee et al. [121] have estimated that the sulfate binding protein represents approximately 1% of the total soluble protein in derepressed cells of S. typhimurium.

Mutants specifically defective in sulfate uptake (and no other enzyme of sulfate assimilation) have been isolated in Salmonella [109,121,127] and a number of fungi [41,111,112,119,128]. In S. typhimurium such mutants map as three separate complementation groups in the cys A region of the chromosome [129]. Using deletions covering the entire cys A region, as well as nonsense point mutations within each of the cys A complementation groups, Ohta et al. [127] showed conclusively that cys A does not represent the structural gene for the sulfate binding protein. Certain mutations in the cys A region do reduce the level of binding activity, however (although, curiously, deletion of the entire cys A region does not), suggesting an interaction between the cys A gene product(s) and the binding protein which affects the properties of the latter. Binding protein is absent in cys B mutants of S. typhimurium [127]; such mutants are deficient in most of the enzymes of the sulfate assimilation pathway (see below). Although cys B clearly is not the structural gene for binding protein [127], the effect of mutation at this regulatory locus on binding activity does indicate that binding protein is involved in some manner in sulfate assimilation in vivo.

Sulfate transport systems have also been reported in E. coli [130], Salmonella pullorum [37], yeast [131], and Euglena [132], although none of these systems has been studied extensively.

E. Regulation of Sulfate Assimilation

Mechanisms of regulation of the pathway of sulfate assimilation have been studied extensively in enterobacteria and yeast, organisms in which the metabolic relationship between the "end products" of the assimilation, cysteine and methionine, differs markedly. As might be anticipated, the regulatory signals utilized by these organisms appear to reflect these differences.

1. REGULATION IN ENTEROBACTERIA

As shown in Fig. 3, cysteine biosynthesis in E. coli and S. typhimurium represents a branched convergent pathway, one arm of which involves production of the sulfur precursor, H_2S, from sulfate, and the other involves production of the activated carbon precursor, O-ace-

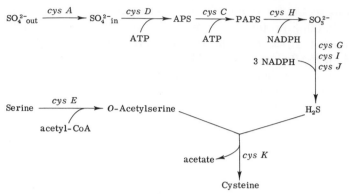

FIG. 3. Pathway of sulfate assimilation in enterobacteria. Genetic loci corresponding to individual enzymatic steps are those defined for *S. typhimurium* [38].

tylserine, from serine and acetyl-CoA, in a reaction catalyzed by serine acetyltransferase (EC 2.3.1.30; acetyl-CoA:L-serine *O*-acetyl-transferase) [133]. The two precursors are then joined by the action of *O*-acetylserine sulfhydrylase [EC 4.2.99.8; *O*-acetyl-L-serine acetate-lyase (adding hydrogen sulfide)] to form cysteine, with release of acetate [133]. The latter reaction has also been demonstrated in extracts of yeast [67,134], *Neurospora* [134], *Aspergillus* [135], spinach [68], and turnip [67]. Structural genes specifying each of the enzymes in Fig. 3, mutation in any of which (except that for *O*-acetylserine sulfhy-drylase*) leads to cysteine auxotrophy, have been identified in *S. typhimurium* and *E. coli* [36,38,49,129,136,140–142]. The *Salmonella* gene designation for each of these enzymes is indicated in Fig. 3.

In the enterobacteria, the sulfur atom of methionine is derived from that of cysteine, so that cysteine, and not H_2S, represents the starting point for methionine biosynthesis. Since two key enzymes involved in

* There are two immunologically unrelated species of *O*-acetylserine sulfhydrylase activity in *S. typhimurium,* termed OASS-A and OASS-B, respectively [136–138]. OASS-A is by far the major species in sulfur starved cells [139]. While the level of OASS-A is markedly repressed in cells grown with cysteine as sulfur source [38,133,137], that of OASS-B is not significantly altered by growth on different sulfur sources [136,137]. All of the serine acetyltransferase activity normally found in *S. typhimurium* extracts is bound in a complex with OASS-A; OASS-B does not complex with serine ace-tyltransferase [133,138]. These results suggest that OASS-A should play a role in cys-teine biosynthesis in *S. typhimurium.* However, mutants (termed *cys K* [136]) which lack OASS-A do not require cysteine for growth, a result which indicates that OASS-B, together with serine acetyltransferase, is capable of supplying the needs of the cell for cys-teine.

converting methionine to cysteine in higher organisms, cystathionine β-synthase [EC 4.2.1.22; L-serine hydro-lyase (adding homocysteine)] and γ-cystathionase [EC 4.4.1.1; L-cystathionine cysteine-lyase (deaminating)], are not present in these bacteria [143], methionine cannot give rise to cysteine in these organisms. The end product suitable for feedback regulation of the pathway of sulfate assimilation would appear then to be cysteine in these bacteria. Indeed, with the exception of serine acetyltransferase, all of the enzymes concerned with cysteine biosynthesis in *S. typhimurium* and *E. coli* are subject to derepression by sulfur starvation and to repression by growth of the organisms on cysteine as sulfur source [36,38,71,109,133,144–147]. The level of repression varies with the intracellular cysteine concentration [148]. The effect, however, is not simple repression by cysteine, since under conditions of sulfur starvation, *O*-acetylserine, the direct organic precursor of cysteine, is necessary for derepression, as shown by the fact that *cys E* mutants (which lack serine acetyltransferase [133,137]) lack the enzymes of sulfate uptake, activation, and reduction and have low levels of *O*-acetylserine sulfhydrylase activity, unless exogenous *O*-acetylserine is added [38,140,141,149]. *O*-Acetylserine, then, acts as an internal inducer for the sulfate assimilation pathway. However, when cells are grown in the presence of both *O*-acetylserine and cysteine (or H_2S), the inductive effect of *O*-acetylserine is blocked, indicating a competition between internal inducer and end product at the site of enzyme synthesis [38,140].

Thus, production of the enzymes necessary for synthesis of the immediate sulfur precursor of cysteine, H_2S, is controlled, at least in part, by the level of the immediate carbon precursor, *O*-acetylserine. Since serine acetyltransferase does not seem to be subject to repression by cysteine, how is the production of *O*-acetylserine regulated? Kredich and Tomkins [133,138] showed that cysteine, the end product, strongly feedback inhibits serine acetyltransferase ($K_i = 6 \times 10^{-7}\ M$, competitive with respect to acetyl-CoA), so that production of *O*-acetylserine is rigidly controlled by the intracellular level of cysteine. (Ellis [130] has presented evidence indicating direct feedback inhibition of the sulfate transport system by cysteine in *E. coli*. No further reports on this subject have appeared, and Dreyfuss and Pardee [109,150] could find no inhibition of the *S. typhimurium* sulfate transport system by either cysteine or H_2S.)

A survey of mutants blocked in the pathway of cysteine biosynthesis showed that most mutations at the *cys B* locus, including deletions, lead to repression of all the enzymes of the pathway (except serine acetyltransferase) even under conditions of sulfur starvation and an excess of

O-acetylserine [38,49,141,146,149,151]. Certain mutations mapping in or near the *cys B* locus, however, lead to partial constitutivity of all or some of the sulfate assimilation enzymes in the presence of cysteine and absence of O-acetylserine [38,149]. A few mutations in *cys B* lead to more pronounced repression (or constitutive synthesis) of some of the enzymes of sulfate assimilation than of others, a result which suggests that genes specifying these enzymes are under the control of independent "operators," i.e., recognition sites for the *cys B* gene product. In keeping with this idea, repression of the various enzymes of the sulfate assimilation pathway does not appear to be coordinate [38,71], despite the fact that at least some of the structural genes involved (*cys C, D, H, I,* and *J*) map on a common transducing fragment [129]. Studies with diploids containing both a normal and a mutant *cys B* allele have shown that *cys B* control is "positive" [141,152], i.e., the *cys B* gene specifies a diffusible product required for derepressed expression of the "target" gene loci. Jones-Mortimer [151] has postulated that O-acetylserine and the *cys B* gene product must cooperate to induce the enzymes of sulfate assimilation. This induction is in some fashion antagonized by cysteine or H_2S [38]. The enzymes of sulfate activation and reduction are also repressed by growth of *Bacillus subtilis* cells on cysteine [71]; the mechanism of regulation in this organism is not known.

2. REGULATION IN YEAST

As shown in Fig. 4, the metabolic relationships between H_2S, cysteine, and methionine are more complex in yeast than in the enterobacteria. Thus, while yeast, like the bacteria, possesses enzymes to convert H_2S and O-acetylserine to cysteine [67,134] and cysteine to methi-

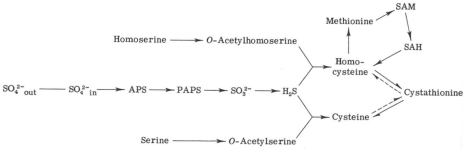

FIG. 4. Pathways of sulfate assimilation and methionine-cysteine interconversion in yeast and other fungi. The reactions marked with dashed lines are of questionable significance in yeast, although they are demonstrably important in the metabolism of other fungi. SAM, S-Adenosylmethionine; SAH, S-adenosylhomocysteine.

onine, via cystathionine and homocysteine [153], it can also convert methionine to cysteine [143]. Methionine, in fact, is a better sulfur source for growth of yeast cells than is sulfate (generation time 70 minutes versus 90 minutes, in sulfur-supplemented minimal medium), while cysteine is a poorer sulfur source than either (generation time 120 minutes) [154]. These results suggest that not only can cysteine be formed efficiently from methionine *in vivo,* but methionine is formed more efficiently from sulfate than it is from cysteine. While these results may merely reflect permeability phenomena, they also suggest that there may be a mechanism for incorporation of the sulfur atom of methionine in yeast which does not involve the intermediate formation of cysteine. This possibility was strengthened by the finding of DeVito and Dreyfuss [154] that, while H_2S and SO_3^{2-} completely inhibit the incorporation of radioactivity from $^{35}SO_4^{2-}$ into yeast protein *in vivo* and methionine inhibits 75%, cysteine inhibits only a small proportion (less than 30% at 1×10^{-3} *M* cysteine) of the $^{35}SO_4^{2-}$ incorporation.

Indeed, yeast extracts contain high activities of an enzyme, termed *O*-acetylhomoserine sulfhydrylase (or homocysteine synthase), which catalyzes the reaction of *O*-acetylhomoserine with H_2S to form homocysteine [134,153,155,156]. Similar enzymes have been described in enterobacteria [134,157], *Neurospora* [134,158,159], and spinach [68,160], although sometimes *O*-succinyl- or *O*-phosphorylhomoserine is the preferred H_2S acceptor. In the enterobacteria and *Neurospora* there is evidence that the sulfhydrylase activity leading to homocysteine formation is not of major physiological significance for methionine biosynthesis, since mutants blocked in β-cystathionase [EC 4.4.1.8; cystathionine L-homocysteine-lyase (deaminating)] which cleaves cystathionine to homocysteine, pyruvate, and NH_3, are methionine auxotrophs [161,162]. Indeed, in yeast, as in *S. typhimurium, O*-acetylhomoserine sulfhydrylase activity and cystathionine-γ-synthase (EC 4.2.99.9) activity may be catalyzed by a common enzyme, since a single mutation (*met-8*) leads to loss of both activities (and methionine auxotrophy) [153]. No yeast mutants blocked in β-cystathionase have been found, however, and the cystathionine-γ-synthase activity is present at only 2–4% of the level of *O*-acetylhomoserine sulfhydrylase activity in crude extracts (unlike the situation in *Salmonella* and *Neurospora,* which organisms show comparable levels of the two activities in extracts) [153,162]. Such considerations, among others, led Cherest *et al.* [155] to favor the scheme shown in Fig. 4 as the pathway for biosynthesis of cysteine and methionine from SO_4^{2-} in yeast. Yeast accumulates large quantities of *S*-adenosylmethionine (SAM) when grown on exogenous methionine [163–165]. The free methionine pool, on the other hand, remains relatively constant even when large excesses of methionine are present in the growth medium [166].

The enzymes of sulfate assimilation appear to be regulated as follows in this complicated pathway: sulfate permease, ATP-sulfurylase, APS-kinase, sulfite reductase, homoserine acetyltransferase (EC 2.3.1.31; acetyl-CoA-L-homoserine O-acetyltransferase), and O-acetylhomoserine sulfhydrylase (PAPS-reductase, serine acetyltransferase, and O-acetyl-serine sulfhydrylase were not examined) are all repressed when yeast cells are grown on media containing excess methionine [131,155,167]. Growth on cysteine does not repress ATP-sulfurylase (the only enzyme examined) [154]. The repression appears to be exerted through two systems. One may involve charged met-tRNA, since modified methionine mediated repression was observed in a strain of yeast with an impaired met-tRNA synthetase [167]. Mutation at the independent loci *eth-2, eth-3,* and *eth-10* (all of which endow the mutant cell with resistance to the toxic effects of the methionine analogue ethionine, probably due to overproduction of methionine by these strains) leads to loss of the methionine mediated repression system [167–170]. Cherest *et al.* [166] have suggested that these mutations may involve alterations in sequence or maturation of one or more species of met-tRNA, although there is no direct evidence for this. A second repressive effect is exerted by SAM, or a metabolite derived from SAM (possibly homocysteine?), and is not due to free methionine or met-tRNA [166,171]. This SAM mediated repression is not affected by mutations at the *eth-2, eth-3,* or *eth-10* locus [169]. It has been suggested that the met-tRNA and SAM mediated repression mechanisms may represent independent means of regulating enzyme synthesis at the levels of translation and transcription, respectively [166].

Repression of the enzymes of sulfate assimilation by methionine and/or SAM may be a general phenomenon in the fungi, Thus, repression of sulfate permease and sulfite reductase by growth of *Neurospora* on methionine containing media has been reported [172–174]. Loss of this repression in a mutant (*eth-1^r*) which is deficient in SAM-synthetase [162], the enzyme that converts methionine to SAM, suggests that SAM may function as a corepressor in this organism. Interestingly, ATP-sulfurylase is not repressed by methionine in *Neurospora* [114]. The sulfate permease [111,113] and ATP-sulfurylase [28] activities of *Penicillium* and *Aspergillus* are repressed by growth on exogenous methionine but not cysteine. Growth on SAM has been shown to repress sulfate permease, sulfite reductase, and O-acetylserine sulfhydrylase in *A. nidulans* [175].

At least two genetic loci, termed *cys-3* and *scon^c*, are involved (in addition to *eth-1*[r]) in regulation of the sulfate permease in *Neurospora* [173,174,176,177]. The system is apparently under positive control in that the *cys-3* gene product is required for derepressed synthesis of the permease. Sulfite reductase, on the other hand, does not appear to be under control of the *cys-3* gene system.

The pathway of sulfate assimilation is also controlled by feedback inhibition in yeast and other fungi, since ATP-sulfurylases from both yeast and *Penicillium* are strongly inhibited by H_2S, the end product of the inorganic pathway [28,154]. Homoserine acetyltransferase and *O*-acetylhomoserine sulfhydrylase, which together catalyze conversion of homoserine to homocysteine in yeast, are inhibited, respectively, by SAM [178] and methionine [155,156]. Regulation of cysteine synthesis has not been studied. Feedback inhibition of the sulfate uptake mechanism by SO_4^{2-} and cysteine or some metabolite readily derived from these two compounds has been suggested for *Aspergillus* and *Penicillium* [111]; on the other hand, no such inhibition was observed with the sulfate permease of *Neurospora* [114].

III.   DISSIMILATORY SULFATE REDUCTION

Two bacterial genera, *Desulfovibrio* and *Desulfotomaculum* [11,179], are capable of utilizing sulfate (and reduction products thereof) as a terminal electron acceptor for anaerobic cellular respiration. The sulfate reducing bacteria are heterotrophic obligate anaerobes [180]; they may satisfy their metabolic requirement for ATP either by substrate-level phosphorylation (e.g., via phosphoroclastic cleavage of pyruvate) or by respiratory chain phosphorylation associated with reduction of sulfate or fumarate [180–183]. Simple three and four carbon compounds (e.g., lactate, pyruvate) are generally used by these bacteria as hydrogen donors, although complex organic material can be used for growth [11,179]. H_2 also serves as a good electron donor for SO_4^{2-} reduction [184], and the sulfate reducing bacteria have active hydrogenase enzymes [185–187]. Some strains of sulfate reducing bacteria can grow with pyruvate or fumarate as carbon source in the absence of sulfate [188,189], electrons generated by pyruvate oxidation being used to reduce protons and evolve H_2.

Kluyver [190] classified nitrate reduction in microorganisms into two types: "assimilatory" reduction to satisfy the biosynthetic requirements of the cell for nitrogen at the oxidation level of NH_3, and "dissimilatory" reduction associated with the use of nitrate as terminal electron acceptor for cellular respiration. The latter process is associated with the excretion of large quantities of nitrogen in oxidation states intermediate between NO_3^{2-} and NH_3. Postgate [191], in analogy with Kluyver's classification, has termed the large scale reduction of sulfate conducted by *Desulfovibrio* and *Desulfotomaculum* "dissimilatory" sulfate reduction. We shall use the terms dissimilatory and respiratory sulfate reduction interchangeably.

The pathway of dissimilatory sulfate reduction, determined through the classical studies in the laboratories of Peck and Ishimoto, can be outlined as follows:

$$SO_4^{2-} \xrightarrow{ATP} APS \xrightarrow{H_2} SO_3^{2-} \xrightarrow{3 H_2} H_2S$$

In vitro studies generally have utilized H_2 as ultimate reductant, with MV added to serve as an electron carrier between the hydrogenase present in crude extracts (or added to more purified preparations) and the reductive enzymes of sulfate metabolism. On occasion, chemically reduced methyl viologen has been used. It is of interest to note that, although dissimilatory sulfate reduction is associated with phosphorylation *in vivo* and *in vitro*, the enzymes of the pathway are found largely in soluble form in extracts of the sulfate reducing bacteria [194,195].

A. Sulfate Activation

Sulfate reduction in extracts of sulfate reducing bacteria is dependent on the presence of ATP and Mg^{2+} [192–194]. Ishimoto and Fujimoto [195] and Peck [193,194] showed that this requirement was due to reaction of SO_4^{2-} with ATP to form APS. Akagi and Campbell [17] partially purified the ATP-sulfurylase from *Desulfovibrio* and *Desulfotomaculum* species. The enzyme from these organisms catalyzes SO_4^{2-} and Mg^{2+}-dependent $^{32}PP_i$-ATP exchange [194], as well as production of PP_i from ATP in the presence of MoO_4^{2-}, CrO_4^{2-}, or WO_4^{2-} [17,193,194], Group VI anions which competitively inhibit sulfate reduction *in vivo* and *in vitro*. These results suggest that the formation of APS occurs by the same type of mechanism in both assimilatory and dissimilatory sulfate reduction.

Extracts of sulfate reducing bacteria do not contain APS-kinase [193,194], and added PAPS is neither degraded nor reduced by these extracts [80,194,196]. APS is therefore the immediate substrate for sulfate reduction in these organisms [194] as it is in *Chlorella*. However, since the sulfate reducing bacteria, unlike *Chlorella*, lack APS-kinase, they are dependent solely upon the action of inorganic pyrophosphatase to drive APS accumulation via the unfavorable ATP-sulfurylase reaction. It is not surprising then, that the pyrophosphatase of *Desulfovibrio* may be subject to metabolic regulation *in vivo*. This enzyme, the activity of which is stimulated by either Mg^{2+}, Mn^{2+}, or Co^{2+} [197], has been purified to homogeneity [198] and found to have a molecular weight of 42,000. Ware and Postgate [198,199] reported that the pyrophosphatase activity, as normally isolated, could be stimulated up to 130-fold by treatment with reducing agents such as $S_2O_4^{2-}$, BH_4^-, or S^{2-}. They termed this phenomenon "reductant-dependent activation." The treatment with reducing agents was associated with an increase in titratable sulfhydryl groups from 3 to 9 in the inactive versus reductant activated enzyme. There was no change in molecular weight of the enzyme following reductant treatment.

The interconversion between active and inactive pyrophosphatase forms could be readily demonstrated *in vivo*. Exposure of the bacteria to small amounts of O_2 in the absence of oxidizable carbon source and sulfate yielded enzyme which was almost completely inactive unless reductant was added to the extract. Addition of lactate and sulfate to aerated bacteria led to reversal of the inactivation *in vivo*. Both inactivation and reactivation are rapid processes which do not require protein synthesis. The mechanism of the inactivation process is unknown, since it could not be duplicated *in vitro* upon the addition of O_2 or other oxidizing agents to active enzyme. Ware and Postgate [199] have suggested that the reversible inactivation of pyrophosphatase in sulfate reducing bacteria is a control mechanism which ensures that sulfate activation, a drain on the cellular pool of ATP, will occur only when intracellular conditions are such as to permit APS reduction. A summary of their scheme is presented in Fig. 5. In keeping with this scheme, these authors were able to demonstrate reductant-stimulated formation of AMP from sulfate and ATP by crude extracts derived from *Desulfovibrio* cells exposed to O_2 *in vivo*; the results were not obtained consistently, however. Since a number of electron carriers in *Desulfovibrio*, including reduced ferredoxin and cytochrome c_3, may be autooxidizable by O_2, it is possible that one of these carriers, in its reduced form, activates the pyrophos-

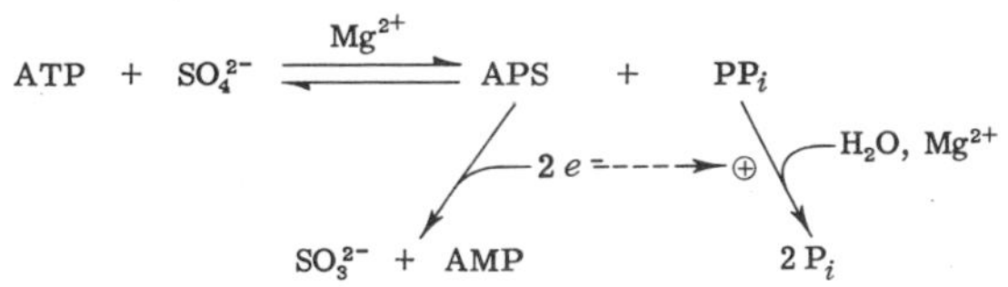

FIG. 5. Relationships between sulfate activation, pyrophosphate hydrolysis, and APS reduction in sulfate reducing bacteria. Reductant-dependent activation of pyrophosphatase is indicated by the dashed arrow.

phatase *in vivo*. Reductant stimulation of inorganic pyrophosphatase may not be limited to sulfate reducing bacteria, since this phenomenon has also been reported in several species of *Clostridium,* an obligate anaerobe which obtains energy solely by fermentation [200].

B. Reduction of APS

The reduction of APS to SO_3^{2-} by extracts of *Desulfovibrio* proceeds with the following stoichiometry when H_2, in the presence of MV, is used as electron donor:

$$APS^{2-} + H_2 \rightarrow AMP^{2-} + 2\ H^+ + SO_3^{2-}$$

Chemically reduced MVH can serve as reductant, but NADPH or NADH cannot [194]. GPS, UPS, and CPS can substitute for APS, but are reduced more slowly [201]. Reduction of all nucleoside phosphosulfates is strongly inhibited by AMP, but not other nucleoside monophosphates [201]. APS-reductase [EC 1.8.99.2; AMP, sulfite:(acceptor) oxidoreductase] has been highly purified by Ishimoto and Fujimoto [201] and Peck *et al.* [202]; it comprises 1–2% of the soluble protein of *Desulfovibrio vulgaris*. APS-reductase has a molecular weight of 220,000 [203] and contains, per molecule, one FAD and 7–9 Fe/S prosthetic groups. As isolated, it contains 1 mole of tightly bound AMP per mole of flavin. The enzyme undergoes polymerization in phosphate buffer, 15 S and 19 S forms appearing in addition to the monomeric 10 S form; the polymerization is reversed upon addition of AMP.

Since the E_0' of the APS/SO_3^{2-}, AMP redox couple is -0.06 V, one might expect SO_3^{2-} and AMP to react to form APS in the presence of an electron acceptor of positive E_0' capable of interacting with APS-reductase. Indeed, Peck *et al.* [202] have shown that purified *Desulfovibrio* APS-reductase catalyzes the following reaction:

$$AMP^{2-} + SO_3^{2-} + 2\ Fe(CN)_6^{3-} \rightarrow APS^{2-} + 2\ Fe(CN)_6^{4-}$$

O_2 can substitute for ferricyanide as electron acceptor, although the turnover number with ferricyanide is 600 times that with O_2 [203]. No other compound tested, including cytochrome c, was effective as an electron acceptor for sulfite oxidation. GMP, IMP, and dAMP can substitute for AMP [202], while pyrimidine nucleotides are inactive.

Desulfovibrio APS-reductase (studied in the direction of sulfite oxidation) loses activity on storage at 0°. The activity can be restored by incubating enzyme with thiol compounds for several minutes prior to assay [202]. The enzyme is strongly inhibited by PCMB or $HgCl_2$ at low concentrations; however, it is not inhibited by arsenite, nor is the MVH-APS-reductase activity stimulated by lipoate, results which contrast with those reported for the thiol-dependent PAPS-reductase system of yeast [201,202]. Peck *et al.* [204] have shown that PCMB, at low concentration, causes removal of Fe/S, but not FAD, from the enzyme. Such PCMB treated enzyme can no longer utilize ferricyanide as an electron acceptor; it can, however, still utilize O_2 as efficiently as the native enzyme as an electron acceptor for AMP-dependent sulfite oxidation [204].

The reaction mechanism of *Desulfovibrio* APS-reductase has been studied only in the direction of AMP-dependent SO_3^{2-} oxidation to APS. Although this reaction is not the physiologically important one in the sulfate reducing bacteria, it does represent the formation of an adenylate anhydride of a compound capable of undergoing oxidation and reduction, and, as such, might serve as a model useful to our understanding of respiratory chain phosphorylation. Furthermore, the AMP-dependent oxidation of SO_3^{2-} is of true physiological significance in the sulfur oxidizing bacteria (see below). For these reasons, we shall consider the mechanism proposed by Peck *et al.* [203,204] for this enzyme in some detail.

Michaels *et al.* [203] showed that addition of SO_3^{2-} to AMP-free APS-reductase caused a bleaching of the enzyme FAD spectrum in the visible wavelength region. However, the resulting spectrum was not simply that of a flavin hydroquinone (FH_2) in that SO_3^{2-} treatment, unlike BH_4^- reduction of the enzyme, led to an increase in absorption, compared to that of oxidized enzyme, in the 300–330 nm region. Since covalent adducts between flavins and SO_3^{2-} (believed to involve bond formation at the N-5 position of the isoalloxazine ring) are reported to have absorption spectra similar to that of 1,5-dihydroflavin (FH_2) except for a new broad absorption band in the 310–330 nm region [205,206], Michaels *et al.* looked directly for the presence of an FAD-sulfite adduct in trichloroacetic acid extracts of SO_3^{2-} treated APS-reductase. The extracted flavin, following centrifugation to remove the protein, showed the prominent 320 nm absorption band expected for such an adduct. The spectrum changed to that of oxidized FAD on allowing the extracted flavin to incubate in the presence of air for 15–30 minutes; a similar time of dissociation of model flavin-sulfite adducts was reported by Muller and Massey [205]. A control using BH_4^- treated enzyme (borohydride reduces the flavin to the hydroquinone level) gave no indication of the 320 nm absorbing species following trichloroacetic acid extraction, and the spectrum quickly reverted in air to that of oxidized FAD, as would be expected for a dihydroflavin. Addition of sulfite to FAD or to the trichloroacetic acid extract of APS-reductase did not yield the 320 nm peak. The increase in 320 nm absorption and bleaching of visible wavelength absorbance in the enzyme still occurred when

PCMB treated enzyme, which has lost its Fe/S, was reacted with sulfite [204]. These results suggest that the interaction of SO_3^{2-} with APS-reductase results in formation of a covalent adduct between sulfite and the enzyme FAD:

$$E\Big\langle{}^{Fe/S}_{FAD} + SO_3^{2-} \rightleftharpoons E\Big\langle{}^{Fe/S}_{FAD-SO_3^{2-}}$$

Addition of AMP to sulfite treated enzyme results in formation of APS. While AMP does not affect the spectrum of oxidized enzyme, its addition to SO_3^{2-} treated enzyme, whether PCMB treated or native, results in partial loss of the 320 nm absorbance, both in the native enzyme itself and in trichloroacetic acid extracts of the enzyme [203,204]. These results suggest the reaction:

$$2H^+ + E\Big\langle{}^{Fe/S}_{FAD-SO_3^{2-}} + AMP^{2-} \rightleftharpoons APS^{2-} + E\Big\langle{}^{Fe/S}_{FADH_2}$$

Addition of AMP to sulfite treated native enzyme can result in further bleaching of the visible absorption spectrum, indicating electron transfer to the Fe/S groups of the enzyme. In keeping with this suggestion, no such additional bleaching of visible absorbance was observed on addition of AMP to a mixture of PCMB treated enzyme and SO_3^{2-}:

$$E\Big\langle{}^{Fe/S}_{FADH_2} \rightleftharpoons E\Big\langle{}^{(Fe/S)_{red}}_{FAD} + 2\,H^+$$

The enzyme can presumably react with additional sulfite and AMP to yield the forms

$$E\Big\langle{}^{(Fe/S)_{red}}_{FAD-SO_3^{2-}} \quad \text{and} \quad E\Big\langle{}^{(Fe/S)_{red}}_{FADH_2}$$

in keeping with the further bleaching of the absorption spectrum.

The loss in ability of PCMB treated enzyme to reduce ferricyanide, coupled with retention of the ability to reduce O_2, suggests that the reaction cycle may be completed by either of the following reactions:

$$E\Big\langle{}^{(Fe/S)_{red}}_{FAD} \xrightarrow[\text{fast}]{2\,Fe(CN)_6^{3-}} E\Big\langle{}^{Fe/S}_{FAD}$$

or

$$E\Big\langle{}^{Fe/S}_{FADH_2} \xrightarrow[\text{slow}]{O_2} E\Big\langle{}^{Fe/S}_{FAD}$$

Inspection of the proposed reaction mechanism suggests that MVH (or the natural reductant for APS) may react with the enzyme through the

Fe/S groups, while APS should react only with the reduced enzyme FAD.

In an attempt to test the catalytic significance of the proposed mechanism, Peck *et al.* [204] have shown that the rate of bleaching of the enzyme flavin absorption spectrum in the presence of AMP and sulfite (studied in the stopped-flow apparatus) is equal to the turnover number of the enzyme in the ferricyanide assay. However, in the absence of AMP, the rate of bleaching was only 10% of the turnover number, indicating that reactions between enzyme and sulfite in the absence of AMP may not be of catalytic significance. On the other hand, AMP may serve as an allosteric effector of the enzyme (in keeping, perhaps, with its ability to reverse enzyme polymerization in phosphate buffer and to specifically inhibit APS reduction when other nucleotides do not). Furthermore, Peck *et al.* [204] found that sulfite treatment of APS-reductase results in formation of a substantial EPR signal at $g = 1.94$, characteristic of reduced Fe/S groups, in the absence of AMP. Reduction of the Fe/S by SO_3^{2-} in the absence of AMP is not in keeping, of course, with the proposed reaction mechanism. The effect of AMP on the Fe/S EPR signal of sulfite treated enzyme has not been examined.

C. Reduction of Sulfite

Although crude extracts of *Desulfovibrio* and *Desulfotomaculum* can catalyze the stoichiometric reduction of SO_3^{2-} to H_2S with MVH as electron donor [207–212], deviations from this stoichiometry have been reported by Akagi *et al.* [213–215], who showed that $S_3O_6^{2-}$ and $S_2O_3^{2-}$, ions of oxidation state intermediate between sulfite and sulfide, could be detected as additional products of SO_3^{2-} reduction in these extracts. Kobayashi *et al.* [209] and Suh and Akagi [213] found that the *Desulfovibrio* system catalyzing SO_3^{2-} reduction to H_2S could be separated into at least two components by chromatography and salt fractionation. One fraction catalyzed SO_3^{2-}-dependent H_2 uptake (in the presence of MV and added hydrogenase), but the ratio of $H_2:SO_3^{2-}$ consumed was less than one. The products of the reaction were chiefly $S_3O_6^{2-}$ and $S_2O_3^{2-}$. The second fraction catalyzed reduction of both $S_3O_6^{2-}$ and $S_2O_3^{2-}$ to H_2S with MVH as electron donor. The kinetics of $S_3O_6^{2-}$ and $S_2O_3^{2-}$ formation and disappearance were such as to suggest that they might represent successive intermediates in the reduction of SO_3^{2-} to H_2S:

$$3\ SO_3^{2-} \xrightarrow[\substack{2\ e^-,\ 3\ OH^- \\ 3\ H^+}]{\substack{\text{sulfite} \\ \text{reductase}}} S_3O_6^{2-} \xrightarrow[\substack{2\ e^-\quad SO_3^{2-}}]{\substack{\text{trithionate} \\ \text{reductase}}} S_2O_3^{2-} \xrightarrow[\substack{2e,\ SO_3^{2-} \\ 2H^+}]{\substack{\text{thiosulfate} \\ \text{reductase}}} H_2S$$

This mechanism predicts a recycling pool of SO_3^{2-} and experiments by Findley and Akagi [216] with radioactive sulfite and thiosulfate were in keeping with this prediction. Thus, incubation of $^{35}SO_3^{2-}$ with the "thio-sulfate forming fraction" of *Desulfovibrio* extracts in the presence of MVH yielded double-labeled [^{35}S]thiosulfate. Using crude extracts, Findley and Akagi showed that while both $^{35}SSO_3^{2-}$ and $S^{35}SO_3^{2-}$ yielded $H_2^{35}S$, the rate of $H_2^{35}S$ formation was more rapid from the former than from the latter form of thiosulfate. When extracts were in-cubated with $S^{35}SO_3^{2-}$ and MVH, the ratio of double-labeled to sul-fonate-labeled [^{35}S]thiosulfate increased with time.

The sequential scheme received strong support from Lee *et al.* [211,217] who purified an enzyme to homogeneity that could catalyze sulfite-dependent H_2 uptake (in the presence of MV and hydrogenase) from *D. gigas* and *D. vulgaris* extracts. The sole product of the reduc-tion was reported to be $S_3O_6^{2-}$. The enzyme copurified with "desul-foviridin," a green protein originally described by Ishimoto *et al.* [218] and Postgate [219] as one of the major pigments characteristic of the genus *Desulfovibrio*.

Thiosulfate reductase, the third enzyme in the postulated sequential mechanism for sulfite reduction, was originally detected in *Desulfovibrio* extracts by Ishimoto *et al.* [208,220], who demonstrated that the reac-tion products, with MVH as electron donor, were SO_3^{2-} and H_2S. Haschke and Campbell [221] have reported purification of this enzyme to homogeneity. The enzyme, molecular weight 16,000, has no visible wavelength absorbing material essential for its activity, and is active after treatment with 6 M guanidine hydrochloride. Haschke and Camp bell state that the enzyme can reduce both SO_3^{2-} and $S_2O_3^{2-}$ to H_2S; however, their demonstration that the radioactivity of $^{35}SSO_3^{2-}$, but not $S^{35}SO_3^{2-}$, is converted to $H_2^{35}S$ in the presence of MVH by the enzyme, is not consistent with reduction of sulfite by their enzyme prep-aration. MVH-dependent $S_2O_3^{2-}$-reductase has also been partially puri-fied from extracts of *Desulfotomaculum* by Nakatukasa and Akagi [222]. Using [^{35}S]thiosulfate, it was shown that the sulfane atom was reduced to H_2S with this enzyme, while the sulfonate atom accumulated as SO_3^{2-}. The K_m for $S_2O_3^{2-}$, 0.13 M, was very high with this enzyme. The enzyme did not catalyze SO_3^{2-} reduction; the reduction of $S_2O_3^{2-}$, however, was strongly inhibited by SO_3^{2-}. Although the *Desulfo-tomaculum* thiosulfate reductase is strongly inhibited by mercurials, it is not stimulated by thiol compounds (GSH, cysteine, or 2-mercap-toethanol), indicating that its reaction mechanism may be different from that of the GSH-dependent $S_2O_3^{2-}$ reductase reported in yeast by Kaji and McElroy [223]. Reduced pyridine nucleotides could not replace

MVH as reductant. No report on isolation of trithionate reductase has appeared as yet.

The sequential mechanism of Kobayashi et al. [209] suggests that the processes of assimilatory and dissimilatory sulfite reduction proceed by quite different mechanisms. Recent work, however, has shown that the three enzyme scheme does not represent an obligatory mechanism of sulfite reduction to H_2S in the sulfate reducing bacteria.

Thus, Kobayashi et al. [224] have now reported the isolation of an electrophoretically homogeneous sulfite reductase from D. vulgaris and confirmed the identity of this enzyme with the pigment desulfoviridin. When the products of sulfite reduction by H_2 (in the presence of MV and hydrogenase) were examined, they were found to be a mixture of $S_3O_6^{2-}$, $S_2O_3^{2-}$, and H_2S. Yet, the purified enzyme could not catalyze reduction of either $S_3O_6^{2-}$ or $S_2O_3^{2-}$, and neither ion affected the rate of SO_3^{2-}-dependent H_2 consumption [224,225]. Addition of nonradioactive $S_3O_6^{2-}$ or $S_2O_3^{2-}$ to enzyme catalyzing reduction of $^{35}SO_3^{2-}$ did not result in significant dilution of the radioactivity of the $H_2^{35}S$ formed [225]. The proportions of $S_3O_6^{2-}$, $S_2O_3^{2-}$, and H_2S formed from sulfite depended on the relative concentrations of MVH and SO_3^{2-} in the reaction mixture. Thus, at fixed $[SO_3^{2-}]$, increasing levels of reductant favored production of relatively large amounts of $S_2O_3^{2-}$ and H_2S, as compared to $S_3O_6^{2-}$, while at fixed $[MVH]$, increasing concentrations of sulfite favored production of relatively larger amounts of $S_3O_6^{2-}$ as compared to H_2S. Akagi and Adams [215] have shown that a highly purified sulfite reductase from Desulfotomaculum nigrificans, like the D. vulgaris enzyme, catalyzes the reduction of sulfite by MVH to $S_3O_6^{2-}$, $S_2O_3^{2-}$, and H_2S. The enzyme was shown to be free of trithionate reductase activity. The D. nigrificans sulfite reductase had been originally described as a CO-binding pigment from this organism by Trudinger [210]; the protein comprises one of the major pigments of Desulfotomaculum crude extracts. It has been termed "P-582" because the principle visible wavelength absorption band of the oxidized pigment exhibits a maximum at 582 nm. Similarly, Lee et al. [226] have isolated a sulfite reductase from the "Norway" strain of Desulfovibrio desulfuricans that lacks the pigment desulfoviridin [227], and showed that this enzyme, homogeneous in the ultracentrifuge, produces both $S_3O_6^{2-}$ and H_2S as products of sulfite reduction by MVH. The reddish brown sulfite reductase from this organism has been termed "desulforubidin." The reduction of SO_3^{2-} to H_2S by sulfite reductases from a large number of Desulfovibrio and Desulfotomaculum strains, which enzymes had been completely separated from any thiosulfate reductase activity by gel electrophoresis, has been reported by Skyring and Trudinger [228].

Recently, Saks and Siegel [229], working with highly purified preparations of desulfoviridin, P-582, and desulforubidin, have been able to show
that each of these enzymes is capable of catalyzing the stoichiometric reduction of SO_3^{2-} to H_2S with MVH as reductant when low
sulfite concentrations (approximately 1×10^{-5} M) are used. As the
$[SO_3^{2-}]$ is raised, first $S_2O_3^{2-}$ and then $S_3O_6^{2-}$ are formed in increasing
proportion, until at the sulfite concentrations used by all other workers
(greater than 1×10^{-3} M), $S_3O_6^{2-}$ becomes the predominant product. In
agreement with the results of Kobayashi *et al.* [225], the proportion of
H_2S can be raised, at constant $[SO_3^{2-}]$, by raising the concentration of
MVH.

These results demonstrate, then, that all of the dissimilatory sulfite
reductases purified to date can catalyze the complete six electron reduc-
tion of SO_3^{2-} to H_2S, and that $S_3O_6^{2-}$ and $S_2O_3^{2-}$ are not intermediates in
this process, but are rather alternative products of sulfite reduction.
Kobayashi *et al.* [225] have now suggested that $S_3O_6^{2-}$ and $S_2O_3^{2-}$ represent adducts between sulfite and enzyme-bound monosulfur compounds intermediate in oxidation state between sulfite and sulfide. A
plausible scheme, based on this suggestion, is shown in Fig. 6. If the
reaction proceeds according to this mechanism, involving enzyme-bound
intermediates, the dissimilatory and assimilatory sulfite reductases may
then differ principally in the lifetimes of the enzyme-bound intermediates
"X" and "Y" (the turnover numbers of the dissimilatory sulfite reductases are all considerably less than those of the assimilatory sulfite
reductases, with MVH as electron donor [229]) and perhaps in the
accessibility of the enzyme-bound intermediates to reaction with nucleophiles (such as SO_3^{2-}). It is interesting, in this regard, that Saks and

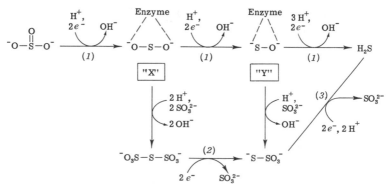

FIG. 6. Mechanism for dissimilatory sulfite reduction involving intermediates bound to a
single enzyme (modified from Kobayashi *et al.* [224, 225]). (*1*) Sulfite reductase; (*2*) trithionate reductase; and (*3*) thiosulfate reductase.

Siegel [229] have found that incubation of desulfoviridin with MVH and SO_3^{2-} in the presence of CN^- leads to the formation of SCN^- in addition to $S_3O_6^{2-}$, $S_2O_3^{2-}$, and H_2S.

While all sulfite reductases from sulfate assimilating organisms are similar in their spectral properties, the enzymes from sulfate reducing bacteria show marked divergence in their absorption spectra. P-582 and desulforubidin exhibit optical spectra characteristic of hemoproteins. The spectrum of P-582, with principal maxima at 392 and 582 nm [95,210,215,230] is similar to that of the assimilatory sulfite reductases, while the spectrum of desulforubidin, with principal maxima at 392 and 545 nm, is somewhat different [96,226]. Both proteins, however, react with CO or pyridine, in the presence of reducing agents, to yield complexes quite similar in spectrum to those of *E. coli* sulfite reductase with the same ligands [96,210,226,230]. EPR spectra of P-582 and desulforubidin show strong signals in the region characteristic of high spin ferriheme [95,96]. Upon extraction with acetone-HCl, both enzymes release a chromophore identical in its spectral and chromatographic properties to siroheme prepared from *E. coli* sulfite reductase [95,96]. The molecular weight of P-582, by gel filtration, is reported to be about 150,000 [210]; desulforubidin, with a sedimentation coefficient of 10 S, is probably similar in size [226]. P-582 is reported to contain Fe/S in addition to the siroheme prosthetic group [210].

Desulfoviridin has a molecular weight of 225,000 [217] and yields two protein bands of approximately equal intensity upon gel electrophoresis in the presence of sodium dodecyl sulfate. The molecular weights of these bands, 42,000–45,000 and 50,000–55,000, respectively, suggest an $\alpha_2\beta_2$ structure for the enzyme [217,224]. The absorption spectrum of desulfoviridin, with a split Soret band (principle maxima at 408 and 390 nm) and a major band in the visible region at 628 nm, is quite unlike that of any other sulfite reductase, and is, in fact, reminiscent of that of the cationic form of sirohydrochlorin, the tetrahydroporphyrin derived from siroheme by removal of the iron [95]. The absorption spectrum of desulfoviridin is not altered by incubation with oxidizing or reducing agents nor by CO [95,217,219], and the enzymatic activity is not inhibited by either CN^- or CO [208,225]. Extraction of desulfoviridin with acetone-HCl, or treatment with any of a number of mild denaturants, yields a fluorescent compound shown to be sirohydrochlorin [95], rather than siroheme. Although these results are in keeping with the suggestion of Postgate [219] that the prosthetic group of desulfoviridin may be a reduced porphyrin rather than a heme, EPR spectra of the enzyme show an EPR signal characteristic of high spin ferriheme [96]. Murphy and Siegel [95] have suggested that the unusual properties of desulfoviridin may be explained if the iron is substantially out of plane with respect to the porphyrin ring in this siroheme enzyme. Hoard [231] has shown that the metal in all high spin ferriheme compounds must be slightly out of plane with respect to the pyrrole nitrogen atoms, and it is anticipated that this effect is markedly exaggerated in desulfoviridin to the point where many of the properties of the tetrapyrrole are more those of a porphyrin than a heme. Whether the metal atom can function in catalysis under such conditions remains, of course, a prime subject for future investigation.

The association of the siroheme prosthetic group with sulfite reduction has possible implications for our understanding of heme evolution [95]. As reviewed by Peck [2], geological evidence indicates that the process of sulfate reduction by microorganisms is an ancient one. Thus, it has been possible to find sulfur with an $^{32}S/^{34}S$ ratio indicative of biological sulfate reduction [5,232,233] in rocks formed as early as 3.5 billion years ago [4], an era earlier in time than that generally proposed for the appearance of photosynthesis and respiration based on O_2 consumption [2]. The presence of the siroheme prosthetic group in both assimilatory and dissimilatory sulfite reductases suggests that this unusual reduced porphyrin compound is essential for the process of biological sulfite, and hence sulfate, reduction. The recent observation that *Clostridium bifermentans,* an obligate anaerobe previously thought to be devoid of heme compounds, possesses a sulfite reductase with a hemclike absorption spectrum (H.D. Peck, Jr. and J. LeGall, personal communication) lends support to this view. It may be inferred that the unique siroheme molecule, associated with one of the most ancient biological processes known, represents an ancient and possibly ancestral type of heme compound. In this regard, the remarkable stability of sirohydrochlorin metal complexes to photoreduction [94] (tetrahydroporphyrins are often the products of photoreduction of porphyrins) suggests that siroheme may have been particularly well suited to survive in a prebiotic photoreducing environment.

In addition to desulfoviridin, MVH-sulfite reductases of relatively low molecular weight have been purified from *D. vulgaris* by Haschke and Campbell [234] and Lee *et al.* [217]. The preparation of Haschke and Campbell has a molecular weight of 40,000 and reportedly exhibits no visible absorption. The Lee *et al.* [217] enzyme, on the other hand, exists as a mixture of monomer and dimer of 27,000 molecular weight subunits, and exhibits an absorption spectrum similar to that of assimilatory sulfite reductases. The Lee *et al.* enzyme catalyzes the stoichiometric reduction of SO_3^{2-} to H_2S with MVH as electron donor under conditions where the sole product of SO_3^{2-} reduction with desulfoviridin as catalyst was $S_3O_6^{2-}$; Lee *et al.* have suggested that the low molecular weight sulfite reductase, present in *D. vulgaris* in quantities much smaller than is desulfoviridin, may serve a role in sulfate assimilation rather than sulfate respiration in that organism. The pH optimum of the assimilatory sulfite reductase of *Desulfovibrio* is more alkaline than that of desulfoviridin; for that reason, the former enzyme has been referred to as a "sulfite reductase," while the latter has been termed "bisulfite reductase" [214,217].

D. Electron Transport and Phosphorylation

The sulfate reducing bacteria can couple respiratory chain phosphorylation to sulfate reduction to H_2S. The necessity for these organisms to do so is evident when one considers that *Desulfovibrio* is able to assimilate acetate plus CO_2 to form pyruvate in the presence of H_2 and SO_4^{2-} [180]. The energy to drive the condensation of CO_2 and acetate can only be derived from respiratory chain phosphorylation under these conditions. Similarly, *Desulfovibrio* can grow with lactate as sole carbon source, but only in the presence of sulfate. The stoichiometry of the reaction *in vivo* is [235]

$$2\ CH_3CH(OH)COO^- + SO_4^{2-} + 2\ H^+ \rightarrow 2\ CH_3COO^- + 2\ CO_2 + H_2S + 2\ H_2O$$

The overall reaction is made up of the following steps [236]:

(a) $2\ CH_3CH(OH)COO^- \rightarrow 2\ CH_3COCOO^- + 4\ H^+ + 4\ e^-$

(b) $2\ CH_3COCOO^- + 2\ HPO_4^{2-} \rightarrow 2\ CH_3C(O)OPO_3^{2-} + 2\ CO_2 + 2\ H^+ + 4\ e^-$

(c) $2\ CH_3C(O)OPO_3^{2-} + 2\ ADP^{3-} \rightarrow 2\ CH_3COO + 2\ ATP^{4-}$

(d) $SO_4^{2-} + 8\ e^- + 8\ H^+ + ATP^{4-} \rightarrow H_2S + 2\ H_2O + AMP^{2-} + 2\ HPO_4^{2-}$

(e) $AMP^{2-} + ATP^{4-} \rightarrow 2\ ADP^{3-}$

The enzyme systems utilized are, successively: (a) lactate dehydrogenase, (b) the pyruvate phosphoroclastic system, (c) acetokinase, (d) the sulfate reduction system, and (e) adenylate kinase. It is evident from inspection of these equations that lactate oxidation coupled to sulfate reduction can yield no net ATP by substrate-level phosphorylation alone. Vosjan [237] has calculated that *Desulfovibrio* must be able to conduct respiratory chain phosphorylation during electron transfer from lactate to sulfate with an average $P/2e^-$ ratio of approximately 0.5 to account for the cellular growth yields observed with this carbon source; i.e., *Desulfovibrio* must be able to produce up to 2 high energy phosphate bonds during the reduction of one molecule of SO_4^{2-} to H_2S.

Peck [236] provided support for the existence of respiratory chain phosphorylation in *Desulfovibrio* by showing that the ability of cells to reduce sulfate (which requires ATP for sulfate activation) with H_2 was strongly inhibited by 2,4-dinitrophenol. The ability to reduce SO_3^{2-} or $S_2O_3^{2-}$ (which does not require ATP) *in vivo* was not affected by the presence of the uncoupling agent. In 1966, Peck (184) directly demonstrated formation of ATP from ADP and P_i in *D. gigas* extracts conducting SO_3^{2-} reduction by H_2. The $P/2e^-$ ratio was low (0.1 to 0.2), but phosphorylation was inhibited *in vitro* by uncouplers such as dinitrophenol and gramicidin. Both soluble and particulate fractions from the bacterium were required; one of the soluble components is probably ferredoxin. A dinitrophenol- and Mg^{2+}-stimulated ATPase activity has been reported by Guarraia and Peck [238] in both the soluble and particulate fractions of the phosphorylating system; its role, if any, in the energy coupling process remains to be determined. No phosphorylation has as yet been found to be associated with the reductions of APS, $S_3O_6^{2-}$, or $S_2O_3^{2-}$. A particulate system from *D. gigas* which catalyzes phosphorylation associated with fumarate reduction by lactate or H_2 has been reported [183]. Fumarate can serve as an alternative terminal electron acceptor to sulfate in some *Desulfovibrio* strains [189].

The minimum sequence of electron transfer between pyruvate, H_2, and SO_3^{2-} compatible with existing data on electron transport in *Desulfovibrio* extracts [239–244] is shown in Fig. 7. Ferredoxin and flavodoxin are Fe/S and FMN containing proteins, respectively, with E_0' values for single electron transfer similar to that of the H_2 electrode (−0.42 V). Cytochrome c_3, as defined by DerVertanian and LeGall [239], is in reality a class of hemoproteins which contain multiple covalently bound mesohemes and exhibit a negative E_0'. There appear to be at least two types of cytochrome c_3 in *Desulfovibrio* extracts, one of molecular weight 13,000, which contains 4 hemes, and a second of molecular weight 26,000, which contains 8 hemes. The hemes are not equivalent within a given molecule of cytochrome c_3, and reduction and

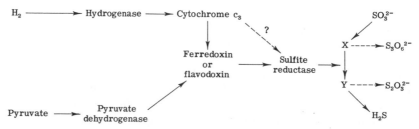

FIG. 7. Pathways of electron transfer between pyruvate, H_2, and sulfite in *Desulfovibrio* extracts.

oxidation proceed through a number of intermediate oxidation and reduction states [239]. The E_0' generally reported for cytochromes c_3,
about -0.25 V, clearly must represent an average potential. The ability
of cytochrome c_3 to transfer electrons between H_2 and ferredoxin, flavodoxin, or $S_2O_4{}^{2-}$, all the E_0' more negative than -0.4 V, indicates that
the cytochrome can operate between at least one set of oxidation states
at such a negative E_0'. Suh and Akagi [213,240] showed that both cytochrome c_3 and ferredoxin were required to couple H_2 oxidation to
$S_2O_3{}^{2-}$ reduction with partially purified *Desulfovibrio* thiosulfate reductase and hydrogenase, a result which indicates that an electron transfer
sequence similar to that of Fig. 7 can operate in $S_2O_3{}^{2-}$ as well as $SO_3{}^{2-}$
reduction. Cytochrome c_3, ferredoxin, and flavodoxin have individually
been reported to stimulate H_2-dependent reduction of $SO_4{}^{2-}$ plus ATP in
crude *Desulfovibrio* extracts, but the specificity of this stimulation remains open to question [245].

Since the average E_0' for $SO_3{}^{2-}$ reduction to H_2S is -0.12 V [20],
while the E_0' for oxidation of pyruvate (to acetyl phosphate and CO_2) or
H_2 is < -0.4 V, it is evident that there is sufficient energy available to
couple synthesis of ATP to $SO_3{}^{2-}$ reduction when either pyruvate or H_2
serves as ultimate electron donor (a $\Delta E_0'$ of 0.3 V corresponds to a $\Delta G_0'$
of -14 kcal/mole of electron pairs transferred). When lactate serves as
carbon source, however, the electrons derived from oxidation of lactate
to pyruvate are at an E_0' (-0.19 V) too positive to be used to drive both
$SO_3{}^{2-}$ reduction to H_2S and ATP synthesis. The electrons derived from
lactate could be used, of course, to reduce APS to sulfite ($E_0' = -0.06$
V). Since two moles of lactate and two moles of pyruvate are oxidized
by *Desulfovibrio in vivo* to yield the eight moles of electrons needed for
reduction of a mole of $SO_4{}^{2-}$ to H_2S, it is clear that at least two of the six
electrons required for $SO_3{}^{2-}$ reduction to H_2S must be derived from lactate itself, while four of the electrons may be derived from pyruvate. It
seems plausible that the low potential electrons from the latter source
would be used to reduce sulfite, via ferredoxin or flavodoxin (Fig. 7),
with concomitant phosphorylation. However, since sufficient electrons
from this source are not available to permit stoichiometric reduction of
sulfite to H_2S, the phenomenon described by Kobayashi *et al.* [225] and
Siegel and Saks [229] for sulfite reduction in the presence of MVH may
come into play; i.e., with low steady state levels of reduced ferredoxin or
flavodoxin available and a relatively high concentration of sulfite
(formed from APS using lactate electrons), the lifetime of the intermediates "X" and "Y" may be so prolonged that they can react with $SO_3{}^{2-}$
to form free $S_3O_6{}^{2-}$ and $S_2O_3{}^{2-}$. The release of $S_3O_6{}^{2-}$ under these circumstances could be of advantage to the organism, since the E_0' for

reduction of $S_3O_6^{2-}$ to $S_2O_3^{2-}$ and SO_3^{2-} is quite positive (+0.35 V [20]). Such a reduction could provide a means for utilizing the "excess" lactate electrons, of relatively positive E_0', during the overall process of sulfite reduction, perhaps even with phosphorylation if the appropriate coupling machinery were available. The reduction of $S_2O_3^{2-}$ to H_2S and SO_3^{2-}, on the other hand, exhibits an E_0' of -0.42 V [20]; electrons derived from lactate could not drive this reaction. Pyruvate, via reduced ferredoxin or flavodoxin [213], could, of course serve as ultimate reductant for $S_2O_3^{2-}$, although no phosphorylation would be expected to accompany this reduction. Since $S_2O_3^{2-}$ reduction represents a potential diversion of electrons from a phosphorylating to a nonphosphorylating pathway, it would be expected that this reduction would occur only when sulfite itself is not readily available for reduction. The high K_m of thiosulfate reductase for $S_2O_3^{2-}$ and the strong inhibition exerted by SO_3^{2-} on this reaction with partially purified enzyme [222] are in keeping with this hypothesis. Since some accumulation of $S_2O_3^{2-}$ would be expected to accompany lactate oxidation in *Desulfovibrio*, it is interesting to recall the old observation of Ishimoto *et al.* [235] that when *Desulfovibrio* cells are incubated with sulfate in the presence of lactate, a pronounced lag period in H_2S production (not seen when pyruvate or H_2 serve as electron donors) is observed.

In summary, then, in the steady state, the model that we have proposed suggests that the electrons of lactate are used predominantly for reduction of APS to SO_3^{2-} and of $S_3O_6^{2-}$ to $S_2O_3^{2-}$, while electrons of lower E_0' are utilized predominantly for reduction of SO_3^{2-} to $S_3O_6^{2-}$ and $S_2O_3^{2-}$ to H_2S. Phosphorylation may be associated with both sulfite and trithionate reductases. With H_2 or pyruvate as electron donors, it is anticipated that SO_3^{2-} reduction would proceed directly to H_2S, via sulfite reductase, without the formation of substantial quantities of either $S_3O_6^{2-}$ or $S_2O_3^{2-}$ *in vivo*.

IV. CHEMOLITHOTROPHIC OXIDATION OF REDUCED SULFUR COMPOUNDS

A. Oxidation of Sulfide and Sulfur

Although several bacterial genera can use the oxidation of reduced sulfur compounds as a source of electrons and energy for biosynthetic reactions, e.g., CO_2 assimilation, the mechanism of these oxidations has been studied to a significant extent only in the genus *Thiobacillus*. The biology and general metabolism of this group of bacteria have been reviewed elsewhere [8,13,180,246].

Extracts that can catalyze the O_2-dependent oxidation of H_2S to SO_4^{2-} have been prepared from a number of thiobacilli [247–252]. One species, *T. denitrificans,* is capable of oxidizing sulfide anaerobically, with nitrate as terminal electron acceptor [252,253]. Elemental sulfur (S^0) frequently precipitates during sulfide oxidation [249,254,255], and at least one preparation has been described in which S^0 was the sole product detected [256]. SO_3^{2-} [252] and $S_2O_3^{2-}$ [248–251,254] have also been detected as products of H_2S oxidation.

In several strains, H_2S-dependent O_2 uptake is distinctly biphasic, both *in vivo* [250,256] and in extracts [250–252,257,258]. The rapid phase is catalyzed by a particulate enzyme system and in *T. concretivorus* extracts is associated with the disappearance of 2.2±0.1 moles of H_2S per mole of O_2 consumed [251], a result that suggests that the product of this phase of H_2S oxidation is at an oxidation level close to that of S^0. No precipitation of S^0 was detected, however, in these experiments. When $H_2{}^{35}S$ was incubated for a brief period with *T. concretivorus* crude extract, 95% of the ^{35}S was converted to a form which was excluded from a column of Sephadex G-50. The "excluded" ^{35}S was not dialyzable and was associated with particulate components of the extract following centrifugation. Boiling of the extract prevented conversion of $H_2{}^{35}S$ to the particle-bound form. Incubation of sulfide with the crude extract was accompanied by appearance of an absorption band at approximately 315 nm [251,252,257]. Since the increased absorbance was largely abolished upon incubation of the sulfide treated extract with CN^- or SO_3^{2-} (nucleophiles which cleave RS—SH bonds), while addition of I_2 led to precipitation of elemental sulfur, it was concluded that the product of H_2S oxidation by the *T. concretivorus* extract is a hydropolysulfide [251,257]. A mass spectrogram of the Sephadex G-50 excluded ^{35}S indicated the presence of molecules containing at least eight covalently linked sulfur atoms, and reactions of the excluded ^{35}S with nucleophiles led to products expected for decomposition of a linear hydropolysulfide. The nature of the particle to which the hydropolysulfide appears to be bound has not been elucidated. Nicholas *et al.* [251,252] have shown that the Sephadex G-50 excluded hydropolysulfide can be oxidized in air by *T. concretivorus* or *T. denitrificans* extracts to yield SO_3^{2-} and SO_4^{2-}, a result that is consistent with a role for particle-bound hydropolysulfide as an intermediate in H_2S oxidation. Although direct formation of elemental sulfur from the hydropolysulfide has not been demonstrated, such a reaction might well be expected as a result of the spontaneous slow decomposition of such a compound, thus accounting for the formation of S^0 during H_2S oxidation by a number of thiobacilli.

A soluble system which catalyzes the aerobic oxidation of elemental sulfur to $S_2O_3^{2-}$ has been purified from a number of *Thiobacillus* species [259–262]. Suzuki and Silver [261,262] have shown that in the presence of the sulfite trapping agent formaldehyde, the initial product of S^0 oxidation is in fact SO_3^{2-}, and that $S_2O_3^{2-}$ is formed as the result of a rapid nonenzymatic reaction between S^0 (or, more likely, a hydropolysulfide intermediate derived from S^0) and sulfite. Adair, by preparing an extract from *T. thiooxidans* which could oxidize both S^0 and SO_3^{2-}, but not $S_2O_3^{2-}$, to SO_4^{2-}, has demonstrated that $S_2O_3^{2-}$ cannot be an intermediate in S^0 oxidation [263]. Suzuki [260] has suggested that the S^0 oxidizing enzyme is an oxygenase, since small amounts of $^{18}O_2$ were incorporated into the $S_2O_3^{2-}$ formed from S^0 in extracts. However, no $^{18}O_2$ incorporation was detected during S^0 oxidation *in vivo,* and the oxygenase hypothesis is opposed by the fact that *T. denitrificans* can oxidize S^0 to SO_4^{2-} anaerobically with NO_3^{2-} as electron acceptor [253].

The purified S^0 oxidizing enzymes require small amounts of GSH for activity [259–262]. Suzuki [259] has shown that GSH can rapidly react with S^0 to form a hydropolysulfide, and that preformed glutathione hydropolysulfide (GS_nSH) can serve as substrate for the soluble S^0 oxidizing enzyme [262,264]. Since several crude S^0 oxidizing systems do not require GSH, but are strongly inhibited by thiol binding reagents [249,263,265], it is probable that GSH is replacing a carrier thiol (RSH), possibly particle-bound, that acts in such extracts to convert S^0 into the more reactive hydropolysulfide. Taylor [249] has shown that

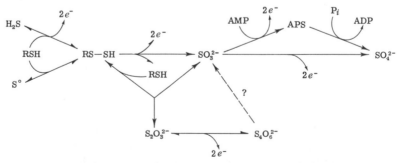

FIG. 8. Proposed scheme for oxidation of inorganic sulfur compounds in the thiobacilli. RSH represents a particle-bound thiol which may be replaced *in vitro* by GSH.

incubation of S^0 with a crude extract capable of S^0 oxidation leads to production of some H_2S under anaerobic conditions. Sulfide is not an obligatory intermediate in S^0 oxidation, however, since partially purified S^0 oxidizing enzyme preparations (GSH-dependent) will not catalyze H_2S oxidation [262]. These results have led to the suggestion [12], for which there is admittedly little evidence, that a common intermediate, probably a membrane-bound hydropolysulfide (RS_nSH), may be involved in both H_2S and S^0 oxidation in the thiobacilli (Fig. 8).

B. Oxidation of Sulfite

There are two pathways for the oxidation of SO_3^{2-} to SO_4^{2-} in the thiobacilli. One, originally proposed by Peck [266], is dependent on AMP and involves participation of the following reactions, catalyzed by APS-reductase, ADP-sulfurylase (EC 2.7.7.5; ADP:sulfate adenylyl-transferase), and adenylate kinase (EC 2.7.4.3; ATP:AMP phospho-transferase), respectively [80,202,247,266–270]:

$$SO_3^{2-} + AMP^{2-} \rightarrow APS^{2-} + 2\ e^-$$

$$APS^{2-} + P^{2-} \rightarrow SO_4^{2-} + ADP^{3-} + H^+$$

$$2\ ADP^{3-} \rightarrow AMP^{2-} + ATP^{4-}$$

This pathway conserves some of the free energy of oxidation of SO_3^{2-} as ATP, via substrate-level phosphorylation. The second pathway, cata-lyzed by sulfite oxidase (EC 1.8.2.1; sulfite: ferricytochrome c oxido-reductase), is not dependent on the presence of AMP [248, 252,256,263,270–278], and is probably associated with respiratory chain phosphorylation *in vivo:*

$$SO_3^{2-} + H_2O \rightarrow SO_4^{2-} + 2\ H^+ + 2\ e^-$$

Extracts of *T. thioparus* and *T. denitrificans* have been shown to possess simultaneously both APS-reductase and sulfite oxidase [252,270], although in some thiobacilli only the latter activity has been detected [256,274].

APS-reductase represents 3–5% of the soluble protein in extracts of *T. thioparus* and *T. denitrificans,* a result that indicates a major role for this enzyme in the energy metabolism of these organisms. The enzyme has been highly purified from both species, and is similar in structure to the APS-reductase of *Desulfovibrio;* it has a molecular weight of 170,000 [279] and contains FAD and Fe/S prosthetic groups

[269,279]. Both the *T. thioparus* and *T. denitrificans* APS-reductases can utilize ferricyanide as an electron acceptor for AMP-dependent SO_3^{2-} oxidation; the *T. thioparus* enzyme, however, is far more active if yeast cytochrome c is used as electron acceptor. Both enzymes catalyze a very slow oxidation of sulfite in the presence of O_2. Analysis of the steady state kinetics of SO_3^{2-} oxidation by the *T. thioparus* APS-reductase in the presence of AMP and cytochrome c led Lyric and Suzuki [280] to propose a reaction mechanism in which SO_3^{2-} and two molecules of oxidized cytochrome c react with the enzyme prior to AMP, while reduced cytochrome c is released prior to APS. This mechanism is at variance with that proposed by Peck *et al.* [203,204] for the *Desulfovibrio* APS-reductase.

AMP-independent SO_3^{2-} oxidation activity is associated with particulate material in extracts of several thiobacilli [252,263,272,278]; however, a soluble enzyme catalyzing SO_3^{2-} oxidation with cytochrome c or ferricyanide as electron acceptor has been purified [256,270,274–276]. The *T. thioparus* sulfite oxidase appeared to be more than 80% pure by gel electrophoresis following a 160-fold purification [270]. Unlike APS-reductase and the sulfite oxidase of animals, purified *Thiobacillus* sulfite oxidase does not seem to utilize O_2 as an electron acceptor at all. When native cytochrome c and particles containing cytochrome c oxidase activity are added to the purified sulfite oxidase, however, an activity which catalyzes electron transfer from SO_3^{2-} to O_2 can be reconstituted [256,274]. Like the animal sulfite oxidase, the *T. thioparus* enzyme has a molecular weight of about 54,000 [270], and contains sulfite reducible molybdenum [281]. Unlike the animal enzyme, however, the *Thiobacillus* sulfite oxidase does not appear to contain a heme prosthetic group [270], a fact which may correlate with the inability of the latter enzyme to utilize O_2. Steady state kinetic analysis of sulfite oxidation by cytochrome c catalyzed by the *T. thioparus* sulfite oxidase suggests that SO_3^{2-} initially reduces the enzyme, following which the enzyme is reoxidized by cytochrome c in two separate univalent steps [280]. The reason for having two separate mechanisms for sulfite oxidation in the thiobacilli is not apparent. Lyric and Suzuki [282] have suggested that the two enzymes may function *in vivo* at different pH values.

C. Oxidation of Thiosulfate

The ability to oxidize $S_2O_3^{2-}$ to SO_4^{2-} is a property of all species and strains of thiobacilli studied to date [250,256,264,277,283–292]. $S_4O_6^{2-}$ and other polythionates are generally formed transiently during $S_2O_3^{2-}$

oxidation *in vivo* and in extracts, and kinetic studies have generally shown that $S_4O_6^{2-}$ is the first product of $S_2O_3^{2-}$ oxidation to be formed *in vivo* [285,287,288,293], both sulfur atoms of the thiosulfate appearing in the tetrathionate formed. It is natural, then, that $S_4O_6^{2-}$ should have been proposed as an obligatory intermediate in $S_2O_3^{2-}$ oxidation to SO_4^{2-} in the thiobacilli [248,294–296]. Charles and Suzuki [255] and Lee *et al.* [271] have shown that this is not the case, however, by preparing extracts which were able to oxidize $S_2O_3^{2-}$ to SO_4^{2-}, but which were unable to convert either $S_2O_3^{2-}$ to $S_4O_6^{2-}$ or $S_4O_6^{2-}$ to SO_4^{2-}. It seems likely that $S_2O_3^{2-}$ can be oxidized by two different routes in the thiobacilli: (a) it may be oxidized to $S_4O_6^{2-}$, and the latter compound converted to other polythionates and ultimately oxidized to SO_4^{2-} (possibly by disproportionation and/or rereduction of the polythionates to $S_2O_3^{2-}$ and/or SO_3^{2-} [12]) and (b) it may be cleaved into its sulfane and sulfonate moieties, and the two resulting compounds independently oxidized to SO_4^{2-}. $S_4O_6^{2-}$ formation is favored under conditions of high concentration of $S_2O_3^{2-}$ and O_2 relative to cell density [297]; when either of these compounds becomes limiting, $S_2O_3^{2-}$ becomes increasingly converted to SO_4^{2-}.

Enzymes that catalyze oxidation of $S_2O_3^{2-}$ to $S_4O_6^{2-}$ with ferricyanide as electron acceptor have been partially purified from several thiobacilli [264,270,282,298,299]. Although these $S_2O_3^{2-}$-ferricyanide reductases generally react poorly with mammalian cytochrome c, a *T. neapolitanus* cytochrome $c_{553.5}$ serves as an excellent electron acceptor for the enzyme from that organism [298]. In the presence of $S_2O_3^{2-}$-ferricyanide reductase, cytochrome $c_{553.5}$, and a particulate fraction from *T. neapolitanus* containing cytochrome oxidase activity $S_2O_3^{2-}$ is rapidly and quantitatively converted to $S_4O_6^2$ with O_2 as electron acceptor [298,300]. The thiosulfate-ferricyanide reductase from *T. thioparus* is reported to have a molecular weight of 115,000; the 150-fold purified preparation contains 1–2 atoms of nonheme iron per enzyme equivalent weight [282]. Pyridine nucleotides, O_2, and diaphorase acceptors were inactive as electron acceptors for the enzyme. The $S_2O_3^{2-}$-ferricyanide reductase activity is relatively insensitive to inhibition by thiol binding agents, iron chelating agents, or CN^- [264]. On the other hand, SO_3^{2-} irreversibly inactivates the enzyme, 50% inactivation being observed with 5×10^{-6} M sulfite. Since SO_3^{2-} is a product of the "cleavage" pathway of $S_2O_3^{2-}$ oxidation, inhibition by this ion could be of physiological significance in preventing oxidation of $S_2O_3^{2-}$ to $S_4O_6^{2-}$ under conditions where reducing power (and ATP) are available from alternative sources, Trudinger [301] reported that the *in vivo* oxidation of $S_2O_3^{2-}$ to

$S_4O_6^{2-}$ by *T. neapolitanus* cells is inhibited 80% in the presence of $1 \times 10^{-3} M$ SO_3^{2-}.

The mechanism of cleavage of the sulfane and sulfonate moieties of $S_2O_3^{2-}$ is unclear. *In vivo* studies with [^{35}S] thiosulfate showed that although both sulfur atoms give rise to SO_4^{2-}, there is frequently a preferential conversion of the sulfonate S into SO_4^{2-} [247,294,297, 302–305]. Santer [306] showed that when [^{18}O]phosphate is incubated with *T. thioparus* cells oxidizing $S_2O_3^{2-}$, ^{18}O is incorporated into the SO_4^{2-} formed; the ^{18}O incorporation is insensitive to 2,4-dinitrophenol. Peck and Stulberg [268] interpreted this to mean that SO_3^{2-} is an intermediate in $S_2O_3^{2-}$ oxidation, and that SO_3^{2-} oxidation under the conditions of Santer's experiment proceeds largely via the APS pathway. Peck and Fisher [247,266] were able to prepare extracts from *T. thioparus* that oxidized $S_2O_3^{2-}$ to SO_4^{2-}; the oxidation was dependent upon the presence of stoichiometric quantities of GSH and was markedly stimulated by AMP. During the *in vitro* $S_2O_3^{2-}$ oxidation, ^{18}O was transferred from [^{18}O]AMP (labeled in the phosphate moiety) into SO_4^{2-} [268], and ^{32}P$_i$ was esterified into ADP in a reaction insensitive to dinitrophenol [247], both results being consistent with oxidation of $S_2O_3^{2-}$ by a pathway involving APS. The *T. thioparus* extracts were shown to contain an enzyme similar to one described in yeast by Kaji and McElroy [223], which catalyzes the reduction of $S_2O_3^{2-}$ to SO_3^{2-} and H_2S by GSH:

$$S_2O_3^{2-} + 2 \text{ GSH} \rightleftharpoons \text{GSSG} + SO_3^{2-} + H_2S$$

Pcck suggested that this reaction represents the mechanism of $S_2O_3^{2-}$ cleavage in thiobacilli, and that H_2S is therefore an obligatory intermediate in thiosulfate oxidation [247,266].

The Peck mechanism has been challenged by other workers [255,256,275] who have found that (a) *Thiobacillus* extracts can be prepared which do not oxidize H_2S to SO_4^{2-}, yet rapidly catalyze $S_2O_3^{2-}$ oxidation to SO_4^{2-}, in the absence of GSH, and (b) GSH, in stoichiometric quantities, is generally strongly inhibitory to $S_2O_3^{2-}$ oxidation in *Thiobacillus* extracts. Charles and Suzuki [255,256] suggested, instead, that $S_2O_3^{2-}$ is cleaved to SO_3^{2-} and elemental S^0. Indeed S^0 is frequently observed as a product of $S_2O_3^{2-}$ oxidation *in vivo,* and any S^0 which is formed arises exclusively from the sulfane atom [247,296,297,302,303]. When Peck and Fisher [247] added a large pool of nonradioactive S^0 to *T. thioparus* extracts oxidizing ^{35}SSO_3^{2-}, at least half of the radioactivity was recovered in ^{35}S^0, a result which suggests that, at the least, S^0 is in isotopic equilibrium with a product derived from the sulfane atom of thiosulfate. On the other hand, Kelly and Syrett [305] could find no

[35]S appearing in S^0 during simultaneous oxidation of [35]SSO_3^{2-} and nonlabeled S^0 by *Thiobacillus* strain C, and Charles [256] has described an extract of *T. intermedius* which can oxidize $S_2O_3^{2-}$, but not S^0, to SO_4^{2-}. These results suggest that neither S^0 nor H_2S is an obligatory intermediate in oxidation of the sulfane atom of $S_2O_3^{2-}$. The data are compatible, however, with cleavage of thiosulfate by transfer of the sulfane moiety to a carrier thiol, possibly the particulate RSH believed to be involved in H_2S and S^0 oxidation, to form a hydropolysulfide (Fig. 8). Such a reaction could be catalyzed by the thiosulfate sulfurtransferase (EC 2.8.1.1; rhodanese, thiosulfate:cyanide sulfurtransferase) commonly found in *Thiobacillus* extracts [262,274,307,308]. It should be emphasized that even the existence of the RSH carrier is speculative at this point.

D. Electron Transport and Phosphorylation

Most thiobacilli are chemautotrophs, i.e., they can grow with CO_2 as sole carbon source, with the ATP and reducing equivalents necessary for CO_2 assimilation (via the ribulose-1,5-diphosphate carboxydismutase pathway) being provided by the oxidation of inorganic sulfur compounds [180,246]. Most of the thiobacilli are obligate aerobes, O_2 serving as ultimate electron acceptor for this oxidation. At least one species, *T. denitrificans,* however, can assimilate CO_2 under anaerobic conditions, with nitrate serving as terminal electron acceptor for sulfur compound oxidation. Although ATP can be formed by substrate level phosphorylation during SO_3^{2-} oxidation via the APS pathway, Hempfling and Vishniac [291,309] have calculated that the amount of ATP which can be obtained by this mechanism is insufficient to account for growth yields of thiobacilli observed when $S_2O_3^{2-}$ oxidation to SO_4^{2-} serves as sole source of reducing equivalents and energy. These calculations indicate that thiobacilli must be capable of respiratory chain phosphorylation, and indeed, formation of ATP sensitive to inhibition by uncoupling agents has been demonstrated during the aerobic oxidation of H_2S [250,251,310–312], $S_2O_3^{2-}$ [250,310,311,313,314], and SO_3^{2-} [251,274,315] by extracts of several *Thiobacillus* species. The following P/O ratios have been observed *in vitro:* H_2S, 1.0–1.4 [311]; $S_2O_3^{2-}$, 0.4–0.9 [311,313,314]; and SO_3^{2-}, 0.1 [274].

Thiobacilli contain a large number of cytochromes (a, b, c, d, and o types have been observed), as well as other potential respiratory carriers such as flavins and ubiquinone. Although it is possible that separate electron transport chains exist for the oxidation of different sulfur compounds, available evidence, based on (a) studies of the effect of site-specific inhibitors (the specificity of which has been rigorously established for animal respiratory systems, but not for thiobacilli) on the oxidation of individual sulfur compounds in extracts, (b) the ability of individual sulfur compounds to reduce specific respiratory components as determined by difference spectroscopy, and (c) limited attempts to reconstitute electron transport from sulfur compounds to O_2 using partially purified enzymes and respiratory components, has led to the minimum sequence of electron flow from H_2S, $S_2O_3^{2-}$, and SO_3^{2-} to O_2 shown in Fig. 9. The entry of $S_2O_3^{2-}$ electrons into the respiratory chain at the level of cytochrome c (E_0' for native *Thiobacillus* cytochrome c is approximately $+0.2$ V [316,317]), seems well established [250,272,298,299,313] and is in accord with the E_0' of $+0.08$ V for the $S_2O_3^{2-}/S_4O_6^{2-}$ redox couple [318]. On the other hand, it is not clear whether H_2S interaction with the chain involves a flavoprotein, since amytal is not inhibitory to H_2S oxidation in two *Thiobacillus* species [250,312], while it does inhibit in a third [311]. The observed inhibitions, the ability to obtain P/O ratios greater than one for H_2S oxidation by O_2, and the E_0' for the H_2S/S^0 redox couple (-0.24 V [319]) are all consistent with entry of H_2S electrons at a carrier prior to cytochrome b [250,252,257,292,311]. Inhibitor data [278,292] also indicate that SO_3^{2-} electrons enter the respiratory chain at an early point, probably at the flavoprotein level, in keeping with the very negative E_0' of the SO_3^{2-}/SO_4^{2-} redox couple (-0.48 V [20]). However, considerable data do suggest a direct interaction of sulfite oxidase with cytochrome c in extracts [256,274], and only very small P/O ratios have as yet been obtained for SO_3^{2-} oxidation *in vitro* [274]. Although S^0 oxidation in *Thiobacillus* extracts is inhibited by CO and CN^- [251,274,315,320–326], there are no data concerning the site of entry of S^0 electrons into the respiratory chain, if indeed such electrons do enter

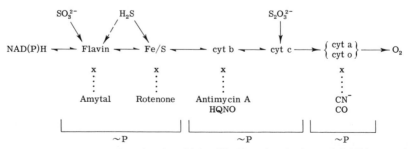

FIG. 9. Electron transport in the thiobacilli. Postulated sites of inhibitor action ($\cdots$x) and phosphorylation ($\sim P$) are indicated. HQNO, 2-*n*-Heptyl-4-hydroxy-quinoline-*N*-oxide.

the chain at all. Adair [323] has suggested that ubiquinone may be involved in S^0 oxidation. The relationship between APS and the *Thiobacillus* respiratory chain has not been demonstrated.

Since the E_0' for $S_2O_3^{2-}/S_4O_6^{2-}$ is considerably more positive than that of $NAD(P)H/NAD(P)^+$, the reduction of pyridine nucleotides, necessary for CO_2 assimilation, by electrons derived from $S_2O_3^{2-}$ is an energy requiring process. Aleem *et al.* [324–326] have shown that electrons can flow uphill from $S_2O_3^{2-}$ (or reduced cytochrome c) to $NAD(P)^+$ in *Thiobacillus* extracts under anaerobic conditions if ATP is added; 2–3 ATP are utilized per molecule of pyridine nucleotide reduced. ATP-dependent reduction of flavins by $S_2O_3^{2-}$ has also been demonstrated [313,326]. The ATP-dependent reverse electron flow is inhibited by uncoupling agents, oligomycin, flavoprotein inhibitors, and inhibitors of electron flow between cytochromes b and c [326]. In the presence of O_2, reverse electron flow from $S_2O_3^{2-}$ to $NAD(P)^+$ in extracts does not require ATP addition [324,325], suggesting that either ATP or a high energy intermediate or state generated *in situ* during electron flow from $S_2O_3^{2-}$ to O_2 is utilized to drive the reverse electron flow. In agreement with these results, Roth *et al.* [312] have reported that $NAD(P)^+$ is reduced by either $S_2O_3^{2-}$ or H_2S *in vivo* by *Thiobacillus* cells under aerobic (but not anaerobic) conditions; the $NAD(P)^+$ reduction is strongly inhibited by uncoupling agents and amytal, while ATP formation accompanying $S_2O_3^{2-}$ or H_2S oxidation by O_2 is sensitive to uncouplers but not amytal. The capacity to couple ATP formation to the exergonic flow of electrons from inorganic sulfur compounds to O_2, while utilizing a portion of the ATP formed to drive reverse electron flow from these compounds to pyridine nucleotides, explains the ability of the thiobacilli to function as lithotrophic chemautotrophs. The thermodynamic efficiency of energy utilization by these organisms for CO_2 assimilation when oxidizing S^0 has been estimated as better than 20% [327–329].

V. PHOTOLITHOTROPHIC OXIDATION OF REDUCED SULFUR COMPOUNDS

Bacteria of the families Chlorobiaceae (green sulfur bacteria) and Chromatiaceae (purple sulfur bacteria) [330] are capable of anaerobic light-dependent oxidation of inorganic sulfur compounds such as H_2S, S^0, $S_2O_3^{2-}$, and SO_3^{2-} to SO_4^{2-} [14,331]. The electrons derived from these oxidations may be used for reductive assimilation of CO_2 (making these organisms photolithotrophic autotrophs), reduction of N_2, or evolution of H_2 [332,333]. There appears to be a single bacteriochlorophyll photocenter in these bacteria [334–337] that is involved in (a) cyclic photophosphorylation, via a pathway involving cytochrome c_{555} in *Chlorobium thiosulfatophilum* and *Chromatium vinosum* strain D [338–341], and (b) a noncyclic pathway of electron flow from H_2S to $NAD(P)^+$, which involves cytochrome c_{553} in the same organisms [341–343]. There is evidence that cytochromes c_{553} and c_{555}, in addition to donating electrons directly to the photochemical center [335,336,344], can transfer electrons between themselves [337,345]. The photosynthetic sulfur bacteria contain a number of other electron carriers in addition to these cytochromes; the roles of most of these carriers is not clear, although schemes incorporating them have been devised [339,341,344,346–348].

The primary acceptor of electrons from the photocenter bacteriochlorophyll ($E_0' = +0.3$–0.5 V [339,344,349]) appears to be an Fe/S carrier ($E_0' = -0.13$ V [344,349,350]) with a characteristic EPR signal at $g = 1.82$. Cytochromes c_{553} and c_{555} have E_0' values of 0–0.1 V [351,352] and $+0.15$–0.3 V [344,353] respectively. Kusai and Yamanaka [345] have found that a purified cytochrome c_{553} from *C. thiosulfatophilum*, a protein of 50,000 molecular weight that contains one molecule each of covalently bound mesoheme and FMN [353], can be reduced directly by H_2S and the electrons transferred in the dark to purified cytochrome c_{555}. A similar flavocytochrome has been purified from *Chromatium D* [354].

A cytochrome c_{551} ($E_0' = +0.14$ V [352]) has been implicated as an electron acceptor for the oxidation of $S_2O_3^{2-}$ catalyzed by a partially purified enzyme from *C. thiosulfatophilum* [345]. Curiously, the enzyme catalyzed transfer of electrons from $S_2O_3^{2-}$ to cytochrome c_{551} requires the presence of small amounts of cytochrome c_{555}, despite the fact that the latter cytochrome does not appear to serve directly as an electron acceptor for the enzyme. The $S_2O_3^{2-}$ oxidation is highly sensitive to inhibition by CN^- and SO_3^{2-}; the latter inhibition, as in the thiobacilli, may be of physiological significance.

The mechanism of reduction of low potential compounds such as ferredoxin and $NAD(P)^+$ by the photoreducible primary electron acceptor, with its relatively positive E_0' of -0.13 V, is unresolved. Gest and others [347,355,356] have suggested that ATP or a high energy intermediate or state generated by cyclic photophosphorylation drives this reaction by reverse electron flow. Others have presented data which opposes this view, however [357]. A scheme which incorporates the data currently available is presented in Fig. 10.

The limited information available on the pathways of H_2S and $S_2O_3^{2-}$

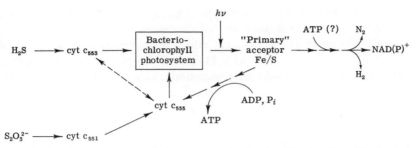

FIG. 10. Postulated sequence of electron transfer in photolithotrophic sulfur bacteria.

oxidation in the photosynthetic bacteria suggests that the reactions involved are similar to those in the thiobacilli. Thus, during the photooxidation of either compound, elemental S^0 is rapidly formed [351, 358,359] and may accumulate either within or outside the bacterial cell [330,360] prior to its slower oxidation to sulfate. The S^0 is derived exclusively from the sulfane atom of $S_2O_3^{2-}$, the sulfonate sulfur being rapidly converted to SO_4^{2-} *in vivo* [359,361]. Both a thiosulfate reductase, which catalyzes reduction of $S_2O_3^{2-}$ to H_2S and SO_3^{2-} in the presence of either GSH, dihydrolipoamide, or MVH as electron donor [362], and a thiosulfate sulfurtransferase [359,362] have been detected in *Chromatium* extracts. Although there is no direct evidence that SO_3^{2-} is an intermediate in the oxidation of H_2S, S^0, or $S_2O_3^{2-}$ by photolithotrophic bacteria, its involvement is likely, since high levels of APS-reductase activity have been detected in several strains of these bacteria [352,363]. The APS-reductase from *Thiocapsa roseopersicina* has been purified to homogeneity by Truper and Rogers [364]; it has a molecular weight of 180,000 and contains 1 flavin, 4–6 Fe/S, and 2 heme c per enzyme molecule. Both cytochrome c and ferricyanide can serve as electron acceptors for AMP-dependent sulfite oxidation to APS. In some photosynthetic sulfur bacteria, the APS-reductase appears to be bound to the chromatophores rather than free in solution [363]. ADP-sulfurylase activity has been detected in extracts of *Thiocapsa fluoridana* [352]. No AMP-independent SO_3^{2-} oxidase has been detected in any of the photosynthetic bacteria examined [352], a result which suggests that these organisms may utilize only the APS pathway of sulfite oxidation. Both *Chromatium* and *Chlorobium* species contain an enzyme which will catalyze the oxidation of $S_2O_3^{2-}$ to $S_4O_6^{2-}$ [365,366], and it is presumably this enzyme which Kusai and Yamanaka [345] have isolated as an $S_2O_3^{2-}$-cytochrome c_{551} reductase. Tetrathionate seems to be produced from $S_2O_3^{2-}$ *in vivo* in *Chromatium* only at pH values below the optimum for growth [365], a result which makes this reaction of questionable physiological significance in that organism.

The Rhodospirillaceae, or purple nonsulfur bacteria, are characterized by an inability to grow well with H_2S or S^0 as photosynthetic electron donors. Instead, simple organic compounds, as well as H_2, can serve this function [330]. These bacteria are microaerophilic and can grow, in the presence of O_2, in the dark as well as the light. A few strains of the Rhodospirillaceae are able to grow photoautotrophically with $S_2O_3^{2-}$ as reductant. Eley *et al.* [367,368] have prepared an extract of photolithotrophically grown *Rhodopseudomonas palustris* which catalyzes S_2O_3 oxidation with cytochrome c or O_2 as electron acceptors. The active fraction contained c, a, and o type cytochromes and was virtually devoid of bacteriochlorophyll. The extract catalyzed an anaerobic reduction of NAD^+ dependent upon addition of $S_2O_3^{2-}$ and ATP. This reverse electron flow was inhibited by uncouplers, Amytal, rotenone, antimycin, and HQNO, indicating a pathway of electron transfer similar to that found in the thiobacilli. It seems likely that the $S_2O_3^{2-}$ oxidizing system studied by Eley *et al.* is more related to the aerobic dark metabolism of *R. palustris* than to its photometabolism.

VI. OXIDATION OF REDUCED SULFUR COMPOUNDS IN ANIMALS

A. Oxidation of Sulfide and Thiosulfate

Animals can oxidize H_2S, $S_2O_3^{2-}$, and SO_3^{2-} to sulfate. When any of these compounds is administered to rats or dogs, either by injection or ingestion (or, in the case of SO_2, by inhalation), a large proportion of the input sulfur is eventually excreted in the urine as free or esterified sulfate [369–375]. The oxidations can be catalyzed *in vitro* [376–379], with the highest activities being observed in liver and kidney [380,381], although other organs may be capable of oxidizing some of these compounds at lower rates. The enzyme systems involved appear to be localized in the mitochondria [380–382].

Haggard [383] found that oxygenated blood plasma could oxidize H_2S and that this oxidation could account for the ability of a dog to survive a slow venous infusion of H_2S, the administration of which in a single dose would have killed the animal [384]. Curtis *et al.* [375] have shown that the rate of H_2S oxidation by blood is far too slow to account for the rate of its oxidation in the rat *in vivo*, and that a rapid oxidation of H_2S occurs if blood containing this compound is perfused through the liver. Baxter *et al.* [380,385] and Sörbo [386] reported that extracts of rat liver and kidney could catalyze an O_2-dependent oxidation of H_2S to $S_2O_3{}^{2-}$. Elemental S^0 could not substitute for H_2S in this system [380]. Both hydropolysulfide [385,387] and $SO_3{}^{2-}$ were implicated as intermediates in this oxidation, and it is likely that the $S_2O_3{}^{2-}$ observed is formed via a side reaction between $SO_3{}^{2-}$ and polysulfide. The pathway of H_2S oxidation in the rat thus appears to be similar to that in the thiobacilli. It is not at all clear, however, that the rat liver H_2S oxidizing system studied by these workers (which was only partially heat stable) represents a physiological process, since rates of H_2S oxidation to $S_2O_3{}^{2-}$ similar to those observed by Baxter *et al.* could be obtained with any of a number of low molecular weight iron chelates and iron–protein complexes (including ferritin, hemoglobin, and iron serum albumin) [386,388], at least some of which were probably present in the extracts examined. It is possible that iron chelates catalyze H_2S oxidation by producing the reactive $O_2{}^-$ radical upon reaction of the ferrous form with O_2; the report by Ichihara [389] that hypoxanthine (which in the presence of xanthine oxidase, undoubtedly present in the extracts studied, and O_2, generates $O_2{}^-$) stimulates H_2S oxidation by beef liver extracts is in accord with this interpretation.

As in bacteria, the oxidation of $S_2O_3{}^{2-}$ to $SO_4{}^{2-}$ by animals involves cleavage of the sulfane and sulfonate moieties. *In vivo,* sulfate is formed preferentially from the sulfonate moiety, with at least part of the sulfane atom being incorporated into protein [390]. In homogenates, the radioactivity of $S^{35}SO_3{}^{2-}$ rapidly appears in $^{35}SO_4{}^{2-}$, while the radioactivity of $^{35}SSO_3{}^{2-}$ appears in the sulfonate moiety of thiosulfate prior to its conversion to $^{35}SO_4{}^{2-}$ [391]. The production of $SO_4{}^{2-}$ from $S_2O_3{}^{2-}$ in extracts is greatly stimulated by addition of thiol compounds, the most effective being glutathione and dihydrolipoate [381,392,393]. Koj [393] has shown that both thiosulfate reductase [223] and thiosulfate sulfurtransferase contribute significantly to the cleavage observed with liver extracts at pH 7.4 in the presence of either thiol. The thiosulfate is reduced to $SO_3{}^{2-}$ and H_2S [381,392]. Since the reactions catalyzed by thiosulfate sulfurtransferase all appear to involve sulfane transfer, it has been postulated [394,395] that the reduction involves intermediate formation of a persulfide between the thiol and the sulfane atom of thiosulfate, this persulfide linkage being cleaved in the presence of excess thiol:

$$^-S—SO_3{}^- + RSH \rightleftharpoons RS—SH + SO_3{}^{2-}$$

$$RS—SH + RSH \rightleftharpoons RSSR + H_2S$$

It is possible, of course, that it is the persulfide (associated with a more physiological carrier thiol?) rather than free H_2S that is the substrate for oxidation of the sulfane atom of $S_2O_3{}^{2-}$ in animal tissues.

B. Oxidation of Sulfite

Sulfite oxidase (EC 1.8.3.1; sulfite:oxygen oxidoreductase), which catalyzes the oxidation of SO_3^{2-} to SO_4^{2-} with O_2, cytochrome c, or ferricyanide as electron acceptor, has been detected in a number of animal [281,396–400] and plant [401–403] tissues. The enzyme has been highly purified from beef [396–399], rat [396,400], and chicken [281] liver. It contains one molecule each of molybdenum and protoheme per 55,000 molecular weight subunit, and the enzyme may exist in solution as a monomer or dimer of such units [281,399,400]. The absorption spectrum of sulfite oxidase is similar [396,399], but not identical [382,404], to that of microsomal cytochrome b_5. However, unlike the cytochrome b_5 of microsomes or outer mitochondrial membrane, the heme component of purified sulfite oxidase is not reducible by NADH [399]. Neither sulfite oxidase nor a 12,000 molecular weight fragment obtained from it by tryptic hydrolysis is related immunologically to microsomal cytochrome b_5 [405].

The molybdenum component of purified sulfite oxidase is reduced to the EPR detectable Mo(V) state on addition of SO_3^{2-} to the enzyme [281,400,406]. Sulfite reducible molydenum has also been observed, by EPR, in partially purified preparations of human liver, wheat germ, and T. thioparus sulfite oxidase [281]. Since reduction of cytochrome c and ferricyanide by SO_3^{2-} catalyzed by the enzyme is sensitive to inhibition by anions, while O_2 reduction is not, it has been concluded that cytochrome c (and ferricyanide) and O_2 interact with sulfite oxidase at different sites [281,397,398]. Examination of the turnover of the molybdenum and protoheme prosthetic groups in the presence of different acceptors led Cohen et al. [406] to propose the following sequence of electron flow in the enzyme:

$$SO_3^{2-} \rightarrow Mo \rightarrow heme \rightarrow O_2$$
$$\downarrow$$
cytochrome c,
$$Fe(CN)_6^{3-}$$

Since cytochrome c is probably an acceptor of physiological importance for sulfite oxidation in mitochondria (see below), it is of interest to note that the K_m for cytochrome c of purified beef liver sulfite oxidase, in the absence of inhibitory anions, is reported to be 2×10^{-8} M [397].

Rat liver sulfite oxidase has been localized to the intermembranal space of the mitochondrion [382,404,407]. Under anaerobic conditions, the cytochromes of intact rat liver mitochondria were reduced by SO_3^{2-} to the same extent as by succinate [382]. The reduction was not inhibited by antimycin A or rotenone, suggesting that the site of entry of sulfite electrons into the respiratory chain of the rat is at the level of cytochrome c. Seligman *et al.* [408] have indicated that cytochrome c can be present on the outer surface of the inner mitochondrial membrane, so that it is in position to accept electrons from sulfite oxidase in the intermembranal space, and pass these electrons to cytochrome c oxidase and O_2 in the inner membrane. In the presence of O_2, sulfite oxidation in mitochondria proceeds with a stoichiometry indicating reduction of O_2 to H_2O [382]; with purified enzyme, however, the product of aerobic SO_3^{2-} oxidation (via the heme component of the enzyme) is H_2O_2 [396,397].

Mitochondrial oxidation of SO_3^{2-} by O_2 is accompanied by esterification of P_i into ATP, with a $P/2e^-$ ratio of 0.8 observed [382,409]. ATP formation was strongly inhibited by uncoupling agents or CN^-, but not by rotenone, results strongly indicative of respiratory chain phosphorylation. Sulfite oxidation in plant mitochondria is also accompanied by esterification of P_i into ATP with a $P/2e^-$ of approximately one [401,410]. These results indicate that in all organisms capable of oxidizing SO_3^{2-} to SO_4^{2-}, whether via APS or the direct oxidative pathway, the reaction is accompanied by energy conservation in the form of phosphate anhydride bond formation. The role of this reaction in the overall energy metabolism of heterotrophic organisms is not clear.

The physiological importance of sulfite oxidation to normal human metabolism is suggested by the discovery of a human infant lacking hepatic and renal sulfite oxidase [411–414]. The urine of this child contained large amounts of SO_3^{2-} and $S_2O_3^{2-}$, but virtually no SO_4^{2-}. The child exhibited pronounced neurological abnormalities at birth, and his condition deteriorated until he died, virtually decorticate, at nine months of age. No alteration in the levels of a number of sulfate esters could be detected in the tissue and urine of the patient, a result that suggests that it was the accumulation of sulfite and/or its by-products, rather than a deficiency of SO_4^{2-}, which was responsible for the pathological effects.

A potentially important tool for further study of the physiological role of sulfite oxidase in animals derives from the recent findings of Johnson *et al.* [415–417] that treatment of rats, kept on a low Mo diet, with tungstate, an antagonist of Mo uptake in the rat, leads to depletion of sulfite oxidase activity, with accumulation in the liver of an Mo-free apoenzyme partially substituted with tungsten. The sulfite oxidase deficient rats were shown to be more susceptible to the toxic effects of large amounts of intraperitoneally injected HSO_3^- or inspired SO_2 than were normal rats. These results indicate that sulfite oxidase should be viewed, at least in part, as an enzyme of detoxification. The efficiency of this enzyme in normal animals is indicated by the fact that amounts of sulfite approaching 0.2% of the total weight of the diet can be ingested by rats or pigs for up to two years without noticeable impairment to health [418,419].

C. Oxidation of Cysteine

Inorganic sulfur compounds can arise in animal tissues as products of cysteine catabolism. Since the routes of cysteine metabolism are discussed elsewhere in this volume (Chapter 14), only an outline of the principal processes leading to SO_4^{2-} production from cysteine will be presented here.

The capacity to produce H_2S from cysteine is a property of many animal tissues [420–424] as well as bacteria [425]. Production of H_2S by intestinal bacteria may serve as a significant source of reduced sulfur in ruminants. In animal tissues, however, there appears to be no specific cysteine desulfhydrase similar to that isolated from bacteria [426–431], which can catalyze the direct cleavage of cysteine to pyruvate, NH_3, and H_2S, probably via 2-aminoacrylate as an intermediate. Instead, desulfuration of cysteine can be catalyzed indirectly by a number of enzymes present in animal tissues, including γ-cystathionase [432–438], cystathionine β-synthase [439,440], and cysteine aminotransferase (EC 2.6.1.3) in conjunction with 3-mercaptopyruvate transsulfurase [441, 442]. The physiological function of the first two enzymes lies in cysteine biosynthesis rather than its cleavage. Yamaguchi et al. [424, 443] have reported that total hepatic cysteine desulfhydrase activity (in contrast to cysteine oxygenase) is not elevated when rats are fed large quantities of cysteine. These authors have provided evidence that the cleavage of sulfur from cysteine prior to its oxidation is not a pathway of major significance in vivo in the rat [424].

Thiosulfate, a normal constituent of urine [444,445], can arise in animals via sulfane transfer reactions involving 3-mercaptopyruvate (formed by transamination between cysteine and oxalacetate or α-ketoglutarate) or thiocysteine (formed via action of γ-cystathionase on cystine) and SO_3^{2-} as substrates. Transfer of the sulfane moiety is catalyzed by 3-mercaptopyruvate sulfurtransferase (EC 2.8.1.2; 3-mercaptopyruvate:cyanide sulfurtransferase) [442,446] and thiosulfate sulfurtransferase [447–449], respectively. Other, more complicated, routes of thiosulfate formation, involving alanine thiosulfonate [450,451], thiotaurine [451], or cysteine S-sulfonate [452,453] have been suggested. It seems likely that in animals, $S_2O_3^{2-}$ represents part of a labile pool of sulfane sulfur which is available for either oxidation or incorporation into organic linkage according to the needs of the organism [454].

Quantitatively, the most important route of thiol oxidation in animals

[424] involves the direct oxidation of cysteine itself to cysteine sulfinic acid [376,455–457], a reaction catalyzed by the cytoplasmic enzyme cysteine dioxygenase (EC 1.13.11.20; L-cysteine:oxygen oxidoreductase) [443,458–461]:

$$HSCH_2CH(NH_2)COO^- + O_2 \rightarrow {}^-O_2SCH_2CH(NH_2)COO^- + H^+$$

Both atoms attached to the sulfur atom of cysteine sulfinate are derived exclusively from molecular oxygen and not H_2O [460]. The enzyme has been purified 200-fold from rat liver by Sakakibara et al. [461]; it is strongly inhibited by iron chelating agents, suggesting an involvement of nonheme Fe groups in catalysis. The purified enzyme, unlike cruder fractions [443,458,459], does not require pyridine nucleotides for activity nor is it stimulated by Fe^{2+} if the enzyme has been preincubated with cysteine [461]. Cysteine dioxygenase activity, located primarily in liver and kidney, is inducible in rats by cysteine feeding [443]. The reaction that it catalyzes appears to be rate limiting in the overall oxidative catabolism of cysteine in the rat [424].

Cysteine sulfinic acid may be decarboxylated to yield hypotaurine and the latter oxidized to taurine, a major product of cysteine oxidation in animals. Alternatively, cysteine sulfinate may undergo transamination with oxalacetate or α-ketoglutarate in a reaction catalyzed by aspartate aminotransferase [453], or oxidative deamination in an NAD^+-dependent reaction catalyzed by rat liver mitochondria, to yield the unstable compound 3-sulfinylpyruvate [462,463], which, in the presence of Mn^{2+}, rapidly decomposes to SO_3^{2-} and pyruvate [464]. The latter reaction may be enzyme catalyzed [463]. SO_3^{2-} is oxidized to SO_4^{2-} by mitochondrial sulfite oxidase, as described previously. The overall route of cysteine oxidation to SO_4^{2-} in animals, then, is:

$$Cysteine + O_2 \rightarrow cysteine\ sulfinate$$

$$Cysteine\ sulfinate + \begin{Bmatrix} \alpha\text{-ketoglutarate} \\ oxalacetate \\ NAD^+ + H_2O \end{Bmatrix} \rightarrow 3\text{-sulfinylpyruvate} + \begin{Bmatrix} glutamate \\ aspartate \\ NADH + NH_4^+ \end{Bmatrix}$$

$$3\text{-Sulfinylpyruvate} \rightarrow pyruvate + SO_3^{2-}$$

$$SO_3^{2-} + \tfrac{1}{2} O_2 \rightarrow SO_4^{2-}$$

The existence of a different route of cysteine oxidation to SO_4^{2-}, not involving cysteine sulfinate and localized in rat liver mitochondria, has been suggested by Wainer [465]; the status of this alternate pathway is uncertain at present, however.

The ability to oxidize organic or inorganic sulfur compounds, with the ultimate production of sulfate, is a characteristic of many heterotrophic soil organisms, including fungi and bacteria [7,466–471], as well as higher plants [472] and animals. Despite the ubiquity of the process and the fact that oxidation of sulfur compounds by heterotrophs appears to be quantitatively more important in many soils than the oxidation due to autotrophic sulfur oxidizing bacteria [466,473], the enzymatic mechanism of the oxidations in these organisms remains largely unexplored.

REFERENCES

1. J. R. Postgate, *in* "Inorganic Sulfur Chemistry" (G. Nickless, ed.), p. 259. Elsevier, Amsterdam, 1968.
2. H. D. Peck, Jr., *in* "Lectures on Theoretical and Applied Aspects of Modern Microbiology," p. 1. Univ. of Maryland Press, University Park, 1967.
3. J. R. Postgate, *Proc. Roy. Soc., Ser. B* **171**, 67 (1968).
4. R. M. Klein and A. Cronquist, *Quart. Rev. Biol.* **42**, 105 (1967).
5. W. V. Ault and J. L. Kulp, *Geochim. Cosmochim. Acta* **16**, 201 (1959).
6. H. D. Peck, Jr., *Soc. Gen. Microbiol.* **24**, 241 (1974).
7. H. D. Peck, Jr., *Bacteriol. Rev.* **26**, 67 (1962).
8. A. B. Roy and P. A. Trudinger, "The Biochemistry of Inorganic Sulfur Compounds." Cambridge Univ. Press, London and New York, 1970.
9. P. A. Trudinger, *Advan. Microbial Physiol.* **3**, 111 (1969).
10. J. A. Schiff and R. C. Hodson, *Annu. Rev. Plant Physiol.* **24**, 381 (1973).
11. J. LeGall and J. R. Postgate, *Advan. Microbial Physiol.* **10**, 81 (1973).
12. P. A. Trudinger, *Rev. Pure Appl. Chem.* **17**, 1 (1967).
13. H. D. Peck, Jr., *Annu. Rev. Microbiol.* **22**, 489 (1968).
14. N. Pfennig, *Annu. Rev. Microbiol.* **21**, 285 (1967).
15. E. Kun, *in* "Metabolic Pathways" (D. M. Greenberg, ed.), 3rd ed., Vol. 3, p. 375. Academic Press, New York, 1969.
16. P. W. Robbins and F. Lipmann, *J. Biol. Chem.* **233**, 686 (1958).
17. J. M. Akagi and L. L. Campbell, *J. Bacteriol.* **84**, 1194 (1962).
18. W. H. Shaw and J. W. Anderson, *Biochem. J.* **139**, 27 (1974).
19. R. W. Gwynn, L. T. Webster, Jr., and R. L. Veech, *J. Biol. Chem.* **249**, 3248 (1974).
20. G. C. Wagner, R. J. Kassner, and M. D. Kamen, *Proc. Nat. Acad. Sci. U. S.* **71**, 253 (1974).
21. B. Lindberg, *Acta Chem. Scand.* **9**, 1323 (1955).
22. T. Harada and B. Spencer, *J. Gen. Microbiol.* **22**, 520 (1960).
23. P. Nissen and A. A. Benson, *Science* **134**, 1759 (1961).
24. I. H. Segel and M. J. Johnson, *Biochim. Biophys. Acta* **69**, 443 (1963).
25. W. G. McGuire and G. A. Marzluff, *Arch. Biochem. Biophys.* **161**, 570 (1974).
26. E. Percival and R. H. McDowell, "Chemistry and Enzymology of Marine Algal Polysaccharides." Academic Press, New York, 1967.
27. A. K. Varma and D. J. D. Nicholas, *Biochim. Biophys. Acta* **227**, 373 (1971).
28. J. W. Tweedie and I. H. Segel, *J. Biol. Chem.* **246**, 2438 (1971).
29. W. H. Shaw and J. W. Anderson, *Biochem. J.* **127**, 237 (1972).
30. C. S. Hawes and D. J. D. Nicholas, *Biochem. J.* **133**, 541 (1973).
31. W. H. Shaw and J. W. Anderson, *Biochem. J.* **139**, 37 (1974).
32. P. W. Robbins, *in* "The Enzymes" (P. D. Boyer, H. Lardy, and K. Myrbäck, eds.), 2nd ed., Vol. 6, p. 469. Academic Press, New York, 1962.
33. P. W. Robbins and F. Lipmann, *J. Biol. Chem.* **233**, 681 (1958).
34. L. G. Wilson and R. S. Bandurski, *J. Biol. Chem.* **233**, 975 (1958).

35. H. G. Wood, J. J. Davis, and H. Lochmuller, *J. Biol. Chem.* **241**, 5692 (1966).
36. J. F. Wheldrake and C. A. Pasternak, *Biochem. J.* **96**, 276 (1965).
37. B. C. Kline and D. E. Schoenhard, *J. Bacteriol.* **102**, 142 (1970).
38. N. M. Kredich, *J. Biol. Chem.* **246**, 3474 (1971).
39. A. K. Varma and D. J. D. Nicholas, *Arch. Mikrobiol.* **78**, 99 (1971).
40. B. Spencer and T. Harada, *Biochem. J.* **77**, 305 (1960).
41. B. Spencer, E. C. Hussey, B. A. Orsi, and J. M. Scott, *Biochem. J.* **106**, 461 (1968).
42. G. J. E. Balharry and D. J. D. Nicholas, *Biochim. Biophys. Acta* **220**, 513 (1970).
43. J. A. Schiff and R. C. Hodson, *Ann. N. Y. Acad. Sci.* **175**, 555 (1970).
44. R. C. Hodson, J. A. Schiff, A. J. Scarsella, and M. Levinthal, *Plant Physiol.* **43**, 563 (1968).
45. E. I. Mercer and G. Thomas, *Phytochemistry* **8**, 2281 (1969).
46. A. Schmidt and A. Trebst, *Biochim. Biophys. Acta* **180**, 529 (1969).
47. J. N. Burnell and J. W. Anderson, *Biochem. J.* **134**, 565 (1973).
48. E. C. Brunngraber, *J. Biol. Chem.* **233**, 472 (1958).
49. J. Dreyfuss and K. J. Monty, *J. Biol. Chem.* **238**, 1019 (1963).
50. A. Schmidt, *Z. Naturforsch. B* **27**, 183 (1972).
51. A. Schmidt, *Arch. Mikrobiol.* **84**, 77 (1972).
52. W. R. Abrams and J. A. Schiff, *Arch. Mikrobiol.* **94**, 1 (1973).
53. A. Schmidt, *Arch. Mikrobiol.* **93**, 29 (1973).
54. A. Schmidt and J. D. Schwenn, *Proc. Int. Congr. Photosyn. 2nd, 1971* p. 507 (1972).
55. A. Schmidt, W. R. Abrams, and J. A. Schiff, *Plant Physiol.* **51**, S53 (1973).
56. R. C. Hodson, J. A. Schiff, and J. P. Mather, *Plant Physiol.* **47**, 306 (1971).
57. R. C. Hodson and J. A. Schiff, *Plant Physiol.* **47**, 300 (1971).
58. M. Tsang, E. Goldschmidt, and J. A. Schiff, *Plant Physiol.* **47**, S20 (1971).
59. M. Tsang and J. A. Schiff, *Plant Physiol.* **51**, S53 (1973).
60. L. M. Siegel, M. J. Murphy, and H. Kamin, *J. Biol. Chem.* **248**, 251 (1973).
61. E. Saito and G. Tamura, *Agr. Biol. Chem.* **35**, 491 (1971).
62. G. Tamura, *J. Biochem. (Tokyo)* **57**, 207 (1965).
63. K. Asada, G. Tamura, and R. S. Bandurski, *J. Biol. Chem.* **244**, 4904 (1969).
64. G. Tamura, *Bot. Mag.* **77**, 239 (1964).
65. G. Tamura, K. Asada, and R. S. Bandurski, *Plant Physiol.* **42**, S36 (1967).
66. A. Trebst and A. Schmidt, *Progr. Photosyn. Res.* **3**, 1510 (1969).
67. J. F. Thompson and D. P. Moore, *Biochem. Biophys. Res. Commun.* **31**, 281 (1968).
68. J. Giovanelli and S. H. Mudd, *Biochem. Biophys. Res. Commun.* **27**, 150 (1967).
69. D. Fujimoto and M. Ishimoto, *J. Biochem. (Tokyo)* **50**, 533 (1961).
70. L. G. Wilson, T. Asahi, and R. Bandurski, *J. Biol. Chem.* **236**, 1822 (1961).
71. C. A. Pasternak, R. J. Ellis, M. C. Jones-Mortimer, and C. E. Chrichton, *Biochem. J.* **96**, 270 (1965).
72. H. Hilz, M. Kittler, and G. Knappe, *Biochem. Z.* **332**, 151 (1959).
73. J. Mager, *Biochim. Biophys. Acta* **41**, 553 (1960).
74. A. Yoshimoto, T. Nakamura, and R. Sato, *J. Biochem. (Tokyo)* **50**, 553 (1961).
75. J. D. Kemp, D. E. Atkinson, A. Ehret, and R. A. Lazzarini, *J. Biol. Chem.* **238**, 3466 (1963).
76. R. J. Ellis, *Biochim. Biophys. Acta* **85**, 335 (1964).
77. N. Naiki, *Plant Cell Physiol.* **6**, 179 (1965).
78. F. J. Leinweber and K. J. Monty, *J. Biol. Chem.* **240**, 782 (1965).
79. R. J. Ellis, *Nature (London)* **211**, 1266 (1966).
80. H. D. Peck, Jr., *J. Bacteriol.* **82**, 933 (1961).
81. H. Hilz and M. Kittler, *Biochem. Biophys. Res. Commun.* **3**, 140 (1960).

82. R. Hodson and J. A. Schiff, *Plant Physiol.* **47**, 296 (1970).
83. T. Asahi, R. S. Bandurski, and L. G. Wilson, *J. Biol. Chem.* **236**, 1830 (1961).
84. P. G. Porque, A. Baldesten, and P. Reichard, *J. Biol. Chem.* **245**, 2371 (1970).
85. K. Torii and R. S. Bandurski, *Biochim. Biophys. Acta* **136**, 286 (1967).
86. L. M. Siegel, H. Kamin, D. C. Rueger, R. P. Presswood, and Q. H. Gibson, *in* "Flavins and Flavoproteins" (H. Kamin, ed.), p. 523. Univ. Park Press, Baltimore, Maryland, 1971.
87. A. Yoshimoto and R. Sato, *Biochim. Biophys. Acta* **153**, 555 (1968).
88. A. Yoshimoto, T. Nakamura, and R. Sato, *J. Biochem. (Tokyo)* **62**, 756 (1967).
89. L. M. Siegel, P. S. Davis, and H. Kamin, *J. Biol. Chem.* **249**, 1572 (1974).
90. E. J. Laishley, P. M. Lin, and H. D. Peck, Jr., *Can. J. Microbiol.* **17**, 889 (1971).
91. M. J. Murphy, L. M. Siegel, and H. Kamin, *J. Biol. Chem.* **249**, 1610 (1974).
92. L. M. Siegel and H. Kamin, *in* "Flavins and Flavoproteins" (K. Yagi, ed.), p. 15. Univ. Park Press, Baltimore, Maryland, 1968.
93. G. S. Kennedy and R. S. Bandurski, *Fed. Proc., Fed. Amer. Soc. Exp. Biol.* **29**, 892 (1970).
94. M. J. Murphy, L. M. Siegel, H. Kamin, and D. Rosenthal, *J. Biol. Chem.* **248**, 2801 (1973).
95. M. J. Murphy and L. M. Siegel, *J. Biol. Chem.* **248**, 6911 (1973).
96. M. J. Murphy, L. M. Siegel, H. Kamin, D. V. DerVertanian, J. P. Lee, J. LeGall, and H. D. Peck, Jr., *Biochem. Biophys. Res. Commun.* **54**, 82 (1973).
97. M. J. Murphy, L. M. Siegel, S. R. Tove, and H. Kamin, *Proc. Nat. Acad. Sci. U. S.* **71**, 612 (1974).
98. J. Vega, R. Garrett, and L. M. Siegel, unpublished data.
99. A. Yoshimoto and R. Sato, *Biochim. Biophys. Acta* **220**, 190 (1970).
100. K. Asada, *J. Biol. Chem.* **242**, 3646 (1967).
101. A. Yoshimoto and R. Sato, *Biochim. Biophys. Acta* **153**, 576 (1968).
102. L. M. Siegel and H. Kamin, *Fed. Proc., Fed. Amer. Soc. Exp. Biol.* **30**, 1261 (1971).
103. L. M. Siegel and P. S. Davis, *J. Biol. Chem.* **249**, 1587 (1974).
104. L. M. Siegel and K. J. Monty, *Biochim. Biophys. Acta* **112**, 346 (1966).
105. E. J. Faeder, P. S. Davis, and L. M. Siegel, *J. Biol. Chem.* **249**, 1599 (1974).
106. W. H. Orme-Johnson, *Annu. Rev. Biochem.* **42**, 159 (1973).
107. L. M. Siegel, E. J. Faeder, and H. Kamin, *Z. Naturforsch. B* **27**, 1087 (1972).
108. H. Bothe and B. Falkenberg, *Z. Naturforsch. B* **27**, 1090 (1972).
109. J. Dreyfuss, *J. Biol. Chem.* **239**, 2292 (1964).
110. B. Spencer and B. G. Moore, *Trans. Biochem. Soc.* **1**, 304 (1973).
    G. Bradfield, P. Sommerfield, T. Meyn, M. Holby, D. Babcock, D. Bradley, and I. H. Segel, *Plant Physiol.* **46**, 720 (1970).
112. J. W. Tweedie and I. H. Segel, *Biochim. Biophys. Acta* **196**, 95 (1970).
113. L. A. Yamamoto and I. H. Segel, *Arch. Biochem. Biophys.* **114**, 523 (1966).
114. G. A. Marzluff, *Arch. Biochem. Biophys.* **138**, 254 (1970).
115. R. T. Wedding and M. K. Black, *Plant Physiol.* **35**, 72 (1960).
116. M. Vallee and R. Jeanjean, *Biochim. Biophys. Acta* **150**, 599 (1968).
117. K. R. Roberts and G. A. Marzluff, *Arch. Biochem. Biophys.* **142**, 651 (1971).
118. M. Vallee, *Biochim. Biophys. Acta* **173**, 486 (1969).
119. G. A. Marzluff, *J. Bacteriol.* **102**, 716 (1970).
120. J. Dreyfuss and A. B. Pardee, *Biochim. Biophys. Acta* **104**, 308 (1965).
121. A. B. Pardee, L. S. Prestidge, M. B. Whipple, and J. Dreyfuss, *J. Biol. Chem.* **241**, 3962 (1966).
122. A. B. Pardee, *J. Biol. Chem.* **241**, 5886 (1966).
123. A. B. Pardee, *Science* **156**, 1627 (1967).

124. R. Langridge, H. Shinagawa, and A. B. Pardee, *Science* **169**, 59 (1970).
125. T. Imagawa and A. Tsugita, *J. Biochem.* (*Tokyo*) **72**, 889 (1972).
126. A. B. Pardee and K. Watanabe, *J. Bacteriol.* **96**, 1049 (1968).
127. N. Ohta, P. R. Galsworthy, and A. B. Pardee, *J. Bacteriol.* **105**, 1053 (1971).
128. C. Hussey, B. Orsi, J. Scott, and B. Spencer, *Nature* (*London*) **207**, 632 (1965).
129. K. Mizobuchi, M. Demerec, and D. H. Gillespie, *Genetics* **47**, 1617 (1962).
130. R. J. Ellis, *Biochem. J.* **93**, 19P (1964).
131. J. Antoniewski and H. deRobichon-Szulmajster, *Biochimie* **55**, 529 (1973).
132. A. Abraham and B. Bachhawat, *Indian J. Biochem.* **1**, 192 (1964).
133. N. M. Kredich and G. M. Tomkins, *J. Biol. Chem.* **241**, 4955 (1966).
134. J. L. Wiebers and H. R. Garner, *J. Biol. Chem.* **242**, 5644 (1967).
135. N. J. Pieniazek, P. P. Stepien, and A. Paszewski, *Biochim. Biophys. Acta* **297**, 37 (1973).
136. M. D. Hulanicka, N. M. Kredich, and D. M. Treiman, *J. Biol. Chem.* **249**, 867 (1974).
137. M. A. Becker and G. M. Tomkins, *J. Biol. Chem.* **244**, 6023 (1969).
138. N. M. Kredich, M. A. Becker, and G. M. Tomkins, *J. Biol. Chem.* **244**, 2428 (1969).
139. M. A. Becker, N. M. Kredich, and G. M. Tomkins, *J. Biol. Chem.* **244**, 2418 (1969).
140. M. C. Jones-Mortimer, J. F. Wheldrake, and C. A. Pasternak, *Biochem. J.* **107**, 51 (1968).
141. M. C. Jones-Mortimer, *Biochem. J.* **110**, 589 (1968).
142. M. C. Jones-Mortimer, *Heredity* **31**, 213 (1973).
143. C. Delavier-Klutchko and M. Flavin, *J. Biol. Chem.* **240**, 2537 (1965).
144. C. A. Pasternak, *Biochem. J.* **81**, 2P (1961).
145. C. A. Pasternak, *Biochem. J.* **85**, 44 (1962).
146. J. Dreyfuss and K. J. Monty, *J. Biol. Chem.* **238**, 3781 (1963).
147. R. J. Ellis, S. K. Humphries, and C. A. Pasternak, *Biochem. J.* **92**, 167 (1964).
148. J. F. Wheldrake, *Biochem. J.* **105**, 697 (1967).
149. H. T. Spencer, J. Collins, and K. J. Monty, *Fed. Proc., Fed. Amer. Soc. Exp. Biol.* **26**, 677 (1967).
150. J. Dreyfuss and A. B. Pardee, *J. Bacteriol.* **91**, 2275 (1966).
151. M. C. Jones-Mortimer, *Biochem. J.* **110**, 594 (1968).
152. I. Ino and M. Demerec, *Genetics* **59**, 167 (1968).
153. M. A. Savin and M. Flavin, *J. Bacteriol.* **112**, 299 (1972).
154. R. DeVito and J. Dreyfuss, *J. Bacteriol.* **88**, 1341 (1964).
155. H. Cherest, F. Eichler, and H. deRobichon-Szulmajster, *J. Bacteriol.* **97**, 328 (1969).
156. S. Yamagata, *J. Biochem.* (*Tokyo*) **70**, 1035 (1971).
157. M. Flavin and C. Slaughter, *Biochim. Biophys. Acta* **132**, 400 (1967).
158. D. S. Kerr and M. Flavin, *Biochem. Biophys. Res. Commun.* **31**, 124 (1968).
159. D. S. Kerr, *J. Biol. Chem.* **246**, 95 (1971).
160. A. H. Datko, J. Giovanelli, and S. H. Mudd, *J. Biol. Chem.* **249**, 1139 (1974).
161. M. M. Kaplan and M. Flavin, *J. Biol. Chem.* **241**, 5781 (1966).
162. D. S. Kerr and M. Flavin, *J. Biol. Chem.* **245**, 1842 (1970).
163. F. Schlenk and R. E. DePalma, *J. Biol. Chem.* **229**, 1037 (1957).
164. F. Schlenk, J. L. Dainko, and F. S. Stanford, *Arch. Biochem. Biophys.* **83**, 28 (1959).
165. G. Spihla and F. Schlenk, *J. Bacteriol.* **100**, 198 (1960).
166. H. Cherest, Y. Surdin-Kerjan, J. Antoniewski, and H. deRobichon-Szulmajster, *J. Bacteriol.* **114**, 928 (1973).
167. H. Cherest, Y. Surdin-Kerjan, and H. deRobichon-Szulmajster, *J. Bacteriol.* **106**, 758 (1971).
168. M. Masselot and H. deRobichon-Szulmajster, *Genetics* **71**, 535 (1972).

169. H. Cherest, Y. Surdin-Kerjan, J. Antoniewski, and H. deRobichon-Szulmajster, *J. Bacteriol.* **115**, 1084 (1973).
170. Y. Surdin-Kerjan, H. Cherest, and H. deRobichon-Szulmajster, *J. Bacteriol.* **113**, 1156 (1973).
171. A. J. Ferro and K. D. Spence, *J. Bacteriol.* **116**, 812 (1973).
172. R. L. Metzenberg and J. W. Parson, *Proc. Nat. Acad. Sci. U.S.* **55**, 629 (1966).
173. G. A. Marzluff and R. L. Metzenberg, *J. Mol. Biol.* **33**, 423 (1968).
174. E. G. Burton and R. L. Metzenberg, *J. Bacteriol.* **109**, 140 (1972).
175. N. J. Pieniazek, I. M. Kowalska, and P. P. Stepien, *Mol. Gen. Genet.* **126**, 367 (1973).
176. R. L. Metzenberg and S. K. Algren, *Genetics* **68**, 369 (1970).
177. P. S. Dietrich and R. L. Metzenberg, *Biochem. Genet.* **8**, 73 (1973).
178. H. deRobichon-Szulmajster and H. Cherest, *Biochem. Biophys. Res. Commun.* **28**, 256 (1967).
179. J. R. Postgate and L. L. Campbell, *Bacteriol. Rev.* **30**, 732 (1966).
180. S. C. Rittenberg, *Advan. Microbial Physiol.* **3**, 159 (1969).
181. J. M. Akagi, *J. Bacteriol.* **88**, 813 (1964).
182. H. D. Peck, Jr., *Biochem. Biophys. Res. Commun.* **22**, 112 (1966).
183. L. L. Barton, J. LeGall, and H. D. Peck, Jr., *Biochem. Biophys. Res. Commun.* **41**, 1036 (1970).
184. J. R. Postgate, *J. Gen. Microbiol.* **5**, 725 (1951).
185. T. Yagi, M. Hanya, and N. Tamiya, *Biochim. Biophys. Acta* **153**, 699 (1968).
186. R. L. Haschke and L. L. Campbell, *J. Bacteriol.* **105**, 249 (1971).
187. J. LeGall, M. Bruschi-Heriaud, and D. V. DerVertanian, *Biochim. Biophys. Acta* **234**, 499 (1971).
188. J. R. Postgate, *J. Bacteriol.* **85**, 1450 (1963).
189. J. D. A. Miller and D. S. Wakerley, *J. Gen. Microbiol.* **43**, 101 (1966).
190. A. J. Kluyver, *in* "Microbial Metabolism," p. 71. Symp. Ist. Superiore di Sanita, Rome, 1953.
191. J. R. Postgate, *Annu. Rev. Microbiol.* **13**, 505 (1959).
192. M. Ishimoto, *J. Biochem.* (*Tokyo*) **46**, 105 (1959).
193. H. D. Peck, Jr., *Proc. Nat. Acad. Sci. U.S.* **45**, 701 (1959).
194. H. D. Peck, Jr., *J. Biol. Chem.* **237**, 198 (1962).
195. M. Ishimoto and D. Fujimoto, *Proc. Jap. Acad. Sci.* **35**, 243 (1959).
196. M. Ishimoto and D. Fujimoto, *Symp. Enzyme Chem.* **14**, 48 (1960).
197. J. M. Akagai and L. L . Campbell, *J. Bacteriol.* **86**, 563 (1963).
198. D. A. Ware and J. R. Postgate, *J. Gen. Microbiol.* **67**, 145 (1971).
199. D. A. Ware and J. R. Postgate, *Nature* (*London*) **226**, 1250 (1970).
200. A. J. D'Eustachio, E. Knight, Jr., and R. W. F. Hardy, *J. Bacteriol.* **90**, 288 (1965).
201. M. Ishimoto and D. Fujimoto, *J. Biochem.* (*Tokyo*) **50**, 299 (1961).
202. H. D. Peck, Jr., T. E. Deacon, and J. T. Davidson, *Biochim. Biophys. Acta* **96**, 429 (1965).
203. G. B. Michaels, J. T. Davidson, and H. D. Peck, Jr., *in* "Flavins and Flavoproteins" (H. Kamin, ed.), p. 555. Univ. Park Press, Baltimore, Maryland, 1971.
204. H. D. Peck, Jr., R. Bramlett, and D. V. DerVertanian, *Z. Naturforsch. B* **27**, 1084 (1972).
205. F. Muller and V. Massey, *J. Biol. Chem.* **244**, 4007 (1969).
206. L. Hevesi and T. C. Bruice, *Biochemistry* **12**, 290 (1973).
207. J. Millet, *in* "Colloque sur la Biochimie du Soufre," p. 77. CNRS, Paris, 1956.
208. M. Ishimoto and T. Yagi, *J. Biochem.* (*Tokyo*) **49**, 103 (1961).
209. K. Kobayashi, S. Tachibana, and M. Ishimoto, *J. Biochem.* (*Tokyo*) **65**, 155 (1969).

210. P. A. Trudinger, *J. Bacteriol.* **104,** 158 (1970).
211. J. P. Lee and H. D. Peck, Jr., *Biochem. Biophys. Res. Commun.* **45,** 583 (1971).
212. E. C. Hatchikian, J. LeGall, M. Bruschi, and M. Dubourdieu, *Biochim. Biophys. Acta* **258,** 701 (1972).
213. B. Suh and J. M. Akagi, *J. Bacteriol.* **99,** 210 (1969).
214. J. E. Findley and J. M. Akagi, *Biochem. Biophys. Res. Commun.* **36,** 266 (1969).
215. J. M. Akagi and V. Adams, *J. Bacteriol.* **116,** 392 (1973).
216. J. E. Findley and J. M. Akagi, *J. Bacteriol.* **103,** 741 (1973).
217. J. P. Lee, J. LeGall, and H. D. Peck, Jr., *J. Bacteriol.* **115,** 529 (1973).
218. M. Ishimoto, J. Koyama, and Y. Nagai, *J. Biochem. (Tokyo)* **41,** 763 (1954).
219. J. R. Postgate, *J. Gen. Microbiol.* **14,** 545 (1956).
220. M. Ishimoto, J. Koyama, and Y. Nagai, *J. Biochem. (Tokyo)* **42,** 41 (1955).
221. R. Haschke and L. L. Campbell, *J. Bacteriol.* **106,** 603 (1971).
222. W. Nakatukasa and J. M. Akagi, *J. Bacteriol.* **98,** 429 (1969).
223. A. Kaji and W. D. McElroy, Jr., *J. Bacteriol.* **77,** 630 (1959).
224. K. Kobayashi, E. Takahashi, and M. Ishimoto, *J. Biochem. (Tokyo)* **72,** 879 (1972).
225. K. Kobayashi, Y. Seki, and M. Ishimoto, *J. Biochem. (Tokyo)* **75,** 519 (1974).
226. J. P. Lee, C. S. Yi, J. LeGall, and H. D. Peck, Jr., *J. Bacteriol.* **115,** 453 (1973).
227. J. D. A. Miller and A. M. Saleh, *J. Gen. Microbiol.* **37,** 419 (1964).
228. G. W. Skyring and P. A. Trudinger, *Can. J. Biochem.* **50,** 1145 (1972).
229. J. Saks and L. M. Siegel, unpublished data.
230. P. A. Trudinger and L. A. Chambers, *Biochim. Biophys. Acta* **293,** 26 (1973).
231. J. L. Hoard, *in* "Structural Chemistry and Molecular Biology" (A. Rich and N. Davidson, eds.), p. 573. Freeman, San Francisco, California, 1968.
232. I. R. Kaplan and S. C. Rittenberg, *J. Gen. Microbiol.* **34,** 195 (1954).
233. N. Nakai and M. L. Jensen, *Geochim. Cosmochim. Acta* **28,** 1893 (1964).
234. R. Haschke and L. L. Campbell, *Fed. Proc., Fed. Amer. Soc. Exp. Biol.* **27,** 989 (1968).
235. M. Ishimoto, J. Koyama, T. Omura, and Y. Nagai, *J. Biochem. (Tokyo)* **41,** 537 (1954).
236. H. D. Peck, Jr., *J. Biol. Chem.* **235,** 2734 (1960).
237. J. H. Vosjan, *Antonie van Leeuwenhoek; J. Microbiol. Serol.* **36,** 585 (1970).
238. L. Guarraia and H. D. Peck, Jr., *J. Bacteriol.* **106,** 890 (1971).
239. D. V. DerVertanian and J. LeGall, *Biochim. Biophys. Acta* **346,** 79 (1974).
240. J. M. Akagi, *J. Biol. Chem.* **242,** 2478 (1967).
241. J. LeGall and N. Dragoni, *Biochem. Biophys. Res. Commun.* **23,** 145 (1966).
242. J. LeGall and E. C. Hatchikian, *C. R. Acad. Sci.* **264,** 2580 (1967).
243. E. C. Hatchikian and J. LeGall, *Ann. Inst. Pasteur, Paris* **118,** 288 (1970).
244. K. Irie, K. Kobayashi, M. Kobayashi, and M. Ishimoto, *J. Biochem. (Tokyo)* **73,** 353 (1973).
245. L. J. Guarraia, E. J. Laishley, N. Forget, and H. D. Peck, Jr., *Bacteriol. Proc.* p. 130 (1968).
246. D. P. Kelley, *Annu. Rev. Microbiol.* **25,** 177 (1971).
247. H. D. Peck, Jr. and E. Fisher, *J. Biol. Chem.* **237,** 190 (1962).
248. J. London and S. C. Rittenberg, *Proc. Nat. Acad. Sci. U.S.* **52,** 1183 (1964).
249. B. F. Taylor, *Biochim. Biophys. Acta* **170,** 112 (1968).
250. J. Saxena and M. I. H. Aleem, *Can. J. Biochem.* **51,** 560 (1973).
251. D. J. W. Moriarty and D. J. D. Nicholas, *Biochim. Biophys. Acta* **197,** 143 (1970).
252. M. Aminuddin and D. J. D. Nicholas, *Biochim. Biophys. Acta* **325,** 81 (1973).
253. K. Baalsrud and K. S. Baalsrud, *Arch. Mikrobiol.* **20,** 34 (1954).
254. I. Suzuki and C. H. Werkman, *Proc. Nat. Acad. Sci. U.S.* **45,** 239 (1959).

255. A. M. Charles and I. Suzuki, *Biochim. Biophys. Acta* **128,** 510 (1966).
256. A. M. Charles, *Arch. Biochem. Biophys.* **129,** 124 (1969).
257. D. J. W. Moriarty and D. J. D. Nicholas, *Biochim. Biophys. Acta* **184,** 114 (1969).
258. D. J. W. Moriarty and D. J. D. Nicholas, *Biochim. Biophys. Acta* **216,** 130 (1970).
259. I. Suzuki, *Biochim. Biophys. Acta* **104,** 359 (1965).
260. I. Suzuki, *Biochim. Biophys. Acta* **110,** 97 (1965).
261. I. Suzuki and M. Silver, *Biochim. Biophys. Acta* **122,** 22 (1966).
262. M. Silver and D. G. Lundgren, *Can. J. Biochem.* **46,** 457 (1968).
263. F. W. Adair, *J. Bacteriol.* **92,** 899 (1966).
264. M. Silver and D. G. Lundgren, *Can. J. Biochem.* **46,** 1215 (1968).
265. T. Tano and K. Imai, *Agr. Biol. Chem.* **32,** 51 (1968).
266. H. D. Peck, Jr., *Proc. Nat. Acad. Sci. U.S.* **46,** 1053 (1960).
267. H. D. Peck, Jr., *Biochim. Biophys. Acta* **49,** 621 (1961).
268. H. D. Peck and M. P. Stulberg, *J. Biol. Chem.* **237,** 1648 (1962).
269. T. J. Bowen, F. C. Happold, and B. F. Taylor, *Biochim. Biophys. Acta* **118,** 566 (1966).
270. R. M. Lyric and I. Suzuki, *Can. J. Biochem.* **48,** 334 (1970).
271. H. B. LeJohn, L. VanCaeseele, and H. Lees, *J. Bacteriol.* **94,** 1484 (1967).
272. G. Milhaud, J. P. Aubert, and J. Millet, *C. R. Acad. Sci.* **246,** 1766 (1958).
273. A. M. Charles and I. Suzuki, *Biochem. Biophys. Res. Commun.* **19,** 686 (1965).
274. A. M. Charles and I. Suzuki, *Biochim. Biophys. Acta* **128,** 522 (1966).
275. W. P. Hempfling, P. A. Trudinger, and W. Vishniac, *Arch. Mikrobiol.* **59,** 149 (1967).
276. J. R. Vestal and D. G. Lundgren, *Can. J. Biochem.* **49,** 1125 (1971).
277. C. A. Adams, G. M. Warnes, and D. J. D. Nicholas, *Biochim. Biophys. Acta* **235,** 398 (1971).
278. A. Kodama, T. Kodama, and T. Mori, *Plant Cell Physiol.* **11,** 701 (1970).
279. R. M. Lyric and I. Suzuki, *Can. J. Biochem.* **48,** 344 (1970).
280. R. M. Lyric and I. Suzuki, *Can. J. Biochem.* **48,** 594 (1970).
281. D. L. Kessler and K. V. Rajagopalan, *J. Biol. Chem.* **247,** 6566 (1972).
282. R. M. Lyric and I. Suzuki, *Can. J. Biochem.* **48,** 355 (1970).
283. R. L. Starkey, *J. Bacteriol.* **28,** 365 (1934).
284. R. L. Starkey, *J. Bacteriol.* **28,** 387 (1934).
285. W. Vishniac, *J. Bacteriol.* **64,** 363 (1952).
286. C. D. Parker and J. Prisk, *J. Gen. Microbiol.* **8,** 344 (1953).
287. P. A. Trudinger, *Biochim. Biophys. Acta* **31,** 270 (1959).
288. M. Santer, M. Margulies, N. Klinman, and R. Kaback, *J. Bacteriol.* **79,** 213 (1960).
289. G. L. Jones and F. C. Happold, *J. Gen. Microbiol.* **26,** 361 (1961).
290. D. Wooley, G. L. Jones, and F. C. Happold, *J. Gen. Microbiol.* **26,** 361 (1962).
291. W. P. Hempfling and W. Vishniac, *J. Bacteriol.* **93,** 874 (1967).
292. J. Peeters and M. I. H. Aleem, *Arch. Mikrobiol.* **71,** 319 (1970).
293. M. Okuzumi and T. Kita, *Agr. Biol. Chem.* **29,** 1063 (1965).
294. D. P. Kelley, *Aust. J. Sci.* **31,** 165 (1968).
295. W. Vishniac and M. Santer, *Bacteriol. Rev.* **21,** 195 (1957).
296. W. Vishniac and P. A. Trudinger, *Bacteriol. Rev.* **26,** 168 (1962).
297. P. A. Trudinger, *Aust. J. Biol. Sci.* **17,** 738 (1964).
298. P. A. Trudinger, *Biochem. J.* **78,** 680 (1961).
299. M. I. H. Aleem, *J. Bacteriol.* **90,** 95 (1965).
300. P. A. Trudinger, *Biochim. Biophys. Acta* **30,** 211 (1958).
301. P. A. Trudinger, *Biochem. J.* **90,** 640 (1964).
302. B. Skarzynski, W. Ostrowski, and A. Krawczyk, *Bull. Acad. Pol. Sci.* **4,** 159 (1957).
303. W. Ostrowski and A. Krawczyk, *Acta Biochim. Pol.* **4,** 249 (1957).

304. P. A. Trudinger, *Aust. J. Biol. Sci.* **17,** 577 (1964).
305. D. P. Kelley and P. J. Syrett, *Biochem. J.* **98,** 537 (1966).
306. M. Santer, *Biochem. Biophys. Res. Commun.* **1,** 9 (1959).
307. T. J. Bowen, P. J. Butler, and F. C. Happold, *Biochem. J.* **95,** 5P (1965).
308. T. J. Bowen, P. J. Butler, and F. C. Happold, *Biochem. J.* **97,** 651 (1965).
309. W. P. Hempfling and W. Vishniac, *Biochem. Z.* **342,** 272 (1965).
310. D. P. Kelley and P. J. Syrett, *J. Gen. Microbiol.* **34,** 307 (1964).
311. J. S. Cole and M. I. H. Aleem, *Proc. Nat. Acad. Sci. U.S.* **70,** 3571 (1973).
312. C. W. Roth, W. P. Hempfling, J. N. Conners, and W. Vishniac, *J. Bacteriol.* **114,** 592 (1973).
313. A. J. Ross, R. L. Schoenhoff, and M. I. H. Aleem, *Biochem. Biophys. Res. Commun.* **32,** 301 (1968).
314. J. S. Cole and M. I. H. Aleem, *Biochem. Biophys. Res. Commun.* **38,** 736 (1970).
315. E. A. Davis and E. J. Johnson, *Can. J. Microbiol.* **13,** 873 (1967).
316. P. A. Trudinger, *Biochem. J.* **78,** 673 (1961).
317. T. Tano, H. Kagawa, and K. Imai, *Agr. Biol. Chem.* **32,** 279 (1968).
318. "Handbook of Biochemistry," 2nd ed., p. J-36. Chem. Rubber Publ. Co., Cleveland, Ohio, 1970.
319. "Handbook of Biochemistry," 2nd ed., p. J-38. Chem. Rubber Publ. Co., Cleveland, Ohio, 1970.
320. H. Iwatsuka and T. Mori, *Plant Cell Physiol.* **1,** 163 (1960).
321. A. Kodama and T. Mori, *Plant Cell Physiol.* **9,** 725 (1968).
322. A. Kodama, *Plant Cell Physiol.* **10,** 645 (1969).
323. F. W. Adair, *J. Bacteriol.* **95,** 147 (1968).
324. M. I. H. Aleem, *J. Bacteriol.* **91,** 729 (1966).
325. M. I. H. Aleem, *Biochim. Biophys. Acta* **128,** 1 (1966).
326. J. Saxena and M. I. H. Aleem, *Arch. Mikrobiol.* **84,** 317 (1972).
327. M. P. Silverman and D. G. Lundgren, *J. Bacteriol.* **78,** 326 (1959).
328. N. N. Lyalikova, *Mikrobiologiya* **27,** 556 (1958).
329. G. I. Karavaiko and T. A. Pivovarova, *Mikrobiologiya* **42,** 389 (1973).
330. N. Pfennig and H. G. Truper, *Int. J. Syst. Bacteriol.* **21,** 17 (1971).
331. E. N. Kondratieva, "Photosynthetic Bacteria." Israel Program Sci. Transl., Jerusalem, 1965.
332. D. I. Arnon, M. Losada, M. Nozaki, and K. Tagawa, *Nature (London)* **190,** 601 (1961).
333. H. G. Truper and H. G. Schlegel, *Antonie van Leeuwenhoek; J. Microbiol. Serol.* **30,** 225 (1964).
334. J. P. Thornberg, *Biochemistry* **9,** 2688 (1970).
335. W. W. Parson and G. D. Case, *Biochim. Biophys. Acta* **205,** 232 (1970).
336. M. Siebert and D. DeVault, *Biochim. Biophys. Acta* **205,** 220 (1970).
337. G. D. Case and W. W. Parson, *Biochim. Biophys. Acta* **325,** 441 (1973).
338. M. A. Cusanovich and M. D. Kamen, *Biochim. Biophys. Acta* **153,** 418 (1968).
339. M. A. Cusanovich, R. G. Bartsch, and M. D. Kamen, *Biochim. Biophys. Acta* **153,** 397 (1968).
340. P. L. Dutton, *Biochim. Biophys. Acta* **226,** 63 (1971).
341. G. Hind and J. M. Olson, *Annu. Rev. Plant Physiol.* **19,** 249 (1968).
342. J. M. Olson and B. Chance, *Arch. Biochem. Biophys.* **88,** 26 (1960).
343. S. Morita, M. Edwards, and J. Gibson, *Biochim. Biophys. Acta* **109,** 45 (1965).
344. P. L. Dutton and J. S. Leigh, *Biochim. Biophys. Acta* **314,** 178 (1973).
345. K. Kusai and T. Yamanaka, *Biochim. Biophys. Acta* **325,** 304 (1973).
346. I. P. Vernon, *Bacteriol. Rev.* **32,** 243 (1968).

347. T. Horio and M. D. Kamen, *Annu. Rev. Microbiol.* **24**, 399 (1970).
348. H. Gest, *Advan. Microbial Physiol.* **7**, 243 (1972).
349. D. B. Knaff, B. B. Buchanan, and R. Malkin, *Biochim. Biophys. Acta* **325**, 94 (1973).
350. P. L. Dutton, J. S. Leigh, and D. W. Reed, *Biochim. Biophys. Acta* **292**, 654 (1973).
351. V. D. Fedorov and V. M. Maksimov, *Dokl. Akad. Nauk SSSR* **162**, 100 (1965).
352. H. H. Thiele, *Antonie van Leeuwenhoek; J. Microbiol. Serol.* **34**, 350 (1968).
353. T. E. Meyer, R. G. Bartsch, M. A. Cusanovich, and J. H. Mathewson, *Biochim. Biophys. Acta* **153**, 854 (1968).
354. R. G. Bartsch, T. E. Meyer, and A. B. Robinson, *in* "Structure and Function of Cytochromes" (K. Okunuki, M. D. Kamen, and I. Sezuki, eds.), p. 443. Univ. of Tokyo Press, Tokyo, 1968.
355. H. Gest, *Nature (London)* **209**, 879 (1966).
356. D. L. Keister and N. Y. Yike, *Arch. Biochem. Biophys.* **121**, 415 (1967).
357. B. B. Buchanan and M. C. W. Evans, *Biochim. Biophys. Acta* **180**, 123 (1969).
358. H. G. Truper, *Antonie van Leeuwenhoek; J. Microbiol. Serol.* **30**, 385 (1964).
359. A. J. Smith and J. Lascelles, *J. Gen. Microbiol.* **42**, 357 (1966).
360. H. G. Truper and J. C. Hathaway, *Nature (London)* **215**, 435 (1967).
361. H. G. Truper and N. Pfennig, *Antonie van Leeuwenhoek; J. Microbiol. Serol.* **32**, 361 (1966).
362. F. Hashwa and N. P. Fennig, *Arch. Mikrobiol.* **81**, 36 (1972).
363. H. G. Truper and H. D. Peck, Jr., *Arch. Mikrobiol.* **73**, 125 (1970).
364. H. G. Truper and L. A. Rogers, *J. Bacteriol.* **108**, 1112 (1971).
365. A. J. Smith, *J. Gen. Microbiol.* **42**, 371 (1966).
366. J. Mathewson, L. J. Burger, and H. G. Millstone, *Fed. Proc., Fed. Amer. Soc. Exp. Biol.* **27**, 774 (1968).
367. J. H. Eley, K. Knobloch, and M. I. H. Aleem, *Arch. Biochem. Biophys.* **147**, 419 (1971).
368. K. Knobloch, J. H. Eley, and M. I. H. Aleem, *Arch. Mikrobiol.* **80**, 97 (1971).
369. Z. Nyiri, *Biochem. Z.* **141**, 160 (1923).
370. W. Denis and L. Reed, *J. Biol. Chem.* **72**, 385 (1972).
371. D. D. Dziewjakowski, *J. Biol. Chem.* **161**, 723 (1945).
372. S. Skarzynski, T. W. Szczepkowski, and M. Weber, *Nature (London)* **184**, 994 (1959).
373. B. Bhaghat and M. F. Lockett, *J. Pharm. Pharmacol.* **12**, 690 (1960).
374. E. Yokoyama, R. E. Yoder, and N. R. Frank, *Arch. Environ. Health* **22**, 389 (1971).
375. C. G. Curtis, T. C. Bartholemew, F. A. Rose, and K. S. Dodgson, *Biochem. Pharmacol.* **21**, 2313 (1972).
376. N. W. Pirie, *Biochem. J.* **28**, 305 (1934).
377. M. A. DerGarabedian and C. Fromageot, *C. R. Acad. Sci.* **216**, 216 (1943).
378. C. V. Smyth, *Arch. Biochem.* **2**, 259 (1943).
379. M. Heimberg, I. Fridovich, and P. Handler, *J. Biol. Chem.* **204**, 913 (1953).
380. C. F. Baxter, R. VanReen, P. B. Pearson, and C. Rosenberg, *Biochim. Biophys. Acta* **27**, 584 (1958).
381. A. Koj and J. Frendo, *Folia Biol. (Warsaw)* **15**, 49 (1967).
382. H. J. Cohen, S. Betcher-Lange, D. L. Kessler, and K. V. Rajagopalan, *J. Biol. Chem.* **247**, 7759 (1972).
383. H. W. Haggard, *J. Biol. Chem.* **49**, 519 (1921).
384. C. Lovatt Evans, *Quart J. Exp. Physiol. Cog. Med. Sci.* **52**, 231 (1967).
385. C. F. Baxter and R. VanReen, *Biochim. Biophys. Acta* **28**, 567 (1958).
386. B. Sörbo, *Biochim. Biophys. Acta* **27**, 324 (1958).
387. B. Sörbo, *Biochim. Biophys. Acta* **38**, 349 (1960).

388. C. F. Baxter and R. VanReen, *Biochim. Biophys. Acta* **29,** 573 (1958).
389. A. Ichihara, *Mem. Inst. Protein Res., Osaka Univ.* **1,** 177 (1959).
390. B. Skarzynski, J. W. Szczepkowski, and M. Weber, *Acta Biochim. Pol.* **7,** 105 (1960).
391. A. Koj, J. Frendo, and Z. Janik, *Biochem. J.* **103,** 791 (1967).
392. B. Sörbo, *Acta Chem. Scand.* **18,** 821 (1964).
393. A. Koj, *Acta Biochim. Pol.* **15,** 161 (1968).
394. B. Sörbo, *Acta Chem. Scand.* **16,** 243 (1962).
395. M. Villarejo and J. Westley, *J. Biol. Chem.* **238,** 4016 (1963).
396. R. M. MacLeod, W. Farkas, I. Fridovich, and P. Handler, *J. Biol. Chem.* **236,** 1841 (1961).
397. L. G. Howell and I. Fridovich, *J. Biol. Chem.* **243,** 5941 (1968).
398. H. J. Cohen and I. Fridovich, *J. Biol. Chem.* **246,** 359 (1971).
399. H. J. Cohen and I. Fridovich, *J. Biol. Chem.* **246,** 367 (1971).
400. D. L. Kessler, J. L. Johnson, H. J. Cohen, and K. V. Rajagopalan, *Biochim. Biophys. Acta* **334,** 86 (1974).
401. J. M. Tager and N. Rautanen, *Biochim. Biophys. Acta* **18,** 111 (1955).
402. O. Arrigoni, *Ital. J. Biochem.* **8,** 181 (1959).
403. R. G. Strickland, *Nature (London)* **190,** 648 (1961).
404. A. Ito, *J. Biochem. (Tokyo)* **70,** 1061 (1971).
405. K. Fukushima, A. Ito, T. Omura, and R. Sato, *J. Biochem. (Toyko)* **71,** 447 (1972).
406. H. J. Cohen, I. Fridovich, and K. V. Rajagopalan, *J. Biol. Chem.* **246,** 374 (1971).
407. S. Wattiaux-DeConinck and R. Wattiaux, *Eur. J. Biochem.* **19,** 552 (1971).
408. A. M. Seligman, M. J. Karnovsky, and H. L. Wasserkrug, *J. Cell Biol.* **38,** 1 (1968).
409. F. E. Hunter and L. Ford, *Fed. Proc., Fed. Amer. Soc. Exp. Biol.* **13,** 234 (1954).
410. O. Arrigoni and G. Rossi, *G. Biochem.* **10,** 463 (1961).
411. S. H. Mudd, F. Irreverre, and L. Laster, *Science* **156,** 1599 (1967).
412. F. Irreverre, S. H. Mudd, W. D. Heizer, and L. Laster, *Biochem. Med.* **1,** 187 (1967).
413. A. K. Percy, S. H. Mudd, F. Irreverre, and L. Laster, *Biochem. Med.* **2,** 198 (1968).
414. W. I. Rosenblum, *Neurology* **18,** 1187 (1968).
415. J. L. Johnson, K. V. Rajagopalan, and H. J. Cohen, *J. Biol. Chem.* **249,** 859 (1974).
416. H. J. Cohen, R. T. Drew, J. L. Johnson, and K. V. Rajagopalan, *Proc. Nat. Acad. Sci. U.S.* **70,** 3655 (1974).
417. J. L. Johnson, H. J. Cohen, and K. V. Rajagopalan, *J. Biol. Chem.* **249,** 5046 (1974).
418. H. P. Til, V. J. Feron, and A. P. DeGroot, *Food Cosmet. Toxicol.* **10,** 291 (1972).
419. V. J. Feron, A. P. DeGroot, and P. vanderWal, *Food Cosmet. Toxicol.* **10,** 463 (1972).
420. C. Fromageot, E. Wookey, and P. Chaix, *C. R. Acad. Sci.* **209,** 1019 (1939).
421. C. V. Smyth, *J. Biol. Chem.* **142,** 387 (1942).
422. C. Fromageot, *Advan. Enzymol.* **7,** 369 (1947).
423. G. Barboza, E. Fernandez, and J. Pulgar, *Comp. Biochem. Physiol.* **45,** 1 (1973).
424. K. Yamaguchi, S. Sakakibara, J. Asamizu, and I. Ueda, *Biochim. Biophys. Acta* **297,** 48 (1973).
425. J. C. Andrews, *J. Biol. Chem.* **122,** 687 (1937).
426. D. E. Metzler and E. E. Snell, *J. Biol. Chem.* **198,** 353 (1952).
427. G. Guarneros and M. V. Ortega, *Biochim. Biophys. Acta* **198,** 132 (1970).
428. N. M. Kredich, B. S. Keenan, and L. J. Foote, *J. Biol. Chem.* **247,** 7157 (1972).
429. N. M. Kredich, L. J. Foote, and B. S. Keenan, *J. Biol. Chem.* **248,** 6187 (1973).
430. J. M. Collins and K. J. Monty, *J. Biol. Chem.* **248,** 3769 (1973).
431. J. M. Collins and K. J. Monty, *J. Biol. Chem.* **248,** 5943 (1973).

432. Y. Matsuo and D. M. Greenberg, *J. Biol. Chem.* **230,** 545 (1954).
433. D. Cavallini, C. DeMarco, and A. Scioscia-Santoro, *Arch. Biochem. Biophys.* **96,** 456 (1962).
434. M. Flavin, *J. Biol. Chem.* **237,** 768 (1962).
435. J. Loiselet and F. Chatagner, *Biochim. Biophys. Acta* **89,** 330 (1964).
436. D. Deme and F. Chatagner, *Biochim. Biophys. Acta* **258,** 643 (1972).
437. M. T. Costa, A. M. Wolf, and D. Giarnieri, *Enzymologia* **43,** 271 (1972).
438. T. E. Pascal, H. H. Tallan, and B. M. Gillam. *Biochim. Biophys. Acta* **285,** 48 (1972).
439. A. Sentenac and P. Fromageot, *Biochim. Biophys. Acta* **81,** 258 (1964).
440. A. E. Braunstein, E. V. Goryachenkova, E. A. Tolosa, I. H. Willhardt, and L. L. Yefremova, *Biochim. Biophys. Acta* **242,** 247 (1971).
441. A. Meister, P. E. Fraser, and S. V. Tice, *J. Biol. Chem.* **206** 561 (1954).
442. D. Fanshier and E. Kun, *Biochim. Biophys. Acta* **58,** 266 (1962).
443. K. Yamaguchi, S. Sakakibara, K. Koga, and I. Ueda, *Biochim. Biophys. Acta* **237,** 502 (1971).
444. C. Fromageot and A. Royer, *Enzymologia* **11,** 361 (1945).
445. J. H. Gast, K. Arai, and F. L. Aldrich, *J. Biol. Chem.* **196,** 875 (1952).
446. B. Sörbo, *Biochim. Biophys. Acta* **24,** 324 (1957).
447. T. W. Szczepkowski, *Acta Biochim. Pol.* **8,** 251 (1961).
448. C. DeMarco, M. Coletta, and D. Cavallini, *Experientia* **13,** 117 (1962).
449. T. W. Szczepkowski and J. L. Wood, *Biochim. Biophys. Acta* **139,** 469 (1967).
450. B. Sörbo, *Biochim. Biophys. Acta* **23,** 412 (1957).
451. C. DeMarco, M. Coletta, B. Mondovi, and D. Cavallini, *Ital. J. Biochem.* **9,** 3 (1960).
452. B. Sörbo, *Acta Chem. Scand.* **12,** 1990 (1958).
453. P. Luchi and C. DeMarco, *Ital. J. Biochem.* **18,** 451 (1969).
454. J. F. Schneider and J. Westley, *J. Biol. Chem.* **244,** 5735 (1969).
455. G. Medes, *Biochem. J.* **31,** 1330 (1937).
456. F. Chapeville and P. Fromageot, *Biochim. Biophys. Acta* **17,** 275 (1955).
457. J. Awapara and V. M. Doctor, *Arch. Biochem. Biophys.* **58,** 506 (1955).
458. L. Ewetz and B. Sörbo, *Biochim. Biophys. Acta* **128,** 296 (1966).
459. J. B. Lombardini, P. Turini, D. R. Biggs, and T. P. Singer, *Physiol. Chem. Phys.* **1,** 1 (1969).
460. J. B. Lombardini, T. P. Singer, and P. D. Boyer, *J. Biol. Chem.* **244,** 1172 (1969).
461. S. Sakakibara, K. Yamaguchi, I. Ueda, and Y. Sakamoto, *Biochem. Biophys. Res. Commun.* **52,** 1093 (1973).
462. T. P. Singer and E. B. Kearney, *Biochem. Biophys. Acta* **14,** 570 (1954).
463. T. P. Singer and E. B. Kearney, *in* "Amino Acid Metabolism" (W. D. McElroy and H. B. Glass, eds.), p. 558. Johns Hopkins Press, Baltimore, Maryland, 1955.
464. E. B. Kearney and T. P. Singer, *Biochim. Biophys. Acta* **11,** 276 (1953).
465. A. Wainer, *Biochim. Biophys. Acta* **141,** 466 (1967).
466. M. Alexander, "Introduction to Soil Microbiology." Wiley, New York, 1961.
467. J. R. Freney, *in* "Soil Biochemistry" (A. D. McClaren and G. H. Peterson, eds.), p. 229. Dekker, New York, 1967.
468. J. R. Freney and F. J. Stevenson, *Soil. Sci.* **101,** 307 (1966).
469. G. Guittonneau, *C. R. Acad. Sci.* **184,** 45 (1927).
470. G. Guittonneau and J. Keiling, *Acad. Sci.* **184,** 898 (1927).
471. M. R. Hall and R. S. Beck, *Can. J. Microbiol.* **18,** 235 (1972).
472. N. I. Sheryakova and T. G. Leonova, *Fiziol. Rast.* **18,** 1188 (1971).
473. M. I. Vitolins and R. J. Swaby, *Aust. J. Soil. Res.* **7,** 171 (1969).

CHAPTER 8

Sulfate Activation and Transfer

Romano Humberto De Meio

I. INTRODUCTION*

In his textbook, "The Elements of Experimental Chemistry" (1815), Henry [1] noted the presence of two forms of sulfate in urine. One that

* The following abbreviations will be used in this chapter: ATP, adenosine triphosphate; ADP, adenosine diphosphate; p-NP, p-nitrophenol; p-NPS, p-nitrophenyl sulfate; APS, adenosine 5'-phosphosulfate or adenylyl sulfate; PAPS, 3'-phosphoadenosine 5'-phosphosulfate or 3'-phosphoadenylyl sulfate; PAP, adenosine 3',5'-diphosphate or 3'-phosphoadenosine 5'-phosphate; PP_i, inorganic pyrophosphate; GSH, reduced glutathione; APS-kinase, adenylyl sulfate kinase.

precipitated from acidified urine by addition of $BaCl_2$ was considered to be inorganic sulfate; the other required acid hydrolysis of the urine to precipitate. Baumann [2], almost 50 years later, demonstrated that this second fraction consisted primarily of sulfate esters. He further suggested that inorganic sulfate was the precursor of these compounds produced in the mammalian body. Direct proof of this assumption was not obtained until radioactive tracer methods were available. Borsook et al. [3] first used radioactive sulfate to study the excretion of radioactivity in the urine after oral administration to a human subject. The excretion was followed for 48 hours and 47% of the administered dose was found in the urine collected during the first 24 hours. Dziewiatkowski [4], in similar studies in rats, found a high rate of excretion, with 95% of the sulfate accounted for in the urine and feces collected during the first 120 hours after intraperitoneal administration of the radioactive sulfate.

That exogenous inorganic sulfate was the source of the ester-bound sulfate was clearly indicated by the close agreement among the specific activities of the sulfur containing fractions of urine (inorganic sulfate, ester sulfate, and total sulfate) found by Laidlaw and Young [5] and Dziewiatkowski [6]. This was a confirmation and extension of the assumption by early in vivo work [2,7,8] and in vitro work with liver tissue slices [9].

An extensive literature on autoradiographic studies on $^{35}SO_4^{2-}$ incorporation has appeared [10–12]. Though most of the incorporation of ^{35}S is due to uptake of sulfate into sulfated polysaccharides, there is no doubt that in some instances the sulfate is taken up into other compounds.

From the foregoing, it appears that exogenous preformed sulfate may be an important source for the formation of sulfated compounds. It is well known that sulfate is the final product of catabolism of sulfur in animal organisms. We can expect therefore that part of the sulfate for sulfation may have this origin (see Chapter 14).

Knowledge concerning the biological mechanism of sulfation had its origin in the in vitro studies on the fate of phenol in mammalian liver. Arnolt and De Meio [13] studied phenol conjugation and came to the conclusion that it was an enzymatic process that required a coupled oxidative reaction that provided the energy for the formation of the conjugate (conjugation did not take place under anaerobic conditions or in presence of cyanide). In a following paper De Meio and Arnolt [14] found that tissue brei did not conjugate phenol.

In a preliminary note De Meio and Tkacz [15] were able to demonstrate that rat liver homogenate conjugated phenol with sulfate, provided

α-ketoglutarate, adenylic acid, and sulfate were added to the incubation medium. This showed the need for a coupled oxidation to provide the energy for the synthetic reaction. The role of the mitochondria and oxygen in the homogenate, in the presence of an oxidizable substrate and catalytic amounts of adenylic acid, obviously consists of the generation of ATP for the sulfation reaction. De Meio and Tkacz [16] found that 2, 4-dinitrophenol inhibited sulfation of phenol, indicating that phosphorylation is involved in this process. Further support for this assumption was obtained by De Meio et al. [17] from the inhibition of sulfation by methylol gramicidin. This study dealt with the activity of the fractions obtained from rat liver homogenate and led the authors to the conclusion that the particle-free fraction required only SO_4^{2-}, ATP, and Mg^{2+} to sulfate phenol. Bernstein and McGilvery [18], using m-aminophenol as a substrate, reported the same findings.

II.   SULFATE ACTIVATION

A. Mechanism of Activation

Bernstein and McGilvery [19] showed that the high speed supernatant fraction incubated with inorganic sulfate, ATP, and Mg^{2+}, but without m-aminophenol, led to the accumulation of an intermediate. This was interpreted as evidence for a preliminary step of enzymatic activation of inorganic sulfate by ATP, prior to conjugation of m-aminophenol with sulfate.

The formation of active sulfate was confirmed by De Meio et al. [20]. These authors were able to separate two activities from an ammonium sulfate fraction (1.7–2.3 M) of the particle-free supernate of rat liver homogenate. The "activating enzyme" remains after heating the preparation for 15 minutes at 52°, and the "transferring enzyme" can be demonstrated in preparations kept in contact for 4 hours with 0.03 M glycylglycine. Thus, it was clearly demonstrated that sulfation required the activation of sulfate, followed by its transfer to the substrate. Segal [21] carried out a kinetic analysis of the system and studied the inhibition by iodoacetate in detail; this contributed to a better understanding of the problem.

Hilz and Lipmann [22] undertook the study of the activation of sulfate using preparations obtained from lamb liver and Neurospora sitophila. Incubation of these preparations with ATP and $^{35}SO_4^{2-}$ led to the appearance of a labeled compound that would transfer its sulfate to p-NP. The compound appeared to be adenylyl sulfate (APS), as the stoichiometry of the reaction suggested that inorganic pyrophosphate (PP_i) was formed from ATP, and the labeled compound had an absorption peak at 260 nm. De Meio and Wizerkaniuk [23] also concluded that active sulfate was an adenylic derivative. They visualized the activation step (formation of active sulfate intermediate) as the common first step in the biosynthesis of all sulfuric acid esters. They suggested that this first step was followed by the transfer of the sulfate from the active sulfate to the acceptor (phenols, steroid phenols and alcohols, amino sugar containing compounds, and others) probably mediated by specific transferring enzymes.

Robbins and Lipmann [24] isolated the active sulfate by Dowex-1 chromatography and found that the ratio of adenosine to phosphate was $1:2$. By the behavior of the active sulfate on acid hydrolysis and with the use of 3'-nucleotidase, the second phosphate was identified as located in the 3' position. On the basis of this, and other, information, they characterized active sulfate as 3'-phosphoadenosine 5'-phosphosulfate (PAPS). Final identification of the active sulfate as such a compound was reported by the same authors [25]. They found that the phosphosulfate link is unstable at acid pH but stable at neutral and higher pH. These authors identified APS by comparison with the synthetic compound, prepared according to the method of Baddiley et al. [26], which had been shown not to behave as active sulfate. This method gave about 5% yield by sulfation of adenylic acid with pyridine-SO_3 in sodium bicarbonate solution. Reichard and Ringertz [27] used sulfuric acid and carbodiimide in pyridine as the sulfating agent with a yield of 20–25%, but the disadvantage of this technique is the formation of a number of derivatives of adenosine sulfated on the ribose. Triethylamine-N-sulfonic acid in a mixture of pyridine, dioxane, and dimethylformamide as the sulfating agent was used by Cherniak and Davidson [28] with a yield of 60–75% of APS from adenylic acid. This method appears to be the best for the preparation of APS. A 40% yield of APS was obtained by a method based on an anion-exchange reaction [29] that also proved useful for the synthesis of uridylyl sulfate.

The structure of PAPS (Fig. 1) was confirmed by chemical synthesis by Baddiley et al. [30] and the synthetic compound was shown to be active as a sulfate donor in the process of sulfation. The chemical synthesis of PAPS becomes more difficult because the starting material (PAP) is not readily available. Baddiley et al. [31] synthesized PAP by the phosphorylation of adenosine with dibenzyl phosphoryl chloride, followed by sulfation with pyridine-SO_3 in sodium bicarbonate solution to give PAPS [32,33]. The yield was about 10%. Fogarty and Rees [34] prepared PAP using 2-cyanoethyl phosphate. The use of adenosine

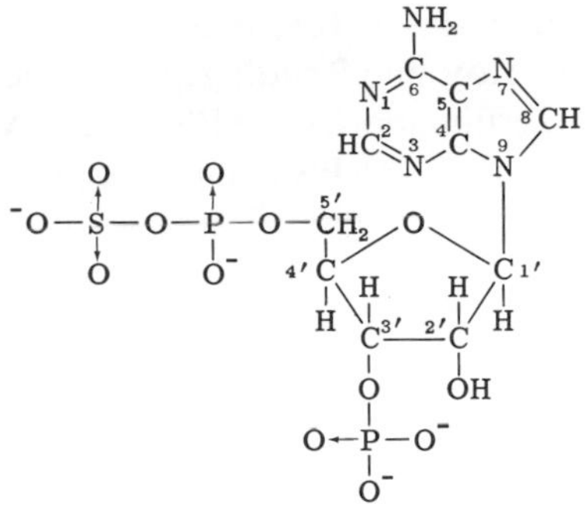

FIG. 1. 3'-Phosphoadenosine 5'-phosphosulfate or 3'-phosphoadenylyl sulfate.

2'(3'),5'-diphosphate, prepared from adenosine treated with dibenzyl phosphoryl chloride, as the starting material for the preparation of PAPS by Cherniak and Davidson [28] eliminates some of the side reactions. This nucleotide was sulfated with triethylamine-N-SO_3 and the cyclic phosphate ester hydrolyzed with T_2 ribonuclease from Taka diastase. This appears to be the method of preference as the yield of PAPS from the diphosphate was from 40 to 65%.

PAPS may be prepared enzymatically. Brunngraber [35] has developed an improved method for the isolation of substrate amounts of PAPS produced by biosynthesis from rat liver and Banerjee and Roy [36] have used a modification of this method. The latter method and that described by Robbins [37] from yeast, both using the sulfate activating system with a yield of about 5% from the ATP, which is the starting material, can easily provide PAPS in 0.1 mmole quantities. Balasubramanian et al. [38] developed a similar method for the sulfate activating system of Furth mouse mastocytoma. Analogous techniques have been described for *Chlorella pyrenoidosa* [39] and for chicken liver by Sass and Martin [40]. All these methods can be adapted to prepare PAP^{35}S [37].

APS and PAPS, when separated from other nucleotides, can be determined measuring the sulfate freed by short hydrolysis in 0.1 N HCl or by measuring their absorption at 260 nm. Accurate values for the extinction coefficients of APS and of PAPS appear not to be available. Baddiley et al. [26] report an ϵ_{259} of 15,200 for APS; and an ϵ_{260} of 14,500 for PAPS has been used by both Robbins and Lipmann [25] and Cherniak and Davidson [28].

APS can be determined by the ATP produced through the action of ATP-sulfurylase [25]. PAPS has been determined using it as sulfate donor with an appropriate acceptor in a reaction catalyzed by phenol sulfotransferase. Jansen and van Kempen [41] have used 4-methylumbelliferone as acceptor. Roy and Trudinger [42] deal with the methods of preparation and determination of APS and PAPS, and describe a technique of determination of PAPS using 2-naphthol as acceptor.

B. The Sulfate-Activating System: The Biosynthesis of Adenylyl Sulfate (APS) and 3'-Phosphoadenylyl Sulfate (PAPS)

Hilz and Lipmann [22] arrived at the tentative conclusion that activation of sulfate was a one-step reaction producing APS (assumed to be the active sulfate) with the probable formation of inorganic pyrophosphate. But the APS synthesized by Baddiley *et al.* [26] did not behave as sulfate donor.

Wilson and Bandurski [43] showed that the sulfate activating system was present in bakers' yeast, and were able to demonstrate, by protamine fractionation, that two separate protein fractions are required for sulfate activation. Segal [44] observed a sulfate-dependent exchange of pyrophosphate with nucleotide phosphate and an inhibition of the esterification of sulfate by pyrophosphate. These results supported the assumption that pyrophosphate is released in a reversible reaction during the process of sulfate activation.

Bandurski *et al.* [45], using enzyme fractions obtained from yeast, confirmed earlier findings [43] and, from the evidence obtained, concluded that two steps were involved in the activation of sulfate. The first, catalyzed by ATP-sulfurylase, leads to the formation of APS [Eq. (1)].

$$ATP + SO_4^{2-} \rightleftharpoons APS + PP_i \tag{1}$$

This is followed by the conversion of APS into PAPS with the participation of APS-kinase [Eq. (2)].

$$APS + ATP \rightarrow PAPS + ADP \tag{2}$$

The same sequence for the biosynthesis of PAPS was reported, at the same time, by Robbins and Lipmann [46]. They gave later a detailed account of their work on the biosynthesis of APS [47] and of the activating system [48]. These authors turned to yeast as a source of the enzymatic system because of the sensitivity of the liver system. Yeast is furthermore a convenient source because it is lacking in any known sulfotransferase (the sulfate-transferring enzyme of De Meio *et al.* [20]), as is the mold *N. sitophila* [22]. The enzyme system that synthesizes PAPS catalyzing the overall reaction represented by Eq. (3) is widely distributed in nature.

$$2 \, ATP + SO_4^{2-} \rightarrow ADP + PAPS + PP_i \tag{3}$$

If we assume that the component enzymes are present in all tissues that form PAPS (only in a few cases have they been isolated, or even identified) and that this is the only known path for the formation of sulfate esters (see Section V), the distribution of this enzymatic system would be as follows. It is present in most mammalian tissues including adrenal, brain, cartilage, cornea, heart muscle and valves, intestinal mucosa, kidney, liver, lung, mast cells, muscle, ovary, pancreas, placenta, retina, skin, spleen, and a variety of tumors. It has also been found in chick embryo [49], hen oviduct [50], and frog liver [51]. Among the invertebrates it occurs in snails [52,53] some sea urchins [54], and the clam *Spisula solidissima* [55]. It has been shown to be present in higher plants [56], yeast [43], *Fusarium solani* [23], *Euglena gracilis* [57,58], and marine algae [52]. The system seems to be generally present in microorganisms.

Though first found in mammalian tissues, the quantitative distribution of the activating system and the variation of its activity under normal and pathological conditions is poorly understood. In fetal rat liver the biosynthesis of PAPS is carried out at almost adult rates, but it brings about sulfation only of acid mucopolysaccharides as acceptors, because of the absence of other sulfotransferases [59]. The lower activity of the activating system compared to that found in adult rats was also reported by Wengle [60]. These findings were attributed to the low vitamin A content of fetal rat liver by Carroll and Spencer [61] (Section II, D).

Sasaki [62] found that 6 days after wounding, the granulation tissue showed the highest values for sulfate activation. According to Balasubramanian and Bachhawat [63] PAPS formation in rat brain is maximal 12 days after birth, coinciding with the maximum of myelination. Jansen et al. [64] found a maximum activity at birth, decreasing gradually thereafter. Increased activity related to growth of the tissue was observed by Gerlach [65] in the heart muscle of young rats. The sulfate activation dropped rather sharply when the body weights of the animals reached values between 50 gm and 150 gm.

In vitro, at least, the addition of NAD^+ stimulates the biosynthesis of PAPS by a mitochondria-free preparation from a mouse mastocytoma [66]. In a representative experiment, the PAPS/APS ratio was 15 after NAD^+ addition and 3 without added NAD^+. These results may be explained by the increased glycolysis producing rather high concentrations of ATP, known to stimulate the production of PAPS from APS [50, p. 257].

APS and PAPS are known to be dialyzable compounds, but there have been claims of bound forms that do not dialyze. This is supposed to be the case with PAPS formed by rat or guinea pig skin *in vitro,* associated with amino acids possibly as a peptide [67], and with APS formed by a preparation of ATP-sulfurylase from sheep liver [68], assumed to be partially protein-bound. The evidence presented in the latter case has been questioned by Roy and Trudinger [42, p. 99]. Rat skin appears to contain a sulfated cytidine nucleotide according to Barker *et al.* [67].

Control of sulfate activation is known to take place in some of the microorganisms that reduce sulfate. In *Escherichia coli* the formation of PAPS is repressed by cysteine, and in *Bacillus subtilis* by both cysteine and glutathione [69]. The two steps of activation seem to be repressed simultaneously [70], and there is an inverse relationship between the specific activity of the activating system and the intracellular concentration of cysteine [71]. This type of control is not present in all microorganisms; for instance in *Desulfovibrio*, ATP-sulfurylase is not repressed by either cysteine or sulfite. This behavior could be explained by the fact that in this microorganism APS is the only active form of sulfate produced and sulfate serves as final electron acceptor, with formation of hydrogen sulfide (see Chapter 7). The sulfate reducing pathway in yeast appears to be regulated at the ATP-sulfurylase step by feedback inhibition by sulfide, and methionine as a repressor [72].

1. First Step: ATP-Sulfurylase (EC 2.7.7.4; ATP:Sulfate Adenylyltransferase)

As discussed above, the first step in the biosynthesis of PAPS [Eq. (1)], which leads to the formation of APS, is catalyzed by the enzyme ATP-sulfurylase. This enzyme has been purified extensively from bakers' yeast by Wilson and Bandurski [73] and Robbins and Lipmann [47,48]. The outstanding characteristic of the sulfurylase reaction is that the reaction as written is actually greatly favored energetically in the backward direction. That is, that the sulfuryl potential in APS is considerably higher than the pyrophosphoryl potential in ATP. The apparent equilibrium constant, at pH 8 and 37°, was found by Robbins and Lipmann [47] to be about 10^{-8} and Wilson and Bandurski [73] obtained a value of 4×10^{-8} at pH 7.5 and 30°. Later, Akagi and Campbell [74] found a value of the same order of magnitude. Since the equilibrium of the sulfurylase reaction is so unfavorable, the forward reaction proceeds to a reasonable extent only when the products of the reaction, APS and PP_i, are removed by subsequent reactions. Pyrophosphate is removed by the quite ubiquitous pyrophosphatase, and APS is removed by phosphorylation with the participation of APS-kinase [Eq. (2)] in the second step of the sulfate activation process. Both of these reactions are strongly exergonic. The standard free energy of the reaction is approximately $+11,000$ calories, meaning that the sulfate group potential in APS must be of the order of 19,000 calories compared with a phosphate group potential of about 8000 calories in ATP.

Because of the extremely unfavorable equilibrium for APS formation, the assay of ATP-sulfurylase is difficult due to the lack of accumulation of significant amounts of this compound. The forward reaction may be used for this purpose; measuring the incorporation of $^{35}SO_4^{2-}$ into PAPS, when the sulfurylase reaction is coupled with the APS-kinase reaction [Eq. (2)]. The reaction [Eq. (1)] has been run in the reverse direction for measuring sulfurylase activity by the disappearance of pyrophosphate or the appearance of ATP, observed when APS is incubated with pyrophosphate and ATP-sulfurylase. The interesting work of Wilson and Bandurski [73] on the effect of the SeO_4^{2-} ion, and other similar group VI anions, on the activity of ATP-sulfurylase led to the development of an easier method of assay. This method is based on the formation of orthophosphate, derived from the pyrophosphate produced when ATP-sulfurylase is incubated with ATP and, for instance, MoO_4^{2-} ions. The basis for this method of assay is discussed below. ATPase activity or a molybdate catalyzed breakdown of ATP, as well as displacement of phosphate ions bound to proteins by MoO_4^{2-} ions [70] are possible sources of error that require carefully controlled techniques. ATP can be determined in picomole amounts by the luciferin-luciferase technique [75,76]. Balharry and Nicholas [77] have developed an assay of ATP-sulfurylase for plant tissues using the luciferin-luciferase method. They determine the ATP produced in the reaction [Eq. (1)] run in the reverse direction, by the luciferin-luciferase enzyme reaction in a liquid scintillation spectrometer, based on the sensitive chemiluminescent assay of Stanley and Williams [78] for detecting picomole amounts of ATP.

ATP-sulfurylase has been purified from bakers' yeast, with about a thousandfold increase in specific activity when compared with the crude extract [47]. The purified enzyme gave a single sharp peak in the ultracentrifuge, but it was not shown whether it was due to the enzyme or to a contaminating protein. The enzyme had a rather broad pH optimum between 7.5 and 9, and was stimulated by Mg^{2+} and Co^{2+}. Hawes and Nicholas [79], for a 40-fold purified enzyme from the same source, also found a broad optimum pH from 7.0 to 9.0. The substrate appeared to be the $Mg–ATP^{2-}$ complex, and free ATP and APS acted as inhibitors. A homogeneous protein by disc gel electrophoresis, gel filtration, and analytical ultracentrifugation, corresponding to 945-fold purification, was prepared by Tweedie and Segel [80,81] from *Penicillium chrysogenum*. It showed a broad pH optimum between 7.5 and 9.0 and required a divalent cation for activity (Mg^{2+}, Mn^{2+}, Co^{2+}). The actual substrate is the ATP–Mg^{2+} complex, free ATP being a competitive inhibitor to ATP–Mg^{2+} and MoO_4^{2-}. The enzyme is inhibited by APS and sulfide. The MW is in the neighborhood of 425,000–440,000, the enzyme being an octamer with subunits of an MW of 56,000.

ATP-sulfurylase has been found in *N. sitophila* [22], but is missing in sulfiteless mutants of this mold [82]. The enzyme has been detected in several microorganisms; a partial purification was carried out from *Desulfotomaculum nigrificans* and from *Desulfovibrio desulfuricans* [74]. An 820-fold purification was achieved by Varma and Nicholas [83] for ATP-sulfurylase from *Nitrobacter agilis*. The enzyme preparation produced a single protein band on starch-electrophoresis, showed a pH optimum of 7.4, and an approximate MW of 700,000. *p*-Chloromercuribenzoate inhibited this enzyme and the inhibition was reversed by thiols. Inhibition was also produced by Group VI anions, and the enzyme was stimulated by thiols. The enzyme occurs in extracts of higher plants tissues, such as spinach chloroplasts [84, 85]. Direct evidence for its presence in animal tissues has been provided only for rat liver and colon by Sundaresan [86], though it is presumed to have a wide distribution. He demonstrated that, at least in those tissues, the enzyme is partly associated with the particulate fraction. It is generally accepted that the enzymes responsible for sulfate activation are present only in the soluble fraction of the cell.

Little is known of the physicochemical properties of ATP-sulfurylases of animal tissues; they do appear, however, to show considerably less stability than the microbial enzymes. De Meio *et al.* [20] obtained an activating enzyme fraction by destroying the sulfotransferase (heating the preparation at 52° for 15 minutes). These results indicate that the ATP-sulfurylase is somewhat heat stable. On the other hand, the enzyme from the thermophile *D. nigrificans* can withstand a temperature of 60° for 1 hour, while the sulfurylase from *Desulfovibrio* is fully inactivated in 3 minutes under the same conditions.

ATP-sulfurylase has also been purified from animal tissues. Levi and Wolf [87] purified it about 1000-fold, compared to the original homogenate from rat liver. The molecular weight, determined by gel filtration, is around 800,000–900,000. A 200-fold purification of the enzyme from sheep liver has been achieved [68]. The K_m values for ATP-sulfurylases from several sources are shown in Table I [68,79,81,83,85,87–89]. The rat liver enzyme is not stabilized by sulfate ions, in spite of earlier claims, but is by phosphate ions. Its instability in Tris-HCl buffers is undoubtedly one of the reasons for previous failures in attempts at purification of the enzyme. An ATP-sulfurylase has been

TABLE I

K_m (mM) Values for the Substrates of Purified ATP-Sulfurylases

Source of enzyme	ATP	ATP-Mg^{2+}	APS	Deoxy-APS	PP$_i$	SO$_4^{2-}$	MgCl$_2$	MoO$_4^{2-}$	Reference
Rat liver	1.6	—	0.25	—	0.037	0.1	—	—	[87]
Sheep liver	—	—	2.0	—	1.7	—	—	—	[68]
Furth mouse mastocytoma									
E-I	—	—	0.05	1	0.15	—	—	—	[88]
E-II	—	—	0.1	1.2	0.2	—	—	—	[88]
Spinach leaves	—	—	0.00047	—	0.003	—	—	—	[85]
Saccharomyces cerevisiae	—	0.07	—	—	—	5	—	—	[89]
Saccharomyces cerevisiae	—	0.046	—	—	—	—	—	0.17	[79]
Penicillium chrysogenum	—	—	0.0071	—	0.077	—	—	0.15	[81]
Nitrobacter agilis	1.4	—	0.025	—	0.12	—	0.35	—	[83]

purified 545-fold from Furth mouse mastocytoma by ammonium sulfate fractionation, hydroxylapatite column chromatography, and Geon resin electrophoresis [88,90–92]. The purified enzyme has a pH optimum of 8.5 and is free of pyrophosphatase, ATP-phosphohydrolase, APS-kinase, and APS-sulfohydrolase. Two forms of the enzyme with similar properties were obtained. There was no absolute requirement for metal ions, but the activity was increased by Mn^{2+}, Mg^{2+}, and Zn^{2+}, and EDTA was a strong inhibitor. The enzyme was activated by 0.1–0.05 mM p-hydroxymercuribenzoate, and it appears not to require –SH groups for its activity. It reacted only with APS or deoxy-APS, and it formed an enzyme–ATP complex converted to enzyme-bound APS upon addition of sulfate. A sequential mechanism is proposed for the reaction based on these observations. These interpretations do not agree with those of Hawes and Nicholas [79] and Tweedie and Segel [80,81], who consider Mg^{2+} part of the complex, and Levi and Wolf [87] who postulate an enzyme–AMP complex as intermediate. A more stable ATP-sulfurylase was obtained from bakers' yeast [93] and purified to a higher degree, the enzyme being homogeneous according to acrylamide disc gel electrophoresis and ultracentrifugation. It differed in some properties from that of mouse mast cell tumor. Maximal formation of the enzyme–ATP complex required Mn^{2+}, but exogenous Mg^{2+} was not needed. Sulfate appeared not to be bound to the enzyme even in the presence of ATP, probably because of the undetectable amounts of APS formed.

Sulfate activation decreases in vitamin A deficiency (see Section II,D) but the nature of this decrease and the cause of it are not yet clear. Levi and Wolf [87] have prepared the enzyme from livers of rats treated with [^{14}C]vitamin A after rendering them deficient in this vitamin and found no radioactivity in their most highly purified preparation. This is a definite proof that vitamin A, or any derivative containing the labeled carbon atoms 6 and 7, is not part of the structure of ATP-sulfurylase.

The studies of specificity carried out with ATP-sulfurylase from yeast [48,73] and $D.$ $nigrificans$ or $Desulfovibrio$ [74] led to the conclusion that they show an absolute specificity for ATP. This is not true for the sulfate ion. Wilson and Bandurski [73] presented evidence for two other types on reactions with group VI anions, other than the SO_4^{2-} ion. The second reaction type was encountered with selenate as anion. With the help of radioactive selenate they were able to demonstrate the formation of APSe, the selenate analogue of APS. The small amounts of APSe formed were detected by charcoal adsorption and electrophoresis. The ^{75}Se-labeled compound had approximately the same mobility as APS. ^{32}PP$_i$ exchange into ATP was observed, but the amount of PP$_i$ produced was much greater than the APSe formed. The reaction slowly led to the complete breakdown of ATP, and the final products were only AMP, PP_i, and selenate.

The third reaction type was observed with sulfite, chromate, molybdate, and tungstate as anions. In this case ATP was irreversibly changed to AMP and PP_i. $^{32}PP_i$ exchange into ATP was not detected, and an AMP-anion anhydride could not be demonstrated. The mechanism of these reactions has not been clarified. It could be interpreted as a catalytic cleavage of PP_i from ATP without intermediate anhydride formation or formation of AMP-anion anhydride of insufficient half-life to permit PP_i exchange. ATP could not be replaced by adenosine diphosphate (ADP) or the triphosphates of uridine, cytosine, and guanosine. Phosphate formation in the presence of sulfite and molybdate was inhibited by addition of sulfate. APS also acted as an inhibitor of this process.

Robbins [89] found that the breakdown of ATP by molybdate catalyzed by the ATP-sulfurylase of yeast reaches half of its maximum velocity in 0.5 mM MoO_4^{2-} ions. The velocity of the reaction gradually increases in going from pH 7 to pH 9. In extracts of *D. nigrificans* an optimum pH of about 7 is observed and in *D. desulfuricans* the rate of the reaction increased up to pH 8 and then remained constant up to pH 9.5, the highest pH tested.

The behavior of those other Group VI anions differs from that of sulfate. With this anion the reaction [Eq. (1)] involves the formation of small amounts of APS and PP_i in equilibrium with ATP and sulfate. The amount of APS recovered from reaction mixtures approximates the amount predicted by PP_i formation. $^{32}PP_i$ exchange into ATP is observed, and the reaction equilibrium is shifted by the addition of inorganic pyrophosphatase.

At this point it may be of interest to consider the mechanism proposed for the ATP-sulfurylase reaction [Eq. (1)]. It is generally interpreted as a result of a nucleophilic displacement by SO_4^{2-} on the inner phosphorus atom of ATP with release of pyrophosphate. Kosower [94] prefers a mechanism that would involve the formation of an enzyme–AMP complex which would react with SO_4^{2-} ions to give APS. He believes that the other mechanism is improbable, due to the low nucleophilicity of the SO_4^{2-} ion coupled with the expected repulsion between doubly negative sulfate and the negatively charged ATP (in spite of a Mg^{2+} ion chelated to the nucleotide). A sequential mechanism for the reaction catalyzed by this enzyme has been proposed by several authors [79,80,92]. On the other hand, Levi and Wolf [87] suggested that the mechanism is ping-pong.

2. SECOND STEP: APS-KINASE (EC 2.7.1.25; ATP:ADENYLYLSULFATE 3'-PHOSPHOTRANSFERASE)

We have seen that APS, at first thought to be the active sulfate, does not behave as sulfate donor in the sulfation process. The actual donor was identified as adenosine 3'-phosphate-5'-phosphosulfate (PAPS), and therefore APS requires a phosphorylation to be converted to PAPS.

The reaction involved [Eq. (2)] is catalyzed by APS-kinase. Robbins and Lipmann [48] are the only workers who have achieved a certain degree of purification of this type of enzyme from yeast. It was found to have a very high affinity for APS, giving the highest reaction rate at the lowest measurable concentration (5 μM), and this characteristic of the enzyme helps to drive the first step [Eq. (1)] in the forward direction by eliminating APS from equilibrium. An additional factor, working in the same direction, is the essential irreversibility of the second step which was estimated to have a ΔG^0 of -6000 at pH 8 and 37° [47]. Such are the conditions that may lead to an accumulation of PAPS as high as 1 mM in reaction mixtures [35]. The reaction has a pH optimum between 8.5 and 9 and Mg^{2+} acts as a cofactor which may be replaced by Mn^{2+}. Co^{2+} is also active, while Ca^{2+} is inactive. Inosine 5'-phosphosulfate was not a substrate for APS-kinase and did not inhibit the reaction if present in concentrations similar to APS. There is a possibility that the specificity toward the phosphate acceptor may not be absolute.

Wilson and Bandurski [73] have observed increased production of phosphate upon incubating ATP-sulfurylase, ATP, and SeO_4^{2-} ions with APS-kinase. They interpreted these results as the consequence of the formation of APSe in the the first step, the APSe being removed in the second step with the assumed formation of PAPSe, driving forward the ATP-sulfurylase reaction to form pyrophosphate, which changes to phosphate. This interpretation cannot be definitely accepted, as these workers were not able to detect PAPSe or its degradation product PAP.

Robbins and Lipmann [48] have shown that cytidine triphosphate has approximately the same activity as ATP, whereas the triphosphates of guanosine, inosine, and uridine are somewhat more active. The significance of this finding is in doubt, because the presence of nucleotide diphosphokinases and traces of ADP or ATP were not excluded.

Adams and Rienits [95] found that galactosamine inhibited rather strongly the APS-kinase of chick embryo cartilage, but no inhibition was observed with glucosamine or the two corresponding N-acetyl sugars.

As stated above this enzyme has been isolated from yeast, but its distribution in nature is not well known. Being responsible for the formation of PAPS from APS, one can assume a wide distribution of this enzyme, and that it should be present in all animal tissues which form

PAPS. Formation of PAPS is not required for *D. desulfuricans* and in this organism ATP-sulfurylase was found in the absence of APS-kinase [70]. In this case the sulfate is activated to APS, which is utilized in a dissimilatory reduction process leading to the formation of hydrogen sulfide with the concomitant liberation of energy. Here the sulfate is serving as a terminal electron acceptor for energy yielding electron transfers (see Chapter 7). The presence of APS-kinase activity associated with chloroplasts was found in extracts of spinach obtained by sonication [96] and in bean and corn leaves [97].

C. ADP-Sulfurylase (EC 2.7.7.5; ADP:Sulfate Adenylyltransferase)

Another sulfurylase, of not well understood function, was found and partially purified from yeast by Robbins and Lipmann [46,48]; it catalyzes the reaction of Eq. (4).

$$ADP + SO_4^{2-} \rightleftharpoons APS + P_i \tag{4}$$

The forward reaction has never been studied and the presence of the enzyme was detected by the formation of ADP, constituting the reverse reaction. Purified ATP-sulfurylase does not catalyze the reaction and the two enzymes can be easily separated. The reaction can be distinguished from a reaction between APS and P_i catalyzed by a combination of ATP-sulfurylase and pyrophosphatase yielding ATP, because pyrophosphatase requires high concentrations of Mg^{2+} while the ADP-sulfurylase reaction proceeds without added Mg^{2+}. Peck [98] has shown that ADP-sulfurylase does not catalyze a reaction between ADP and Group VI anions. This enzyme is unstable at 37° and catalyzes the arsenolysis of APS to AMP and sulfate. Adams and Nicholas [99] have also studied this enzyme from bakers' yeast and suggest that it may be important in controlling the production of sulfated and reduced sulfur compounds.

ADP-sulfurylase has been detected in microorganisms, in *Thiobacillus thioparus* [100] and in *D. desulfuricans* [101] by Peck, and by Thiele [102] in Thiorhodaceae. Peck feels that this enzyme may function indirectly in the formation of ATP in *T. thioparus* by generating ADP from APS and phosphate ions. Peck and Stulberg [103] report that ADP-sulfurylase appears to be involved in the oxidation of thiosulfate to sulfate with concomitant generation of ADP, in the same microorganism.

Grunberg-Manago *et al.* [104] have found an enzyme in yeast claimed to catalyze the exchange between the terminal phosphate of ADP and orthophosphate ions. The enzyme showed a low degree of specificity toward the nucleotide. Ribotide and deoxyribotide also acted as substrates and catalyzed the exchange with SO_4^{2-}. Because of this behavior, they identified the enzyme with ADP-sulfurylase. This conclusion requires further confirmation. ADP-sulfurylase was freed by sonication from spinach chloroplasts [105] and its activity assayed by the sulfate-dependent $^{32}P_i$-ADP exchange, which is inhibited by ATP. The enzyme is activated by Ba^{2+} and Ca^{2+}, Mg^{2+} being ineffective. The purified yeast and spinach enzymes are sensitive to SH reagents and fluoride, and show a pH optimum of 8.

D. Vitamin A and Sulfate Activation

It seems to be definitely established that vitamin A deficiency causes changes in the metabolism of sulfate. It is not clear, however, what the nature of these changes is, or the specific role vitamin A plays in them [106]. Several authors have reported that vitamin A markedly influences sulfate metabolism [107–109]. The study of the relation of vitamin A to specific metabolic reactions started with the observation of Wolf and Varandani [110] that in vitamin A deficiency, the incorporation of $^{35}SO_4^{2-}$ and [^{14}C]glucose into mucopolysaccharides *in vitro,* by colonic segments from rats, was greatly decreased. Retinol, retinal, or retinoic acid added *in vitro* restored the activity to normal. The defect was found related to a very low rate of formation of PAPS, which could be restored to normal by addition of retinol *in vitro* [111]. These observations were essentially confirmed by the work of Subba Rao *et al.* [112], and Subba Rao and Ganguly [113,114]. Very low ATP-sulfurylase activity in vitamin A deficiency was found in rat liver and colon by Sundaresan [86]. The activity could be restored to normal by an acidic lipid-soluble factor prepared from liver of normal animals, but not by addition of retinol or retinoic acid *in vitro*. Perumal and Cama [115] reported that the low ATP-sulfurylase activity could be enhanced by addition of 5,6-monoepoxyretinoic acid and a lipid extract of normal enzyme. An enhancement of PAPS synthesizing activity of fetal liver by vitamin A derivatives supported the idea of a specific cofactor role for the vitamin or a derivative [59]. This would imply that the activity of ATP-sulfurylase is dependent upon the presence of vitamin A or one of its metabolites. The work of Levi and Wolf [87] showed that the ATP-sulfurylase from rat liver can be prepared free from vitamin A or any of its derivatives still retaining carbon atoms 6 and 7. As a consequence some explanation, other than a structural part of the enzyme, must be sought for this influence of vitamin A on ATP-sulfurylase activity.

Other authors have been unable to find a decrease in the biosynthesis of PAPS in a variety of tissues from vitamin A deficient rats and rabbits

[116–119]. These contradictory results probably may be explained by the degree of avitaminosis and the general nutritional state of the animals used. Protein deprivation alone results in a considerable decrease in the rat tissues level of ATP-sulfurylase. But here again, Levi *et al.* [120] claim that an effect of the vitamin deficiency can be demonstrated particularly in the adrenals, followed by the liver and other tissues, while Geison *et al.* [121] find that when such nutritional effects are minimized, the levels of the enzyme in normal and vitamin A deficient rats are the same, within experimental error. In conclusion, it appears that the nature of the effect of vitamin A on the biosynthesis of PAPS is far from being understood and requires more well-controlled experimentation.

E. Enzymatic Degradation of APS and PAPS

The knowledge of the mechanisms involved in the breakdown of APS and PAPS is of physiological importance for the understanding of the control of sulfate activation and in the study of their biosynthesis and utilization in tissues. The practical significance of these breakdown processes lies in the biological preparation of APS and PAPS.

As mentioned before, the hydrolysis of PAPS, by purified 3'-nucleotidase of rye grass, to APS was used to determine the position of the phosphate introduced in the second step of activation [25]. In the same paper it was shown that bull semen 5'-nucleotidase hydrolyzed APS with the formation of adenosine, phosphate and SO_4^{2-}. The venom of *Crotalus atrox* hydrolyzed APS with the same results [26]. This type of hydrolysis is the result of the activity of the many phosphatases and nucleotidases present in the venom.

Rat liver and other tissue preparations degrade APS and PAPS to yield a variety of products [35,122,123] such as adenosine, sulfate, phosphate, inosine, and hypoxanthine. These findings indicate the participation of 3'- and 5'-nucleotidases and of various deaminases, in addition to the possible role of sulfohydrolase enzymes.

The sulfohydrolase activity is represented by the so-called PAPS-sulfatases which split the sulfate–phosphate bond.

Enzymes able to catalyze the hydrolysis of PAPS to PAP and SO_4^{2-} are widely distributed in nature. This activity has been found in hen oviduct [50, p. 257], in several mammalian tissues [38,124–126], in the mucus gland of the marine mollusk *Charonia lampas* [53], and in *E. gracilis* [127]. Generally, a low degree of purification was obtained. PAPS-sulfatase from sheep brain was shown to be activated by Co^{2+} or Mn^{2+} and inhibited by sulfhydryl compounds, fluoride, and ADP [125]. Koizumi *et al.* [128] reported that in rat liver the activity is localized in both the lysosomal and the supernatant fractions of the cell. Whenever it is necessary to control the degradation of PAPS one can take advantage of the rather strong inhibition of PAPS-sulfatase produced by fluoride or phosphate ions [50, p. 257].

Another type of sulfohydrolase hydrolyzes APS to AMP and SO_4^{2-}. The properties and cellular distribution of two of these sulfatases have been studied by Armstrong *et al.* [129]. Austin *et al.* [130] have studied the subcellular distribution of a sulfatase and a phosphatase that degrade PAPS. In rat liver, Bailey-Wood *et al.* [131] showed that high speed supernatant preparations can desulfate PAPS in two different ways. The direct path involves a PAPS-sulfohydrolase, the other is preceded by formation of APS, catalyzed by a Co^{2+} activated 3'-nucleotidase, followed by desulfation of the product by a specific APS-sulfohydrolase. The latter enzyme has been considerably purified from the high speed supernatant fraction (137-fold) and the lysosomes (56-fold). The authors present evidence that suggests that the two enzymes are not identical. The lysosomal enzyme retains the ability to change PAPS to APS, but Co^{2+} has no effect on this process. The enzyme shows an optimum pH of 5.2, and the enzyme from the high speed supernate is strongly inhibited by ADP, ATP, GTP, PP_i, and P_i. Further studies on these enzymes from bovine liver by Denner *et al.* [132] have shown that PAPS-sulfohydrolase is strongly inhibited or not affected by Co^{2+}, and the pH optimum is 8.9 or 9.4, in both cases depending on the buffer used. Farooqui and Balasubramanian [133] described a dephosphorylation of PAPS to APS not related to 3'-nucleotidase and PAPS-sulfohydrolase activities. This enzymatic reaction had an optimum pH of 5 and was inhibited by Mg^{2+}, Co^{2+}, Mn^{2+}, and most potently by PAP. The APS-sulfohydrolase from bovine liver was purified 1200-fold by ammonium sulfate fractionation, DEAE-cellulose chromatography, and Sephadex G-100, giving a single homogeneous active component on polyacrylamide disc gel electrophoresis [134]. The enzyme is specific for APS, free from PAPS-sulfohydrolase and 3'-nucleotidase, has an MW of 68,000–69,000, and is distributed between the lysosomal and high speed supernate fractions of the bovine liver cell. The kinetic properties were reported by Stokes *et al.* [135]. The enzyme had a pH optimum of 5.4 and showed its maximum activity with 4 mM APS. Substrate inhibition was observed above this concentration, which is not the case with the rat liver enzyme. AMP and SO_4^{2-} inhibit the enzyme noncompetitively and competitively, respectively. The K_m of 0.95 mM for APS is about the same as that of the rat liver enzyme ($K_m = 1$ mM). The pattern of inhibition is consistent with an ordered uni-bi reaction sequence with SO_4^{2-} as the last released product.

F. Other Natural and Synthetic Sulfate Containing Nucleotides

A very interesting nucleotide was isolated from extracts of cod liver by Tsuyuki and Idler [136] using chromatography on refrigerated columns. The refrigeration prevented the decomposition of the compound to 6-succinyladenylic acid. The nucleotide was shown to be a compound of 6-succinyladenosine 5'-sulfatophosphate and a peptide formed by an equal number of serine and glutamic acid residues. The nature of the linkage between the nucleotide and the peptide is not known. The peptide could be removed by 1 hour treatment at room temperature with 0.01 N HCl, leaving the rest of the molecule intact, including the sulfatophosphate link. It is regrettable that no further work has been published on this compound.

Ishimoto and Fujimoto [137] have chemically prepared synthetic analogues of APS and studied their biochemical behavior. These compounds were the sulfates of guanosine, cytidine, and uridine, which were all substrates for the APS-reductase of *Desulfovibrio* (see Chapter 7).

Egami and Yagi [138] have studied the effects of adenosine monosulfate and riboflavin monosulfate on D-amino acid oxidase, and Niwa *et al.* [139], the effects of 5'-adenosine monosulfate as a substrate analogue on 5'-adenylic acid deaminase and 5'-nucleotidase. Adenosine monosulfate acted as a competitive inhibitor of D-amino acid oxidase and 5'-nucleotidase, but did not act either as a substrate or inhibitor of 5'-adenylic acid deaminase. Riboflavin monosulfate competitively inhibits D-amino acid oxidase. Adenosine 5'-sulfatopyrophosphate, an analogue of ATP, has been prepared by Yount *et al.* [140,141]. They have shown that it does not serve as an energy source for the contraction of myosin fibers *in vitro* and is not a substrate for myosin, actomyosin, or myosin ATPase.

III.   SULFATE TRANSFER: THE SULFOTRANSFERASES

A. Introduction

From the discussion about the enzymes involved in the activation of sulfate in microorganisms, and the distribution of sulfotransferases, which will be considered below, it will become apparent that evolutionary changes have taken place in the relative functional significance of the active forms of sulfate. Bacteria use these compounds (APS and PAPS) mainly in sulfate reduction processes (see Chapter 7). In fungi, algae, and plants, sulfation is more common, though the ability to reduce

PAPS for assimilatory processes persists. Sulfation becomes prevalent in higher organisms (reductive mechanisms involving PAPS virtually disappear), and reaches more complex patterns in mammals.

The transfer of sulfate to ordinary phenol has historic interest because it was Baumann [2] in 1876 who isolated for the first time potassium phenyl sulfate from the urine of dogs that had been fed phenol. In 1941 Arnolt and De Meio [13], studying phenol conjugation, discovered the characteristics of this reaction that led to the recognition of the mechanism of sulfate activation and transfer.

We have already discussed the process of sulfate activation, with the formation of PAPS. The sulfate of PAPS is transferred to a suitable acceptor and the reaction is catalyzed by an enzyme belonging to a group known as sulfotransferases (wrongly designated sometimes as sulfokinases). As a result of this type of reaction, when alcoholic or phenolic functions are involved, a sulfate ester and PAP are produced [Eq. (5)].

$$PAPS + R \cdot OH \rightleftharpoons R \cdot OSO_3^- + PAP \qquad (5)$$

This reaction is practically irreversible with the exception of nitrophenyl and dinitrophenyl sulfates, which will be considered below.

A wide variety of sulfate esters are found in nature. For instance, phenolic compounds normally present in the diet or originating in metabolic processes within the tissues or in the intestinal tract give rise to many and varied arylsulfates [142]. Among them are those derived from simple phenols and phenolic acids, phenolic hormones, phenolic steroids such as estrone, and heterocyclic phenols such as indoxyl and skatoxyl. Few have been isolated and identified due to their presence in small amounts, high degree of solubility, and relative instability [143]. The most significant biologically, of those definitely identified from normal urine, were estrone sulfate [144,145], L-tyrosine O-sulfate [146,147], 2-amino-3-hydroxyacetophenone O-sulfate [148], estriol 3-sulfate [149], skatoxyl sulfate [150,151], and serotonin O-sulfate [152]. Sulfotransferase activity is present in most ocular tissues [153]. Biogenic amines and their metabolites have been assumed to be substrates for phenol sulfotransferase in rat brain and liver [154].

Systematic studies investigating the presence of arylsulfates in tissue fluids other than urine and in mammalian tissues have not been pursued, although a few have been identified from these sources. Many of these compounds not yet identified have been found in living organisms. Paper chromatography has been applied to the search in urine and bile for sulfated compounds after administration of $^{35}SO_4^{2-}$ to experimental animals [124,155,156]. Boström and Vestermark [157], using more indirect methods, have shown that human urine contains a complex pattern of sulfate acceptors.

There is no doubt that the sulfation of such diversified types of substrates, alkyl and aryl alcohols, phenols and phenolic steroids, carbohydrates, and glycolipids calls for a variety of sulfotransferases with differing degrees of specificity. The knowledge on this subject is rather meager because of the difficulties in purifying the sulfotransferases. They are extremely unstable enzymes and their mixtures are difficult to resolve. Kinetic characteristics and similar evidence are used in their differentiation rather than a clear-cut separation by the usual enzyme fractionation techniques.

A considerable amount of work on the formation of sulfate esters has been carried out incubating unfractionated tissue preparations with ATP and SO_4^{2-}. Sulfotransferases are certainly involved in these processes, but in most instances nothing is known about the enzymes participating in those reactions.

The study of the kinetics of these types of enzymes would require preparations devoid of the activating system and of the rather widespread PAPS degrading enzymes. PAPS then has to be added as a sulfate donor, taking into consideration the fact that crude preparations may contain PAP and ADP as contaminants and these substances are strong inhibitors of sulfotransferases. As a consequence, only purified preparations of these enzymes and of PAPS would allow a proper study of the kinetic properties. Unfortunately these conditions have been met only a few times, and with some of the crude preparations only the data concerning specificity are of some value. For instance, Adams and Poulos [158] claim to have obtained highly specific estrogen sulfotransferase from beef adrenal glands, only showing a very weak activity toward the synthetic estrogens, stilbestrol, and hexestrol. On the other hand, Roy and Banerjee [159] claimed that guinea pig estrogen sulfotransferase can also use simple phenols and arylamines as substrates. The formation of sulfate esters from catecholamine metabolites and pyrogallol have been studied in the rat brain *in vivo* with the aid of $^{35}SO_4^{2-}$ [160]. With this technique it was found that 5-hydroxytryptamine and normetanephrine were not sulfate acceptors. The inhibition of the activation reaction by the use of EDTA is not to be recommended because many sulfotransferases require the presence of Mg^{2+} ions for their activity. A few sulfotransferases have been partially purified, and some of their properties are presented in Table II.

The determination of sulfotransferase activity may be carried out with methods of general application, or with specific substrates when one wishes to study a particular type of sulfotransferase.

Chromatography has been used by Vestermark and Boström [161] and Spencer [124] in a rather simple technique based on the use of PAP^{35}S.

TABLE II

PROPERTIES OF PARTIALLY PURIFIED SULFOTRANSFERASES

Sulfotransferase	Source	Optimum pH	K_m (mM) for sulfate acceptor	K_m (mM) for PAPS	K_m (mM) for PAP	Mg^{2+} requirement	SH enzyme	Reference
Phenol	Guinea pig liver	5.6	0.08	0.036	0.014	−	+	[184]
Phenol	Rabbit liver	7.2	—	—	—	+	?	[224]
p-NP	Rat liver	7.5, 8.0	0.0015, 0.051	0.014	—	?, −	+, ?	[188,211]
Tyrosine methyl ester	Rat liver	7.5	2.9	—	—	?	+	[188]
Serotonin	Rabbit liver	8.2	2	—	—	+	?	[224]
Estrone	Beef adrenal	8.0	0.014	0.07	—	+	+	[158]
Estrone	Guinea pig liver	6.2	0.025	—	—	+	?	[36]
Androstenolone	Guinea pig liver	7.5	0.02	0.043	—	+	?	[248]
Androstenolone	Human liver	7.4	0.0067	0.004	—	?	+	[253]
Choline	Aspergillus nidulans	7.8	12	0.022	—	−	+	[296]
Chondroitin-6-sulfotransferase	Mouse liver	8.2	0.06,[a] 0.018[b]	0.018	—	?	?	[351]

[a] Chondroitin.
[b] Chemically desulfated chondroitin sulfate.

Wengle [162] has taken advantage of the insolubility of the barium salts of PAPS and SO_4^{2-} (the barium salts of sulfate esters are generally soluble) to devise an excellent method with the use of $PAP^{35}S$. Of more limited application is the method of Roy [42, p. 66; 163] based on the solubility in chloroform of the methylene blue salts of some sulfate esters. This method cannot be used for the determination of simple alkyl sulfates, aminophenyl, or carbohydrate sulfates. It has been applied successfully to the study of the biosynthesis of aryl and steroid sulfates.

In other methods a particular substrate is used, therefore limiting their application. p-Nitrophenol was used by De Meio [164] and Gregory and Lipmann [165], m-aminophenol by Bernstein and McGilvery [18], serotonin by Hidaka et al. [166], and 4-methylumbelliferone by van Kempen and Jansen [167].

B. Aryl (or Phenol) Sulfotransferase (EC 2.8.2.1; 3'-Phosphoadenylylsulfate : Phenol Sulfotransferase)

The transfer of sulfate to phenolic compounds is one of the first sulfation processes studied and a great deal of attention has been devoted to it. In spite of this, remarkably little information is available concerning the properties of the individual purified enzymes.

Judging by the presence of this particular type of sulfotransferase activity (i.e., by the ability of biological material to produce esters of phenols), we can conclude that they are among the most widespread of this group of enzymes. We have direct evidence of their presence in many mammalian tissues [168,169]. If the presence of ester sulfate in the excreta is considered as indirect evidence, then phenol sulfotransferases must be present in all mammals, including several species of whales [170]. The activity of these enzymes vary in different species. According to Stekol [171], the conjugation of isobarbituric acid with sulfate is very low in the pig, whereas it is rapidly sulfated in many other mammals. Other species, including birds [172], amphibians [173], and mollusks [52] have the ability to sulfate phenols, and among the latter, those of the genus Murex were the source of the Tyrian Purple (derived from the substituted indoxyl sulfates found in them). Though birds, reptiles, and amphibia form sulfuric esters of phenols, the fish do not appear to be able to [174–176]. Among invertebrates, insects and arachnids have been shown to be able to form aryl sulfates [174,175,177–179]. Aryl sulfates have rarely been found in microorganisms, but Ruelius and Gauhe [180] isolated fusarubin sulfate (hydroxynaphthoquinone sulfate) from the culture medium of the mold Fusarium solani. In this case it appeared that the phenol sulfotransferase, presumably present, must have had a certain degree of specificity, because cell-free preparations of the mold required the addition of rat liver sulfotransferase to synthesize p-nitrophenyl sulfate [23]. In the alga *Polysiphonia lanosa,* Hodgkin *et al.* [181] found 2,3-dibromobenzylalcohol 4,5-disulfate, which indicates the presence of a phenol sulfotransferase presumably responsible for its formation. The only indications of the probable occurrence of phenol sulfotransferases in higher plants has been the isolation of rhamnazin 3-sulfate from *Polygonum hydropiper* by Hörhammer and Hänsel [182], and of similar flavonoid sulfates from two species of eel grass [183]. We do not know if the flavonols are substrates for phenol sulfotransferase. Though the presence of phenol sulfotransferases in microorganisms and plants is questionable, they are able to produce PAPS which plays a role in sulfate reduction (see Chapter 7).

Most of the sulfotransferase activity, according to what is known at present, is found in the soluble fraction of cell preparations where the enzymatic system responsible for the biosynthesis of PAPS has been found.

Phenol sulfotransferases catalyze the type of reaction represented by Eq. (5), where R · OH is a phenol. In the case of simple aryl sulfates, such as phenyl or the naphthyl sulfates, the reaction is essentially irreversible. If the "sulfate group potential" is high, as in the case of p-NPS or the dinitrophenyl sulfates, the reaction is reversible. Gregory and Lipmann [165] have devised an assay for phenol sulfotransferase based on the reverse reaction that can be used provided some precautions are observed.

The quantitative distribution of phenol sulfotransferase has not, unfortunately, been studied properly. Frequently the total synthetic process, including the formation of PAPS, has been measured. In other instances PAPS was added to the system, but in both cases the possible interference of APS and PAPS degrading enzymes, present in crude tissue preparations, has often been disregarded.

The method of assay mentioned above [165] is based on the following type of reaction:

$$p\text{-NPS} + \text{PAP} \xrightarrow{\text{phenol sulfotransferase}} p\text{-NP} + \text{PAPS} \qquad (6)$$

A closely related and particularly effective reaction has been used by Gregory and Lipmann [165] and Brunngraber [35]. The transfer of sulfate was shown to be catalyzed by the phenol sulfotransferase from rabbit liver to some suitable phenolic acceptor such as phenol or *m*-aminophenol, with PAP acting as a cofactor [Eq. (7)].

$$p\text{-NPS} + \text{phenol} \xrightarrow[\text{PAP}]{\text{phenol sulfotransferase}} p\text{-NP} + \text{phenyl sulfate} \qquad (7)$$

Because the phenyl sulfate has a very low sulfate group potential, the essential irreversibility of the reaction is assured and the released p-nitrophenol, determined spectrophotometrically, is a measure of the enzymatic activity. Other types of compounds, such as steroids, have been found inactive in this system. It must be assumed that PAP accepts sulfate forming PAPS as an intermediary in this reaction. Because the enzyme preparations contain other sulfotransferases, the PAPS must remain bound to the enzyme; otherwise it could be transferred to other acceptors. A very low concentration of PAP is required in the test. According to Brunngraber [35], 2 μM PAP saturates the phenol sulfotransferase from a crude preparation of rabbit liver and concentrations greater than 10 μM strongly inhibit the reaction. This method may be used for the determination of either the sulfotransferase or PAP. Unfortunately, its general validity is doubtful following the failure of Banerjee and Roy [184] to achieve a similar transfer with a purified phenol sulfotransferase from guinea pig liver.

Is there one or several individual phenol sulfotransferases in liver and generally in other mammalian tissues? From the scant information available it appears that there are certainly more than one. Attempts have also been made to study the activity of these enzymes, in crude tissue preparations, under different physiological conditions. Wengle [185] found low activity in fetal human tissue extracts, using PAP³⁵S as sulfate donor and phenol and various steroids as acceptors. Carroll and Spencer [59] in rats, and Spencer and Raftery [186] in guinea pigs, reported similar findings and attributed the low results to fetal deficiency of an essential cofactor related to vitamin A. The deficiency appeared to result from low maternal transfer of the vitamin to the fetus. This interpretation is open to question and requires confirmation. In rat brain the activity is minimal at birth and increases gradually to the adult value [64]. Because 4-methylumbelliferone was used as sulfate acceptor, the values may not represent phenol sulfotransferase activity. The fact that fetal skin was able to transfer sulfate from PAPS to endogenous acidic glycosaminoglycans [59]—a basic process for biosynthesis of all connective tissue in the developing fetus—led to the view that the low phenol sulfotransferase activity is a reflection of the low requirement for "detoxication" in the fetus. The observation of extremely high phenol sulfotransferase levels in the enclosed system of the developing egg [186] lends some indirect support to the above interpretation.

It appears that at least three distinct sulfotransferases transferring the sulfate to phenols or phenolic steroids exist in mammalian tissues, namely, phenol, estrogen, and L-tyrosine-methyl-ester sulfotransferases.

The papers published on phenol sulfotransferase in the early period disagree in the general and kinetic properties reported. Roy [187] has critically reviewed this work.

The only contributions that have dealt properly with this problem are those of Roy and co-workers. Banerjee and Roy [36] obtained, by chromatography on DEAE-Sephadex, from extracts of guinea pig liver three fractions showing phenol sulfotransferase activity. One appears to be a true phenol sulfotransferase, because it did not catalyze the transfer of the sulfuryl group of PAPS to any other acceptor tested. Later on, all the fractions were shown to catalyze the sulfation of L-tyrosine methyl ester, in addition to the sulfation of p-nitrophenol, by Jones and Roy (cited by Barford and Jones [188, p. 432]). The other two phenol sulfotransferases also showed steroid sulfotransferase and arylamine sulfotransferase activities. Indirect evidence suggested that these three activities were due to one enzyme, but only further work can help in reaching a decision.

Banerjee and Roy [184] provide the most reliable data on the reaction kinetics of the true phenol sulfotransferase isolated and partly purified from guinea pig liver. The optimum pH is 5.6 in acetate buffer, the enzyme being very unstable, and at 37°, under test conditions, only gives zero-order kinetics for 10 minutes. In contrast with some other sulfotransferases, it does not require Mg^{2+}, it is a sulfhydryl enzyme sensitive to metals or oxidation, and is not inhibited by EDTA. The estimated molecular weight is 70,000. The kinetic studies of the forward reaction and the inhibition produced by the products indicated that there are independent binding sites for the phenol and PAPS, and that the reaction catalyzed by the enzyme is a rapid equilibrium random bi bi reaction with one dead-end ternary complex of enzyme-3'-phosphoadenylic acid–p-nitrophenol (Cleland's nomenclature [189]. The reaction can be represented by Eq. (8).

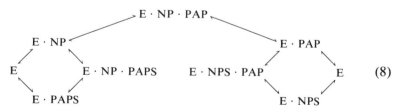

$$(8)$$

The dissociation constants of the enzyme–substrate complexes are given in Table III.

The specificity of the phenol sulfotransferases requires further study. We know that phenol sulfotransferase activity from mammalian tissues is responsible for the sulfation of a variety of simple phenols. Some of them are artificial substrates like p-nitrophenol, others of physiological

TABLE III

DISSOCIATION CONSTANTS OF COMPLEXES OF PHENOL SULFOTRANSFERASE OF
GUINEA PIG LIVER AT pH 5.6 AND 37° WITH DIFFERENT SUBSTRATES[a]

Substrate	K_s (mM)[b]
p-Nitrophenol	0.070 ± 0.012
PAPS	0.036 ± 0.010
p-Nitrophenyl sulfate	0.068 ± 0.008
PAP	0.024 ± 0.002

[a] From Roy and Trudinger [42].
[b] Values given with their 95% confidence limits.

importance such as adrenaline [190], serotonin [191], or the thyroid
hormones [192]. A large variety of aryl sulfates have been found
in nature; these include bufothionine in toad venom [193], 2,3-
dibromobenzyl alcohol 4,5-disulfate in the marine alga *Polysiphonia
lanosa* [181], hydroxynaphthoquinone sulfate in a mold [180], ommatin
D in insects [194], and tyrosine *O*-sulfate in mammalian fibrinogens
[195]. The variety of chemical structures involved must lead us to the
conclusion that either the responsible enzymes are of very low specificity or there are many specific phenol sulfotransferases.

The work of Banerjee and Roy [184] with their purified enzyme from
guinea pig liver also has provided the most detailed information about
the specificity of a phenol sulfotransferase. Their results are summarized
in Table IV.

TABLE IV

SPECIFICITY OF THE PHENOL SULFOTRANSFERASE FROM GUINEA PIG LIVER[a]

Substrate	K_m (mM)	V
Phenol	2.5	0.30
p-Nitrophenol	0.070	1.00
1-Naphthol	0.025	1.02
2-Naphthol	0.025	0.95
5,6,7,8-Tetrahydro-2-naphthol	0.056	0.63
4-Nitro-1-naphthol	0.018	0.39
2-Phenanthrol	0.017	0.51
15,16-Dihydro-3-hydroxy-17- ketocyclopentena[a]phenanthrene	0.015	0.13
Equilin	0.020	0.21
Equilenin	0.015	0.10
Estrone	—	0.00

[a] The reaction was in all cases studied at pH 5.6 in acetate buffer at 37° with PAPS as
the sulfate donor. The maximum velocities are expressed relative to that with p-nitrophenol.

We see that many phenols can act as substrates. Only estrone is not sulfated and it was found that it does not inhibit the formation of p-nitrophenyl sulfate. Steric hindrance preventing the combination of estrone to the enzyme probably accounts for this behavior.

The sulfation of 4-methylumbelliferone has been reported by van Kempen and Jansen [196] using sheep brain and rat liver as sources of enzyme. Later [167] they proposed a method of determination of phenol sulfotransferase activity based on this sulfation, but the specificity of the reaction was not determined.

The studies of Williams [197] of in vivo sulfoconjugation in the rabbit after oral administration of phenols with different substituent groups in the benzene ring have some bearing on the specificity of sulfotransferases. It was found that the electrophilic ortho substituent groups tended to depress sulfation, whereas nucleophilic ortho substituent groups had the opposite effect. Meta substituent groups had a similar, but lower, effect while para substituents had relatively little effect. The "ortho effect" was attributed to the lack of sulfation of o-hydroxybenzoic acid by the rabbit. It would be of interest to study the in vitro behavior of these substrates with purified sulfotransferases, as feeding experiments of this type are open to many questions at the enzymatic level. Some of the drawbacks have been discussed by Bray et al. [198,199].

Of further interest in connection with enzyme specificity is the behavior of compounds having more than one phenolic hydroxyl group available for sulfation. The monosulfates of orally administered polyphenols were isolated from the urine. Garton and Williams [200,201] studied catechol and quinol and Dodgson et al. [143] used 4-chloro-2-hydroxyphenol. But Vestermark and Boström [202] indicate that in vitro catechol, resorcinol, phloroglucinol, and pyrogallol may be polysulfated by properly fortified liver preparations. This reaction apparently did not occur to a significant extent in vivo.

Since the isolation of L-tyrosine O-sulfate by Tallan et al. [146] from human urine, the ester has been found in other mammalian urines. L-tyrosine O-sulfate may be routinely determined by an autoanalytical technique [147]. The biosynthesis of this sulfate of tyrosine has posed intriguing problems. The early attempts to obtain sulfation of free L-tyrosine with properly fortified preparations from liver and other sources under conditions which would produce sulfation of other substrates ended in failure [50, p. 257; 203,204]. Segal and Mologne [205] investigated this problem and concluded, from indirect evidence, that only tyrosine derivatives in which the carboxyl group is blocked or absent and the amino group is free (and uncharged) were sulfated by PAPS-phenol sulfotransferase system. These findings were confirmed and firmly established by Dodgson and co-workers. Tyrosine methyl and ethyl esters were sulfated by an unfractionated sulfotransferase preparation from rat liver. The sulfation products gave free L-tyrosine O-sulfate *in vivo* or *in vitro* [206]. L-Tyrosylglycine and L-tyrosylalanine acted as a sulfate acceptor, but glycyltyrosine did not [207]. Tyramine was also a substrate, but estrone was not sulfated. Mixed substrate studies and other evidence indicated that a different enzyme is involved in the sulfation of dehydroisoandrosterone. The enzyme responsible for the formation of L-tyrosine O-sulfate (EC 2.8.2.9; 3'-phosphoadenylylsulfate:L-tyrosine-methyl-ester sulfotransferase) has been partially purified about 450-fold from rat liver, but no clear cut separation from all other sulfotransferases has been achieved [208]. The three fractions of guinea pig liver, mentioned above, that show phenol sulfotransferase activity can also sulfate tyrosine methyl ester. It appears, though, that this substrate is not sulfated by phenol sulfotransferase. A partial separation from the activity toward *p*-nitrophenol and some difference in their behavior points to two different enzymes. The reaction catalyzed by the enzyme appears to be of the rapid equilibrium random bi-bi type, the binding of PAPS and L-tyrosine methyl ester taking place at independent sites. The enzyme is specific for a nonprotonated amino group and probably a protonated phenolic hydroxyl group, at least in the case of L-tyrosine methyl ester and tyramine. Mattock and Jones [209] found that GSH, 2-mercaptoethanol, and dithiothreitol inhibit sulfation of *p*-nitrophenol, but enhance sulfation of derivatives of L-tyrosine by the rat liver enzymes. The stimulating effect of GSH and cysteine on phenol sulfotransferase of rabbit and rat livers was observed by Gregory and Lipmann [165] and Subba Rao *et al.* [112]. Recently Barford and Jones [188] separated two enzymes (A and B) from female rat livers. Both catalyzed the sulfation of *p*-nitrophenol and L-tyrosine methyl ester, but enzyme A also used dehydroepiandrosterone as a sulfate acceptor. Enzyme B undergoes oxidation when stored in the cold at 0°, and the K_m and V_{max} for *p*-nitrophenol sulfation increase about 200-fold and 4-fold, respectively. This form of enzyme did not catalyze the sulfation of L-tyrosine methyl ester. The K_m for PAPS was not affected under these conditions. Prolonged storage in the cold and low ionic strength led to a considerable degree of polymerization of both enzyme activities. The changes in kinetic properties and molecular size of enzyme B during storage were reversed by dithiothreitol. This thiol did not affect the enzyme activity of freshly prepared liver supernatant which was interpreted to mean that the enzymes existed *in vivo* in the reduced state. Is there a single enzyme responsible for the sulfation of phenol and tyrosine? According to Barford and Jones [210] the best substrates are L-tyrosine amide, L-tyrosine methyl ester, tyramine, and 5-hydroxytryptamine (serotonin). They believed that L-tyrosine-methyl-ester sulfotransferase was not likely to be involved in the biosynthesis of proteins containing L-tyrosine O-sulfate residues, but could be important in the sulfation of physiologically active amines. A phenol sulfotransferase was purified 2000-fold from male rat livers [211], that did not catalyze sulfation of dehydroepiandrosterone, butan-1-ol, L-tyrosine methyl ester, 1-naphthylamine, or serotonin and is inhibited by thiol compounds. The transfer of sulfate from PAPS to dopamine forming the 3-O-sulfate was demonstrated by Jenner and Rose [212] in rat liver and brain (brain also formed the 4-O-sulfate). They concluded that the L-tyrosine-methyl-ester sulfotransferase was the enzyme involved. The activity was doubled in the presence of dithiothreitol and L-dopa was not sulfated when the dopa decarboxylase present was inhibited.

L-Tyrosine O-sulfate, though isolated as such from urine, is not found free as a component of mammalian body tissues. It has been detected, however, in the tissues of some rather primitive organisms such as shrimps, sea urchins, and starfishes [213]. As part of fibrinopeptide B, L-tyrosine O-sulfate is present in mammalian and other fibrinogens [195,214,215]. The distribution of L-tyrosine O-sulfate residues differs in fibrinogens from different species. Fibrinopeptides A and B arise during the conversion of fibrinogen to fibrin by the action of thrombin and are released before the polymerization that produces the insoluble fibrin gel. Single L-tyrosine O-sulfate residues are found in the fibrinopeptides B of most mammalian species, but in man, some other primates, rat, and guinea pig no such residues are found. In rare cases (dog and cat) two such residues lying adjacent to each other have been detected [216]. L-Tyrosine O-sulfate has been also found in hog gastrin II [217] and two hypotensive peptides, phyllokinin isolated from the skin of a South American tree frog *Phyllomedusa rohdei* [218] and caerulein from the skins of various Australian frogs of the genus *Hyla,* particularly the frog *Hyla caerulea* [219,220].

The structures of these peptides are compared below with that of bradykinin.

Ala-Asp-Asp-Tyr(SO₃H)(Asp,Glu,Pro,Leu,Asp,Val)-Asp-Ala-Arg
Rabbit fibrinopeptide B

Pyr-Gly-Pro-Try-Met-Glu-Glu-Glu-Glu-Glu-Ala-Tyr(SO₃H)-Gly-Try-Met-Asp-Phe·NH₂
Hog gastrin II

Pyr-Gln-Asp-Tyr(SO₃H)-Thr-Gly-Try-Met-Asp-Phe·NH₂
Caerulein

Arg-Pro-Pro-Gly-Phe-Ser-Pro-Phe-Arg-Ile-Tyr(SO₃H)
Phyllokinin

Arg-Pro-Pro-Gly-Phe-Ser-Pro-Phe-Arg
Bradykinin

The similarity between caerulein and gastrin II is quite remarkable. Both peptides have a blocked N-terminus of pyrrolidone carboxylic acid and the C-terminal protion consists of the same pentapeptide amide. In addition, phyllokinin is a bradykinin derivative with the peptide residue isoleucyltyrosyl O-sulfate attached to the chain.

If all the tyrosine O-sulfate present in human urine was derived from that type of residue in proteins, then other proteins besides fibrinogen must be the source. The lack of information to back this assumption may be due to the fact that the acid hydrolysis used in studies of protein structure brings about a destruction of the residue. The method of Jevons [221], based on the stability of the ester to alkali, should be of use in the study of the distribution of tyrosyl O-sulfate. Little is known about the function of this ester in proteins, but there is no doubt that its presence must change the whole charge pattern of the protein. The sulfate group in gastrin II does not appear to have any obvious function. However, the desulfation of phyllokinin and caerulein causes an appreciable decrease of their biological activity [218].

Our present knowledge tells us that free tyrosine cannot be sulfated by the PAPS-phenol sulfotransferase system, but yet tyrosine O-sulfate residues are found in proteins. It could be assumed that sulfation takes place at an early stage during the formation of the polypeptide chain, but this would require a specific code for the incorporation of sulfated tyrosine residues, or that sulfation follow the incorporation of tyrosine into the polypeptide chain.

The enzymatic O-sulfation of serotonin was reported by Goldberg and Delbrück [52]. Hidaka et al. [222,223] confirmed the formation of serotonin O-sulfate by rat liver preparations and later showed that serotonin lost its biological activity on sulfation. The sulfotransferase from rabbit liver soluble fraction was purified about 65-fold and still retained full phenol sulfotransferase activity [224]. The pH optimum for the sulfation of serotonin was 8.2 and for phenol 7.2 and the K_m for serotonin was 2 mM. This enzyme was found in rabbit and human brain [225,226]. Hidaka and co-workers found the highest activity in liver, followed by kidney, lung, intestine, brain, heart, and spleen. The enzyme was similar to phenol sulfotransferase in some properties, but requires Mg^{2+} for its activity. Further work is needed to definitely establish the specificity of this enzyme. Methods of purification and assay are described by Yagi [227].

The general mechanism of the phenol sulfotransferase reaction is not understood. Mayers and Kaiser [228] proposed that a compound similar to the SO_3 adduct of imidazole might be involved in the transfer of sulfate, as they are powerful sulfating agents in the laboratory. There is no evidence for this type of reaction, and the same applies to the suggestion of Weidman *et al.* [229] based on the formation of methyl sulfate by oxidation of hydroquinone monosulfate in the presence of methanol. The proposal of Ford and Ruoff [230] that sulfated ascorbic acid may act as a sulfating agent may have some merit in view of recent findings (see Section V).

C. Estrone Sulfotransferase (EC 2.8.2.4; 3'-Phosphoadenylylsulfate : Estrone 3-Sulfotransferase)

The designation of estrone sulfotransferase applies to the enzyme that forms estrone sulfate using PAPS as a sulfate donor. This type of enzyme, presumably able to form other estrogen sulfates, was first described by Nose and Lipmann [204] in rat liver. Later it was discovered in human and bovine adrenal gland [231,232].

As in the case of phenol sulfotransferase, the distribution of this enzyme appears to be confined to the animal world. Estrone sulfotransferase does not appear to be widely distributed in adult mammalian tissues. For instance, in beef, only placenta, adrenal, and liver are good sources [168], while the adrenal, jejunal mucosa, and liver are important sources in the human [169]. The enzyme is found in the human fetus [185,233] with a rather widespread distribution. Of the tissues examined, only the brain did not show detectable activity [234]. Here a species difference may be noted; the fetal and neonatal rat livers show low or no activity when compared to that of the adults [59,235]. The hormonal nature of the acceptors have also led to investigations for its presence in the placenta. Here again a species difference is apparent; a high enzyme activity was found in bovine and guinea pig placentas, whereas the activity was quite low in human placenta [168,236,237]. The enzyme has been detected in the hen [238] and it also occurs in embryonic chick, but not in calf cartilage [239]. Creange and Szego [54] reported that the gut of the sea urchin *Strongylocentrotus franciscanus* showed quite a high rate of formation of estradiol 3-sulfate, which indicated the presence of estrone sulfotransferase in this species.

Estradiol and estriol have been shown to form disulfates [240,241] but in these instances we must assume that estrone sulfotransferase was only responsible for the sulfation of the phenolic hydroxyl [158].

The steroid sulfotransferase of guinea pig liver has been studied in some detail by Banerjee and Roy [36,159]. It was separated only from phenol sulfotransferase, but not from other sulfotransferases. The authors were able to conclude that it is different from others that transfer sulfate to steroid alkyl hydroxyl groups. The optimum pH was in the region of 6.0 if estrone is the sulfate acceptor. The enzyme activity depended on the presence of Mg^{2+}, differing in this respect from phenol sulfotransferase. The sulfation rates of the estrogens by the partially purified enzyme decreased in the following order: estrone > 17-deoxy-estrone > estradiol > 2-methoxyestrone > estriol.

An estrone sulfotransferase obtained from beef adrenal gland was shown to be free from most other types of sulfotransferase activity (L-tyrosine methyl ester was not tested as an acceptor with the purified enzyme preparation). The enzyme could not use simple phenols, 17β-estradiol 3-methyl ether, androstenolone, or 2-naphthylamine as substrates; it could utilize the synthetic estrogens stilbestrol and hexestrol but not dienestrol [158,242]. By DEAE-cellulose chromatography and electrophoresis on Geon resin two forms (A and B) of the enzyme have been found. Form B is changed to A on standing and only B is obtained if the enzyme is isolated in the presence of mercaptoethanol. Form B is related to A as a trimer or tetramer to a monomer, the latter having a molecular weight of about 67,000. Mercaptoethanol appeared to maintain the thiol groups of the enzyme in the reduced state thus helping polymerization. Both A and B forms gave multiple bands on gel electrophoresis, corresponding to individual isoenzymes. The kinetic constants have been obtained with the monomeric form, and are presented in Table II for estrone as acceptor. The rate of sulfation of estrone, estradiol, and estriol decreases in the reverse order as that found for the guinea pig enzyme. The enzyme activity was considerably stimulated by Mg^{2+} ions, and by certain other divalent cations. Since the enzyme is not inhibited by EDTA, an absolute requirement for these cations could not be determined. The steroid sulfotransferase of guinea pig liver showed a similar behavior [36]. A more complex type of kinetics can be expected for the polymeric form. The kinetic studies suggested that the mechanism of action of the enzyme is of the sequential type, catalyzing a rapid equilibrium random bi-bi reaction.

Adams [243] obtained a sulfotransferase preparation containing estrone that accompanied the enzyme throughout the purification procedure. The estrone was released as estrone sulfate by incubating the enzyme with PAPS. This finding has lent support to the idea that estrone is the true substrate for the enzyme.

We have indicated that estrone sulfotransferase can catalyze the sulfation of other estrogens. With this restricted type of specificity it is difficult to believe that this same enzyme is responsible for the formation of the 3-sulfate of estriol 16-glucuronide, reported to take place with a crude sulfotransferase from guinea pig liver [244].

D. Steroid Sulfotransferases

The generic name heading this section is supposed to cover all the sulfotransferases that catalyze the sulfation of alcoholic hydroxyls and do not participate in the formation of sulfuric esters of phenolic groups in steroid compounds. This type of activity was noted in crude preparations of rabbit, rat, and beef liver [245–247] at an early stage in the study of sulfation. A variety of steroids served as sulfate acceptors, including 17α-, 17β-, and 21-hydroxysteroids as well as 3α- and 3β-hydroxysteroids of the 5α, 5β, and Δ^5 series. Nose and Lipmann [204] were the first to point out the possibility that these activities were due to several enzymes. This view gained support from the work of Holcenberg and Rose [168], who showed that many bovine tissues sulfated dehydroepiandrosterone, but few were able to sulfate testosterone to any significant degree. Relatively few detailed studies of these enzymes have been carried out. We again find some information in the study of sulfotransferases of guinea pig liver by Banerjee and Roy [36]. From their work they clearly differentiated at least two, if not three, steroid sulfotransferases. One of these sulfated dehydroepiandrosterone, androsterone, and cholesterol (i.e., the secondary cyclic alcohol group at C-3) [159,248]. All information available at present seems to indicate that a single enzyme is responsible for this type of activity. The other enzyme transfers sulfate to testosterone and deoxycorticosterone, in this case to either a C-17 secondary or C-21 primary alcohol grouping. Because of the fact that a partial separation of the two types of activity can be obtained on DEAE-Sephadex, further work may show that two separate enzymes are responsible for the sulfation of testosterone and deoxycorticosterone. The enzyme involved in the sulfation of 3β-hydroxysteroids is (EC 2.8.2.2) 3'-phosphoadenylylsulfate:3β-hydroxysteroid sulfotransferase. A case for the existence of a specific etiocholanolone sulfotransferase can be made based on the following evidence. In rat [245] and rabbit [204] liver the two etiocholanolones (3-hydroxy-5β-androstan-17-ones) are sulfated at a slower rate than androstenolone (dehydroepiandrosterone or 3β-hydroxyandrost-5-en-17-one). On the other hand, the rate of sulfation is high for both acceptors in human adrenal [231,249] and human liver [190]. The steroid sulfotransferases seem to be of limited distribution in the adult, as seen for estrone sulfotransferase. Significant concentrations are found only in the adrenal, the liver, and the jejunal mucosa in the human [169] and beef tissues [168]. Wallace and Silberman [250], using ^{14}C- and ^{3}H-labeled steroids as substrates, were able to detect steroid sulfotransferase activity in the human ovary. Adams [251] and Boström and Wengle [169] failed to detect any such activity in this tissue, probably because of the lower sensitivity of their technique. The presence of androstenolone sulfate in blood from the spermatic vein and the testes of the pig [252] indicated that androstenolone sulfotransferase must be the sulfation catalyst in the testes of the pig.

Wengle and Boström [240] and Carroll and Spencer [59] have reported that alkyl steroid sulfotransferases are practically absent from rat fetal liver, but their full activity develops rapidly after birth. Wengle [185,234] on the other hand, has found that steroid sulfotransferases are present in the human fetus in the same tissues as in the adult with the addition of the kidney. The adrenal gland is a very rich source of the enzyme, while the other tissues show a lower activity than in the adult. A 22-fold purified steroid sulfotransferase was obtained from human liver by Gugler et al. [253]. The MW by gel filtration was about 50,000, it was inhibited by –SH blocking agents and by several cations, among them Cu^{2+} and Hg^{2+}, and stimulated by Co^{2+}, Ni^{2+}, Mn^{2+}, and Ca^{2+}. Other steroids were sulfated to a lesser degree, and testosterone and cholesterol were not sulfate acceptors. Ryan and Carroll [254] purified a sulfotransferase 60-fold from rat liver that apparently specifically transferred sulfate to dehydroepiandrosterone and had negligible activity with p-NP, serotonin, or tyrosine methyl ester.

Some interesting observations were made by Adams and Edwards [255] on the behavior of androstenolone sulfotransferase from human adrenal gland. They noted unusual kinetic properties when dehydroepiandrosterone was used as the acceptor. They suggested that the enzyme was present in several polymeric forms of a monomer having a molecular weight of 65,000. They were led to this conclusion by experiments with gel filtration and sucrose density gradient centrifugation. The separated components reformed an equilibrium mixture on standing for about 72 hours, but the position of this equilibrium depended on the presence of substrates and cofactors. Androstenolone, cysteine, and Mg^{2+} favored association to polymeric forms and PAPS favored formation of the monomer. A high speed supernatant from adrenal gland homogenate was used in the experiments, and no attempt was made to purify the enzyme. Further work with purified preparations is needed since these observations may have resulted from the formation of complexes with other proteins or sulfotransferases.

An indirect biosynthesis of steroid sulfates results from metabolic changes where cholesterol sulfate serves as an intermediate. The compounds formed in this manner, or by direct sulfation, may undergo extensive metabolism without prior removal of the sulfate group (see reviews by Döllefeld and Breuer [256] and Wang and Bulbrook [257]).

The sexual differences in hepatic sulfation of deoxycorticosterone in rats were studied by Carlstedt-Duke and Gustafsson [258]. They found that adult females sulfated five times more efficiently than males.

We have mentioned above that many monohydroxysteroids undergo sulfation in presence of tissue preparations. Wengle and Boström [240] have shown that polyhydroxysteroids can form disulfates, and this is particularly true of the 3β-, 17β-, and 3β, 20β-dihydroxysteroids. The behavior of these steroids as substrates for steroid sulfotransferase depends on some structural characteristics. For example, the sulfation of a 17β-hydroxysteroid is inhibited by the presence of a methyl or ethyl group on the 17α position. A double bond in the 4 → 5 position inhibits sulfation of a 3β-hydroxysteroid.

From the quite different types of steroids sulfated by enzyme preparations from mammalian livers, and the isolation of double conjugates from body fluids, one is led to believe that a considerable variety of alkyl steroid sulfotransferases exist. Herrmann and Repke [259] have shown that a liver preparation sulfates several cardioactive genins such as sarmentogenin, digitoxigenin, and uzarigenin. Androstenolone sulfotransferase could be involved in this case as the steroids contain 3β-hydroxy groups. The double conjugate pregn-5-ene-3β, 20α-diol-20-(2'-acetamido-2'-deoxy-α-D-glucoside) 3-sulfate was found in urine by Arcos and Lieberman [260], and Palmer [261] and Palmer and Bolt [262] isolated the 3-sulfates of glycolithocholic and taurolithocholic acids from bile. The presence of sulfated bile acids in human serum and urine was reported by Makino et al. [263,264] and Palmer [265] found them in rat urine and feces after administration of [24-14C]lithocholic acid. An interesting sulfate diester has been found in plasma after administration of ACTH to humans [266]. The compound was considered to be sulfatidyl 17-ketosteroid complex in which sulfate acts as a diester bridge between the 17-ketosteroid moiety and glycerol. Other similar complexes involving androsterone, dehydrocpiandrosterone, and etiocholanolone were later detected [267]. Oertel et al. [268] believe that 80% of the plasma sulfated steroids are found in association with the α-lipoprotein fraction. Oertel and co-workers have studied the biosynthesis and metabolism of these steroid sulfatides in some detail [269–271], but their meaning is still a matter of speculation.

Laatikainen et al. [272] have detected the presence in human bile of a variety of disulfated steroids involving the C-3, C-17, and C-20 hydroxyl groups. Similar disulfates occur in humans after administration of several steroid acceptors [273,274].

From the point of view of specificity and mechanism of action, the observation of Levitz et al. [275] pose many questions. They found that estriolsulfoglucosiduronate could be biosynthesized by incubating estriol with human liver homogenate in the presence of UDP-glucosiduronic acid, and then incubating the estriol 16-glucosiduronate formed with guinea pig liver homogenate with added ATP. If estriol 3-sulfate was incubated with liver preparations and UDP-glucosiduronic acid, no double conjugate was formed.

It is of interest that of other natural sources, the steroid alkyl sulfates appear to be restricted to the bile of lower vertebrates. A primary or secondary alcohol function in the C-17 side chain is generally the site of sulfation. Ranol sulfate $(3\alpha,7\alpha,12\alpha,24\epsilon,26$-pentahydroxy-27-nor-5$\alpha$- and 5$\beta$-cholestane 24-sulfate) was obtained from the bile of the frog *Rana temporaria* [276] and scymnol sulfate $(3\alpha,7\alpha,12\alpha,24,26,27$-hexahydroxy-5$\beta$-cholestane 26- or 27-sulfate) was found in the bile of some fishes of the Elasmobranchii group [277]. The bile of the hagfish was the source of myxinol disulfate $(3\beta,27$-disulfate of $3\beta,7\alpha,16\alpha,26(27)$-tetrahydroxy-5$\alpha$-cholestane) [278,279]. Similar compounds were isolated and identified from the biles of the carp (cyprinol 27-sulfate) [280], of the coelacanth (latimerol 27-sulfate) [281], and of the *Chimaera monstrosa* (chimaerol sulfate) [282]. These findings are of interest in relation to the problem of bile salt evolution, especially after the recent discovery of numerous steroid sulfates in human bile [272]. Few contributions have been made to the study of the biosynthesis and metabolism of these rare steroid sulfates. Bridgwater and Ryan [51] have shown that frog liver preparations can sulfate ranol, scymnol, and cholan-24-ol when incubated with ATP and SO_4^{2-}. Scully *et al.* [283] have reported the sulfation of bile alcohols (scymnol and β-cyprinol) by the liver of the toad *Xenopus laevis* and separated this activity from that involved in the sulfation of simple aliphatic alcohols by chromatography on DEAE-cellulose. The specifity of the sulfotransferases involved may be only determined by further work. Steroid sulfates have been recently isolated from an insect, the tobacco hornworm *Manduca sexta* Johannson [284]. Tritium-labeled ecdysone (insect molting hormone) injected into the grasshopper (*Locusta migratoria*) is inactivated by sulfation and sulfates of some of its metabolites are formed [285]. The larval gut tissues of the southern armyworm (*Prodenia eridania* Cramer) sulfate *p*-NP, cholesterol, α-ecdysone, and several other steroids [286]. The same authors [287] found similar types of sulfation in several insects. A tabulation of steroids sulfated by various animal tissues has been provided by Wang and Bulbrook [257].

Recently, the sulfation of vitamin D_2 and D_3 has been shown to occur in rat liver and various other tissues [288], while vitamin D sulfate was isolated from human and cow's milk [289]. As vitamin D is not a steroid, the sulfotransferase involved in this process may be of another type.

A technique to measure picomole amounts of radioactive steroid sulfate has been developed using ^{35}S-labeled sulfate and dehydroepiandrosterone or 17β-estradiol as sulfate acceptors to study steroid sulfation [290].

E. Choline Sulfotransferase (EC 2.8.2.6; 3'-Phosphoadenylylsulfate:Choline Sulfotransferase)

Choline O-sulfate is found in appreciable amounts in the higher fungi [291–294]. It is also found in red algae [295] and several *Pseudomonas* species [296]. According to the work of Spencer and co-workers, this ester is used as a convenient store of sulfate [297–299]. When there is an adequate supply of sulfur, sulfate is converted to choline sulfate through the action of choline sulfotransferase, and in times of sulfur deprivation, inorganic sulfate is released by the activity of choline sulfohydrolase and can be used for cysteine production. Choline sulfotransferase uses only PAPS as a sulfate donor in an irreversible reaction [300]. It was wrongly reported earlier [301] that APS could be the sulfate donor.

This enzyme was found difficult to purify because of its instability. Obtained from *Aspergillus nidulans* [297], it was shown to be a sulfhydryl enzyme not requiring the addition of Mg^{2+} ions for maximum activity. This latter observation disagrees with the findings of Kaji and Gregory [300] for *Aspergillus sydowi*. Many compounds structurally related to choline were tested as possible substrates for choline sulfotransferase but only choline, dimethylethylaminoethanol, and dimethylaminoethanol were sulfated, with K_m values 0.012, 0.020, and 0.025 M, respectively. The K_m value for PAPS was 0.022 mM, and was independent of the concentration of the other substrates. Here again, as in the cases of phenol, estrone, and tyrosine-methyl-ester sulfotransferases reactions, the kinetics appear to follow a rapid equilibrium random bi-bi reaction. The enzyme was inhibited by some analogues of choline such as thiocholine. Choline sulfotransferase has not been so widely found in nature as the other sulfotransferases so far considered. Choline sulfate was reported to occur in higher plants [56]. Later on, Benson and Atkinson [302] showed that of fifteen species of mangrove only those excreting salt could biosynthesize choline sulfate. They believe that in those species choline sulfate may play a role in salt transport.

F. Arylamine Sulfotransferase (EC 2.8.2.3; 3'-Phosphoadenylylsulfate:Arylamine Sulfotransferase)

The discovery of the end product of the reaction catalyzed by this enzyme was the result of work on drug metabolism. Boyland *et al.* [303]

and Parke [304] showed that simple arylamines such as aniline and 2-naphthylamines are excreted to a small extent in rat or rabbit urine as phenyl sulfamate and naphthyl 2-sulfamate, respectively. In this case the acceptor molecule is an aromatic amine, which differs from the other known sulfotransferases. The activity was first found by Roy [305] in extracts of rat and guinea pig liver which catalyzed the formation of naphthyl 2-sulfamate. Equation (9) expresses the general reaction involved.

$$R \cdot NH_2 + PAPS \rightarrow R \cdot NH \cdot SO_3^- + PAP \tag{9}$$

The rat liver enzyme is able to transfer sulfate from PAPS to acceptors such as aniline and 1- and 2-naphthylamines, but not to glucosamine or to benzylamine, where a $-CH_2$ grouping is between the nitrogen and the aromatic ring. The studies of Roy [306–308] have led to the conclusion that, as in the case of several other sulfotransferases, the activity of the enzyme depends on SH groups and is stimulated by Mg^{2+} ions. In addition, he observed some interesting effects of the 17-ketosteroids (3β-methoxyandrost-5-en-17-one was used). These substances acted as partially competitive activators of the rat enzyme and partially competitive inhibitors of the guinea pig enzyme. Further work by Banerjee and Roy [36] showed that arylamine sulfotransferase was not a separate enzyme, but appeared to be, at least in guinea pig liver, an additional transfer activity of estrone and androstenolone sulfotransferase activities. These activities in turn were associated with phenol sulfotransferase. It would be of great importance to clarify this situation, particularly since Adams and Poulos [158] reported that a highly purified estrone sulfotransferase of bovine adrenal glands entirely lacked transfer activity toward 2-naphthylamine.

The natural distribution of arylamine sulfotransferase is not well known. Dodgson and Rose [309] indicated that toad liver is a potent source (as well as of many other sulfotransferases). The opposite was found in the opossum by Roy [310], little or no activity toward 2-naphthylamine, estrone, testosterone, androstenolone, and p-NP being found. The ability to form aryl sulfamates has also been found in some birds [174,175] and arachnids [178,311].

The sulfation of the amino groups of heparin does not seem to be related to arylamine sulfotransferase activity. While the latter activity is found in the soluble fraction of the cell, the enzyme catalyzing the transfer of sulfate to the amino groups of heparin is mainly associated with the endoplasmic reticulum.

G. Transfer in the Formation of Sulfated Glycosaminoglycans

1. INTRODUCTION

An indirect indication of the use of sulfate in the formation of glycosaminoglycans is found in the studies of the fate of $^{35}SO_4^{2-}$ *in vivo* or *in vitro*. We will only deal with the work that definitely established the relation of sulfate incorporation to the transfer of sulfate to glycosaminoglycans. The interested reader will find quite extensive lists of references to this type of work in the review by Boström and Rodén [12].

Layton [312] found the highest uptake of $^{35}SO_4^{2-}$ *in vitro* in cartilage, blood vessels, heart, and other mesenchymal tissues. The work of Dziewiatkowski *et al.* [313] provided support for the idea that inorganic sulfate was incorporated into sulfated glycosaminoglycans by showing that ^{35}S activity was incorporated into the knee joint cartilage of suckling rats. Later, Dziewiatkowski [314] was able to isolate ^{35}S-labeled chondroitin sulfate after addition of nonradioactive carrier chondroitin sulfate from articular cartilage. Additional evidence was supplied by the work of Boström [10,315] with the preparation of radioactive chondroitin sulfate from rib cartilage of rats (without using carrier) after administration of $^{35}SO_4^{2-}$. It was this work that opened the way for the study of the sulfation of glycosaminoglycans beginning with the transfer of sulfate to chondroitin sulfate.

The sulfated polysaccharide was always found combined with a protein forming a macromolecule, generally known as proteoglycan. Protein synthesis appears to be required for the initiation of polysaccharide chain formation. Thus polysaccharide synthesis is inhibited by puromycin in cartilage mince [316,317], in cultured fibroblasts by puromycin or cycloheximide [318], and in other studies by puromycin [319]. The main proteoglycan components of embryonic chick cartilage have been identified as chondroitin 4- and 6-sulfate linked to protein (termed "proteochondroitin sulfates").

The polysaccharide moieties are joined to the protein core by a linkage region consisting of glucuronosylgalactosylgalactosylxylosylserine. Okayama *et al.* [320] have shown that *p*-nitrophenyl β-D-xyloside added to surviving cartilage slices, in appropriate media, acts as initiator of growth of polysaccharide chains, thus taking the place of the protein core. The formation of the polysaccharide chain, producing a mixture of protein-free chondroitin sulfates, took place even in presence of puromycin and both $^{35}SO_4^{2-}$ and [U-^{14}C]glucose were incorporated. We cannot deal here with the structural details of the complex polysaccharide structure of the various glycosaminoglycans (formerly known as mucopolysaccharides). Jeanloz [321] has written an excellent chapter on the subject, and Rodén [322] another on the biosynthesis of acidic glycosaminoglycans.

Considering the complexity of the structure, added to the presence of protein as part of the macromolecule, we realize how intricate the biosynthesis must be. This poses several problems in relation to the incorporation of sulfate into the macromolecule which will be considered below. Some of the difficulties arise from the lack of fine detail on the chemical structures of the polymers themselves, particularly in the cases of heparin and heparan sulfate. The experimental approach is also hampered by the relative metabolic inertness of connective tissues, particularly in the adult animal, which have a slow turnover of their macromolecules. Boström [315] and Boström and Gardell [323] have determined the half-life times of the chondroitin sulfates of costal cartilage and the dermatan sulfate of the skin of adult rats. The estimated values were 16 days and 9–10 days, respectively, using $^{35}SO_4^{2-}$. Essentially the same results were obtained for dermatan sulfate of rabbit skin by Schiller *et al.* [324] by simultaneously labeling with ^{14}C-labeled glucose and acetate and $^{35}SO_4^{2-}$. An increase in the rate of synthesis of glycosaminoglycans (chondroitin 4- and 6-sulfate and dermatan sulfate) has been demonstrated in transformed fibroblast cells using $^{35}SO_4^{2-}$, by addition of dibutyryl cyclic AMP and theophylline [325]. A problem is faced in finding a proper source of the sulfation system, and separating the sparse cells responsible for the biosynthesis from the mass of extracellular material in which they are embedded. In many instances the sulfate is transferred to endogenous acceptors of not too well understood structure, and frequently more than one proteoglycan exists in the same tissue. In addition, the actual amount of sulfate incorporated, corresponding to the measured radioactivity, is so low that it is extremely difficult to isolate and purify the resulting labeled compound. Another aspect of this picture concerns the enzymatic systems involved. Leaving aside for the moment the enzymes and cofactors required for the biosynthesis of the polysaccharide and the attached protein, and assuming the use of PAPS as a sulfate donor, we find that we do not generally have methods available for the preparation of specific sulfotransferases. In spite of all these difficulties, it is amazing to see how much progress, though slow, has been made in this field.

2. Chondroitin Sulfate

The biosynthesis of sulfated glycosaminoglycans in general, and of chondroitin sulfate in particular, raises the problem of the mechanism of sulfation and of the nature of the sulfate acceptors. This process may be visualized as taking place at the level of one of the monosaccharide intermediates prior to polymerization, or alternatively, during or after polymerization. When Strominger [326] reported the isolation of UDP-N-acetylgalactosamine sulfate from hen oviduct it was naturally believed that sulfation took place at the monosaccharide level, and that polymerization of this type of compound formed the polysaccharide. This view was particularly strengthened by the identification of the same compound in embryonic rat epiphyses [327]. Strangely enough, up to the present, all attempts to demonstrate the participation of this compound in glycosaminoglycan synthesis have failed. The compound incubated together with UDP-glucuronic acid in cell-free systems obtained from hen oviduct or cartilage [50, p.257; 328] did not lead to the incorporation of the sulfated nucleotide into a macromolecular product. The function of this compound remains unknown. It has been suggested that it may be a product of degradation of the glycosaminoglycans.

On the other hand, Davidson and Meyer [329] isolated chondroitin sulfate fractions with low sulfate content, and without sulfate (chondroitin) from cornea. On the basis of these findings they suggested that a polysaccharide rather than a monosaccharide was the sulfate acceptor. Perlman *et al.* [330] showed that the 105,000 g particulate enzyme preparation from epiphyseal embryo cartilage catalyzed the formation of polysaccharide in the absence of sulfation, thus apparently reinforcing that opinion. Chondroitin was tried as a sulfate acceptor by several workers with contradictory results.

The work of D'Abramo and Lipmann [49,331] opened the way for a more direct study of the problem of polysaccharide sulfation. These authors showed that inorganic $^{35}SO_4^{2-}$ and $^{35}SO_4^{2-}$ from PAP^{35}S was incorporated into chondroitin sulfate in a cell-free extract of embryonic chick or beef cartilage. It is the result of work along these lines, on the biosynthesis of this, and other, sulfated glycosaminoglycans that has led to the present interpretation of the stage at which sulfation takes place. It is generally held that this process occurs at the polymer stage, or at least concurrent with polymerization, rather than at a monosaccharide containing precursor stage. The problem had been that the crude sulfotransferase preparations contained endogenous acceptors which made difficult a more detailed study of the acceptor specificity. The removal of the endogenous acceptors was achieved by Delbrück and Lipmann [332] by fractionation of the sulfotransferase on DEAE-cellulose. The preparation obtained from chick embryo cartilage catalyzed the transfer of sulfate to added exogenous acceptors such as chondroitin sulfates, heparan sulfate, and their desulfated derivatives, but not to heparin and hyaluronic acid. Adams [333–335] freed the same type of preparation by precipitating the polysaccharide acceptor with protamine, and found that it did not sulfate chondroitin and dermatan sulfate, but it did sulfate chondroitin 4-sulfate and chondroitin 6-sulfate. Suzuki and Strominger [50, pp. 257 and 267; 336] using a soluble extract of oviduct made a thorough study of sulfate acceptors. The transfer of sulfate from PAP[35]S to a number of glycosaminoglycans (including chondroitin 4-sulfate, chondroitin 6-sulfate, chondroitin sulfate from shark cartilage, chondroitin prepared by chemical desulfation of chondroitin 4-sulfate, chondroitin from beef cornea, dermatan sulfate, and a heparan sulfate-like heptasaccharide) was catalyzed at an optimum pH of 6.6 by the preparation. They noted that the naturally occurring chondroitin was a far better acceptor than the chemically desulfated one. Hyaluronic acid, heparin, keratan sulfate, and charonin sulfate were inert. The sulfate was also transferred to numerous model acceptors, such as N-acetylgalactosamine, N-acetylgalactosamine monosulfate, and several nonsulfated and sulfated oligosaccharides obtained from chondroitin and chondroitin sulfate by the action of hyaluronidase and β-glucuronidase. The larger molecules were generally sulfated more rapidly than the smaller ones, with the exception of the nonsulfated trisaccharide and pentasaccharide, which were sulfated at the same rate as chondroitin 4-sulfate. The two monosaccharides yielded derivatives tentatively characterized as N-acetylgalactosamine 4-sulfate and N-acetylgalactosamine 4,6-disulfate, respectively. It was also found that fully sulfated compounds could serve as acceptors. The physiological meaning of the latter observation is not clear, as it is not known whether avian tissues contain oversulfated chondroitin sulfate disaccharide units. Ringertz [337] also found, for heparin fractions, that the sulfation proceeded more readily when the sulfate content was low. The specificity of the sulfotransferases was tested again in a preparation of hen oviduct by Suzuki *et al.* [338]. They found that the apparent Michaelis constant and maximum velocity differed for chondroitin 4- and 6-sulfate and heparan sulfate as substrates. This indicated the presence of different sulfotransferases. Partial separation of the three activities was obtained by chromatography on DEAE-cellulose. Heparan sulfate sulfotransferase (EC 2.8.2.12; 3'-phosphoadenylylsulfate: heparitin N-sulfotransferase), being more stable on storage than the others, was obtained completely free of other activities. These findings seem to indicate the presence in tissues of a number of rather specific sulfotransferases. Hasegawa *et al.* [339] showed that the sulfotransferase of a human chondrosarcoma extract was specific for chondroitin 6-sulfate, but had practically no activity towards chondroitin 4-sulfate or dermatan sulfate. On the contrary, Meezan and Davidson [340] found a relative lack of specificity for the cartilage sulfotransferase. In this case, the enzyme used chondroitin sulfates as acceptors, and in addition dermatan sulfate, keratan sulfate, heparin, and most efficiently, heparan sulfate. Keratan sulfate was present in minute amounts in the preparation used, whereas the other three compounds have never been found in cartilage. In all the work that has been carried out only two substrates have shown consistent results. Hyaluronic acid has never acted as sulfate acceptor, and on the contrary heparan sulfate has always proved to be a substrate for sulfate transfer.

The location of the sulfate groups transferred to exogenous acceptors has presented a difficult problem in the past. Meezan and Davidson [341] distinguished the 4- and 6-sulfate groups from the kinetics of a hydrolysis curve; they differ on their stability. A very useful technique appears to be the degradation of the polysaccharides by the chondroitinases from *Proteus vulgaris* developed by Suzuki *et al.* [50,336,338]. The digest obtained can easily be separated into nonsulfated, 4- and 6-sulfated disaccharide products. Adequate changes of the technique would allow the differentiation of even oversulfated products with extra sulfate groups on either the hexosamine or the uronic acid component.

Chondroitin 4- and 6-sulfate appear to be the most prevalent components, but there is a strong possibility that chondroitin sulfate may occur in mammalian tissues as a hybrid molecule containing both 4- and 6-sulfated hexosamine residues in the same molecule. In the chondroitin sulfate known as E obtained from squid cartilage, most of the hexosamine residues are actually sulfated in both positions [342,343]. This was confirmed by Suzuki *et al.* [344] who isolated the disulfated monosaccharide, 2-acetamido-2-deoxy-4,6-di-*O*-sulfo-D-galactose. In the case of the "oversulfated" chondroitin sulfate D from elasmobranch cartilage, the hexosamine units contain only one sulfate in position 6, and the other is in position 2 or 3 of the uronic acid residue [345–350].

With the exception of the experiments where the sulfate uptake of tissues was investigated, in all the other instances exogenous acceptors have been added to the sulfotransferase preparations. There is no doubt that the physiological acceptor is the endogenous one, and studies concerning its behavior on sulfation will give more information about the course of this process *in vivo*. With the idea of studying the properties of the naturally occurring endogenous sulfate acceptors, Meezan and Davidson [341] studied the compounds present in a boiled supernate from a homogenate of embryonic chick cartilage. The boiled supernate was fractionated on Sephadex G-50 after pronase treatment. The determination of acceptor activity of the fractions showed that the molecular size was of little or no importance, since the retarded fractions had the same activity as the fractions emerging with the void volume. A different picture was found after chromatography on Ecteola. In this case, rather active components were present in the early part of the elution, whereas materials of lower activity appeared in the more retarded fractions. These results were interpreted on the basis of charge density, therefore the early fractions would have material of lower sulfate content. These different fractions could not be considered as representing material containing actual physiological intermediates in the biosynthesis of chondroitin sulfate, but their study has added more information on the relation of the degree of sulfation and the activity of a compound as an acceptor. These authors made the interesting observation that the protein-free chondroitin sulfate was mainly sulfated in position 6 of the hexosamine moiety, whereas when the protein portion was intact, position 4 was preferred. This difference in behavior has been explained by a change in conformation of the protein polysaccharide molecule brought about by the presence of the protein, making the hydroxyl groups in position 4 more accessible to sulfation. Okayama et al. [320] have demonstrated the sulfation in position 4 and 6 of a protein-free polysaccharide by slices of embryonic chick cartilage. They conclude that the attachment of a protein core to the polysaccharide moiety is not a prerequisite for introduction of sulfate residues. The enzyme named chondroitin sulfotransferase (EC 2.8.2.5; 3'-phosphoadenylylsulfate: chondroitin 4-sulfotransferase) is supposed to catalyze the transfer of sulfate from PAPS to chondroitin, but other oligo- and polysaccharides containing 2-acetyl-D-galactosamine can act as acceptor. Momburg et al. [351] have achieved a 4000-fold purification of a PAPS: chondroitin 6-sulfotransferase from a 150,000 g supernatant of mouse liver homogenate. The enzyme still showed a low phenol sulfotransferase activity, and chondroitin and some glycosaminoglycans acted as acceptors. The optimum pH was at 8.2 and the K_m was 0.018 mM for PAPS, 0.06 mM for chondroitin, and 0.018 mM for chemically desulfated chondroitin sulfate.

Kimata et al. [352] have shown that the sulfotransferase responsible for the 4-sulfation in the 105,000 g supernatant of cartilage homogenate is more heat labile than that involved in 6-sulfation and that the latter can be obtained essentially free from it by chromatography on Sephadex G-200 [353].

The nature of the actual physiological sulfate acceptors remains somewhat obscure for the following reasons. The rate of PAPS synthesis, as well as concentrations of sulfotransferase and acceptors, has influenced the observations made in the many studies carried out utilizing $^{35}SO_4^{2-}$. Most of the work was done with homogenates which contain polysaccharides of the extracellular matrix, whereas sulfation must occur intracellularly on protein-bound polysaccharide [317,354,355]. Only partially purified sulfotransferases have been used and the amount of sulfate incorporated is extremely low, if one considers the high specific activity of the $^{35}SO_4^{2-}$ used in the experiments.

From what has been presented and discussed above it is evident that only further research into many aspects of the problem of sulfation will give us a more coherent and harmonious picture of the whole. The suggestion has been made that sulfation *in vivo* occurs concomitantly with the elongation of the polysaccharide chains. Sulfation must lag behind elongation of the chain, because of the observation that the 4-sulfated pentasaccharide does not act as an acceptor for glucuronosyl transfer.

We have seen that the biosynthesis of PAPS is brought about by an enzymatic system present in the cytosol. The sulfotransferases which catalyze the transfer of sulfate in the formation of glycosaminoglycans are particle-bound enzymes and are found in the microsomal fraction. Their distribution in microsomal subfractions, in the case of the biosynthesis of chondroitin sulfate, has been studied by Dorfman [356]. The specific activity of the smooth subfraction is four or five times that of the rough and about twice that of the 20,000 g pellet. The assay, however, is dependent on an endogenous acceptor of unknown nature. The smooth subfraction contains about four times as much uronic acid containing material as does the rough and presumably four times as much endogenous acceptor. They observed that the presence of Mg^{2+} during the preparation as well as during the assay enhanced the sulfotransferase activity.

Dorfman [356] believes, on the basis of the information available, that the biosynthesis of chondromucoprotein follows a path within the cell organelles with the following sequence of events. The starting point is the formation of the core protein at the ribosome. The linkage sugars are added primarily at the rough endoplasmic reticulum and the molecule is elongated and completed as it proceeds through the smooth endoplasmic reticulum to final secretion. The addition of N-acetylgalactosamine and glucuronic acid units occurs in both the rough and smooth fractions. The higher incorporation of sulfate in the smooth endoplasmic reticulum results from the higher concentration of appropriate polysaccharide acceptor, as mentioned above. With the aid of electron microscope autoradiography after injecting ^{35}S-labeled sulfate into rats and mice, Young [357] detected sulfate initially in the smooth membranes and vesicles of the Golgi complex and concluded that the enzymes required for sulfate transfer to a variety of acceptors were probably located in the Golgi complex. The same findings were reported for the Golgi fraction from 12-day-old chick embryo cartilage by Kimata *et al.*

[353]. The recent work of Silbert and co-workers [358–360] with a microsomal preparation of chick embryo epiphyseal cartilage adds valuable information to the timing and degree of polymerization and sulfation. At pH 6.5 the incorporation of SO_4^{2-} from PAPS was 60–70% into chondroitin 6-sulfate and 30–40% into chondroitin 4-sulfate, while at pH 7.8 the incorporation was exclusively into chondroitin 6-sulfate. All acceptor molecules were linked to protein in both cases. Incorporation took place into occasional nonsulfated galactosamine units in predominantly sulfated molecules with no oversulfation to form 4,6-disulfated molecules. The final product of polymerization on primer (endogenous acceptor) consisted of 2 distinct types of glycosaminoglycans: 25–50% newly formed nonsulfated chondroitin chains (MW ~ 35,000) on small primer (MW ~ 8000), and the remaining 50–75% chondroitin sulfate primer (MW ~ 25,000) plus addition of newly formed chondroitin (MW ~ 5000–15,000). There were no primers of intermediate size, no molecules of low sulfate content, and no endogenous nonsulfated chondroitin. Sulfation during chondroitin polymerization occurred essentially in an "all or nothing" rather than a random fashion, so that maximally 75% of the final chondroitin chains was 96% sulfated, while only approximately 5% were free of sulfate. The data suggested that the sulfating enzymes were located together with polymerizing enzymes in an enzyme complex so that sulfation of the heteropolysaccharide chain proceeded during or immediately following polymerization. The data obtained by these authors are inconsistent with the suggestion [361] that in vivo sulfation is a prerequisite for further polysaccharide formation since it is clearly shown that in vitro extensive polymerization can precede extensive sulfation. Furthermore it does not seem likely that sulfation occurs as the growing polysaccharide moves along the reticuloendothelial system. In this case sulfation would occur after the completed polysaccharide has moved to a separate site as proposed by Horwitz and Dorfman [362].

Mohanram and Reddy [363] have found decreased incorporation of $^{35}SO_4^{2-}$ into acid glycosaminoglycans in the colon of vitamin A deficient children and the level was restored to normal after treatment with vitamin A.

3. DERMATAN SULFATE

Davidson and Riley [364] partially purified a sulfotransferase from extracts of rabbit skin. This enzyme exhibited an acceptor specifity for dermatan (desulfated dermatan sulfate) with a considerable specific activity, resulting in one sulfate group being incorporated per five to six disaccharide units. Heparan sulfate also acted as an acceptor, whereas chondroitin, chondroitin 4-sulfate, chondroitin 6-sulfate, dermatan sulfate, hyaluronic acid, and keratan sulfate were inactive. This indicated that different sulfotransferases were responsible for the sulfation of the galactosamine residue in dermatan sulfate and chondroitin 4-sulfate, in spite of the fact that in both instances the sulfate group is located at the C-4 position. Sulfation of dermatan by this system was strongly stimulated by UTP. This finding is not yet understood, but is assumed to be an indication of the close relationship between sulfation and the polymerization process.

A sulfotransferase specific for dermatan sulfate has been reported for a leiomyosarcoma by Adams and Meaney [365]. The problem of incorporation of sulfate in the case of dermatan sulfate appears to be more complex, because the sulfate groups are located not only at C-4 of the galactosamine, but also to a certain extent on uronic acid residues. It is not yet clear whether both glucuronic and iduronic acid are involved.

4. HEPARIN AND HEPARAN SULFATE

Heparin and heparan sulfate differ in several characteristics from the other glycosaminoglycans. They have α-glycosidic linkages rather than the β linkages that predominate in other polysaccharides of this group. They are found intracellularly, and the process of their sulfation involves not only the hydroxyl groups of the hexosamines, but also their amino groups and the hydroxyl groups in their uronic acids components.

Heparin is found in mast cells, which has been the experimental material used for the study of sulfate incorporation. $^{35}SO_4^{2-}$ was shown to be introduced into the mast cells of normal tissues [366], mastocytomas in dogs and mice [367–369], dog liver heparin *in vivo* [370], and heparin from slices and homogenates of rat liver [371]. The introduction of two transplantable mouse mast cell tumors, the solid type of Furth *et al.* [372] and the ascites type of Dunn and Potter [373], contributed a suitable source of heparin synthesizing tissue, and opened a new era for the study of its biosynthesis.

The incorporation of $^{35}SO_4^{2-}$ into heparin in slices of mast cell tumors of the Dunn–Potter type was reported by Day and Green [374] and Korn [375]. The sulfation of heparin in tumor homogenates from inorganic $^{35}SO_4^{2-}$ or PAP^{35}S was first found by Spolter and Marx [376,377]. This observation was confirmed and extended in more detailed studies on particle-free preparations of mast cell solid type tumors by Pasternak [378], Ringertz [379], and Korn [380], and of ascitic tumors by Pasternak [378]. That two separate sulfotransferases are involved in the formation of O-sulfate esters and N-sulfate groups was shown by Korn [375,380]. He found that mast cell tumor slices incorporated $^{35}SO_4{}^{2-}$ almost equally into the O-sulfate and the N-sulfate groups of heparin, while extracts of the tumor formed mainly O-sulfate groups. These preparations were shown to form PAPS [378–380], as well as APS [379]. With the particle-free mast cell tumor extract the heparin or heparinlike fractions with lower sulfate content behaved as better acceptors of sulfate [379]. Most of the activity of the homogenate was found in the particles sedimenting at 11,000 g, and only a small fraction of the total activity was present in the particle-free supernate [381].

The biosynthesis of the carbohydrate skeleton of heparin has been studied in considerable detail by Silbert [382–385] with a microsomal preparation of the Dunn–Potter [373] type of mast cell tumor. This preparation catalyzed the transfer of N-acetylglucosamine and glucuronic acid from their UDP derivatives to an endogenous acceptor. The compound formed in the absence of PAPS was a low sulfated polymer which behaved similar to hyaluronic acid on ion-exchange chromatography. The product was not changed by treatment with testicular hyaluronidase, but was degraded by bacterial heparinase to low molecular weight fragments retarded on gel filtration. These results indicated that it is a heparin like substance. When PAPS was present during incubation, the sulfated polysaccharide formed amounted to approximately half of the total polysaccharide synthesized. The relative proportions of sulfate and glucuronic acid incorporated varied between 0.5 and 1.3 sulfate groups per disaccharide unit. N-Sulfate represented 40–50% of the total radioactive sulfate incorporated.

As indicated above, the sulfation of amino groups is one of the problems in the formation of heparin and heparan sulfate. The process may occur after introduction of glucosamine into the molecule (from UDP-glucosamine) followed by sulfation of the free amino group. Another possibility is the formation of the sulfated precursor UDP-glucosamine N-sulfate. But Silbert and Brown [386] have shown that UDP-glucosamine does not act as precursor in the polymerizing system, although it is produced by the mast cell tumor preparations from glucosamine 1-phosphate and UTP, and there is no evidence for the formation of UDP-glucosamine N-sulfate. The mechanism proposed by Silbert [385] is the only one supported by experimental findings. He incubated the polymerase preparation from mast cell tumor with UDP-N-acetylglucosamine, labeled with tritium in the acetyl group, and UDP-glucuronic acid. After $1\frac{1}{2}$ hours, PAPS was added, which partially removed the labeled acetyl groups. Evidently the sulfate groups replaced the acetyl residues by an exchange reaction. The mechanism of this reaction and the fate of the liberated acetyl groups are not yet known. This mechanism would explain the occurrence of a variety of heparan sulfate fractions which differ in their relative proportions of N-sulfate and N-acetyl groups. The difference could be attributed to a different degree of replacement of acetyl groups by sulfate during biosynthesis. The formation of the N-sulfate group was studied in some detail by Eisenman et al. [387] and Rice et al. [388] with a postmicrosomal particulate fraction from a mouse mast cell tumor. The reaction run with N-desulfoheparin as acceptor and PAPS as donor had a pH optimum between 6.7 and 7.2, needed added Mg^{2+} ions, and was not saturated at 0.04 mM PAPS. Heparin was not an acceptor, but other polysaccharides such as heparan sulfate, chondroitin 4-sulfate, and dermatan sulfate were. The enzymatic activity is linked to the presence of (EC 2.8.2.8.) 3'-phosphoadenylylsulfate:N-desulfoheparin N-sulfotransferase.

Undoubtedly there are other unknown factors that determine the degree of sulfation, that is whether the molecule remains incompletely sulfated or is completely sulfated and thus becomes heparin. Heparin appears to be the prevailing polysaccharide, and heparan sulfate, if present at all, is in very low concentration in mast cells from peritoneal washings [389–392]. On the contrary, mast cell tumors produce appreciable quantities of heparan sulfate and other polysaccharides [337, 381,393–395]. Heparan sulfate sulfotransferase activity, sulfating predominantly free NH_2, has been partially purified from beef lung [396]. Desulfated heparin and heparan sulfate were better acceptors, and the authors believe that N-deacetylation and N-sulfation are not obligatorily linked as Silbert has assumed. Further research is needed to clarify the origin of heparan sulfate, as this compound does not seem to be present in mast cells from peritoneal washings, while it is produced with heparin by neoplastic mast cells. From what has been presented above, it appears that the initial stages of heparin and heparan sulfate biosynthesis are identical. Cifonelli [397] presented chemical data, based on the occurrence of heparan sulfate fractions having variable contents of N-acetylated and N-sulfated hexosamines, to support the suggestion of an interrelationship between heparin and heparan sulfate biosynthesis.

The biosynthesis of heparin presents other complications in view of recent findings. According to Lindahl [398] and Lindahl and Axelsson [399], the uronic acid-bound sulfate is located mainly on iduronic acid residues. With heparin having sulfate groups located in carbons 3 and 6 of the glucosamine, the amino group, and the uronic acid, we may assume that four different sulfotransferases are required for its biosynthesis.

The isolation of a sulfotransferase preparation from mast cell tumor [400] that preferentially sulfated amino groups has opened the way for the study of the problem with soluble purified enzymes. The enzyme was found in a postmicrosomal particulate fraction and was solubilized by treatment with snake venom (*Ancistrodon p. piscivorus*) and fractionated on DEAE-cellulose. N-Desulfated heparan sulfate was the best acceptor, followed by heparan sulfate and N-desulfated heparin. Heparin and chondroitin 4-sulfate were nearly completely inactive and dermatan sulfate had some acceptor activity. The apparent K_m value for PAPS was 0.022 mM with either N-desulfoheparin or heparan sulfate as sulfate acceptor. The products of incubation with N-desulfoheparin and heparan sulfate released 75–79% of the incorporated sulfate on mild acid hydrolysis, an indication of N-sulfate groups. Dermatan sulfate released only 30% of sulfate under similar treatment. Acetylation of N-desulfoheparin decreased sulfate incorporation to 20% of the initial level. This observation appears to contradict the mechanism proposed by Silbert of exchange of sulfate for acetyl groups. From this work it can be assumed that the enzyme preparation has at least two enzymes catalyzing sulfate transfer to amino groups and to hydroxyl groups.

A sulfotransferase catalyzing the sulfation of heparan sulfate was separated from other sulfotransferases present in an extract of hen oviduct by Suzuki *et al.* [338]. The location of the sulfate groups introduced was not determined, and the relation of this sulfotransferase to the similar activities demonstrated in mast cell tumor has not been investigated.

Johnson and Baker [401] reported enzymatic sulfation of N-desulfated heparin, dermatan sulfate, heparan sulfate, and its N-desulfated derivative by hen uterus using PAPS as a SO_4^{2-} donor. Heparin, chondroitin sulfate, hyaluronic acid, glucosamine, and its N-acetyl derivative were not sulfated.

5. KERATAN SULFATE

The structure of keratan sulfate is poorly understood. This stems from the difficulties encountered in the isolation of the compound. The lack of proper techniques for the identification of metabolic intermediates in its biosynthesis have greatly delayed progress in the study of this process.

Rodén [402] isolated labeled keratan sulfate from nucleus pulposus after incubation with $^{35}SO_4^{2-}$ or [^{14}C]glucose. The incubation of extracts of beef corneal epithelium with $^{35}SO_4^{2-}$ in the presence of ATP yielded a radioactive compound having an electrophoretic behavior similar to chondroitin sulfate. The sulfotransferase catalyzing this sulfate transfer appears to be able to work with the phenol sulfotransferase also present in the extracts to catalyze the transfer of sulfate from p-nitrophenyl sulfate to polysaccharides with PAP as a cofactor. The phenol sulfotransferase of liver has not shown this ability. Wortman [403,404] also showed that sulfate transfer in this system could take place with keratan sulfate, when it was added as an acceptor.

6. OTHER SULFATED GLYCOSAMINOGLYCANS AND SULFATED CARBOHYDRATES

Unfortunately there is very little we can say about the sulfotransferases and their specificity in the process of formation of the compounds described below.

Sulfated polysaccharides are found in high concentrations in seaweeds and are present in most, if not all, animal organisms. In primitive animal organisms they are frequently found in mucous secretions, usually as polyhexose polysulfates. In *Buccinum undatum L.* polysaccharide sulfate [405] and charonin sulfate [406], the structure is mainly that of a $\beta(1 \rightarrow 4)$ and $\alpha(1 \rightarrow 4)$ linked polysulfated glucose. In the chondroid odontophore of the marine snail *Busycon caniculatum* a polyglucose sulfate was identified by Lash and Whitehouse [407]. In the jelly coats of sea urchins sulfated polygalactose or polyfucose, assumed to be important factors in the process of fertilization, have been found [408,409]. The horatin sulfates of *C. lampas* are more complex, since they contain several sugars as well as sialic acid [410]. The mactins from mollusks are anticlotting agents like heparin and of complex nature [411]. The participation of PAPS in the sulfation of mactin has been demonstrated by De Meio *et al.* [55]. Other recently found polysaccharide sulfates are the "lorenzan sulfates" of elasmobranch fish [412] and "spisulan" from clam tissues [413]. A polysaccharide sulfate of unknown structure occurs in the limpet *Patella vulgata* [414]. Katzman and Jeanloz [415] have called a polysaccharide isolated from the dermis of the sea cucumber *Thyone briareus* thyonatan 4-sulfate. On hydrolysis it gives mainly L-fucose 4-sulfate and contains about 3% galactose and 6.5% firmly bound peptide.

Simple carbohydrate sulfate esters have been detected in tissues of some animals. UDP-N-acetyl-D-galactosamine 4-sulfate and other sulfated sugar nucleotides have been isolated from hen oviduct by Strominger [326], Suzuki and Strominger [50, p. 257] and Nakanishi *et al.* [416]. UDP-N-acetyl-D-galactosamine 4-sulfate was also found in ossifiable cartilage and aorta of rats following injection of $^{35}SO_4^{2-}$ [417]. This compound can be formed by sulfation of UDP-N-acetyl-D-galactosamine by hen oviduct preparations [50, p. 267; 418]. A different sulfotransferase from the same source appears to be responsible for the sulfation of UDP-N-acetyl-D-galactosamine 4-sulfate in position 6. This enzyme had no activity toward the desulfated UDP derivative [419]. A low optimum pH at 4.8 is reported for this sulfotransferase (EC 2.8.2.7; 3'-phosphoadenylylsulfate: UDP-2-acetamido-2-deoxy-D-galactose-4-sulfate 6-sulfotransferase) and the K_m values found by the authors are presented in Table V.

The last acceptor mentioned in Table V showed a V_{max} that was about 5% of the first; the other two showed intermediate values. The same type of activity was found for a mucopolysaccharide sulfotransferase from squid cartilage by Habuchi et al. [420]. The enzyme was purified about ninefold with DEAE Sephadex A-50, and was only active when exogenous acceptors were added. Using PAPS as sulfate donor, the enzyme sulfated at C-6 (besides acetylgalactosamine 4-sulfate) various mono-, di-, and polysaccharides provided the acetylgalactosamine residue was already sulfated at C-4. The role, if any, that these two sulfated UDP derivatives play is not yet known. Other interesting carbohydrate esters, but also of unknown function, are neuramin lactose 6-sulfate and lactose 6-sulfate found in rat mammary glands by Barra and Caputto [421]. They showed that extracts of rat mammary gland produce both compounds, using $PAP^{35}S$ or $Na_2{}^{35}SO_4$ as sulfate donors. Neuramin lactose 6-sulfate has also been isolated from rat mammary gland by Carubelli et al. [422] and Ryan et al. [423]. The presence of two different sulfotransferases in rat brain showing varying degrees of activity toward two distinct types of acceptor was reported by Cumar et al. [424]. One of them catalyzed the transfer of sulfate to galactose, lactose, neuramin lactose, and p-nitrophenyl β-D-galactoside. The other formed sulfatides from galactosylceramide, galactosyl sphingosine, and lactosyl ceramide. The two enzyme activities were not separated, but were clearly differentiated by kinetic, inhibition, and cell distribution studies. Another sulfotransferase is likely to be involved in the biosynthesis of holothurin A, present in the tropical sea cucumber Actinopyga agassizi Selenka [425]. A closely related compound, holothurin B, has been recognized in two other sea cucumber species

TABLE V

SPECIFICITY OF A SULFOTRANSFERASE FROM HEN OVIDUCT USING SIMPLE SUGAR DERIVATIVES AS SULFATE ACCEPTORS

Acceptor	K_m (mM)
UDP-N-acetylgalactosamine 4-sulfate	0.05
N-Acetylgalactosamine 4-sulfate	1.4
N-Acetylgalactosamine 1-phosphate 4-sulfate	0.13
$\Delta^{4,5}$-Glucuronido-N-acetylgalactosamine 4-sulfate	2.0

[426]. These two compounds are steroidal glycosides in which a sulfate group is present on a sugar residue. The holothurins are potent neurotoxins. Their toxicity, like that of phyllokinin, decreases considerably on desulfation [427].

H. Formation of Sulfolipids or Cerebroside Sulfates

The glycolipid sulfates are generally found in mammalian tissues. The only nonmammalian source seems to be the halophilic bacterium, *Halobacterium cutirubrum*, which synthesizes a complex glycolipid containing galactose 3-sulfate residues [428]. A phytanyl derivative of a phosphatidyl glycerosulfate has been obtained from the same bacterium [429]. This glycolipid sulfate was identified also as a phytanyl derivative of glycerol, containing in addition galactose 3-sulfate, mannose, and glucose in equimolar proportions [430]. In mammals they seem to be of particular importance in nervous tissue, where they are abundant in both white and gray matter of brain and in the myelin sheaths of nerves. As part of the cell membrane they occur in other tissues.

The transfer of sulfate from PAPS to lipid material was reported for the first time by Goldberg and Delbrück [52] and Goldberg [431]. The enzyme preparation from rat liver or rat brain contained the sulfotransferase in a particulate fraction, as in the case of the sulfotransferases participating in the sulfation of glycosaminoglycans. The resulting sulfated product was lipid of undetermined nature, probably N-acetylsphingosine sulfate, and it was rather unlikely to be cerebroside sulfate.

The formation of cerebroside sulfate appears to take place as follows:

$$\text{PAPS} + \text{ceramide galactoside} \rightarrow \text{ceramide galactoside 3-sulfate} + \text{PAP} \quad (10)$$

This reaction was described as occurring in sonicated extracts of rat brain microsomes [432]. Naturally occurring ceramide galactoside, as well as synthetic N-palmitoyl sphingosyl galactoside, but not ceramide glucoside, stimulated incorporation of $^{35}SO_4^{2-}$ into a lipid which had the properties of sulfatide. This incorporation was dependent upon the presence of ATP and so presumably upon the formation of PAPS. Galactocerebroside sulfotransferase (EC 2.8.2.11; 3'-phosphoadenylylsulfate:galactocerebroside 3'-sulfotransferase) was characterized in rat brain by Farrell and McKhann [433]. This enzyme catalyzed the reaction represented by Eq. (10) and is primarily located in microsomes but may also be present in myelin of 17-day-old rat brain. Virtually all the activity is extracted with Triton X-100 and has a K_m of 0.08 mM for galactosylceramide as exogenous substrate. Other sulfate acceptors are α-hydroxy and nonhydroxy fatty acid galactosyl sphingosine, lactosyl ceramide, and galactosyl sphingosine. Glucosyl ceramide and other lipids examined were not acceptors. The activity of the cerebroside sulfotransferase of rat kidney homogenate Golgi fraction was found to be about 80-fold greater than that of the homogenate, while other cell fractions showed very low activity [434]. Balasubramanian and Bachhawat [435] used an ammonium sulfate fraction obtained from sheep brain white matter that is activated by EDTA and cysteine for the study of sulfatide biosynthesis. The preparation could utilize only endogenous galactocerebrosides as acceptors; added ceramide galactoside was inert. It was concluded that the endogenous sulfate acceptor was tightly bound to protein and by indirect evidence was considered to be a protein-bound galactosyl ceramide. Cumar *et al.* [424] found two different sulfotransferases in rat brain, one forming sulfatides from galactosyl ceramide, galactosyl sphingosine, and lactosyl ceramide and the other sulfating simple sugars (see Section III,G,6). The activity of galactocerebroside sulfotransferase in developing rat brain was found to be maximal at the peak of the myelination period (twentieth day) by Saxena *et al.* [436]. An enzymatic activity catalyzing the transfer of sulfate from PAPS to galactosylsphingosine with a pH optimum of 6.8 was demonstrated in kidney, brain, and liver of young mice [437]. In a study of the sulfatides biosynthesized *in vitro,* Stoffyn *et al.* [438] came to the conclusion that the sulfate is introduced at position C-3 of the galactose. This corresponded to the position of the sulfate found in sulfatides isolated from tissues, and was an indication of the specificity of the sulfotransferase involved in the *in vivo* synthesis. They showed that exogenous cerebrosides can act as acceptors by using galactose ^{14}C-labeled phrenosine and PAPS as sulfate donor.

A novel sulfoglycolipid, named "seminolipid," obtained from boar testis and spermatozoa, was shown to contain equimolar amounts of fatty acid (mostly palmitic) and galactose 3-sulfate, bound to an alkyl (mainly chimyl) ether of glycerol [439]. Goren [440] reported the abundant production of a glycolipid sulfate by surface grown cultures of *Mycobacterium tuberculosis.* This was characterized as complex 2,3,6,6'-tetraacyl-α,α'trehalose 2'-sulfate; the purified ammonium salt undergoes rapid quantitative desulfation on mere dissolution in reagent grade anhydrous ether (a mechanism to explain this behavior is proposed by the author). These sulfolipids are found in virulent strains, while the avirulent and attenuated strains of *M. tuberculosis* var. *hominis* produce little or none. The phytoflagellate *Ochromonas danica* produces large amounts of chlorosulfolipids [441]. These compounds are derivatives of docosane 1,14-disulfate and tetracosane 1,15-disulfate. The chlorination apparently occurs after sulfation and six chlorine atoms arc substituted for H.

I. Other Sulfations

Alkyl sulfates have not generally been found in nature, with one exception. Isopropyl sulfate occurs in appreciable quantities with other simple alkyl sulfates in the allantois of the hen egg [442]. Extracts of embryonic chick liver were shown by Spencer and Raftery [186] to synthesize alkyl sulfates. Sulfotransferases are certainly responsible for the formation of all of these alkyl sulfates and for the *in vitro* sulfation of methanol, ethanol, propanol, and butanol by preparations from rat liver [59,61,124,443]. Scully *et al.* [283] found an optimum pH of 8 for sulfate activation by toad liver. Sulfate from PAPS was transferred to phenols, primary and secondary alcohols, steroids, and aromatic amines. In rat and chick liver only simple aliphatic alcohols were sulfated. In toad liver they separated (by chromatography on DEAE-cellulose) the activity toward simple aliphatic alcohols from that toward bile alcohols (scymnol and β-cyprinol). It is of interest that the major metabolite of Dimetridazole in turkeys is the alkyl sulfate derived from it [444]. The *in vitro* formation of ester sulfates of aliphatic polyols has been reported by Vestermark and Boström [445]. Bilirubin disulfate occurs in bile, but the location of the sulfate groups has not been clarified [446]. The sulfation of bilirubin was reported earlier by Isselbacher and McCarthy [447].

Cysteine and glutathione form S-sulfoconjugates when incubated with scrapings of rat intestinal mucosa in the presence of inorganic sulfate [448,449]. This reaction requires PAPS, but it is not known if the transfer takes place directly from PAPS or whether PAPS is reduced to inorganic sulfite as a preliminary step to the formation of the S-sulfoconjugate. The biological role of these compounds is obscure; one of them, *S*-sulfoglutathione, occurs in the lens of the eye [450].

Cormier *et al.* [451] obtained an enzyme from the sea pansy (*Renilla reniformis*) which sulfates luciferin. This enzyme (EC 2.8.2.10; 3'-phosphoadenylylsulfate:luciferin sulfotransferase) does not exhibit any phenol sulfotransferase activity, appears to be specific, and requires the presence of Ca^{2+}. The biological role of the enzyme may be to regulate the level of luciferin available for bioluminescence.

Ascorbic acid 2-sulfate has been found in the eggs of the brine shrimp *Artemia salina* [452], rat liver, and spleen after administration of $^{35}SO_4^{2-}$ and [1-^{14}C]ascorbic acid. It appears also to be present in rat urine and adrenals [453] and is a significant metabolite of ascorbic acid excreted in human urine [454]. Evidently ascorbic acid is probably sulfated by the same general mechanism. In any event, ascorbic acid sulfate may be a sulfate reservoir and act as a sulfating agent (see Section V).

King and Phillips [455] showed that the soluble fraction of rat liver synthesizes acetylaminofluorene N-sulfate from N-hydroxy-2-acetyl-aminofluorene and PAPS. The enzyme responsible for this activity has been named N-hydroxy-2-acetylaminofluorene sulfotransferase and its substrate becomes carcinogenic on sulfation [456]. This enzyme has not been purified and requires further work to determine its specificity.

Sulfated glycopeptides have been isolated and characterized from eggshell membranes and hen oviduct [457].

IV. ROLE OF SULFATE ACTIVATION IN TAURINE BIOSYNTHESIS

Taurine is known to be formed from cysteic acid [458] and cysteine sulfinic acid [459], as well as from the catabolism of the sulfur amino acids [460] (see Chapters 13 and 14). The incorporation of sulfate sulfur into taurine was observed by several workers. Boström and Åqvist [461] reported it in rat liver following $^{35}SO_4^{2-}$ injection with no appreciable radioactivity in cystine or methionine. A clear demonstration of its direct incorporation in chicken embryo was given by Machlin et al. [462]. The injection of $^{35}SO_4^{2-}$ into the egg led to the incorporation of 65% of the radioactivity into taurine. No trace of radioactivity was found in cystine or methionine. The same results were obtained by Lowe and Roberts [463]. These observations eliminated the possibility of an indirect utilization of sulfate for the biosynthesis of taurine through the formation of sulfur amino acids.

The studies of Martin and co-workers have clarified the mechanism of the incorporation of sulfate into taurine in chick and rat liver preparations. From their work with the chick liver Sass and Martin [464] proposed the following sequence of events. PAPS formed in this tissue interacts with α-aminoacrylic acid. This reaction, catalyzed by a sulfotransferase and requiring pyridoxal phosphate, yields enzyme bound cysteic acid (as no free [^{35}S]cysteic acid was observed) and PAP. The enzyme bound cysteic acid is decarboxylated to taurine by cysteic acid decarboxylase, requiring pyridoxal phosphate as a coenzyme. The α-aminoacrylic acid is produced from the dehydration of L-serine through the activity of L-serine dehydratase (EC 4.2.1.13) which requires pyridoxal phosphate. A similar pathway appears to lead to the formation of taurine in rat liver [465]. The specific activity of the sulfotransferase system was 19.8 cpm [^{35}S]taurine formed/μg protein/5 minutes for chick liver and 21 cpm for rat liver. According to some observations of these authors, the enzymes responsible for taurine biosynthesis from PAPS may be present in a sulfur metabolizing complex capable of dehydrating and sulfonating serine and decarboxylating cysteic acid. The enzyme complex appeared to require free SH groups for activity and the pH optimum was at 8.5. Cysteine was found to inhibit the formation of taurine. This was interpreted as resulting from the repression of ATP-sulfurylase [466]. In the presence of optimal dietary cysteine, the formation of taurine from PAPS and serine is inhibited and taurine is formed by the degradation of the sulfur amino acids. DL-Methionine enhances the *in vivo* and *in vitro* biosynthesis of taurine from sulfate by the chick liver enzymes [467]. Addition of DL-methionine to the diet of chicks did not enhance the enzymatic activation of sulfate [468]. As a consequence, it appeared to bring about the increase in taurine formation by increasing the availability of the sulfate acceptor, dehydrated serine. According to Martin [469], in chick and rat, the heart had higher activity than the liver and several other tissues in the formation of taurine by this pathway. The activity was definitely present in heart and liver of dog, cat, rat, mouse, sheep, guinea pig, hamster, and chicken.

V.   OTHER MECHANISMS OF SULFATION

The transfer of sulfate to different acceptors takes place mainly in reactions catalyzed by sulfotransferases in which the sulfate donor is PAPS. A closely related route, but of unknown physiological significance, is sulfate transfer from an aryl sulfate of high sulfate potential to a phenol or other acceptors, by a reaction requiring PAP as a cofactor, resulting in the formation of PAPS as an intermediate. This possible path for sulfation is based on the study of the transfer of sulfate among phenolic compounds with PAP as coenzyme [165]. Equation (7) shows this type of reaction in the case of *p*-NPS as a sulfate donor and phenol as acceptor. This reaction, besides indicating another path for the formation of sulfated phenols, represents a sensitive method of assay for PAP and phenol sulfotransferase.

It has been shown for human arylsulfatases that the cleavage takes place at the S–O bond. These enzymes can therefore be classified as sulfate transferring enzymes with a high specificity for water as acceptor. The possibility of sulfotransferase activity for the glucosulfatase and the associated arylsulfatase of the mollusk *C. lampas* has been suggested by Suzuki *et al.* [470]. Acetone dried preparations of the mucus gland of this mollusk could not incorporate $^{35}SO_4^{2-}$ into charonin sulfate (a sulfated polyglucose found in the mucus gland of *C. lampas*), but could incorporate the sulfate from *p*-NP^{35}S. Because the purified preparations of sulfatases of the mucus gland did not catalyze the transfer of sulfate [471], while the partly purified preparation was effective, the authors concluded that a dialyzable cofactor was needed. The p-NPS acting as a sulfate donor in the presence of the arylsulfatase, transferred sulfate to charonin. The latter reaction was catalyzed by glucosulfatase and formed a sulfated cofactor intermediate. The process was inhibited by phosphate and fluoride and suggested the participation of sulfatases in the sulfate transfer. No further work has appeared, and the fact that neither a natural aryl sulfate donor of high sulfate group potential, nor a cofactor have been identified make this mechanism questionable. On the other hand, Spencer [472] found no sulfation of glucose, galactose, galactosamine, choline, or simple alcohols with p-NPS and nitrocatechol sulfate as donors and human arylsulfatases A and B as enzymes.

Ford and Ruoff [230] have proposed that ascorbic acid may be involved in sulfation. Synthetic ascorbic acid 3-sulfate was shown to be capable of nonenzymically sulfating certain steroids [473] and alcohols [474] in the presence of adequate oxidizing agents. Verlangieri and Mumma [475] found that the amount of cholesterol [^{35}S]sulfate, excreted primarily with the feces, was increased 50-fold when L-[^{35}S]ascorbic acid 2-sulfate and 2-fold when L-ascorbic acid and ^{35}SO$_4{}^{2-}$ were injected into rats, as compared with the single injection of ^{35}SO$_4{}^{2-}$. These observations acquire greater importance now that we know that sulfation of ascorbic acid occurs *in vivo* [452–454]. A similar process of oxidative transfer of sulfate has been suggested by Butenandt *et al.* [194] in the case of ommatin D. This reaction is identical to the transfer of sulfate from hydroquinone monosulfate to methanol [229].

Another nonenzymatic process was reported by Adams [476] in the sulfation of hexosamines and N-acetylhexosamines by PAP^{35}S (of high specific activity) in the presence of charcoal. Mayers and Kaiser [??8] have shown that sulfur analogues of 1-phosphoimidazole can participate in sulfate transfer reactions. These type of reactions probably do not have any biological significance.

It appears to be definitely established that sulfate esters *in vivo*, as well as *in vitro*, are formed mainly through the process that has been discussed, i.e., activation of inorganic sulfate to PAPS followed by transfer of the sulfate group to the acceptor. Wellers [477] has put forward another possible pathway involving the utilization of cysteine sulfur without its prior conversion to SO$_4{}^{2-}$ in the sulfation of indoxyl and p-nitrophenol. His views are based on detailed and careful sulfur balance studies in rats maintained on various artificial diets with and without the addition of indole as sulfate acceptor [478]. Their results clearly show that with regard to the formation of sulfate esters *in vivo*, a more efficient utilization of sulfur from dietary cysteine and methionine compared to the sulfur from dietary sulfate occurs. In feeding experiments it is important to consider gastrointestinal absorption which has been definitely demonstrated to be relatively inefficient for sulfate ions. In addition, the work of Bray *et al.* [198] has shown that in the rabbit, the endogenous level of SO_4^{2-} is normally rate limiting in the formation of aryl sulfates *in vivo*. On these grounds it appears difficult to accept Wellers' conclusion that direct utilization of cysteine sulfur for the biosynthesis of sulfate esters occurs.

VI. SOMATOMEDIN (SULFATION FACTOR)

Salmon and Daughaday [479] reported a stimulation of sulfate uptake by rat cartilage produced by a factor present in normal rat serum, which they called sulfation factor. Hypophysectomy of the rats halved the sulfate incorporation by the cartilage, and injection of growth hormone restored the sulfate uptake to normal. Addition of growth hormone to the incubation medium, in the presence or absence of hypophysectomized rat plasma, did not increase the $^{35}SO_4^{2-}$ uptake by hypophysectomized rat cartilage. Incubation of hypophysectomized rat cartilage with normal rat serum, however, more than doubled the sulfate uptake as compared to cartilage incubated in hypophysectomized rat serum.

In their studies of sulfate incorporation by embryonic chick or embryonic rat cartilage, Herington *et al.* [480] found that extracellular glucose stimulated sulfate uptake, while stimulation of amino acid transport and incorporation (apparently due to the same factor) was independent of extracellular glucose. Stimulation of protein synthesis in the absence of glucose resulted in a subsequent stimulation of sulfate incorporation provided the tissue was transferred to a second medium containing glucose but not the sulfation factor. These results are interpreted as suggesting that in embryonic cartilage the limiting step in sulfated proteoglycan synthesis is protein synthesis in the presence of extracellular glucose and polysaccharide synthesis in the absence of glucose. Evidently these results seem to indicate that the stimulation of sulfation depends on the stimulation of the biosynthesis of the sulfate acceptor molecule.

Daughaday *et al.* [481] proposed renaming the sulfation factor to somatomedin. This proposed change stems from the fact that the factor stimulates other processes besides sulfation. Among the other activities, we may mention an induced widening of the tibial epiphyseal cartilage, insulinlike actions on isolated rat diaphragm, stimulation of conversion of [^{14}C]glucose to $^{14}CO_2$ by rat adipose tissue, and stimulation of thymidine incorporation into DNA in rat costal cartilage, of proline into collagen, and of uridine into RNA. The factor has been purified approximately 25,000-fold by gel filtration, ion exchange chromatography, and electrophoresis. The biological activity could be attributed to a neutral peptide with a molecular weight of about 8000. It appears that at least some of the activities may be related to different factors. Uthne [482] achieved a higher level of purification and on the basis of the behavior of his preparations he suggests calling somatomedin A a fraction stimulating sulfation and somatomedin B the one that stimulates DNA synthesis in human glialike cells.

REFERENCES

1. W. Henry, "The Elements of Experimental Chemistry," 7th ed., Vol. II, p. 352. Baldwin, Cradock, & Joy, London 1815.
2. E. Baumann, *Arch. Gesamte Physiol. Menschen Thiere* **12**, 69 (1876); *Ber. Deut. Chem. Ges.* **9**, 54 (1876).
3. H. Borsook, G. Keighley, D. M. Yost, and E. McMillan, *Science* **86**, 525 (1937).
4. D. D. Dziewiatkowski, *J. Biol. Chem.* **178**, 197 (1949).
5. J. C. Laidlaw and L. Young, *Biochem. J.* **42**, Proc. 1 (1948); **54**, 142 (1953).
6. D. D. Dziewiatkowski, *J. Biol. Chem.* **178**, 389 (1949).
7. T. S. Hele, *Biochem. J.* **18**, 110 (1924).
8. T. S. Hele, *Biochem. J.* **25**, 1736 (1931).
9. F. Bernheim and M. L. C. Bernheim, *J. Pharmacol.* **78**, 394 (1943).
10. H. Boström, *Ark. Kemi* **6**, 43 (1953).
11. D. D. Dziewiatkowski, *Int. Rev. Cytol.* **7**, 159 (1958).
12. H. Boström and L. Rodén, in "The Amino Sugars" (E. A. Balazs and R. W. Jeanloz, eds.), Vol. 2B, p. 45. Academic Press, New York, 1966.
13. R. I. Arnolt and R. H. De Meio, *Rev. Soc. Argent. Biol.* **17**, 570 (1941).
14. R. H. De Meio and R. I. Arnolt, *J. Biol. Chem.* **156**, 577 (1944).
15. R. H. De Meio and L. Tkacz, *Arch. Biochem.* **27**, 242 (1950).
16. R. H. De Meio and L. Tkacz, *J. Biol. Chem.* **195**, 175 (1952).
17. R. H. De Meio, M. Wizerkaniuk, and E. Fabiani, *J. Biol. Chem.* **203**, 257 (1953).
18. S. Bernstein and R. W. McGilvery, *J. Biol. Chem.* **198**, 195 (1952).
19. S. Bernstein and R. W. McGilvery, *J. Biol. Chem.* **199**, 745 (1952).
20. R. H. De Meio, M. Wizerkaniuk, and I. Schreibman, *J. Biol. Chem.* **213**, 439 (1955).
21. H. L. Segal, *J. Biol. Chem.* **213**, 161 (1955).
22. H. Hilz and F. Lipmann, *Proc. Nat. Acad. Sci. U. S.* **41**, 880 (1955).
23. R. H. De Meio and M. Wizerkaniuk, *Biochim. Biophys. Acta* **20**, 428 (1956).
24. P. W. Robbins and F. Lipmann, *J. Amer. Chem. Soc.* **78**, 2652 (1956).
25. P. W. Robbins and F. Lipmann, *J. Biol. Chem.* **229**, 837 (1957).
26. J. Baddiley, J. G. Buchanan, and R. Letters, *J. Chem. Soc., London* p. 1067 (1957).
27. P. Reichard and N. R. Ringertz, *J. Amer. Chem. Soc.* **81**, 878 (1959).
28. R. Cherniak and E. A. Davidson, *J. Biol. Chem.* **239**, 2986 (1964).
29. A. M. Michelson and F. Wold, *Biochemistry* **1**, 1171 (1962).
30. J. Baddiley, J. G. Buchanan, and R. Letters, *Proc. Chem. Soc., London* p. 147 (1957).
31. J. Baddiley, J. G. Buchanan, and R. Letters, *J. Chem. Soc., London* p. 1000 (1958).
32. J. Baddiley, J. G. Buchanan, R. Letters, and A. R. Sanderson, *J. Chem. Soc., London* p. 1731 (1959).
33. J. Baddiley and A. R. Sanderson, *Biochem. Prep.* **10**, 3 (1963).

34. L. M. Fogarty and W. R. Rees, *Nature* (*London*) **193**, 1180 (1962).
35. E. G. Brunngraber, *J. Biol. Chem.* **233**, 472 (1958).
36. R. K. Banerjee and A. B. Roy, *Mol. Pharmacol.* **2**, 56 (1966).
37. P. W. Robbins, *in* "Methods in Enzymology" (S. P. Colowick and N. O. Kaplan, eds.), Vol. 6, p. 770 Academic Press, New York, 1963.
38. A. S. Balasubramanian, L. Spolter, L. I. Rice, J. B. Sharon, and W. Marx, *Anal. Biochem.* **21**, 22 (1967).
39. R. C. Hodson and J. A. Schiff, *Arch. Biochem. Biophys.* **132**, 151 (1969).
40. N. L. Sass and W. G. Martin, *Proc. W. Va. Acad. Sci.* **42**, 65 (1970).
41. G. S. I. M. Jansen and G. M. J. van Kempen, *Anal. Biochem.* **51**, 324 (1973).
42. A. B. Roy and P. A. Trudinger, "The Biochemistry of Inorganic Compounds of Sulfur." Cambridge Univ. Press, London and New York, 1970.
43. L. G. Wilson and R. S. Bandurski, *Arch. Biochem. Biophys.* **62**, 503 (1956).
44. H. L. Segal, *Biochim. Biophys. Acta* **21**, 194 (1956).
45. R. S. Bandurski, L. G. Wilson, and C. L. Squires, *J. Amer. Chem. Soc.* **78**, 6408 (1956).
46. P. W. Robbins and F. Lipmann, *J. Amer. Chem. Soc.* **78**, 6409 (1956).
47. P. W. Robbins and F. Lipmann, *J. Biol. Chem.* **233**, 686 (1958).
48. P. W. Robbins and F. Lipmann, *J. Biol. Chem.* **233**, 681 (1958).
49. F. D'Abramo and F. Lipmann, *Biochim. Biophys. Acta* **25**, 211 (1957).
50. S. Suzuki and J. L. Strominger, *J. Biol. Chem.* **235**, 257, 267, and 274 (1960).
51. R. J. Bridgwater and D. A. Ryan, *Biochem. J.* **65**, 24P (1957).
52. I. H. Goldberg and A. Delbrück, *Fed. Proc., Fed. Amer. Soc. Exp. Biol.* **18**, 235 (1959).
53. H. Yoshida and F. Egami, *J. Biochem.* (*Tokyo*) **57**, 215 (1965).
54. J. E. Creange and C. M. Szego, *Biochem. J.* **102**, 898 (1967).
55. R. H. De Meio, Y. C. Lin, and S. Narasimhulu, *Comp. Biochem. Physiol.* **20**, 581 (1967).
56. P. Nissen and A. A. Benson, *Science* **134**, 1759 (1961).
57. A. Abraham and B. K. Bachhawat, *Biochim. Biophys. Acta* **70**, 104 (1963).
58. W. H. Davies, E. I. Mercer, and T. W. Goodwin, *Biochem. J.* **98**, 369 (1966).
59. J. Carroll and B. Spencer, *Biochem. J.* **94**, 20P (1965).
60. B. Wengle, *Acta Soc. Med. Upsal.* **68**, 154 (1963).
61. J. Carroll and B. Spencer, *Biochem. J.* **96**, 79P (1965).
62. S. Sasaki, *Clin. Chim. Acta* **17**, 215 (1967).
63. A. S. Balasubramanian and B. K. Bachhawat, *J. Sci. Ind. Res., Sect. C* **20**, 202 (1961).
64. G. S. I. M. Jansen, R. van Elk, and G. M. J. van Kempen, *J. Neurochem.* **20**, 9 (1973).
65. U. Gerlach, *Klin. Wochenschr.* **41**, 873 (1963).
66. L. Spolter, L. I. Rice, R. Yamada, and W. Marx, *Biochem. Pharmacol.* **16**, 229 (1967).
67. S. A. Barker, C. N. D. Cruikshank, and T. Webb, *Carbohyd. Res.* **1**, 62 (1965).
68. K. R. Panikkar and B. K. Bachhawat, *Biochim. Biophys. Acta* **151**, 725 (1968).
69. C. A. Pasternak, *Biochem. J.* **85**, 44 (1962).
70. J. F. Wheldrake and C. A. Pasternak, *Biochem. J.* **96**, 276 (1965).
71. J. F. Wheldrake, *Biochem. J.* **105**, 697 (1967).
72. P. C. de Vito and J. Dreyfus, *J. Bacteriol.* **88**, 1341 (1964).
73. L. G. Wilson and R. S. Bandurski, *J. Biol. Chem.* **233**, 975 (1958).
74. J. M. Akagi and L. L. Campbell, *J. Bacteriol.* **84**, 1194 (1962).

75. B. L. Strehler, in "Methods of Enzymatic Analysis" (H. U. Bergmeyer, ed.), p. 559. Academic Press, New York, 1965.
76. S. Addanki, J. F. Sotos, and P. D. Rearick, Anal. Biochem. 14, 261 (1966).
77. G. J. E. Balharry and D. J. D. Nicholas, Anal. Biochem. 40, 1 (1971).
78. P. E. Stanley and S. G. Williams, Anal. Biochem. 29, 381 (1969).
79. C. S. Hawes and D. J. D. Nicholas, Biochem. J. 133, 541 (1973).
80. J. W. Tweedie and I. H. Segel, Prep. Biochem. 1, 90 (1971).
81. J. W. Tweedie and I. H. Segel, J. Biol. Chem. 246, 2438 (1971).
82. J. B. Ragland, Arch. Biochem. Biophys. 84, 541 (1959).
83. A. K. Varma and D. J. D. Nicholas, Biochim. Biophys. Acta 227, 373 (1971).
84. T. Asahi, Biochim. Biophys. Acta 82, 58 (1964).
85. G. J. E. Balharry and D. J. D. Nicholas, Biochim. Biophys. Acta 220, 513 (1970).
86. P. R. Sundaresan, Biochim. Biophys. Acta 113, 95 (1966).
87. A. S. Levi and G. Wolf, Biochim. Biophys. Acta 178, 262 (1969).
88. M. Shoyab and W. Marx, Life Sci., Part II 9, 1151 (1970).
89. P. W. Robbins, in "Methods in Enzymology" (S. P. Colowick and N. O. Kaplan, eds.), Vol. 5, p. 964. Academic Press, New York, 1962.
90. M. Shoyab and W. Marx, Arch. Biochem. Biophys. 146, 368 (1971).
91. M. Shoyab and W. Marx, Biochim. Biophys. Acta 258, 125 (1972).
92. M. Shoyab, L. Y. Su, and W. Marx, Biochim. Biophys. Acta 258, 113 (1972).
93. W. Marx, M. Shoyab, and L. Y. Su, Int. J. Biochem. 5, 471 (1974).
94. E. M. Kosower, "Molecular Biochemistry," p. 263. McGraw-Hill, New York, 1962.
95. J. B. Adams and K. G. Rienits, Biochim. Biophys. Acta 51, 567 (1961).
96. J. N. Burnell and J. W. Anderson, Biochem. J. 134, 565 (1973).
97. E. I. Mercer and G. Thomas, Phytochemistry 8, 2281 (1969).
98. H. D. Peck, Jr., Bacteriol. Rev. 26, 67 (1962).
99. C. A. Adams and D. J. D. Nicholas, Biochem. J. 128, 647 (1972).
100. H. D. Peck, Jr., Proc. Nat. Acad. Sci. U. S. 46, 1053 (1960).
101. H. D. Peck, Jr., J. Biol. Chem. 237, 198 (1962).
102. H. H. Thiele, Antoine van Leeuwenhoek; J. Microbiol. Serol. 34, 350 (1968).
103. H. D. Peck, Jr. and M. P. Stulberg, J. Biol. Chem. 237, 1648 (1962).
104. M. Grunberg-Manago, A. del Campillo-Campbell, L. Dondon, and A. M. Michelson, Biochim. Biophys. Acta 123, 1 (1966).
105. J. N. Burnell and J. W. Anderson, Biochem. J. 133, 417 (1973).
106. O. A. Roels, in "The Vitamins" (W. H. Sebrell Jr. and R. S. Harris, eds.), 2nd ed., Vol. 1, pp. 233–241. Academic Press, New York, 1967.
107. S. D. Balakhovski and I. V. Kuznetsova, Dokl. Akad. Nauk SSSR 118, 331 (1958).
108. D. D. Dziewiatkowski, J. Exp. Med. 100, 11 (1954).
109. H. B. Fell, E. Mellanby, and S. R. Pelc, J. Physiol. (London) 134, 179 (1956).
110. G. Wolf and P. T. Varandani, Biochim. Biophys. Acta 43, 501 (1960).
111. P. T. Varandani, G. Wolf, and B. C. Johnson, Biochem. Biophys. Res. Commun. 3, 97 (1960).
112. K. Subba Rao, P. S. Sastry, and J. Ganguly, Biochem. J. 87, 312 (1963).
113. K. Subba Rao and J. Ganguly, Biochem. J. 90, 104 (1964).
114. K. Subba Rao and J. Ganguly, Biochem. J. 98, 693 (1966).
115. A. S. Perumal and H. R. Cama, Indian J. Biochem. 4, 152 (1967).
116. C. A. Pasternak, S. K. Humphries, and A. Pirie, Biochem. J. 86, 382 (1963).
117. C. A. Pasternak and A. Pirie, Exp. Eye Res. 3, 365 (1964).
118. M. O. Hall and B. R. Straatsma, Biochim. Biophys. Acta 124, 246 (1966).
119. B. Mukherji and B. K. Bachhawat, Biochem. J. 104, 318 (1967).

120. A. S. Levi, S. Geller, D. M. Root, and G. Wolf, *Biochem. J.* **109,** 69 (1968).
121. R. L. Geison, W. E. Rogers, Jr., and B. C. Johnson, *Biochim. Biophys. Acta* **165,** 448 (1968).
122. M. H. R. Lewis and B. Spencer, *Biochem. J.* **85,** 18P (1962).
123. T. Fujiwara and B. Spencer, *Biochem. J.* **85,** 19P (1962).
124. B. Spencer, *Biochem. J.* **77,** 294 (1960).
125. A. S. Balasubramanian and B. K. Bachhawat, *Biochim. Biophys. Acta* **59,** 389 (1962).
126. J. B. Adams, *Biochim. Biophys. Acta* **83,** 127 (1964).
127. A. Abraham and B. K. Bachhawat, *Indian J. Biochem.* **1,** 192 (1964).
128. T. Koizumi, T. Suematsu, A. Kawasaki, K. Hiramatsu, and N. Iwabori, *Biochim. Biophys. Acta* **184,** 106 (1969).
129. D. Armstrong, J. Austin, T. Luttenegger, B. K. Bachhawat, and D. Stumpf, *Biochim. Biophys. Acta* **198,** 523 (1970).
130. J. Austin, D. Armstrong, D. Stumpf, T. Luttenegger, and M. Dragoo, *Biochim. Biophys. Acta* **192,** 29 (1969).
131. R. Bailey-Wood, K. S. Dodgson, and F. A. Rose, *Biochim. Biophys. Acta* **220,** 284 (1970).
132. W. H. B. Denner, A. M. Stokes, F. A. Rose, and K. S. Dodgson, *Biochim. Biophys. Acta* **315,** 394 (1973).
133. A. A. Farooqui and A. S. Balasubramanian, *Biochim. Biophys. Acta* **198,** 56 (1970).
134. A. M. Stokes, W. H. B. Denner, F. A. Rose, and K. S. Dodgson, *Biochim. Biophys. Acta* **302,** 64 (1973).
135. A. M. Stokes, W. H. B. Denner, and K. S. Dodgson, *Biochim. Biophys. Acta* **315,** 402 (1973).
136. H. Tsuyuki and D. R. Idler, *J. Amer. Chem. Soc.* **79,** 1771 (1957).
137. M. Ishimoto and D. Fujimoto, *J. Biochem. (Tokyo)* **50,** 299 (1961).
138. F. Egami and K. Yagi, *J. Biochem. (Tokyo)* **43,** 153 (1956).
139. M. Niwa, S. Higuchi, and F. Egami, *J. Biochem. (Tokyo)* **45,** 89 (1958).
140. R. G. Yount, S. Simchuk, I. Yu, and M. Kottke, *Arch. Biochem. Biophys.* **113,** 288 (1966).
141. R. G. Yount, I. Yu, and S. Simchuk, *Arch. Biochem. Biophys.* **113,** 296 (1966).
142. R. T. Williams, "Detoxication Mechanisms." Chapman & Hall, London, 1959.
143. K. S. Dodgson, F. A. Rose, and B. Spencer, *Biochem. J.* **60,** 346 (1955).
144. B. Schacter and G. F. Marrian, *J. Biol. Chem.* **126,** 663 (1938).
145. J. McKenna, E. Menini, and J. K. Norymberski, *Biochem. J.* **79,** 11P (1961).
146. H. H. Tallan, S. T. Bella, W. H. Stein, and S. Moore, *J. Biol. Chem.* 217, 703 (1955).
147. R. A. John, F. A. Rose, F. S. Wusteman, and K. S. Dodgson, *Biochem. J.* **100,** 278 (1966).
148. C. E. Dalgliesch, *Biochem. J.* **61,** 334 (1955).
149. P. Troen, B. Nilsson, N. Wiqvist, and E. Diczfalusy, *Acta Endocrinol. (Copenhagen)* **38,** 361 (1961).
150. R. M. Acheson and A. R. Hands, *Biochim. Biophys. Acta* **51,** 579 (1961).
151. R. A. Heacock and M. E. Mahon, *Can. J. Biochem.* **42,** 813 (1964).
152. V. E. Davis, J. A. Huff, and H. Brown, *Clin. Chim. Acta* **13,** 380 (1966).
153. B. Jacobson, E. A. Balazs, and H. Boström, *Exp. Eye Res.* **10,** 156 (1970).
154. J. L. Meek and N. H. Neff, *J. Neurochem.* **21,** 1 (1973).
155. H. Boström, B. E. Gustafsson, and B. Wengle, *Proc. Soc. Exp. Biol. Med.* **114,** 742 (1963).
156. H. Boström and A. Vestermark, *Nature (London)* **183,** 1593 (1959).
157. H. Boström and A. Vestermark, *Scand. J. Clin. Lab. Invest.* **12,** 323 (1960).

158. J. B. Adams and A. Poulos, *Biochim. Biophys. Acta* **146,** 493 (1967).
159. A. B. Roy and R. K. Banerjee, *Proc. Int. Congr. Horm. 2nd, 1966 Int. Congr. Ser.* No. 132, p. 397 (1967).
160. D. Eccleston and L. M. Ritchie, *J. Neurochem.* **21,** 635 (1973).
161. A. Vestermark and H. Boström, *Acta Chem. Scand.* **13,** 827 (1959).
162. B. Wengle, *Acta Chem. Scand.* **18,** 65 (1964).
163. A. B. Roy, *Biochem. J.* **62,** 41 (1956).
164. R. H. De Meio, *Acta Physiol. Lat. Amer.* **2,** 195, (1952).
165. J. D. Gregory and F. Lipmann, *J. Biol. Chem.* **229,** 1081 (1957).
166. H. Hidaka, T. Nagatsu, and K. Yagi, *Anal. Biochem.* **19,** 388 (1967).
167. G. M. J. van Kempen and G. S. I. M. Jansen, *Anal. Biochem.* **46,** 438 (1972).
168. J. S. Holcenberg and S. W. Rosen, *Arch. Biochem. Biophys.* **110,** 551 (1965).
169. H. Boström and B. Wengle, *Acta Endocrinol. (Copenhagen)* **56,** 691 (1967).
170. S. Schmidt-Nielsen and J. Holmsen, *Arch. Int. Physiol.* **18,** 128 (1921).
171. J. A. Stekol, *J. Biol. Chem.* **113,** 675 (1936).
172. L. L. Layton and D. R. Frankel, *Arch. Biochem. Biophys.* **31,** 161 (1951).
173. R. H. De Meio, *Arch. Biochem.* **7,** 323 (1945).
174. J N. Smith, *Comp. Biochem.* **6,** 403 (1964).
175. J. N. Smith, *Advan. Comp. Physiol. Biochem.* **3,** 173 (1968).
176. R. P. Maickel, W. R. Jondorf, and B. B. Brodie, *Fed. Proc., Fed. Amer. Soc. Exp. Biol.* **17,** 390 (1958).
177. J. N. Smith, *Biol. Rev. Cambridge Phil. Soc.* **30,** 455 (1955).
178. M. Hitchcock and J. N. Smith, *Biochem. J.* **93,** 392 (1964).
179. F. G. Darby, M. P. Heenan, and J. N. Smith, *Life Sci.* **5,** 1499 (1966).
180. H. W. Ruelius and A. Gauhe, *Justus Liebigs Ann. Chem.* **570,** 121 (1950).
181. J. H. Hodgkin, J. S. Craigie, and A. G. McInnes, *Can. J. Chem.* **44,** 74 (1966).
182. L. Hörhammer and R. Hänsel, *Arch. Pharm. (Weinheim)* **286,** 153 (1953).
183. P. Nissen and A. A. Benson, *Biochim. Biophys. Acta* **82,** 400 (1964).
184. R. K. Banerjee and A. B. Roy, *Biochim. Biophys. Acta* **151,** 573 (1968).
185. B. Wengle, *Acta Soc. Med. Upsal.* **69,** 105 (1964).
186. B. Spencer and J. Raftery, *Biochem. J.* **99,** 35P (1966).
187. A. B. Roy, *Advan. Enzymol.* **22,** 205 (1960).
188. D. J. Bartord and J. G. Jones, *Biochem. J.* **123,** 427 (1971).
189. W. W. Cleland, *Biochim. Biophys. Acta* **67,** 104 (1963).
190. H. Boström and B. Wengle, *Acta Soc. Med. Upsal.* **69,** 41 (1964).
191. B. T. Chadwick and J. H. Wilkinson, *Biochem. J.* **76,** 102 (1960).
192. G. L. Cohn, *Nature (London)* **208,** 80 (1965).
193. H. Wieland and F. Vocke, *Justus Liebigs Ann. Chem.* **481,** 215 (1930).
194. A. Butenandt, E. Biekert, N. Koga, and P. Traub, *Hoppe-Seyler's Z. Physiol. Chem.* **321,** 258 (1960).
195. R. F. Doolittle and B. Blombäck, *Nature (London)* **202,** 147 (1964).
196. G. M. J. van Kempen and G. S. I. M. Jansen, *Experientia* **27,** 485 (1971).
197. R. T. Williams, *Biochem. J.* **32,** 878 (1938).
198. H. G. Bray, B. G. Humphris, W. V. Thorpe, K. White, and P. B. Wood, *Biochem. J.* **52,** 419 (1952).
199. H. G. Bray, W. V. Thorpe, and K. White, *Biochem. J.* **52,** 423 (1952).
200. G. A. Garton and R. T. Williams, *Biochem. J.* **43,** 206 (1948).
201. G. A. Garton and R. T. Williams, *Biochem. J.* **44,** 234 (1949).
202. A. Vestermark and H. Boström, *Experientia* **16,** 408 (1960).
203. A. J. Grimes, *Biochem. J.* **73,** 723 (1959).

204. Y. Nose and F. Lipmann, *J. Biol. Chem.* **233**, 1348 (1958).
205. H. L. Segal and L. A. Mologne, *J. Biol. Chem.* **234**, 909 (1959).
206. J. G. Jones and K. S. Dodgson, *Biochem. J.* **94**, 331 (1965).
207. J. G. Jones, S. M. Scotland, and K. S. Dodgson, *Biochem. J.* **98**, 138 (1966).
208. K. S. Dodgson, J. M. Basford, J. G. Jones, and P. Mattock, *Indian J. Biochem.* **4**, Suppl., 7 (1967).
209. P. Mattock and J. G. Jones, *Biochem. J.* **116**, 797 (1970).
210. D. J. Barford and J. G. Jones, *Biochem. J.* **125**, 76P (1971).
211. F. A. McEvoy and J. Carroll, *Biochem. J.* **123**, 901 (1971).
212. W. N. Jenner and F. A. Rose, *Biochem. J.* **135**, 109 (1973).
213. J. S. Kittredge, D. G. Simonsen, E. Roberts, and B. Jelinek, *in* "Amino Acid Pools" (J. T. Holden, ed.), p. 176. Elsevier, Amsterdam, 1962.
214. F. R. Bettelheim, *J. Amer. Chem. Soc.* **76**, 2838 (1954).
215. G. A. Mross and R. F. Doolittle, *Arch. Biochem. Biophys.* **122**, 674 (1967).
216. T. Krajewski and B. Blombäck, *Acta Chem. Scand.* **22**, 1339 (1968).
217. H. Gregory, P. M. Hardy, D. S. Jones, G. W. Kenner, and R. C. Sheppard, *Nature (London)* **204**, 931 (1964).
218. A. Anastasi, G. Bertaccini, and V. Erspamer, *Brit, J. Pharmacol. Chemother.* **27**, 479 (1966).
219. A. Anastasi, V. Erspamer, and R. Endean, *Experientia* **23**, 699 (1967).
220. A. Anastasi, V. Erspamer, and R. Endean, *Arch. Biochem. Biophys.* **125**, 57 (1968).
221. F. R. Jevons, *Biochem. J.* **89**, 621 (1963).
222. H. Hidaka, T. Nagatsu, and K. Yagi, *Arch. Biochem. Biophys.* **117**, 196 (1966).
223. H. Hidaka, T. Nagatsu, K. Takeya, S. Matsumoto, and K. Yagi, *J. Pharmacol.* **166**, 272 (1969).
224. H. Hidaka, T. Nagatsu, and K. Yagi, *Biochim. Biophys. Acta* **177**, 354 (1969).
225. H. Hidaka, T. Nagatsu, and K. Yagi, *J. Neurochem.* **16**, 783 (1969).
226. H. Hidaka and J. Austin, *Biochim. Biophys. Acta* **268**, 132 (1972).
227. K. Yagi, *in* "Methods in Enzymology" (H. Tabor and C. W. Tabor, eds.), Vol. 17B, p. 825. Academic Press, New York 1971.
228. D. F. Mayers and E. T. Kaiser, *J. Amer. Chem. Soc.* **90**, 6192 (1968).
229. S. W. Weidman, D. F. Mayers, O. R. Zaborsky, and E. T. Kaiser, *J. Amer. Chem. Soc.* **89**, 4555 (1967).
230. E. A. Ford and P. M. Ruoff, *Chem. Commun.* p. 630 (1965).
231. J. B. Adams, *Biochim. Biophys. Acta* **82**, 572 (1964).
232. A. Sneddon and G. F. Marrian, *Biochem. J.* **86**, 385 (1963).
233. E. Diczfalusy, O. Cassmer, C. Alonso, and M. de Miquel, *Acta Endocrinol. (Copenhagen)* **38**, 31 (1961).
234. B. Wengle, *Acta Endocrinol (Copenhagen)* **52**, 607 (1966).
235. H. R. Raud and R. Hobkirk, *Can. J. Biochem.* **44**, 657 (1966).
236. M. Levitz, G. P. Condon, and J. Dancis, *Endocrinology* **68**, 825 (1961).
237. E. Bolté, S. Mancuso, G. Eriksson, N. Wiqvist, and E. Diczfalusy, *Acta Endocrinol. (Copenhagen)* **45**, 535 (1964).
238. H. R. Raud and R. Hobkirk, *Can. J. Biochem.* **46**, 749 (1968).
239. J. B. Adams, *Arch. Biochem. Biophys.* **101**, 478 (1963).
240. B. Wengle and H. Boström, *Acta Chem. Scand.* **17**, 1203 (1963).
241. A. H. Payne and M. Mason, *Biochim. Biophys. Acta* **71**, 719 (1963).
242. J. B. Adams and M. Chulavatnatol, *Biochim. Biophys. Acta* **146**, 509 (1967).
243. J. B. Adams, *Biochim. Biophys. Acta* **146**, 522 (1967).
244. M. Levitz, J. Katz, and G. H. Twombly, *Steroids* **6**, 553 (1965).

245. A. B. Roy, *Biochem. J.* **63**, 294 (1956).
246. J. J. Schneider and M. L. Lewbart, *J. Biol. Chem.* **222**, 787 (1956).
247. R. H. De Meio, C. Lewycka, M. Wizerkaniuk, and O. Salciunas, *Biochem. J.* **68**, 1 (1958).
248. R. K. Banerjee and A. B. Roy, *Biochim, Biophys. Acta* **137**, 211 (1967).
249. H. Boström, C. Franksson, and B. Wengle, *Acta Endocrinol. (Copengagen)* **47**, 633 (1964).
250. E. Wallace and N. Silberman, *J. Biol. Chem.* **239**, 2809 (1964).
251. J. B. Adams, *J. Clin. Endocrinol. Metab.* **24**, 988 (1964).
252. E. E. Baulieu, I. Fabre-Jung, and L. G. Huis in't Veld, *Endocrinology* **81**, 34 (1967).
253. R. Gugler, G. S. Rao, and H. Breuer, *Biochim. Biophys. Acta* **220**, 69 (1970).
254. R. Ryan and J. Carroll, *Biochem. J.* **124**, 25P (1971).
255. J. B. Adams and A. M. Edwards, *Biochim. Biophys. Acta* **167**, 122 (1968).
256. E. Döllefeld and H. Breuer, *Z. Vitam.-, Horm.-Fermentforsch.* **14**, 193 (1966).
257. D. Y. Wang and R. D. Bulbrook, *Advan. Reprod. Physiol.* **3**, 113–146 (1968).
258. J. Carlstedt-Duke and J. A. Gustafsson, *Eur. J. Biochem.* **36**, 172 (1973).
259. J. Herrmann and K. Repke, *Naunyn-Schmiedebergs Arch. Exp. Pathol. Pharmakol.* **248**, 370 (1964).
260. M. Arcos and S. Lieberman, *Biochemistry* **6**, 2032 (1967).
261. R. H. Palmer, *Proc. Nat. Acad. Sci. U. S.* **58**, 1047 (1967).
262. R. H. Palmer and M. G. Bolt, *J. Lipid Res.* **12**, 671 (1971).
263. I. Makino, S. Nakagawa, K. Shinozaki, and K. Mashimo, *Lipids* **7**, 750 (1972).
264. I. Makino, K. Shinozaki, and S. Nakagawa, *Lipids* **8**, 47 (1973).
265. R. H. Palmer, *J. Lipid Res.* **12**, 680 (1971).
266. G. W. Oertel, *Biochem. Z.* **334**, 431 (1961).
267. G. W. Oertel and E. Kaiser, *Clin. Chim. Acta* **7**, 463 (1962).
268. G. W. Oertel, E. Kaiser, and P. Brühl, *Biochem. Z.* **336**, 154 (1962).
269. G. W. Oertel, *Biochem. Z.* **339**, 125 (1963).
270. G. W. Oertel and L. Treiber, *Hoppe-Seyler's Z. Physiol. Chem.* **344**, 163 (1966).
271. P. Benes and G. W. Oertel, *J. Steroid Biochem.* **3**, 925 (1972).
272. T. Laatikainen, P. Peltokallio, and R. Vihko, *Steroids* **12**, 407 (1968).
273. J. R. Pasqualini and M. F. Jayle, *J. Clin. Invest.* **41**, 981 (1962).
274. E. E. Baulieu and C. Corpéchot, *Bull. Soc. Chim. Biol.* **47**, 443 (1965).
275. M. Levitz, J. Katz, and G. H. Twombly, *Steroids* **6**, 553 (1965).
276. G. A. D. Haslewood, *Biochem. J.* **90**, 309 (1964).
277. G. A. D. Haslewood, *J. Lipid Res.* **8**, 535 (1967).
278. G. A. D. Haslewood, *Biochem. J.* **100**, 233 (1966).
279. I. G. Anderson, G. A. D. Haslewood, A. D. Cross, and L. Tökés, *Biochem. J.* **104**, 1061 (1967).
280. I. G. Anderson, T. Briggs, and G. A. D. Haslewood, *Biochem. J.* **90**, 303 (1964).
281. I. G. Anderson and G. A. D. Haslewood, *Biochem. J.* **93**, 34 (1964).
282. R. J. Bridgwater, G. A.·D. Haslewood, and J. R. Watt, *Biochem. J.* **87**, 28 (1963).
283. M. F. Scully, K. S. Dodgson, and F. A. Rose, *Biochem. J.* **119**, 29P (1970).
284. R. F. N. Hutchins and J. N. Kaplanis, *Steroids* **13**, 605 (1969).
285. J. Koolman, J. A. Hoffmann, and P. Karlson, *Hoppe-Seyler's Z. Physiol. Chem.* **354**, 1043 (1973).
286. R. S. H. Yang and C. F. Wilkinson, *Biochem. J.* **130**, 487 (1972).
287. R. S. H. Yang and C. F. Wilkinson, *Comp. Biochem. Physiol.* **46B**, 717 (1973).
288. M. Higaki, M. Takahashi, T. Suzuki, and Y. Sahashi, *J. Vitaminol. (Kyoto)* **11**, 261 and 266 (1965).

354 ROMANO HUMBERTO DE MEIO

289. Y. Sahashi, T. Suzuki, M. Higaki, and T. Asano, *J. Vitaminol.* (*Kyoto*) **13**, 33 (1967).
290. B. S. Leung, W. Jack, and D. C. Wood, *Physiol. Chem. Phys.* **4**, 543 (1972).
291. D. W. Woolley and W. H. Peterson, *J. Biol. Chem.* **122**, 213 (1937).
292. T. Harada and B. Spencer, *J. Gen. Microbiol.* **22**, 520 (1960).
293. B. Spencer and T. Harada, *Biochem. J.* **77**, 305 (1960).
294. M. Itahashi, *J. Biochem.* (*Tokyo*) **50**, 52 (1961).
295. B. Lindberg, *Acta Chem. Scand.* **9**, 1323 (1955).
296. J. W. Fitzgerald, *Biochem. J.* **136**, 361 (1973).
297. B. A. Orsi and B. Spencer, *J. Biochem.* (*Tokyo*) **56**, 81 (1964).
298. C. Hussey, B. A. Orsi, J. Scott, and B. Spencer, *Nature* (*London*) **207**, 632 (1965).
299. J. Scott and B. Spencer, *Biochem. J.* **95**, 50P (1965).
300. A. Kaji and J. D. Gregory, *J. Biol. Chem.* **234**, 3007 (1959).
301. A. Kaji and W. D. McElroy, *Biochim. Biophys. Acta* **30**, 190 (1958).
302. A. A. Benson and M. R. Atkinson, *Fed. Proc., Fed. Amer. Soc. Exp. Biol.* **26**, 394 (1967).
303. E. Boyland, D. Manson, and S. F. D. Orr, *Biochem. J.* **65**, 417 (1957).
304. D. V. Parke, *Biochem. J.* **77**, 493 (1960).
305. A. B. Roy, *Biochem. J.* **74**, 49 (1960).
306. A. B. Roy, *Biochem. J.* **79**, 253 (1961).
307. A. B. Roy, *Biochem. J.* **82**, 66 (1962).
308. A. B. Roy, *J. Mol. Biol.* **10**, 176 (1964).
309. K. S. Dodgson and F. A. Rose, in "Metabolic Conjugation and Metabolic Hydrolysis" (W. H. Fishman, ed.), Vol. 1, p. 310. Academic Press, New York, 1970.
310. A. B. Roy, *Aust. J. Exp. Biol. Med. Sci.* **41**, 331 (1963).
311. J. N. Smith, *Nature* (*London*) **195**, 399 (1962).
312. L. L. Layton, Cancer **3**, 725 (1950).
313. D. D. Dziewiatkowski, R. E. Benesch, and R. Benesch, *J. Biol. Chem.* **178**, 931 (1949).
314. D. D. Dziewiatkowski, *J. Biol. Chem.* **189**, 187 (1951).
315. H. Boström, *J. Biol. Chem.* **196**, 477 (1952).
316. L. F. Adamson, S. Gleason, and C. S. Anast, *Biochim. Biophys. Acta* **83**, 262 (1964).
317. A. Telser, H. C. Robinson, and A. Dorfman, *Proc. Nat. Acad. Sci. U. S.* **54**, 912 (1965).
318. R. Matalon and A. Dorfman, *Proc. Nat. Acad. Sci. U. S.* **60**, 179 (1968).
319. G. de la Haba and H. Holtzer, *Science* **149**, 1263 (1965).
320. M. Okayama, K. Kimata, and S. Suzuki, *J. Biochem.* (*Tokyo*) **74**, 1069 (1973).
321. R. W. Jeanloz, *Compr. Biochem.* **5**, 262 (1963).
322. L. Rodén, in "Metabolic Conjugation and Metabolic Hydrolysis" (W. H. Fishman, ed.), Vol. 2, p. 345. Academic Press, New York, 1970.
323. H. Boström and S. Gardell, *Acta Chem. Scand.* **7**, 216 (1953).
324. S. Schiller, M. B. Mathews, J. A. Cifonelli, and A. Dorfman, *J. Biol. Chem.* **218**, 139 (1956).
325. J. F. Goggins, G. S. Johnson, and I. Pastan, *J. Biol. Chem.* **247**, 5759 (1972).
326. J. L. Strominger, *Biochim. Biophys. Acta* **17**, 283 (1955).
327. J. Picard, A. Gardais, and L. Dubernard, *Nature* (*London*) **202**, 1213 (1964).
328. A. Telser, Ph.D. thesis, University of Chicago (1968) (cited by Rodén [322]).
329. E. A. Davidson and K. Meyer, *J. Biol. Chem.* **211**, 605 (1954).
330. R. L. Perlman, A. Telser, and A. Dorfman, *J. Biol. Chem.* **239**, 3623 (1964).
331. F. D'Abramo and F. Lipmann, *Abstr. Int. Congr. Biochem., 4th, 1958* Vol. 15, Sect. 6-41, p. 75 (1960).

332. A. Delbrück and F. Lipmann, cited by A. Delbrück, in "Struktur und Stoffwechsel des Bindegewebes" (W. H. Hauss and H. Losse, eds.), p. 38. Thieme, Stuttgart, 1960.
333. J. B. Adams, Nature (London) 184, 274 (1959).
334. J. B. Adams, Biochim. Biophys. Acta 32, 559 (1959).
335. J. B. Adams, Biochem. J. 76, 520 (1960).
336. S. Suzuki and J. L. Strominger, Biochim. Biophys. Acta 31, 283 (1959).
337. N. R. Ringertz, "Acid Polysaccharides of Mast Cell Tumors." Almqvist & Wiksell, Stockholm, 1960.
338. S. Suzuki, R. H. Trenn, and J. L. Strominger, Biochim. Biophys. Acta 50, 169 (1961).
339. E. Hasegawa, A. Delbrück, and F. Lipmann, Fed. Proc., Fed. Amer. Soc. Exp. Biol. 20, 86 (1961).
340. E. Meezan and E. A. Davidson, J. Biol. Chem. 242, 1685 (1967).
341. E. Meezan and E. A. Davidson, J. Biol. Chem. 242, 4956 (1967).
342. M. B. Mathews, J. Duh, and P. Person, Nature (London) 193, 378 (1962).
343. Y. Kawai, N. Seno, and K. Anno, J. Biochem. (Tokyo) 60, 317 (1966).
344. S. Suzuki, H. Saito, T. Yamagata, K. Anno, N. Seno, Y. Kawai, and T. Furuhashi, J. Biol. Chem. 243, 1543 (1968).
345. T. Soda, F. Egami, and T. Horigome, J. Chem. Soc. Jap., Pure Chem. Sect. 61, 43 (1940).
346. S. Suzuki, J. Biol. Chem. 235, 3580 (1960).
347. T. Furuhashi, J. Biochem. (Tokyo) 50, 546 (1961).
348. T. Furuhashi, Seikagaku 33, 746 (1961).
349. M. B. Mathews, Biochim. Biophys. Acta 58, 92 (1962).
350. B. Anderson and K. Meyer, Fed. Proc., Fed. Amer. Soc. Exp. Biol. 21, 171 (1962).
351. M. Momburg, H. W. Stuhlsatz, R. Kisters, and H. Greiling, Hoppe-Seyler's Z. Physiol. Chem. 353, 1351 (1972).
352. K. Kimata, M. Okayama, and S. Suzuki, Seikagaku 39, 548 (1967); 40, 501 (1968) (cited by Kimata et al. [353]).
353. K. Kimata, M. Okayama, A. Oohira, and S. Suzuki, Mol. Cell. Biochem. 1, 211 (1973).
354. U. Lindahl and L. Rodén, J. Biol. Chem. 240, 2821 (1965).
355. A. Telser, H. C. Robinson, and A. Dorfman, Arch. Biochem. Biophys. 116, 458 (1966).
356. A. Dorfman, Advan. Biol. Skin. 10, 123 (1970).
357. R. W. Young, J. Cell Biol. 57, 175 (1973).
358. M. E. Richmond, S. DeLuca, and J. E. Silbert, Biochem. Biophys. Res. Commun. 46, 263 (1972).
359. M. E. Richmond, S. DeLuca, and J. E. Silbert, Biochemistry 12, 3898 and 3904 (1973).
360. S. DeLuca, M. E. Richmond, and J. E. Silbert, Biochemistry 12, 3911 (1973).
361. J. G. Derge and E. A. Davidson, Biochem. J. 126, 217 (1972).
362. A. L. Horwitz and A. Dorfman, J. Cell Biol. 38, 358 (1968).
363. M. Mohanram and V. Reddy, Int. J. Vitam. Nutr. Res. 43, 56 (1973).
364. E. A. Davidson and J. G. Riley, J. Biol. Chem. 235, 3367 (1960).
365. J. B. Adams and M. F. Meaney, Biochim. Biophys. Acta 54, 592 (1961).
366. E. Jorpes, E. Odeblad, and H. Boström, Acta Haematol. 9, 273 (1953).
367. G. Asboe-Hansen, Cancer Res. 13, 587 (1953); 14, 94 (1954).
368. S. Magnusson and B. Larsson, Acta Chem. Scand. 9, 534 (1955).
369. B. Larsson, Nord Veterinaermed. 8, 581 (1956).

370. H. B. Eiber and I. Danishefsky, *J. Biol. Chem.* **226,** 721 (1957).
371. T. Sato, T. Suzuki, T. Fukuyama, and H. Yoshikawa, *J. Biochem.* (*Tokyo*) **45,** 237 (1958).
372. J. Furth, P. Hagen, and E. I. Hirsch, *Proc. Soc. Exp. Biol. Med.* **95,** 824 (1957).
373. T. B. Dunn and M. Potter, *J. Nat. Cancer Inst.* **18,** 587 (1957).
374. S. M. Day and J. P. Green, *Fed. Proc., Fed. Amer. Soc. Exp. Biol.* **18,** 381 (1959).
375. E. D. Korn, *J. Biol. Chem.* **234,** 1321 (1959).
376. L. Spolter and W. Marx, *Fed. Proc., Fed. Amer. Soc. Exp. Biol.* **17,** 314 (1958).
377. L. Spolter and W. Marx, *Biochim. Biophys. Acta* **32,** 291 (1959).
378. C. A. Pasternak, *J. Biol. Chem.* **235,** 438 (1960).
379. N. R. Ringertz, *Ark. Kemi* **16,** 67 (1960).
380. E. D. Korn, *J. Biol. Chem.* **234,** 1647 (1959).
381. N. R. Ringertz, *Ann. N. Y. Acad. Sci.* **103,** 209 (1963).
382. J. E. Silbert, *J. Biol. Chem.* **238,** 3542 (1963).
383. J. E. Silbert *J. Biol. Chem.* **242,** 2301 (1967).
384. J. E. Silbert, *J. Biol. Chem.* **242,** 5146 (1967).
385. J. E. Silbert, *J. Biol. Chem.* **242,** 5153 (1967).
386. J. E. Silbert and D. H. Brown, *Biochim. Biophys. Acta* **54,** 590 (1961).
387. R. A. Eisenman, A. S. Balasubramanian, and W. Marx, *Arch. Biochem. Biophys.* **119,** 387 (1967).
388. L. I. Rice, L. Spolter, Z. Tokes, R. Eisenman, and W. Marx, *Arch. Biochem. Biophys.* **118,** 374 (1967).
389. S. Schiller and A. Dorfman, *Biochim. Biophys. Acta* **31,** 278 (1959).
390. S. Schiller, *Ann. N. Y. Acad. Sci.* **103,** 199 (1963).
391. G. Bloom and N. R. Ringertz, *Ark. Kemi* **16,** 51 (1960).
392. A. C. Parekh and D. Glick, *J. Biol. Chem.* **237,** 280 (1962).
393. N. R. Ringertz, *Acta Chem. Scand.* **14,** 312 (1960).
394. N. R. Ringertz, *in* "The Amino Sugars" (E. A. Balazs and R. W. Jeanloz, eds.), Vol. 2A, pp. 209–217. Academic Press, New York, 1965.
395. L. Rodén and A. Dorfman, *Acta Chem. Scand.* **13,** 2121 (1959).
396. T. Foley and J. R. Baker, *Biochem. J.* **135,** 187 (1973).
397. J. A. Cifonelli, *in* "Chemistry and Molecular Biology of the Intercellular Matrix" (E. A. Balazs, ed.), Vol. 2, p. 961. Academic Press, New York, 1970.
398. U. Lindahl, *in* "Chemistry and Molecular Biology of the Intercellular Matrix" (E. A. Balazs, ed.), Vol. 2, pp. 943–960. Academic Press, New York, 1970.
399. U. Lindahl and O. Axelsson, *J. Biol. Chem.* **246,** 74 (1971).
400. A. S. Balasubramanian, N. S. Joun, and W. Marx, *Arch. Biochem. Biophys.* **128,** 623 (1968).
401. A. H. Johnson and J. R. Baker, *Biochim. Biophys. Acta* **320,** 341 (1973).
402. L. Rodén, *Ark. Kemi* **10,** 383 (1956).
403. B. Wortman, *J. Biol. Chem.* **236,** 974 (1961).
404. B. Wortman, *Fed. Proc., Fed. Amer. Soc. Exp. Biol.* **22,** 413 (1963).
405. S. Hunt and F. R. Jevons, *Biochem. J.* **98,** 522 (1966).
406. F. Egami and N. Takahashi, *in* "Biochemistry and Medicine of Mucopolysaccharides" (F. Egami and Y. Oshima, eds.), p. 53. Maruzen, Tokyo, 1962.
407. J. W. Lash and M. W. Whitehouse, *Biochem. J.* **74,** 351 (1960).
408. J. Immers, "Investigations on Macromolecular Sulfated Polysaccharides in Sea Urchin Development." Almqvist & Wiksell, Stockholm, 1962.
409. K. Ishihara, *Exp. Cell Res.* **51,** 473 (1968).
410. S. Inoue, *Biochim. Biophys. Acta* **101,** 16 (1965).

411. S. L. Burson Jr., M. J. Fahrenbach, L. H. Frommhagen, B. A. Riccardi, R. A. Brown, J. A. Brockman, H. V. Lewry, and E. L. R. Stockstad, *J. Amer. Chem. Soc.* **78,** 5874 (1956).
412. J. Doyle, *Biochem. J.* **103,** 325 (1967).
413. J. A. Cifonelli and M. B. Mathews, *Fed. Proc., Fed. Amer. Soc. Exp. Biol.* **27,** 808 (1968).
414. P. F. Lloyd and K. O. Lloyd, *Nature (London)* **199,** 287 (1963).
415. R. L. Katzman and R. W. Jeanloz, *J. Biol. Chem.* **248,** 50 (1973).
416. Y. Nakanishi, H. Sonohara, and S. Suzuki, *J. Biol. Chem.* **245,** 6046 (1970).
417. J. Picard and A. Gardais, *Bull. Soc. Chim. Biol.* **49,** 1689 (1967).
418. S. Suzuki and J. L. Strominger, *J. Biol. Chem.* **235,** 2768 (1960).
419. T. Harada, S. Shimizu, Y. Nakanishi, and S. Suzuki, *J. Biol. Chem.* **242,** 2288 (1967).
420. O. Habuchi, T. Yamagata, and S. Suzuki, *J. Biol. Chem.* **246,** 7357 (1971).
421. H. S. Barra and R. Caputto, *Biochim. Biophys. Acta* **101,** 367 (1965).
422. R. Carubelli, L. C. Ryan, R. E. Trucco, and R. Caputto, *J. Biol. Chem.* **236,** 2381 (1961).
423. L. C. Ryan, R. Carubelli, R. Caputto, and R. E. Trucco, *Biochim. Biophys. Acta* **101,** 252 (1965).
424. F. A. Cumar, H. S. Barra, H. J. Maccioni, and R. Caputto, *J. Biol. Chem.* **243,** 3807 (1968).
425. J. D. Chanley, R. Ledeen, J. Wax, R. F. Nigrelli, and H. Sobotka, *J. Amer. Chem. Soc.* **81,** 5180 (1959).
426. T. Yasumoto, N. Nakamura, and Y. Hashimoto, *Agr. Biol. Chem.* **31,** 7 (1967).
427. S. L. Fries, R. C. Durant, J. D. Chanley, and F. J. Fash, *Biochem. Pharmacol.* **16,** 1617 (1967).
428. M. Kates, B. Palameta, M. P. Perry, and G. A. Adams, *Biochim. Biophys. Acta* **137,** 213 (1967).
429. A. J. Hancock and M. Kates, *J. Lipid Res.* **14,** 422 (1973).
430. M. Kates and P. W. Deroo, *J. Lipid Res.* **14,** 438 (1973).
431. I. H. Goldberg, *Fed. Proc., Fed. Amer. Soc. Exp. Biol.* **19,** 220 (1960).
432. G. M. McKhann, R. Levy, and W. Ho, *Biochem. Biophys. Res. Commun.* **20,** 109 (1965).
433. D. F. Farrell and G. M. McKhann, *J. Biol. Chem.* **246,** 4694 (1971).
434. B. Fleischer and F. Zambrano, *Biochem. Biophys. Res. Commun.* **52,** 951 (1973).
435. A. S. Balasubramanian and B. K. Bachhawat, *Biochim. Biophys. Acta* **106,** 218 (1965).
436. S. Saxena, E. George, S. Kokrady, and B. K. Bachhawat, *Indian J. Biochem. Biophys.* **8,** 1 (1971).
437. J. L. Nussbaum and P. Mandel, *J. Neurochem.* **19,** 1789 (1972).
438. P. Stoffyn, A. Stoffyn, and G. Hauser, *J. Lipid Res.* **12,** 318 (1971).
439. I. Ishizuka, M. Suzuki, and T. Yamakawa, *J. Biochem. (Tokyo)* **73,** 77 (1973).
440. M. B. Goren, *Biochim. Biophys. Acta* **210,** 116 (1970); *Lipids* **6,** 40 (1971).
441. C. L. Mooney and T. H. Haines, *Biochemistry* **12,** 4469 (1973).
442. T. Yagi, *Biochim. Biophys. Acta* **82,** 170 (1964); *J. Biochem. (Tokyo)* **59,** 495 (1966).
443. A. Vestermark and H. Boström, *Exp. Cell Res.* **18,** 174 (1959).
444. G. L. Law, G. P. Mansfield, D. F. Muggleton, and E. W. Parnell, *Nature (London)* **197,** 1024 (1963).
445. A. Vestermark and H. Boström, *Acta Chem. Scand.* **13,** 2133 (1959).
446. B. A. Noir, A. T. De Walz, and E. A. Rodriguez Garay, *Biochim. Biophys. Acta* **222,** 15 (1970).

447. K. J. Isselbacher and E. A. McCarthy, *Biochim. Biophys. Acta* **29**, 658 (1958).
448. H. C. Robinson and C. A. Pasternak, *Biochem. J.* **93**, 487 (1964).
449. H. C. Robinson, *Biochem. J.* **94**, 687 (1965).
450. S. G. Waley, *Biochem. J.* **71**, 132 (1959).
451. M. J. Cormier, K. Hori, and Y. D. Karkhanis, *Biochemistry* **9**, 1184 (1970).
452. C. J. Mead and F. J. Finamore, *Biochemistry* **8**, 2652 (1969).
453. R. O. Mumma and A. J. Verlangieri, *Biochim. Biophys. Acta* **273**, 249 (1972).
454. E. M. Baker, III, D. C. Hammer, S. C. March, B. M. Tolbert, and J. E. Canham, *Science* **173**, 826 (1971).
455. C. M. King and B. Phillips, *Science* **159**, 1351 (1968).
456. J. R. De Baun, E. C. Miller, and J. A. Miller, *Cancer Res.* **30**, 577 (1970).
457. J. Picard, A. P. Gardais, and M. Vedel, *Biochim. Biophys. Acta* **320**, 427, (1973).
458. H. Blaschko, S. P. Datta, and H. Harris, *Brit. J. Nutr.* **7**, 364 (1953).
459. D. B. Hope, *Biochem. J.* **59**, 497 (1955).
460. J. G. Jacobsen and L. H. Smith, *Physiol. Rev.* **48**, 424 (1968).
461. H. Boström and S. Åqvist, *Acta Chem. Scand.* **6**, 1557 (1952).
462. L. J. Machlin, P. B. Pearson, and C. A. Denton, *J. Biol. Chem.* **212**, 469 (1955).
463. I. P. Lowe and E. Roberts, *J. Biol. Chem.* **212**, 477 (1955).
464. N. L. Sass and W. G. Martin, *Proc. Soc. Exp. Biol. Med.* **139**, 755 (1972).
465. W. G. Martin, N. L. Sass, L. Hill, S. Tarka, and R. Truex, *Proc. Soc. Exp. Biol. Med.* **141**, 632 (1972).
466. C. A. Pasternak, R. J. Ellis, M. C. Jones-Mortimer, and C. E. Crichton, *Biochem. J.* **96**, 270 (1965).
467. R. J. Miraglia and W. G. Martin, *Proc. Soc. Exp. Biol. Med.* **132**, 640 (1969).
468. W. G. Martin, *Poultry Sci.* **51**, 608 (1972).
469. W. G. Martin, personal communication.
470. S. Suzuki, N. Takahashi, and F. Egami, *Biochim. Biophys. Acta* **24**, 444 (1957).
471. S. Suzuki, N. Takahashi, and F. Egami, *J. Biochem. (Tokyo)* **46**, 1 (1959).
472. B. Spencer, *Proc. Int. Symp. Enzyme Chem. 1957* p. 96 (1958).
473. T. M. Chu and W. R. Slaunwhite, *Steroids* **12**, 309 (1968).
474. R. O. Mumma, *Biochim. Biophys. Acta* **165**, 571 (1968).
475. A. J. Verlangieri and R. O. Mumma, *Atherosclerosis* **17**, 37 (1973).
476. J. B. Adams, *Biochim. Biophys. Acta* **62**, 17 (1962).
477. G. Wellers, *Prod. Pharm.* **15**, 307 (1960).
478. G. Wellers and G. Boelle, *Arch. Int. Physiol. Biochim.* **68**, 299 (1960).
479. W. D. Salmon, Jr. and W. H. Daughaday, *J. Clin. Invest.* **35**, 733 (1956).
480. A. C. Herington, L. F. Adamson, and J. Bornstein, *Biochim, Biophys. Acta* **286**, 164 (1972).
481. W. H. Daughaday, K. Hall, M. S. Raben, W. D. Salmon, Jr., J. L. van den Brande, and J. J. van Wyk, *Nature (London)* **235**, 107 (1972).
482. K. Uthne, *Acta Endocrinol. (Copenhagen)* **73**, Suppl. 175 (1973).

CHAPTER 9

Sulfohydrolases

K. S. Dodgson and F. A. Rose

I. GENERAL INTRODUCTION

Sulfuric acid esters and other related sulfoconjugates constitute a very diverse group of compounds, representatives of which are widely distributed throughout the worlds of microorganisms, plants, and animals. The true sulfuric acid ester, of course, is formed by the esterification of a hydroxyl group with sulfuric acid, the resulting compound containing the C–O–S bond. The practical consequences of such an esterification are profound. The sulfate group is highly acidic; when many such groups are present in the same molecule, polyanionic compounds are formed that can bind counterions and maintain a highly solvated domain, features that may have profound implications for the architecture of tissues. In other instances the introduction of sulfate serves to mask physiologically important hydroxyl groups and the possibility exists that the control of the physiological activity of such groups could be achieved by alternate enzymatic sulfation and desulfation of them. Other possibilities arise from the increased water solubility and decreased lipid solubility conferred by the introduction of the sulfate group, properties reflected in the importance of biological sulfation to the excretion of phenolic and steroidal end products of metabolism. In other cases the sulfate group, in association with a suitable hydrophobic moiety, provides the necessary physical properties for a surface-active agent, a property that finds expression in the bile salts of many relatively primitive animal organisms.

Related to the true sulfate esters are compounds in which sulfate is present in linkage forms other than C–O–S. Clearly these compounds are strictly not sulfate esters, but the relationship is so close in many ways that it is convenient to include some of them as members of the general group. The types of linkages found fall into three types: compounds in which sulfate is linked to phosphate via a P–O–S linkage; sulfate esters of substituted hydroxylamines in which the linkage is N–O–S; and compounds in which sulfate is linked to nitrogen via an N–S linkage.

Liberation of inorganic sulfate from all the different classes of "sulfate ester" is achieved by hydrolytic enzymes that have generally been called "sulfatases." In more recent times the term "sulfohydrolases" has found more general favor, but either term is really a misnomer for those enzymes acting on linkages other than C–O–S. For the sake of simplicity, the term sulfohydrolase will generally be used in this account.

For many years the sulfohydrolases attracted relatively little attention, mainly because the few that were known appeared to be enzymological curiosities without obvious physiological function. For example, the question of why mammals should possess arylsulfohydrolases was difficult to answer. Mammals go to great lengths to form and excrete arylsulfate esters as part of their mechanism of defense against the toxicity of phenolic compounds. Once formed, the subsequent hydrolysis of such esters seems to be pointless and would only reexpose the animal to the toxicity problem. The view was sometimes held that these particular hydrolytic enzymes were actually responsible for the biosynthesis of arylsulfate esters, but evidence against this was not difficult to obtain [1]. Other reasons for the tendency to regard sulfohydrolases as curiosities are not far to seek. For example, some enzymes appeared to have no natural substrates, while some natural substrates appeared to have no enzymes.

The interest of the Cardiff laboratories in these enzymes has extended over a period of about 20 years and has been paralleled by a similar interest in the laboratories of Dr. A. B. Roy. Over this period interest has been sustained by our confidence in nature's aversion to prodigality and by our belief that such a diverse and widespread group of enzymes could certainly not be regarded as vestigial. Confidence that fundamental studies on the enzymes would eventually lead to the solution of function was more than repaid when, as a direct result of four years of basic research in our laboratories, workers in Germany and the United States were able to establish that deficiency of a sulfohydrolase was the explanation for the metabolic disorder metachromatic leukodystrophy. Since then other evidence of function has been emerging, particularly as a result of increasing knowledge of the chemical structures and physiological functions of some of the sulfohydrolase substrates. Interest in the enzymes has become widespread, in consequence, and has been further stimulated by discoveries that other sulfohydrolases are deficient in other serious metabolic diseases.

Much still remains to be accomplished, many enzymes remain to be studied, and others undoubtedly await discovery. Few of the enzymes have yet been isolated in pure form and the mechanism of action of none of them is clearly understood. From time to time attempts to ascribe transferase functions to sulfohydrolases have been made. The probability that most of them act by splitting the O–S bond of the C–O–S linkage [2] according to Eq. (1) would imply that any such transfer would involve the sulfate moiety. No firm evidence of such transfer has yet been obtained and in our view the enzymes are purely hydrolytic.

$$R—O\!\!\mid\!\!-SO_3^-$$
$$+ \qquad\qquad \longrightarrow R—OH + SO_4^{2-} + H^+ \qquad\qquad (1)$$
$$H \mid OH$$

From the point of view of comparative and evolutionary biochemistry there would be much to be gained from a study of the detailed structures of individual enzymes, particularly since there are indications that the numbers of enzymes present and the degree of sophistication of enzyme specificity increase with increasing complexity of the organism. Moreover, some sulfohydrolases have special features of kinetic behavior that are of considerable interest to the enzymologist.

In the present review of the sulfohydrolases we have included only those references that appear to us to be particularly relevant to the topic at its present state of development. For additional detail the reader is referred to a number of earlier reviews that are cited in various sections of the following account.

II. ENZYMES HYDROLYZING C–O–SO$_3$ LINKAGES

A. Arylsulfohydrolases

1. MAMMALIAN ARYLSULFOHYDROLASES

Arylsulfohydrolases, sometimes referred to as "phenolsulfatases" or "arylsulfatases" (EC 3.1.6.1; aryl-sulfate sulfohydrolase), are the most widely distributed of all the sulfohydrolases. They liberate sulfate from a large number of arylsulfate esters and this currently serves as the basis for their classification. However, recent work suggests that this may be an oversimplified (and hence misleading) representation of their activities which has delayed recognition of their true function. The early history and development of this field have been the subject of several recent reviews [3–6].

The observation of enzyme activity toward indoxyl sulfate, first reported in snails (*Murex trunculus*) in 1911 [7], led to searches for arylsulfohydrolases [8,9, for example] and their natural substrates. In the succeeding years, however, little progress was made toward an understanding of the enzymes and indeed confusion arose because it was always assumed that observed activity reflected the presence of one enzyme only. The development of a sensitive assay procedure [10], followed by detailed studies on the arylsulfohydrolases of rat [11] and mouse [12] liver, represented the first major step forward. Briefly, work in Cardiff established that enzyme activity toward potassium *p*-acetylphenyl sulfate (I) was localized in the microsomal fraction of the liver cell, whereas it had been claimed by other workers that activity toward a different substrate, dipotassium 2-hydroxy-5-nitrophenyl sulfate (nitrocatechol sulfate) (II) was localized in the mitochondrial fraction.

(I)   (II)

(III)

Apart from its distinct subcellular localization, the microsomal enzyme (now known as arylsulfohydrolase C) was characterized by its alkaline optimum pH, greater activity and affinity toward simple arylsulfates such as *p*-acetylphenyl sulfate and *p*-nitrophenyl sulfate (III) than toward the more complex nitrocatechol sulfate, and by its inhibition by cyanide but not by sulfate or phosphate. The mitochondrial activity was distinguishable by its acid optimum pH and by its substrate affinity and inhibition behavior that was the converse of that shown by arylsulfohydrolase C. Eventually the mitochondrial activity was shown to reflect the combined activity of two enzymes, arylsulfohydrolases A and B [see 13–15], and subsequently, employing refined experimental techniques, both enzymes were established as lysosomal in origin [16].

From these studies it became clear that mammalian tissues contained at least 3 arylsulfohydrolases and there emerged the concept that mammalian arylsulfohydrolases could be classified as either Type I or Type II according to whether their properties resembled, respectively, those of enzyme C or of A and B [17]. While this principle was useful for the study of the enzymes from mammalian sources, its real significance has never been fully explored. Moreover, it has often been misused by carrying the analogy too far or by regarding it as universally true and inflexible.

Many of the studies on the three enzymes that have since been made are of limited value only, since the tendency has been toward oversimplification and, in spite of more than adequate documentation, the multiplicity of the enzymes and the complexity of their kinetic behavior have been ignored. Nevertheless, several imaginative studies, based on certain disease processes in which sulfohydrolase deficiency was observed (Section II,A,1,e), have given the enzymes a new significance. Indeed, a new classification of the enzymes, based on activity toward naturally occurring sulfate esters, may soon be necessary.

a. Arylsulfohydrolase A. Kinetic studies that followed the partial separation of the A and B enzymes of ox liver [18] revealed that the former enzyme reacted with substrate (nitrocatechol sulfate) in an abnormal manner that manifested itself in the form of anomalous enzyme concentration–activity curves. Similar behavior was subsequently reported for the human liver enzyme [19] and other studies have since established this as a general feature of A enzymes from other sources, including the brains of various species [20–23]. Various theories, based on enzyme polymerization or impure substrate, were advanced as an explanation but were shown to be untenable by workers in Cardiff [19,24–26]. These studies revealed that the basic anomaly was that the reaction was not of zero-order. Incubation of human enzyme and substrate apparently converted the former to a less active or inactive form which could be subsequently isolated and reactivated by sulfate; it was also markedly affected by other anions. Similar conclusions have since been drawn for pure ox arylsulfohydrolase A [27]. A modified and low activity form of the enzyme was separated from enzyme-substrate mixtures and freed from substrate and reaction products at low temperature. This form was reactivated by sulfate, phosphate, and pyrophosphate, all of which inactivated the native enzyme. Similar effects were noted with 4-nitrocatechol, but to a lesser degree. On the basis of these results, the reaction mechanism proposed by Baum and Dodgson [26] was modified slightly (see Fig. 1). The key step is the formation of F (the inactive form of the enzyme) during incubation with substrate. F appears to possess a second binding site which can accept substrate, products, or other anions. The

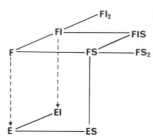

FIG. 1. Proposed scheme for the interconversion of active and inactive forms of mammalian arylsulfohydrolase A in the presence of substrate and other components [26,27]. E native enzyme; S, substrate; F, inactive enzymes; and I, SO_4^{2-}, P_i, PP_i, or 4-nitrocatechol. See text for explanation.

formation of FS_2, FI_2, and FIS (where S is the substrate and I is sulfate, phosphate, pyrophosphate, or 4-nitrocatechol) as well as that of ES and EI (where E is the native enzyme) would account for the observed behavior. Other related studies [28] on arylsulfohydrolase A from human kidney have established that formate and certain other monovalent anions can affect the rate of inactivation by substrate, as well as the subsequent reactivation by sulfate.

The pure ox enzyme is a relatively heat stable, proline rich glycoprotein that can exist as a monomer (MW 107,000) at pH 7.5 or as a tetramer (MW 411,000) at pH 5.0, depending on protein concentration and the ionic strength of the medium [29]. In this respect the rat and human enzymes behave similarly and the property has been exploited in the purification of these particular enzymes [30, and unpublished results, these laboratories]. In the presence of 6 mM dithiothreitol and 8M urea at pH 8.1, the ox enzyme dissociated further into units of 27,000 MW, the change being accompanied by irreversible inactivation [31]. It was concluded that the native enzyme was composed essentially of four such units, linked together in pairs by disulfide bridges and hydrogen bonds. These observations point to the need to maintain clearly defined experimental conditions when studying the enzyme. Recent analytical work on the ox liver enzyme has confirmed its glycoprotein nature and has pinpointed the sialic acid content of the molecule [32], a characteristic common to other lysosomal enzymes [33]. Enzyme that had been desialylated with neuraminidase retained its catalytic properties and was in no way identifiable with arylsulfohydrolase B (Section II,A,1,b), a possibility suggested by other workers [34].

Much attention has been devoted to the active site of aryl-sulfohydrolase A and to the interaction of the enzyme with group-specific reagents [35,36]. However, very little emerged from the studies except that the participation of a histidyl group (and possibly a tyrosyl group) appeared to be essential for enzyme activity. This conclusion is supported in part by studies on the nonenzymatic hydrolysis of 2-[4(5)-imidazolyl]phenyl sulfate, which suggested a possible involvement of the imidazole moiety [37].

The most significant recent discovery in relation to arylsulfohydrolase A has been the finding that it is actually a cerebroside sulfohydrolase (EC 3.1.6.8) [38]. This possibility had previously been suspected [39–43], but its acceptance was not immediate, as it was difficult to understand why an arylsulfohydrolase should be active toward cerebroside sulfate esters (IV). Moreover, the initial studies had indicated that aryl-

(IV)

sulfohydrolase A was not active toward such esters unless a heat stable factor was also present. Even then, the activity of the complete system was unfortunately very low and it proved difficult to make appropriate kinetic studies. However, it was subsequently shown [38] that the cerebroside sulfohydrolase activity of liver enzyme A could readily be measured, provided that the reaction mixture contained sodium taurodeoxycholate and manganese chloride. Under these circumstances, mixed micelles containing enzyme and cerebroside sulfate were formed and enzyme action proceeded at a reasonable rate. The substrate specificity of arylsulfohydrolase A is therefore unusual in that it will hydrolyze nitrocatechol sulfate (and to some extent a number of simpler arylsulfates [25,44]) as well as the D-galactose 3-O-sulfate residues present in cerebroside sulfates. The enzyme had previously been shown to be inactive toward simple carbohydrate sulfate esters [41], but a recent report [45] that ascorbic acid 2-sulfate (see Section II,C,4) can serve as a substrate suggests that the specificity question requires further study.

The recognition that deficiency of arylsulfohydrolase A is responsible for the primary lesion in human metachromatic leukodystrophy (Section II,A,1,e) has focused attention on the isolation and characterization of the enzyme from various human tissues [46–49]. Some conflicting observations have emerged from these studies, including the claim that two catalytically similar and interconvertible forms of the enzyme can be separated from human urine. The chicken brain enzyme also seems to be somewhat anomalous in that some of its properties were typical of an A enzyme, but others were typical of a B enzyme [23].

b. Arylsulfohydrolase B. Relatively few detailed studies have been made on the B enzyme, probably because of the difficulties surrounding its purification on an adequate scale. The precise relationship between the enzymes A and B has occupied the attention of some investigators and even the possibility of their interconversion has been considered. Suggestions that conversion of A into B might result from interaction with pyrophosphate and other ions [26] and the concept that B is a desialylated form of A [34] have been discounted. Both enzymes are inhibited by phosphate, but only enzyme A exhibits the remarkable kinetic behavior described previously. However, some features of aryl-sulfohydrolases B are unusual [50]. For example, the partially purified human liver enzyme was markedly affected by buffer concentration while pH–activity and substrate concentration–activity curves showed anomalies. These effects were attributed to the tendency of the enzyme to interact with inert protein and, in support of this, further purification abolished the effects. A somewhat similar behavior was exhibited by partially purified ox B, but was not abolished by further purification. A further difference between the enzymes from the two sources was re-flected in their behavior toward nitrocatechol sulfate when chloride ions were present. The ox B enzyme was activated, whereas human B was inhibited. The activity of both ox and human enzymes toward *p*-ni-trophenyl sulfate was quite low. However, activity increased sharply in the presence of Cl^- [51,52]. Because Cl^- did not affect the K_m of the enzyme [51] and because the binding of substrate and its subsequent catalytic hydrolysis were independent processes [50], it seems reason-able to conclude that the ion exerted its effects on the latter process. By analogy with known effects of Cl^- on other hydrolytic enzymes [e.g., 53], conformational changes in either the enzyme or enzyme–substrate complex may be effected by the ion. Meanwhile, *p*-nitrophenyl sulfate continues to be employed as a substrate and Cl^- is often a component of the incubation mixtures. The implications of this have rarely been con-sidered by investigators.

Studies on arylsulfohydrolases B are further complicated by its exis-tence in multiple forms. For example, two forms (β_α and β_β) of ox liver enzyme have been separated and purified [54]. Both exhibited the same kinetics, were of similar size (MW 25,000), and differed only in charge. Under conditions of low ionic strength they aggregated to give mixtures of polymers of up to 300,000 MW, presumably as a result of electro-static interaction. Early claims that ox and rabbit cornea [55] and ox brain [56] preparations contained several subfractions may also reflect this property. Three enzyme B fractions have been separated from human placenta but, in contrast to the ox liver subfractions, they showed individual differences in substrate affinity and molecular weight [57]. Investigations on arylsulfohydrolase B activity in ox brain [22,58,59] culminated in the claim that seven subfractions could be detected by subjecting an apparently homogeneous enzyme to ion-exchange chromatography [60]. The possibility that these were artifacts of the purification procedure (which involved autolysis) was apparently eliminated.

The reason for the coexistence in lysosomes of two arylsulfohydrolases that have much the same specificities toward arylsulfate esters has always been puzzling. The observation that ox A and B were not extracted at the same rate from liver lysosomes [61] suggested that the two enzymes were present in different parts of the lysosome or were in different lysosomes or even in different cells. There is no doubt, however, that the last possibility can be disregarded. Both enzymes do coexist in the same type of cell, but the relative amounts of each varies considerably from one cell type to another [62]. For example, in rat liver the ratio of activities of enzyme B to enzyme A is much higher in nonparenchymal cells than in parenchymal cells. The precise significance of this is uncertain, but it suggests that the two enzymes vary in individual importance in maintaining the integrity of different types of cell.

A logical explanation for the coexistence of the A and B enzymes in lysosomes only really becomes possible if one assumes that neither is concerned with the degradation of arylsulfate esters. Under certain experimentally contrived circumstances, it is possible to achieve an appreciable and rapid hydrolysis of nitrocatechol sulfate by intact rat liver [63]. Under such circumstances the ester is obviously entering the lysosomes. However, under normal conditions there is no apparent reason why any simple arylsulfate should enter the lysosomes for the purpose of undergoing degradation and we must accept that the principal function of these organelles is concerned with the intracellular degradation of macromolecules. If this is so then we must assume that, in common with arylsulfohydrolase A, the B enzyme is concerned with the degradation of compounds other than simple arylsulfates. Sulfated glycosaminoglycans or sulfated glycoproteins would be likely contenders for the roles of natural substrates. This possibility has already been raised for the arylsulfohydrolase B of brain tissues [64]. However, none of the seven subfractions of ox brain B showed any ability to liberate sulfate from polymeric chondroitin 4-sulfate [60].

c. Arylsulfohydrolase C and Estrogen Sulfohydrolase. Mention has already been made (Section II,A,1) of the presence of arylsulfohydrolase C in the microsomal fraction of mammalian livers. In reporting on this enzyme the investigators [10,11] advised caution in employing *p*-nitrophenyl sulfate as an assay substrate because liberated p-nitrophenol could not be quantitatively recovered from liver homogenates. A further problem associated with this substrate stems from the fact that both aryl-sulfohydrolases A and B are also active toward it, particularly the latter enzyme, when Cl⁻ is present. Few investigators have paid any attention to these points and have continued to employ the substrate. Similar conditions do not apply to p-acetylphenyl sulfate and the most recent work with this substrate has established precise conditions whereby aryl-sulfohydrolase C can be specifically measured in the presence of enzymes A and B [65].

Early work on arylsulfohydrolase C in rat, human, and ox livers [15,66,67] revealed the enzyme to be insoluble and particulate and resistant to the then known procedures for the extraction of membrane-

(V)

bound proteins. A pseudosolubilization, accompanied by activation, was achieved with rat liver enzyme by treating microsomal preparations with nonionic and cationic surface-active agents. However, both effects were shown to be a function of the presence of detergent micelles, within which particulate material became dispersed, and insolubility returned following removal of detergent. Similar results were obtained with anionic detergents except that apparent solubilization was accompanied by strong competitive inhibition. An enzyme preparation that appeared to be truly soluble was obtained by treating a washed microsomal fraction with crude pancreatic lipase in the presence of a nonionic detergent (Lissapol-N). However, recoveries were very low.

The human and ox liver enzymes have resisted all attempts to obtain them in a truly soluble form, although there has been a claim that two soluble enzyme C subfractions have been obtained from an acetone powder of human placenta by treatment with trypsin or Triton X-100 [68].

The comparative stability of rat arylsulfohydrolase C, its ease of assay, and its specific localization in the membrane of the endoplasmic reticulum of the rat liver cell has led to the suggestion that it could serve as a useful marker for this subcellular component [65,69]. Similar conclusions have been reached for the membranes of rat spleen and kidney (unpublished results, these laboratories) and may well apply to other tissues and species.

The discovery that rat liver, kidney, and placenta possessed sulfohydrolase activity toward a naturally occurring arylsulfate ester (V), estrone sulfate [70], and that activity was localized in the microsomal fraction [71] has led several workers to consider the possibility that arylsulfohydrolase C was responsible. However, kinetic studies on the situation in human placenta [72] and rat kidney [73] did not support this view. A detailed study of the possibility was made recently in Cardiff [74] following studies on the metabolism of estrone sulfate in whole animals and in isolated, perfused livers [75]. The results strongly supported the concept that rat liver arylsulfohydrolase C and estrogen sulfohydrolase were one and the same enzyme. Thus enzyme activities toward estrone sulfate and p-acetylphenyl sulfate were both very low in fetal liver and increased at the same rate after birth. Similar parallel behavior was also observed in heat inactivation, mixed substrate, and competitive inhibition experiments on liver microsomal preparations. The results of work still to be published have further shown that the two activities have the same ultrastructural localization in microsomal subfractions, are elevated to the same extent in the livers of animals receiving phenobarbitone, and are dependent upon the lipid component of the membrane. A considerable weight of evidence has therefore accumulated which supports the concept that, in rat liver at least, both activities are due to one enzyme and the possibility of defining a physiological role for arylsulfohydrolase C now becomes more feasible.

All the arylsulfohydrolases C that have been studied have alkaline pH optima and are unaffected by sulfate and phosphate. Some species differences have been noted; for example, K^+ and Na^+ have little effect on the human and rat enzymes but inhibit that from ox. Further comment on the relationship between arylsulfohydrolase C and steroid sulfohydrolases are made in Section II,B,2,a.

d. Distribution of Mammalian Arylsulfohydrolases. A catalog of references to the occurrence and tissue distribution of these enzymes would probably have occupied the greater part of this chapter and would almost certainly have established that every mammalian tissue that has been examined contained at least one of the enzymes. However, many of the observations are difficult to interpret and are of limited value. Enzyme assays have been performed without regard for the abnormal kinetics of enzymes A and B, the presence of endogenous inhibitors and activators, or the choice of substrates. Attention has repeatedly been drawn to these points [4,5,17], and much effort has been devoted in our laboratories to devising specific methods for the assay of each enzyme.

One such method was developed for the independent assay of the A and B enzymes in human urine [76], and this was later adapted for use with human tissues [21]. However, the method has sometimes been misused and applied to systems other than human although it is applicable only to human sources of the enzymes. A method that permits an approximate determination of the individual A and B activities of rat livers has also been developed [30], but the enzymes have first to be separated by gel-filtration chromatography. Chromatographic separation has also formed the basis of a method for the human serum enzymes [77]. A fluorimetric assay employing 4-methylumbelliferone sulfate as substrate [78] is now in regular use. Improvements in the method of preparation of the substrate seem to have overcome the earlier limitations of its use [79,80] and the greater sensitivity of a fluorimetric method is obviously advantageous for measuring low activities [80]. However, as with earlier work using nitrocatechol sulfate, failure to appreciate the abnormal behavior of the enzymes and the species differences is again leading to some confusion. A thorough investigation of the activity of the various enzymes toward this substrate is urgently needed and would help to clarify the significance of some of the recent studies in which it has been used [e.g., 81,82].

For details of some of the tissue distribution studies that have been made, the reader is referred to the reviews cited previously (Section II,A,1). However, some observations require brief comment from the point of view of the physiological role of the enzyme. Thus, histochemical [83] and enzymological [84] studies indicated that Type II arylsulfohydrolases resembling A and B were present in the acrosomes of spermatozoa and their involvement in the fertilization process has been suggested. Type II enzymes have also been noted in cells from amniotic fluid and attempts have been made to correlate activity with the duration of pregnancy [85] and to use enzyme activity as a diagnostic index of fetal abnormality [86]. Studies on the development of aryl-sulfohydrolase activity in rat and human fetal tissues and the changes in levels that occur subsequently may be of some significance in this respect. The levels of enzyme C [87] gradually increase during the prenatal period and continue to do so for some time after birth. Increases in A and B levels have also been claimed [88], as well as decreases in old animals [89,90]. However, other lysosomal hydrolases behave similarly and the effect is therefore nonspecific.

Most distribution studies have relied on *in vitro* enzyme assays on tissues or extracts, but the need for more precise information has been appreciated by investigators who have attempted to achieve this at the light and electron microscope level. However, the problem of detecting individual arylsulfohydrolases also applies to the field of histochemistry. Among the many studies made there are claims that individual enzymes have been detected [e.g., 91,92] and in one interesting case [93], B was claimed to be present in mitochondria of the proximal convoluted tubules of rat kidney.

Finally, some differences in the distribution of the enzymes between sexes and between species are not easy to understand at the present time. For example, in rats, greatest enzyme C activity is found in the liver, with the levels of activity in males being approximately twice that of females [89,94]. In contrast, guinea pig [89] and some marsupials [95,96] have enzyme levels that are barely detectable or are undetectable in many tissues, presumably indicating that the enzyme is not of major importance to the well being of these particular species.

e. Physiological and Pathological Aspects of Mammalian Arylsulfohydrolases. Undoubtedly, one of the significant landmarks in the progress of our knowledge of the arylsulfohydrolases was the finding that enzyme A was deficient or absent in the tissues and body fluids of patients with metachromatic leukodystrophy [21,97,98]. The disease process is characterized by accumulation of cerebroside sulfates (sulfatides) in brain and other tissues. As noted in Section II,A,1,a, it is now established that arylsulfohydrolase A is a cerebroside sulfohydrolase [38], the activity of which (*in vitro* at least) appears to depend on the presentation of substrate in a micellar form.

The preparation of a monospecific antibody to the human enzyme led to immunological studies that showed that in metachromatic leukodystrophy the enzyme protein was present but was inactive [99]. Surprisingly, the antibody enhanced the activity of the normal enzyme but, of course, did not restore activity to the abnormal protein. Cultured skin fibroblasts from patients with the disease were used as *in vitro* models and were shown to be deficient in enzyme A and to be able to accumulate cerebroside sulfate [100,101]. Supplementation of the culture medium with arylsulfohydrolase A was followed by uptake of the enzyme into the fibroblasts and a return of ability to degrade ingested sulfatide [102]. Another interesting finding was that the naturally occurring factor that early work [43] established as necessary for sulfatide hydrolysis was an activator of enzyme A and could be purified from human tissues [103]. The therapeutic possibilities arising from these various findings are fairly obvious.

Some confusion surrounds the question of the association between sulfohydrolase levels and the onset of the disease since absolute determinations of enzyme activity for the various subgroups (late infantile, juvenile, and adult) have been made under different conditions [104].

The situation may be more complex than suspected, since some forms of the disease are characterized by deficiency of enzymes B and C as well as A [105,106]. These so called "multiple sulfohydrolase deficiency" diseases are represented by accumulation of sulfated glycosaminoglycan and cholesterol sulfate, as well as sulfatide. A recent report suggests that prenatal detection of these and other diseases can be achieved by measuring enzyme A levels in amniotic fluid and cells cultured from it [86,107].

The low enzyme B activity noted in patients where glycosaminoglycan accumulation occurs (e.g., multiple sulfohydrolase deficiency diseases) supports the possibility that this enzyme plays some part in glycosaminoglycan metabolism. Some deficiency of liver arylsulfohydrolase A, together with other lysosomal enzymes, has also been reported in various mucopolysaccharidoses (see Sections II,C,1,c and II,C,8,b) including Hurler, Hunter, and Sanfilippo B syndromes [108,109], but in some of these cases increases in the levels of enzyme B were observed, a finding that is at variance with the possibility that B participates in glycosaminoglycan degradation.

The literature contains numerous references to arylsulfohydrolase levels in various other pathological conditions including cancer, vitamin A deficiency, fatty liver, kwashiorkor, alcoholism, and multiple sclerosis [e.g., see 110–113]. Most of the observations are of doubtful value for reasons explained earlier but, in any case, where changes were observed, they were accompanied by similar changes in the activities of other lysosomal enzymes and are therefore nonspecific.

Pathological variations in the levels of arylsulfohydrolase C, specifically, have rarely been determined, although this is a pity if the enzyme is indeed an estrogen sulfohydrolase. However, "placental sulfatase (sulfohydrolase) deficiency" does sometimes occur in pregnancy and is characterized by a low urinary excretion of estrogens, a circumstance indicating that the well being of the fetus is threatened [114]. The part played by arylsulfohydrolase C in this tissue may then be concerned with interconversion of physiologically inactive estrogen sulfates and physiologically active free estrogens.

It will be apparent that, in spite of many experimental observations, it is still not possible to define precisely the physiological roles of the mammalian arylsulfohydrolases. The probability exists that the A enzyme is concerned with the turnover of tissue sulfatides. The ability of the enzyme to degrade simple aryl sulfates would then be seen as a coincidental catalytic feature of its active site. In the case of arylsulfohydrolase B, the function is still unknown although it is tempting to assume that it is concerned with the turnover of undefined sulfated glycosaminoglycan or glycoprotein entities (see Section II,C). The fact that the multiple forms of B exist would then be interpreted as indicating different sulfohydrolases acting on different polymers, and the ability of each enzyme to hydrolyze nitrocatechol sulfate would again reflect a catalytic site property without physiological significance. One possible type of polymer candidate for the role of substrate for enzyme B would be polypeptides (e.g., fibrinogen) containing sulfated L-tyrosine residues. However, such residues are not desulfated *in vivo* and do not serve as substrates for mammalian arylsulfohydrolases [see 44,115]. It is worth noting that neither enzyme A nor B is significantly active toward other naturally occurring aryl sulfates such as serotonin *O*-sulfate [116] and the 3- and 4-*O*-sulfates of 3,4-dihydroxyphenylethylamine (dopamine) (unpublished results, these laboratories). This gives further point to the view that the prime function of the two enzymes, as with other lysosomal hydrolases, is concerned with the degradation of relatively complex molecules rather than with simple, readily excretable compounds such as arylsulfates.

Finally there is the question of arylsulfohydrolase C. The possibility exists that it is concerned with maintaining a balance between active estrogen and inactive (and possibly transport form) estrogen sulfates. In accord with this is the fact that the estrogens are powerful competitive inhibitors of the sulfohydrolase [74]. Attempts to substantiate this type of role for the enzyme are currently in progress in these laboratories.

2. MICROBIAL ARYLSULFOHYDROLASES

A variety of different arylsulfohydrolases occur in microorganisms. Enzymes that are somewhat similar to those of Type I in mammals seem to occur frequently, but others resemble either Type II enzymes or appear to be intermediate between the two types. Formation of some of these enzymes depends on the sulfur state of the organism.

a. Fungal Arylsulfohydrolases. The presence of arylsulfohydrolase in *Aspergillus oryzae* has long been known [8] and commercial preparations of Taka diastase from this organism have frequently served as crude, but ready, sources of the enzyme. Several workers have studied the activity of this preparation in some detail and have noted general similarity to Type I enzymes in terms of specificity and behavior toward inhibitors. Subsequently, three isoenzymes, designated I, II, and III on the basis of electrophoretic mobility were detected [117–119], and since then several reports on their properties have appeared [120–122]. Isoenzyme II was shown [123] to have a molecular weight of 65,000 [but see 120], to be rich in aspartic and glutamic acids, and to be inhibited by phosphate, sulfite, and fluoride. The effects of pH on K_m and V_{max} suggested that histidine was involved in the catalytic center and other evidence pointed to a functional tyrosine group. Collectively, these results suggest a resemblance to ox liver arylsulfohydrolase A. However, activity was not inhibited by sulfate, chloride, or fluoride and in this respect was unlike that of either Type I or Type II arylsulfohydrolases.

The isolation of isoenzyme I (MW 100,000) has also been claimed [123]. The K_m was similar to that of isoenzyme II but the pH optimum (6.8) was higher (4.8 for II). The various isoenzyme forms were not related to each other by subunit association. Two of them were secreted into the culture medium while the other remained in the mycelium [121]. The production of all three enzymes was repressed by sulfate, sulfite, methionine, cystine, and cysteine [121] and their appearance seemed to relate to circumstances of sulfur insufficiency. Arylsulfohydrolase activity has been observed in other fungi [124]. In *Aspergillus nidulans* activity was repressed by sulfite, sulfate, methionine, and cysteine and two molecular species were detected which were similar in many respects, but differed in their relative substrate affinities, electrophoretic mobilities, and heat stabilities [125]. As with *A. oryzae* enzymes, the ones from *A. nidulans* were inhibited by phosphate, sulfate, and cyanide, behavior that was intermediate to that of enzymes of Types I and II. Some *Neurospora* species can produce arylsulfohydrolases [126,127] and attempts have been made to locate the enzyme activity present in the conidia and mycelia of *Neurospora crassa* [128]. Within the mycelium most activity was associated with particulate material and especially with a lysosomelike particle [129]. No detailed information has emerged concerning the general properties of this enzyme, but it has been studied by Metzenberg's group during investigations on the genetic control of the synthesis of enzymes involved in sulfate assimilation [130–133].

b. Bacterial Arylsulfohydrolases. Many bacterial sources of arylsulfohydrolases are now known and some of them have been examined in some detail [for accounts of the early work see 3,4,17].

The enzyme produced by *Aerobacter aerogenes* [134,135] was repressed by sulfate and cysteine but not by methionine. It was purified to homogeneity (MW 41,000) and shown to have some unusual features including an anomalous temperature-activity profile. Phosphate inhibited the enzyme but sulfate did not; cyanide inhibited irreversibly but only in the presence of substrate [136]. It is therefore difficult to relate its properties to any of the mammalian enzymes.

The arylsulfohydrolase *Proteus vulgaris* [137] shows some resemblance to the Type II enzymes of mammalian tissues, particularly aryl-sulfohydrolases B. The enzyme was inhibited by sulfate and phosphate and its activity toward p-nitrophenyl sulfate was activated by Cl$^-$. Early studies on *Proteus rettgeri* [138] suggested the presence of two aryl-sulfohydrolases which did not appear simultaneously during growth of the organism. Multiple electrophoretically distinct forms of enzyme were detected [139], but further work appears to be necessary before the significance of this finding can be evaluated.

Pseudomonads also produce arylsulfohydrolases that are repressed by sulfate, sulfite, and cysteine but not by methionine. Activity present in *Pseudomonas aeruginosa* [140] has been shown to be due to two aryl-sulfohydrolases, both similar in properties to Type I mammalian enzymes [141]. Both enzymes were active toward nitrocatechol sulfate and p-nitrophenyl sulfate and were inhibited by cyanide. Phosphate and sulfate activated one enzyme but only toward p-nitrophenyl sulfate. The optimum pH of each enzyme varied with increasing substrate concentration, a phenomenon thought to represent a conformational change in the enzyme and previously noted in Cardiff for the arylsulfohydrolase of the snail *Helix pomatia* [142]. *Pseudomonas* C12B, an organism isolated from soil near a sewage outfall, produced considerable aryl-sulfohydrolase activity when grown in the presence of methionine as sole sulfur source [143], but the enzyme has not yet been purified.

In only one instance has a detailed study been made of the mechanism of action of a bacterial arylsulfohydrolase. This study was made in our laboratories with a partially purified enzyme from *Alcaligenes metalcaligenes* [144,145]. The enzyme possessed Type I characteristics, being inhibited by cyanide but not by sulfate or phosphate. The introduction of electrophilic substituents into the benzene ring of the basic arylsulfate substrate structure enhanced enzyme activity, while nucleophilic substituents gave the reverse effect. In both cases there was a good relationship between the substitution constants (Hammet values) and K_m and V_{max} for the enzyme. From studies in which the variation of K_m and V_{max} with pH was measured, curves were obtained that indicated the presence of substrate-binding groups with pK of 8.2 and 9.4, respectively. On the basis of these and other results a reaction mechanism was proposed. Inhibition of the enzyme by carbonyl reagents was of a unique anticompetitive type [146].

c. General Comments. At the present time there is little to be gained from comparing the properties of different microbial arylsulfohydrolases. Moreover, the significance of the multiple forms of the enzyme that have often been noted is far from clear. Unfortunately, little is known about the relative specificities of the different forms, and activities have usually been measured with nitrocatechol sulfate or p-nitrophenyl sulfate, both of which appear to be excellent substrates for the enzymes. It is therefore impossible to say whether the production of multiple forms is related to the need for enzymes that, collectively, would allow the microorganism to deal with any type of arylsulfate that it might encounter. There are already indications that the specificities of individual microbial arylsulfohydrolases are limited. For example, the enzyme from *A. aerogenes* cannot use α-naphthyl sulfate or phenolphthalein disulfate as substrates [136]. In detailed specificity studies [145] on the *A. metalcaligenes* enzyme, the V_{max} values obtained (relative to 1.0 for phenylsulfate) with 21 different substrates varied from 0.1 for *m*-aminophenyl sulfate to 16.7 for p-acetylphenyl sulfate.

Further work carried out in Cardiff (unpublished results) is pertinent to this problem of specificity. The microbial floras present in acidic (podzol, pH 4.0), basic (rendzina, pH 8.4), and neutral (brown earth, pH 6.2) soils were tested for ability to degrade [35S]-labeled aryl sulfate esters. Nitrocatechol [35S]sulfate and p-nitrophenyl [35S]sulfate were hydrolyzed at a rapid rate when added to soils and no such hydrolysis occurred with sterilized soils (see also Section II,C,2,d). With phenyl [35S]sulfate the rate of hydrolysis was very much lower, while with L-tyrosine O-sulfate and L-tyrosylglycine O-sulfate, hydrolysis commenced only after a considerable lag period. The implications of these findings, presumably, are that microorganisms are present in soils that are immediately capable of accomplishing the hydrolysis of some aryl sulfate esters but not others. Additional arylsulfatases must first be "induced" (either by the same or different microorganisms), in order to deal with these other substrates. Hence, a possible explanation for multiple forms of arylsulfohydrolase may be related to the need of the microorganism to deal with all types of aryl sulfate ester that they continually encounter and that enter soils and waters as components of sewage and animal excreta. There is now a great need for detailed studies on this general problem.

Studies on the production of arylsulfohydrolase activity in microorganisms provide strong evidence that the appearance of the enzymes in fungi and bacteria is a reflection of the need to acquire sulfate for growth purposes. Expressed in the most simple terms, in times of sulfur sufficiency it seems probable that the presence of sulfate or sulfur containing intermediates on the pathway to cysteine are sufficient to repress bacterial arylsulfohydrolases. There are good reasons for supposing that cysteine itself is the corepressor [e.g., see 134,135]. In these microorganisms, the conversion of methionine to cysteine appears not to be possible and hence arylsulfohydrolase is not repressed if the sulfur for growth is supplied only by methionine. The situation in fungi is different in that the ability of many of these organisms to form cysteine from methionine means that the latter compound, although probably not the actual corepressor [126], appears to serve in that capacity. In times of sulfur insufficiency derepression occurs and not only arylsulfohydrolases appear, but also a number of other enzymes (including permeases) that are concerned with securing supplies of the element [131].

3. OTHER ARYLSULFOHYDROLASES

In Section II,A,1, the claim was made that the arylsulfohydrolases were the most widely distributed of all sulfohydrolases. In addition to mammalian and microbial sources, activity has been noted in birds, reptiles, amphibians, fish, insects, a host of lower organisms of both marine and terrestial origins, and in some plants [e.g., 95,142,147–153]. In fact, probably all members of the animal kingdom possess arylsulfohydrolases of some kind. Studies on enzymes from these other sources have been few and have generally suffered from the experimental limitations mentioned previously in relation to other arylsulfohydrolases. Enzymes that bear some resemblance to the Type I or Type II mammalian enzymes have been detected, but the majority appear to have properties that are intermediate between these two extremes.

Early reports of arylsulfohydrolase activity in various gastropods, pelecypods, cephalopods, arthropods, and amphineurates common to tropical waters [149,154] stimulated other searches for the enzyme in marine organisms. Considerable activity, accompanied by an equally impressive β-glucuronidase activity, was detected in a number of mollusks [150]. A procedure was developed for the preparation of highly active and stable enzyme concentrates from the viscera of the limpet *Patella vulgata* and the periwinkle *Littorina littorea* and some general properties of the enzymes were described [155]. The collective studies suggested that the enzymes had properties similar to those of the Type II mammalian enzymes. These early studies were not extended, but they did lead to the commercial production of an enzyme concentrate (containing arylsulfohydrolase and β-glucuronidase) from *Patella* that has been employed for the analysis of urinary steroids. Multiple enzyme forms have been claimed to be present in *Patella* preparations [156] and also in extracts of the liver of *Charonia lampas* [157].

High arylsulfohydrolase activity is also present in terrestial mollusks and detailed studies have been made [142,158] on a partially purified preparation from the digestive juice of the snail *H. pomatia*. The enzyme had curious properties that distinguished it from other arylsulfohydrolases. In particular, it showed a marked instability that was considerably enhanced in the presence of the substrate, nitrocatechol sulfate. Even more striking was the marked upward shift in optimum pH that occurred as the concentration of substrate was increased. This type of behavior is not unknown among those hydrolytic enzymes that display some ability to act as transferases. However, the *Helix* enzyme failed to exhibit transfer activity to a wide range series of acceptors and attempts to explain the behavior on the basis of other known enzymological phenomena were unsuccessful. Although the enzyme behaved like a typical Type II enzyme, it did not exhibit either the anomalous kinetic behavior of mammalian arylsulfohydrolase A or the enhanced activity toward *p*-nitrophenyl sulfate shown by arylsulfohydrolase B in the presence of chloride.

It is difficult to assess the physiological significance of the extremely high arylsulfohydrolase activity present in invertebrate organisms. The major sources of the enzyme in mollusks (and insects) seem to be the digestive organs or digestive juices and one might then be tempted to assume that the enzyme must be associated in some way with digestion. If the enzyme is indeed concerned with the desulfation of aryl sulfates it is difficult to see why such high levels of enzyme are required or in what type of food would significant quantities of aryl sulfate esters exist. Why indeed should the organisms want to desulfate such esters in the first place? They would surely have no nutritional significance. However, considerable enzyme activity is present in other tissues, including the hypobranchial glands of many mollusks. Here it seems possible that it may be concerned with the production of the so-called "Tyrian Purple" dyes [159]. A wide variety of colored pigments are produced by a variety of mollusks and have figured significantly in the development of civilizations flourishing on the shores of the Mediterranean and in other parts of the world. These pigments are produced from a number of different sulfate ester precursors. Hydrolysis of the esters by molluscan arylsulfohydrolase is followed by oxidation of the liberated phenol to give complex colored quinoid compounds. Such compounds probably serve the mollusks as warning or defensive agents and the arylsulfohydrolase clearly has an important role to play.

In other organisms similar problems concerning function face the investigator. For example, enzyme activity in the southern armyworm (*Prodenia eridania*) changed during the life cycle of the organism, reaching a maximum during the molt. This was taken to represent its possible participation in the regulation of the activity of insect steroid hormones [160].

The present situation is clearly highly unsatisfactory. Although commercial sources of the enzymes of *Patella* and *Helix* are of some importance in the identification and measurement of aryl sulfates in urines, it is particularly frustrating to have a plethora of enzymes apparently searching for a physiological function. New thoughts and novel approaches to the problem would now be most timely.

4. GENERAL COMMENTS

The principal problems that now require to be resolved in relation to the arylsulfohydrolases concern substrate specificity and hence, physiological function. Until this is done, there seems little point in finding other sources of the enzymes and purifying them in order to collect more kinetic and similar data. In our view it seems possible that arylsulfohydrolases may perhaps be resolved into five major types: (1) microbial enzymes required for the mobilization of sulfur, (2) mammalian enzymes present in insoluble form in smooth endoplasmic reticulum and concerned with the metabolic regulation of estrogen levels, (3) enzymes enclosed in lysosomal particles and concerned with the turnover of macromolecules in which sulfate is not necessarily esterified to phenolic–OH groups, (4) enzymes present in the digestive organs and juices of invertebrates and which are involved in the digestion of food containing sulfated molecules other than aryl sulfates, and (5) enzymes present in certain mollusks that are specifically concerned with the production of protective pigments. Possibly other highly specialized functions of this sort may eventually emerge. When taken as a whole, these views represent a marked departure from those previously held about the enzymes, but certainly some new thinking is required if the large gap in our understanding of function is ever to be bridged.

A final thought about the enzymes concerns their mode of action at molecular level. As mentioned earlier (Section I), the available evidence indicates that they are involved in the scission of the O–S bond of the C–O–S linkage. There are further indications that histidine residues may be important to the activity of some of them. Nevertheless, some members of the group operate at a relatively acid pH (e.g., mammalian A and B and molluscan enzymes, pH range 5–6) while others operate at a relatively alkaline pH (e.g., arylsulfohydrolases C and the *Alcaligenes* enzyme, pH range 8–9). It is hoped that future studies will reveal whether arylsulfohydrolases operating optimally at a pH difference of this magnitude are doing so by the same type of reaction mechanism.

B. Alkylsulfohydrolases

1. SIMPLE ALKYLSULFOHYDROLASES

Enzymatic desulfation of the sulfate esters of simple alcohols by hydrolytic fission of the O–S bond has not been detected in mammals.

This is somewhat unexpected since mammalian tissues are known to be capable of the sulfoconjugation of alcohols such as methanol, ethanol, propan-1-ol, and butan-1-ol [161,162], although such sulfoconjugates have not been detected in the tissues under normal circumstances. The livers of several lower animal organisms such as toad, frog, and carp also possess the ability to sulfate a variety of aliphatic alcohols [163] and the bile salts of many of these animals possess a sulfated aliphatic alkyl side chain attached to a steroid nucleus. In these species, also, there has been no indication of the presence of alkylsulfohydrolase activity. Isopropyl sulfate has been found in the hen's egg [164], but alkylsulfohydrolase activity cannot be detected there.

An enzyme that can liberate sulfate from L-serine O-sulfate (EC 4.3.1.10; L-serine-O-sulfate ammonia-lyase) was discovered some years ago by workers in our laboratories [165]. The enzyme was widely distributed in mammalian tissues but was not a true sulfohydrolase. Pyruvic acid, ammonia, and sulfate were released simultaneously in equimolar amounts as a result of an $\alpha\beta$-elimination reaction [166]. A similar, but nonenzymatic degradation of L-serine O-sulfate could be catalyzed by pyridoxal phosphate in the presence of metal ions [167], but the enzymatic reaction did not require these cofactors. The rate determining step in the enzymatic degradation was the removal, probably by a histidine residue, of the hydrogen attached to the α-carbon of the substrate. The resulting carbonium ion then underwent an electronic rearrangement with the expulsion of the β-substituent (sulfate); aminoacrylic acid was produced and this hydrolyzed spontaneously to pyruvate and ammonia [168,169]. Aspartate aminotransferase [170] and alanine aminotransferase [171] can also degrade L-serine O-sulfate, but both require pyridoxal phosphate as cofactor.

Whole animal studies on the metabolism of the surfactant, sodium dodecyl sulfate, indicated the possible presence of a sulfohydrolase active toward this ester [172]. However, it was subsequently shown that butyric acid 4-sulfate was first formed as a result of the ω- and β-oxidation of the alkyl chain and sulfate was then liberated from this compound as a result of a pH-dependent, nonenzymatic intramolecular rearrangement. No sulfohydrolase activity toward butyric acid 4-sulfate could be detected in rat tissues.

In contrast to the situation in the animal kingdom, alkylsulfohydrolases are formed in a number of microorganisms. Many higher fungi and some bacteria produce an enzyme active toward choline sulfate (VI), an interesting internally compensated sulfate ester that accumulates in considerable amounts in fungal mycelia during times of sulfur sufficiency [173,174]. The ester serves as a readily utilizable store of sulfur that can be called upon when other sources of sulfur are

$$H_3C\diagdown\overset{+}{\underset{\diagup}{N}}\diagup CH_2CH_2OSO_3^-$$
$$H_3C\diagup\diagdown CH_3$$

(VI)

not available [175,176]. The internal compensation of charge allows large quantities to be stored without disturbance of ionic and osmotic balances. The compound can serve as the sole sulfur source for fungal growth but its sulfur becomes available only following the liberation of inorganic sulfate by choline sulfohydrolase (EC 3.1.6.6; choline-sulfate sulfohydrolase). In *A. nidulans* during times of sulfur sufficiency, the sulfohydrolase is repressed, so that during growth on sulfate, sulfite, cysteine, or methionine, the enzyme cannot be detected. However, transfer of the organism to a sulfur deficient medium quickly results in the appearance of the enzyme [177,178]. Cysteine seems to be the corepressor and also serves to control the activity of the enzyme by feedback inhibition. The choline sulfohydrolase of *A. nidulans* has been studied in some detail [179]. The enzyme was temperature sensitive, being inactivated at temperatures greater than 30° and showing instability even at 2°, factors that prevented any appreciable purification of the enzyme. Cysteine inhibited completely at a concentration of 1 mM, and sulfite, phosphate, and cyanide were also good inhibitors. Sulfate and fluoride inhibited only slightly at similar concentrations.

The presence of choline sulfate has also been noted in plants [180], lichens [181], and red algae [182] where again it has been assumed to serve as a sulfur store. However, no attempts appear to have been made to search for choline sulfohydrolase in such organisms.

Until very recently it was believed that bacteria did not produce choline sulfate. It has now been found that *Pseudomonas* C12B (see Section II,A,2,b), when cultured on growth limiting concentrations of sulfate, secretes the ester into the growth medium [183]. Similar activity was noted with other *Pseudomonas* species, but the significance of the process is still far from clear. It seems certain, however, that the ester does not serve as a sulfur store. Attempts to detect a choline sulfohydrolase in *Pseudomonas* C12B were not successful [143]; however, such activity has been noted in other pseudomonads [140,184]. The enzymes present in *P. nitroreducens* and *P. aeruginosa* [140], in contrast to those present in fungi, are induced when the organisms are grown on choline sulfate. The two enzymes are somewhat similar to that of *A. nidulans* in being very temperature sensitive, but there are some differences in their behavior toward inhibitors.

Microorganisms are also capable of producing other sulfohydrolases that are active toward the sulfate esters of long chain primary and secondary alcohols. It has been known for a considerable time that sodium dodecyl sulfate was rapidly removed from sewage by mixed microflora [185] and subsequently an alkylsulfohydrolase, active toward that sulfate ester, was detected in various pseudomonads that had been grown on the ester and that had originated from either sewage [186,187] or soil [188,189].

Further work on one of these organisms, *Pseudomonas* C12B (see Section II,A,2,b), led to the finding that both primary and secondary alkylsulfohydrolases were present in cell extracts [190–192], but some confusion arose concerning the precise number (and the specificities) of enzymes that were present under different growth conditions [193]. This confusion was recently resolved to a considerable extent and the following situation was established for stationary phase cells [194, 195]. During growth on nutrient broth alone, C12B produced one primary alkylsulfohydrolase (designated as P1) active toward dodecyl sulfate (VII) and two secondary alkylsulfohydrolases (designated as S1 and S2) that were active toward decan 5-sulfate (VIII). When hexan 1-sulfate was included in the culture medium the same three enzymes ap-

$$CH_3—(CH_2)_{10}—CH_2—OSO_3^-$$
(VII)

$$CH_3—(CH_2)_4—\underset{\underset{(VIII)}{\overset{|}{OSO_3^-}}}{CH}—(CH_2)_3—CH_3$$

peared, together with an additional primary alkylsulfohydrolase (designated as P2) active toward dodecyl sulfate. When grown on nutrient broth containing Oronite (a commercial detergent consisting of a C_{10}–C_{20} mixture of secondary alkyl sulfates), all four enzymes appeared, together with an additional secondary alkylsulfohydrolase (designated as S3) active toward decan 5-sulfate. A mixture of tetradecan 2-sulfate and tetradecan-2-ol or of hexadecan 2-sulfate and hexadecan-2-ol could substitute for Oronite in the induction of the S3 enzyme. The presence of the various enzymes was readily detected by a novel acrylamide-gel electrophoresis technique [191,196].

The precise specificity of each of these enzymes is not yet defined and work is currently in progress to establish this. The P1 and P2 enzymes will certainly act on primary alkylsulfates of chain length between C_6 and C_{12} while the S1 and S2 enzymes can hydrolyze secondary alkylsulfates with chains containing 7, 8, 9, and 10 carbons. The S3 enzyme can further be distinguished from the S1 and S2 enzymes because of its ability to hydrolyze pentan 3-sulfate. There is no cross-specificity between the primary and secondary members of the group although there are indications that the primary enzymes are inhibited by substrates of the secondary enzymes and vice versa.

These various alkylsulfohydrolases may well be proteins with unusual molecular features in the light of their ability to use powerful anionic detergents such as sodium dodecyl sulfate as substrates. One of the enzymes (S1) present in crude cell extracts can tolerate the presence of 20 mM dodecyl sulfate for 20 hours at room temperature without losing activity (unpublished results, these laboratories). During this time a large proportion of the protein present in the extracts is denatured and precipitated. Clearly, comparative and other studies at the molecular level will be of considerable interest when each enzyme is available in homogeneous form.

In contrast to the arylsulfohydrolase that is produced by *Pseudomonas* C12B during growth on limiting amounts of sulfate [143], the alkylsulfohydrolases were not repressed by sulfate or cysteine [197]. However, the synthesis of the enzymes was terminated or diminished by the presence in media of a number of different carbon sources including some alcohols, some intermediates of the tricarboxylic acid cycle, and by acetate and propionate. These findings suggest catabolite repression and indicate that the enzymes are not concerned with the acquisition of sulfur for growth, but with the removal of the ester sulfate group so that other enzymes present in the organism can proceed with degradation of the carbon chain. Support for this has come from studies on the metabolism of ^{35}S-labeled hexan-1-yl sulfate by the growing organism. The sulfate ester acted as an inducer for the P2 enzyme without the prior need to modify the carbon moiety of the ester [195]. The absence of significant quantities of labeled metabolites in the cells or culture media suggested that desulfation of the ester was a prerequisite for the degradation of the carbon skeleton.

Very recently, alkylsulfohydrolase activity has been detected in another microorganism that was isolated from the same soil as *Pseudomonas* C12B (unpublished results from the Microbiology Department of the University of Georgia and our own laboratories). This organism (identified as a *Comamonas* sp.) produces two secondary alkylsulfohydrolases when grown on nutrient broth. One of the enzymes predominates and neither have electrophoretic mobilities comparable to those of the enzymes of *Pseudomonas* C12B. The enzymes are currently being purified for detailed study.

Finally, mention should be made of a long standing observation that seems not to have been pursued. This concerns the ability of *Bacillus cereus mycoides* to liberate sulfate from the herbicide 3,5-dichlorophenoxyethyl sulfate [198]. This sulfate ester can replace choline sulfate as a sulfur source for growth of some of the higher fungi. This may indicate the possibility that either choline sulfohydrolase or alkylsulfohydrolases analogous to those of *Pseudomonas* C12B are also active toward 3,5-dichlorophenoxyethyl sulfate, or that an entirely different type of alkylsulfohydrolase is involved.

2. STEROID ALKYLSULFOHYDROLASE

a. Dehydroepiandrosterone Sulfohydrolase. One of the most significant mammalian 17-ketosteroid conjugates, dehydroepiandrosterone sulfate (IX), is hydrolyzed by a sulfohydrolase (EC 3.1.6.2; sterol-sulfate sulfohydrolase) present in mammals and certain marine and terrestrial mollusks. The enzyme is sometimes called "steroid sulfatase" or

(IX)

"steroid 3-sulfatase" but this term could equally well be applied to some other enzymes that are currently considered to be distinct and it is therefore somewhat misleading. Dehydroepiandrosterone sulfohydrolase is relatively specific and hydrolyzes only the 3β-sulfates of the 5α and Δ^5 series of steroids [199]. The sulfate esters of epiandrosterone, pregnenolone, and androstene diol can therefore serve as substrates for the enzyme. In contrast, other sulfate esters including androsterone sulfate and the 17-sulfates of the androstane series and the 20-sulfates of the pregnane series cannot.

Several studies on the tissue distribution of the enzyme in various mammals [e.g., 199–203] have established that the reproductive organs, particularly placenta and testes, generally possess greatest activity. In pigs [203], the fetal liver and kidney had greater activity than the placenta, but in other species activity was generally low at birth and then steadily increased with age until a plateau was reached [87,204]. In mammals, the enzyme is present in an insoluble form in the endoplasmic reticulum of the cell, a situation similar to that noted for the aryl-sulfohydrolase C/estrogen sulfohydrolase enzyme (see Section II,A,1,c) in rat liver. The latter enzyme has not been clearly distinguished from the dehydroepiandrosterone sulfohydrolase and no purification of either has been achieved because of their insolubility.

Inhibition studies [205,206] on dehydroepiandrosterone sulfohydrolase from human and rat placentas and testes suggest that the enzyme might be subject to regulatory control through interaction with various free steroids. In turn, the function of the enzyme is considered to be concerned with controlling the availability of free steroids that may be required for further metabolic conversion. This is seen as a means of regulating the production of estrogens and testosterone in these tissues. The molecular structures of those steroids that competitively inhibit the enzyme bear a striking similarity to those of steroids that inhibit rat liver estrogen sulfohydrolase (unpublished results, these laboratories). This may indicate some characteristic feature that is common to both enzymes or may even point to the possibility that a single enzyme is responsible for the sulfohydrolase activity toward arylsulfates, estrogen sulfates, and dehydroepiandrosterone sulfate. However, studies on enzyme activities in human placenta [72] have generally been taken to indicate that this is not so.

The presence of dehydroepiandrosterone sulfohydrolase in digestive organs of the limpet *P. vulgata* [67], the snail *H. pomatia* [207], and several other mollusks [148,208] is puzzling. Various sterols [209] and steroids [210,211] have been reported as present in molluscan tissues, but their significance is not understood.

In contrast to the mammalian enzymes, the molluscan enzymes are soluble. The properties of the mammalian and molluscan enzymes differ in other respects as well; for example, the former enzymes have high pH optima (7–8) and are not inhibited by sulfate and phosphate, while the latter enzymes have low pH optima (4–5) and are noncompetitively inhibited by sulfate and phosphate [212]. The *Patella* enzyme is without activity toward etiocholanolone sulfate (X), whereas *Helix* preparations hydrolyze the ester, although only at about 10% of the rate that dehydroepiandrosterone is hydrolyzed [212]. This may indicate that *Helix* possesses two distinct enzymes and there is some kinetic support for this.

(X)

b. Cortisone Sulfohydrolase. Enzyme concentrates from *Helix* that were active toward dehydroepiandrosterone sulfate also hydrolyzed cortisone 21-sulfate (XI) [213]. In the latter substrate, sulfate is esterified to a primary alcohol group and hence the compound may be regarded as a simple alkyl sulfate ester. The activities toward the two substrates have not been separated and, indeed, substrate competition experiments [212,214] support the possibility that the two activities are a feature of a single enzyme. The structures of the two substrates are so different, however, that it is difficult to accept this view. A suggestion that glyco-sulfohydrolase (Section II,C,2,a) was responsible for the hydrolysis of cortisone 21-sulfate has been disproved [214].

(XI)

Cortisone 21-sulfohydrolase is present in a number of mollusks, but no activity has been detected in mammals. However, appreciable desulfation of cortisone 21-sulfate does occur in rats injected with the ester [215] and the manner in which this occurs is still a mystery.

C. Carbohydrate Sulfohydrolases

1. General Comments

These enzymes constitute a much neglected group of sulfohydrolases which, until recently, have not generally been deemed sufficiently interesting to warrant detailed study. Potential substrates are extremely widely distributed in nature and occur in many diverse molecular forms. With a few exceptions they are polymeric and generally appear to be serving in a protective and/or structural role as ion-binding components of supporting tissues, for example, the sulfated polysaccharides of seaweeds that probably help to cushion the cells against physical stress, as well as provide an ionic barrier between cell contents and the high salinity of the ocean. In relatively primitive animal organisms, polysaccharide sulfates frequently appear as important structural components of mucus secretions, usually in the form of polyhexose polysulfates [e.g., 216–218], but sometimes in more complex forms with a multiplicity of different sugars and sometimes sialic acid [219–221]. Other complex sulfated heteropolysaccharides exist in nonmammalian and mammalian organisms. The most common of these are the glycosaminoglycan sulfates (acid mucopolysaccharides) that constitute a major part of the amorphous extracellular "ground substance" of connective tissues. In those tissues they are principally concerned with supporting, stabilizing, and orientating the collagen and elastin fibers and the cellular elements of the tissue, as well as maintaining a controlled ionic and aqueous environment. With most of the polymers that have been studied, protein is covalently associated with the sulfated polysaccharide moiety, the whole providing a proteoglycan of extreme complexity. Sulfated glycoprotein polymers (protein being predominant) are present in tissues and in mucus secretions [e.g., 222–225] but few studies have been made on these compounds and their significance is obscure.

At the other end of the scale are carbohydrate sulfate esters of relatively simple structure. For example, the lactose 6-sulfate and the neuramin lactose 6-sulfate of mammary glands [226,227], the holothurins of the sea cucumbers [228,229], the sulfated glycolipids of the halophilic bacteria [230] and of mammalian tissues [e.g., 231,232], the uridine diphospho-N-acetyl-D-galactosamine 4-sulfate isolated from hen's oviduct [233], and the ascorbic acid 2-sulfate of mammalian and other tissues [234]. More detailed accounts of the structures of many of these simple and complex carbohydrate sulfate esters have been presented [for example, see 5,235–237].

The distribution of sulfate groups in carbohydrate sulfates at first sight appears to be haphazard. Glucose, galactose, fucose, rhamnose, N-acetylglucosamine, N-acetylgalactosamine, iduronic acid, and possibly glucuronic acid are some of the residues known to exist in sulfated form. There is apparently no single preferred position of sulfation on the sugar molecules; in some instances more than one sulfate group is present on the same sugar residue and in others a sugar amino group is sulfated (N-sulfate or sulfamate). However, there is little doubt that the type of sugar involved, as well as the position and number of sulfate groups, exerts a pronounced influence on the conformation of the polymeric carbohydrate sulfates and this, in turn, may be of vital importance to physiological function.

The existence of this diverse class of compounds, widely distributed in organisms of all stages of evolutionary complexity, reinforces the view that ester sulfate groups have been essential to the continuing development of living organisms. It also requires a singular lack of imagination not to deduce that a variety of sulfohydrolase enzymes must exist that are able to hydrolyze those groups. Unfortunately, the importance of such enzymes is not immediately apparent and the difficulties of studying them are formidable. These two factors have been responsible for the paucity of useful recorded work. Undoubtedly, other carbohydrate sulfohydrolases are therefore still awaiting discovery and this situation might have continued indefinitely had it not been for recent developments concerned with lysosomal storage diseases. At least three such diseases can now be attributed to deficiency of enzymes that participate in the degradation of sulfated proteoglycans or glycolipids by removing sulfate groups from sugar residues. These findings will certainly ensure a quickening interest in carbohydrate sulfohydrolases.

2. GLYCOSULFOHYDROLASE

a. Molluscan Enzymes. More than 40 years ago an enzyme was discovered in snails of the Eulota family that could liberate sulfate from mono-, di-, and tri-sulfated mono- and disaccharides of undefined structure. Other invertebrate organisms, particularly marine mollusks, also possessed a similar enzyme for which the name "glucosulfatase" was coined [for early review, see 154]. Since the enzyme will hydrolyze sulfate esters of sugars other than glucose, this name has been superseded by "glycosulfatase" or, more correctly, "glycosulfohydrolase" (EC 3.1.6.3; sugar-sulfate sulfohydrolase).

Between 1931 and 1950, Japanese workers studied the enzyme present in the liver of the mollusk *C. lampas.* The properties of a partially purified enzyme were described [154,238,239] and these, together with findings by other workers [240–244], served to distinguish it from arylsulfohydrolase, chondrosulfohydrolase, cellulose polysulfohydrolase, and keratansulfohydrolase. In common with many other sulfohydrolases, the *Charonia* enzyme was inhibited by borate, phosphate, and fluoride.

Subsequently, glycosulfohydrolase activity was discovered in other marine mollusks [245] and a partially purified enzyme was separated from the viscera of the periwinkle *L. littorea* [214,246]. The enzyme was distinguished from other sulfohydrolases in the preparation that were active toward cortisone 21-sulfate and androstenolone sulfate. A similar enzyme was present in the limpet *P. vulgata* [247,248]. All the molluscan glycosulfohydrolases exhibited optimal activity toward D-glucose 6-*O*-sulfate at pH 5.5–5.8.

Charonia and *Littorina* glycosulfohydrolases were also active (although somewhat less so) toward D-glucose 3-*O*-sulfate, D-galactose 6-*O*-sulfate, *N*-acetyl-D-glucosamine 6-*O*-sulfate, and *N*-acetyl-D-galactosamine 6-*O*-sulfate. *Littorina* glycosulfohydrolase did not hydrolyze *N*-acetylchondrosin 6-*O*-sulfate, the principal disaccharide repeating unit of chondroitin 6-sulfate, and this distinguished the enzyme from known chondrosulfohydrolases. The *Patella* enzyme was active toward the 2-, 3-, and 6-*O*-sulfates of D-galactose and toward fucose monosulfates [248].

On the whole, it would seem prudent to regard the various specificity claims with some degree of caution. The enzyme preparations that have been employed have not been pure and there are observations on microbial glycosulfohydrolases and chondrosulfohydrolases that lead one to suspect the presence of more than one glycosulfohydrolase in molluscan preparations.

   b. *Microbial Enzymes.* Following brief reports of glycosulfohydrolase activity in bacteria [249] and in the mold *Trichoderma viride* [250], a more detailed study of the *Trichoderma* enzyme was made [251]. The enzyme liberated equimolar amounts of D-glucose and SO_4^{2-} from D-glucose 6-*O*-sulfate and exhibited maximum activity at pH 7.4 and 28°. Activity at 37° was only about 35% of that at 28°. D-Galactose 6-*O*-sulfate also served as a substrate, but D-glucose 3-*O*-sulfate was not attacked. This behavior contrasts with that of molluscan enzyme that can use the 3-*O*-sulfate as a substrate and may indicate that more than one glycosulfohydrolase is present in molluscan preparations.

   Pseudomonas carrageenovora is another microorganism that produces a glycosulfohydrolase of relatively high specificity [252], while rat fecal microorganisms possess enzymes active toward the 6-*O*-sulfates of hexoses [253] and hexosamines [254]. One such organism, responsible for the desulfation of D-glucose 6-*O*-sulfate, was identified as a *Lactobacillus* species and produced the desulfating enzyme only when grown in a medium containing the ester [255].

   c. *Other Sources of Glycosulfohydrolases.* The existence of glycosulfohydrolase-type enzymes in vertebrate organisms ought to be predictable from the existence of such compounds as lactose 6-sulfate, neuramin lactose 6-sulfate, and cerebroside sulfates (containing D galactose 3-*O*-sulfate). However, apart from cerebroside sulfate sulfohydrolase (arylsulfohydrolase A, see Section II,A,1,a) and a glycosulfohydrolase present in the embryos and allantoic and amniotic fluids of embryonated eggs [256], there is no evidence to suggest that such enzymes are common in vertebrates. Indeed, many searches for mammalian sources of the enzyme have been made in our own laboratories, employing the ^{35}S-labeled 6-*O*-sulfate esters of hexoses and N-acctylhexosamines as potential substrates and either direct incubations of tissues and substrates or metabolic degradation studies in intact animals. Although the latter type of experiment frequently indicated some *in vivo* hydrolysis of these compounds, it was not possible to detect enzyme activity directly and unequivocally.

   Searches for glycosulfohydrolases have been made in marine algae but early reports of activity in the red seaweed *Porphyra umbilicalis* [257] were later shown to reflect the presence of an enzyme which cleaved the C–O bond of the C–O–S linkage in the D-galactose 6-*O*-sulfate residues in the polymer to yield 3,6-anhydrogalactose units [258]. A true sulfo-hydrolase is therefore not involved.

d. *Physiological Significance.* Attempts to assign specific functions to glycosulfohydrolases are, for the most part, speculative. In mollusks the enzyme may be a component of multienzyme systems concerned with the overall degradation of sulfated polysaccharides that form constituents of the diet or of mucus secretions. This concept will be further developed in Section II,C,5. The ability of *Charonia* preparations containing polysaccharases and glycosulfohydrolase to degrade the polyfucan polysulfates present in the jelly coats of some sea urchin eggs [216,259] and the presence of glycosulfohydrolase in embryonated hen's eggs may point to a role for the enzyme in the fertilization process, but the view has long been held without receiving adequate substantiation. On the whole, we incline to the view that enzymes are concerned with the degradation of macromolecules, but the inability to be more precise is a sad reflection of progress on a group of enzymes first discovered more than 40 years ago.

The presence of glycosulfohydrolases in microorganisms is more readily explained, although again definitive proof is lacking. However, it seems reasonable to assume that the enzymes are important factors in securing supplies of sulfur for growth requirements. This theme was developed in Section II,A,2,c in relation to bacterial arylsulfohydrolases. Certainly hexose sulfate esters are rapidly degraded when added to a variety of acid, basic, and neutral soils (unpublished results, these laboratories) and sulfate esters of this type when returned to the soil or to the ocean muds are probably rapidly desulfated and liberated SO_4^2 thereby returned to the sulfur cycle. Here is an interesting field that would repay detailed study.

3. D-GLUCONATE-6-SULFATE AND D-GLYCERATE-3-*O*-SULFATE SULFOHYDROLASES

Following the studies on *Trichoderma* glycosulfohydrolase (Section II,C,2,b), a search was made for similar enzymes in a variety of pseudomonads [260,261]. One organism, *P. fluorescens* N.C.I.B. 8248, grew particularly well when D-glucose 6-*O*-sulfate was employed as either the sole source of carbon and sulfur or of sulfur. Under the former circumstances cell extracts liberated SO_4^{2-} from D-gluconate 6-*O*-sulfate and D-glycerate 3-*O*-sulfate, but not from D-glucose 6-*O*-sulfate. During growth D-glucose 6-*O*-sulfate was first oxidized to the corresponding D-gluconate ester by an enzyme which appeared to be analogous to the glucose dehydrogenase present in some members of the genus *Pseudomonas* [262]. This was not altogether surprising since previous work [263,264] had established that extracts of *T. viride* were also able to convert D-glucose 6-*O*-sulfate to D-gluconate 6-*O*-sulfate. In *P. fluorescens*, the latter compound was apparently metabolized further, via the Entner and Doudoroff pathway [265], as though it was D-gluconate 6-phosphate. In this type of pathway the 6-carbon phosphate ester eventually gives rise to two C_3 fragments, pyruvate and D-glycerate 3-phosphate, the latter then being metabolized via the corresponding 2-phosphate. In the case of D-gluconate 6-*O*-sulfate, pyruvate and D-glycerate 3-*O*-sulfate would be the C_3 fragments first formed and there are theoretical reasons for supposing that the latter compound could not be converted enzymatically to the corresponding 2-*O*-sulfate. In support of this the 3-*O*-sulfate accumulated in large amounts in the culture medium during the growth of the organism.

Other evidence strongly suggested that separate sulfohydrolases were responsible for the desulfation of D-gluconate 6-*O*-sulfate and D-glycerate 3-*O*-sulfate, respectively. Maximum activity toward the latter compound was obtained at 34°C and pH 6.5 with a substrate concentration of 0.9 mM.

When *P. fluorescens* was grown under conditions in which D-glucose 6-*O*-sulfate served only as the sole source of sulfur, D-gluconate 6-*O*-sulfate accumulated in the culture medium and appreciable sulfohydrolase activity toward that sulfate ester was present in cell extracts. No sulfohydrolase activity toward D-glucose 6-*O*-sulfate or D-glycerate 3-*O*-sulfate was detected and this supports the view that activity toward the latter, exhibited by extracts of cells grown on the former as sole carbon and sulfur sources, is due to a distinct D-glycerate-3-*O*-sulfate sulfohydrolase.

When *P. fluorescens* was grown on D-glucose 6-*O*-sulfate as sole carbon and sulfur sources, enzyme activities toward D-gluconate 6-*O*-sulfate and D-glycerate 3-*O*-sulfate appeared and disappeared over a comparatively short period during the early exponential growth phase. Activities declined virtually to zero by the late exponential growth. These findings support the view that the function of the enzymes is concerned with the acquisition of sulfur for growth purposes. However, the fact that under certain growth circumstances only one of the enzymes appears indicates separate control mechanisms for each one. It is not without interest that at least one sulfated glycerol-containing compound occurs naturally, namely the 2,3-di-*O*-phytanyl-*sn*-1-glycerophosphoryl-*sn*-3'-glycerol 1'-sulfate of *Halobacterium cutirubrum* [266]. This may point to the possible natural occurrence of other C_3 sulfate esters and the existence of other sulfohydrolases analogous to D-glycerate-3-*O* sulfate sulfohydrolase.

4. ASCORBATE-2-SULFATE SULFOHYDROLASE

Ascorbate 2-sulfate has occupied the attention of several investigators since its discovery in brine shrimps [267] and mammalian tissues [e.g., 234,268]. Principal interest focuses on its potential as a sulfating agent [e.g., 59,269] and hence the possibility that it participates in the biological transfer of ester sulfate groups. At the time of writing there has been one short report [45] on sulfohydrolase activity toward the ester, but the situation appears to be complicated and merits further study. Briefly, a partially purified preparation of arylsulfohydrolase A was able to hydrolyze ascorbate 2-sulfate to give ascorbate and SO_4^{2-}. When subjected to column chromatography on agarose, arylsulfohydrolase, and ascorbate-2-sulfate sulfohydrolase activity coincided. On DEAE-cellulose however, three peaks of ascorbate 2-sulfate sulfohydrolase activity separated and only two of them exhibited arylsulfohydrolase activity. It was suggested that ascorbate-2-sulfate sulfohydrolase had a physiological role in equilibrating the ascorbate/ascorbate 2-sulfate pools in mammalian tissues. No doubt further work on this sulfohydrolase activity will emerge in the near future.

5. CELLULOSE POLYSULFATE SULFOHYDROLASE

The existence of this enzyme was reported [270] following the observation that extracts of the liver of the mollusk C. *lampas* were able to liberate inorganic sulfate from chemically sulfated cellulose polysulfate. A polysulfated cellulose-type polysaccharide [$\beta(1 \rightarrow 4)$-linked polyglucose] had previously been isolated from the mucus gland of the same mollusk [271] and this polymer, charonin sulfate, was also desulfated by enzyme extracts.

The discovery of this enzyme was particularly interesting since several puzzling observations made previously immediately became clear. As early as 1936 it was noted [272] that crude preparations of charonin sulfate (the chemical constitution of which was then obscure) could be partially desulfated by crude *Charonia* extracts, but it was assumed that this activity was due to glycosulfohydrolase. Later, a chondrosulfohydrolase was detected in the extracts [238] and for some time this enzyme was held to be responsible for the desulfation. The discovery of cellulose polysulfate sulfohydrolase clarified the situation and led to the elucidation [241,242,273] of the complete pathway for the enzymatic degradation of cellulose polysulfate and hence, for the degradation of charonin sulfate. Ester sulfate groups at positions 2 and/or 3 on the glucose residues were removed by the cellulose polysulfate sulfohydrolase while those at position 6 were unaffected. Subsequently, a cellulase-like polysaccharase depolymerized the partially desulfated compound to yield D-glucose 6-O-sulfate residues from which the sulfate groups were then removed by glycosulfohydrolase. The latter enzyme was inactive toward the intact polymer. Charonin sulfate is presumed to be degraded in a similar fashion, although absolute proof has not been furnished.

Support for the proposed reaction sequence came from studies [243] in which the polysulfate sulfohydrolase was separated from the polysaccharase and tested for activity toward a number of naturally occurring and synthetic sulfated polysaccharides. Compounds hardly affected were dextran and amylose polysulfates, heparin, glycogen sulfate, chondroitin 4-sulfate, shark chondroitin sulfate, the polyfucan polysulfate from the eggs of *Hemicentrotus pulcherrimus,* a sulfated polysaccharide from a seaweed (a *Chondrus* sp., containing sulfated galactose residues), and the polyglucose polysulfate of *Busycon canniculatum.*

Presumably, the cellulose polysulfate sulfohydrolase in *Charonia* is concerned with the catabolism of charonin sulfate and possibly, sulfated polysaccharides present in the food of the organism. No doubt a search would reveal analogous enzymes in other organisms that either produce polyhexose polysulfates or encounter them as food components.

6. SULFOHYDROLASES ATTACKING ISOMERIC CHONDROITIN SULFATES

In describing these and other sulfohydrolases that participate in the degradation of sulfated glycosaminoglycans, space does not permit a complete description of the structures of the naturally occurring substrates and the reader is referred to reviews previously mentioned (Section II,C,1). Suffice it to say that within the tissues these sulfated glycosaminoglycans are covalently bound to a protein core. The individual carbohydrate chains themselves are generally hybrid in character, the degree of sulfation is not constant, the isolated proteoglycans are very variable in composition depending on the tissues of origin, and often contain more than one type of glycosaminoglycan. Moreover, many proteoglycan subunits, in association with glycoprotein, can aggregate to form giant proteoglycan complexes [for an indication of the structural complexities of some of these macromolecules, see 274–277]. The degradation of such aggregates requires a spectrum of enzymes, including those capable of degrading protein, the protein–polysaccharide linkage regions, and the glycosaminoglycan chains proper. In now describing the involvement of sulfohydrolases in these processes, the chemical structures of the glycosaminoglycan substrates have been deliberately oversimplified and have been regarded as regular linear polymers in which substituted hexosamine residues alternate in regular fashion with either hexuronic acid or hexose residues. Wherever this treatment is inadequate for the understanding of sulfohydrolase action, additional details have been provided in the text.

Three isomeric chondroitin sulfates (and variants thereof) are widely distributed in animal organisms. In the most common repeating disaccharide unit of chondroitin 4-sulfate (XII), D-glucuronic acid residues alternate with N-acetyl-D-galactosamine 4-O-sulfate residues; in dermatan sulfate (XIII), L-iduronic acid replaces D-glucuronic acid and sometimes carries a sulfate group at position 2, while in chondroitin 6-sulfate (XIV), N-acetyl-D-galactosamine 6-O-sulfate replaces the corresponding 4-O-sulfate. Much of the early work on sulfohydrolases that attack this type of compound involved substrates that were chemically ill-defined, but almost certainly they consisted mainly of chondroitin 4-sulfate. It has become the practice to refer to sulfohydrolases

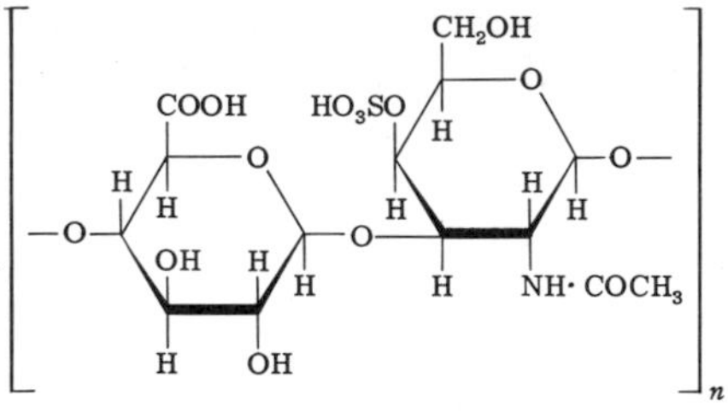

(XII)

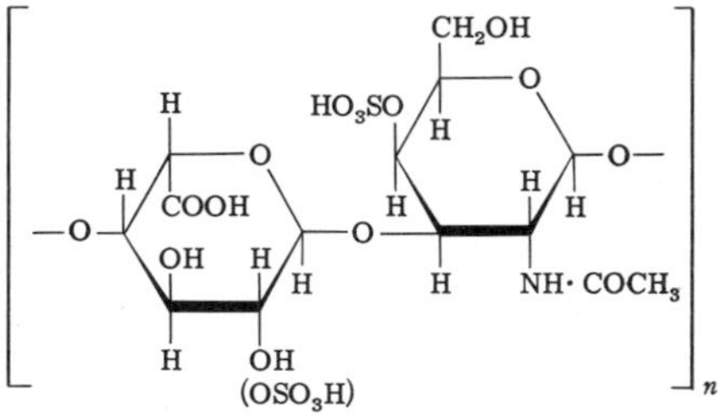

(XIII)

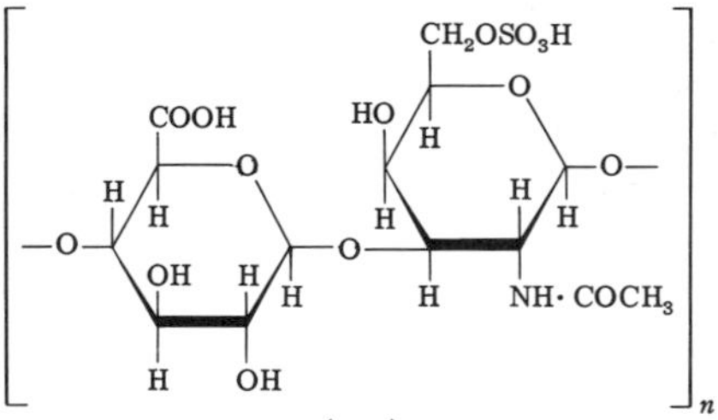

(XIV)

attacking chondroitin sulfates as "chondrosulfohydrolases" (chondrosulfatases) although some of them exhibit no activity toward the intact glycosaminoglycan.

The first reports of chondrosulfohydrolases were provided by Carl Neuberg [278–281] working with various *Pseudomonas* sp. and with *P. vulgaris*. Subsequently, similar enzymes were noted in other bacteria [e.g., 282,283], in fungi [284], and in marine mollusks [e.g., 238,285].

Most studies were superficial in character, but one important general-
ization emerged, namely, that a depolymerizing enzyme (generally
termed a "chondroitinase") was associated with the sulfohydrolase and
both enzymes participated in the degradative process. More recently,
detailed studies have been made on chondrosulfohydrolases from bacterial, molluscan, and mammalian sources and have revealed striking differences between them in terms of the molecular size of their true substrates.

  a. Bacterial Chondrosulfohydrolases. Following earlier reports of enzyme activity in bacteria, workers in our own laboratories [286,287]
searched for activity in *Pseudomonas* sp. and *P. vulgaris* strains. All the
strains of the latter organism were active and one strain (NCTC 4636)
particularly so. A chondroitinase was also present in cell extracts, but
such extracts exhibited no sulfohydrolase activity towards other sulfate
esters including heparin, carrageenin, charonin sulfate, agar, fucoidin,
Chondrus ocellatus mucilage, D-glucose 6-*O*-sulfate, *N*-acetyl-D-galac-
tosamine 6-*O*-sulfate, *N*-acetyl-D-glucosamine 6-*O*-sulfate, and UDP-*N*-
acetyl-D-galactosamine 4-*O*-sulfate [see also 288]. Time–activity curves
for chondroitin 4-sulfate degradation suggested that sulfohydrolase
activity had to be preceded by depolymerization of the glycosaminoglycan chain, and this was confirmed by separating the chondrosulfohydrolase from the chondroitinase by preferentially absorbing the former on
calcium phosphate gel. The separated sulfohydrolase exhibited no activity toward polymeric chondroitin sulfate. However, when the latter was
first degraded with testicular hyaluronidase, SO_4^{2-} was rapidly released
from the digests following the addition of the sulfohydrolase. Sulfate was
also liberated from the sulfated disaccharide isolated from digests of
chondroitin sulfate by *Proteus* chondroitinase and from chemically synthesized *N*-acetylchondrosine 6-*O*-sulfate [see 289]. With this bacterial
system therefore, depolymerization was a prerequisite for removal of
sulfate groups and the name chondrosulfohydrolase is revealed as being
somewhat inadequate.

  A considerable weight of evidence suggested that the depolymerization of chondroitin 4-sulfate by the bacterial extracts proceeded via a
simple hydrolytic attack [290]. However, later studies by other
workers, using cells that had been grown in the presence of chondroitin
sulfate, established that depolymerization proceeded via an elimination
type of mechanism to give sulfated disaccharides unsaturated in the $\Delta^{4,5}$
position of the uronic acid [291]. It has since been generally accepted
that this is the reaction mechanism of all bacterial depolymerases acting
on glycosaminoglycans. However, recent work on the degradation of
heparin by *Flavobacterium heparinum* [292] has indicated the probabil-

ity that the growth conditions may determine the type of depolymerization system predominating in cell extracts and there is clearly a need for further studies on the induction of chondroitinase activity.

In retrospect, it is clear that these findings came several years too early for the potential significance of the enzymes as structural tools to be fully appreciated, and more than 10 years elapsed before Japanese workers took up the problem. Then, with the availability of defined substrates and new enzyme purification techniques, it was established [293–296] that two distinct sulfohydrolases coexisted in *P. vulgaris* NCTC 4636 that had been grown in the presence of chondroitin sulfate. One of these was active toward the saturated disaccharide repeating unit of chondroitin 4-sulfate (*N*-acetylchondrosine 4-*O*-sulfate, XII) and the corresponding $\Delta^{4,5}$-unsaturated disaccharide, 2-acetamido-2-deoxy-3-*O*-(β-D-gluco-4-enepyranosyluronic acid)-4-*O*-sulfo-D-galactose (XV); the other toward the saturated disaccharide repeating unit of chondroitin 6-sulfate (XIV) and the corresponding $\Delta^{4,5}$-unsaturated disaccharide (XVI).

(XV)   (XVI)

The relative specificity of each enzyme was high and the 4-*O*-sulfohydrolase would remove only the 4-*O*-sulfate group from a 4,6-di-*O*-sulfated unsaturated disaccharide prepared by digesting squid cartilage chondroitin sulfate with *Proteus* chondroitinase, while the 6-*O*-sulfohydrolase removed only the 6-*O*-sulfate group from the compound. Neither enzyme hydrolyzed the sulfate group present at either the 2 or 3 position or the uronic acid moiety of an unsaturated disaccharide prepared by digesting oversulfated dermatan sulfate.

According to the Japanese group, all three isomeric chondroitin sulfates are completely degraded to unsaturated disaccharides and, of course, the same unsaturated disaccharide will be obtained from either chondroitin 4-sulfate or dermatan sulfate in spite of the different uronic acids originally present in the molecules. Clearly then, the 4-*O*-sulfohydrolase can participate in the biodegradation of both chondroitin 4-sulfate and dermatan sulfate. This particular sulfohydrolase appears to be very unstable and hence resembles the heparin disaccharide sulfohy-

drolase described in Section II,C,8,a, but differs from other sulfohydrolases, most of which are remarkably stable.

Flavobacterium heparinum is another microorganism that exhibits
chondroitinase and chondrosulfohydrolase activities toward the isomeric
chondroitin sulfates. Either one or two chondroitinases are produced
depending on growth conditions [291,297]. One of these will depolymerize chondroitin 4- and 6-sulfates while the other will depolymerize
dermatan sulfate in addition to the others. Sulfohydrolase activity is
exhibited toward both chondroitin 4- and 6-sulfates (probably at least
two enzymes are involved although this has not been investigated), but
preliminary depolymerization of the glycosaminoglycan is necessary.
Both saturated and $\Delta^{4,5}$-unsaturated sulfated oligosaccharides are attacked suggesting that the system is similar to that of *P. vulgaris*.

Under other growth conditions, *Flavobacterium* produces enzymes
that are able to degrade heparin [298], and at least three separate sulfate
liberating enzymes are involved. These various enzymes will be discussed in a later Section (II,C,8,a), but it is worth recording here that it
is doubtful whether any of them are identical with the sulfohydrolases
participating in the degradation of isomeric chondroitin sulfates.

No doubt other microorganisms can produce chondrosulfohydrolases.
The amounts of sulfated glycosaminoglycans returned to the soil in the
form of dead animal tissue may be enormous and many microorganisms
may have developed the ability to capitalize on this valuable supply of
C, N, and S. Meanwhile, the chondroitinases and sulfohydrolases of
P. vulgaris and *F. heparinum* are proving to be valuable analytical tools
for the study of glycosaminoglycans [295,296,299,300].

b. Molluscan Chondrosulfohydrolase. The digestive gland of
C. lampas was the first reported source of this enzyme [238,239]. The
enzyme was partially purified and distinguished from other sulfohydrolases. A chondroitinase, which was also present in the gland, could
degrade polymeric chondroitin sulfate (of undefined structure) in the absence of sulfohydrolase activity. There was no indication of the molecular size of the substrate for the sulfohydrolase. Later, a similar sulfohy-
drolase/chondroitinase system was noted in the viscera of the limpet,
P. vulgata [301] and, in this case, either polymeric chondroitin 4-sulfate
or sulfated oligosaccharides prepared by treating the polymer with testicular hyaluronidase could serve as substrates for the sulfohydrolase.
Polymer chondroitin 6-sulfate was also desulfated by the enzyme preparation, but no attempt was made to see whether one or a multiplicity of
sulfohydrolases are responsible for the different effects noted.

One further molluscan enzyme must be mentioned, the sulfohydrolase
present in the liver of the squid (*Ommastrephes sloani pacificus*) [302].

As usual there was an associated chondroitinase. In common with other molluscan sulfohydrolases, the squid enzyme had an acid optimum pH (5.0) and was inhibited by phosphate. A partially purified preparation desulfated chondroitin 4-sulfate and (at a much slower rate) chondroitin 6-sulfate and was also active toward sulfated oligosaccharides prepared from chondroitin 4-sulfate by testicular hyaluronidase digestion, and toward the sulfated tetrasaccharide isolated from the digestion mixture. There was negligible activity toward the $\Delta^{4,5}$-unsaturated disaccharide (XV) prepared by degrading chondroitin 4-sulfate with *Proteus* chondroitinase.

Probably other mollusks possess chondrosulfohydrolase, but its importance to the organism is obscure. Possibly it is concerned with the degradation of connective tissue glycosaminoglycans, but there is no evidence to support that view. Much more work is required on the specificity of molluscan chondrosulfohydrolase, but isolation of homogeneous enzyme should be the prerequisite for such studies. In any event it does seem that the term chondrosulfohydrolase is more correctly applicable to enzymes of molluscan origin rather than to those from bacteria.

c. Mammalian Chondrosulfohydrolase. Proteoglycan degradation in mammals has been of growing interest in recent years. It seems likely that degradation of the protein core of proteoglycans may be responsible for some of the damage to cartilage that occurs in degenerative arthritis. Furthermore, it is now emerging that the so-called mucopolysaccharidoses can be attributed to the absence of one or more lysosomal enzymes that participate in proteoglycan degradation. However, there is still some way to go before the nature of these various involvements are completely understood.

Evidence currently available indicates that degradation of proteoglycans proceeds in cell lysosomes. The glycosaminoglycan chains are released from their attachment to protein as a result of proteolysis by cathepsins [see 5,235], protein–glycosaminoglycan linkage regions may be hydrolyzed by β-xylosidase and/or β-galactosidases, and the glycosaminoglycan chain is degraded by a combination of glycosidases and sulfohydrolases. Precisely which cells can accomplish these changes is still uncertain. For example, macrophages can depolymerize chondroitin 4-sulfate proteoglycan by means of lysosomal proteases and hyaluronidase, but there is no evidence of secondary degradation of the depolymerization products [303]. In contrast, rat liver cells appear to be capable of the uptake and complete degradation of chondroitin 4- and 6-sulfates [304]. Relatively little information concerning the *in vivo* fate of the sulfate groups of the isomeric chondroitin sulfates is available. However, ^{35}S-labeled inorganic sulfate appears in urine following the administration of ^{35}S-labeled chondroitin 4-sulfate to rats [305,306], and studies from our laboratories [307] have shown the liver to be particularly effective in accomplishing the desulfation. It was calculated that 30 mg of chondroitin 4-sulfate could be handled by the rat over a period of 10 days (the approximate half-life of chondroitin 4-sulfate proteoglycan). Lysosomes were responsible for the desulfation and for the degradation of the polymer chain that also occurred. It was not possible to say precisely how sulfate was released during degradation.

Up to the time of writing, three sulfohydrolases have been recognized which are presumed to participate in the degradation of one or the other of the isomeric chondroitin sulfates. One of these was separated from rat liver lysosomes [308] and shown to be of unusual specificity. In common with bacterial systems, degradation of the chondroitin 4-sulfate chain was a prerequisite for sulfohydrolase action. However, the substrate for the latter appeared to be an odd numbered oligosaccharide unit (probably heptasaccharide) containing a terminal nonreducing N-acetyl-D-galactosamine 4-O-sulfate residue. This was the residue that was apparently desulfated. More work needs to be done on this enzyme, but it must be presumed that lysosomal hyaluronidase, β-glucuronidase, β-N-acetylhexosaminidase, and the sulfohydrolase cooperate in degrading chondroitin 4-sulfate to readily excretable products.

The second type of sulfohydrolase was detected in bovine aorta and exhibited maximum activity toward polymeric chondroitin 4-sulfate at about pH 4 [309]. The copresence of hyaluronidase in incubation mixtures did not increase the rate of desulfation, which, in the event, was relatively low. The enzyme was apparently incapable of desulfating either chondroitin 6-sulfate or dermatan sulfate. Other proteoglycan degrading enzymes were present in the aortal extracts [310], including hyaluronidase, cathepsin D, β-N-acetylhexosaminidase, β-glucuronidase, and a carboxypeptidase. It was suggested that the proteases were involved in the degradation of the protein core while the others were concerned with the degradation of the glycosaminoglycan chains. Two possible routes (Fig. 2) for the degradation were envisaged, although no unequivocal proof for either exists.

The third mammalian sulfohydrolase concerned with isomeric chondroitin sulfates participates in the degradation of oversulfated, L-iduronic acid containing regions of dermatan sulfate chains. In these regions, L-iduronic acid residues carry an ester sulfate group at position 2 [296]. Recent studies [311–313] have established the existence of a sulfohydrolase that is responsible for hydrolysis of such groups during the degradation of dermatan sulfate. Cultured fibroblasts from patients with Hunter's syndrome [see 314], unlike those from normal subjects, are

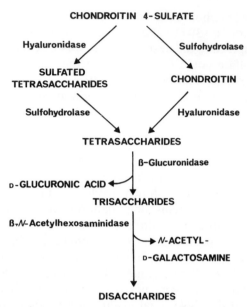

FIG. 2. Alternative pathways suggested for the enzymatic degradation of chondroitin 4-sulfate in aortal tissue [310].

deficient in this sulfohydrolase and hence dermatan sulfate accumulates. Not all the sulfated iduronate residues are available to the enzyme and only a very small amount of the total ester sulfate in the molecule is hydrolyzed. Nevertheless, this desulfation is an essential stage in the degradation of the polymer and the release of L-iduronic acid residues by α-L-iduronidase [for example, see 315] cannot proceed without it. Very little is known about the sulfohydrolase; it is inhibited in the presence of citrate-phosphate buffers, has an optimum pH of about 4.0, and cannot be detected in homogenates of fibroblasts prepared by freeze–thawing techniques. Although much more work remains to be done, it seems likely that the oversulfated L-iduronic acid regions of the hybrid dermatan sulfate chains may be degraded in stepwise fashion by the sequential action of L-iduronate sulfate sulfohydrolase, α-L-iduronidase, a sulfohydrolase acting on N-acetyl-D-galactosamine 4-O-sulfate, and β-N-acetylhexosaminidase.

It will not pass unnoticed that the present state of knowledge about mammalian sulfohydrolases active toward isomeric chondroitin sulfates is unsatisfactory. Above all, there still seems to be no indication how sulfate groups are removed from the 6-O-position of N-acetyl-D-galactosamine residues of chondroitin 6-sulfate (and of some regions of hybrid dermatan sulfate chains). The problems of finding and studying sulfohydrolases that participate sequentially, with other enzymes, in degrading hybrid macromolecular structures need not be spelled out, and there is need of clearly defined substrates of low molecular weight if further progress is to be made.

7. SULFOHYDROLASES ATTACKING KERATAN SULFATES

Keratan sulfates are widely distributed in cartilagenous structures and are also a particularly interesting feature of the cornea of the eye. In common with other glycosaminoglycans, they are covalently bound to protein, usually in association with chondroitin 4- and/or 6-sulfates, the whole forming a giant molecule of great complexity. There is considerable variation between keratan sulfate proteoglycans from different sources and the individual complexes are very heterogeneous. For example, the mode of attachment of glycosaminoglycan chains to protein core varies, some of the chains are hybrid, the chain length and degree of sulfation varies, and some unusual sugar residues are present [space does not permit detailed discussion, but for some indication of complexity see 276,316–318]. However, the major parts of the keratan sulfate chains are made up of alternating D-galactose and N-acetyl D-glucosamine 6-O-sulfate residues (XVII), although in oversulfated regions, an ester sulfate group is also found on the 6-position of the D-galactose residues.

(XVII)

In cartilage, the keratan sulfate content increases with increasing age and the turnover of the glycosaminoglycan is extremely slow in some tissues. This may be one reason why comparatively few searches for keratan sulfate sulfohydrolases have been made. However, enzymes capable of degrading keratan sulfate chains have been noted in microorganisms, mollusks, and mammals.

a. Bacterial Keratansulfohydrolases. Meyer *et al.* [319] were the first to note the presence of a keratan sulfate degrading multienzyme system in a *Coccobacillus* sp. Extracts of cells degraded the glycosaminoglycan to oligosaccharides, D-galactose, N-acetyl D-glucosamine, and inorganic sulfate. A later report [320] described a *Coccobacillus* enzyme preparation from which a sulfohydrolase and exoglycosidases had been removed. This preparation contained an *endo-β*-galactosidase that degraded keratan sulfate to sulfated oligosaccharides. Unfortunately there has been no further mention of the sulfohydrolase, but the *endo-β*-galactosidase was more effective when acting on a chemically desulfated keratan sulfate and this may point to the importance of the former enzyme.

Other investigators have examined aerobic and anaerobic microorganisms and have isolated one from human feces that was able to degrade keratan sulfate from whale nasal cartilage, provided that the glycosaminoglycan was included in the growth medium [321]. The microorganism, identified as a strain of *Escherichia freundii* [322], produced an *N*-acetylhexosaminidase, *β*-galactosidase, a depolymerase, and a sulfohydrolase. Cell extracts degraded keratan sulfate to D-galactose, *N*-acetyl D-glucosamine, and inorganic sulfate. Subsequently, the polymerase was purified to homogeneity [323], but no further studies on the sulfohydrolase appear to have been made.

b. Molluscan Keratansulfohydrolase. To the extensive list of sulfohydrolases known to be present in the liver of *C. lampas* two enzymes that can participate in keratan sulfate degradation must be added [244]. They form part of a multienzyme system that degrades the glycosaminoglycan to D-galactose, *N*-acetyl D-glucosamine, and inorganic sulfate and were separated from cell extracts by DEAE-Sephadex chromatography. Each could release sulfate from oversulfated shark keratan sulfate and from ^{35}S-labeled bovine corneal keratan sulfate. However, only one of the enzymes was able to liberate sulfate from horatin sulfate, a sulfated polysaccharide containing fucose, mannose, glucose, galactose, glucosamine, and galactosamine, present in *Charonia* liver [324]. The two sulfohydrolases therefore show some difference in specificity, but the significance of this is quite obscure. Certainly, neither enzyme appeared to be identical with glycosulfohydrolase or chondrosulfohydrolase, although, in common with those enzymes, the keratan sulfohydrolases had acid optimum pH values (4.5 and 6.0) and were inhibited by phosphate. A *β*-galactosidase and a *β-N*-acetylhexosaminidase were also separated from the *Charonia* extracts and experiments in which combinations of the separated enzymes were employed indicated that the keratan sulfate sulfohydrolases first hydrolyzed the ester sulfate groups present in the glycosaminoglycan. This was followed by alternate action of the *β*-galactosidase and *β-N*-acetylhexosaminidase, probably working from the nonreducing end of the molecule.

The fact that over very long incubation periods, one of the sulfohydrolases was able to release all the sulfate from oversulfated keratan sulfate suggests that the enzyme had a dual specificity toward sulfate groups on position 6 of either D-galactose or N-acetyl D-glucosamine residues. However, because of the close specificity apparently exhibited by other glycosaminoglycan sulfohydrolases, this conclusion should be viewed with caution.

c. *Mammalian Keratansulfohydrolases*. Although it is generally accepted that keratan sulfate has a relatively long half-life within mammalian tissues [325,326], there have been recent indications of more than one metabolic pool for the compound and, in some tissues, a portion of it may turnover at a fairly rapid rate [276]. The characteristic presence of excessive amounts of keratan sulfate in the urines of patients with the Morquio syndrome [314] presumably points to the possibility that normal turnover is reduced because of a lack of some enzyme that is concerned with the degradative process. However, apart from the recognition that a tissue β-galactosidase can remove the terminal nonsulfated galactose residue from the nonreducing end of each chain, information about other mammalian enzymes acting on the polymer is very scanty. Brief reports [327,328] have concerned keratansulfohydrolase activity in enzyme extracts from rabbit kidney and liver and a degradation scheme for the polymer has been proposed in which a sulfohydrolase active toward 6-O-sulfate residues on D-galactose and N-acetyl D-glucosamine is shown working cooperatively with a β-galactosidase and a β-N-acetylhexosaminidase. Presumably further studies on this proposal are currently in progress.

8. SULFOHYDROLASES ATTACKING HEPARIN

Heparin is a most important glycosaminoglycan from the commercial, as well as the physiological, point of view and considerable quantities are marketed annually because of its unique properties as an anticoagulant. Heparin differs from other glycosaminoglycans in several ways, not the least of which relates to its existence in the tissues as an intracellular rather than an extracellular component. In its principal storage sites, mast cells, it is present in proteoglycan form and its importance in those cells may be concerned with the binding and release of histamine.

The glycosaminoglycan portion of heparin proteoglycan is extremely complicated and may show considerable variation in heparins from different sources. Irrespective of source, the polymer contains D-glucosamine, D-glucuronic acid, L-iduronic acid, N-acetyl, N-sulfate, and O-sulfate residues, although in very variable quantities. For example, the variation in the relative amounts of D-glucuronic acid and L-iduronic acid in heparins from different sources is considerable, with L-iduronic acid generally predominating. Quantitative variation of components is also common within an individual glycosaminoglycan chain. The reader is referred to recent studies [329–332] for details but, for the present purpose, the position can be simplified by delineating the positions in which sulfate can be found on the most common disaccharide repeating unit. This unit is internally $\alpha(1 \rightarrow 4)$-linked, each unit being linked to the next in an $\alpha(1 \rightarrow 4)$ manner, and maximum sulfation would show sulfate groups on position 2 of the L-iduronic acid and on position 6 and the amino group of the D-glucosamine (XVIII).

(XVIII)   (XIX)

Bacterial enzymes are known that can remove the sulfate from all these positions but, at the time of writing, only the N-sulfate (sulfamate) group can be liberated by *known* mammalian enzymes. Strictly speaking, those enzymes that release sulfate from its combination with the amino group are sulfamatases (sulfamidases, N-sulfohydrolases) and not true sulfohydrolases. Sulfamatases are further discussed in Section V, but it will be convenient to mention certain aspects at this point.

a. Bacterial Heparin Sulfohydrolases. Only one microorganism is currently known that is capable of producing heparin degrading enzymes [298,333]. The organism was isolated from soil and given the name *F. heparinum.* As outlined previously (Section II,C,6,a), under certain growth conditions extracts of the organism will degrade the isomeric chondroitin sulfates but will not degrade heparin. It is claimed that the organism must be grown in a medium containing heparin before the appropriate degrading systems are produced. In important initial studies, Korn and Payza [334–336] concluded that a depolymerase, a sulfohydrolase, and a sulfamatase were collectively responsible for the degradation. Later [337,338], it was established that depolymerization progressed via an elimination mechanism rather than by hydrolysis, the resulting unsaturated disaccharide units being subsequently hydrolyzed by a further enzyme with the production of α,β-keto acid and hexosamine residues.

Much subsequent work was concerned with elaborating the details of this system [e.g., 339–343] and culminated in the elucidation by Dietrich [344] of the sequence of events. Briefly, five enzymes that collectively could degrade heparin to its basic constituents were separated and characterized. The initial step involved a depolymerase (heparinase) which degraded heparin mainly to a trisulfated $\Delta^{4,5}$-unsaturated disaccharide (XIX). Subsequently, this was converted to a disulfated $\Delta^{4,5}$-unsaturated disaccharide by a disaccharide sulfohydrolase that removed the sulfate group from the 2-position of the uronic acid residues. Following this, an α-glycuronidase degraded the disulfated disaccharide to a 2,6-disulfated D-glucosamine (2-deoxy-2-sulfamino-D-glucose 6-O-sulfate) and an α,β-keto acid. The disulfated D-glucosamine then served as a substrate for a second sulfohydrolase which removed the 6-O-sulfate grouping and, finally, the N-sulfate group was removed by a sulfamatase. The disaccharide sulfohydrolase was unstable and it was not possible to purify it. Heparin, sulfated hexa- and tetrasaccharides, disulfated $\Delta^{4,5}$-unsaturated disaccharide, and the 2,6-disulfated D-glucosamine were not substrates for the enzyme. The monosaccharide 6-O-sulfohydrolase was stable, readily inhibited by phosphate, had an optimum temperature of 45°, an apparent K_m of 8×10^{-5} M, and an optimum pH of about 7.0. The sulfamatase was also stable, had an optimum temperature of 25°, an apparent K_m of 1.3×10^{-4} M, and an optimum pH of about 7.0. Both enzymes acted only on the appropriate sulfated hexosamine.

The close specificities shown by these three sulfate liberating enzymes suggests that none of them is identical with the 4-O- and 6-O-sulfohydrolases that are responsible for degradation of the isomeric chondroitin sulfates and that are present in extracts of nonadapted cells of *F. heparinum* (see Section II,C,6,a).

The precise requirements for the induction of the heparin degrading system in *Flavobacterium* are currently under investigation. Present indications [345] are that the minimum requirement is the presence of a uronic acid $\alpha(1 \rightarrow 4)$ linked to either an N-acetyl or N-sulfated hexosamine. However, in the experiments leading to these conclusions the criterion for degradation was the production of the 2,6-disulfated D-glucosamine and the conclusions really appertain more to chain degradation than to sulfate release. The experiments also suggest that D-glucosamine N-sulfate was only a weak inducer of heparin degradation. In contrast, other workers [346] have established that growth of *Flavobacterium* in media containing ^{35}S-labeled D-glucosamine N-sulfate leads to the liberation of the label as inorganic [35S]sulfate. This illustrates the difficulties inherent in induction experiments in which only parts of a multienzyme system are considered. Clarification of the situation must await the development of techniques that will enable each enzyme of the system to be studied separately.

This particular point is further emphasized as a result of some very recent work [292] which has indicated that *F. heparinum* cells grown on a heparin-free tryptone-soya peptone medium containing glucose produce a heparin depolymerase that appears to operate by a hydrolytic mechanism rather than by an elimination mechanism. This situation is reminiscent of that noted for *P. vulgaris,* where different views have been expressed about the types of enzymes responsible for the depolymerization of chondroitin 4-sulfate (see Section II,C,6,a).

Presumably the functions of the sulfohydrolases and associated enzymes in *Flavobacterium* are related to the environment of the organism in soil and the need to capitalize on whatever sources of C, N, and S are available. Meanwhile, these enzymes, in association with those of *P. vulgaris,* will continue to be valuable aids to the elucidation of the structures of complex proteoglycans [300].

b. Mammalian Heparin Sulfohydrolases. It has been known for many years that heparin does not retain its anticoagulant activity for long after being injected into mammals, and in 1940 Jaques [347] described an enzyme (heparinase) from rabbit liver which appeared to be responsible for the inactivation. Little further has been heard about this enzyme, but there are clear indications from metabolic studies that mammalian tissues can degrade and desulfate heparin to some extent [for brief reviews of some studies, see 20,288,348]. These findings were extended in our laboratories [348,349] as a result of metabolic studies on rats injected with heparin containing ^{35}S-labeled *N*-sulfate groups. Considerable amounts of the label were recovered from urine as inorganic sulfate, findings that led us to search for a rat tissue sulfamatase. Such an enzyme was detected in rat and pig spleens [339,350] and this finding was later confirmed by others [351]. Further comment about this enzyme is made in Section II,C,9,b, but it is discussed in greater detail in Section V in relation to the rupture of the N–S linkage.

It would indeed be surprising if further mammalian enzymes capable of removing sulfate groups from heparin are not discovered. Meanwhile, although there is no direct evidence, the existence of a mammalian iduronate sulfohydrolase that will hydrolyze the ester sulfate groups on position 2 of the iduronic acid residues of heparin may be supposed. The reasons for suggesting this are detailed in the following section.

9. SULFOHYDROLASES ATTACKING HEPARAN SULFATE

The heparan sulfates form a group of closely related polymers that occur in connective tissues as proteoglycans. Other types of glycosaminoglycan can coexist with heparan sulfate in the same proteoglycan. Like all other glycosaminoglycans, heparan sulfates show considerable heterogeneity, including differences in molecular size, total charge, and the manner in which charge is distributed. The general structures of heparin and heparan sulfates are similar; both are α-linked polymers and both contain D-glucosamine, D-glucuronic acid, L-iduronic acid, N-acetyl, N-sulfate, and O-sulfate. Analysis of heparan sulfates from different sources has shown that the ratio of L-iduronic acid to D-glucuronic acid, although variable, is generally appreciably lower than that found for heparins from different sources. In contrast, the content of N-acetyl groups is much higher [332], and these two findings together indicate the presence in heparan sulfate chains of segments containing little or no sulfate where the repeating unit is probably a $\beta(1 \rightarrow 4)$-linked disaccharide containing D-glucuronic acid and N-acetyl D-glucosamine. The general similarity between heparin and heparan sulfates has led to the view that the latter compounds are intermediate biosynthetic precursors of the former, although this view is not held by all investigators. The reader is referred to recent studies [332,352–354] for an impression of the complexity of the field. Meanwhile, it is reasonably certain that sulfate groups can occur on the same positions on L-iduronic acid and D-glucosamine residues as found in heparin. The disaccharide repeating unit presented for heparin in Section II,C,8 (XVIII) will therefore suffice for the present discussion of sulfohydrolase activity toward heparan sulfates.

a. *Bacterial Heparan Sulfohydrolases.* When *F. heparinum* was grown in media containing either heparin or heparan sulfate, extracts of the cells degraded both of the polymers [355]. Two distinct enzymes were present [340], both being eliminases producing sulfated $\Delta^{4,5}$-unsaturated disaccharides that could be degraded further by other enzymes. One of the eliminases (heparinase) was relatively specific for heparin (but showing some activity towards heparan sulfate) while the other (heparanase) was specific for heparan sulfate. The detailed specificity requirements for the heparinase included the presence of O- and N-sulfate groups on the polymer. In contrast, the heparanase required O-sulfate groups to be absent. During degradation of heparan sulfate by extracts of adapted *Flavobacterium,* inorganic sulfate is released from the polymer, although evidence for this is mainly indirect [240,337,356,357]. We are left to assume that desulfation proceeds in an analogous fashion to that observed for heparin and that the same three sulfate liberating enzymes are involved (Section II,C,8,a). Unfortunately, as mentioned above, the specificity claimed for heparanase suggests that its activity depends on the absence of O-sulfate groups. If this is so, then it must be presumed that either the enzyme acts only at regions of the heparan sulfate chains in which O-sulfate groups are absent (in which event the coexisting heparinase enzyme must degrade the remaining parts of the chain), or that extracts contain additional sulfohydrolases that act on O-sulfate groups in the intact polymer. There is obviously a need for further definitive studies similar to those made by Dietrich *et al.* [344] with the heparin degrading multienzyme system.

b. *Mammalian Heparan Sulfohydrolases.* Relatively little is known about the mammalian systems responsible for the degradation of heparan sulfates, but what is known is highly significant. Briefly, one of the heritable disorders affecting connective tissues, the Sanfilippo A syndrome [see 314], is characterized biochemically by an excessive urinary excretion of heparan sulfate. Fibroblasts cultured from the skin of patients accumulate excessive quantities of a heparan sulfate with a high N-sulfate content. The rate of degradation of the polymer appears to be inadequate for normal turnover. A corrective factor was separated from normal urine [358] which, when administered exogenously to Sanfilippo fibroblasts, corrected the abnormal metabolism. This, and other observations, led to the conclusion that the urine factor was a sulfate liberating enzyme. Subsequently, it was shown [359] that skin fibroblasts from normal subjects contained at least two enzymes that participate in the desulfation of heparan sulfate. One of these liberated inorganic sulfate from a de-N-sulfated heparan sulfate containing ^{35}S-labeled O-sulfate groups. The responsible O-sulfohydrolase was optimally active at pH 4.2 and was also present in Sanfilippo A cells. Although not absolutely clear from the experimental results, this enzyme probably acts on polymeric heparan sulfate.

Also present in normal fibroblast extracts was a sulfamatase that was capable of hydrolyzing the N-sulfate groups of polymeric heparan sulfate. This sulfamatase had an optimum pH of 4.9 and was not present in fibroblasts from Sanfilippo A patients. The enzyme was identical with the "corrective factor" isolated from normal urine. In its absence heparan sulfate accumulates and this presumably means that removal of the N-sulfate groups is a prerequisite for degradation of the glycosaminoglycan chain. Further comment on this enzyme is made in Section V.

10. GENERAL COMMENTS

It will be apparent from the contents of Section III that the carbohydrate sulfohydrolases still constitute a largely unexplored group of enzymes from the point of view of the enzymologist. Moreover, it will be surprising if further examples of these enzymes do not come to light. It will not have escaped notice that no mention has been made of sulfohydrolases that act on sulfated glycoproteins. Recently, a brief report [360] described the enzymatic liberation of inorganic [^{35}S]sulfate from ^{35}S-labeled polyanions isolated from rat stomach and duodenum. The enzyme preparation came from gastric mucosa and was maximally active at pH 4.0. There was no indication, from the report, of the precise nature of the polyanionic material, but the known general interests of the authors lead one to suspect that sulfated glycoprotein might have been present. Almost certainly, sulfohydrolase enzymes acting on sulfated glycoproteins must be present in some organisms.

It is difficult to generalize about such a diverse group of enzymes as the carbohydrate sulfohydrolases and there may be little to be gained at the moment from attempting such generalizations. Evidence presently available suggests that most, if not all, of the enzymes are inhibited by phosphate and this may indicate some general similarity in the mechanisms of action of them. Certainly, one of the exciting prospects for the future would be the possibility of comparative molecular and mechanistic studies on purified samples of the different enzymes, particularly because of the evolutionary implications. Unfortunately, in common with many other sulfohydrolases, purification of sufficient quantities of the enzymes for detailed studies is a major problem.

One further point might be made concerning those carbohydrate sulfohydrolases that are involved in the degradation of polymers. A picture seems to emerge of sulfohydrolases of bacterial origin acting only after depolymerization of polysaccharide chains. On the other hand, animal sulfohydrolases appear to exist that act on polymeric and oligomeric forms as well as on dimers and monomers. At the present time it is not possible to account for these variations purely in terms of the molecular conformations adopted by the substrates, for a conformation that might in animals require sulfate removal before internal glycosidic bonds are sufficiently exposed ought surely to present the same problem to the microorganisms.

III. ENZYMES HYDROLYZING P–O–SO$_3^-$ LINKAGES

A. Sulfatophosphate Sulfohydrolases

1. GENERAL COMMENTS ON THE DEGRADATION OF SULFATOPHOSPHATES

The so-called "active" forms of sulfate, adenylyl sulfate (adenosine 5'-sulfatophosphate, APS) (XX) and 3'-phosphoadenylyl sulfate (3'-phosphoadenosine 5'-sulfatophosphate, PAPS) (XXI), occupy a central position in the biochemistry of the sulfate ion in animals, plants, and microorganisms. Their roles in the biotransformation of inorganic forms

(XX)   (XXI)

of sulfur have previously been reviewed [4,361] and further information is presented elsewhere in this volume.

Following the elucidation of the structures of APS and PAPS [362], efforts were made to understand how they were synthesized in living tissues and how they were involved in the biotransformation of inorganic forms of sulfate. Almost no attention was paid to the possible occurrence of enzymes that could participate in the degradation of the sulfated nucleotides, although it slowly became apparent that such enzymes did exist. For example, Brunngraber [363], during studies on the biosynthesis of PAPS by rat liver supernatant preparations, noted that optimum concentrations of PAPS were obtained relatively quickly and thereafter the compound was gradually broken down. Breakdown was attributed to the presence of a magnesium-dependent 3′-nucleotidase that removed the 3′-phosphate group of PAPS to yield APS [cf. 364] and that was also able to remove the 3′-phosphate group from 3′-phosphoadenosine 5′-phosphate (PAP). These observations were later confirmed by Spencer [162], who also indicated that a sulfohydrolase was present that liberated SO_4^{2-} from any APS formed. A later and very brief report [365] also indicated the presence of deaminases, a 5′-nucleotidase, and a PAPS-sulfohydrolase, all of which could participate in PAPS and APS degradation. Unfortunately, a detailed account of this work has not been published.

Other studies with hen oviduct [233], yeast [366], and snake venom [367] preparations provided indirect evidence of the existence of sulfohydrolases active toward PAPS and it gradually became evident [see also 368,369] that their presence in cell preparations was an undesirable complication in experiments in which PAPS was being produced for biosynthetic purposes. Realization of this caused a number of laboratories to begin studies on the enzymes, studies that initially led to a confusion that is still not completely resolved.

The first detailed study [368] was probably responsible for some of the confusion. An extract of sheep's brain was employed as enzyme source, ^{35}S-labeled PAPS as substrate, and incubation of the two was always carried out in the presence of cobalt ions. The work apparently demonstrated the existence of a Co^{2+} activated PAPS-sulfohydrolase (Mn^{2+} gave a similar activation) with a pH optimum in the region of 6.0 and a K_m value of 4.5×10^{-5} M. PAP was claimed to be a product of enzyme action, although the evidence for this was not unequivocal. Various rat tissues were also claimed to possess the enzyme.

From previous work on sulfohydrolases it might have been considered questionable whether Co^{2+} was likely to serve as an activator for a PAPS-sulfohydrolase. On the other hand, it was known that some 5'- and 3'-nucleotidases were markedly activated by certain divalent cations [363,370,371]. In spite of this, Co^{2+} was always included in the assay mixture in most of the investigations that were subsequently made. Such studies included one by Koizumi et al. [372] on preparations from male albino rats of Sprague–Dawley strain. It was concluded that "PAPS-degrading enzyme" (assayed by measuring $^{35}SO_4^{2-}$ liberated from PAP^{35}S) was mainly localized in lysosomes of the normal liver cell, although activity was also present in other particulate cellular components. Chronic damage (CCl_4 or hypervitaminosis A) led to elevated levels of the enzyme in the soluble fraction of the cell and it was suggested that release of lysosomal enzymes was occurring in these cases.

Other work on pig kidney cortex [373] confirmed the wide distribution of PAPS degrading enzymes within the cell. Again Co^{2+} was present in incubation mixtures, which were maintained at a pH favorable to 3'-nucleotidase activity. In this study, enzyme assay involved measurement of the amount of PAP^{35}S that had disappeared that could not be accounted for by the appearance of AP^{35}S. Release of $^{35}SO_4^{2-}$ was ascribed to a PAPS-sulfohydrolase. However, the assay method employed was not unequivocal and no account was taken of the possible degradative route:

$$PAP^{35}S \rightarrow AP^{35}S \rightarrow AMP + {}^{35}SO_4^{2-}$$

In a similar study [374], PAP^{35}S was degraded by cell-free extracts of *Euglena gracilis* in the presence of Mg^{2+}. Liberated $^{35}SO_4^{2-}$ was measured and PAP was also claimed as a product, although no attempts were made to establish the quantitative relationship between the two. Maximum activity was obtained at pH 5.5, but the pH–activity curve suggested the presence of more than one enzyme. Mg^{2+} stimulated enzyme activity about fourfold and Co^{2+} about threefold. From the quantitative point of view, the results were again equivocal and the effects of Mg^{2+} and Co^{2+} are such that one suspects the presence of more than one degrading enzyme (including a 3'-nucleotidase) in the extracts.

Two further studies are relevant to the problem of enzyme multiplic-

412 K. S. DODGSON AND F. A. ROSE

ity. The first [375], described the properties and subcellular distribution of two pig kidney APS-sulfohydrolases while the other [376] reported on the presence of a phosphohydrolase in sheep brain that dephosphorylated PAPS to yield APS. The latter enzyme appeared to be distinct from 3'-nucleotidase.

Collectively, these results present a confusing picture of PAPS degradation and the number and subcellular localization of enzymes involved therein. However, some light has recently been shed as a result of work from our laboratories [377–383]. This established that the ability to degrade PAPS is a feature of many invertebrate and vertebrate organisms and also of a wide variety of different mammalian organs. Investigations were concentrated on enzymes present in the cell sap of rat and ox liver cells and emphasis was placed on the elucidation of the role of Co^{2+} in the degradative process. Initial studies on rat provided clear indications of the presence in the cell sap of at least 3 enzymes that could participate in PAPS degradation, namely, a PAPS-sulfohydrolase, an APS-sulfohydrolase, and a 3'-nucleotidase. Only the last of these was activated by Co^{2+}. APS-sulfohydrolase activity was also present in the particulate components of the cell and the lysosomal enzyme was partially purified and shown to be quite different from the corresponding cell sap enzyme [278].

In more detailed studies with ox liver, the cell sap APS- and PAPS-sulfohydrolases and the 3'-nucleotidase were separated and studied independently. Only the 3'-nucleotidase was activated by Co^{2+}. The APS-sulfohydrolase was purified to homogeneity and its properties are described in Section III,A,2. Collectively, the studies established the existence in the cell saps of bovine and rat livers of at least two alternative routes for the biodegradation of PAPS (Fig. 3). The first of these involves direct desulfation of the sulfatophosphate by a PAPS-

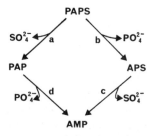

FIG. 3. Alternative pathways suggested for the enzymatic degradation of PAPS by mammalian liver cytosol preparations [382]. Enzymes on the pathways are (a) PAPS-sulfohydrolase, (b) PAPS 3'-nucleotidase and/or PAPS-phosphohydrolase, (c) APS-sulfohydrolase, (d) PAP-phosphohydrolase.

sulfohydrolase that is not activated by Co^{2+} (Fig. 3, step a). The PAP formed can lose its 3'-phosphate group to give AMP (Fig. 3, step d). It is not yet certain whether only a 3'-nucleotidase is involved in this step or whether a PAP-phosphohydrolase [363,365] also participates. Thus the optimum pH of PAPS-sulfohydrolase is in the region of 9 (see Section III,A,3), whereas that of the 3'-nucleotidase (acting on PAPS) is at 5.9, a difference that might be regarded as incompatible with the concept of two enzymes operating sequentially within the same cell compartment.

The second route for PAPS degradation involves the action of the 3'-nucleotidase [and PAPS-phosphohydrolase, see 376] to give APS (Fig. 3, step b) and the subsequent desulfation of the latter by APS-sulfohydrolase to yield AMP (Fig. 3, step c). The APS-sulfohydrolase is not activated by Co^{2+} and the pH optima for the dephosphorylation (pH 5.9) and desulfation steps (pH 5.2) are compatible.

Other results from various laboratories [see also 384] show that PAPS degradation is also a feature of particulate components of mammalian cells but, in order to establish the precise nature of the systems involved, it will be necessary to proceed systematically with each type of particle, as was done with the cell sap preparations of rat and bovine livers. What is certain is that distinct and specific sulfohydrolases exist that use PAPS or APS as substrates and further details of some of these enzymes are given in the following sections.

2. APS-SULFOHYDROLASE

The cell sap and the lysosomal enzymes from rat liver have been studied in some detail in Cardiff [377,378]. Briefly, the cell sap enzyme was purified 137-fold from a high speed supernatant fraction of rat liver. The resulting preparation showed no 3'-nucleotidase or sulfohydrolase activity toward PAPS, although several protein bands separated on gel electrophoresis. The enzyme exhibited maximum activity at pH 5.2 and an APS concentration of 7.5 mM ($K_m = 1.0$ mM). However, although stable at pH 6.5–9.0, the enzyme was relatively unstable at its optimum pH and this necessitated the use of very short incubation times. The products of enzyme action were SO_4^{2-} and AMP which were produced in equimolar amounts. Metals such as Co^{2+}, Mn^{2+}, Cu^{2+}, and Mg^{2+} tended to inhibit the enzyme rather than activate. ATP at a concentration of 0.1 mM was strongly inhibitory, ADP and AMP hardly so at the same concentration. Sulfate ions did not inhibit significantly at 10.0 mM concentration, whereas phosphate and pyrophosphate did.

The lysosomal enzyme was purified 56-fold from whole lysosomes, but the preparation was still very impure and still retained some ability to degrade PAPS. Activity toward APS was maximal at pH 5.0 in the presence of 12.0 mM substrate but, in contrast to the cell sap enzyme, the pH–activity curve showed an extremely broad profile. AMP and SO_4^{2-} were the sole products of enzyme action and were produced in equimolar amounts.

The lysosomal and cell sap enzymes exhibited very different behavior and are clearly distinct enzymes (see Table I).

Because of the difficulty of preparing significant quantities of pure APS-sulfhohydrolase from rat liver, attention was turned toward ox liver. The cell sap enzyme was purified to homogeneity [381] and attempts are currently in progress to purify the lysosomal enzyme. The molecular weight of the cell sap enzyme was 68,000 daltons while the amino acid composition showed that the high proline content characteristic of ox liver arylsulfohydrolase A [29,47] was not a feature of APS-sulfohydrolase. The enzyme exhibited maximum activity at pH 5.4 and 4.0 mM APS, inhibition with excess substrate being observed at higher APS concentrations. AMP and SO_4^{2-}, the sole products of enzymatic hydrolysis, inhibited the enzyme noncompetitively and competitively, respectively, a pattern of product inhibition characteristic of an ordered uni-bi reaction sequence with SO_4^{2-} as the last released product. Although an intermediate enzyme–SO_4^{2-} complex is implicit in the mechanism, attempts to demonstrate the complex were not successful. Photochemical inactivation and other studies indicated the presence of an essential histidine residue at the active site of the enzyme. The enzyme could not utilize PAPS as a substrate, neither was it activated by any of a number of metals that were tested. ATP was an extremely powerful competitive inhibitor ($K_i = 7.5 \times 10^{-6}$ M) while ADP and AMP were less powerful noncompetitive inhibitors (2.5×10^{-5} M and 1.5×10^{-3} M, respectively).

TABLE I

SUMMARY OF THE CONTRASTING PROPERTIES OF THE SUPERNATANT AND LYSOSOMAL APS-SULFOHYDROLASES OF RAT LIVER

Treatment	Supernatant enzyme	Lysosomal enzyme
Alumina C_γ gel	Adsorbed	Not adsorbed
DEAE-Sephadex	Adsorbed	Not adsorbed
pH	Sharp pH optimum at 5.2	Broad pH optimum at 5.0
	Unstable at optimum pH	Stable at optimum pH
0.1 mM Co^{2+}	20% inhibition	8% inhibition
0.1 mM Cu^{2+}	76% inhibition	14% inhibition
10.0 mM P_i	55% inhibition	14% inhibition
10.0 mM F^-	24% inhibition	0% inhibition

The only other detailed studies made on APS-sulfohydrolase [375] involved a 13,000 *g* supernatant from a water homogenate of pig kidney. It is difficult to draw any sound conclusions from the work because of the crude nature of the system and the use of water as homogenizing medium. However, it was suggested that at least two APS-sulfohydrolases were present in such preparations, one with a pH optimum of about 5.0 and the other of about 7.0. Neither enzyme appeared to be activated by Co^{2+} or other divalent metals and ATP was an inhibitor. The method of preparation of the cell extract, together with certain other factors, when considered in the light of the work on the rat liver cell sap and lysosomal APS-sulfohydrolases [378] suggest that both cell sap and lysosomal enzymes were present in the pig kidney preparations.

3. PAPS-SULFOHYDROLASE

Little meaningful work has yet been done on this enzyme. As explained in Section III,A,1, the decision to include Co^{2+} as a component of incubation mixtures in most studies has, in retrospect, been unfortunate and it is difficult to derive much value from the results. Clearly new studies ought to be undertaken, but only on a separated and purified enzyme.

Most of the available information which is of value concerns the bovine liver cell sap enzyme [382]. The enzyme was not highly purified, but was completely free from 3'-nucleotidase and APS-sulfohydrolase activities. Maximum activity was obtained at pH 8.8 in 0.1 *M* Tris-HCl buffer and at pH 9.4 in 0.1 *M* glycine-glycylglycine-NaOH buffer ($K_m = 1.7 \times 10^{-5}$ *M*). Co^{2+}, Mn^{2+}, and Mg^{2+} did not serve as activators for the enzyme and were inhibitory at 0.2 m*M* concentration; AMP, ADP, and ATP inhibited to the extent of 55, 61, and 83%, respectively, at 1.0 m*M* concentration.

4. PHYSIOLOGICAL ROLES

It is apparent from the literature that the story of PAPS degradation is far from complete and that the enzymes involved in this degradation are probably widely distributed within the cell. The APS- and PAPS-sulfohydrolases that have been studied in detail show no cross-specificity regarding their substrates and there are no indications that they are present for purposes other than that suggested by their names. Certainly, the lysosomal APS-sulfohydrolase cannot be identical to either arylsulfohydrolases A and B [385], neither does the cell sap enzyme possess phosphatase activity toward ADP or ATP [381]. Of course, the cell sap also contains ATP-sulfurylase. This enzyme is involved in APS biosynthesis and the equilibrium of the reaction catalyzed by the enzyme certainly lies in favor of APS degradation. However, inorganic pyrophosphate would be required as a reaction component and the product (other than SO_4^{2-}) would be ATP, facts which clearly distinguish the enzyme from APS-sulfohydrolase.

At present, therefore, the APS- and PAPS-sulfohydrolases must be regarded as hydrolytic enzymes participating in the degradation of active forms of sulfate in order to produce AMP. Both types of enzyme appear to be strongly inhibited by ATP and would therefore be inoperative when ATP levels in the cell are high and when PAPS biosynthesis is proceeding in the cell sap. Presumably, when the requirement for PAPS biosynthesis ceases, the cell sap sulfohydrolases participate in the process of regenerating ATP (via AMP) from unwanted excess PAPS and/or APS, and this may well be a useful and important step for the conservation of adenosine compounds. Whether this explanation could be offered for the APS- and PAPS-sulfohydrolases present in other cellular components would seem to be less certain. In any event, the whole field of PAPS degradation and the associated enzymes is clearly ripe for further study.

IV. ENZYMES HYDROLYZING N–O–SO$_3^-$ LINKAGES

This type of linkage occurs naturally in the members of an interesting group of compounds, the mustard oil glycosides. Probably the best known of these compounds is myronate (sinigrin, XXII), a component of black mustard seeds.

$$CH_2{=}CH{-}CH_2{-}C\underset{N \cdot OSO_3K}{\overset{S \cdot C_6H_{11}O_5}{\big<}}$$

(XXII)

A wide variety of analogous sulfated thioglycosides are found in plants of the Cruciferae family [e.g., garlic, onion, horseradish, mustard, see 386,387] and their value as potential medicines and food flavorings has long been recognized. This value depends on the fact that they are readily degraded by coexisting plant enzymes to yield inorganic sulfate, glucose, and complex isothiocyanates known as mustard oils. The medicinal and culinary importance of the parent glycosides can be traced to the enzymatic production of these mustard oils, which are generally characterized by their biting taste and pungent odor.

The existence of enzymes that were able to degrade these glycosides was known before the turn of the century, but it was left to Carl Neuberg to make detailed studies of them [388–390]. As a result of these studies it was generally accepted that two types of enzyme were involved, a glycosidase and a sulfohydrolase. The latter enzyme could apparently be detected in mammalian tissues [but see 391] as well as those of plants.

Space does not permit a detailed account of the confusion that subsequently followed, but much of it could be related to the inadequacy of the general chemical structure assigned to the mustard oil glycosides by early workers. The revision of structure by Ettlinger and Lundeen [392,393] led to the concept that only one enzyme was required to degrade the compounds, namely, a thioglycosidase (EC 3.2.3.1; thioglucoside glucohydrolase) that liberated glucose from the substrate. This would then be followed by a spontaneous, nonenzymatic loss of sulfate from the aglycone as a rearrangement (similar to the Lossen rearrangement of hydroxamic acids) of the C and N atoms occurred to yield the mustard oil.

Following this work, a number of attempts were made [394–397] to establish whether a sulfohydrolase acting on mustard oil glycosides actually existed and, at the present time, the weight of evidence is against this. Indeed, the sinigrin sulfohydrolase listed in the first and second editions of the Enzyme Commission report has been deleted from the recent third edition. Studies now concentrate on the thioglycosidase and reports on the enzyme in rapeseed [398], white mustard seed [399], Aspergillus sydowi [400], and Aspergillus niger [401] have all assumed the absence of a sulfohydrolase.

Meanwhile, the situation is still unsatisfactory. For example, it is claimed [395] that a sulfohydrolase can be separated from yellow mustard seeds that is able to hydrolyze chemically synthesized oxime O-sulfates, as well as mustard oil glycosides. Furthermore, the claims for the existence of an independent sulfohydrolase in mollusks [273,402] have never been adequately refuted. It has previously been stressed [4] that the status of sinigrin sulfohydrolase is currently quite unsatisfactory and we can only restate the necessity for further studies using oxime O-sulfates that are unable to serve as substrates for the thioglycosidase.

V. ENZYMES HYDROLYZING N–SO₃⁻ LINKAGES

Sulfamatases (N-sulfohydrolases, sulfamidases) have been found in the bacterium F. heparinum and in mammalian organisms. The bacterial enzyme was discussed briefly in Section II,C,8,a, particularly from the point of view of its participation in the degradation of heparin and heparan sulfate. The enzyme has been studied in two laboratories [339,341] and both have confirmed that the heparin polymer is not a substrate for the enzyme. Either 2-sulfoamino-2-deoxy-D-glucose or 2-sulfoamino-2-deoxy-D-glucose 6-O-sulfate can serve as substrate and the name sulfoglucosamine sulfamidase has been recommended on this basis (EC 3.10.1.1). This name is in fact not strictly correct as the enzyme substrates have sulfamate rather than sulfamide groups.

The *Flavobacterium* enzyme was inhibited by phosphate and sulfate, but not by the *N*-sulfates of L-serine, D-cycloserine, and aniline.

Comment was made in Section II,C,9,b that a mammalian sulfamatase that participates in the degradation of heparan sulfates is absent from the tissues of patients with the Sanfilippo A syndrome. In its absence, heparan sulfate appears in urine and accumulates in tissues. Loss of *N*-sulfate groups therefore seems to be a prerequisite for polymer degradation in mammals. The question now arises as to whether this enzyme is identical to the one reported some years ago by our own laboratories [350]. The enzyme was found in rat spleen following studies [348,349] on the metabolic fate of re-N-[^{35}S] sulfated heparin [403] and a similar enzyme was later partially purified from pig spleen [339]. The pig spleen enzyme was maximally active toward polymeric heparin at pH 5.0, but was inactive toward 2-sulfoamino-2-deoxy-D-glucose and the corresponding 6-O-sulfate. Sulfate and phosphate were inhibitory and, in contrast to the *Flavobacterium* enzyme, so also were the *N*-sulfates of L-serine, D-cycloserine, and aniline. Subsequently, a similar sulfamatase was noted in lymphoid tissues of dogs, humans, and rats [351]. A partially purified preparation of the enzyme (presumably from rat spleen, but not clear from the report) showed properties analogous to those of the pig spleen enzyme, but two further points of some importance were noted. The first of these indicated that only one sulfate group was liberated from each heparin molecule, and the other that unsaturated tetrasaccharide and higher oligosaccharide fragments, prepared by digesting heparin with *Flavobacterium* enzymes, could also serve as substrates for the sulfamatase. Although no direct evidence is available, it seems possible that the mammalian sulfamatase active towards heparin is the same one that is active toward heparan sulfate and that is deficient in the Sanfilippo A syndrome. Meanwhile, it seems reasonably certain that the mammalian heparin sulfamatase is different from any of the mammalian arylsulfohydrolase enzymes [404,405].

It will be surprising if further microbial and animal sources of this type of enzyme do not come to light in the near future.

VI. GENERAL COMMENTS

It will be apparent from this account that the sulfohydrolases are likely to form a subject of interest for some years to come and there is little doubt that a number of new enzymes will emerge as a result of future studies. Probably several lines of research will become particularly prominent; for example, the nature and extent of the roles played by microbial sulfohydrolases in the general recycling of sulfur. The sulfur cycle is vital to life on this planet and the amounts of the element that are returned to soils and waters in the form of sulfate esters must be enormous. A rough estimate suggests that in terms of human excreta alone, almost 50 tons of sulfur are daily returned to the sulfur cycle in the form of sulfate esters. Indeed, Freney [406] has shown that ester linked sulfate can be detected in significant quantities in a variety of soils. However, we have confirmed his conclusion that this sulfate is present as a component of certain complex soil colloids including the so-called humic acids and, as such, is remarkably resistant to attack by microorganisms. We are led to conclude (see also Section II,A,2,c) that the majority of the sulfate esters that are returned to the soil from living organisms and industrial sources (e.g., detergents) are degraded almost immediately. The inorganic sulfate so formed may then serve as a sulfur source for other microorganisms and for plants. It must therefore be presumed that many microorganisms are capable of producing sulfohydrolases of one type or another. The induction and repression of these enzymes and their genetic relationship to each other and to other sulfur metabolizing enzymes when present in the same organism ought to be of great interest to the microbial geneticist and is already beginning to at tract the attention of a number of workers. Apart from this genetic interest, the possibility exists that some of the sulfohydrolases produced by microorganisms may be of considerable potential value as analytical or structural tools. This is already well recognized for the sulfohydrolases present in preparations from *P. vulgaris* and *F. heparinum* (see Sections II,C,6,a and II,C,8,a) but, for example, there is still a need for highly specific enzymes that might be used for the estimation of sulfated steroids in urine and other body fluids.

When methods for the isolation of the sulfohydrolases in quantity have been improved, the enzymologist and the student of biochemical evolution should find the enzymes to be a particularly appealing field for their researches. For example, evidence is slowly accumulating that histidine residues may be important features of the active sites of several sulfohydrolases and the possibility arises that the enzymes may share a common catalytic site, but vary in the manner in which the substrate molecules are accommodated on the enzyme molecule. The question then arises as to whether the enzymes have developed from a common ancestral form and, if so, what form. Was it, for example, an APS-sulfohydrolase—a possibility suggested by the vital role that APS must have played in primitive microorganisms that flourished before molecular oxygen appeared on earth?

Most intriguing of all is the fact that, of all the many different types of hydrolases present in living organisms, the sulfohydrolases are probably the most mysterious from the point of view of physiological function. Much of the future work may well be directed toward the elucidation of function and, in particular cases, to establishing which sulfate esters really serve as the natural substrates. In mammals the possible involvement of sulfohydrolases in disease processes is particularly relevant at the moment and we may expect to see further interesting developments on these lines in the future.

Ultimately, the aim must be to understand how these enzymes and their substrates fit into the overall picture of sulfur biochemistry. This is certainly far from clear at the moment, but at least both enzymes and substrates are no longer regarded as curiosities of nature, long overtaken by the events of evolution.

VII. ADDENDUM

A number of studies have been reported since the completion of the present chapter, some of which provide important additional information.

In relation to the arylsulfohydrolases, it has been shown [407] that human arylsulfohydrolase A (cerebroside sulfohydrolase) is able to hydrolyze lactosyl sulfatide isolated from human kidney and [408] the sulfated glycerogalactolipid present in human testes. In both of these compounds the ester sulfate group is present at position 3 on the galactose residue as is the case with cerebroside sulfate. Other studies [409] on arylsulfohydrolase A appear to substantiate the claims [45] that the enzyme is also capable of hydrolyzing ascorbic acid 2-sulfate (see Section II,C,4). Enzyme preparations from certain invertebrate organisms that are rich in arylsulphohydrolase activity are also capable of hydrolyzing cerebroside sulfate [410] or ascorbic acid 2-sulfate [411]. However, it is too early to say whether the relationship between the activities noted is more than coincidental.

Significant developments appear to be imminent in relation to human arylsulphohydrolase B. Thus, tissues [412,413] and fibroblasts [414] from certain patients with the clinical picture of Maroteaux–Lamy syndrome have been shown to be markedly deficient in arylsulphohydrolase B. The possibility exists that the enzyme is actually concerned with the enzymatic desulfation of the N-acetylgalactosamine 4-sulfate residues present in the dermatan sulfate [415] that accumulates in fibroblasts and other tissues in Maroteaux-Lamy patients. This possibility would be in accord with the predictions made earlier in Sections II,A,1,b and II,C,6,c.

In relation to choline sulfohydrolase, further studies on the regulation of the N. crassa enzyme have been reported [416]. The enzyme undergoes a rapid turnover during development of the organism in sharp contrast to arylsulfohydrolase, which is a stable species. Although the two enzymes are similarly metabolically and genetically regulated, it seems probable that there is superimposed a developmental control system that can override the other regulatory mechanisms at appropriate stages in the life cycle of the organism. A new bacterial source (Pseudomonas V-A, isolated from sewage) of choline sulfohydrolase has been reported [417]. In common with arylsulfohydrolases of Pseudomonas sp. (Section II,A,2,b), but in contrast to the alkylsulfohydrolases of such species (Section II,B,1), formation of the enzyme is subject to inhibition by inorganic sulfate and L-cysteine.

It has been known for some years that the isthmus region of the hen oviduct contains a unique series of sulfated sugar nucleotides, including UDP-N-acetyl-D-galactosamine 4-O-sulfate and UDP-N-acetyl-D-galactosamine 4,6-di-O-sulfate. Recent work by Suzuki and his colleagues [418] has shown that the isthmus region also contains a sulfohydrolase which hydrolyzes ester sulfate groups at position 4 on these nucleotides. High specificity is claimed for the enzyme and there appears to be good reason to suppose that it may be an entirely new sulfohydrolase.

ACKNOWLEDGMENT

The authors of this chapter wish to express their appreciation to The Wellcome Trust for support extending over several years which has enabled our own laboratories to contribute to progress in the field described.

REFERENCES

1. H. Baum and K. S. Dodgson, Nature (London) 181, 115 (1958).
2. B. Spencer, Biochem. J. 69, 155 (1958).

3. K. S. Dodgson and B. Spencer, *Rep. Progr. Chem.* **53,** 318 (1957).
4. A. B. Roy and P. A. Trudinger, "The Biochemistry of Inorganic Compounds of Sulphur." Cambridge Univ. Press, London and New York, 1970.
5. K. S. Dodgson and F. A. Rose, *in* "Metabolic Conjugation and Metabolic Hydrolysis" (W. H. Fishman, ed.), Vol. 1, p. 239. Academic Press, New York, 1970.
6. R. G. Nicholls and A. B. Roy, *in* "The Enzymes" (P. D. Boyer, ed.), 3rd ed., Vol. 5, p. 21. Academic Press, New York, 1971.
7. M. Derrien, *Bull. Soc. Chim. Fr.* **9,** 110 (1911).
8. C. Neuberg and K. Kurono, *Biochem. Z.* **140,** 295 (1923).
9. C. Neuberg and E. Simon, *Ergeb. Physiol.* **34,** 896 (1932).
10. K. S. Dodgson and B. Spencer, *Biochem. J.* **53,** 444 (1953).
11. K. S. Dodgson, B. Spencer, and J. Thomas, *Biochem. J.* **56,** 117 (1954).
12. A. B. Roy, *Biochim. Biophys. Acta* **14,** 149 (1954).
13. A. B. Roy, *Biochem. J.* **53,** 12 (1953).
14. K. S. Dodgson, B. Spencer, and J. Thomas, *Biochem. J.* **59,** 29 (1955).
15. K. S. Dodgson, B. Spencer, and C. H. Wynn, *Biochem. J.* **62,** 500 (1956).
16. R. Viala and R. Gianetto, *Can. J. Biochem. Physiol.* **33,** 839 (1955).
17. K. S. Dodgson and B. Spencer, *Methods Biochem. Anal.* **4,** 211 (1957).
18. A. B. Roy, *Biochem. J.* **55,** 653 (1953).
19. K. S. Dodgson and B. Spencer, *Biochem. J.* **62,** 30P (1956).
20. A. S. Balasubramanian and B. K. Bachhawat, *J. Neurochem.* **10,** 201 (1963).
21. J. H. Austin, D. McAfee, D. Armstrong, M. O'Rourke, L. Shearer, and B. K. Bachhawat, *Biochem. J.* **95,** 150 (1964).
22. W. S. Bleszynski and L. M. Działoszyński, *Biochem. J.* **97,** 360 (1965).
23. A. A. Farooqui and B. K. Bachhawat, *Biochem. J.* **126,** 1025 (1972).
24. K. S. Dodgson and B. Spencer, *Biochim. Biophys. Acta* **21,** 175 (1956).
25. H. Baum, K. S. Dodgson, and B. Spencer, *Biochem. J.* **69,** 567 (1958).
26. H. Baum and K. S. Dodgson, *Biochem. J.* **69,** 573 (1958).
27. R. G. Nicholls and A. B. Roy, *Biochim. Biophys. Acta* **242,** 141 (1971).
28. K. Stinshoff, *Biochim. Biophys. Acta* **276,** 475 (1972).
29. L. W. Nichol and A. B. Roy, *Biochemistry* **4,** 386 (1965).
30. M. Worwood, K. S. Dodgson, G. E. R. Hook, and F. A. Rose, *Biochem. J.* **134,** 183 (1973).
31. A. B. Roy and A. Jerfy, *Biochim. Biophys. Acta* **207,** 156 (1970).
32. E. R. B. Graham and A. B. Roy, *Biochim. Biophys. Acta* **329,** 88 (1973).
33. A. Goldstone and H. Koenig, *Life Sci.* **9,** 1341 (1970).
34. A. Goldstone, P. Konecny, and H. Koenig, *FEBS (Fed. Eur. Biochem. Soc.) Lett.* **13,** 68 (1971).
35. A. Jerfy and A. B. Roy, *Biochim. Biophys. Acta* **175,** 355 (1969).
36. A. B. Roy, *Biochim. Biophys. Acta* **198,** 76 (1970).
37. S. J. Benkovic and L. K. Dunikoski, *Biochemistry* **9,** 1390 (1970).
38. A. Jerfy and A. B. Roy, *Biochim. Biophys. Acta* **293,** 178 (1973).
39. H. Jatzkewitz, *Hoppe-Seyler's Z. Physiol. Chem.* **318,** 265 (1960).
40. E. Mehl and H. Jatzkewitz, *Hoppe-Seyler's Z. Physiol. Chem.* **331,** 292 (1963).
41. E. Mehl and H. Jatzkewitz, *Hoppe-Seyler's Z. Physiol. Chem.* **339,** 260 (1964).
42. E. Mehl and H. Jatzkewitz, *Biochem. Biophys. Res. Commun.* **19,** 407 (1965).
43. E. Mehl and H. Jatzkewitz, *Biochim. Biophys. Acta* **151,** 619 (1968).
44. K. S. Dodgson, F. A. Rose, and N. Tudball, *Biochem. J.* **71,** 10 (1959).
45. B. M. Tolbert, W. W. Bullen, M. Downing, and E. M. Baker, *Fed. Proc., Fed. Amer. Soc. Exp. Biol.* **32,** 931 (1973).
46. E. Neuwelt, D. Stumpf, J. Austin, and P. Kohler, *Biochim. Biophys. Acta* **236,** 333 (1971).

47. J. L. Breslow and H. R. Sloan, *Biochem. Biophys. Res. Commun.* **46,** 919 (1972).
48. R. L. Stevens, M. Hartman, A. L. Fluharty, and H. Kihara, *Biochim. Biophys. Acta* **302,** 338 (1973).
49. R. K. Draper and J. Edmond, *Fed. Proc., Fed. Amer. Soc. Exp. Biol.* **32,** 553 (1973).
50. K. S. Dodgson and C. H. Wynn, *Biochem. J.* **68,** 387 (1958).
51. E. C. Webb and P. F. W. Morrow, *Biochem. J.* **73,** 7 (1959).
52. E. C. Webb and P. F. W. Morrow, *Biochim. Biophys. Acta* **39,** 542 (1960).
53. R. C. Davies, A. Neuberger, and B. M. Wilson, *Biochim. Biophys. Acta* **178,** 294 (1969).
54. E. Allen and A. B. Roy, *Biochim. Biophys. Acta* **168,** 243 (1968).
55. B. Wortman, *Arch. Biochem. Biophys.* **97,** 70 (1962).
56. W. Bleszynski, *Enzymologia* **32,** 169 (1967).
57. J. Gniot-Szulżycka, *Acta Biochim. Pol.* **19,** 181 (1972).
58. W. Bleszynski, *Zesz. Nauk. Uniw. Mikolaja Kopernika Toruniu, Nauki Mat.-Pyzr.* **15,** Biol. IX, 33 (1966).
59. W. S. Bleszynski and A. Leznicki, *Enzymologia* **33,** 373 (1967).
60. W. S. Bleszynski and A. B. Roy, *Biochim. Biophys. Acta* **317,** 164 (1973).
61. A. B. Roy, *Biochem. J.* **77,** 380 (1960).
62. G. E. R. Hook, K. S. Dodgson, F. A. Rose, and M. Worwood, *Biochem. J.* **134,** 191 (1973).
63. T. G. Flynn, K. S. Dodgson, G. M. Powell, and F. A. Rose, *Biochem. J.* **105,** 1003 (1967).
64. A. A. Farooqui and B. K. Bachhawat, *J. Neurochem.* **18,** 635 (1971).
65. D. W. Milsom, F. A. Rose, and K. S. Dodgson, *Biochem. J.* **128,** 331 (1972).
66. K. S. Dodgson, F. A. Rose, and B. Spencer, *Biochem. J.* **66,** 357 (1957).
67. A. B. Roy, *Biochem. J.* **62,** 41 (1956).
68. J. Gniot-Szulżycka and M. Komoszyński, *Enzymologia* **42,** 11 (1972).
69. D. W. Milsom and F. A. Rose, *Biochem. J.* **118,** 22P (1970).
70. D. J. Hanahan, N. B. Everett, and C. D. Davis, *Arch. Biochem. Biophys.* **23,** 501 (1949).
71. M. O. Pulkkinen and I. Paunio, *Ann. Med. Exp. Biol. Fenn.* **41,** 283 (1963).
72. A. P. French and J. C. Warren, *Biochem. J.* **105,** 233 (1967).
73. N. G. Zuckerman and D. D. Hagerman, *Arch. Biochem. Biophys.* **135,** 410 (1969).
74. J. O. Dolly, K. S. Dodgson, and F. A. Rose, *Biochem. J.* **128,** 337 (1972).
75. J. O. Dolly, C. G. Curtis, K. S. Dodgson, and F. A. Rose, *Biochem. J.* **123,** 261 (1971).
76. H. Baum, K. S. Dodgson, and B. Spencer, *Clin. Chim. Acta* **4,** 453 (1959).
77. J. Gniot-Szulżycka, *Clin. Chim. Acta* **32,** 17 (1971).
78. W. R. Sherman and E.F. Stanfield, *Biochem. J.* **102,** 905 (1967).
79. G. C. Guilbault and J. Hieserman, *Anal. Chem.* **41,** 2006 (1969).
80. H. Rinderknecht, M. C. Geokas, C. Carmack, and B. J. Haverback, *Clin. Chim. Acta* **29,** 481 (1970).
81. H. E. Hirsch, *J. Neurochem.* **16,** 1147 (1969).
82. A. S. Perumal and E. Robins, *J. Neurochem.* **21,** 501 (1970).
83. A. C. Allison and E. F. Hartree, *J. Reprod. Fert.* **21,** 501 (1970).
84. C. H. Yang., P. N. Srivastava, and W. L. Williams, *Fed. Proc., Fed. Amer. Soc. Exp. Biol.* **32,** 310 (1973).
85. W. R. Den Tandt, *Clin. Chim. Acta* **40,** 199 (1972).
86. C. B. van der Hagen, A. L. Borresen, K. Molne, G. Oftedal, K. Bjøro, and K. Berg, *Clin. Genet.* **4,** 256 (1973).
87. M. O. Pulkkinen, *Acta Physiol. Scan.* **52,** Suppl. 180, P9 (1961).
88. J. Kuczynski, T. Prydzik, and A. Wenclewski, quoted in *Chem. Abstr.* **61,** 13687e (1964).

89. A. B. Roy, *Biochem. J.* **68**, 519 (1958).
90. G. Ugazio, *Ital. J. Biochem.* **9**, 98 (1960).
91. A. Leznicki and W. S. Bleszynski, *Histochemie* **24**, 251 (1970).
92. G. Rappay, L. Kondics, and E. Bácsy, *Histochemie* **34**, 271 (1973).
93. T. Makita and E. B. Sandborn, *Experientia* **27**, 187 (1971).
94. K. S. Dodgson, B. Spencer, and J. Thomas, *Biochem. J.* **53**, 452 (1953).
95. A. B. Roy, *Aust. J. Biol. Med. Sci.* **41**, 331 (1963).
96. A. B. Roy, *Biochim. Biophys. Acta* **227**, 129 (1971).
97. J. H. Austin, A. S. Balasubramanian, T. N. Pittabiraman, S. Saraswathi, D. K. Basu, and B. K. Bachhawat, *J. Neurochem.* **10**, 805 (1963).
98. J. H. Austin, D. McAfee, and L. Shearer, *Arch. Neurol.* (*Chicago*) **13**, 593 (1965).
99. D. Stumpf, E. Neuwelt, J. H. Austin, and P. Kohler, *Arch. Neurol.* (*Chicago*) **25**, 427 (1971).
100. M. T. Porter, A. L. Fluharty, and H. Kihara, *Proc. Nat. Acad. Sci. U.S.* **62**, 887 (1969).
101. M. T. Porter, A. L. Fluharty, S. E. Harris, and H. Kihara, *Arch. Biochem. Biophys.* **138**, 646 (1970).
102. M. T. Porter, A. L. Fluharty, and H. Kihara, *Science* **172**, 1263 (1971).
103. H. Jatzkewitz and K. Stinshoff, *FEBS* (*Fed. Eur. Biochem. Soc.*) *Lett.* **32**, 129 (1973).
104. M. T. Porter, A. L. Fluharty, J. Trammell, and H. Kihara, *Biochem. Biophys. Res. Commun.* **44**, 660 (1971).
105. K. Harzer, K. Stinshoff, W. Mraz, and H. Jatzkewitz, *J. Neurochem.* **20**, 279 (1973).
106. J. H. Austin, *Arch. Neurol.* (*Chicago*) **28**, 258 (1973).
107. F. Huijing, R. J. Warren, and A. G. W. McLeod, *Clin. Chim. Acta* **44**, 453 (1973).
108. V. Patel, A. L. Tappel, and J. S. O'Brien, *Biochem. Med.* **3**, 447 (1970).
109. J. A. Kint, G. Dacremont, D. Carton, E. Orye, and C. Hooft, *Science* **181**, 352 (1973).
110. L. M. Działoszyński and J. Gniot-Szulżycka, *Clin. Chim. Acta* **15**, 381 (1967).
111. M. C. Gcokas and H. Rinderknecht, *Clin. Chim. Acta* **46**, 27 (1973).
112. A. Begum and T. R. Ittyerah, *Clin. Chim. Acta* **28**, 263 (1970).
113. M. L. Cuzner and A. N. Davison, *J. Neurol. Sci.* **19**, 29 (1973).
114. J. R. H. Fliegner, I. Schindler, and J. B. Brown, *J. Obstet. Gynaecol. Brit. Commis.* **79**, 810 (1972).
115. K. S. Dodgson, G. M. Powell, F. A. Rose, and N. Tudball, *Biochem. J.* **79**, 209 (1961).
116. F. A. Rose and W. S. Bleszynski, *Biochem. J.* **122**, 601 (1971).
117. J. D. Cherayil and H. Van Kley, *Fed. Proc., Fed. Amer. Soc. Exp. Biol.* **20**, 235 (1961).
118. J. D. Cherayil and H. Van Kley, *Fed. Proc., Fed. Amer. Soc. Exp. Biol.* **21**, 230 (1962).
119. J. D. Cherayil and H. Van Kley, *Fed. Proc., Fed. Amer. Soc. Exp. Biol.* **22**, 241 (1963).
120. J. F. Drnec, Ph.D. thesis, St. Louis University, St. Louis, Missouri (1968).
121. J. E. Presslitz, Ph.D. thesis, St. Louis University, St. Louis, Missouri (1968).
122. S. J. Benkovic, E. V. Vergara, and R. C. Hevey, *J. Biol. Chem.* **246**, 4926 (1971).
123. M. Rasburn and C. H. Wynn, *Biochim. Biophys. Acta* **293**, 191 (1971).
124. T. Harada and B. Spencer, *Biochem. J.* **82**, 148 (1962).
125. B. N. Apte and O. Siddiqi, *Biochim. Biophys. Acta* **242**, 129 (1971).
126. R. L. Metzenberg and J. W. Parson, *Proc. Nat. Acad. Sci. U.S.* **55**, 629 (1966).
127. R. L. Metzenberg and S. K. Ahlgren, *Genetics* **64**, 409 (1970).

128. W. A. Scott and R. L. Metzenberg, *J. Bacteriol.* **104**, 1254 (1970).
129. W. A. Scott, K. D. Munkres, and R. L. Metzenberg, *Arch. Biochem. Biophys.* **142**, 623 (1971).
130. R. L. Metzenberg and S. K. Ahlgren, *Genetics* **68**, 369 (1971).
131. R. L. Metzenberg, G. S. Chen, and S. K. Ahlgren, *Genetics* **68**, 359 (1971).
132. E. G. Burton and R. L. Metzenberg, *J. Bacteriol.* **109**, 140 (1972).
133. E. G. Burton and R. L. Metzenberg, *J. Bacteriol.* **113**, 519 (1973).
134. T. Harada and B. Spencer, *Biochem. J.* **93**, 373 (1964).
135. D. H. Rammler, C. Grado, and L. R. Fowler, *Biochemistry* **3**, 224 (1964).
136. L. R. Fowler and D. H. Rammler, *Biochemistry* **3**, 230 (1964).
137. K. S. Dodgson, *Enzymologia* **20**, 301 (1959).
138. F. H. Milazzo and J. W. Fitzgerald, *Can. J. Microbiol.* **13**, 659 (1967).
139. J. W. Fitzgerald and F. H. Milazzo, *Can. J. Microbiol.* **16**, 1109 (1970).
140. T. Harada, *Biochim. Biophys. Acta* **81**, 193 (1964).
141. G. J. Delisle and F. H. Milazzo, *Can. J. Microbiol.* **18**, 561 (1972).
142. K. S. Dodgson and G. M. Powell, *Biochem. J.* **73**, 666 (1959).
143. J. W. Fitzgerald and W. J. Payne, *Microbios* **6**, 147 (1972).
144. K. S. Dodgson, B. Spencer, and K. Williams, *Biochem. J.* **61**, 374 (1955).
145. K. S. Dodgson, B. Spencer, and K. Williams, *Biochem. J.* **64**, 216 (1956).
146. K. S. Dodgson, B. Spencer, and K. Williams, *Nature (London)* **177**, 432 (1956).
147. K. H. Ney and R. Ammon, *Hoppe-Seyler's Z. Physiol. Chem.* **315**, 145 (1959).
148. E. D. S. Corner, Y. A. Leon, and R. D. Bulbrook, *J. Mar. Biol. Ass. U.K.* **39**, 51 (1960).
149. T. Soda and F. Egami, *J. Chem. Soc. Jap., Pure Chim. Sect.* **54**, 1069 (1933).
150. K. S. Dodgson, J. I. M. Lewis, and B. Spencer, *Biochem. J.* **55**, 253 (1953).
151. P. Jarrige and R. Henry, *Bull. Soc. Chim. Biol.* **34**, 872 (1952).
152. H. Baum and K. S. Dodgson, *Nature (London)* **179**, 312 (1957).
153. N. Poux, *J. Histochem. Cytochem.* **14**, 932 (1967).
154. T. Soda, *J. Fac. Sci., Univ. Tokyo* **3**, Sect. V, 149 (1963).
155. K. S. Dodgson and B. Spencer, *Biochem. J.* **55**, 315 (1953).
156. B. Wortman and A. Schneider, *Anat. Rec.* **137**, 403 (1960).
157. S. Suzuki, N. Takahashi, and F. Egami, *J. Biochem. (Tokyo)* **46**, 1 (1959).
158. K. S. Dodgson and G. M. Powell, *Biochem. J.* **73**, 672 (1959).
159. J. T. Baker, *Endeavour* **33**, 11 (1974).
160. R. S. H. Yang, J. G. Pelliccia, and C. F. Wilkinson, *Biochem. J.* **136**, 817 (1973).
161. A. Vestermark and H. Boström, *Exp. Cell Res.* **18**, 174 (1959).
162. B. Spencer, *Biochem. J.* **77**, 294 (1960).
163. M. F. Scully, K. S. Dodgson, and F. A. Rose, *Biochem. J.* **119**, 29P (1970).
164. T. Yagi, *J. Biochem. (Tokyo)* **59**, 495 (1966).
165. K. S. Dodgson, A. G. Lloyd, and N. Tudball, *Biochem. J.* **79**, 111 (1961).
166. N. Tudball and J. H. Thomas, *Abstr. 2nd Meet. Fed. Eur. Biochem. Soc., 1965* p. 201 (1965).
167. J. H. Thomas, K. S. Dodgson, and N. Tudball, *Biochem. J.* **110**, 687 (1968).
168. N. Tudball and P. Thomas, *Biochem. J.* **126**, 187 (1972).
169. N. Tudball and P. Thomas, *Biochem. J.* **128**, 41 (1972).
170. R. A. John and P. Fasella, *Biochemistry* **8**, 4477 (1969).
171. N. Tudball, J. H. Thomas, and J. A. Fowler, *Biochem. J.* **114**, 299 (1969).
172. W. H. B. Denner, A. H. Olavesen, G. M. Powell, and K. S. Dodgson, *Biochem. J.* **111**, 43 (1969).
173. D. W. Woolley and W. H. Peterson, *J. Biol. Chem.* **122**, 213 (1937).
174. T. Harada and B. Spencer, *J. Gen. Microbiol.* **22**, 520 (1960).

175. M. Itahashi, *Bot. Mag.* **72**, 275 (1959).
176. I. Takebe, *J. Gen. Appl. Microbiol.* **6**, 83 (1960).
177. J. M. Scott and B. Spencer, *Biochem. J.* **95**, 50P (1965).
178. C. Hussey, B. A. Orsi, J. M. Scott, and B. Spencer, *Nature (London)* **207**, 632 (1965).
179. J. M. Scott and B. Spencer, *Biochem. J.* **106**, 471 (1968).
180. P. Nissen and A. A. Benson, *Science* **134**, 1759 (1961).
181. B. Lindberg, *Acta Chem. Scand.* **9**, 917 (1955).
182. B. Lindberg, *Acta Chem. Scand.* **9**, 1323 (1955).
183. J. W. Fitzgerald, *Biochem. J.* **136**, 361 (1973).
184. I. Takebe, *J. Biochem. (Tokyo)* **50**, 245 (1961).
185. R. H. Bogan and C. N. Sawyer, *Sewage Ind. Wastes* **27**, 917 (1955).
186. Y. -C. Hsu, *Nature (London)* **200**, 1091 (1963).
187. Y. -C. Hsu, *Nature (London)* **207**, 385 (1965).
188. W. J. Payne and V. E. Feisal, *Appl. Microbiol.* **11**, 339 (1963).
189. J. Williams and W. J. Payne, *Appl. Microbiol.* **12**, 360 (1964).
190. W. J. Payne, J. P. Williams, and W. R. Mayberry, *Appl. Microbiol.* **13**, 698 (1965).
191. W. J. Payne and B. G. Painter, *Microbios* **3**, 199 (1971).
192. W. J. Payne, J. P. Williams, and W. R. Mayberry, *Nature (London)* **214**, 623 (1967).
193. J. W. Fitzgerald and W. J. Payne, *Microbios* **5**, 87 (1972).
194. K. S. Dodgson, J. W. Fitzgerald, and W. J. Payne, *Biochem. J.* **138**, 53 (1974).
195. J. W. Fitzgerald, K. S. Dodgson, and W. J. Payne, *Biochem. J.* **138**, 63 (1974).
196. W. J. Payne, J. W. Fitzgerald, and K. S. Dodgson, *Appl. Microbiol.* **27**, 154 (1974).
197. J. W. Fitzgerald and W. J. Payne, *Microbios* **6**, 55 (1972).
198. A. J. Vlitos, *Contrib. Boyce Thompson Inst.* **17**, 127 (1953).
199. A. B. Roy, *Biochem. J.* **66**, 700 (1957).
200. S. Burstein and R. I. Dorfman, *J. Biol. Chem.* **238**, 1656 (1963).
201. J. C. Warren and A. P. French, *J. Clin. Endocrinol. Metab.* **25**, 278 (1965).
202. A. L. Verde and W. D. Drucker, *Endocrinology* **90**, 138 (1972).
203. L. Ainsworth, *Steroids* **19**, 741 (1972).
204. S. Burstein and C. Westort, *Endocrinology* **80**, 1102 (1967).
205. A. H. Payne, *Biochim. Biophys. Acta* **258**, 473 (1972).
206. J. D. Townsley, *Endocrinology* **93**, 172 (1973).
207. P. Jarrige, *Bull. Soc. Chim. Biol.* **45**, 761 (1963).
208. K. Savard, E. Bagnoli, and R. I. Dorfman, *Fed. Proc., Fed. Amer. Soc. Exp. Biol.* **13**, 289 (1954).
209. W. Bergman, in "Comparative Biochemistry" (M. Florkin and H. S. Mason, eds.), Vol. 3, Part A, p. 103. Academic Press, New York, 1962.
210. H. Gottfried and O. Lusis, *Nature (London)* **212**, 1488 (1966).
211. H. Gottfried, R. I. Dorfman, and P. E. Wall, *Nature (London)* **215**, 409 (1967).
212. P. Jarrige, J. Yon, and M. F. Jayle, *Bull. Soc. Chim. Biol.* **45**, 783 (1963).
213. P. Jarrige, "Purification et propriétiés des sulfatase du suc digestif d'helix pomatia." R. Foulon, Paris, 1962.
214. K. S. Dodgson, *Biochem. J.* **78**, 324 (1961).
215. K. S. Dodgson, P. W. Gatehouse, A. G. Lloyd, and G. M. Powell, *Biochem. J.* **95**, 18P (1965).
216. F. Egami and N. Takahashi, in "Biochemistry and Medicine of Mucopolysaccharides" (F. Egami and Y. Oshima, eds.), p. 53. Maruzen, Tokyo, 1962.
217. S. Hunt and F. R. Jevons, *Biochem. J.* **98**, 522 (1966).
218. K. Ishihara, *Exp. Cell Res.* **51**, 473 (1968).

219. S. Inoue and F. Egami, *J. Biochem.* (*Tokyo*) **54**, 557 (1963).
220. R. L. Katzman and R. W. Jeanloz, *J. Biol. Chem.* **248**, 50 (1973).
221. J. Doyle, *Biochem. J.* **103**, 325 (1967).
222. L. Robert and Z. Dische, *Biochem. Biophys. Res. Commun.* **10**, 209 (1963).
223. P. W. Kent, J. Ackers, and J. C. Marsden, *Biochem. J.* **105**, 24p (1967).
224. J. Schrager and M. D. Oates, *Biochem. J.* **106**, 523 (1968).
225. B. L. Slomiany and K. Meyer, *J. Biol. Chem.* **247**, 5062 (1972).
226. H. S. Barra and R. Caputto, *Biochim. Biophys. Acta* **101**, 367 (1965).
227. L. C. Ryan, R. Carubelli, R. Caputto, and R. E. Trucco, *Biochim. Biophys. Acta* **101**, 252 (1965).
228. R. F. Nigrelli, J. D. Chanley, S. K. Kohn, and H. Sobotka, *Zoologica* (*New York*) **40**, 47 (1955).
229. T. Yasumoto, T. K. Nakamura, and Y. Hashimota, *Arg. Biol. Chem.* **31**, 7 (1967).
230. M. Kates, B. Palameta, M. P. Perry, and G. A. Adams, *Biochim. Biophys. Acta* **137**, 213 (1967).
231. I. Ishizuka, M. Suzuki, and T. Yamakawa, *J. Biochem.* (*Tokyo*) **73**, 77 (1973).
232. L. Svennerholm and S. Stalberg-Stenhagen, *J. Lipid Res.* **9**, 215 (1968).
233. S. Suzuki and J. L. Strominger, *J. Biol. Chem.* **235**, 257 (1960).
234. R. O. Mumma and J. Verlangiera, *Biochim. Biophys. Acta* **273**, 249 (1972).
235. K. S. Dodgson and A. G. Lloyd, *in* "Carbohydrate Metabolism and its Disorders" (F. Dickens, P. J. Randle, and W. J. Whelan, eds.), Vol. 1, p. 169. Academic Press, New York, 1968.
236. S. Hunt, "Polysaccharide-Protein Complexes in Invertebrates." Academic Press, New York, 1970.
237. L. Roden, *in* "Metabolic Conjugation and Metabolic Hydrolysis" (W. H. Fishman, ed.), Vol 2, p. 345. Academic Press, New York, 1970.
238. T. Soda and F. Egami, *J. Chem. Soc. Jap., Pure Chem. Sect.* **59**, 1202 (1938).
239. T. Soda, T. Katsura, and O. Yoda, *J. Chem. Soc. Jap., Pure Chem. Sect.* **61**, 1227 (1940).
240. T. Soda and F. Egami, *J. Chem. Soc. Jap., Pure Chem. Sect.* **55**, 256 (1934).
241. N. Takahashi, *J. Biochem.* (*Tokyo*) **48**, 508 (1960).
242. N. Takahashi, *J. Biochem.* (*Tokyo*) **48**, 691 (1960).
243. N. Takahashi and F. Egami, *Biochem. J.* **80**, 384 (1961).
244. M. Nishida-Fukuda and F. Egami, *Biochem. J.* **119**, 39 (1970).
245. K. S. Dodgson and B. Spencer, *Biochem. J.* **57**, 310 (1954).
246. K. S. Dodgson and A. G. Lloyd, *Biochem. J.* **78**, 319 (1961).
247. P. F. Lloyd, K. O. Lloyd, and O. Owen, *Biochem. J.* **85**, 193 (1962).
248. P. F. Lloyd and C. H. Stuart, *Biochem. J.* **99**, 37p (1966).
249. B. Tanko, *Biochem. Z.* **247**, 248 (1932).
250. I. Yamashina, *J. Chem. Soc. Jap. Pure Chem. Sect.* **72**, 124 (1951).
251. A. G. Lloyd, P. J. Large, M. Davies, A. H. Olavesen, and K. S. Dodgson, *Biochem. J.* **108**, 393 (1968).
252. J. Weigl and W. Yaphe, *Can. J. Microbiol.* **12**, 874 (1966).
253. A. G. Lloyd, P. J. Large, A. M. James, and K. S. Dodgson, *J. Biochem.* (*Tokyo*) **55**, 669 (1964).
254. A. G. Lloyd, *Biochim. Biophys. Acta* **58**, 1 (1962).
255. P. J. Large, A. G. Lloyd, and K. S. Dodgson, *Biochem. J.* **90**, 12p (1963).
256. Y. Nonami, *Nippon Nogei Kagaku Kaishi* **33**, 1000 (1959).
257. S. Peat and D. A. Rees, *Biochem. J.* **79**, 7 (1961).
258. D. A. Rees, *Biochem. J.* **81**, 347 (1961).
259. H. Numanoi, *Sci. Pap. Coll. Gen. Educ., Univ. Tokyo* **3**, 55 (1953).

260. J. W. Fitzgerald and K. S. Dodgson, *Biochem. J.* **121,** 521 (1971).
261. J. W. Fitzgerald and K. S. Dodgson, *Biochem. J.* **122,** 277 (1971).
262. J. G. Hange, *in* "Methods in Enzymology" (W. A. Wood, ed.), Vol. 9, p. 92 Academic Press, New York, 1966.
263. A. G. Lloyd, P. J. Large, N. Tudball, M. Davies, A. H. Olavesen, and K. S. Dodgson, *Biochem. J.* **97,** 41P (1965).
264. A. G. Lloyd, P. J. Large, N. Tudball, F. S. Wusteman, and K. S. Dodgson, *Biochem. J.* **97,** 43P (1965).
265. N. Entner and M. Doudoroff, *J. Biol. Chem.* **169,** 853 (1952).
266. A. J. Hancock and M. Kates, *J. Lipid Res.* **14,** 422 (1973).
267. C. G. Mead and F. J. Finamore, *Biochemistry* **8,** 2652 (1969).
268. B. M. Tolbert, D. J. Isherwood, R. W. Atchely, and E. M. Baker, *Fed. Proc., Fed. Amer. Soc. Exp. Biol.* **30,** 529 (1971).
269. A. J. Verlangieri and R. O. Mumma, *Atherosclerosis* **17,** 37 (1973).
270. N. Takahashi and F. Egami, *Biochim. Biophys. Acta* **38,** 375 (1960).
271. F. Egami, T. Asahi, N. Takahashi, S. Suzuki, S. Shikata, and K. Nisizawa, *Bull. Chem. Soc. Jap.* **28,** 685 (1955).
272. T. Soda, C. Hattori, and H. Teraski, *J. Chem. Soc. Jap., Pure Chem. Sect.* **57,** 981 (1936).
273. N. Takahashi, *J. Biochem. (Tokyo)* **47,** 230 (1960).
274. L.-A. Fransson and B. Havsmark, *J. Biol. Chem.* **245,** 4770 (1970).
275. L. C. Rosenberg, S. Pal, and R. J. Beale, *J. Biol. Chem.* **248,** 3681 (1973).
276. S. Lohmander, C. A. Antonopoulos, and U. Friberg, *Biochim. Biophys. Acta* **304,** 430 (1973).
277. H. Habuchi, T. Yamagata, H. Iwata, and S. Suzuki, *J. Biol. Chem.* **248,** 6019 (1973).
278. C. Neuberg and O. Rubin, *Biochem. Z.* **67,** 82 (1914).
279. C. Neuberg and E. Hofmann, *Biochem. Z.* **234,** 345 (1931).
280. C. Neuberg and W. Cahill, *Biochem. Z.* **275,** 328 (1934).
281. C. Neuberg and W. Cahill, *Enzymologia* **1,** 22 (1936).
282. H. J. Buehler, P. A. Katzman, and E. A. Doisy, *Proc. Soc. Exp. Biol. Med.* **78,** 3 (1951).
283. R. L. Hartles and W. D. McLean, *Brit. Dent. J.* **93,** 147 (1952).
284. P. Pincus, *Nature (London)* **166,** 187 (1950).
285. Y. Horiguchi and M. Mikaya, *Bull. Jap. Soc. Sci. Fish.* **19,** 957 (1954).
286. K. S. Dodgson, A. G. Lloyd, and B. Spencer, *Biochem. J.* **65,** 131 (1957).
287. K. S. Dodgson and A. G. Lloyd, *Biochem. J.* **66,** 532 (1957).
288. K. S. Dodgson, *in* "The Amino Sugars" (R. W. Jeanloz and E. A. Balazs, eds.), Vol. 2B, p. 201. Academic Press, New York, 1966.
289. A. G. Lloyd, A. H. Olavesen, P. A. Woolley, and G. Embery, *Biochem. J.* **102,** 37P (1967).
290. K. S. Dodgson and A. G. Lloyd, *Biochem. J.* **68,** 88 (1958).
291. A. Linker, P. Hoffman, K. Meyer, P. Sampson, and E. D. Korn, *J. Biol. Chem.* **265,** 3061 (1960).
292. G. Embery and J. M. Day, *Biochem. Soc. Trans.* **1,** 271 (1973).
293. T. Yamagata, Y. Kawamura, and S. Suzuki, *Biochim. Biophys. Acta* **115,** 250 (1966).
294. T. Yamagata, H. Saito, O. Habuchi, and S. Suzuki, *J. Biol. Chem.* **243,** 1523 (1968).
295. H. Saito, T. Yamagata, and S. Suzuki, *J. Biol. Chem.* **243,** 1536 (1968).
296. S. Suzuki, H. Saito, T. Yamagata, K. Anno, N. Seno, Y. Kawai, and T. Furuhashi, *J. Biol. Chem.* **243,** 1543 (1968).
297. Y. M. Michelacci and C. P. Dietrich, *Biochimie* **55,** 893 (1973).
298. E. D. Korn and A. N. Payza, *Biochim. Biophys. Acta* **20,** 596 (1956).

299. Z. Yosizawa, *Biochim. Biophys. Acta* **141**, 600 (1967).
300. C. P. Dietrich and S. M. C. Dietrich, *Anal. Biochem.* **46**, 209 (1972).
301. P. F. Lloyd and R. J. Fielder, *Biochem. J.* **109**, 14ᴘ (1968).
302. K. Atsumi, Y. Kawai, N. Seno, and K. Anno, *Biochem. J.* **128**, 983 (1972).
303. W. E. Huffer, *Fed. Proc., Fed. Amer. Soc. Exp. Biol.* **32**, 829 (1973).
304. N. N. Aronson and E. A. Davidson, *J. Biol. Chem.* **242**, 437 (1967).
305. C. H. Dohlman, *Acta Physiol. Scand.* **37**, 220 (1956).
306. D. D. Dziewiatkowski, *J. Biol. Chem.* **223**, 239 (1956).
307. K. M. Wood, F. S. Wusteman, and C. G. Curtis, *Biochem. J.* **134**, 1009 (1973).
308. N. Tudball and E. A. Davidson, *Biochim. Biophys. Acta* **171**, 113 (1969).
309. E. Held and E. Buddecke, *Hoppe-Seyler's Z. Physiol. Chem.* **348**, 1047 (1967).
310. E. Held, O. Hoefele, G. Reich, U. Stein, E. Werries, and E. Buddecke, *Z. Klin. Chem. Klin. Biochem.* **4**, 244 (1968).
311. G. Bach, F. Eisenberg, M. Cantz, and E. F. Neufeld, *Proc. Nat. Acad. Sci. U.S.* **70**, 2134 (1973).
312. G. V. Coppa, J. Singh, B. L. Nichols, and N. Di Ferrante, *Anal. Lett.* **6**, 225 (1973).
313. I. Sjoberg, L.-A. Fransson, R. Matalon, and A. Dorfman, *Biochem. Biophys. Res. Commun.* **54**, 1125 (1973).
314. V. A. McKusick, "Heritable Disorders of Connective Tissues." Mosby, St. Louis, Missouri, 1972.
315. B. Weismann and R. Santiago, *Biochem. Biophys. Res. Commun.* **46**, 1430 (1972).
316. V. P. Bhavanandan and K. Meyer, *J. Biol. Chem.* **242**, 4352 (1967).
317. V. P. Bhavanandan and K. Meyer, *J. Biol. Chem.* **243**, 1052 (1968).
318. V. C. Hascall and R. L. Riolo, *J. Biol. Chem.* **247**, 4529 (1972).
319. O. Rosen, P. Hoffman, and K. Meyer, *Fed. Proc., Fed. Amer. Soc. Exp. Biol.* **19**, 147 (1960).
320. S. Hirano and K. Meyer, *Biochem. Biophys. Res. Commun.* **44**, 1371 (1971).
321. M. Kitamikado, R. Ueno, and T. Nakamura, *Bull. Jap. Soc. Sci. Fish.* **36**, 592 (1970).
322. M. Kitamikado, R. Ueno, and T. Nakamura, *Bull. Jap. Soc. Sci. Fish.* **36**, 1172 (1970).
323. M. Kitamikado and R. Ueno, *Bull. Jap. Soc. Sci. Fish.* **36**, 1175 (1970).
324. S. Inoue, *Biochim. Biophys. Acta* **101**, 16 (1965).
325. E. A. Davidson and W. Small, *Biochim. Biophys. Acta* **69**, 459 (1963).
326. N. Katsura and E. A. Davidson, *Biochim. Biophys. Acta* **121**, 135 (1966).
327. H. Greiling, H. W. Stuhlsatz, and R. Kisters, *Hoppe-Seyler's Z. Physiol. Chem.* **350**, 669 (1969).
328. H. Greiling and M. Kaneko, *Arzneim. Forsch.* **23**, 593 (1973).
329. A. S. Perlin and D. M. Mackie, *Carbohyd. Res.* **18**, 185 (1971).
330. J. A. Cifonelli and J. King, *Carbohyd. Res.* **21**, 173 (1972).
331. S. Inoue and M. Miyawaki, *Biochim. Biophys. Acta* **320**, 73 (1973).
332. R. L. Taylor, J. E. Shively, H. E. Conrad, and J. A. Cifonelli, *Biochemistry* **12**, 3633 (1973).
333. A. N. Payza and E. D. Korn, *Nature (London)* **177**, 88 (1956).
334. A. N. Payza and E. D. Korn, *J. Biol. Chem.* **223**, 853 (1956).
335. E. D. Korn and A. N. Payza, *J. Biol. Chem.* **223**, 859 (1956).
336. E. D. Korn, *J. Biol. Chem.* **226**, 841 (1957).
337. A. Linker and P. Sampson, *Biochim. Biophys. Acta* **43**, 366 (1960).
338. A. Linker and P. Hovingh, *J. Biol. Chem.* **240**, 3724 (1965).
339. A. G. Lloyd, L. J. Fowler, G. Embery, and B. A. Law, *Biochem. J.* **110**, 54ᴘ (1968).
340. P. Hovingh and A. Linker, *J. Biol. Chem.* **245**, 6170 (1970).

341. C. P. Dietrich, *Biochem. J.* **111**, 91 (1969).
342. A. Linker and P. Hovingh, *Biochemistry* **11**, 563 (1972).
343. C. P. Dietrich, *Biochemistry* **8**, 2089 (1969).
344. C. P. Dietrich, M. E. Silva, and Y. M. Michelacci, *J. Biol. Chem.* **248**, 6408 (1973).
345. M. E. Silva and C. P. Dietrich, *Biochimie* **55**, 1101 (1973).
346. A. G. Lloyd, B. A. Law, L. J. Fowler, and G. Embery, *Biochem. J.* **110**, 54P (1968).
347. L. B. Jaques, *J. Biol. Chem.* **133**, 445 (1940).
348. A. G. Lloyd, G. Embery, F. S. Wusteman, P. J. Large, and K. S. Dodgson, *in* "Structure and Function of Connective and Skeletal Tissues" (S. Fitton-Jackson *et al.*, eds.), p. 459. Butterworth, London, 1965.
349. A. G. Lloyd, G. Embery, F. S. Wusteman, and K. S. Dodgson, *Biochem. J.* **98**, 33P (1966).
350. A. G. Lloyd, G. Embery, G. M. Powell, C. G. Curtis, and K. S. Dodgson, *Biochem. J.* **98**, 33P (1966).
351. C. P. Dietrich, *Can. J. Biochem.* **48**, 725 (1970).
352. A. Linker and P. Hovingh, *Carbohy. Res.* **29**, 41 (1973).
353. C. P. Dietrich, H. B. Nader, L. R. G. Britto, and M. E. Silva, *Biochim. Biophys. Acta* **237**, 430 (1971).
354. L. Jansson and U. Lindahl, *Biochem. J.* **117**, 699 (1970).
355. A. Linker, P. Hoffman, P. Sampson, and K. Meyer, *Biochim. Biophys. Acta* **43**, 366 (1960).
356. A. Linker, P. Hoffman, P. Sampson, and K. Meyer, *Biochim. Biophys. Acta* **29**, 443 (1958).
357. A. Linker and P. Hovingh, *Biochim. Biophys. Acta* **165**, 89 (1968).
358. H. Kresse and E. F. Neufeld, *J. Biol. Chem.* **247**, 2164 (1972).
359. H. Kresse, *Biochem. Biophys. Res. Commun.* **54**, 1111 (1973).
360. Y. H. Liau and M. I. Horowitz, *Fed. Proc., Fed. Amer. Soc. Exp. Biol.* **32**, 676 (1973).
361. K. S. Dodgson and F. A. Rose, *Nutr. Abstr. Rev.* **36**, 327 (1966).
362. H. Hilz and F. Lipmann, *Proc. Nat. Acad. Sci. U.S.* **41**, 880 (1955).
363. E. G. Brunngraber, *J. Biol. Chem.* **233**, 472 (1958).
364. R. W. Robbins and F. Lipmann, *J. Biol. Chem.* **233**, 686 (1958).
365. M. Lewis and B. Spencer, *Biochem. J.* **85**, 18P (1962).
366. L. G. Wilson, T. Asahi, and R. Bandurski, *J. Biol. Chem.* **236**, 1822 (1961).
367. N. R. Ringertz and P. Reichard, *Acta Chem. Scand.* **11**, 1081 (1157).
368. A. S. Balasubramanian and B. K. Bachhawat, *Biochim. Biophys. Acta* **59**, 389 (1962).
369. J. E. Silbert, *J. Biol. Chem.* **242**, 5146 (1967).
370. L. A. Heppel, *in* "The Enzymes" (P. D. Boyer, H. Lardy, and K. Myrbäck, eds.), 2nd ed., Vol. 5, p. 49. Academic Press, New York, 1961.
371. R. Itoh, A. Mitsui, and K. Tsushima, *Biochim. Biophys. Acta* **146**, 151 (1967).
372. T. Koizumi, T. Suematsu, A. Kawasaki, K. Hiramatsu, and N. Iwabori, *Biochim. Biophys. Acta* **184**, 106 (1969).
373. J. Austin, D. Armstrong, D. Stumpf, T. Luttenegger, and M. Dragoo, *Biochim. Biophys. Acta* **192**, 29 (1969).
374. A. Abraham and B. K. Bachhawat, *Indian J. Biochem.* **1**, 192 (1964).
375. D. Armstrong, J. Austin, T. Luttenegger, B. Bachhawat, and D. Stumpf, *Biochim. Biophys. Acta* **198**, 523 (1970).
376. A. A. Farooqui and A. S. Balasubramanian, *Biochim. Biophys. Act* **198**, 56 (1970).
377. R. Bailey-Wood, K. S. Dodgson, and F. A. Rose, *Biochem. J.* **112**, 257 (1969).
378. R. Bailey-Wood, K. S. Dodgson, and F. A. Rose, *Biochim. Biophys. Acta* **220**, 284 (1970).

379. A. M. Stokes, W. H. B. Denner, and K. S. Dodgson, *Biochem. J.* **123,** 134P (1972).
380. W. H. B. Denner, A. M. Stokes, and K. S. Dodgson, *Biochem. J.* **128,** 133P (1972).
381. A. M. Stokes, W. H. B. Denner, F. A. Rose, and K. S. Dodgson, *Biochim. Biophys. Acta* **302,** 64 (1973).
382. W. H. B. Denner, A. M. Stokes, F. A. Rose, and K. S. Dodgson, *Biochim. Biophys. Acta* **315,** 394 (1973).
383. A. M. Stokes, W. H. B. Denner, and K. S. Dodgson, *Biochim. Biophys. Acta* **315,** 402 (1973).
384. J. M. Fry and S. B. Koritz, *Proc. Soc. Exp. Biol. Med.* **140,** 1275 (1972).
385. R. Bailey-Wood, Ph.D. thesis, University of Wales, (1970).
386. A. Kjaer, *in* "Organic Sulfur Compounds" (N. Karasch, ed.), Vol. 1, p. 409. Pergamon, Oxford, 1961.
387. A. I. Virtanen, *Phytochemistry* **4,** 207 (1965).
388. C. Neuberg and J. Wagner, *Biochem. Z.* **174,** 457 (1926).
389. C. Neuberg and J. Wagner, *Z. Gesamte Exp. Med.* **56,** 334 (1927).
390. C. Neuberg and O. Schoenbeck, *Biochem. Z.* **265,** 223 (1933).
391. H. Baum and K. S. Dodgson, *Nature (London)* **179,** 312 (1957).
392. M. G. Ettlinger and A. J. Lundeen, *J. Amer. Chem. Soc.* **78,** 4172 (1956).
393. M. G. Ettlinger and A. J. Lundeen, *J. Amer. Chem. Soc.* **79,** 1764 (1957).
394. R. D. Gaines and K. J. Goering, *Biochem. Biophys. Res. Commun.* **2,** 207 (1960).
395. R. D. Gaines and K. J. Goering, *Arch. Biochem. Biophys.* **96,** 13 (1962).
396. Z. Nagashima and M. Uchiyama, *Nippon Nogei Kagaku Kaishi* **33,** 881 (1959).
397. I. Tsuruo, M. Yoshida, and T. Hata, *Agr. Biol. Chem.* **31,** 18 (1967).
398. B. Lönnerdal and J.-C. Janson, *Biochim. Biophys. Acta* **315,** 421 (1973).
399. R. Björkman and J.-C. Janson, *Biochim. Biophys. Acta* **276,** 508 (1972).
400. M. Ohtsuru, I. Tsuruo, and T. Hata, *Agr. Biol. Chem.* **33,** 1309 (1969).
401. M. Ohtsuru, I. Tsuruo, and T. Hata, *Agr. Biol. Chem.* **37,** 976 (1973).
402. M. Ishimoto and I. Yamashina, *Symp. Enzymes Chem.* **2,** 36 (1949).
403. A. G. Lloyd, G. Embery, and L. J. Fowler, *Biochem. Pharmacol.* **20,** 637 (1971).
404. Y. Friedman and C. Arsenis, *Biochem. Biophys. Res. Commun.* **48,** 1133 (1972).
405. Y. Friedman and C. Arsenis, *Biochem. J.* **139,** 699 (1974).
406. J. R. Freney, G. E. Melville, and C. H. Williams, *J. Sci. Food Agr.* **20,** 440 (1969).
407. K. Harzer and H. U. Benz, *Hoppe-Seyler's Z. Physiol. Chem.* **335,** 744 (1974).
408. A. L. Fluharty, R. L. Stevens, D. L. Sanders and H. Kihara, *Fed. Proc., Fed. Amer. Soc. Exp. Biol.* **33,** 1233 (1974).
409. R. W. Carlson, M. Downing, and B. M. Tolbert, *Fed. Proc., Fed. Amer. Soc. Exp. Biol.* **33,** 1377 (1974).
410. W. Mraz and H. Jatzkewitz, *Hoppe-Seyler's Z. Physiol. Chem.* **355,** 33 (1974).
411. H. Hatanaka, Y. Ogawa, and F. Egami, *J. Biochem. (Tokyo)* **75,** 861 (1974).
412. D. A. Stumpf, J. H. Austin, A. C. Crocker, and M. LaFrance, *Amer. J. Dis. Child.* **126,** 744 (1973).
413. N. DiFerrante, B. H. Hyman, W. Klish, P. V. Donnelly, D. L. Nichols, R. V. Dutton, and J. Gniot-Szulżycka, *Johns Hopkins Med. J.* **135,** 42 (1974).
414. A. L. Fluharty, R. L. Stevens, D. L. Sanders, and H. Kihara, *Biochem. Biophys. Res. Commun.* **59,** 455 (1974).
415. J. F. O'Brien, M. Cantz, and J. Spranger, *Biochem. Biophys. Res. Commun.* **60,** 1170 (1974).
416. W. G. McGuire and G. A. Marzluf, *Arch. Biochem. Biophys.* **161,** 360 (1974).
417. J. W. Fitzgerald and C. L. Scott, *Microbios* **10,** 121 (1974).
418. M. Tsuji, M. Hamano, Y. Nakanishi, K. Ishihara, and S. Suzuki, *J. Biol. Chem.* **249,** 879 (1974).

CHAPTER 10

Thiosulfate Sulfurtransferase and Mercaptopyruvate Sulfurtransferase

Bo Sörbo

I. INTRODUCTION

Thiosulfate and mercaptopyruvate sulfurtransferases were originally implicated in the biosynthesis of thiocyanate, but later studies have demonstrated that they also may participate in the formation of other compounds containing divalent sulfur. The first of these enzymes to be described was thiosulfate sulfurtransferase (EC 2.8.1.1; thiosulfate: cyanide sulfurtransferase). Lang [1] thus reported in 1933 that certain tissues of higher animals contained an enzyme that catalyzed the reaction

$$CN^- + S_2O_3^{2-} \rightarrow CNS^- + SO_3^{2-} \tag{1}$$

High activities of the enzyme were detected in the liver and kidney and the biological function of the enzyme was thought to be the detoxification of cyanide. Lang named the enzyme "rhodanese" from the German name for thiocyanate ("rhodanid") with the ending "ese" indicating that this compound was formed through the action of the enzyme. Rhodanese is still often used as a trivial name in the literature, but according to existing nomenclature rules the unfortunately longer name, thiosulfate sulfurtransferase, is preferred. Twenty years later, Wood and Fiedler [2] reported that cyanide when incubated with a rat liver extract and β-mercaptopyruvate (which is a transamination product of cysteine [3]) was converted to thiocyanate according to the reaction

$$HSCH_2COCOO^- + CN^- \rightarrow CH_3COCOO^- + CNS^- \qquad (2)$$

They assumed that the enzyme responsible for this conversion was thiosulfate sulfurtransferase. However, this enzyme had, at that time, been isolated and crystallized [4] and it was quickly found [5] that crystalline thiosulfate sulfurtransferase did not catalyze Eq. (2). Another enzyme must be responsible for the catalytic effect of liver extracts on this reaction, an enzyme later named mercaptopyruvate sulfurtransferase (EC 2.8.1.2; 3-mercaptopyruvate:cyanide sulfurtransferase). Further studies of thiosulfate sulfurtransferase and mercaptopyruvate sulfurtransferase (for which we shall, in the following, use the abbreviations TST and MST, respectively) demonstrated that compounds other than cyanide may function as sulfur acceptors in reactions catalyzed by these enzymes and their physiological role is still a matter of discussion. As other reviews are available on TST [6–9] and MST [8,10], which contain extensive references to earlier (and sometimes outdated) work in this field, the present communication will concentrate on more recent findings, although references to older publications will be given when necessary.

II.   THIOSULFATE SULFURTRANSFERASE

A. Assay

TST activity is most conveniently [11] followed by the production of thiocyanate from thiosulfate and cyanide. Thiocyanate gives a red colored complex with ferric ions in acid solution, which allows for its easy determination. This method of assay is applicable to TST preparations of all degrees of purity, but it must be kept in mind that highly purified TST is inactivated by dilution. This can be prevented by including thiosulfate and bovine serum albumin [11] or high concentrations of glycine [12] in the diluting medium. TST activity can also be measured from the formation of sulfite, the other product in the cyanolysis of thiosulfate. Sulfite can be determined colorimetrically with *p*-rosaniline [13] or by its ability to reduce 2',6'-dichloroindophenol in the presence of *N*-methylphenazonium methosulfate [14]. The latter method allows for the continuous recording of TST activity, but it should be noted that the dyes used in this assay inhibit the enzyme to some extent [14]. With highly purified TST it is also possible to follow the enzyme catalyzed reaction between thiosulfate and cyanide spectrophotometrically in the UV region [12]. If the thiosulfate is replaced by an aromatic thiosulfonate in this assay method, the latter becomes more sensitive [15]. Another spectrophotometric assay method has been reported [16] in which thiosulfate and dihydrolipoate are used as substrates and the formation of oxidized lipoate measured. Other methods not involving colorimetry or spectrophotometry are available. Thus a polarographic method, in which the dropping mercury electrode may be used for measuring both the two substrates and the two products of Eq. (1), has been introduced by Westley and co-workers [17,18]. This technique is of special value for the study of TST catalyzed reactions involving stoichiometric amounts of enzyme or sulfur acceptors other than cyanide. Recently, methods for the assay of TST with the aid of ion-selective electrodes have been described [19,20]. So far, only cyanide sensitive electrodes appear to yield acceptable results, although thiocyanate sensitive electrodes have been described. These methods, however, do not appear to offer any particular advantages over the other, simpler techniques previously described. Finally, it should be mentioned that special techniques are available for locating TST activity in polyacrylamide gels after disc electrophoresis. A simple method was thus reported by Gullbault *et al.* [21], in which the gel is incubated in a medium containing thiosulfate, cyanide, and calcium ions. The sulfite formed by the action of the enzyme precipitates as the insoluble calcium sulfite, resulting in an opaque white zone, where the enzyme is located. Another recently reported method [22] for the same purpose is a modification of the dichloroindophenol method of Smith and Lascelles [14].

B. Occurrence and Distribution

Earlier, Lang [1] observed that TST was widely distributed in nature, from man down to bacteria. An examination of the enzyme activity of different dog tissues revealed that suprarenals had the highest activity followed by liver, brain, and salivary glands. The enzyme was not detected in muscle and blood. Lang also compared the enzyme activities of liver from different animal species and found pronounced differences: frog liver was thus 50-fold, and rabbit liver 30-fold more active than dog liver. Although Lang performed his determinations on acetone-dried material (with the inherent risk of enzyme inactivation), later studies employing homogenates of fresh [23,24] or frozen [25] tissues have, in general, confirmed Lang's findings. However, the distribution of TST in dog tissues is different from that of other mammalian species, in which liver is usually the most active tissue, followed by kidney. Fairly detailed distribution studies of the enzyme in tissues from rat [24,26], mouse [24], guinea pig [26,27], cow [25], horse [25], pig [25], sheep [25], pigeon [26], and frog [26,28] have been published. Nonmammalian tissues, with the exception of frog liver, generally contain very little TST activity [28]. This applies to insects as well [29]. Of special interest is the careful study of Reinwein [25] on the distribution of TST in human tissues. As the determinations had to be performed on autopsy material, the effect of autolysis on the enzyme activity was evaluated, but found to be of minor significance. The human tissue with highest activity was liver (with about the same activity as bovine liver) followed by kidney, suprarenals, and thyroid gland.

Studies of the intracellular distribution of TST in mammalian liver have revealed that the enzyme is confined to the mitochondria [30–32]. On the other hand, the fairly small amounts of the enzyme detected in bovine cardiac muscle were apparently not associated with the sarcosomes [33], although these particles show a certain structural and functional resemblance to mitochondria. It should be noted that TST is a "latent" enzyme in mitochondria [31,34], meaning that the enzyme activity of intact mitochondria is low, but increases when the mitochondrial membrane is disrupted by freezing, hypotonicity, or the action of detergents. This phenomenon (which should be kept in mind when tissue homogenates are assayed for TST activity) has been attributed [31,34] to a restricted permeability of thiosulfate through the mitochondrial membrane.

It was earlier claimed [35] that TST is present in certain thiocyanate rich plants, such as beets, white cabbage, and stinging nettle, but subsequent studies [6] did not confirm these findings. The enzyme may, however, according to recent reports, be present in plants as well. Thus the leaves from the tapioca plant, also known as cassava (*Manihot utilissima*), were found to contain TST [36]. This is of certain interest as this plant contains high amounts of cyanogenic glucosides. The enzyme was also detected in chloroplasts from spinach, parsley, and turnip [37], although only trace amounts of activity appear to be present.

A number of studies have been devoted to the presence of TST in microorganisms. The enzyme is apparently absent from yeast [6] and *Neurospora crassa* [18], but the presence of a fungal TST in culture filtrates from *Trametes sanguinea* has recently been reported [38]. Certain bacteria contain TST. Lang [1] and later Stearns [39] and more recently Schiewelbein *et al.* [28] found the enzyme in *Escherichia coli,* whereas Villarejo and Westley [18] found their strain of this bacterium to be inactive. On the other hand, the latter authors detected significant activities in *Bacillus subtilis, Bacillus coagulans,* and *Bacillus stearo- thermophilus* and the enzyme is also present in *Pseudomonas aerugi- nosa* and *Alcaligenes* [40] as well as in many photosynthetic bacteria [14,41,42]. Considerable attention has been paid to the presence of TST in sulfur bacteria and its possible role in their peculiar metabolism of inorganic sulfur compounds. The thiobacilli species have been studied especially from this point of view [18,43–51]. TST appears to be a con- stitutive enzyme in most of the bacteria studied [14,18,40,42], but was reported to be inducible with cyanide in *Thiobacillus denitrificans* [44].

C. Purification

TST was first isolated in crystalline form from bovine liver [4,11] by a procedure consisting of treatment of a liver homogenate with basic lead acetate, ammonium sulfate fractionation at weakly acid and weakly alkaline pH's, acetone fractionation at −5°, and ammonium sulfate fractionation at a weakly acid pH followed by crystallization from am- monium sulfate at pH 7.8. The acetone fractionation step used in this procedure has apparently presented difficulties to other workers in the field (strict temperature control of this step is necessary according to the reviewer's experience). Improved purification procedures were later worked out by Westley and co-workers [12,52], in which the acetone fractionation step was replaced by batchwise fractionation with DEAE- cellulose. An apparently very simple preparation method for the crys- talline enzyme has been described by Horowitz and De Toma [53], which in essence consists of extraction of the liver at a faintly acid pH followed by ammonium sulfate fractionations at different pH's. This method of purification was later used in another laboratory [54] to ob- tain material for the preparation of large TST crystals suitable for X-ray studies. On the other hand, the method of Horowitz and De Toma may apparently present difficulties and in a following paper from Westley's laboratory [55] the DEAE-cellulose step was reintroduced in their puri- fication method. TST has also been purified to crystallinity from bovine kidney [56,57]. As the kidney enzyme was found to be identical with the liver enzyme in many respects [56,58] and much higher yields are obtained from bovine liver, the latter appears to be the preferable source of TST. Attempts to purify the enzymes of microbiological origin have resulted in apparently rather inhomogenous preparations [14,38,44,50,59].

D. Molecular Properties

1. PHYSICOCHEMICAL PROPERTIES

The crystalline liver enzyme was found to behave as a homogenous protein on electrophoresis and in the ultracentrifuge [4], and a molecular weight of 37,000 daltons was determined by sedimentation velocity-diffusion methods. The isoelectric point could not be accurately determined, as it was found in an acid region, where the enzyme was unstable. Subsequent work by Westley and Green [56] confirmed the original sedimentation data, but Volini *et al.* [60] later obtained evidence from gel filtration studies of the enzyme on Sephadex G-100, indicating that TST is a dimer that under certain conditions dissociates into monomers of 19,000 molecular weight. Further evidence for the existence of the enzyme in a rapid dimer–monomer equilibrium was obtained by the same group from kinetic studies of the enzyme reaction [61] and fluorescence polarization measurements [62]. On the other hand, Blumenthal and Heinrikson [55] have reported that the molecular weight of the enzyme, as determined by gel chromatography on Sephadex G-75 or by electrophoresis in SDS-polyacrylamide, was 35,000—a finding difficult to reconcile with the data of Volini *et al.* [60]. Blumenthal and Heinrikson [55] also observed that crystalline TST could be separated by polyacrylamide gel electrophoresis or by chromatography on DEAE-cellulose into two equally active components, which furthermore appeared to be identical with respect to kinetic parameters, molecular weight, amino acid composition, and other analytical characteristics. The ratio between the two components was dependent upon the method used for preparing the enzyme and it was suggested that one component arose by deamidation of the other during the purification of the enzyme.

Although microbial TST has not been obtained in a pure form, the results obtained by Bowen *et al.* [44] deserve some consideration in this connection. These authors obtained an enzyme preparation from *T. denitrificans* that on Sephadex chromatography gave a main component with an estimated molecular weight of about 38,000, but also an active "subunit" with a molecular weight of about 9000 and an enzymatically active tetramer of about 150,000 molecular weight. Treatment of the latter with mercaptoethanol produced small (but still enzymatically active) fragments with molecular weights on the order of 2000 and 7000. It was suggested that the components of higher molecular weights were comprised of the 2000 and 7000 MW fragments, bound together by disulfide bonds.

2. COMPOSITION

The crystalline enzyme shows a typical protein ultraviolet absorption spectrum [4] with an absorption peak at 280 nm. Its amino acid composition has been reported from two laboratories [55,63] and the results are in essential agreement. The N-terminal amino acid is lysine [55], whereas the C-terminal amino acid has apparently not been identified. The enzyme contains 4 sulfhydryl groups [64,65] per 37,000 MW and their functional role will be discussed later. Peptide mapping studies [60] of tryptic peptides obtained from TST revealed 18–20 peptides. As the sum of the number of lysine and arginine residues per 37,000 MW of the enzyme is 35–37 [55,63], the number of tryptic peptides obtained would conform with the assumption that the enzyme molecule is made up of two identical monomers. The amino acid composition of some of these peptides was later determined [66] and more recently the amino acid sequences of two tryptic peptides obtained from the S-carboxymethylated enzyme have been reported [67]. Probably the complete amino acid sequence of TST will be established in the near future, and a preliminary X-ray crystallographic study [54] has been reported.

An important discovery, of fundamental importance to our understanding of the reaction mechanism of TST, was made by Westley and co-workers [16,68] when they found that the crystalline enzyme contained labile sulfur that directly participated in the enzyme catalyzed reaction. The amount of bound sulfur corresponded to 1.9 atoms per 37,000 MW as determined by a polarographic technique [16] and to 1.5 atoms per molecule as determined with the aid of ^{35}S [68]. Sörbo [69] confirmed the presence of labile sulfur in the enzyme by a colorimetric technique based on the conversion of this sulfur to thiocyanate. However, he detected only 1.1–1.3 atoms of sulfur per 37,000 MW [63,69]. A similar colorimetric method was later used by Blumenthal and Heinrikson [55] who obtained a value corresponding to 1.5 sulfur atoms per 37,000 MW. A possible explanation for these discrepancies (which are of certain importance, as a value of 2 atoms of labile sulfur per molecule must be expected if the enzyme consists of two identical monomers) was recently offered by Westley [9], who pointed out that sulfur tends to dissociate from the sulfur containing enzyme during recrystallization or other purification steps.

Volini et al. [60] reported that TST contained zinc, corresponding to 1 atom per monomer (19,000 MW) and postulated a functional role for this metal in the enzyme catalyzed reaction. However, B. Vallee, in collaboration with the reviewer, had earlier found by spectrographic analysis of crystalline TST that the enzyme does not contain any heavy metal, including zinc. These results were briefly reported in 1962 [70] and subsequent work in other laboratories [57,71] has confirmed that the pure enzyme is free from zinc, if suitable precautions are taken to avoid its contamination from extraneous sources.

E. Catalytic Properties

1. Substrate Specificity

It is now evident that TST is a less specific enzyme than originally thought by Lang [1]. The enzyme may, in fact, be considered to catalyze the transfer of divalent sulfur from a donor molecule (DS) to an acceptor (A) according to the general scheme

$$DS + A \rightleftarrows AS + D \tag{3}$$

Certain other sulfur compounds may thus substitute for thiosulfate in TST catalyzed reactions. Especially active in this respect are thiosulfonates [15,72–77], which may be considered as derivatives of thiosulfate (see Table I) in which one of the ionized oxygens atoms in thiosulfate is replaced by an organic residue. Persulfides as well (Table I), both inorganic [78,79] and organic [18,80], are efficient sulfur donors in TST catalyzed reactions. Thiocystine, an unusually stable trisulfide (Table I), has furthermore been reported [80] to possess substrate activity. Lang reported in his classic paper [1] that colloidal sulfur was a substrate for TST, but his results were obtained with very impure preparations of the enzyme. It was later shown [72], however, that although crude tissue preparations (including blood serum that is devoid of TST

TABLE I

COMPOUNDS ACTIVE AS SUBSTRATES FOR TST

Compound	Structure
Sulfur acceptors	
Cyanide	CN^-
Sulfite	SO_3^{2-}
Sulfinates	RSO_2^-
Thiols	RSH
Sulfur donors	
Thiosulfate	$^-SO_3S^-$
Thiosulfonates	RSO_2S^-
Inorganic persulfide	HS_nH
Organic persulfide	$RSSH$
Thiocystine	$RSSSR$

activity) had a promoting effect on the reaction between colloidal sulfur and cyanide, purified TST was inactive in this respect. It was therefore suggested that an enzyme, different from TST, was responsible for the "catalytic" effect of crude tissue preparations on the cyanolysis of sulfur. Further studies revealed [81] that the catalytic activity of blood serum was confined to serum albumin, which is usually not considered as an enzyme. As sulfur, when dissolved in an organic solvent, reacts very rapidly with cyanide [81,82] and serum albumin is endowed with surfactant properties, its catalytic effect was attributed to a facilitation of the contact possibilities between the colloidal sulfur particles and cyanide. It may be mentioned that Blumenthal and Heinriksen [55] recently claimed that TST could use mercaptopyruvate as a sulfur donor, although with an efficiency only about 1% of that of thiosulfate. The reviewer hesitates to accept this claim from his own experience of the difficulties in preparing mercaptopyruvate free from traces of thiosulfate and persulfides.

In regard to the acceptor specificity of TST, it was assumed for a long time after its discovery that only cyanide could function as an acceptor substrate. Sörbo [83] then demonstrated that sulfite and sulfinates (which may be considered as derivatives of sulfite obtained in a manner similar to thiosulfonates from thiosulfate, see Table I), were also very efficient sulfur acceptors in TST catalyzed reactions. Certain thiol compounds must be included among the acceptor substrates as well. Sörbo [15] thus observed that mercaptoethanol and thioglycollate were weakly active as acceptors when toluene thiosulfonate was used as the donor, whereas inorganic sulfide, cysteamine, and glutathione were inactive. Villarejo and Westley [18] then reported that certain dithiols, notably dihydrolipoic acid and dihydrolipoamide, could participate in TST catalyzed reactions with thiosulfate as the donor. The stoichiometry of the reactions was shown to be

$$S_2O_3{}^{2-} + \text{Lip (SH)}_2 \rightleftarrows H_2S + \text{LipS}_2 + SO_3{}^{2-} \qquad (4)$$

This reaction proceeds, however, in two steps, with the persulfide of lipoate as an intermediate, according to

$$S_2O_3{}^{2-} + \text{Lip (SH)}_2 \rightleftharpoons \text{Lip} \begin{smallmatrix} \text{SSH} \\ \text{SH} \end{smallmatrix} + SO_3{}^{2-} \qquad (5)$$

followed by

$$\text{Lip} \begin{smallmatrix} \text{SSH} \\ \text{SH} \end{smallmatrix} \rightleftharpoons \text{Lip} \begin{smallmatrix} \text{S} \\ | \\ \text{S} \end{smallmatrix} + H_2S \qquad (6)$$

Dimercaptopropanol, but not mercaptoethanol, also was found by Villarejo and Westley to be an acceptor. Koj [13] compared the acceptor ability of cyanide and three thiol compounds. The most active among the latter was dihydrolipoate, followed by cysteine and glutathione, but the relative activity of dihydrolipoate in comparison with cyanide was only 0.06

The compounds of biological interest so far established as substrates for TST are summarized in Table I. It should be noted that all compounds active as acceptor substrates are strongly nucleophilic for sulfur, and all donor substrates contain a divalent planetary sulfur atom exclusively bound to another sulfur atom in the molecule. Thus disulfides (RSSR) or thiosulfate esters (RSSO$_3^-$) are not substrates for TST [18,70,72] and thiocyanate is a product, but not a sulfur donor for the enzyme. The TST catalyzed cyanolysis of thiosulfate is thus irreversible [1,73] in contrast to the reactions between a thiosulfonate and a sulfinate [15] or between thiosulfate and dihydrolipoate [18].

2. INHIBITION AND INACTIVATION

The crude enzyme from bovine liver is stable from pH 4.5 to 10.0, the stability being somewhat extended at lower temperatures [31]. Temperatures above 45° inactivate the liver enzyme, but the temperature stability is increased in the presence of thiosulfate [31]. The tapioca enzyme appears to be unusually thermoresistant [36]. TST from animal sources is inactivated if incubated with cyanide [31,84,85] or sulfite [31] in the absence of thiosulfate; these inactivations may also be prevented by cysteine [31]. Cyanide is a compound with a high affinity for heavy metals, but other metal enzyme inhibitors have no inactivating or inhibitory action on TST, even at high concentrations [31]. Cyanide and sulfite also react with carbonyl groups, but inhibition experiments with other carbonyl group reagents seem to exclude the presence of an active carbonyl group in the enzyme [31]. Cyanide and sulfite may also attack disulfide bonds in proteins, however, which was the main reason why Sörbo [73] erroneously inferred that the enzyme contained an active disulfide bond. This interpretation was justly criticized by Green and Westley [17], who noted that the dilute enzyme was more sensitive to cyanide and sulfite than the concentrated enzyme, the opposite of what should be expected if these compounds attacked a disulfide bond in the enzyme. These authors, as earlier mentioned, unequivocally demonstrated that the isolated enzyme contained labile sulfur, which was rapidly transferred to cyanide or sulfite, and suggested that the free enzyme thus obtained was unstable and rapidly denatured. The situation turned out to be more complicated, however. Sörbo [69] later noted that the inactivation of TST required the presence of oxygen and that the inactivated enzyme could be reactivated by treatment with thiol compounds. Moreover, experiments with ^{35}S-labeled sulfite revealed that this compound became bound to the enzyme during the inactivation and was liberated again during the reactivation by a thiol compound. Determinations of the sulfhydryl content of TST before and after inactivation with sulfite revealed that one sulfhydryl group per enzyme molecule had disappeared during the inactivation. These results suggested that the enzyme contained a reactive sulfhydryl group that bound the labile sulfur as a persulfide. When the enzyme-bound sulfur was removed by sulfite or cyanide in the presence of oxygen, the active sulfhydryl group became autooxidizable and a reaction with sulfite ensued to give a thiosulfate ester derivative of the enzyme according to

$$ESH + O + SO_3^{2-} \rightarrow ESSO_3^- + OH^- \qquad (7)$$

The reactivation of the inactivated enzyme by a thiol compound could then be explained by the well-known reaction between a thiosulfate ester and a thiol according to

$$ESSO_3^- + 2\,RSH \rightarrow ESH + HSO_3^- + (RS)_2 \qquad (8)$$

It may be noted that a similar mechanism was found by Nichol and Creeth [86] to account for the inactivating effect of sulfite on urease, a well-known sulfhydryl enzyme. That the presence of sulfhydryl groups in TST was necessary for the enzyme activity had in fact been deduced from inhibition experiments in an early publication by Sörbo [31]. Further evidence for their presence was later obtained from experiments in which a reaction between TST and p-hydroxymercuribenzoate was demonstrated spectrophotometrically together with a simultaneous inhibition of enzyme activity [64]. The presence of an essential sulfhydryl group in the enzyme has also been confirmed by Wang and Volini [65] from inhibition experiments of a similar type.

TST is furthermore inhibited by certain aromatic compounds (pyridylpyridinium chloride [12], benzenesulfonates [65], and iodothyronines [87]) indicating the presence of an apolar active site region in the enzyme. As N-bromosuccinimide has been found to destroy tryptophan residues in the enzyme with a concomitant loss of enzyme activity [12], tryptophan residues may play an important role in the hydrophobic active site.

3. MECHANISM OF ACTION

Sörbo [73] proposed a reaction mechanism for the enzyme catalyzed cyanolysis of thiosulfate in which thiosulfate was postulated to attack the supposed active disulfide bond in the enzyme with formation of a sulfenylthiosulfate derivative of the enzyme. The latter was then assumed to react with cyanide, giving thiocyanate, sulfite, and the free enzyme. This mechanism was disproved when Westley and co-workers [17,68] established the formation of an enzyme–sulfur intermediate in the enzyme reaction. Green and Westley [17] first demonstrated by polarographic measurements that the crystalline enzyme contained bound sulfur, which in the presence of stoichiometric amounts of cyanide or sulfite reacted to give thiocyanate and thiosulfate, respectively. Furthermore, the transfer of the divalent planetary sulfur atom of thiosulfate to the enzyme during the enzymatic reaction was established by Westley and Nakamoto [68] with the aid of ^{35}S-labeled thiosulfate; no transfer of the inner hexavalent sulfur occurred. The enzyme catalyzed reaction between thiosulfate and cyanide may then be written as

$$E + S_2O_3^{2-} \rightleftarrows ES + SO_3^{2-} \tag{9}$$

$$ES + CN^- \rightarrow E + CNS^- \tag{10}$$

and is evidently of the double displacement type (ping-pong mechanism). Kinetic studies by Westley and Heyse [77] have, in fact, verified that the TST catalyzed cyanolytic reactions of thiosulfate and thiosulfonates rigorously adhere to this type of mechanism. It should be noted that thiosulfate is added during the purification of TST in order to stabilize the enzyme, which probably explains why the crystalline TST is obtained in the form of the sulfur containing intermediate.

The identity of the enzyme-bound sulfur must now be discussed. Sörbo postulated [69], after abandoning his earlier theory, that an active sulfhydryl group in the free enzyme reacted with the sulfur atom of the donor substrate to give an enzyme-bound persulfide group. This was disputed by Westley and co-workers [12,18], primarily on the ground that they were unable to detect any increased ultraviolet absorption of the crystalline enzyme around 340 nm, where low molecular weight persulfides absorb maximally. However, by using more concentrated solutions of TST, Finazzo Agro et al. [88] have recently demonstrated that the crystalline enzyme has, in fact, a low intensity absorption band at 330 nm that disappears on addition of cyanide or sulfite and is then restored by the addition of thiosulfate. They attributed this absorption band to the presence of a persulfide group, which now appears to be unanimously accepted [9]. This persulfide is of an unusual stability in comparison with low molecular weight persulfides, and its possible stabilization by interaction with a hydrophobic site in the enzyme has been suggested [65,69]. Recent fluorescence studies of TST in its sulfur containing and free states [88] neither confirm nor disprove this hypothesis. The participation of a hydrophobic site in the enzyme activity appears nevertheless highly plausible, as previously discussed in Section II,E,2.

The excellent kinetic studies by Westley and co-workers have greatly contributed to our understanding of the enzyme mechanism. Thus Davidson and Westley [12] noticed that high ionic strengths reduced the velocity of the enzyme catalyzed cyanolysis of thiosulfate when the latter was present at a nonsaturating concentration. This suggested that charge neutralization by a cationic group was involved in the initial binding of thiosulfate to the enzyme. Further support for this interpretation was supplied by Mintel and Westley [89], who found that reducing the dielectric constant of the reaction medium had an effect on the reaction velocity opposite to that given by high ionic strength, as predicted by theory if charge neutralization was involved. The chemical nature of the postulated cationic site in the enzyme remains to be established. Davidson and Westley [12] suggested that an ϵ-amino group of lysine or a guanidino group of arginine could fulfill this function. This was later refuted by Mintel and Westley [89], however, from rather indirect evidence. Leininger and Westley [52] and Volini et al. [60] implied that the cationic site was a zinc atom, but this appears very unlikely for reasons previously presented (Section II,D,2). Mintel and Westley [74] also compared the kinetic parameters of thiosulfate and thiosulfonates in the enzyme catalyzed reactions with cyanide, and found that the maximal velocity varied greatly. This signified that the scission of the sulfur–sulfur bond of the donor molecule was the rate limiting step in the overall reaction. From the observed relation between the inductive effect of electron withdrawing substituents in thiosulfonates and enzyme activity, they postulated that an electrophilic site in the enzyme, possibly identical with the cationic site, would facilitate the cleavage of the sulfur–sulfur bond of the donor substrate. No further information on this electrophilic site in TST is apparently available.

Westley and his co-workers have also paid due attention to another important aspect of the enzyme mechanism, protein conformational changes. Leininger and Westley [52] thus studied the TST catalyzed cyanolysis of thiosulfate with respect to the thermodynamic parameters for thiosulfate binding and scission of the sulfur–sulfur bond. The results demonstrated a large change of entropy in the binding step, suggesting a conformational change of the enzyme. Further investigations on such conformational changes have recently been reported by Wang and Volini [90–92].

F. Biological Function

Lang [1] assumed that the function of TST in mammals (including man) was to detoxify cyanide by conversion to thiocyanate with thiosulfate as the donor substrate. As a more detailed discussion of the toxicological aspects of this reaction is beyond the scope of this treatise, the reader is referred to a recent review [93] for this purpose. It should only be stated that thiosulfate is, not unexpectedly, an antidote to cyanide, but its efficiency is limited by its slow penetration through the cellular and mitochondrial membranes. If purified TST is injected together with thiosulfate [94] or thiosulfonates [95], these permeability barriers are circumvented and a much better antidote effect against cyanide is obtained. This probably represents the first case where an enzyme has been used successfully as an antidote. Thiocyanate formation is, however, not restricted to the cyanide poisoned animal. Convincing evidence for a *de novo* formation of thiocyanate in normal animals and human beings [96] was obtained in 1933, but the thiocyanate excreted in mammals may, to a certain extent, be derived from preformed thiocyanate taken in with the food [97]. The formation of its presumed precursors, cyanide [98,99] and thiosulfate [100], also has been convincingly established in normal animals and human beings in support of the assigned role of TST. There have been objections to this hypothesis because there is a discrepancy between the enormous enzyme activity present in many mammalian tissues, and the rather small amounts of thiocyanate formed under normal conditions [6,9,89]. We must keep in mind, however, that such comparisons are usually made on the basis of enzyme activities measured at optimal substrate concentrations that probably do not exist *in vivo*. Furthermore, a similar discrepancy is noted in case of the enzyme acetylcholinesterase, an enzyme with a well-established biological function, and the *in vivo* turnover of its physiological substrate (acetylcholine) [101], which illustrates the fallacy of such comparisons. Nevertheless, other biological functions of TST have been sought. It has been proposed [26,79,80] that a major role of the enzyme is to prevent the formation of inorganic sulfide, a highly toxic compound. The latter may be formed *in vitro* from cysteine-cystine through the action of cysteine desulfhydrase [EC 4.4.1.1; L-cystathionine cysteine-lyase (deaminating)], an enzyme now known to be identical with cystathionase. The persulfide of cysteine (thiocysteine) is an intermediate in this reaction and an effective sulfur donor for TST, as previously noted (Section II,E,1). It was, in fact, demonstrated [80] that cystathionase may be coupled *in vitro* with TST in the presence of cyanide or sulfite to produce thiocyanate and thiosulfate, respectively, from cysteine. Furthermore, a similar distribution of TST and cysteine desulfhydrase (cystathionase) in various animal tissues has also been noted [26], but more evidence for the proposed function of TST is lacking.

TST has also been postulated to participate in the oxidation of thiosulfate to sulfate, both in mammals [6,15,102–104] and in bacteria

[14,47–51]. Thiosulfate, when injected into mammals, is oxidized to sulfate. Isotope experiments [105] revealed that most of the sulfate formed was derived from the inner (hexavalent) sulfur atom of thiosulfate, whereas the outer (divalent) sulfur atom entered into unknown pathways of tissue metabolism. A possible route for this oxidation of thiosulfate involves an initial reductive cleavage of the compound to sulfite followed by an oxidation of the latter, catalyzed by the enzyme sulfite oxidase (EC 1.8.3.1; sulfite:oxygen oxidoreductase). As previously mentioned (Section II,E,1), TST may accomplish a reductive cleavage of thiosulfate in the presence of certain thiol compounds, but evidence has been presented [102–104] which suggests that the initial step of thiosulfate oxidation in mammals is, to a major extent, catalyzed by a different enzyme. This enzyme, thiosulfate reductase (not named by the Enzyme Commission) catalyzes the reaction between thiosulfate and glutathione (GSH) as follows:

$$S_2O_3^{2-} + 2\ GSH \rightarrow SO_3^{2-} + H_2S + (GS)_2 \tag{11}$$

It was originally discovered in yeast [106], but later shown to be present also in the mammalian liver. With regard to the postulated role of TST in the oxidation of thiosulfate in sulfur bacteria [14,47–51], no more tangible evidence is available than the fact that these organisms contain the enzyme.

The theories discussed above assign a catabolic function to TST, but the enzyme has also been suggested to participate in anabolic reactions. Schneider and Westley [107] thus demonstrated the incorporation of the divalent sulfur atom of thiosulfate into cysteine in the presence of lysed rat liver mitochondria and these in vitro studies were followed by in vivo experiments in which rats, injected with labeled thiosulfate, were found to incorporate the label into cystine of the hair [108]. The authors suggested that TST may provide a mechanism for the synthesis of sulfur containing amino acids from inorganic sulfur compounds, but the functional importance of this metabolic route remains to be assessed. Another anabolic function of TST has recently been implicated from the finding [109] that the enzyme may catalyze the formation of the "labile sulfur" of ferredoxin from thiosulfate in the presence of apoferredoxin, ferric ions, and a reducing agent (dithiothreitol).

In conclusion, TST is an enzyme that has been found to catalyze a number of reactions of uncertain physiological significance. The widespread distribution of the enzyme suggests an important metabolic function, but more experimental work is obviously required to clarify the situation.

448

III. MERCAPTOPYRUVATE SULFURTRANSFERASE

A. Assay

MST is conveniently assayed by the formation of thiocyanate from β-mercaptopyruvate and cyanide. Directions for the assay of the enzyme from animal sources are given by Hylin [110], and for the bacterial enzyme by Vachek and Wood [111]. Other, somewhat more complicated methods are based on measurements of pyruvate formed from mercaptopyruvate in the presence of mercaptoethanol [112] or of thiosulfate formed from mercaptopyruvate and sulfite [113].

B. Occurrence and Distribution

MST was originally discovered in rat liver [2,3,5] and studies of the tissue distribution of MST in this animal [3,112,114] demonstrate that the highest activities are present in liver and kidney. It is noteworthy that erythrocytes contain significant amounts of MST [112,114], whereas they are practically devoid of TST activity [26]. Little is known about the occurrence of MST in animal species other than rat, but mouse liver [112,115] and human erythrocytes [121] appear to contain enzyme activities of a magnitude similar to the corresponding rat tissues. Differential centrifugation studies of rat liver homogenates [112] indicate that most of the MST is confined to the cytosol and very little activity is found in the mitochondria, which on the other hand contain most of the TST activity (Section II,B). MST has not been reported to occur in plants, but has been detected in certain bacteria, *E. coli* and *Aerobacter aerogenes,* but not in *Clostridium welchii* [3]. *Escherichia coli* appears, in fact, to be a good source of the enzyme [111,116,117].

C. Purification

Attempts to purify MST from rat liver [112,118] and rat erythrocytes [119] have been described, but were apparently hampered by the instability of the enzyme. No homogenous preparations of MST from animal sources have apparently been achieved, despite claims to the contrary [112,118]. On the other hand, Vachek and Wood [111] recently reported the isolation of the enzyme from *E. coli* in an apparently homogenous form. Their method utilized treatment of an extract from sonicated cells with protamine, chromatography on DEAE-cellulose and Sephadex G-100, followed by preparative electrophoresis on polyacrylamide. The enzyme was stabilized during the purification by the presence of 0.8 *M* KCl, as it rapidly lost its activity in solutions of low ionic strength due to dissociation into monomers.

D. Molecular Properties

Only the bacterial enzyme [111], which as noted is apparently homogenous, will be described in this respect. The absorption spectrum of the purified enzyme was that of a typical protein. Its isoelectric point was determined by cellulose acetate membrane electrophoresis to be pH 4.5. The molecular weight (at high ionic strength), as determined by sedimentation equilibrium ultracentrifugation, was 23,800 daltons; values obtained from gel filtration studies and amino acid analysis were in satisfactory agreement. In solutions of low ionic strength, the enzyme was found to dissociate into two fragments of about 12,000 molecular weight, each apparently endowed with enzymatic activity, but less stable than the parent dimer. Amino acid analysis demonstrated that some amino acids occurred in an odd number per dimer, indicating that the monomers were not identical. In line with this interpretation is the finding that only one of the two cysteine residues in the dimer molecule was found to react with 5,5'-dithiobis(2-nitrobenzoic acid). Metal analysis revealed the presence of 0.5 atom of copper and 1 atom of zinc per molecule of enzyme. Whether these metals have any functional role in the enzyme action is unsettled. It is pertinent to note that Kun and Fanshier [112,118] reported that their purified MST from rat liver contained copper and assigned a functional role for this metal. Van Den Hamer *et al.* [119] were, on the other hand, able to remove all copper from their purified enzyme preparations without any impairment of enzyme activity and concluded that MST was not a copper enzyme.

E. Catalytic Properties

1. SUBSTRATE SPECIFICITY

The only known sulfur donor substrate for MST is β-mercaptopyruvate [3,120]. It is of certain interest that mercaptopyruvate may be enzymatically reduced by NADH in the presence of lactate dehydrogenase (EC 1.1.1.27; L-lactate:NAD$^+$ oxidoreductase) to yield β-mercaptolactate [121]. The latter is not a substrate for MST [120,121], but as the lactate dehydrogenase reaction is reversible, it may have a metabolic control function for the transsulfuration of mercaptopyruvate. With regard to the acceptor specificity of MST, the enzyme shows strong resemblance to TST. Meister *et al.* [3] reported that if mercaptopyruvate was incubated with mercaptoethanol in the presence of a rat liver preparation, stoichiometric amounts of pyruvate and hydrogen sulfide were formed, presumably according to

$$HSCH_2COCOO^- + RSH \rightarrow CH_3COCOO^- + RSSH \tag{12}$$

followed by

$$RSSH + RSH \rightarrow (RS)_2 + H_2S \qquad (13)$$

where RSH denotes mercaptoethanol and RSSH and $(RS)_2$, the corresponding persulfide and disulfide, respectively. It should be noted that the actual formation of the two latter compounds has not been verified. No systematic survey of other thiol compounds with respect to their ability to function as sulfur acceptors for MST has been reported, but the bacterial enzyme was found by Vachek and Wood [111] to use cysteamine and thioglycollate for this purpose, although less efficiently than mercaptoethanol.

Sörbo discovered [114] that rat liver homogenates catalyzed the transsulfuration between mercaptopyruvate and sulfite according to

$$HSCH_2COO^- + SO_3^{2-} \rightarrow CH_3COCOO^- + S_2O_3^{2-} \qquad (14)$$

Later work in another laboratory [112,122] demonstrated that this reaction was also catalyzed by a partly purified preparation of MST, indicating that the same enzyme is responsible for the transsulfuration between mercaptopyruvate and sulfite, cyanide, or mercaptoethanol, although this has not been strictly proved. Sörbo [114] furthermore reported that rat liver homogenates catalyzed a similar transsulfuration between mercaptopyruvate and two sulfinates of biological occurrence, alanine sulfinate (cysteine sulfinate), and hypotaurine. The corresponding thiosulfonates, alanine thiosulfonate (thiocysteate) and thiotaurine, were demonstrated as products:

$$HSCH_2COCOO^- + RSO_2^- \rightarrow CH_3COCOO^- + RSO_2S^- \qquad (15)$$

This was the first demonstration of an enzymatic reaction by which thiosulfonates could be formed from sulfinates, previously shown to be present in animal tissues [123,124].

To summarize, cyanide, sulfite, sulfinates, and thiols function as sulfur acceptors for MST, as well as for TST, but the substrate specificity of the two enzymes with respect to thiol compounds appears to be somewhat different.

2. INHIBITION AND INACTIVATION

Limited data are available on the stability of MST, but the bacterial enzyme appears to be more thermostable, but less acid stable, than the mammalian enzyme [117]. The bacterial enzyme was (similar to TST) inactivated by incubation with cyanide, in contrast to the rat liver enzyme [117]. The crude enzyme from *E. coli* was observed to be moderately sensitive to inactivation from certain sulfhydryl group reagents [117], whereas the pure enzyme was insensitive to these compounds [111]. No inhibition of the pure bacterial enzyme was observed with heavy metal reagents, speaking against the presence of an active heavy metal group in MST. Van Den Hamer *et al.* [119] also noted that diethyldithiocarbamate, a copper reagent, did not inhibit partly purified MST from rat liver or rat erythrocytes. On the other hand, Kun and Fanshier [125] reported earlier that three heavy metal reagents (hydroxyquinoline, *o*-phenanthroline, and bathocuproine) were inhibitory to their partly purified rat liver enzyme, supporting their view that the enzyme contained copper as a catalytically active group.

Of certain interest is the observation that cysteine behaves as a competitive inhibitor of the enzyme [126]. This must apparently complicate studies of *in vitro* systems, where the transamination of cysteine to mercaptopyruvate is coupled to an MST catalyzed transsulfuration of the latter compound. It has furthermore been reported [125] that two other thiol compounds, *o*- and *m*-thiolbenzoate, are potent MST inhibitors, possibly due to some structural relationship to mercaptopyruvate. Another inhibitory compound of interest is pyruvate [3,111]. Being a product of the enzyme reaction, a product type of inhibition can be expected. Apparently the mechanism is more complicated, as the inhibition given by pyruvate is augmented if this compound is preincubated with mercaptopyruvate [111]. To account for these results, it was suggested that pyruvate and mercaptopyruvate reacted to yield a product, possibly of a thiomercaptol structure. The necessity of taking such effects into accounts in future kinetic studies of MST must be noted.

3. MECHANISM OF ACTION

Available, although meager, evidence indicates that MST operates by a double displacement mechanism, similar to that of TST. Sörbo [114] suggested that MST reacted with mercaptopyruvate to form an enzyme–persulfide intermediate; further support for this hypothesis was later provided by Hylin and Wood [126]. Kun and Fashier [112] have proposed another reaction mechanism, assuming that the enzyme contains copper (and thiol groups) in its active site. As this assumption now appears to be rather questionable, no further details of their hypothesis will be presented here. Although studies of the overall kinetics of MST have been reported [111,112], they provide little information on the action mechanism of the enzyme. As MST now appears to be available in a very pure form, studies of its reaction mechanism by methods so successfully applied to TST appear feasible and should bring more light on this interesting topic.

F. Biological Function

Meister *et al.* [3] suggested that the action of MST could provide an alternative pathway of cysteine degradation by which cysteine is converted to mercaptopyruvate by transamination, followed by desulfuration of this keto acid to pyruvate. The ultimate fate of the sulfur was not considered. It may be noted here that mercaptopyruvate has so far not been detected in biological material, but indirect evidence is available for its formation *in vivo*. As earlier mentioned, mercaptopyruvate may exist in a metabolic equilibrium with mercaptolactate, and the mixed disulfide of this compound and cysteine has been demonstrated in normal human urine [127,128].

There is thus some evidence that mercaptopyruvate may function as the sulfur donor of MST in the body, but the identity of the natural acceptor substrate has not been established. It is probably not a thiol compound, as the only thiol compounds so far found to be acceptor substrates of MST are not naturally occurring. Although cyanide must be considered as a possible candidate, there is little evidence speaking in its favor. It may be noticed in this context that mercaptopyruvate was found to have an antidotal action against cyanide in rats, but not in rabbits [129]. Sörbo proposed [114] that sulfite is the natural acceptor substrate of MST, which would account for the formation of thiosulfate normally occurring in the animal body [100]. In support of this hypothesis, Fasth and Sörbo [130] recently showed that the administration of mercaptopyruvate to mice results in a considerable increase of urinary thiosulfate excretion. It should not be concealed from the reader, however, that other pathways for the biosynthesis of thiosulfate have been proposed [131,132]. One of these is of particular interest for this discussion. De Marco *et al.* [132] reported that alanine thiosulfonate and thiotaurine were efficiently converted to thiosulfate when given to rats, and similar results were obtained by Fasth and Sörbo [130] in experiments on mice. As earlier noted, alanine thiosulfonate and thiotaurine are products of MST catalyzed reactions between the corresponding sulfinates and mercaptopyruvate. Thiotaurine has furthermore been isolated from the urine of rats given a cystine load [133], implying that this thiosulfonate is a normal metabolite in the mammalian body. Thiosulfate may thus be formed from mercaptopyruvate through two MST-dependent pathways. One is a direct transsulfuration to sulfite, the other involves a thiosulfonate as an intermediate. Concerning the details of the latter, very little is known. De Marco and Coletta [133] have reported that rat liver mitochondria catalyze transamination reactions of alanine thiosulfonate that yield thiosulfate as an ultimate product. However, a similar transamination of thiotaurine appears very unlikely from our present knowledge of the substrate specificity of transaminases. Fasth and Sörbo [130] pointed out that thiosulfate may be formed from thiosulfonates and sulfite through the action of TST (Section II,E,1), but more substantial proof for the implied coupling of MST and TST *in vivo* is obviously required.

In conclusion, there are good reasons to believe that MST participates in the biosynthesis of thiosulfate in higher animals. Concerning the function of the enzyme in bacteria, nothing is known and speculations will, at this stage, be fruitless.

REFERENCES

1. K. Lang, *Biochem. Z.* **259**, 243 (1933).
2. J. L. Wood and H. Fiedler, *J. Biol. Chem.* **205**, 231 (1953).
3. A. Meister, P. E. Fraser, and S. V. Tice, *J. Biol. Chem.* **206**, 561 (1954).
4. B. H. Sörbo, *Acta Chem. Scand.* **7**, 1129 (1953).
5. B. Sörbo, *Acta Chem. Scand.* **8**, 694 (1954).
6. B. H. Sörbo, *Sv. Kem. Tidskr.* **65**, 169 (1953).
7. K. Lang in "Handbuch der physiologisch- und pathologisch-chemischer Analyse," Vol. 6B, pp. 780–785. Springer-Verlag, Berlin and New York, 1966.
8. A. R. Roy and P. A. Trudinger, "The Biochemistry of Inorganic Compounds of Sulphur," pp. 190–206. Cambridge Univ. Press, London and New York, 1970.
9. J. Westley, *Advan. Enzymol.* **39**, 327 (1973).
10. E. Kun in "Metabolic Pathways" (D. M. Greenberg, ed.), 3rd ed., Vol. 3, pp. 382–390. Academic Press, New York, 1969.
11. B. H. Sörbo, in "Methods in Enzymology" (S. P. Colowick and N. O. Kaplan, eds.), Vol. 2, p. 334. Academic Press, New York, 1955.
12. B. Davidson and J. Westley, *J. Biol. Chem.* **240**, 4463 (1965).
13. A. Koj, *Acta Biochim. Pol.* **15**, 161 (1968).
14. A. J. Smith and J. Lascelles, *J. Gen. Microbiol.* **42**, 357 (1966).
15. B. Sörbo, *Acta Chem. Scand.* **16**, 243 (1962).
16. M. Volini and J. Westley, *J. Biol. Chem.* **241**, 5168 (1966).
17. J. R. Green and J. Westley, *J. Biol. Chem.* **236**, 3047 (1961).
18. M. Villarejo and J. Westley, *J. Biol. Chem.* **238**, 4016 (1963).
19. R. A. Llenado and G. A. Rechnitz, *Anal. Chem.* **44**, 1366 (1972).
20. R. A. Llenado and G. A. Rechnitz, *Anal. Chem.* **45**, 826 (1973).
21. G. G. Guilbault, S. S. Kuan, and R. Cochran, *Anal. Biochem.* **43**, 42 (1971).
22. J. R. Murphy, J. Frankenfeld, J. Calvert, and R. C. Tilton, *Appl. Microbiol.* **24**, 283 (1972).
23. W. A. Himwich and J. P. Saunders, *Amer. J. Physiol.* **153**, 348 (1948).
24. F. M. Gal, F.-H. Fung, and D. M. Greenberg, *Cancer Res.* **12**, 574 (1952).
25. D. Reinwein, *Hoppe-Seyler's Z. Physiol. Chem.* **326**, 94 (1961).
26. A. Koj and J. Frendo, *Acta Biochim. Pol.* **9**, 373 (1962).
27. H. Schiewelbein, E. Werle, E. K. Schulz, and R. Baumeister, *Naunyn-Schmiedeberg's Arch. Pharmacol. Exp. Pathol.* **262**, 358 (1969).
28. H. Schiewelbein, R. Baumeister, and R. Vogel, *Naturwissenschaften* **56**, 416 (1969).
29. J. Parsons and M. Rotschild, *J. Insect Physiol.* **8**, 285 (1962).
30. S. Ludewig and A. Chanutin, *Arch. Biochem.* **29**, 441 (1950).

31. B. H. Sörbo, *Acta Chem. Scand.* **5,** 724 (1951).
32. C. De Duve, B. C. Pressman, R. Gianetto, R. Watteaux, and F. Appelmans, *Biochem. J.* **60,** 604 (1955).
33. J. M. Moyle, *Nature* (*London*) **172,** 508 (1953).
34. G. D. Greville and J. B. Chappell, *Biochim. Biophys. Acta* **33,** 267 (1959).
35. K. Gemeinhardt, *Ber. Deut. Bot. Ges.* **56,** 275 (1938).
36. M. Y. Chew and C. G. Boey, *Phytochemistry* **11,** 167 (1972).
37. V. Tomati, G. Federici, and C. Cannella, *Physiol. Chem. Phys.* **4,** 193 (1972).
38. S. Oi, *Agr. Biol. Chem.* **37,** 629 (1973).
39. R. N. Stearns, *J. Cell. Comp. Physiol.* **41,** 163 (1953).
40. R. M. Hall and R. S. Berk, *Can. J. Microbiol.* **14,** 515 (1968).
41. D. C. Yoch and E. S. Lindstrom, *J. Bacteriol.* **106,** 700 (1971).
42. F. Hashwa and N. Pfennig, *Arch. Mikrobiol.* **81,** 36 (1972).
43. C. A. McChesney, *Nature* (*London*) **181,** 347 (1958).
44. T. J. Bowen, P. J. Butler, and F. C. Happold, *Biochem. J.* **97,** 651 (1965).
45. K. Sargeant, P. W. Buck, J. W. S. Ford, and R. G. Yeo, *Appl. Microbiol.* **14,** 998 (1966).
46. I. Suzuki and M. Silver, *Biochim. Biophys. Acta* **122,** 22 (1966).
47. A. M. Charles and I. Suzuki, *Biochim. Biophys. Acta* **128,** 510 (1966).
48. H. B. LeJohn, L. Van Caeseele, and H. Lees, *J. Bacteriol.* **94,** 1484 (1967).
49. D. P. Kelly, *Aust. J. Sci.* **31,** 165 (1968).
50. R. Tabita, M. Silver, and D. G. Lundgren, *Can. J. Biochem.* **47,** 1141 (1969).
51. A. M. Charles, *Arch. Biochem. Biophys.* **129,** 124 (1969).
52. K. R. Leininger and J. Westley, *J. Biol. Chem.* **243,** 1892 (1968).
53. P. Horowitz and F. De Toma, *J. Biol. Chem.* **245,** 984 (1970).
54. J. Drenth and J. D. G. Smit, *Biochem. Biophys. Res. Commun.* **45,** 1320 (1971).
55. K. M. Blumenthal and R. L. Heinrikson, *J. Biol. Chem.* **246,** 2430 (1971).
56. J. Westley and J. R. Green, *J. Biol. Chem.* **234,** 2325 (1959).
57. C. Cannella, L. Pecci, and G. Federici, *Ital. J. Biochem.* **21,** 1 (1972).
58. J. Westley, *J. Biol. Chem.* **234,** 1857 (1959).
59. M. Villarejo and J. Westley, *Biochim. Biophys. Acta* **117,** 209 (1966).
60. M. Volini, F. De Toma, and J. Westley, *J. Biol. Chem.* **242,** 5220 (1967).
61. M. Volini and J. Westley, *J. Biol. Chem.* **241,** 5168 (1966).
62. P. Horowitz and J. Westley, *J. Biol. Chem.* **245,** 986 (1970).
63. B. Sörbo, *Acta Chem. Scand.* **17,** 2205 (1963).
64. B. Sörbo, *Acta Chem. Scand.* **17,** S107 (1963).
65. S. F. Wang and M. Volini, *J. Biol. Chem.* **243,** 5465 (1968).
66. F. De Toma and J. Westley, *Biochim. Biophys. Acta* **207,** 144 (1970).
67. K. M. Blumenthal and R. L. Heinrikson, *Biochim. Biophys. Acta* **278,** 530 (1972).
68. J. Westley and T. Nakamoto, *J. Biol. Chem.* **237,** 547 (1962).
69. B. Sörbo, *Acta Chem. Scand.* **16,** 2455 (1962).
70. B. Sörbo, *Proc. Int. Pharmacol. Meet., 1st, 1961* Vol. 6, pp. 121–128 (1962).
71. R. C. Bryant and S. Rajender, *Biochem. Biophys. Res. Commun.* **45,** 532 (1971).
72. B. H. Sörbo, *Acta Chem. Scand.* **7,** 32 (1953).
73. B. H. Sörbo, *Acta Chem. Scand.* **7,** 1137 (1953).
74. R. Mintel and J. Westley, *J. Biol. Chem.* **241,** 3381 (1966).
75. B. Sörbo, *Bull. Soc. Chim. Biol.* **40,** 1859 (1958).
76. B. Eriksson and B. Sörbo, *J. Biol. Chem.* **21,** 958 (1967).
77. J. Westley and D. Heyse, *J. Biol. Chem.* **246,** 1468 (1971).
78. B. Sörbo, *Biochim. Biophys. Acta* **38,** 349 (1960).
79. T. W. Szczepkowski, *Acta Biochim. Pol.* **8,** 251 (1961).

80. T. W. Szczepkowski and J. L. Wood, *Biochim. Biophys. Acta* **139**, 469 (1967).
81. B. Sörbo, *Acta Chem. Scand.* **9**, 1656 (1955).
82. P. D. Bartlett and R. E. Davies, *J. Amer. Chem. Soc.* **80**, 2513 (1958).
83. B. Sörbo, *Acta Chem. Scand.* **11**, 628 (1957).
84. K. Lang, *Z. Vitam.-, Horm.- Fermentforsch.* **2**, 288 (1950).
85. J. P. Saunders and W. A. Himwich, *Amer. J. Physiol.* **163**, 404 (1950).
86. L. W. Nichol and J. M. Creeth, *Biochim. Biophys. Acta* **71**, 509 (1963).
87. D. Reinwein, *Klin. Wochenschr.* **39**, 1216 (1961).
88. A. Finazzi Agro, G. Federici, C. Giovagnoli, C. Cannella, and D. Cavallini, *Eur. J. Biochem.* **28**, 89 (1972).
89. R. Mintel and J. Westley, *J. Biol. Chem.* **241**, 3386 (1966).
90. S. F. Wang and M. Volini, *J. Biol. Chem.* **248**, 7376 (1973).
91. M. Volini and S. F. Wang, *J. Biol. Chem.* **248**, 7386 (1973).
92. M. Volini and S. F. Wang, *J. Biol. Chem.* **248**, 7392 (1973).
93. B. Sörbo, *in* "Sulfur in Organic and Inorganic Chemistry" (A. Senning, ed.), Vol. 2, pp. 143–169. Dekker, New York, 1972.
94. C.-J. Clemedson, H. I:son Hultman, and B. Sörbo, *Acta Physiol. Scand.* **32**, 245 (1954).
95. C.-J. Clemedson, H. I:son Hultman, and B. Sörbo, *Acta Physiol. Scand.* **35**, 31 (1955).
96. B. Stuber and K. Lang, *Deut. Arch. Klin. Med.* **176**, 213 (1933/1934).
97. P. Langer and N. Michajlovskij, *Hoppe-Seyler's Z. Physiol. Chem.* **312**, 31 (1958).
98. G. E. Boxer and J. C. Richards, *Arch. Biochem. Biophys.* **39**, 7 (1952).
99. G. E. Boxer and J. C. Richards, *Arch. Biochem. Biophys.* **39**, 287 (1952).
100. J. H. Gast, K. Arai, and F. L. Aldrich, *J. Biol. Chem.* **196**, 875 (1952).
101. H. McIlwain, "Biochemistry and the Central Nervous System," p. 308. Churchill, London, 1966.
102. B. Sörbo, *Acta Chem. Scand.* **18**, 821 (1964).
103. A. Koj and J. Frendo, *Folia Biol. (Prague)* **18**, 49 (1967).
104. A. Koj, J. Frendo, and Z. Janik, *Biochem. J.* **103**, 791 (1967).
105. B. Skarzynski, T. W. Szczepkowski, and M. Weber, *Nature (London)* **184**, 994 (1959).
106. A. Kaji and W. D. McElroy, *J. Bacteriol.* **77**, 630 (1959).
107. J. E. Schneider and J. Westley, *J. Biol. Chem.* **238**, PC3516 (1963).
108. J. F. Schneider and J. Westley, *J. Biol. Chem.* **244**, 5735 (1969).
109. A. Finazzo Agro, C. Cannella, M. T. Graziani, and D. Cavallini, *FEBS (Fed. Eur. Biochem. Soc.) Lett.* **16**, 172 (1971).
110. J. W. Hylin, *in* "Methods in Enzymology" (S. P. Colowick and N. O. Kaplan, eds.), Vol. 5, p. 987. Academic Press, New York, 1962.
111. H. Vachek and J. L. Wood, *Biochim. Biophys. Acta* **258**, 133 (1972).
112. E. Kun and D. W. Fanshier, *Biochim. Biophys. Acta* **32**, 338 (1959).
113. B. H. Sörbo, *in* "Methods in Enzymology" (S. P. Colowick and N. O. Kaplan, eds.), Vol. 5, p. 990. Academic Press, New York, 1962.
114. B. Sörbo, *Biochim. Biophys. Acta* **24**, 324 (1957).
115. E. Kun, C. Klausner, and D. W. Fanshier, *Experientia* **16**, 55 (1960).
116. Y. Kondo, T. Kameyama, and N. Tamiya, *J. Biochem. (Tokyo)* **43**, 749 (1956).
117. J. W. Hylin, H. Fiedler, and J. L. Wood, *Proc. Soc. Exp. Biol. Med.* **100**, 165 (1959).
118. D. W. Fanshier and E. Kun, *Biochim. Biophys. Acta* **58**, 266 (1962).
119. C. J. A. Van Den Hamer, A. G. Morell, and I. H. Scheinberg, *J. Biol. Chem.* **242**, 2514 (1967).
120. H. Fiedler and J. L. Wood, *J. Biol. Chem.* **222**, 387 (1956).

121. E. Kun, *Biochim. Biophys. Acta* **25**, 135 (1957).
122. E. Kun and D. W. Fanshier, *Biochim. Biophys. Acta* **33**, 26 (1959).
123. B. Bergeret and F. Chatagner, *Biochim. Biophys. Acta* **14**, 297 (1954).
124. B. Bergeret and F. Chatagner, *Biochim. Biophys. Acta* **14**, 543 (1954).
125. E. Kun and D. W. Fanshier, *Biochim. Biophys. Acta* **48**, 187 (1961).
126. J. W. Hylin and J. L. Wood, *J. Biol. Chem.* **234**, 2141 (1959).
127. T. Ubuka, K. Kobayashi, K. Yao, H. Koduma, K. Fuji, K. Hirayama, T. Kuwaki, and S. Mizuhara, *Biochim. Biophys. Acta* **138**, 493 (1968).
128. A. Niederwieser, P. Giliberti, and K. Baerlocher, *Clin. Chim. Acta* **43**, 405 (1973).
129. C.-J. Clemedson, T. Fredriksson, B. Hansen, H. Hultman, and B. Sörbo, *Acta Physiol. Scand.* **42**, 41 (1958).
130. A. Fasth and B. Sörbo, *Biochem. Pharmacol.* **22**, 1337 (1973).
131. D. Cavallini and F. Stirpe, *Atti Accad. Naz. Lincei, Cl. Sci. Fis., Mat. Natur., Rend.* [8] **20**, 378 (1956).
132. C. De Marco, M. Coletta, B. Mondovi, and D. Cavallini, *Ital. J. Biochem.* **9**, 3 (1960).
133. C. De Marco and M. Coletta, *Biochim. Biophys. Acta* **47**, 257 (1961).

CHAPTER 11

Methionine Biosynthesis

Martin Flavin

I. INTRODUCTION

Methionine was isolated from protein in 1922 [1] and identified in 1928 [2]. Its function in transmethylation was inferred in 1939, when it was found that homocysteine + choline could replace methionine as an essential nutrient for animals [3,4], but was only confirmed in 1953 with the identification [5] of S-adenosylemethionine (AMe) as a virtually universal biological methylating agent. Its role as a precursor of polyamines was not established until 1958 [6], and as the initiator of protein synthesis until 1966 [7,8].

Toward the end of a decade of active interest in amino acid biosynthesis, Davis wrote in 1955: "The exploration of this previously inaccessible area has been made possible by the development of several experimental approaches, including use of: (a) isotopically labeled compounds; (b) extracted and purified enzyme systems; (c) growth inhibiting analogues of metabolites; and (d) autotrophic microbial mutants" [9]. The elucidation of methionine biosynthesis took place not during, but in two intervals before and after, this period of active interest. The general outline was established by studies of mammalian metabolism in the 1930's, but the specific reactions and intermediates involved in the several alternate pathways now recognized were not discovered until the 1960's, and other variants of these pathways probably still remain to be identified, particularly in vascular plants. Thus, considerable new information has been added since the subject was reviewed in this series in 1969 [10]. This article will deal with the recent advances, but I shall first outline the early work of the 1930's, and then comment on factors involved in the long period of inactivity that followed.

A. Historical Development of Ideas about Methionine Biosynthesis

1. EARLY STUDIES RELATED TO THE OVERALL PATHWAY OF METHIONINE BIOSYNTHESIS: 1930–1945

The early results, reviewed in references [4,11], were obtained largely by studying animal nutrition and reflect some virtuosity. Mammals cannot synthesize methionine, but they can carry out the last step in the pathway, the methylation of homocysteine. Since homocysteine is not present in a natural diet, the utility of this mammalian reaction does not lie in making more methionine for new protein, but in a cyclic regeneration of methionine for transmethylations. The reaction itself was not discovered in this early work. However, when it was found that homocysteine could replace methionine in animal diets if they were also supplemented with choline or betaine, a few animals were noted to grow even in the absence of the latter [3]. It was suspected that this might have been due to a variable availability of unknown vitamins [12] and ten years later, after many efforts, three laboratories reported that folate and vitamin B_{12} could replace choline in a homocysteine diet [13]. Insight into the relation between these vitamins and methionine came later and from another direction.

A second clue to methionine biosynthesis stemmed from nutritional studies bearing on the pathway of its degradation in animals. A series of experiments in the period 1932–37 [4,11] led to the discovery that cysteine can spare the requirement for methionine, and that homocysteine can replace cysteine for this purpose, i.e., in the absence of choline or a vitamin fraction. In a paper reporting that dietary homocysteine, as well as methionine, led to increased cystine excretion by human cystinurics, Brand et al. postulated in 1936 [14] that the transfer of sulfur from homocysteine to cysteine might involve the intermediate formation of cystathionine (Fig. 1). The latter had been tentatively identified [15],

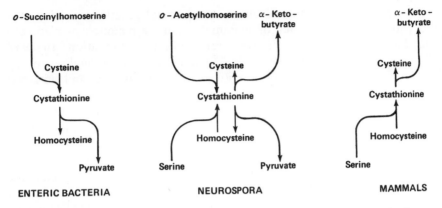

FIG. 1. Patterns of biological transsulfuration. The structure of cystathionine is

$$HOOC—\underset{\underset{NH_2}{|}}{CH}—CH_2—S—CH_2—CH_2—\underset{\underset{NH_2}{|}}{CH}—COOH$$

but was not shown to occur naturally until 1940, when it was isolated from seleniferous grain in an isomorphous combination with its selenium analogue [16]. In the meantime, nutritional experiments with synthetic cystathionine were consistent with the hypothesis of Brand [4]. Eventually these results were shown to be due to the presence of two pyridoxal-P mammalian enzymes [4]: cystathionine β-synthase [17], catalyzing the replacement of the β-hydroxyl of serine by the sulfur of homocysteine, and γ-cystathionase [18], catalyzing the elimination of cysteine from the γ-position of the 4-carbon chain of cystathionine (Fig. 1). These reactions, jointly called "transsulfuration" [4], function in mammals to conserve the sulfur of methionine, as cysteine, while allowing the carbon chain to be metabolized via α-ketobutyrate for generation of energy. Before turning to the implication of this work for methionine biosynthesis, it should be noted that, as indicated in Fig. 1, enzymes catalyzing the mammalian type of transsulfuration, from homocysteine to cysteine, have also been found in *Neurospora crassa* [19–21], *Saccharomyces cerevisiae* [21], and *Aspergillus nidulans* [22], but have never been reported in prokaryotic bacteria [21,23]. A mutant lacking cystathionine β-synthase has been isolated from *Aspergillus* [22] (an earlier report of a *Neurospora* mutant lacking γ-cystathionase [19] could not be confirmed [20]), and has made it possible to show that transsulfuration is the *only* route for cysteine synthesis from methionine in this organism [22]. It is not clear why this pathway has been widely retained in eukaryotic fungi, but is dispensable for bacteria [21].

In one of the earliest applications of microbial genetics to amino acid biosynthesis, these studies of mammalian sulfur metabolism were extrapolated to postulate a pathway for methionine formation. In 1947 Horowitz and his collaborators [24,25] studied the nutritional responses of a collection of methionine auxotrophs of *Neurospora,* and obtained evidence supporting the following pathway: homoserine $\rightarrow$ cystathionine $\rightarrow$ homocysteine $\rightarrow$ methionine. They proposed reactions homologous to those found in animals (Fig. 1), namely, replacement of the homoserine hydroxyl by sulfur of cysteine (cystathionine γ-synthase), and elimination of homocysteine from cystathionine (β-cystathionase). At this point, the overall path of methionine biosynthesis had been correctly surmised. It is interesting that the tool that enabled confirmation of this pathway 20 years later was already available in 1944. This was the fact that three unlinked genes were involved in the postulated step, homoserine $\rightarrow$ cystathionine. However, this information (H. R. Buss, dissertation, cited in Fling and Horowitz [26]) was not published until Murray began a systematic study of methionine genetics in 1960 [27].

2. OBSTACLES AND CLUES IN THE FURTHER ELUCIDATION OF METHIONINE BIOSYNTHESIS: 1945–1960

The difficulties that delayed further progress relate to the postulated Eqs. (1) and (2)

$$\text{Homoserine} + \text{cysteine} \rightarrow \text{cystathionine} \tag{1}$$

$$\text{Homocysteine} + \text{``C}_1\text{''} \xrightarrow[\text{folate}]{B_{12}} \text{methionine} \tag{2}$$

(β-cystathionase was easily detected in microbial extracts [19,20]). There is no record of how many laboratories tried unsuccessfully between 1947 and 1964 to demonstrate the first reaction [20]. My own firm impression is that many biochemists in the 1950's believed the reaction to be an established fact, which may have dissuaded them from looking for it, but I have not found the source of this misconception. Certainly during this time, reaction mixtures were also supplemented with acetyl-CoA and succinyl–CoA or systems generating them, in shotgun type experiments [28,29], with negative results.

The solution came from biochemical genetics, which in this case provided not only the tool, but also the answer. In 1959 Davis and collaborators [30] briefly reported the discovery of an *Escherichia coli* mutant that required succinate for aerobic growth due to the absence of α-ketoglutarate dehydrogenase. In the absence of succinate the mutant would respond to a mixture of lysine + methionine + threonine (at a later time the threonine requirement could not be confirmed [31]). A succinylated intermediate had already been identified in the bacterial biosynthesis of lysine [32]. News of this development, which was never published except in the abstract, was brought by one of the coauthors to D. D. Woods at Oxford, who soon began to investigate whether a succinylated derivative of homoserine was involved in the bacterial biosynthesis of methionine (Section II,A,1,a).

To understand the difficulties that delayed elucidation of the formation of methionine from homocysteine, it is necessary to anticipate some later results (Section II,B). Homocysteine methylation occurs in nature by two different reactions that are very similar, except that one involves vitamin B_{12} as a coenzyme (for example, the mammalian reaction mentioned above). Most organisms have only one or the other, but E. coli, which has been the principal organism used in this field, has both reactions. In both reactions the source of the methyl group is an N^5-methyl tetrahydropteroylglutamate derivative (CH_3—$H_4PteGlu_n$). These derive, via reduction of CH_2=$H_4PteGlu_n$ (Fig. 2), from formate or serine (Fig. 3), the latter being the chief precursor in E. coli [for reviews of folic acid, see 33,34]. Elucidation of the role of folate (pteroylglutamate) was one factor in the delay in characterizing methionine biosynthesis; formation of methionine, by cell-free extracts, from homocysteine and formate was first observed in 1953 [35], and from homocysteine and serine in 1955 [36].

The non-B_{12} homocysteine transmethylase was not detected until 1961 [37] due to another problem connected with folic acid. It was recognized in the mid-1940's [33] that naturally occurring forms of the vitamin were poly-γ-glutamate derivatives (Fig. 2), but the original purification of the vitamin was followed by bioassay with microorganisms impermeable to polyglutamates. Consequently, the vitamin fraction had to be treated with peptidase during the isolation, which removed all but the proximal glutamate [33]. Many enzymes of folate metabolism are somewhat more active with polyglutamate than with the vitamin form, but until now the *only* enzymatic reaction that has been found com-

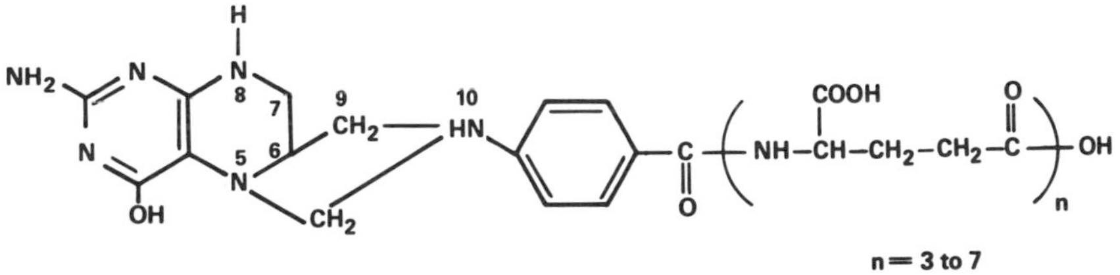

FIG. 2. Structure of 5,10-methylene-5,6,7,8-tetrahydropteroylpolyglutamate (CH_2= $H_4PteGlu_n$).

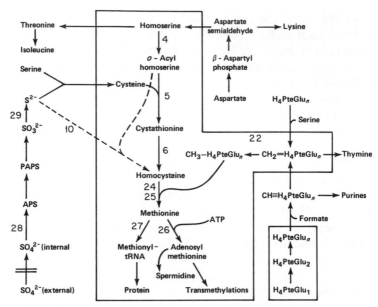

FIG. 3. An outline of methionine biosynthesis and related branch pathways. Reactions wholly enclosed in boxes are essential only for methionine synthesis and utilization. The numbers adjacent to arrows indicate numbered reactions in the text.

pletely inactive with the monglutamate derivative is the non-B_{12} homocysteine transmethylase (Section II,B,2). A single synthesis of a triglutmate was performed in 1948 [38], but was apparently never repeated until very recently (by more effective procedures [39,40]), and the material never became generally available. This curious neglect of coenzyme chemistry hampered studies of methionine biosynthesis until the past few years.

The B_{12}-dependent homocysteine transmethylase is active with the monoglutamate derivative, CH_3—$H_4PteGlu_1$, so studies of this reaction were not handicapped by the unavailability of polyglutamate derivatives. Although vitamin B_{12} (cyanocobalamin; cobalamin = 5,6-dimethylbenzimidazole cobamide) was first implicated in homocysteine methylation in animal tissues in 1949 [13], progress thereafter depended almost entirely on an *E. coli* mutant isolated in 1950 by Davis and Mingioli [41]. The mutant would grow if supplied with either methionine or vitamin B_{12} and is now known to lack the non-B_{12} transmethylase (met-E, Section II,B,3). From 1952 to 1959 this mutant was studied, in relative isolation, by D. D. Woods and his collaborators [42,43]. Using organisms grown on methionine they were able to show, first, that meth-

ionine synthesis by cell suspensions was stimulated by adding cyanocobalamin [42], and second, that addition of the vitamin to cell-free
extracts increased methionine synthesis from homocysteine and serine
[43]. During this decade the intimidating complexity of vitamin B_{12} apparently inhibited any thought that it might be functioning as a coenzyme in the reaction, something that was rather easily shown once the
idea was abroad. Thus in 1955 Davis wrote: "The role of B_{12} is completely obscure" [9].

The situation was transformed in 1958 when, in the course of studying
some bizarre enzymatic rearrangement reactions [44,45], Barker an-
nounced the discovery of a coenzyme form of vitamin B_{12} [46] that was
soon identified as 5'-deoxyadenosylcobalamin [47,48]. The presence
of a new carbon–cobalt bond in the latter at once suggested the possibility that methyl cobalamin (i.e., with a methyl–cobalt bond) might be an
intermediate in methionine biosynthesis, and by the following year several laboratories had purified a B_{12} containing transmethylase protein
from the met-E *E. coli* mutant [49,50]. The B_{12}-dependent transmethylase reaction has since been very actively studied (Section II,B,3)
in these and other [51,52] laboratories.

B. Scope of This Review: 1960–1973

Methionine biosynthesis is interlinked at a great many converging and
diverging branch points with other biosynthetic pathways (Fig. 3). The
following section will describe those reactions that are uniquely required
for methionine biosynthesis (enclosed in boxes in Fig. 3) and the properties of the enzymes involved, including allosteric controls. The pathway
from homoserine to homocysteine will be stressed because it has not
been reviewed as much as homocysteine transmethylation. Homocysteine biosynthesis was studied first, and is now best understood, in
N. crassa and in *E. coli* and *Salmonella typhimurium*. For economy, the
two bacteria will be referred to as "enteric bacteria" wherever they are
the same; small, but not necessarily trivial, differences are occasionally
found between them, in gene location and in the regulation of enzyme
synthesis (comparable differences also occur between different strains of
E. coli). The bacterial and fungal pathways are distinctively different,
however. Still other pathways may well exist in plants and other microorganisms, but have not yet been established (Section II,A,3).

The pathway from homocysteine to methionine (Section II,B) will be
described in three parts: (1) methylene tetrahydrofolate reductase, (2)
non-B_{12} homocysteine transmethylase, and (3) B_{12} transmethylase. In
Section II,B,2 I will also include available information (which is scanty

and recent) on the polyglutamylation of folate. The latter reactions are not part of, but are needed only for, the methionine biosynthetic pathway, and *Neurospora* mutants lacking these reactions score as methionine auxotrophs.

Section III will survey the overall control by methionine of its biosynthesis, including allosteric controls of enzyme activities in branch pathways and all aspects of the control of enzyme synthesis.

I will not discuss methionine yielding reactions that are not part of *de novo* biosynthetic pathways. Due to the delay in establishing the biosynthetic pathways (Section I,B,2), some reactions of this type came to be promoted as possible candidates to mediate *de novo* biosynthesis. One of these was the reaction [53]:

$$\text{AMe} + \text{homocysteine} \rightarrow \text{methionine} + \text{adenosylhomocysteine} \qquad (3)$$

however, no source of the AMe methyl group other than methionine could be identified [54–56]. The same is true for the methyl group of *S*-methylcysteine [57,58], which had been proposed as a possible intermediate in methionine biosynthesis [59–61]. More recently, various bacteria were found, when grown on ethanol as the sole carbon source, to accumulate *O*-ethylhomoserine, and the latter was postulated to be an intermediate in methionine biosynthesis [62]. Later results indicated that the *O*-alkylhomoserine was formed from *O*-acetylhomoserine by an inducible enzyme unrelated to methionine biosynthesis [63].

The following topics (with references to reviews or recent original articles) could not be included: utilization of methionine for protein synthesis and its initiation [64–66], for transmethylations [67,68], and for polyamines [69]; methionine transport [70,70a]; methionine antagonists [11]; and selenomethionine [71].

II. ENZYMES AND CONTROL OF ENZYME ACTIVITY IN REACTIONS UNIQUE TO METHIONINE BIOSYNTHESIS

A. Homoserine to Homocysteine

1. ENTERIC BACTERIA

In enteric bacteria homocysteine is synthesized from homoserine by three sequential reactions, catalyzed by homoserine transsuccinylase [Eq. (4)], cystathionine γ-synthase [Eq. (5)], and β-cystathionase [Eq. (6)].

$$\text{Succinyl-CoA} + \text{L-homoserine} \rightarrow \textit{O}\text{-succinyl-L-homoserine} + \text{CoA} \qquad (4)$$

$$\text{Succinylhomoserine} + \text{L-cysteine} \rightarrow \text{L-cystathionine} + \text{succinic acid} \qquad (5)$$

$$\text{Cystathionine} + H_2O \rightarrow \text{L-homocysteine} + \text{pyruvate} + NH_3 \qquad (6)$$

The genes coding enzymes 4–6 are designated *met-A, met-B*, and *met-C*, respectively (Table I) [72–78]. In *Salmonella*, systematic isolation and mapping of genes involved in methionine biosynthesis and its regulation was started in 1961 by D. A. Smith in the Demerec laboratory [79,80]. In *E. coli*, methionine auxotrophs were isolated sporadically by many different investigators over a longer period of time. However, the organisms now appear to be essentially isogenic and the same gene symbols are used for both.

    a. O-Succinylhomoserine. Shortly after genetic studies that unexpectedly implicated succinate in methionine biosynthesis (Section I,A,2), Rowbury, in the laboratory of D. D. Woods, reported that extracts of an *E. coli* met-C mutant formed cystathionine from homoserine and cysteine if supplemented with succinate, CoA, and ATP [81]. Extracts of met-B mutants incubated with homoserine, succinate, and ATP formed an intermediate that was subsequently converted to cystathionine on addition of cysteine and met-A extract [82]. The intermediate was ninhydrin positive and alkali labile, and was postulated to be the succinic ester of homoserine [83]. At this point further progress was facilitated by the development of procedures for organic synthesis of *O*-succinyl- and *N*-succinylhomoserine [28]. Although neither compound supported the growth of *Salmonella* met-A mutants, probably due to impermeability, *O*-succinyl-L-homoserine was very rapidly converted to cystathionine in the presence of cysteine and a pyridoxal-P enzyme absent from met-B; in contrast the reaction, reported by Rowbury, starting with succinyl-CoA in met-C extracts was actually not detectable in *Salmonella* [28]. If cysteine was omitted, the same enzyme (i.e., absent from met-B) decomposed succinylhomoserine by Eq. (7), which provides a particularly convenient assay for cystathionine γ-synthase.

$$\text{Succinyl-L-homoserine} + H_2O \rightarrow \alpha\text{-ketobutyrate} + NH_3 + \text{succinic acid} \qquad (7)$$

Subsequently, Rowbury and Woods confirmed the identity of the intermediate [84] and reported that met-B mutants of both *E. coli* and *Salmonella* could accumulate quite large amounts of succinylhomoserine [85].

    The synthesis of succinylhomoserine [86], which is prerequisite to an effective study of reaction (4) (Section II,A,1,b), as well as of reaction (5), can be used to introduce radioactive labels into either moiety [87]. Titration curves, infrared spectrum, and optical rotatory dispersion of the L-isomer have been reported. The compound is quite stable at pH values below 7.5, but is rapidly converted to *N*-succinylhomoserine

TABLE I

GENES AND ENZYMES OF METHIONINE BIOSYNTHESIS IN ENTERIC BACTERIA AND *Neurospora*

Equation number	Enzyme	Structural gene	Enzyme specific activity in extracts[a] (nmoles/min × mg)			Reference
			Fully repressed and/or inhibited	Wild-type grown in minimal medium	Maximum observed derepressed	
Enteric bacteria						
4	Homoserine transuccinylase	met-A	0	10	300	[72]
5	Cystathionine γ-synthase	met-B	5	30	200	[72]
6	β-Cystathionase	met-C	15	20	100	[72]
22	Methylene tetrahydrofolate reductase: forward[b]	met-F	0.1	1.0	2	[73]
	reverse[b]		0.06		6	[74]
24	Non-B$_{12}$ homocysteine transmethylase (EC 2.1.1)	met-E	0.2	3	12	[74]
25	B$_{12}$-dependent homocysteine transmethylase (EC 2.1.1)	met-H		0.5	3.5	[74]
26	Adenosylmethionine synthase (EC 2.4.2.13)	met-K		2	7	[72]
27	Methionyl-tRNA^{Met} synthase (tRNA charging assay)	met-G		6	13	[75]

Neurospora					
15	Homoserine transacetylase	*met-5*	4	4	4	[76]
17	Cystathionine γ-synthase	*met-3, met-7*	0	20	20	[76]
19	O-Acetylhomoserine sulfhydrylase		8	8	8	[76]
6	β-Cystathionase	*met-2*	2	2	2	[76]
	Folate diglutamate synthase	*mac*				[77]
	Folate polyglutamate synthase	*met-6*				[77]
22	Methylene tetrahydrofolate reductase: reverse[b] (EC 1.1.1.68)	*met-1*		0.1		c
24	Non-B$_{12}$ homocysteine transmethylase (EC 2.1.1)	*met-8*		7		[78]
26	Adenosylmethionine synthase (EC 2.4.2.13)	*eth-1*		1		c
27	Methionyl-tRNA^{Met} synthase					

[a] For various reasons the numerical values for specific activities are approximations in some cases. Those interested in the quantitative validity of the data should consult the listed references.

[b] Measured in the forward direction with FADH$_2$ as reductant and in the reverse direction with menadione as oxidant.

[c] D. Kerr and M. Flavin, unpublished.

between pH 7.5 and 9.5 [86]; chromatography before and after alkali treatment provides a definitive analytical procedure [87]. Several more rapid assays can also be used, based on the same rearrangement in alkali; the most effective utilizes small Dowex columns [87].

With regard to the long delay in the discovery of succinylhomoserine (Section I,A,2), it should be noted that esterification of homoserine prior to reaction with cysteine was not anticipated because it is not necessary on thermodynamic or kinetic grounds. The formation of a thioether from an alcohol and a mercaptan is strongly exergonic [67]. Although an ester is a better leaving group than an unprotonated hydroxyl, enzymatic acceleration is clearly of overriding importance, as witnessed by the mammalian cystathionine β-synthase [17].

b. Homoserine Transsuccinylase. The first enzyme of homocysteine biosynthesis is coded by the *met-A* gene. Met-A mutants cannot be scored by growth response to succinylhomoserine, which is presumably not permeable [28,85]. However, they will grow if provided *O*-acetylhomoserine [88]; this compound (Section II,A,2,a) is not synthesized by enteric bacteria (29) but can be utilized by bacterial cystathione γ-synthase at $\frac{1}{10}$ the rate of succinylhomoserine [88,89].

The difficulty of assaying homoserine transsuccinylase in crude extracts is documented by the fact that it is often omitted from surveys of all the other enzymes of homocysteine biosynthesis, or else activities are reported as percent of a control rather than actual rates. The obvious assay based on the appearance of the sulfhydryl group of free CoA is satisfactory after a modest (30-fold) purification [90,91]. Lawrence [92] has also reported reasonable rates using this assay with crude extracts in a range of very low protein concentration, but others have found this could not be done due to enzymatic and spontaneous hydrolysis of succinyl-CoA, and spontaneous succinylation of the amino group of homoserine [72]. The transsuccinylase can be reliably assayed by virtue of an exchange reaction [Eq. (8)] that it catalyzes; reaction (8) is totally absent from met-A and is therefore not catalyzed by any other enzyme

[³H]Homoserine + *O*-succinyl-L-homoserine

$$\rightleftarrows \text{homoserine} + O\text{-succinyl-L-[}^3\text{H]homoserine} \quad (8)$$

[29]. Labeled *O*-succinylhomoserine is easily isolated, after conversion by alkali to *N*-succinylhomoserine, by use of ion-exchange columns [87]. More accurate results can be obtained, over a wide (30-fold) range of crude extract protein concentration, by a second method in which the extract is supplemented with a large excess of purified cystathionine γ-synthase, and the α-ketobutyrate formed according to Eq. (9) [Eqs. (4) + (7)] is assayed with lactic dehydrogenase [72]. Specific activ-

ities of Eq. (4) measured this way in extracts are at least 10 times higher

$$\text{Succinyl-CoA} + \text{homoserine} \rightarrow \alpha\text{-ketobutyrate} + NH_3 + \text{succinate} + \text{CoA} \qquad (9)$$

than reported by any other procedure [29,93]. This is evidently due to a marked product inhibition that is relieved when succinylhomoserine is continuously removed [72].

The transsuccinylase has not been purified very much [90,91] and the mechanism of the reaction has not been studied. The exchange reaction [Eq. (8)] suggests the intermediary formation of a succinylated enzyme. The molecular weight is about 65,000 daltons (W. Shive, personal communication, 1969), and K_m values of 5×10^{-4} M for homoserine and 1.5×10^{-3} M for succinyl-CoA have been reported [92].

As expected for the first enzyme of the pathway, the transsuccinylase is the target for allosteric control. Methionine [94] and AMe inhibit enzyme activity separately, but much more effectively when present together [90]. This synergistic end product inhibition is not observed for the exchange reaction [Eq. (8)] [72]. α-Methylmethionine mimics methionine as a feedback inhibitor [93] and resistant mutants have been isolated that overproduce methionine, contain a transsuccinylase insensitive to inhibition by methionine or α-methylmethionine, and map in the met-A gene [80]. Genetic evidence suggests both catalytic and regulatory sites are in a single polypeptide species [80], but protein subunits have not been studied.

No allosteric controls have been found for the other two enzymes of homocysteine biosynthesis. In view of the fact that CH_3—$H_4PteGlu_n$ is an allosteric activator of homocysteine synthesis in Neurospora (Section 11,A,2,c), the effects of various folate derivatives on the corresponding bacterial enzymes were examined. No evidence of folate control over enzyme activity or synthesis was found [72].

At elevated temperatures, growth of E. coli is limited by ability to synthesize methionine [95], which is due to a rapid and immediately reversible inhibition of homoserine transsuccinylase rather unlike conventional denaturation [91]. Similar results were obtained with a Bacillus [96] homoserine transacetylase (Section 11,A,2,b). The utility of special sensitivity of methionine biosynthesis to temperature is unclear [95], but any mechanisms for unusually rapid depletion or restitution of methionine may be worth noting in relation to its role in initiation of protein synthesis. Transsuccinylase activity also selectively declines in Salmonella entering resting phase [72].

c. Cystathionine γ-Synthase of Enteric Bacteria. The met-B gene, which codes the second enzyme of homocysteine biosynthesis, maps near, but is not contiguous with [80], met-F (Section 11,B,1) and is located on different sides of *met-F* in *E. coli* and *Salmonella* [97]. No specific assay for thioethers is available, but reaction (5) can be assayed, after the enzyme is slightly purified, by measuring cysteine sulfhydryl disappearance during a brief anaerobic incubation [98]. It is more convenient to assay cystathionine γ-synthase by measuring the amount of α-ketobutyrate formed in Eq. (7) [98], which the enzyme catalyzes in the absence of cysteine. Cystathionine has also been isolated from reaction mixtures, crystallized, and unequivocally identified [99].

This enzyme has been isolated in 20% yield as a pure yellow protein (MW 160,000) after a 250-fold purification from a derepressed *Salmonella* mutant [98,100]; it constitutes 0.03% of the extractable protein of wild-type. It contains 4 residues of tightly bound pyridoxal-P and has an absorption maximum of 422 nm with a molar absorbance of 39,000 in the pH range 5.5–9.2. In guanidine it dissociates into 4 identical subunits. Each subunit contains two reactive cysteine residues, at least one of which is required for catalytic activity, two additional unexposed sulfhydryl groups, and probably one intramonomeric disulfide bridge [100]. The optical rotatory dispersion spectrum of the native enzyme shows a positive extrinsic cotton effect centered at 422 nm; the α-helix content was calculated to be 43% [101].

Cystathionine γ-synthase is not completely specific in regard to substrate structure and can utilize acetylhomoserine, though much less effectively (Table II) [102,103] in Eqs. (5) and (7). With either homoserine ester, sulfide or methylmercaptan can replace cysteine in Eq. (5) [88,89]. These acylhomoserine sulfhydrylase and methylthiolase reactions [Eqs. (10) and (11)] clearly provide potential alternate pathways to

$$\text{Succinylhomoserine} + H_2S \rightarrow \text{homocysteine} + \text{succinate} \qquad (10)$$

$$\text{Succinylhomoserine} + CH_3SH \rightarrow \text{methionine} + \text{succinate} \qquad (11)$$

methionine, bypassing homocysteine and/or cystathionine. The later identification of similar reactions in eukaryotic fungi has provoked some controversy as to whether Eq. (10) might provide a major alternate pathway (Section II,A,2,d). In enteric bacteria, reactions (10) and (11) are catalyzed by cystathionine γ-synthase and by no other enzyme [88]. Reaction (11) cannot provide an alternate pathway because there is no mechanism of synthesis of methyl mercaptan that does not require prior synthesis of methionine. Reaction (10) cannot provide a major alternative because if it did, met-C mutants, lacking only β-cystathionase [Eq. (6)], would score as wild-types. It probably does provide an ineffective alternative because met-C mutants are slightly leaky and can grow in

TABLE II

KINETIC CONSTANTS FOR SOME REACTIONS CATALYZED BY CYSTATHIONINE γ-SYNTHASE OF *Salmonella*

Elimination and Replacement Reactions

Reaction	Substrate with leaving group	K_m	Substrate introduced	K_m	V_{max} (μmole/min $\times$ μmole enzyme)
γ-Replacement	Succinylhomoserine	3×10^{-4} M^a	Cysteine	7×10^{-5}	15,000
γ-Replacement	Succinylhomoserine	3×10^{-4} M	H_2S	3×10^{-3}	8,000
γ-Replacement	Succinylhomoserine	3×10^{-4} M	CH_3SH	10^{-1}	2,500
γ-Elimination	Succinylhomoserine	3×10^{-4} M	—		3,000
γ-Elimination	Acetylhomoserine	3×10^{-2} M	—		500
β-Elimination	Succinylserine		—		2,800[b]
β-Elimination	Succinylserine	7×10^{-4} M	—		1,200[c]
β-Replacement	Succinylserine		Homocysteine		~30

Hydrogen Exchange Reactions

Exchange into	K_m	Exchange from	V_{max} (μatom H exchange/min $\times$ μmole enzyme)			
			4 carbon chain		3 carbon chain	
			α H	β H	α H	β H
L-Cystathionine	4×10^{-3} M	^3H_2O	4,400	28,000	200,000	11,000
L-Alanine	67×10^{-3} M	^3H_2O			8,000	11,000
L-Alanine	—	^2H_2O				32,000

[a] These K_m values are uncertain [102].
[b] Initial rate [103].
[c] Final inhibited rate [103].

minimal medium with a doubling time of five hours [72]. The ineffectiveness may be due to the high K_m for H_2S (Table II), which might be toxic if it were present at the required concentration. Cystathionine γ-synthase also catalyzes β-elimination from succinylserine, which is not naturally occurring, and very slow synthesis of cystathionine from the latter and homocysteine (Table II). It becomes reversibly inhibited during the β-elimination reaction [103].

Since cystathionine γ-synthase is the only pyridoxal-P enzyme catalyzing elimination and replacement of substituents on C-4 of an amino acid, the mechanisms of Eqs. (5) and (7) have been of special interest to those interested in this coenzyme. The enzyme is unique in catalyzing rapid and stereospecific exchanges of β– as well as α-hydrogens [104,105] in many amino acids not otherwise substrates (Table II). Transient ultraviolet spectral changes during catalysis of the various reactions of Table II suggest that pyridoxamine forms of the coenzyme are present during β-hydrogen exchange and γ-elimination, but not β-elimination, reactions [101,105]. These results indicate [105] labilization of a γ-substituent by:

$$A\text{—}CH_2CH_2\text{—}\overset{\overset{\displaystyle COOH}{|}}{CH}\text{—}N{=}CH\text{—}PyrP \rightarrow A\text{—}CH_2\text{—}CH{=}\overset{\overset{\displaystyle COOH}{|}}{C}\text{—}NH\text{—}CH_2\text{—}PyrP \quad (12)$$

Equation (7) involves a series of intramolecular proton transfers [106,107] by which the new γ-hydrogen of α-ketobutyrate derives in part from both α- and β-hydrogens of the substrate:

$$\text{Succinyl—O—}CH_2\text{—}\overset{\boxed{H}}{CH}\text{—}\overset{\textcircled{H}}{\underset{\underset{\displaystyle NH_2}{|}}{C}}\text{—COOH} \rightarrow \boxed{H}\text{—}CH_2\text{—}CH_2\text{—}\overset{\overset{\displaystyle}{\underset{\underset{\displaystyle O}{\|}}{C}}}{}\text{—COOH} \quad (13)$$

This suggests that a single polyhydric basic group of the enzyme may be involved in labilizing both α- and β-hydrogens [107]. Cystathionine γ-synthase catalyzes an unexpected, and not yet fully explained, reaction in the presence of maleimide [108] in which the product

$$
\begin{array}{c}
\overset{\displaystyle H}{\overset{|}{}}\ \overset{\displaystyle O}{\overset{\|}{}}\\
CH_3-C-C-COOH
\end{array}
$$

is formed instead of α-ketobutyrate. The formation of only one diastereoisomer of this product [108] was the first indication that the last step in elimination reactions

$$CH_3\text{—}CH\text{=}\underset{\underset{NH_2}{|}}{C}\text{—}COOH \rightarrow CH_3\text{—}CH_2\text{—}\underset{\overset{\|}{O}}{C}\text{—}COOH \qquad (14)$$

is enzyme catalyzed [107, footnote 2].

d. β-Cystathionase of Enteric Bacteria. This pyridoxal-P enzyme, catalyzing Eq. (6), is coded by the *met-C* gene [28,109]. Reaction (6) is easily assayed in crude extracts [110] and the enzyme has been partially purified in several laboratories [110–112]. The *Salmonella* enzyme can be obtained as a by-product in the preparation of cystathionine γ-synthase, as it is derepressed (7-fold) with the latter and is purified together with it in the initial heat treatment [110]. Like other cystathionases, it catalyzes β-elimination reactions from a number of amino acids besides cystathionine [110,112].

The enzyme is inhibited by low concentrations of rhizobitoxine, an amino acid produced by the soybean root nodule bacterium *Rhizobium japonicum* that causes chlorosis in the host plant [113]. The inhibition of a spinach β-cystathionase can be overcome by adding additional pyridoxal-P [114], and should be of interest in relation to the mechanism of action of this coenzyme, since rhizobitoxine has recently been identified as an unsaturated oxygen ether analogue of cystathionine [114a]. Inhibition of growth of *Salmonella* by rhizobitoxine is reversed by methionine [113], indicating again that Eqs. (5) and (6) are more important in methionine biosynthesis than Eq. (10) [113,115]. This test would be especially useful in organisms where the more decisive evidence provided by isolation of met-C auxotrophs lacking β-cystathionase is not available (Section II,A,3)

2. *Neurospora crassa*

a. O-Acetylhomoserine. Two years after identification of succinyl-homoserine as a bacterial intermediate, its counterpart in *Neurospora* was discovered as follows. In this organism mutation in any of three unlinked genes (*met-3, met-5, met-7;* each in a different chromosome) results in a methionine requirement that can also be met by cystathionine, but not by homoserine or cysteine. Although this was known in 1944 (Section I,A,2), it was not reported until 1960 [27], shortly after which Murray sent us cultures of representative mutants. A procedure was devised for synthesis of radioactive homoserine [116] and each of the mutants, and wild-type, was then allowed to grow in the presence of unlabeled methionine and [^{14}C]homoserine. The radioactive components in the culture filtrates were fractionated by chromatography and electrophoresis. The principle difference among them was the absence from met-5 of one major and one minor component present in all other strains [117]. The major compound was identified as an ester of 2,4-dihydroxybutyrate and an unknown acid. On the slender grounds that acetylhomoserine had been isolated from a plant source [118], this derivative was synthesized [29,119] and shown to be identical with the minor component. The major component, clearly derived from acetylhomoserine, was shown to be 4-O-acetyl-2,4-dihydroxybutyrate [117]. Acetylhomoserine also supported the growth of met-5, but not met-3 or met-7, mutants [29,119]. These results showed that met-5 mutants were blocked in the acetylation of homoserine and that this reaction was essential for methionine biosynthesis in *Neurospora*.

Acetylhomoserine is even more alkali labile than succinylhomoserine [29]. Procedures for assaying it [119] are similar to those for succinylhomoserine (Section II,A,1,a).

b. Homoserine Transacetylase. The *Neurospora* enzyme, catalyzing reaction (15), can best be assayed in crude extracts by measuring the labeled product formed in the exchange reaction [Eq. (16)] that it also

$$\text{Acetyl-CoA} + \text{L-homoserine} \rightarrow O\text{-acetyl-L-homoserine} + \text{acetic acid} \qquad (15)$$

$$[^3\text{H}]\text{Homoserine} + O\text{-acetylhomoserine} \rightleftarrows \text{homoserine} + O\text{-acetyl-}[^3\text{H}]\text{homoserine} \qquad (16)$$

catalyzes. The procedure is the same [120] as for succinylhomoserine. Reaction (15) can also be measured directly by using [^{3}H]homoserine and isolating the labeled product after conversion to N-acetyl-[^{3}H]homoserine [120], but this assay is not very reliable. Acetyl-CoA has been identified as the acetyl donor [29], and the *Neurospora* transacetylase has been purified 60-fold [29,120]. Extracts of three different met-5 mutants were clearly deficient in the ability to catalyze Eqs. (15) and (16) [76].

Extracts of three other fungi (*S. cerevisiae* [29], *Candida utilis,* and *Cunninghamella blakesleeana* [76,121]) also lacked homoserine transsuccinylase and had similar low levels of transacetylase when assayed by the exchange reaction, the specific activities being about 1% of that of the transsuccinylase in bacteria [76]. Recently, a homoserine transacetylase has been purified from *Bacillus polymyxa* that has a molecular weight of 37,000 and is extraordinarily unstable [96]. Although instability could be the cause of the above apparent low activities, the *Bacillus* enzyme may be more related to the transsuccinylase of enteric bacteria because it is also subject to synergistic end product inhibition by methionine + AMe [96].

The most interesting feature of the *Neurospora* transacetylase is that it is not subject to such inhibition [76] and does not seem to be under methionine control of any kind [122].

   c. *Cystathionine γ-Synthase of Neurospora.* Synthesis of acetylhomoserine was prerequisite to, but did not at once allow, detection of reaction (17) in *Neurospora* extracts. The involvement of two gene

$$O\text{-Acetylhomoserine} + \text{cysteine} \rightarrow \text{cystathionine} + \text{acetic acid} \qquad (17)$$

products (met-3 and met-7), rather than one alone, suggested that there might be differences from enteric bacteria, and for some time cystathionine synthesis could not be detected, using the following, a cumbersome assay that is still the only method available: isolation of labeled cystathionine by high voltage paper electrophoresis after incubations with [^{35}S]cysteine or acetyl-[^{14}C]homoserine [76]. *Neurospora* cystathionine γ-synthase does not catalyze the analytically useful γ-elimination analogous to Eq. (7) [76], nor has any other "partial reaction" been found by comparing the products formed from cysteine or acetylhomoserine by extracts of met-3 and met-7 mutants with those formed by wild-type.

   Factors permitting eventual detection of reasonable activities (5–10 nmoles min^{-1} mg^{-1}) of cystathionine γ-synthase were extraction by brief ultrasonic treatment; stabilization by 10 mM dithiothreitol and EDTA (increasing the half-life of the activity in frozen extracts from 1 hour to several weeks); gel filtration of extracts; and utilization of concentrated extracts for assay [76,121]. Specific activity declined with dilution, and plateaued only when the extract protein concentration was greater than 15 mg/ml [76]. The reaction was specific for acetylhomoserine, was completely absent from extracts of met-3 and met-7 mutants, and was reconstituted by mixing the mutant extracts in stoichiometric proportions.

   When *Neurospora* was cultivated in the presence of 5 mM methionine, extracts were inactive (≤1%) until the small molecule fraction was removed by Sephadex G-25 gel filtration. The inhibitor was identified [122] as AMe (50% inhibition by 10^{-5} M), and its concentration in cells grown on methionine, 0.5 μmole/gm wet weight of mycelium [76], was more than enough to account for complete inhibition. Methionine did not inhibit, or augment the inhibition by AMe. A striking lag in the onset of inhibition by AMe was shown not to be due to its conversion to some other compound [122]. Confirmation that this allosteric inhibition does limit methionine synthesis *in vivo* was obtained with an ethionine resistant mutant shown by Metzenberg [123] to overproduce methionine, and to be osmotically remediable; i.e., to become a phenocopy of wild-type when grown in the presence of 1 M glycerol. Eth^r-1 differed from wild-type in that it failed to accumulate AMe when grown on methionine unless the medium also contained 1 M glycerol [76], and it was shown to have a defect in AMe synthase (the first identification of a structural gene for this enzyme). Also confirmatory is the fact that wild-type grown on methionine + [^{14}C]homoserine accumulated nearly as much acetyl-[^{14}C]homoserine as did met-3 and met-7 mutants that are genetically blocked in its utilization [117].

When extracts of *all* methionine auxotrophs were examined, cystathionine γ-synthase was found to be absent from met-1 and met-6, as well as met-3 and met-7, and to be reconstituted by pairwise mixing of gel filtered extracts of the four mutants in all combinations, except met-1 + met-6 [76]. Met-1 had been shown to lack methylenetetrahydrofolate reductase, and met-6 to be blocked in the synthesis of polyglutamate forms of folic acid [124]. The clue to this puzzling result came from the observation by Sakami that gel filtration did not completely separate polyglutamate derivatives from the protein fraction [125], which suggested to him that met-1 and met-6 extracts might lack an allosteric activator for cystathionine γ-synthase. It was indeed found that full activity could be restored to these mutant extracts by adding 5—CH$_3$H$_4$PteGlu$_n$ [125]. Activation was specific for the N^5—CH$_3$ derivative, and at low concentrations (10^{-6}–10^{-7} M) was much more effective with increasing number of glutamate residues (up to 7). The monoglutamate derivative had some activity, however. Therefore met-6 mutants score as methionine auxotrophs because polyglutamate derivatives are needed for homocysteine transmethylation (Section II,B,2), and extracts lacked cystathionine γ-synthase activity only because monoglutamates were removed by the gel filtration required to remove the allosteric inhibitor, AMe. Addition of CH$_3$—H$_4$PteGlu$_n$ also reactivated aged extracts of wild-type, and antagonized inhibition by AMe [125]. Reactivation began after a lag period similar to that for inhibition, but the cumbersome assay has not allowed any further kinetic analysis.

Thus in *Neurospora* the *second* enzyme in the pathway is the target for allosteric control, both activator and inhibitor being active methylating agents. The enzyme is completely inactive in the absence of N^5-methyl folate derivatives. Stabilization by dithiothreitol may be due to protection of the activator against air oxidation [125], and EDTA may prevent enzyme oxidation of methyl to methylene derivatives. However, inactivation by dilution still occurs in the presence of excess activator (M. A. Savin and M. Flavin, unpublished).

The nature of the above reaction remains an interesting problem for the future. It appears to be catalyzed by two proteins that are very weakly associated; catalysis of a 1-step reaction by such a loose "aggregate" is known but rare [126]. The question whether these are catalytic and allosteric proteins can be approached by genetics as well as by enzyme purification, which has progressed slowly. It is remotely possible that Eq. (17) is not a "γ-replacement" analogous to Eq. (5) in enteric bacteria, but proceeds by some other unforeseen mechanism, since it has not been shown unequivocally that pyridoxal-P is required [76]. Earlier tests for such a requirement, based on inhibition by reagents reactive with aldehydes [76], were probably complicated by trapping of formaldehyde from the activator, formed via

$$CH_3H_4PteGlu_n \leftrightharpoons CH_2=H_4PteGlu_n \rightarrow HCHO + H_4PteGlu_n \qquad (18)$$

Activation of cystathionine γ-synthase by CH_3—$H_4PteGlu_n$ clearly serves to coordinate the formation of methyl groups and of homocysteine for methionine synthesis, and to prevent an undesirable [125] overproduction of methylated folates in the absence of homocysteine. I am not aware of an exact precedent for this type of allosteric activation. It resembles "compensatory reversal of end product inhibition," of which the only example appears to be carbamyl phosphate synthase [127], except that activator is still required in the absence of inhibitor.

Thus, whereas the first enzyme of homocysteine biosynthesis is controlled by synergistic end product inhibition in enteric bacteria, in *Neurospora* neither the activity not the synthesis (Section III,B,1) of this enzyme is under any methionine control. Instead, the second enzyme is regulated by allosteric activation and inhibition. This suggests that either acetylhomoserine or the transacetylase might have another vital function unrelated to methionine biosynthesis. Acetylserine, the precursor of cysteine, is thought to be a coinducer of the cysteine biosynthetic enzymes [128,129], but there is no evidence for a similar function for acetylhomoserine [76].

Of course, from the standpoint of sulfur one might consider cystathionine γ-synthase the first enzyme of homocysteine biosynthesis. Perhaps more relevant is the fact, discussed in the following section, that in *Neurospora* the alternate reaction yielding homocysteine directly from acetylhomoserine is catalyzed by an enzyme distinct from cystathionine γ-synthase.

d. O-Acetylhomoserine Sulfhydrylase. Reaction (19) was discovered in 1966 in *Salmonella* [89, footnote 5], and shortly afterward postu-

$$O\text{-Acetyl-L-homoserine} + H_2S \rightarrow \text{L-homocysteine} + \text{acetate} \qquad (19)$$

lated [88] and identified [130] in *Neurospora*. The sulfhydrylase reaction is very easy to measure and was soon confirmed by several other laboratories [131–133]. Wiebers and Garner had also independently described a sulfhydrylation of homoserine itself [134] but this reaction has not been confirmed [57]. It was thought that reactions (17) and (19) might be catalyzed by the same enzyme in *Neurospora,* since this is so for reactions (5) and (10) in enteric bacteria (Section II,A,1,c). As a result, when difficulties were encountered in detecting cystathionine synthesis, the sulfhydrylase was purified, thinking that this might unmask cystathionine γ-synthase activity; the purified enzyme remained completely inactive when sulfide was replaced with cysteine [57,130]. The recognition that cystathionine γ-synthase was composed of two different proteins in *Neurospora* then made it tempting to think that on dissociation of the weak complex, one of the subunits might acquire sulfhydrylase activity, i.e., in the absence of a cysteine site on the other subunit. However, cystathionine γ-synthase activity was not restored by adding purified sulfhydrylase to extracts of met-3 or met-7 mutants [76]. Although final judgment cannot be made, these two enzymes so far seem to be completely distinct in *Neurospora* [57].

O-Acetylhomoserine sulfhydrylase has been partially (500-fold) purified [57]. It requires pyridoxal-P, does not catalyze γ-elimination (α-ketobutyrate formation) in the absence of H_2S, and cannot utilize acetylserine, or succinyl- or phosphorylhomoserine. It can utilize CH_3SH and the V_{max} for Eq. (20) is 1.8 times that for Eq. (19). K_m values are 7×10^{-3} M for acetylhomoserine, 7×10^{-4} M for H_2S, and 8×10^{-4} M for CH_3SH. Reaction (20) is catalyzed much more effectively in *Neurospora* than the corresponding reaction (11) is in enteric bacteria, in which the enzyme has little or no affinity for CH_3SH (Table I), and this may be the real function of the sulfhydrylase in fungi. *S*-Methylcysteine accumulates in *Neurospora* and can be utilized by reaction (21), catalyzed by γ-cystathionase, coupled with reaction (20). Both γ-cys-

$$O\text{-Acetyl-L-homoserine} + CH_3SH \rightarrow \text{methionine} + \text{acetate} \qquad (20)$$

$$S\text{-Methylcysteine} + H_2O \rightarrow CH_3SH + \text{pyruvate} + NH_3 \qquad (21)$$

tathionase [135] and, to a lesser extent, the sulfhydrylase [57] are derepressed by sulfur deprivation. These reactions may function only in a sulfur storage and retrieval system. That they can so function is suggested by the fact that *S*-methylcysteine supports growth of all *Neurospora* methionine auxotrophs except met-5 mutants, which cannot make acetylhomoserine [60,136].

Neither the activity nor the synthesis of acetylhomoserine sulfhydrylase is under any significant methionine control [57], and the enzyme is present at wild-type levels in all known methionine auxotrophs [76]. If Eq. (19) functioned effectively in methionine biosynthesis, not only met-2 mutants (lacking β-cystathionase like met-C in enteric bacteria), but met-3 and met-7 should grow on minimal medium. For all these reasons it seems unlikely that the sulfhydrylase provides a major alternative route to methionine. A minor alternative role is possible since met-2, met-3, and met-7 mutants are leaky, whereas met-5 is not [60,136].

The possibility that homocysteine formed by Eq. (19) is channeled separately from that formed by Eqs. (17) and (6) has not been investigated, nor has the distribution of the respective enzymes between cytoplasm and mitochondria or other organelles been determined. Transmethylation reactions (potentially) require homocysteine only catalytically; it is unequivocally required in bulk for synthesis of protein methionine and of polyamines. Conceivably homocysteine formed by the sulfhydrylase could be channeled to polyamines. This enzyme might then be under spermidine [69] rather than methionine control, and mutants lacking it might score as spermidine auxotrophs. Neurospora mutants with defects in tRNA^{Met} or methionyl-tRNA^{Met} synthase, which have not been reported, might help to clarify whether homocysteine formed by one of the reactions is channeled into protein synthesis.

e. β-Cystathionase of Neurospora. This pyridoxal-P enzyme, catalyzing reaction (6), is coded by the *met-2* gene [19,20]. It has been separated from γ-cystathionase by DEAE-cellulose fractionation, was rather unstable, and appeared to catalyze β-elimination reactions from several amino acids besides cystathionine [20]. In extracts containing both cystathionases it can be assayed, though not very accurately, by procedures distinguishing between pyruvate and α-ketobutyrate, or homocysteine and cysteine [20,127]. Cystathionine cleaving enzymes of *Neurospora* merit further investigation in view of reports that there may be more than one γ- [138] or β-cystathionase [139]. Pyridoxal-P enzymes catalyzing β-elimination reactions are characteristically not specific with regard to the substituent eliminated, and the physiological function of such enzymes is sometimes unclear.

3. OTHER ORGANISMS

Information on homocysteine biosynthesis in other microorganisms and in plants is still very fragmentary (Table III) [21,22,29,63,76,96, 133,139–150] in relation to the questions: (a) Is acetyl- or succinylhomoserine an intermediate? (b) Is cystathionine an intermediate in the major pathway?

a. O-Acylhomoserine Intermediate. O-Succinylhomoserine is so far confined to one bacterial family, whereas some evidence has been reported for O-acetylhomoserine in members of three bacterial families and in all fungi examined. Here some reservation must be noted about

TABLE III

ACYLHOMOSERINE DERIVATIVES, AND ENZYMES OF HOMOCYSTEINE METABOLISM, IDENTIFIED IN VARIOUS ORGANISMS

Organism	Acylhomoserine isolated	Methionine auxotroph response to acyl-homoserine	Acylhomoserine identified as substrate for enzyme of homocysteine biosynthesis			Other enzymes of homocysteine metabolism			References
			Homoserine transacylase	Acylhomoserine sulfhydrylase	Cystathionine γ-synthase	β-Cystathionase	Cystathionine β-synthase	γ-Cystathionase	
Enteric bacteria									
Salmonella typhimurium	Succinyl	Acetyl	Succinyl	Succinyl	Succinyl	+	−	−	Section II,A,1
Escherichia coli	Succinyl		Succinyl	Succinyl	Succinyl	+	−	−	
Aerobacter aerogenes	Succinyl	Acetyl							[140]
Other bacteria									
Arthrobacter paraffineus	Acetyl	Acetyl							[141]
Bacillus polymyxa	Acetyl		Acetyl						[96,142]
Bacillus sp									[141]
Bacillus subtilis			Acetyl			+			[139]
Brevibacterium flavum			Acetyl						[143]
Corynebacterium acetophilum	Acetyl	Acetyl							[63]

								Reference
Fungi								
Neurospora crassa	Acetyl	Acetyl	Acetyl	Acetyl	+	+	+	Section II,A,2
Saccharomyces cerevisiae	Acetyl	Acetyl	Acetyl	(Acetyl)	+	+	+	[21,29,76, 133,144]
Aspergillus nidulans	Acetyl	Acetyl	Acetyl	(Acetyl)		+	+	[22,145,145a]
Candida albicans					+			[146]
Candida utilis	Acetyl	Acetyl	Acetyl					[29,76]
Cunninghamella blakesleeana	Acetyl	Acetyl	Acetyl					[29,76]
Plants								
Lathyrus sativus	Oxalyl	Oxalyl						[147]
Pisum sativum	Acetyl	Acetyl						[147]
Chlorella	Phosphoryl	Phosphoryl		Phosphoryl				[147]
Spinach	Phosphoryl	Phosphoryl	Acetyl	Phosphoryl	+		−	[148–150]

the conclusions that can be drawn from different types of evidence. The results with enteric bacteria show (Table III) that the presence of enzymes that can utilize a given acylhomoserine, or of mutants that show a growth response to it, is inconclusive, since acetylhomoserine serves in both cases although it is not formed by these organisms. The homoserine transacylases are more specific, but are difficult to assay in crude extracts in which the apparent rates may be very low (Section II,A,2,b). The best evidence would seem to be the accumulation of a specific acylhomoserine by mutants blocked in its utilization. Most relevant here is the current status of homocysteine biosynthesis in vascular plants, in which a serious effort has been made to unequivocally identify the intermediary acylhomoserine [149,150]. Spinach extracts were found to catalyze cystathionine synthesis equally well from acetyl- or succinylhomoserine [151], and to have a sulfhydrylase reaction specific for acetylhomoserine [132]; the rates of these reactions were less than 1/1000 of those in extracts of enteric bacteria or *Neurospora*. Homoserine transacylase reactions (Table III) could not be detected except in species that naturally accumulate acetyl- or oxalylhomoserine, in which they were specific for acetyl- or oxalyl-CoA, respectively [150]. However, recent preliminary results suggest that O-phosphorylhomoserine may be the true substrate for cystathione γ-synthase in plants [147]. Phosphorylhomoserine has previously been known only as an intermediate in the conversion of homoserine to threonine [31]. Thus, no one method suffices to establish whether acetyl- or succinylhomoserine is an intermediate in a new organism, and indeed, the possible involvement of still other acyl groups must be considered.

It seems quite possible that homoserine transacylase specificity may not correlate with evolution [21]; a single amino acid substitution in this enzyme might suffice to switch an organism from one acylhomoserine to another, since cystathionine γ-synthase is sometimes nonspecific. It appears that some bacteria utilizing acetylhomoserine have a transacetylase uncontrolled by feedback inhibition [143], like the *Neurospora* enzyme, whereas in others the enzyme is synergistically inhibited [142] like the transsuccinylase of enteric bacteria. Allosteric sites might have more evolutionary significance than substrate sites.

b. Cystathionine γ-Synthase and Acetylhomoserine Sulfhydrylase. In enteric bacteria both reactions are catalyzed by the same enzyme, composed of identical subunits; the sulfhydrylase reaction does not function effectively because (a) mutants lacking only β-cystathionase require methionine (Section II,A,1,c), and (b) growth is inhibited by a specific inhibitor of β-cystathionase (Section II,A,1,d). The same argument applies to the two reactions in *Neurospora* (Section II,A,2,d), in which they are catalyzed by different enzymes, and in addition: (a) the cystathionine pathway is under methionine control and the sulfhydrylase is not; and (b) mutation in any of the three genes coding the enzymes of the cystathionine pathway results in methionine auxotrophy, but no known methionine auxotrophs have a defect in the sulfhydrylase.

Homocysteine biosynthesis differs in several respects in these two well-studied organisms, so one must anticipate the probability of still other variations. The enteric bacterial type of cystathionine γ-synthase is easy to detect if present. The *in vivo* ineffectiveness of the alternate utilization of sulfide instead of cysteine in these organisms might have analogies to some ammonia transfer reactions where ammonia itself can be utilized but, for various reasons, glutamine is a more effective ammonia donor. Examples are anthranilate synthase [70] and carbamyl phosphate synthase [152].

In the *Neurospora* type of pathway the simplest evidence for ineffectiveness of the sulfhydrylase reaction is obtained by showing that a β-cystathionase is present, and particularly, is absent from a methionine auxotroph. This has recently been shown to be the case for *Aspergillus* [145a]. A series of ingenious and decisive experiments [22,145,145a] has also shown that homocysteine biosynthesis in this organism is mediated by Eqs. (15), (17), and (6), as in *Neurospora*. Acetylhomoserine sulfhydrylase [Eq. (19)] is a nonessential enzyme in this fungus, with a potential dual function in providing alternative pathways for synthesis of cysteine and methionine; however, in the latter case the function can be realized only when the level of the sulfhydrylase is derepressed [145a]. The only piece of evidence still missing appears to be the demonstration of cystathionine γ-synthase activity in extracts.

There are at least two reasons why the *Neurospora* type of cystathionine γ-synthase might be hard to detect. First, the activity is completely dependent on an activator, CH_3—$H_4PteGlu_n$ that is very unstable once the cells are broken. In fact the enzyme cannot be detected in extracts of some species, such as *Neurospora tetrasperma*, until additional activator is added [144]. Second, the enzyme is comprised of two proteins so weakly associated that activity is lost on any dilution of the crude extract. If the association were any weaker, activity might be lost by extraction itself.

Aside from *Aspergillus* and plants (Section II,A,3,a), the only other organism, of those listed in Table III, that has been examined for the presence of cystathionine γ-synthase is *S. cerevisiae*. Suitable extracts of yeast catalyze this reaction at $\frac{1}{10}$ the rate of *Neurospora* extracts, but the activity is not stimulated by added CH_3—$H_4PteGlu_n$ or decreased by dilution or addition of AMe [144]. Yeast extracts also contain an acetylhomoserine sulfhydrylase with specific activity 50 times that of cystathionine γ-synthase [133,144] that is 65% repressible (Section III) by growth on methionine [153]. The purified [144,154] sulfhydrylase resembles that from *Neurospora* since it cannot utilize cysteine and is not subject to significant allosteric inhibition. However, it is absent from a methionine auxotroph [133]. Cystathionine γ-synthase is also absent from extracts of the same mutant [144]. At the time of writing, then, yeast appears to have two alternate pathways to homocysteine, the relation between the enzymes catalyzing them is obscure, and there is no evidence inconsistent with the conclusion that in this organism the sulfhydrylase reaction participates in the major pathway of methionine biosynthesis [153]. Further study, particularly a search for methionine auxotrophs lacking β-cystathionase, may change this conclusion.

B. Homocysteine to Methionine

1. METHYLENE TETRAHYDROFOLATE REDUCTASE

The methyl group of methionine originates, in different organisms, from either formate or the hydroxymethyl group of serine. Methylene tetrahydrofolate (Fig. 2, $CH_2\!\!=\!\!H_4PteGlu_n$) derives either directly from serine, or by reduction of various formyl or forminino derivatives of folate (Fig. 3). These reactions are not unique to methionine biosynthesis, since $CH_2\!\!=\!\!H_4PteGlu_n$ is also an intermediate in the synthesis of thymine and hydroxymethyl cytosine (Fig. 3). The reduction of the 1-carbon moiety from the formaldehyde to the methanol level [Eq. (22)] is required only for methionine biosynthesis. Mutants lacking this reac-

$$CH_2\!\!=\!\!H_4PteGlu_n + FADH_2 \rightleftarrows 5\text{-}CH_3\!\!-\!\!H_4PteGlu_n + FAD \qquad (22)$$

tion, met-F in enteric bacteria [80] and met-1 in *Neurospora* [124], score as methionine auxotrophs. $5\text{-}CH_3\!\!-\!\!H_4PteGlu_n$ can only be metabolized by reversal of Eq. (22) or by homocysteine transmethylation; one possible exception has recently been reported in which a brain enzyme transmethylated dopamine directly with $5\text{-}CH_3\!\!-\!\!H_4PteGlu$, rather than with AMe [154].

The enzyme is a flavoprotein probably containing tightly bound FAD [34,155] that utilizes mono- as well as polyglutamate forms of folate [156]. It is usually assayed in extracts in the reverse direction, by measuring the HCHO spontaneously liberated from $CH_2\!\!=\!\!H_4PteGlu_n$; this requires addition of a relatively high potential artificial electron acceptor [157a] such as menadione ($E_0' = -0.5$ V). With liver enzyme no reverse reaction occurred if NADP replaced menadione, but the forward reaction utilized NADPH even better than substrate amounts of FADH$_2$ [157a]. A flavin reductase (i.e., an NADPH specific menadione reductase) copurified with the enzyme catalyzing reaction (22) [155], indicating that a native enzyme complex catalyzed the overall reaction:

$$NADPH + H^+ + CH_2{=}H_4PteGlu_n \rightarrow NADP^+ + CH_3{-}H_4PteGlu_n \qquad (23)$$

However, $CH_2{=}H_4PteGlu_n$ reductase purified from *E. coli* did not have flavin reductase activity and could not utilize NADPH in the forward reaction unless a separate flavin reductase fraction was added [73]. Reversal of Eq. (22) with the bacterial enzyme could be obtained with menadione or, to a very small extent, with FAD ($E_0' = -0.2$ V); from the latter result the equilibrium constant for Eq. (22) was estimated to be 3×10^3 in the forward direction. If, as seems likely, the bacteria also utilize pyridine nucleotides ($F_0' = -0.3$ V) as physiological electron carriers for Eq. (22), it is obvious that the reaction would be essentially irreversible *in vivo*, and that $CH_3{-}H_4PteGlu_n$ should accumulate if homocysteine transmethylation were blocked. Such accumulation has been shown in animals [158] on a diet deficient in methionine (i.e., deprived of homocysteine), or vitamin B$_{12}$ (thereby lacking transmethylase activity). The fact that B$_{12}$ deficiency might thus cause a secondary folic acid deficiency by tying up the latter as $CH_3{-}H_4PteGlu_n$ has stimulated some controversy about a "methyl trap" theory of pernicious anemia [158,159].

The regulation of $CH_2{=}H_4PteGlu_n$ reductase therefore deserves more attention that it has received. The *E. coli* enzyme is not subject to allosteric control, but enzyme synthesis is partially repressed (Section III) by either methionine or vitamin B$_{12}$ [73], probably by different mechanisms [160]. The activity of the rat liver enzyme, in both directions, is inhibited 50% by 10^{-4} M AMe; the enzyme is easily desensitized to inhibition [161]. The *Neurospora* enzyme is similarly inhibited by AMe [125]. The activity in undialyzed extracts of yeast was inhibited by high concentrations of methionine [157]; here the actual allosteric inhibitor might also be AMe.

2. Non-B$_{12}$ Homocysteine Transmethylase and Polyglutamate Forms of Folic Acid

The utilization of $CH_3{-}H_4PteGlu_n$ for homocysteine transmethylation occurs by two reactions in nature, which are similar except that one requires a coenzyme form of vitamin B$_{12}$ and the other requires polyglutamate forms of the methyl folate substrate [162]. The latter requirement is not very demanding since polyglutamate forms naturally predominate in all cells, as far as is known. Cells that can synthesize vitamin B_{12} utilize only the B_{12}-dependent transmethylase (Table IV) [51,78,154,156,163–171] as do mammals that obtain the vitamin from intestinal flora or diet. Plants, fungi, and some bacteria that do not contain vitamin B_{12} utilize only the non-B_{12} transmethylase (Table IV). The three enteric bacteria at the bottom of Table IV are exceptional in having both transmethylases [172]. These bacteria cannot synthesize adequate amounts of the vitamin and can survive in its absence by utilizing the non-B_{12} enzyme, but in the enriched intestinal habitat utilize the more efficient B_{12}-transmethylase; the turnover numbers of the purified *E. coli* enzymes are 14 and 800, respectively [159]. Utilization of the non-B_{12} enzyme is costly in that, when fully derepressed by restriction of both methionine and vitamin B_{12} (Section III), the enzyme constitutes 5% of the soluble protein [51]. Thermodynamically, methionine synthesis is virtually irreversible, reflecting the low free energy of thioethers [67] mentioned in Section I,A,1,a. The inefficiency of the non-B_{12} transmethylase, presumably, is due to the fact that a tertiary amine is quite unreactive with respect to transmethylating an attacking nucleophile; the monovalent cobalt of the reactive form of B_{12} is a powerful nucleophile whose intervention accelerates the reaction [51,52].

The non-B_{12} homocysteine transmethylase [Eq. (24)] is coded by the

$$5\text{-CH}_3\text{---H}_4\text{PteGlu}_{3-7} + \text{L-homocysteine} \rightarrow \text{L-methionine} + \text{H}_4\text{PteGlu}_{3-7} \quad (24)$$

met-E gene in enteric bacteria [41,79] and *met-8* in *Neurospora* [78,124], and has been obtained nearly pure from *E. coli* [173] and yeast [174]. A prerequisite to assaying the enzyme is the synthesis, from folic acid, of $\text{CH}_3\text{---H}_4\text{PteGlu}_3$, which at the time of writing is still a rather formidable undertaking [173,174]. The reaction is followed by measuring the formation of [^{14}C]methionine from [^{14}C]methyl-labeled substrate [173], or $\text{H}_4\text{PteGlu}_3$ from unlabeled substrate [174].

The *E. coli* enzyme was obtained in pure form after 20-fold purification from a methionine auxotroph in which it had been derepressed (fivefold) by growth on D-methionine [173]. The molecular weight is 84,000. The K_m for $\text{CH}_3\text{---H}_4\text{PteGlu}_3$ is 2.4×10^{-6} M and the enzyme is completely inactive with the corresponding di- or monoglutamate derivatives. The reaction requires phosphate and is stimulated by magnesium ions. No allosteric control has been reported for the *E. coli* or yeast enzymes.

The enzyme was obtained nearly pure from commercial bakers' yeast after only 25-fold purification [174], again indicating the large invest-

TABLE IV

DISTRIBUTION OF VITAMIN B_{12}-DEPENDENT AND -INDEPENDENT HOMOCYSTEINE TRANSMETHYLASES IN DIFFERENT ORGANISMS

Organism	Ability to synthesize adequate amounts of vitamin B_{12}	B_{12}-dependent homocysteine transmethylase	Non-B_{12} homocysteine transmethylase	Transmethylase activity with monoglutamate folate derivative	References
Bacillus subtilis	−	−	+	−	[156]
Candida utilis	−	−	+	−	[163]
Coprinus lagopus	−	−	+	−	[163]
Neurospora crassa	−	−	+	−	[78]
Saccharomyces cerevisiae	−	−	+	−	[78]
Vascular plants	−	−	+	15%[a]	[154,164,165]
Bacillus megaterium	+	+	−	15%[a]	[163]
Ochromonas malhemensis	+	+	−	+	[166]
Pseudomonas denitrificans	++	+	−	+	[167]
Rhodopseudomonas spheroides	+	+	−	+	[168]
Streptomyces olivaceus	+	+	−	+	[169]
Mammals	−	+	−	+	[51]
Aerobacter aerogenes	±	+	+	+	[170]
Escherichia coli	−	+	+	[b]	[51]
Salmonella typhimurium	−	+	+	[b]	[171]

[a] The monoglutamate derivative is utilized about 15% as well as the triglutamate.
[b] The monoglutamate derivative is utilized only by the B_{12}-dependent enzyme.

ment of protein necessary for this step of methionine biosynthesis. The molecular weight of the yeast enzyme is about 75,000 and K_m values are 2.2×10^{-5} M for L-homocysteine and 4×10^{-4} M for CH_3—$H_4PteGlu_3$ [174]. The activity requires phosphate ions, but is not stimulated by Mg^{2+}. $CH_3H_4PteGlu_1$ is not utilized, but the diglutamate is as effective a substrate as the triglutamate [78].

Polyglutamate forms of folate have so far been found to be essential in nature only for Eq. (24), with one remarkable exception. A penta- or hexaglutamate derivative of dihydrofolate occurs in association with dihydrofolate reductase in the tail plates of *E. coli* T-even bacteriophages, and the ability to attach to the host cell is lost after peptidase treatment [175]. The otherwise unique requirements for polyglutamate derivatives of Eq. (24) has suggested that the α-carboxyl of the second glutamate might participate in the catalysis of this reaction [173]. This interpretation has been weakened [51] by the finding that the enzyme from vascular plants is exceptional (Table IV) in being able to use the monoglutamate $\frac{1}{6}$ as well as the triglutamate. A converse exception is also provided by the B_{12}-dependent transmethylase of *Bacillus megaterium* (Table IV), which utilizes polyglutamates better than the monoglutamate.

Confirmation that the polyglutamylation of folate is uniquely required for methionine biosynthesis was provided by the discovery that the met-6 methionine auxotrophic mutants of *Neurospora* [27] lacked polyglutamate derivatives [124,125]. Very little is known about the reactions and enzymes involved in the polyglutamylation. *E. coli* extracts have been reported to catalyze a stepwise synthesis of the triglutamate when supplemented with $H_4PteGlu_1$ + ATP and glutamate [176]. Preliminary evidence indicates that at least two enzymes are involved in *Neurospora* [177]. One catalyzes only diglutamate formation and is absent from a not fully characterized mutant, "mac" (methionine-adenine-cysteine), and a second enzyme converting di- to polyglutamates is absent from met-6. The reactions require ATP and are stimulated by coenzyme A. Glutamylation utilized $H_4PteGlu_1$ but not folate, $H_2PteGlu_1$, CH_3—$H_4PteGlu_1$, or 10-formyl-$H_4PteGlu_1$ [177].

In organisms utilizing the non-B_{12} homocysteine transmethylase, the polyglutamylation reactions offer another point at which methionine biosynthesis could be regulated. In *Coprinus* (Table IV) nutrient methionine was shown to evoke a reduction in the proportion of folates in polyglutamate form [178]. Similar results were observed with yeast [157], but dietary methionine had the opposite effect in rats, causing an increase in the proportion of liver polyglutamates [179].

3. Vitamin B_{12}-Dependent Homocysteine Transmethylase

Reaction (25) is catalyzed by a cobalamin containing enzyme that is absent from met-H mutants of enteric bacteria. The latter are isolated in the form of double mutants, met-E and met-H, which score as methi-

$$5\text{-CH}_3\text{—H}_4\text{PteGlu}_1 + \text{L-homocysteine} \xrightarrow[\text{AMe}]{\text{reducing system}} \text{L-methionine} + \text{H}_4\text{PteGlu}_1 \quad (25)$$

onine auxotrophs unresponsive to vitamin B_{12}. Met-H mutants alone are nutritionally indistinguishable from wild-type, due to the presence of the non-B_{12} transmethylase [80,180]. All the materials needed to assay reaction (25) are commercially available [181]. The assay is based on isolation of labeled methionine formed from $^{14}\text{C—CH}_3\text{—H}_4\text{PteGlu}_1$ [181], or bioassay of methionine formed from unlabeled substrate [182].

The distribution of the enzyme is shown in Table IV. It has been purified, but not to homogeneity, from animal tissues [182,183] and from *E. coli* grown in the presence of vitamin B_{12} [52,181]. Since the properties are similar, and the bacterial activities are 30 times higher, I will confine this discussion to the latter.

Catalysis of reaction (25) by purified *E. coli* enzyme requires, in addition to the two substrates, catalytic amounts of AMe [184] and a "reducing system." The rate of catalysis varies with the reducing system, and is highest with FMNH_2 + dithiothreitol [51]. Mercaptans alone are also effective, but reduced pyridine nucleotide coenzymes are active only if a separate flavin reductase enzyme fraction is added [51]. Oxygen markedly inhibits the reaction and the inhibition may be useful in distinguishing it from Eq. (24) in crude extracts [159]. Reaction (25) involves the transfer, from nitrogen to sulfur, of a methyl group without its bonding electrons with the intermediary formation of a methyl–cobalt bond to the enzyme-bound cobalamin. The role of cobalamin in the reaction has been the subject of a great many ingenious experiments [51,52] and still presents challenging questions.

The enzyme has a molecular weight of 140,000 [51], or possibly 250,000 when purified by a different procedure [52]. The most purified fractions contain 0.3–0.6 moles of tightly bound cobalamin per 140,000 gm [51,52]. The latter can be reversibly dissociated by 6 M urea, yielding a labile colorless apoenzyme [51]. The precise structure of the bound cobalamin is still uncertain. It has the spectrum of divalent cob(II)alamin, i.e., with the cobalt one electron more reduced than in the vitamin cyanocob(III)alamin, but does not appear to have an unpaired electron [51]. When purified by other procedures the enzyme may be isolated with coenzyme in the form of methylcob(II)alamin [52].

At high concentrations other mercaptans can replace homocysteine as substrate for Eq. (25) [51]. CH_3—$H_4PteGlu_1$ is more reactive than the triglutamate [52]. The equilibrium constant for Eq. (25) is about 1.4×10^5 M in the forward direction [52]. The apparent K_m values for substrates and AMe are low, but vary with the reducing system used and with other conditions [51].

An early clue to the role of the B_{12} coenzyme was the finding that synthetic free methylcobalamin could serve in place of CH_3—$H_4PteGlu_1$ as a substrate to transmethylate homocysteine [185]. The intermediary formation of methylcobalamin in Eq. (25) has since been fully confirmed [52]. The requirements for AMe and a reducing system relate to the supposition that the coenzyme species that abstracts a methyl group from CH_3—$H_4PteGlu_1$ must be the strongly nucleophilic monovalent cob(I)alamin, with an unshared electron pair [51]. Since the E_0' for cob(II)/cob(I) is -1.07 V, the $FMNH_2$ reducing system would be able to produce only trace amounts of cob(I)alamin. AMe, a more reactive methyl donor than CH_3—$H_4PteGlu_1$, might then react with these trace amounts and displace the equilibrium. Transfer of this methyl to homocysteine would regenerate cob(I)alamin that could now accept a new methyl from CH_3—$H_4PteGlu_1$. With extracted enzyme it is impossible to exclude molecular oxygen completely. Consequently, the coenzyme is eventually reoxidized to cob(II)alamin, and the "catalytic" AMe is gradually consumed. Under maximally anaerobic conditions a turnover of 500 CH_3—$H_4PteGlu_1$ has been observed per one AMe [51].

The coenzyme species that exist *in vivo* and the reducing system that maintains monovalent cobalt are not known. Perhaps AMe activation is required only in extracts, since it might constitute a "horror autotoxicus," if depletion of end product AMe could shut down methionine biosynthesis [52].

III. REGULATION OF ENZYME SYNTHESIS AND OVERALL METHIONINE CONTROL OF ITS BIOSYNTHESIS AND RELATED BRANCH PATHWAYS

Methionine biosynthetic enzymes are far removed from the forefront of knowledge in the area of mechanisms regulating protein synthesis. Elucidation of the methionine pathway occurred after interest had already shifted from metabolic pathways to control mechanisms (Section I,A,2), and the resultant temptation to study regulation before ascertaining what it was that was being regulated may have delayed progress.

The information available so far should be evaluated in relation to the current status of the regulation of amino acid biosynthetic enzymes in general. In enteric bacteria regulatory mechanisms are currently best understood in relation to the biosynthetic enzymes for histidine and tryptophan, in the case of clustered gene "operons," and arginine in the case, like that of methionine, of scattered genes. In the so far inescapable terminology of the *lac* operon, control of synthesis of the histidine enzymes has been shown to involve an operator segment of DNA and a repressor complex that may include one of the biosynthetic enzymes, and definitely includes charged histidyl-tRNA [70]. A consequence of the latter is that a variety of mutants blocked in the synthesis or modification of tRNA[His] have been scored as analogue resistant "regulatory" mutants that overproduce the histidine enzymes [186]. Control of the synthesis of tryptophan enzymes does not involve tryptophanyl-tRNA [70,187], but a repressor protein, present to the extent of about ten molecules per cell, has been isolated [188]. The relevance of the *lac* operon model is much less sure in the case of the arginine enzymes. There is some evidence for a repressor protein and an operator gene for one of the enzymes [189]; the possibility that repression is translational still appears speculative [190].

Finally, it should be emphasized that the mechanism of regulation of protein synthesis in eukaryotes, such as *Neurospora* and yeast, is completely obscure; no operator–constitutive mutant has ever been detected [191], nor is there any other evidence relating control to the operator–repressor model [127].

A. Enteric Bacteria

As described in Section II,A,1,b, allosteric control of methionine biosynthetic enzymes is exerted only on homoserine transsuccinylase, which is subject to synergistic inhibition by methionine + AMe.

1. REGULATION OF SYNTHESIS OF ENZYMES OF METHIONINE SYNTHESIS AND UTILIZATION

Methionine represses the synthesis of the enzymes catalyzing reactions (4), (5), (6), (22), and (24) (Table I), and of AMe synthase [Eq. (26)].

$$\text{L-Methionine} + \text{ATP} \rightarrow S\text{-adenosyl-L-methionine} + PP_i + P_i \qquad (26)$$

This group of enzymes is constitutive, or not repressible by methionine, in regulatory mutants designated met-J [72,74,80,192,193] that define the principal repression system. The latter probably also controls the synthesis of the high affinity methionine permease, coded by the *met-P* gene [70a]. Methionine repression of these enzymes is not coordinate (Table I) and may involve more than one mechanism [72]. The enzymes catalyzing reactions (25) and (27) appear not to be constitutive in met-J

$$\text{L-Methionine} + \text{tRNA}^{Met} + \text{ATP} \rightarrow \text{methionyl-tRNA}^{Met} + \text{AMP} + \text{PP}_i \qquad (27)$$

mutants [74]. Met-J mutants [194] are scored as specifically resistant to ethionine and map near, but are not contiguous with, met-B [80]. They are recessive in partial diploids [195] and revertible by nonsense supressors [196], indicating that the gene product is a protein; it may be a repressor protein, although this is not proved [193].

The corepressor for the met-J control system appears not to be methionine itself and is probably also not methionyl-tRNA^{Met}. Mutants defective in the ability to form the latter score as methionine auxotrophs when the defect results from low affinity of methionine [197] for methionyl-tRNA^{Met} synthase [Eq. (27)]. Met-G mutants (Table I) of this type are not derepressed for methionine biosynthetic enzymes [72,80,197]. An ethionine resistant *E. coli* mutant has also been reported to have a low affinity for tRNA^{Met} in Eq. (27); the authors propose that methionyl-tRNA^{Met} may be a corepressor for its cognate synthase [198]. The latter appears not to be under met-J control.

The corepressor for the met-J control system is probably AMe or something derived from it. Mutants blocked in AMe synthase [Eq. (26)] cannot be scored as auxotrophs, since AMe is not permeable in enteric bacteria [192] and no attempts have been reported to isolate conditional lethal mutants. Met-K (Table I) mutants have, instead, thus far been scored as analogue resistant, i.e., on the basis of the fact that they are constitutive for methionine biosynthesis and overproduce it. Such mutants must be assumed to have only a partial block in AMe synthesis since AMe is needed for many vital functions.

Met-K mutants are not linked to met-J or to genes coding methionine biosynthetic enzymes, but may map close to several genes concerned with polyamine synthesis [199,200]. They are resistant to ethionine, α-methylmethionine, and norleucine [80], and are recessive [195] and amber suppressible [196]. Three classes of met-K mutant have been described among the small number available [193] and each of these poses a puzzling question. Met-K^{nx} mutants have an AMe synthase with a low affinity for methionine; they do not overproduce methionine and it is unclear why they are analogue resistant. Met-K^x mutants overproduce methionine and are constitutive for methionine biosynthetic enzymes, and extracts have no detectable AMe synthase. Since the mutants are viable, it must be assumed that the enzyme has some activity *in vivo* or that there is some other, unknown AMe yielding reaction. That this must be so is also shown by the fact that these mutant cells contain normal amounts of AMe and spermidine [193]. Met-K (721) extracts appear to have a wild-type AMe synthase activity, but the mutant is constitutive for methionine biosynthetic enzymes, suggesting that it might be blocked in the conversion of AMe to some product that is the true corepressor [72,193]. The fact that Met-K (721) complemented all the other met-K mutants also suggests that it might code a different protein [193]. The AMe synthase of enteric bacteria has been partially purified [201] but little studied, and further study of the enzyme and the met-K mutants should soon resolve these questions. It should be noted that no operator-constitutive mutants have been reported even for this principle "JK" repressor system [80], so that the use of the operator-repressor terminology is arbitrary.

A second, independent repression system has been reported to supplement methionine control of the synthesis of the enzymes catalyzing reactions (22) and (24) [160,202]. These two enzymes are repressed by vitamin B_{12} and there is evidence that the B_{12}-dependent homocysteine transmethylase holoenzyme may be part of the repressor complex, i.e., the repression occurs in met-J mutants, but not in met-H [97,180,203]. The utility of B_{12} repression of the non-B_{12} transmethylase is clear, but the nearly complete repression of $CH_2{=}H_4PteGlu_n$ reductase is puzzling, as is the apparently complete repression of homoserine transsuccinylase by vitamin B_{12} in met-E mutants [72].

There are several indications that still other secondary repression systems, responding to precursors or products of methionine [180], remain to be discovered. First, homoserine transsuccinylase is super-repressed in met-E mutants if grown on methionine, but not if grown on vitamin B_{12} [72]; met-E mutants also have high levels [204] of serine transhydroxymethylase (Fig. 1), and (in the case of one of two complementation groups) low levels of $CH_2{=}H_4PteGlu_n$ reductase [170]. These results could reflect direct participation of the met-E protein in control, but they could also be mediated by a low molecular weight effector produced by the transmethylase, i.e., if the enzyme could methylate something other than homocysteine, which then functioned as coinducer or corepressor [72]. Second, met-A mutants have high levels of cystathionine γ-synthase and β-cystathionase [72]; there are other indications of a met-J independent system modulating the latter [196]. Third, as mentioned above, methionyl-tRNA^{Met} may control its cognate synthase [198].

In *Neurospora* the syntheses of methyl groups and homocysteine are coordinated by allosteric activation of cystathionine γ-synthase by CH_3—$H_4PteGlu_n$ (Section I,A,2,c). Folate derivatives have not been shown to control the activity or synthesis of the homocysteine enzymes in enteric bacteria [72], and it is not yet clear how the converging pathways are coordinated.

2. Methionine Control of Branch Pathways

In enteric bacteria no methionine control has been reported for enzymes catalyzing the synthesis of cysteine or serine, or the reduction of sulfate to sulfide (Fig. 3). In the methyl group biosynthetic pathway, serine transhydroxymethylase (Fig. 3) has been reported to be derepressed by limitation of nutrient methionine or vitamin B_{12} in a met-E auxotroph of E. coli [204]; since the product of this reaction is required for thymine as well as methionine, this finding requires further study.

Regulation of the three enzymes (Fig. 3) mediating the synthesis of homoserine from aspartate [205] by the amino acid end products threonine, isoleucine, lysine, and methionine involves varied and complex mechanisms in different bacteria [127,206]. In most species, including even some strains of E. coli [206], the pathway is controlled only by lysine and threonine. Methionine control has been studied almost exclusively in E. coli K_{12}. In this organism the first step is catalyzed by three different isofunctional β-aspartokinase enzymes, each exhibiting different end product inhibition and repression control. Two of these occur as complexes with isofunctional enzymes (homoserine dehydrogenase) catalyzing the third step. The synthesis of the complex designated II is specifically repressed by methionine. Complex II is constitutive in met-J mutants [206]. The activity of complex II is so low as to be virtually undetectable in extracts of E. coli K_{12} wild-type, and even in the presence of a met-J mutation becomes measurable only when there is a genetic block in the threonine + isoleucine controlled complex I [206]. This suggests that there is little channeling, and that intermediates formed by one complex may be available to another [127]. However, some degree of channeling is indicated by other results with a Salmonella mutant in which homoserine produced by complex II is unavailable for threonine at 28° but does overflow the channel at 37° [207]. The fact that cystathionine γ-synthase was undectable in extracts of some Neurospora species until the requirement for a labile activator was recognized (Section II,A,2,c), suggests that other explanations are not excluded for the remarkably low activity of complex II in extracts of wild-type E. coli K_{12}.

In E. coli B, homoserine has been reported to partially inhibit the glutamate dehydrogenase reaction, a step in aspartate biosynthesis. The authors suggest this may compensate for the lack of methionine control over aspartate $\rightarrow$ homoserine in this organism [208], but the concentration of homoserine required is probably too high to effect significant control.

B. Eukaryotic Microorganisms

Eukaryotic regulation of methionine biosynthesis has been studied principally in *Neurospora* and yeast, although at the time of writing rapid progress is also being made with studies of *Aspergillus nidulans* [22,145,145a]. A recent report [208a] describes mutants of the latter partially defective in AMe synthase, and implicates AMe in repression of sulfate permease, sulfite reductase, and acetylhomoserine sulfhydrylase.

1. *Neurospora*

Allosteric control of methionine biosynthetic enzyme activity is exerted at two points in *Neurospora*. AMe inhibits $CH_2{=}H_4PteGlu_n$ reductase (Section II,B,1) and cystathionine γ-synthase (Section II,A,2,c). The latter activity is also completely dependent on allosteric activation by $CH_3{-}H_4PteGlu_n$ (Section II,A,2,c).

No significant methionine control has been reported over the synthesis of enzymes of its biosynthesis or utilization. Because *Neurospora* grows as a syncytial mycelium, the intracellular concentration of metabolic end products cannot be manipulated by culturing auxotrophs in a chemostat. Thus, if an enzyme is partially repressed under the condition of wild-type growing in minimal medium, repression control can be detected by adding end product to the medium, but if it is fully repressed under these conditions, repression control can only be detected by isolating analogue resistant mutants that are constitutive for enzyme synthesis.

Using the latter approach, one group of methionine repressible enzymes has been identified [191] whose common denominator seems to be the potential to mobilize sulfur in times of deprivation of this element. However, since one of the enzymes is sulfate permease, the results are relevant to methionine control over the synthesis of branch pathway enzymes (Fig. 3). It is noteworthy that methionine control over the pathway from sulfate to cysteine has been found only in organisms [*Neurospora, Aspergillus,* and yeast (Section III,B,2)] that have the enzymes for transsulfuration from homocysteine to cysteine (Fig. 1). Elevated sulfate permease levels were discovered in an ethionine resistant mutant, eth^r-1, that was later shown to have a defective AMe synthase (Section II,A,2,c), suggesting that AMe was the corepressor [76]. However, there is independent evidence that the corepressor is related to cysteine, and Metzenberg has pointed out [191] that in *Neurospora* AMe is an obligatory intermediate in the conversion of methionine to cysteine, i.e., by the following pathway: methionine → AMe → adenosylhomocysteine → homocysteine → cystathionine → cysteine. Two other genes affect the regulatory system identified by the eth^r-1 mutant. Cys-3 mutants are superrepressed and score as cysteine auxotrophs due to the absence of sulfate permease (as well as the other enzymes controlled by eth^r-1 [191]. In Scon^c mutants all the enzymes are constitutive; this gene is thought to code a diffusible control element that is unable to escape the nuclear membrane [191].

No evidence of methionine control over other branch pathways has been reported [209].

2. Yeast

In contrast to *Neurospora,* methionine biosynthesis in *S. cerevisiae* appears to be controlled by repression of enzyme synthesis. There is little evidence of significant allosteric control of enzyme activity [210]. Methionine represses the synthesis of a number of enzymes unique to its biosynthesis and also of many enzymes catalyzing steps in converging branch pathways. Studies of enzyme levels in ethionine resistant mutants that map in at least three different genes suggest [211] that a common repression system affects the following four enzymes (the ratio, maximum derepressed/minimum repressed specific enzyme activity observed in extracts is given in parenthesis): homoserine transacetylase (ratio 20) catalyzing reaction (15); acetylhomoserine sulfhydrylase (ratio 30) catalyzing reaction (19); ATP sulfurylase (ratio 120) catalyzing (Fig. 3) reaction (28); and sulfite reductase (ratio 25) catalyzing (Fig. 3) reaction (29). Methionine control over cysteine biosynthesis [Eqs. (28) and

$$\text{ATP} + \text{SO}_4^{-2} \rightarrow \text{adenosine 5'-phosphosulfate} + \text{PP}_i \qquad (28)$$

$$\text{SO}_3^{2-} + 3\,\text{NADPH} + 4\,\text{H}^+ \rightarrow \text{H}_2\text{S} + 3\,\text{NADP}^+ + 3\,\text{H}_2 \qquad (29)$$

(29)] is reasonable in yeast because of the presence of enzymes that can form cysteine from methionine (Fig. 1). As noted in Section II,A,3,a, homoserine transacetylase is difficult to assay in yeast, and the actual reported specific activities are very low [211].

These four enzymes are constitutive, or not repressible by methionine, in the ethionine resistant mutants eth^r-2, eth^r-3, and eth^r-10 [212]. The *eth^r-2* gene appears to code a protein [213]. The corepressor may be

AMe, rather than methionine. In *Neurospora* and enteric bacteria a regulatory role for AMe, or something derived from it, was established with the aid of mutants having a partial block in AMe synthase [Eq. (26)]. Similar mutants have not yet been identified in yeast. However, unlike the other organisms, yeast is permeable to AMe [214] and exogenous AMe expands the internal pool of AMe selectively, whereas methionine expands both the methionine and AMe pools [215]. Repression of enzyme synthesis correlates somewhat with the AMe pool size [215]. Some of the ethionine resistant mutants are still repressible by exogenous AMe, although not by methionine [212]. Although these mutants seem to have an unaltered AMe synthase [212], the complexities associated with the met-K mutants of enteric bacteria (Section III,A,1) suggest that they might nevertheless be involved in coding this enzyme.

Other investigations [211] have shown that the same four enzymes are less repressible by methionine in a temperature sensitive mutant with an impaired methionyl-tRNA^{Met} synthase [Eq. (27)], but the evidence that methionyl-tRNA^{Met} participates in the repressor system is inconclusive [211,212].

Some degree of methionine repressibility has been reported for several other yeast enzymes that are apparently not subject to the repression system specified by the above ethionine resistant mutants: aspartate kinase and homoserine dehydrogenase [211], sulfate permease [210], AMe synthase [212], and homocysteine transmethylase [157]. An equivocal allosteric inhibition of activity has been described for $CH_2{=}H_4PteGlu_n$ reductase [157] and serine transhydroxymethylase [216].

REFERENCES

1. J. H. Mueller, *Proc. Soc. Exp. Biol. Med.* **19**, 161 (1922).
2. G. Barger and F. P. Coyne, *Biochem. J.* **22**, 1417 (1928).
3. V. du Vigneaud, J. P. Chandler, A. W. Moyer, and D. M. Keppel, *J. Biol. Chem.* **131**, 57 (1939).
4. V. du Vigneaud, "A Trail of Research in Sulfur Chemistry and Metabolism." Cornell Univ. Press, Ithaca, New York, 1952.
5. G. L. Cantoni, *J. Biol. Chem.* **204**, 403 (1953).
6. H. Tabor, S. M. Rosenthal, and C. W. Tabor, *J. Biol. Chem.* **233**, 907 (1958).
7. K. Marcker and F. Sanger, *J. Mol. Biol.* **8**, 833 (1964).
8. J. M. Adams and M. R. Capecchi, *Proc. Nat. Acad. Sci. U.S.* **55**, 147 (1966).
9. B. D. Davis, *Advan. Enzymol.* **16**, 247 (1955).
10. D. M. Greenberg, *in* "Metabolic Pathways" (D. M. Greenberg, ed.), 3rd ed., Vol. 3, p. 259. Academic Press, New York, 1969.
11. A. Meister, "Biochemistry of the Amino Acids," Vol. 2, p. 757. Academic Press, New York, 1965.
12. M. A. Bennett, G. Medes, and G. Toennies, *Growth* **8**, 59 (1944).

13. M. A. Bennett; J. A. Stekol, and K. Weiss; T. H. Jukes and E. L. R. Stokstad; *Papers presented at 116th Meet., Amer. Chem. Soc.* (1949).
14. E. Brand, R. J. Block, B. Cassell, and G. F. Cahill, *Proc. Soc. Exp. Biol. Med.* **35**, 501 (1936).
15. W. Küster and W. Irion, *Hoppe-Seyler's Z. Physiol. Chem.* **184**, 225 (1929).
16. M. J. Horn and D. B. Jones, *J. Amer. Chem. Soc.* **62**, 234 (1940).
17. M. Suda, H. Nakagawa, and H. Kimura, *in* "Methods in Enzymology" (H. Tabor and C. W. Tabor, eds.), Vol. 17B, p. 454. Academic Press, New York, 1971.
18. D. M. Greenberg, *in* "Methods in Enzymology" (S. P. Colowick and N. O. Kaplan, eds.), Vol. 5, p. 936. Academic Press, New York, 1962.
19. G. A. Fisher, *Biochim. Biophys. Acta* **25**, 55 (1957).
20. M. Flavin and C. Slaughter, *J. Biol. Chem.* **239**, 2212 (1964).
21. C. Delavier-Klutchko and M. Flavin, *J. Biol. Chem.* **240**, 2537 (1965).
22. N. J. Pieniazek, P. P. Stepien, and A. Paszweski, *Biochim. Biophys. Acta* **297**, 37 (1973).
23. S. Simmonds, *J. Biol. Chem.* **174**, 717 (1948).
24. N. H. Horowitz, *J. Biol. Chem.* **171**, 255 (1947).
25. H. J. Teas. N. H. Horowitz, and M. Fling, *J. Biol. Chem.* **172**, 651 (1948).
26. M. Fling and N. H. Horowitz, *J. Biol. Chem.* **190**, 277 (1951).
27. N. E. Murray, *Heredity* **15**, 199 (1960).
28. M. Flavin, C. Delavier-Klutchko, and C. Slaughter, *Science* **143**, 50 (1964).
29. S. Nagai and M. Flavin, *J. Biol. Chem.* **242**, 3884 (1967).
30. B. D. Davis, H. L. Kornberg, A. Nagler, P. Miller, and E. Mingioli, *Fed. Proc., Fed. Amer. Soc. Exp. Biol.* **18**, 211 (1959).
31. M. M. Kaplan and M. Flavin, *J. Biol. Chem.* **240**, 3928 (1965).
32. C. Gilvarg, *Fed. Proc., Fed. Amer. Soc. Exp. Biol.* **16**, 186 (1957).
33. J. C. Rabinowitz, *in* "The Enzymes" (P. D. Boyer, H. Lardy, and K. Myrbäck, eds.), 2nd ed., Vol. 2, Part A, p. 185. Academic Press, New York, 1960.
34. R. L. Blakley, "The Biochemistry of Folic Acid and Related Pteridines." Wiley, New York, 1969.
35. P. Berg, *J. Biol. Chem.* **205**, 145 (1953).
36. A. Nakao and D. M. Greenberg, *J. Amer. Chem. Soc.* **77**, 6715 (1955).
37. K. M. Jones, J. R. Guest, and D. D. Woods, *Biochem. J.* **79**, 566 (1961).
38. J. H. Boothe, J. H. Mowat, B. L. Hutchings, R. B. Angier, C. W. Waller, E. L. R. Stokstad, J. Semb, A. L. Gazzola, and Y. SubbaRow, *J. Amer. Chem. Soc.* **70**, 1099 (1948).
39. C. L. Krumdieck and C. M. Baugh, *Biochemistry* **8**, 1568 (1969).
40. H. A. Godwin, I. H. Rosenberg, C. R. Ferenz, P. M. Jacobs, and J. Meienhofer, *J. Biol. Chem.* **247**, 2266 (1972).
41. B. D. Davis and E. Mingioli, *J. Bacteriol.* **60**, 17 (1950).
42. F. Gibson and D. D. Woods, *Biochem. J.* **51**, V (1952).
43. C. W. Helleiner and D. D. Woods, *Biochem. J.* **62**, 26P (1956).
44. M. Flavin, *Fed. Proc., Fed. Amer. Soc. Exp. Biol.* **14**, 211 (1955).
45. J. T. Wachsman, *J. Biol. Chem.* **223**, 19 (1956).
46. H. A. Barker, H. Weissbach, and R. D. Smythe, *Proc. Nat. Acad. Sci. U.S.* **44**, 1093 (1958).
47. H. A. Barker, R. D. Smythe, H. Weissbach, A. Munch-Peterson, J. I. Toohey, J. N. Ladd, B. Volcani, and R. M. Wilson, *J. Biol. Chem.* **235**, 181 (1960).
48. P. G. Lenhert and D. C. Hodgkin, *Nature (London)* **192**, 937 (1961).
49. R. L. Kisliuk and D. D. Woods, *Fed. Proc., Fed. Amer. Soc. Exp. Biol.* **18**, 261 (1959).

50. F. T. Hatch, S. Takeyama, and J. M. Buchanan, *Fed. Proc., Fed. Amer. Soc. Exp. Biol.* **18,** 243 (1959).
51. R. T. Taylor and H. Weissbach, *in* "The Enzymes" (P. D. Boyer, ed.), 3rd ed., Vol. 9, p. 121. Academic Press, New York, 1973.
52. H. Rüdiger and L. Jaenicke, *Mol. Cell. Biochem.* **1,** 157 (1973).
53. J. A. Duerre and F. Schlenk, *Arch. Biochem. Biophys.* **96,** 575 (1962).
54. J. L. Botsford and L. W. Parks, *J. Bacteriol.* **94,** 966 (1967).
55. J. A. Duerre, *Arch. Biochem. Biophys.* **124,** 422 (1968).
56. S. K. Shapiro and D. J. Ehninger, *Biochim. Biophys. Acta* **177,** 67 (1969).
57. D. S. Kerr, *J. Biol. Chem.* **246,** 95 (1971).
58. C. M. Chow, S. N. Nigam, and W. B. McConnell, *Biochim. Biophys. Acta* **273,** 91 (1972).
59. J. B. Ragland and J. A. Liverman, *Arch. Biochem. Biophys.* **65,** 574 (1956).
60. S. Tokuno, B. Strauss, and Y. Tsuda, *J. Gen. Microbiol.* **28,** 481 (1962).
61. J. L. Wiebers and H. R. Garner, *J. Bacteriol.* **88,** 1798 (1964).
62. Y. Murooka and T. Harada, *Biochim. Biophys. Acta* **215,** 333 (1970).
63. Y. Murooka, K. Seto, and T. Harada, *Biochem. Biophys. Res. Commun.* **41,** 407 (1970).
64. J. Lucas-Lenard and F. Lipmann, *Annu. Rev. Biochem.* **40,** 409 (1971).
65. D. H. Gauss, F. von der Haar, A. Maeliche, and F. Cramer, *Annu. Rev. Biochem.* **40,** 1045 (1971).
66. M. Simsek, J. Ziegenmeyer, J. Heckman, and U. L. Rajbhandary, *Proc. Nat. Acad. Sci. U.S.* **70,** 1041 (1973).
67. G. L. Cantoni, *Comp. Biochem.* **1,** 220 (1960).
68. J. B. Lombardini and P. Talalay, *Advan. Enzyme Regul.* **9,** 349 (1971).
69. H. Tabor and C. W. Tabor, *Advan Enzymol.* **36,** 203 (1972).
70. P. Truffa-Bachi and G. N. Cohen, *Annu. Rev. Biochem.* **42,** 113 (1973).
70a. P. D. Ayling and E. S. Bridgeland, *J. Gen Microbiol.* **73,** 127 (1972).
71. T. C. Stadtman, *Science* **183,** 915 (1974).
72. M. A. Savin, M. Flavin, and C. Slaughter, *J. Bacteriol.* **111,** 547 (1972).
73. H. M. Katzen and J. M. Buchanan, *J. Biol. Chem.* **240,** 825 (1965).
74. H.-F. Kung, C. Spears, R. C. Greene, and H. Weissbach, *Arch. Biochem. Biophys.* **150,** 23 (1972).
75. F. R. Archibold and L. S. Williams, *J. Bacteriol.* **114,** 1007 (1973).
76. D. S. Kerr and M. Flavin, *J. Biol. Chem.* **245,** 1842 (1970).
77. W. Sakami, S. J. Ritari, C. W. Black, and J. Rzepka, *Fed. Proc., Fed. Amer. Soc. Exp. Biol.* **32,** 471 (1973).
78. E. Burton, J. Selhub, and W. Sakami, *Biochem. J.* **111,** 793 (1969).
79. D. A. Smith, *J. Gen. Microbiol.* **24,** 335 (1961).
80. D. A. Smith, *Advan. Genet.* **16,** 141 (1971).
81. R. J. Rowbury, *Biochem. J.* **81,** 42P (1961).
82. R. J. Rowbury, *Biochem. J.* **82,** 24P (1962).
83. R. J. Rowbury, *J. Gen. Microbiol.* **28,** V (1962).
84. R. J. Rowbury and D. D. Woods, *J. Gen. Microbiol.* **36,** 341 (1964).
85. R. J. Rowbury, *J. Gen. Microbiol.* **37,** 171 (1964).
86. M. Flavin and C. Slaughter, *Biochemistry* **4,** 1370 (1965).
87. M. Flavin, *in* "Methods in Enzymology" (H. Tabor and C. W. Tabor, eds.), Vol. 17B, p. 418. Academic Press, New York, 1971.
88. M. Flavin and C. Slaughter, *Biochim. Biophys. Acta* **132,** 400 (1967).
89. M. M. Kaplan and M. Flavin, *J. Biol. Chem.* **241,** 4463 (1966).
90. L. W. Lee, J. M. Ravel, and W. Shive, *J. Biol. Chem.* **241,** 5479 (1966).

91. E. Z. Ron and M. Shain, *J. Bacteriol.* **107,** 397 (1971).
92. D. E. Lawrence, *J. Bacteriol.* **109,** 8 (1972).
93. S. Schlesinger, *J. Bacteriol.* **94,** 327 (1967).
94. R. J. Rowbury and D. D. Woods, *J. Gen. Microbiol.* **42,** 155 (1966).
95. E. Z. Ron and B. D. Davis, *J. Bacteriol.* **107,** 391 (1971).
96. A. W. Brush and H. Paulus, *Fed. Proc., Fed. Amer. Soc. Exp. Biol.* **32,** 463 (1973).
97. C. -H. Su and R. C. Greene, *Proc. Nat. Acad. Sci. U.S.* **68,** 367 (1971).
98. M. M. Kaplan and S. Guggenheim, *in* "Methods in Enzymology" (H. Tabor and C. W. Tabor, eds.), Vol. 17B, p. 425. Academic Press, New York, 1971.
99. M. M. Kaplan and M. Flavin, *Biochim. Biophys. Acta* **104,** 390 (1965).
100. M. M. Kaplan and M. Flavin, *J. Biol. Chem.* **241,** 5781 (1966).
101. S. Guggenheim and M. Flavin, *J. Biol. Chem.* **246,** 3562 (1971).
102. B. I. Posner, *Biochim. Biophys. Acta* **276,** 277 (1972).
103. S. Guggenheim and M. Flavin, *J. Biol. Chem.* **244,** 3722 (1969).
104. S. Guggenheim and M. Flavin, *J. Biol. Chem.* **244,** 6217 (1969).
105. B. I. Posner and M. Flavin, *J. Biol. Chem.* **247,** 6402 (1972).
106. S. Guggenheim and M. Flavin, *Biochim. Biophys. Acta* **151,** 664 (1968).
107. B. I. Posner and M. Flavin, *J. Biol. Chem.* **247,** 6412 (1972).
108. M. Flavin and C. Slaughter, *J. Biol. Chem.* **244,** 1434 (1969).
109. S. Wijesundera and D. D. Woods, *J. Gen. Microbiol.* **29,** 353 (1962).
110. S. Guggenheim, *in* "Methods in Enzymology" (H. Tabor and C. W. Tabor, eds.), Vol. 17B, p. 439. Academic Press, New York, 1971.
111. F. Binkley, *in* "Methods in Enzymology" (S. P. Colowick and N. O. Kaplan, eds.). Vol. 2, p. 314. Academic Press, New York, 1955.
112. C. Delavier-Klutchko and M. Flavin, *Biochim. Biophys. Acta* **99,** 375 (1965).
113. L. D. Owens, S. Guggenheim, and J. L. Hilton, *Biochim. Biophys. Acta* **158,** 219 (1968).
114. J. Giovanelli, L. D. Owens, and S. H. Mudd, *Biochim. Biophys. Acta* **227,** 671 (1971).
114a. L. D. Owens, J. F. Thompson, J. F. Pitcher, and T. Williams, *Chem. Commun.* p. 714 (1972).
115. J. Giovanelli, L. D. Owens, and S. H. Mudd, *Plant Physiol.* **51,** 492 (1973).
116. M. Flavin and C. Slaughter, *Biochemistry* **3,** 885 (1964).
117. S. Nagai and M. Flavin, *J. Biol. Chem.* **241,** 3861 (1966).
118. H. Grobbelaar and F. C. Steward, *Nature (London)* **182,** 1358 (1958).
119. S. Nagai and M. Flavin, *in* "Methods in Enzymology" (H. Tabor and C. W. Tabor, eds.), Vol. 17B, p. 423. Academic Press, New York, 1971.
120. S. Nagai and D. Kerr, *in* "Methods in Enzymology" (H. Tabor and C. W. Tabor, eds.), Vol. 17B, p. 442. Academic Press, New York, 1971.
121. D. Kerr and M. Flavin, *Biochem. Biophys. Res. Commun.* **31,** 124 (1968).
122. D. Kerr and M. Flavin, *Biochim. Biophys. Acta* **177,** 177 (1969).
123. R. L. Metzenberg, *Arch. Biochem. Biophys.* **125,** 532 (1968).
124. J. Selhub, E. Burton, and W. Sakami, *Fed. Proc., Fed. Amer. Soc. Exp. Biol* **28,** 352 (1969).
125. J. Selhub, M. A. Savin, W. Sakami, and M. Flavin, *Proc. Nat. Acad. Sci. U.S.* **68,** 312 (1971).
126. T. Chase, Jr. and J. C. Rabinowitz, *J. Bacteriol.* **96,** 1065 (1968).
127. H. E. Umbarger, *Annu. Rev. Biochem.* **38,** 323 (1969).
128. H. T. Spencer, J. Collins, and K. J. Monty, *Fed. Proc., Fed. Amer. Soc. Exp. Biol.* **26,** 677 (1967).

129. M. C. Jones-Mortimer, J. F. Wheldrake, and C. A. Pasternak, *Biochem. J.* **107**, 51 (1968).
130. D. Kerr and S. Nagai, *Fed. Proc., Fed. Amer. Soc. Exp. Biol.* **26**, 387 (1967).
131. J. L. Wiebers and H. R. Garner, *J. Biol. Chem.* **242**, 5644 (1967).
132. J. Giovanelli and S. H. Mudd, *Biochem. Biophys. Res. Commun* **27**, 150 (1967).
133. H. Cherest, F. Eichler, and H. de Robichon-Szulmajster, *J. Bacteriol.* **97**, 328 (1969).
134. J. L. Wiebers and H. R. Garner, *J. Biol. Chem.* **242** 12 (1967).
135. M. Flavin and C. Slaughter, *Biochim. Biophys. Acta* **132**, 406 (1967).
136. D. P. Moore, J. F. Thompson, and I. K. Smith, *Biochim. Biophys. Acta* **184**, 124 (1969).
137. M. Flavin, *in* "Methods in Enzymology" (H. Tabor and C. W. Tabor, eds.), Vol. 17B, p. 450. Academic Press, New York, 1971.
138. E. G. Burton and R. L. Metzenberg, *J. Bacteriol* **109**, 140 (1972).
139. T. Sakata, S. Hiroishi, and H. Kadota, *Agr. Biol. Chem.* **36**, 333 (1972).
140. H. Kase, N. Nakayama, and S. Kinoshita, *Agr. Biol. Chem.* **34**, 274 (1970).
141. K. Nakayama, H. Kase, and S. Kinoshita, *Agr. Biol. Chem.* **33**, 1664 (1969).
142. A. Brush and H. Paulus, *Biochem. Biophys. Res. Commun.* **45**, 735 (1971).
143. R. Miyajima and I. Shiio, *J. Biochem. (Tokyo)* **73**, 1061 (1973).
144. M. A. Savin and M. Flavin, *J. Bacteriol.* **112**, 299 (1972).
145. A. Paszewski and J. Grobski, *Acta Biochim. Pol.* **20**, 159 (1973).
145a. A. Paszewski, personal communication.
146. D. N. Mardon, *Can. J. Microbiol.* **19**, 155 (1973).
147. A. H. Datko, J. Giovanelli, and S. H. Mudd, *Plant Physiol.* **51**, Suppl., 50 (1973).
148. J. Giovanelli and S. H. Mudd, *Biochem. Biophys. Res. Commun.* **31**, 275 (1968).
149. J. Giovanelli and S. H. Mudd, *Biochim. Biophys. Acta* **227**, 654 (1971).
150. J. Giovanelli, S. H. Mudd, and A. H. Datko, *Plant Physiol.* **51**, Suppl., 50, (1973).
151. J. Giovanelli and S. H. Mudd, *Biochem. Biophys. Res. Commun.* **25**, 366 (1966).
152. H. R. Mahler and E. H. Cordes, "Biological Chemistry," p. 827. Harper, New York, 1966.
153. H. de Robichon-Szulmajster and Y. Surdin-Kerjan, *in* "The Yeasts" (A. H. Rose and J. S. Harrison, eds.), Vol. 2, p. 235. Academic Press, New York, 1971.
154. P. Laduron, *Nature (London) New Biol.* **238**, 212 (1972).
155. C. Kutzbach and E. L. R. Stokstad, *Fed. Proc., Fed. Amer. Soc. Exp. Biol.* **26**, 559 (1967).
156. A. R. Salem, J. R. Pattison, and M. A. Foster, *Biochem. J.* **126**, 993 (1972).
157. K. L. Lor and E. A. Cossins, *Biochem. J.* **130**, 773 (1972).
157a. K. O. Donaldson and J. C. Keresztasy, *J. Biol. Chem.* **237**, 1298 (1962).
158. K. U. Buehring, K. K. Batra, and E. L. R. Stokstad, *Biochim. Biophys. Acta* **279**, 498 (1972).
159. H. Weissbach and R. T. Taylor, *Vitam. Horm. (New York)* **28**, 415 (1970).
160. L. Milner, C. Whitfield, and H. Weissbach, *Arch. Biochem. Biophys.* **133**, 413 (1969).
161. C. Kutzbach and E. L. R. Stokstad, *Biochim. Biophys.* **139**, 217 (1967).
162. K. M. Jones. J. R. Guest, and D. D. Woods, *Biochem. J.* **79**, 566 (1961).
163. A. R. Salem and M. A. Foster, *Biochem. J.* **127**, 845 (1972).
164. E. G. Burton and W. Sakami, *Biochem. Biophys. Res. Commun.* **36**, 228 (1969).
165. W. A. Dodd and E. A. Cossins, *Biochim. Biophys. Acta* **201**, 461 (1970).
166. J. M. Griffiths and L. J. Daniel, *Arch. Biochem. Biophys.* **134**, 463 (1969).
167. B. D. Lago and A. L. Demain, *J. Bacteriol.* **99**, 347 (1969).
168. S. F. Cauthen, J. R. Pattison, and J. Lascelles, *Biochem. J.* **102**, 774 (1967).
169. H. Ohmori, K. Sato, K. Shimizu, and S. Fukui, *Agr. Biol. Chem.* **35**, 338 (1971).

170. J. F. Morningstar, Jr. and R.L. Kisliuk, *J. Gen. Microbiol.* **39,** 43 (1965).
171. S. E. Cauthen, M. A. Foster, and D. D. Woods, *Biochem. J.* **98,** 630 (1966).
172. M. A. Foster, G. Tejerina, J. R. Guest, and D. D. Woods, *Biochem. J.* **92,** 476 (1964).
173. C. D. Whitfield, E. J. Steers, Jr., and H. Weissbach, *J. Biol. Chem.* **245,** 390 (1970).
174. E. Burton and W. Sakami, *in* "Methods in Enzymology" (H. Tabor and C. W. Tabor, eds.), Vol. 17B, p. 388. Academic Press, New York, 1971.
175. L. M. Kozloff, M. Lute, L. K. Crosby, N. Rao, V. A. Chapman, S. S. DeLong, *J. Virol.* **5,** 726 (1970).
176. M. J. Griffin and G. M. Brown, *J. Biol. Chem.* **239,** 310 (1964).
177. W. Sakami, S. J. Ritari, C. W. Black, and J. Rzepka, *Fed. Proc., Fed. Amer. Soc. Exp. Biol.* **32,** 471 (1973).
178. A. R. Salem and M. A. Foster, *Biochim. Biophys. Acta* **252,** 597 (1971).
179. S. W. Thenen and E. L. R. Stokstad. *J. Nutr.* **103,** 363 (1973).
180. J. M. Whitehouse and D. A. Smith, *Mol. Gen. Genet.* **120,** 341 (1973).
181. R. T. Taylor and H. Weissbach, *in* "Methods in Enzymology" (H. Tabor and C. W. Tabor, eds.), Vol. 17B, p. 379. Academic Press, New York, 1971.
182. J. M. Buchanan, *in* "Methods in Enzymology" (H. Tabor and C. W. Tabor, eds.), Vol. 17B, p. 371. Academic Press, New York, 1971.
183. J. H. Mangum, B. A. Stewart, and J. A. North, *Arch. Biochem. Biophys.* **148,** 63 (1972).
184. J. H. Mangum and K. G. Scrimgeour, *Fed. Proc., Fed Amer. Soc. Exp. Biol.* **21,** 242 (1962).
185. J. R. Guest, S. Friedman, and D. D. Woods, *Nature (London)* **195,** 340 (1962).
186. M. Brenner and B. N. Ames, *J. Biol. Chem.* **247,** 1080 (1972).
187. C. L. Squires, J. K. Rose, C. Yanofsky, H.-L. Yang, and G. Zubay, *Nature (London), New Biol.* **245,** 131 (1973).
188. G. Zubay, D. E. Morse, W. J. Schrenk, and J. H. M. Miller, *Proc. Nat. Acad. Sci. U.S.* **69,** 1100 (1972).
189. G. A. Jacoby and L. Gorini, *J. Mol. Biol.* **39,** 73 (1969).
190. R. H. Vogel, W. L. McLellan, A. P. Hirvonen, and H. J Vogel, *in* "Metabolic Pathways" (H. J. Vogel, ed.), Vol. 5, p. 463. Academic Press, New York, 1971.
191. R. L. Metzenberg, *Annu. Rev. Genet.* **6,** 111 (1972).
192. C. T. Holloway, R. C. Greene, and C. -H. Su, *J. Bacteriol.* **104,** 734 (1970).
193. A. C. Hobson and D. A. Smith, *Mol. Gen. Genet.* **126,** 7 (1973).
194. G. N. Cohen and F. Jacob, *C. R. Acad. Sci.* **248,** 3490 (1959).
195. K. F. Chater, *J. Gen. Microbiol.* **63,** 95 (1970).
196. A. C. Minson and D. A. Smith, *J. Gen. Microbiol.* **70,** 471 (1972).
197. T. S. Gross and R. J. Rowbury, *Biochim. Biophys. Acta* **184,** 233 (1969).
198. E. R. Archibold and L. S. Williams, *J. Bacteriol.* **114,** 1007 (1973).
199. W. K. Mass, *Mol. Gen. Genet.* **119,** 1 (1972).
200. A. Ahmed, *Mol. Gen. Genet.* **123,** 299 (1973).
201. H. Tabor and C. W. Tabor, *in* "Methods in Enzymology" (H. Tabor and C. W. Tabor, eds.), Vol. 17B, p. 393. Academic Press, New York, 1971.
202. J. Dawes and M. A. Foster, *Biochim. Biophys. Acta* **237,** 455 (1971).
203. R. C. Greene, R. D. Williams, H. -F. Kung, C. Spears, and H. Weissbach, *Arch. Biochem. Biophys.* **158,** 249 (1973).
204. A. Mansouri, J. B. Decter, and R. Silber, *J. Biol. Chem.* **247,** 348 (1972).
205. S. Black and N. G. Wright, *in* "Amino Acid Biosynthesis" (W. D. McElroy and B. Glass, eds.), p. 591. Johns Hopkins Press, Baltimore, Maryland, 1955.

206. J. C. Patte, G. Le Bras, and G. N. Cohen, *Biochim. Biophys. Acta* **136**, 245 (1967).
207. R. L. Cafferata and M. Freudlich, *Biochim. Biophys. Acta* **222**, 671 (1970).
208. A. M. Kotre, S. J. Sullivan, and M. A. Savageau, *J. Bacteriol.* **116**, 663 (1973).
208a. N. J. Pieniazek, I. M. Kowalska, and P. P. Stepien, *Mol. Gen. Genet.* **126**, 367 (1973).
209. M. B. Jenkins and V. W. Woodward, *Biochim. Biophys. Acta* **212**, 21 (1970).
210. J. Antoniewski and H. de Robichon-Szulmajster, *Biochimie* **55**, 529 (1973).
211. H. Cherest, Y. Surdin-Kerjan, J. Antoniewski, and H. de Robichon-Szulmajster, *J. Bacteriol.* **106**, 758 (1971).
212. H. Cherest, Y. Surdin-Kerjan, J. Antoniewski, and H. de Robichon-Szulmajster, *J. Bacteriol.* **115**, 1084 (1973).
213. M. Masselot and H. de Robichon-Szulmajster, *Genetics* **71**, 535 (1972).
214. J. T. Murphy and K. D. Spence, *J. Bacteriol* **109**, 499 (1972).
215. H. Cherest, Y. Surdin-Kerjan, J. Antoniewski, and H. de Robichon-Szulmajster, *J. Bacteriol.* **114**, 928 (1973).
216. J. L. Botsford and J. W. Parks, *J. Bacteriol.* **97**, 1176 (1969).

CHAPTER 12

Biosynthesis of Cysteine and Cystine

David M. Greenberg

I. INTRODUCTION

The biosynthesis of cysteine from methionine via cystathionine was the first pathway to be discovered. It appears to be the sole pathway for the formation of cysteine in vertebrates. Subsequently, a second pathway of cysteine formation from serine or O-acyl serine (predominantly O-acetylserine) and sulfide was discovered in microorganisms and plants. The enzyme catalyzing this reaction occurs in mammalian tissues, but is nonoperative because of the absence of a sulfate-sulfite reducing system in the mammal.

II.   SYNTHESIS OF CYSTEINE BY THE
CYSTEINE SYNTHASE SYSTEM (EC 4.2.99.8)

A. Synthesis from Serine and H_2S

The enzyme cysteine synthase, also named L-serine hydrolase(adding H_2S), was discovered by Schlossman and Lynen [1] using an enzyme preparation from yeast. The enzyme was shown to be pyridoxal phosphate-dependent.

In subsequent studies cysteine synthase was purified about 50-fold from bakers' yeast [2] and 100-fold from chicken liver [3].

This enzyme was shown to catalyze the reversible transsulfuration of L-serine to L-cysteine as represented by Eq. (1) below.

$$
\begin{array}{cc}
H_2COH & H_2C-SH \\
| & | \\
HC-NH_2 + H_2S \rightleftharpoons HC-NH_2 + H_2O \\
| & | \\
COOH & COOH \\
\text{L-Serine} & \text{L-Cysteine}
\end{array}
\tag{1}
$$

Investigation of the distribution of cysteine synthase revealed that it was present in higher concentrations in various species of bacteria, plants (spinach), liver, kidney, and brain than in yeast [4].

Proof that sulfide is the reactant for the biosynthesis of cysteine and that sulfate and sulfite serve only as precursors of sulfide was obtained by several investigators. Waldschmidt [5] found that the proteins of rat liver homogenates incubated with $^{35}S^{2-}$ had a 20– to 50–fold higher radioactivity than the corresponding proteins from $^{35}SO_4^{2-}$ incubations.

In intracardially injected rats a considerable fraction of the ^{35}S was found in the intestine 30 minutes after injection [5]. Huovinen and Gustafsson [6] demonstrated that there was no incorporation of ^{35}S in the sulfur amino acids of germfree rat tissue upon injection of $^{35}SO_4^{2-}$ or $^{35}SO_3^{2-}$. On injection of $^{35}S^{2-}$ the sulfur amino acids were about equally labeled in both germfree and conventional rats. These findings confirm the lack of a sulfate reducing system in mammalian tissues.

A highly purified preparation of cysteine synthase has obtained from *Pasteurella multocida* [7]. The enzyme, purified 375-fold by a seven-step process, was estimated to have no more than 7% impurity. The sedimentation coefficient ($s_{20,w}$) in the ultracentrifuge was 3.2–3.6. Optimum activity, estimated from a shallow pH–activity curve, was at pH 8.0 in Tris buffer. The reaction followed simple Michaelis kinetics with a K_m of 1.3 mM for L-serine. Reagents that react with the aldol group of pyridoxal-P (hydroxylamine, hydrazine, and cyanide) inhibited the enzyme significantly at low concentrations (0.2–1 mM). The enzyme activity was enhanced about 50% by 20 mM K^+.

B. Synthesis from O-Acyl Serine and H₂S

Further investigation revealed that the substrate specificity of cysteine synthase was much broader than the reaction represented by Eq. (1). O-Acetylserine was observed to be a superior substrate for the enzyme with extracts of spinach [8], bakers' yeast, turnip leaves [9], *Neurospora* [9], *Escherichia coli* [10], and *Salmonella typhimurium* [11]. These same enzyme preparations exhibited a weaker activity for the formation of homocysteine from O-acetylhomoserine and sulfide [9,10]. The systematic name accordingly was changed to O-acetyl-L-serine acetate-lyase(adding H₂S). Giovanelli and Mudd [11] separated the enzyme activity of spinach extracts into two fractions, one of which was active only with O-acetylserine, whereas the other utilized both O-acetylhomoserine and O-acetylserine. Not only did the reaction take place with sulfide, but also with methyl and ethyl mercaptan, forming the corresponding S-methyl and S-ethyl sulfur amino acids.

The enzymatic formation of S-methylcysteine from methyl mercaptan and serine had been reported in 1956 by Wolff *et al.* [12].

C. Cysteine Biosynthesis in *Salmonella typhimurium* and *Escherichia coli*

Kredich and Tomkins [13,14] demonstrated a two-step pathway from L-serine to L-cysteine in *E. coli* and *S. typhimurium* in which serine transacetylase (EC 2.3.1.30) catalyzes the formation of O-acetyl-L-serine and O-acetyl-L-serine acetate-lyase catalyzes the formation of L-cysteine. The pathway of L-cysteine biosynthesis in *S. typhimurium*, including the reduction of sulfate to sulfide is shown in Fig. 1. The

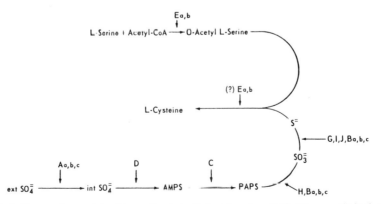

FIG. 1. Pathway of L-cysteine biosynthesis in *S. typhimurium* [14]. The genetic loci concerned with each step are indicated as in Fig. 2. AMPS, adenosine 5′-phosphosulfate; PAPS, 3′-phosphoadenosine 5′-phosphosulfate.

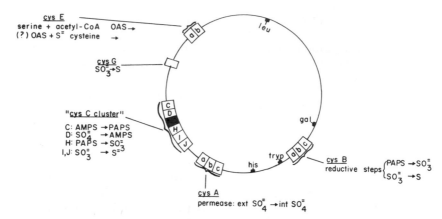

FIG. 2. Map of *S. typhimurium* chromosome showing the genetic areas concerned with the biosynthesis of L-cysteine [14]. OAS = *O*-acetyl-L-serine.

genetics of the reaction scheme in Fig. 1 is illustrated in Fig. 2. Genetic analysis has established the participation of 14 cistrons located in five clusters on the bacterial chromosome [15,16].

Serine transacetylase was purified about 1000-fold from frozen cells of *E. coli* [13]. The purified enzyme is yellow in color, but the color in later work with *S. typhimurium* was shown to be due to contamination of the enzyme with cysteine synthase [17,18]. The purified enzyme preparation exhibited multiple components both by disc gel electrophoresis and analytical centrifugation.

Salmonella typhimurium contains two separate fractions of cysteine synthase activity that can readily be resolved by gel filtration on Sephadex G-200 [17,18]. The first eluted fraction, containing 5% of the enzyme activity, is composed of a complex of both transacetylase and synthase activity. It has been designated by the authors as *O*-acetylserine sulfhydrylase-STA. The remaining 95% *O*-acetylserine sulfhydrase activity, free of transacetylase, is referred to as *O*-acetylserine sulfhydrylase-A.

The *O*-acetylserine sulfhydrylase-A has been obtained as a single component, yellow protein (absorption peak, 415 nm) with a molecular weight of 68,000 daltons. It contains two moles of pyridoxal-P per mole of enzyme. Dissociation in 6 *M* guanidine-HCl yielded a single component with a molecular weight of 32,000 daltons whereas the carboxymethylated protein had a molecular weight of 34,000 daltons. Serine was established to be the only NH_2-terminal amino acid of the undissociated enzyme.

The purified enzyme has a specific activity of 1100 IU/mg. The enzyme is stable at room temperature for several days and retains most of its activity at $-20°$ for many months. It is comparatively heat stable, losing only 5% activity on heating for 10 minutes at 62°.

Substrates for O-acetyl-L-serine sulfhydrylase were determined essentially to be limited to O-acetyl-L-serine and sulfide. Methylmercaptide was found to react and yield S-methylcysteine.

The K_m for O-acetyl-L-serine, determined kinetically, was 5 mM, whereas the equilibrium constant for the substrate binding reaction, determined spectroscopically, was 6×10^{-7} M. The reason for this disparity is not known.

The serine transacetylase–O-acetyl-L-serine sulfhydrylase complex, given the trivial name cysteine synthetase by the authors [17,18], is a multifunctional complex of molecular weight 309,000 daltons. O-Acetyl-L-serine (0.1–1 mM) causes this complex to dissociate reversibly into one molecule of serine transacetylase (MW 160,000) and two molecules of O-acetylserine sulfhydrylase (MW 68,000). The complex is a yellow protein with an absorption at 410–415 nm. The complex was estimated to contain 4.4 moles of pyridoxal-P per 309,000 gm protein. The resolved serine transacetylase is colorless.

The resolved O-acetylserine sulfhydrylase (cysteine synthase) of the complex had properties identical with that of the free sulfhydrylase.

The K_m values for L-serine and acetyl-CoA are 0.69 mM and 0.12 mM, respectively, for the transacetylase in the resolved enzyme, whereas these values are 0.74 mM for L-serine and 0.12 mM for acetyl-CoA in the cysteine synthetase complex.

L-Cysteine inhibits the transacetylase reaction of both forms of the enzyme with a K_i of 6×10^{-7} M.

The K_m for O-acetyl-L-serine is 5 mM for both the free and resolved O-acetylserine sulfhydrylase; the K_m is 20 mM for the reaction in the complex. The participation of O-acetyl-L-serine sulfhydrylase in the cysteine synthetase complex results in a decrease in its catalytic capacity and a decrease in the apparent affinity of the enzyme for O-acetyl-L-serine.

In a study of the regulation of L-cysteine biosynthesis, Kredich [19] confirmed that the enzymes concerned with the biosynthesis of L-cysteine, with the exception of serine transacetylase, are subject to depression by sulfur starvation and repression by sulfide and cysteine. Under conditions of sulfur starvation O-acetyl-L-serine is necessary for depression. Therefore it is an internal inducer. Sulfide and cysteine block the effect of exogenous O-acetylserine. Cells that are starved for sulfide and cysteine presumably accumulate O-acetylserine, since sulfide is required for the utilization of this cysteine precursor. Cysteine decreases the rate of O-acetylserine synthesis by feedback inhibition of serine trans-acetylase. Kredich reasons that increased intracellular levels of O-acetylserine could constitute a signal of sulfur starvation and thus play a role in the derepression or induction of the biosynthetic pathway. Conversely, increased intracellular concentrations of sulfide and cysteine should depress the levels of O-acetylserine and act to repress the biosynthetic pathway.

The ability of cysteine to repress the biosynthetic pathway may, be due, in part, to liberation of sulfide (catalyzed by cysteine desulfhydrase), since sulfide also prevents induction by exogenous O-acetylserine of cysteine desulfhydrase.

Very recently, the structure gene for O-acetylserine sulfhydrylase-A has been identified from a mutant of $S.$ $typhimurium$ resistant to the growth inhibitor, 1,2,4-triazole [19a]. The triazole mutants, which completely lack O-acetylserine sulfhydrylase-A, show no growth requirement for cysteine. This indicates that synthesis of cysteine from O-acetyl-L-serine and sulfide can be catalyzed by another enzyme in $S.$ $typhimurium.$

The cys B region of the genetic map is indicated to be regulatory (Fig. 2). A similar scheme of regulation of the cysteine biosynthetic pathway for cysteine occurs in $E.$ $coli$ [20].

D. Catalysis of β-Carbon Polar Group Replacement Reactions by Cysteine Synthase and Related Enzymes

The specificity of cysteine synthase has been shown to be much broader than was initially anticipated. This enzyme catalyzes replacement reactions of polar groups in the β-position of L-serine and some of its β-substituted analogues and also of L-cysteine and its thioethers upon incubation with mercapto compounds [21,22]. These reactions can be represented by the equations below:

$$HOCH_2 \cdot CHNH_2COOH + H_2S \rightleftharpoons HS \cdot CH_2 \cdot CHNH_2COOH + H_2O \quad (2)$$

$$HOCH_2 \cdot CHNH_2COOH + RSH \rightleftharpoons RSCH_2 \cdot CHNH_2COOH + H_2O \quad (3)$$

$$HSCH_2 \cdot NH_2COOH + RSH \rightleftharpoons RCH_2 \cdot CHNH_2 \cdot COOH + H_2S \quad (4)$$

$$X \cdot CH_2 \cdot CHNH_2 \cdot COOH + RSH \rightleftharpoons RSCH_2 \cdot CHNH_2 \cdot COOH + XH \quad (5)$$

$$(R = \text{alkyl}; X = S\text{-alkyl}, {}^-Cl, {}^-OSO_3H, {}^-CN)$$

Thus, in experiments with a highly purified preparation of chicken liver cysteine synthase it was possible to obtain synthesis of cystathionine in the presence of homocysteine with L-serine, β-chloro-DL-alanine, DL-

TABLE I

SUBSTRATES AND COSUBSTRATES OF CYSTEINE SYNTHASE
AND SIMILAR ENZYMES[a]

Substrates	Cosubstrates	Cysteine thioethers
L-Serine	DL-Homocysteine	S-Carboxymethyl-L-cysteine
O-Acetyl-L-serine	β-Mercaptoethanol	S-Methyl-L-cysteine
β-Chloro-DL-alanine	Cysteamine	S-Hydroxyethyl-L-cysteine
DL-Serine-O-sulfate	N-Acetylcysteamine	N^w-Acetyl-L-thiolysine
β-Cyano-DL-alanine	Methylmercaptan	
	Ethylmercaptan	
	Thioglycolic acid	
	β-Mercaptopropionic acid	
	D-Penicillamine	

[a] From Braunstein et al. [22].

serine-O-sulfate, and β-cyano-DL-alanine [19,20–23]. The reactions of
O-acetylserine with S-methyl and S-ethyl mercaptan have been described above [11,12].

Cysteine and S substituents of cysteine also are substrates of cysteine
synthase. A list of substrates is recorded in Table I.

A previously discovered enzyme, cysteine lyase (EC 4.4.1.10) of the
yolk sac chicken embryos [23–26] had been shown to have a similar
replacement function for the —SH of cysteine, forming lanthionine from
a second molecule of cysteine, and cysteic acid by reaction with sulfite.

Comparison of the substrate activities of the purified chicken liver
cysteine synthase and of highly purified cystathionine β-synthase (EC
4.2.1.22) of rat liver [21,22] has revealed a marked similarity in substrates utilized.

TABLE II

K_m VALUES FOR CYSTEINE SYNTHASE (I) AND
CYSTATHIONINE β-SYNTHASE (II)[a]

	Reaction substrates			
	L-Cysteine (1) + β-mercaptoethanol (2)		L-Serine (3) + DL-homocysteine (4)	
	K_m (mM) (1)	K_m (mM) (2)	K_m (mM) (3)	K_m (mM) (4)
I	22	13	34	83
II	36	24	0.08	18

[a] From Braunstein et al. [22].

The latter enzyme catalyzed a variety of β-replacement reactions described by Eqs. (2–5), listed above, in which the substrate and cosubstrate specificities and the relative reaction rates were similar to that of cysteine synthase.

A kinetic study of several reactions catalyzed by cysteine synthase and of cystathione β-synthase yielded the K_m values shown in Table II. Table II shows that, in the cystathionine synthesis reaction, the K_m values of the two substrates are substantially lower for cystathionine β-synthase than for serine sulfhydrase. In the S-hydroxyethylcysteine synthesis reaction the K_m values are lower for cysteine synthase.

Additional evidence that single enzyme proteins in chicken and rat liver have the total catalytic activities for the reactions described above was obtained by pH-electrofocusing purification experiments [22]. Only one protein fraction was present in extracts of liver of the two species with coincident activities for the synthesis of S-hydroxyethylcysteine and cystathionine. In these experiments, and in experiments with high purity preparations, the isoelectric points were observed to be pH 6.0 for the chicken enzyme and pH 5.5 for the rat liver enzyme.

Since vertebrates do not possess a sulfate reducing system, it has been suggested that the primary biological function of cysteine synthase in these organisms is for the synthesis of cystathionine [21]. Braunstein et al. [22] suggest that the observed differences in kinetic (reaction velocities, affinities) and physical (molecular weights, isoelectric points) properties of the different enzymes are attributable to genetically determined, species-specific variation in the structure of serine sulfhydrase.

III. THE CYSTATHIONINE PATHWAY OF CYSTEINE BIOSYNTHESIS

A. Historical Development

The early studies establishing the transfer of methionine sulfur, via cystathionine, to L-cysteine sulfur have been reviewed by a number of authors [27–31]. These and later findings led to the overall sequence of reactions for the transsulfuration pathway represented by the following equations:

$$\text{ATP} + \text{L-methionine} + H_2O \xrightarrow{1^*} S\text{-adenosyl methionine} + PP_i + P_i \quad (6)$$

* Key to numbers used over arrows in Eqs. (6)–(10):
1. ATP: L-methionine S-adenosyltransferase (EC 2.5.1.6).
2. S-Adenosyl-L-methionine: L-homocysteine S-methyltransferase (EC 2.1.1.10).
3. Adenosylhomocysteinase (EC 3.3.1.1).
4. Cystathionine β-synthase (EC 4.2.1.22).
5. Cystathionase (cystathionine γ-lyase) (EC 4.4.1.1).

S-Adenosyl methionine + dimethylglycine* $\xrightarrow{2}$ S-adenosyl-L-homocysteine + betaine

$$S\text{-Adenosyl-L-homocysteine} + H_2O \xrightarrow{3} \text{adenosine} + \text{L-homocysteine} \qquad (8)$$

$$
\begin{array}{c}
\underset{\text{L-Homocysteine}}{
\begin{array}{l}
H_2C\text{—SH} \\
| \\
HCH \\
| \\
HC\text{—NH}_2 \\
| \\
COOH
\end{array}}
\;+\;
\underset{\text{L-Serine}}{
\begin{array}{l}
H_2COH \\
| \\
H\text{—C—NH}_2 \\
| \\
COOH
\end{array}}
\;\xrightarrow{4}\;
\underset{\text{L-Cystathionine}}{
\begin{array}{l}
H_2C\text{——S——CH}_2 \\
| \qquad\qquad | \\
H\text{—CH} \quad HC\text{—NH}_2 + H_2O \\
| \qquad\qquad | \\
HC\text{—NH}_2 \quad COOH \\
| \\
COOH
\end{array}}
\end{array}
\qquad (9)
$$

$$
\begin{array}{c}
\begin{array}{l}
H_2C\text{——S——CH}_2 \\
| \qquad\qquad | \\
HCH \qquad HC\text{—NH}_2 \\
| \qquad\qquad | \quad + H_2O \\
HC\text{—NH}_2 \quad COOH \\
| \\
COOH
\end{array}
\;\xrightarrow{5}\;
\underset{\text{L-Cysteine}}{
\begin{array}{l}
H_2C\text{—SH} \\
| \\
HC\text{—NH}_2 \\
| \\
COOH
\end{array}}
\;+\;
\underset{\alpha\text{-Ketobutyrate}}{
\begin{array}{l}
CH_3 \\
| \\
H_2C \\
| \\
C\text{=}O \\
| \\
COOH
\end{array}}
\;+ NH_3
\end{array}
\qquad (10)
$$

The transfer of methionine sulfur to cysteine sulfur was demonstrated by Tarver and Schmidt [32] who isolated ^{35}S-labeled cysteine from rat hair after feeding ^{35}S-labeled methionine. DuVigneaud and co-workers [33,34] synthesized the four diastereoisomers of S-(β-amino-β-carboxyethyl)homocysteine (cystathionine) (see below for formulas) and tested their ability to replace the sulfur-containing amino acids in the diet of young rats. L-Cystathionine supported the growth of rats on a diet lacking cysteine and low in methionine; L-allocystathionine promoted growth in the absence of methionine only in the presence of choline. D-Cystathionine and D-allocystathionine were inactive. It was concluded, and subsequently established, that L-cystathionine was cleaved to cysteine and homoserine, whereas L-allocystathionine was cleaved to homocysteine and serine.

$$
\begin{array}{cccc}
\underset{\text{L-Cystathionine}}{
\begin{array}{l}
COOH \\
| \\
H_2N\text{—C—H} \\
| \\
H\text{—C—H} \\
| \\
H\text{—C—H} \\
| \\
S \\
| \\
H\text{—C—H} \\
| \\
HC\text{—NH}_2 \\
| \\
COOH
\end{array}}
&
\underset{\text{D-Cystathionine}}{
\begin{array}{l}
COOH \\
| \\
H\text{—C—NH}_2 \\
| \\
H\text{—C—H} \\
| \\
H\text{—C—H} \\
| \\
S \\
| \\
H\text{—C—H} \\
| \\
H_2N\text{—C—H} \\
| \\
COOH
\end{array}}
&
\underset{\text{L-Allocystathionine}}{
\begin{array}{l}
COOH \\
| \\
H\text{—C—NH}_2 \\
| \\
H\text{—C—H} \\
| \\
H\text{—C—H} \\
| \\
S \\
| \\
H\text{—C—H} \\
| \\
H\text{—C—NH}_2 \\
| \\
COOH
\end{array}}
&
\underset{\text{D-Allocystathionine}}{
\begin{array}{l}
COOH \\
| \\
H_2N\text{—C—H} \\
| \\
H\text{—C—H} \\
| \\
H\text{—C—H} \\
| \\
S \\
| \\
H\text{—C—H} \\
| \\
H_2N\text{—C—H} \\
| \\
COOH
\end{array}}
\end{array}
$$

* Dimethyl glycine is just one of the numerous methyl group acceptors that can function as a substrate for S-methyltransferase.

The nutritional results were corroborated by *in vitro* experiments with liver slices and saline extracts of liver. Incubation of mixtures containing DL-homocysteine and DL-serine yielded L-cystathionine [34]. D-Homocysteine and D-serine did not substitute for the DL forms.

B. Formation of Homocysteine from Methionine

As shown in Eq. (8), methionine is the source of L-homocysteine. The latter is formed by decomposition of S-adenosylmethionine. Considerable evidence has now accumulated [35–39] that shows that after formation of S-adenosylhomocysteine subsequent to transmethylation of the methyl group, the latter is decomposed to L-homocysteine and adenosine. De la Haba and Cantoni [35] found an enzyme in liver that reversibly catalyzes the condensation of adenosine with homocysteine to form adenosylhomocysteine. Although this enzyme can cause the hydrolysis of adenosylhomocysteine, the equilibrium lies far in the direction of synthesis.

An S-adenosyl-L-homocysteine hydrolase has been discovered in yeast [36,37] and germinating peas [38]. Isotopic tracer experiments [36–39] have established that adenosylhomocysteine is cleaved to homocysteine prior to the latter being methylated to methionine.

Study of the enzymatic splitting of the diastereoisomers of cystathionine confirmed that L-cystathionine was cleaved to L-cysteine, whereas L-allocystathionine was predominantly split to D-homocysteine [40–43]. D-Allocystathionine was cleaved to D-cysteine, but much more slowly than was L-cystathionine to L-cysteine.

IV. CYSTATHIONINE β-SYNTHASE OF RAT LIVER (EC 4.2.1.22)

In vertebrates L-cystathionine is formed by condensation of L-homocysteine and L-serine, catalyzed by the enzyme cystathionine β-synthase. The enzymatic synthesis of cystathionine was first reported by Binkley [44]. Selim and Greenberg [45] obtained results from which they concluded that this enzyme also catalyzed the deamination of L-serine. Nagabhushanam and Greenberg [46] obtained a purified enzyme preparation from rat liver that appeared to have L-serine and L-threonine deaminating activity and cystathionine synthesizing activity as well.

However, a number of observations were reported that cast doubt on the supposition that cystathionine synthase had hydroxyamino acid deaminating activity.* These observations were that serine dehydratase

* Porter *et al.* [46a] reported the isolation of a partially purified enzyme from human fetal liver that catalyzes the condensation of sulfide with serine to form cysteine (cysteine synthase), as well as the condensation of homocysteine with serine (cystathionine β-synthase).

was induced by dietary serine or during increased gluconeogenesis whereas cystathionine synthase was not affected [47,48]. Brown *et al.* [49] found that the cystathionine synthase assay employed by Greenberg and co-workers was not valid due to the formation of an adduct of pyruvic acid (derived from serine) and homocysteine. Subsequently, serine dehydratase was crystallized by Nakagawa *et al.* [50] and by Nakagawa and Kimura [51] and was shown to possess no cystathionine synthase activity.

Cystathionine synthase has now been prepared in a high state of purity in a number of laboratories [52–56] and shown to be distinct from serine dehydratase.

A. Assay Methods for Cystathionine Synthase

A widely used unequivocal assay for the enzyme, devised by Mudd *et al.* [57] is one in which [^{14}C]serine is reacted with homocysteine to form [^{14}C]cystathionine and the latter is isolated by column chromatography on Dowex 50-H$^+$ after removal of the residual [^{14}C]serine. Kashiwamata and Greenberg [54] observed that cystathionine yielded a comparatively specific ninhydrin chromogen, absorbing at 455 nm in a strongly acidic medium. Cysteine, formed from cystathionine catalyzed by cystathionase, interferes strongly with the measurement of cystathionine. Cystathionase can be inactivated by the addition of 0.5 mM Cu^{2+} to the incubation medium [58]. Brown and Gordon [56] developed a method suitable for the purified enzyme in which the formed cystathionine is decomposed to serine and homocysteine by an appropriate enzyme in the presence of 5,5'-dithiobis(2-nitrobenzoic acid) (DTNB) (see Section VI,A).

B. Purification and Properties of β Cystathionine Synthase

The purification procedure employed by the several laboratories generally utilized acidification to pH 5.5 to eliminate serine dehydratase, several ammonium sulfate precipitations (40% saturation), calcium phosphate gel treatment, DEAE-cellulose chromatography, and Sepharose-6B chromatography. Certain other steps were used by Kimura and Nakagawa [53] and by Brown and Gordon [56]. The preparation of Kashiwamata and Greenberg represented a 540-fold purification and that of Kimura and Nakagawa, a 1000-fold purification.

It would appear, however, from comparison of the data for the cystathionine synthase preparation of Nakagawa and Kimura [52, Table 1] with that of Kimura and Nakagawa [53, Table 1], that the factor 10^3 has been omitted from the values of the specific activity in the latter and the initial crude extract of this preparation had only half the specific activity of the Nakagawa and Kimura preparation. If this is correct, the specific activity of the purified Kimura and Nakagawa preparation was 2.26 IU, the specific activity of the preparation of Kashiwamata and Greenberg was 8.1 IU, and that of Brown and Gordon was 5.4 IU.

1. PURITY AND MOLECULAR WEIGHT

A summary of the physical, chemical, and kinetic properties of cystathionine synthase is given in Table III.

Employing a Sepharose-4B column, Nakagawa and Kimura [52] obtained a value of 250,000 daltons for the molecular weight of cystathionine synthase. This value was confirmed by Kashiwamata et al. [55] by a similar elution experiment on a Sepharose-6B column. Kimura and Nakagawa [53], however, obtained the value of 112,000 daltons for the molecular weight of their highly purified enzyme preparation estimated from Bio-Gel P-300 chromatography in contrast to the 250,000 daltons for cruder preparations. On polyacrylamide-gel disc electrophoresis, a single protein band was obtained with the cystathionine synthase preparation of Kashiwamata and Greenberg; the preparation of Kimura and Nakagawa yielded five protein bands. Acrylamide-gel disc electrophoresis of sodium dodecyl sulfate treated enzyme yielded two distinct protein bands [55]. Comparison of the rates of migration of these bands with the migration of proteins of known molecular weight indicated that the two bands had molecular weight of 51,000 and 73,000, respectively [55]. It would appear that the 250,000 dalton intact enzyme consists of two units each of the above polypeptides.

2. COENZYME OF CYSTATHIONINE SYNTHASE

The purified preparations of this enzyme have a yellow color, which is characteristic of pyridoxal-P proteins. Kashiwamata and Greenberg's

TABLE III

PHYSICAL, CHEMICAL, AND KINETIC PROPERTIES OF CYSTATHIONINE SYNTHASE

Molecular weight (daltons):	250,000 [52,54]
Number of polypeptide subunits:	$2 \times 51,000$; $2 \times 73,000$ [55]
pH for optimum activity:	8.6 [52], 8.3 [54], 8.0–8.6 [56]
Coenzyme:	Pyridoxal 5-phosphate
Absorption spectra peaks (nm):	278, 360–365, 428–430 [52,54,56]
K_m (mM) for L-serine:	8.3 [54], 0.67 [53], 6.0 [56]
K_m (mM) for L-homocysteine:	2.0 [54], 1.1 [53], 12 [56]
SH inhibitors:	PCMB, Hg^{2+} (protected by L-serine) [52,56]
Heat resistance:	Comparatively stable up to 50°C

preparation [54] exhibited absorption peaks in the visible region at 350 and 430 nm at pH 5.5–9.5; the Kimura and Nakagawa preparation [53] had absorption peaks at 365 and 428 nm, with a small shoulder at 550 nm; Brown and Gordon's preparation [56] had peaks at 360 and 428 nm. The absorption peaks at 350, 365, and 430 nm are due to nonprotonated and protonated Schiff bases, respectively, formed between the coenzyme and apoenzyme molecule.

Additional evidence that pyridoxal-P is the coenzyme of cystathionine synthase is the following: (1) incubation enzyme with cysteine causes a loss of activity that can be largely restored by addition of pyridoxal-P; (2) reduction with sodium borohydride also causes inactivation of the enzyme. This gives only a slight increment in activity upon adding pyridoxal-P. Kashiwamata and Greenberg observed a loss of the characteristic 430 nm peak on reduction with sodium borohydride, Kimura and Nakagawa obtained a considerable decrease in absorbance of the peaks characteristic of pyridoxal-P on reduction, and Brown and Gordon found a loss in enzyme activity, but no change in the absorption peaks.

Kashiwamata and Greenberg hydrolyzed the reduced enzyme and demonstrated the presence of ε-pyridoxyllysine by paper chromatography by comparison with an authentic sample of this compound.

3. KINETIC PROPERTIES AND INHIBITORS

The Michaelis constants obtained for cystathionine synthase for its major substrates in several laboratories are shown in Table III. Kashiwamata and Greenberg and Brown and Gordon obtained figures of the same order of magnitude. The values of Kimura and Nakagawa [53] are considerably lower. These do not agree with the values previously reported by Nakagawa and Kimura [52], which are much closer to the values obtained by the other investigators.

Kashiwamata and Greenberg [54] and Brown and Gordon [56] observed inhibition of the enzyme by PCMB and Hg^{2+}, indicating that cystathionine synthase is an SH enzyme. Kimura and Nakagawa obtained no inhibition of enzyme activity by the above SH reagents. Inhibition by the above have been consistently found by other investigators with both crude and purified enzyme preparation.

As previously mentioned (Section II,D), Braunstein et al. [22] concluded that rat liver cystathionine β-synthase and cysteine synthase are identical proteins.

Brain cystathionine synthase has been studied by a number of investigators [58a,58b]. The brain enzyme was found to a large extent to be bound to the particulate material, presumably the mitochondria. Rats on a pyridoxine deficient diet exhibited about a 50% decrease in cystathionine synthase activity. This enzyme activity could be restored to the control level by the intraperitoneal injection of pyridoxine hydrochloride 20 hours before the rats were killed [58a]. The enzyme was present at about half the maximal levels in the rat fetus and newborn animals and increased to the adult level at about 20 hours after birth.

V.   CYSTATHIONINE γ-SYNTHETASE OF BACTERIA (EC 4.2.99.9)

A cystathionine γ-synthase has been identified in such bacteria as *E. coli* and *S. typhimurium*. This enzyme catalyzes the synthesis of L-cystathionine from the succinyl ester of L-homoserine and L-cysteine. This enzyme is one of the sequence of enzymes concerned with the biosynthesis of methionine in bacteria. Its isolation, properties, and role in methionine biosynthesis are discussed in Chapter 11.

In green plants, in addition to the succinyl ester, *O*-malonyl-, *O*-oxalyl-, and *O*-phosphorylhomoserine have been found to be active substrates for cystathionine synthesis [58c].

VI.   γ-CYSTATHIONASE OF RAT LIVER (EC 4.4.1.1)

The early *in vitro* studies have established that α-ketobutyrate was a product of the enzymatic cleavage of cystathionine [59] and the partial purification of both the cystathionine cleavage enzyme and the synthesizing enzyme had been reported [60,61].

Subsequent studies in clarifying the cystathionine pathway came from studies with purified preparations of the synthesizing and cleavage enzyme in a number of laboratories.

Binkley and co-workers [62–65] obtained considerably purified preparations of the enzyme catalyzing the decomposition of cystathionine and homoserine from rat and pig liver. Binkley [65] reported that L-lanthionine, L-djenkolic acid, *meso*-lanthionine, and D-allocystalhionine may be used as alternative substrates. Binkley and Okeson [62] also observed the liberation of H_2S from cysteine. Binkley [65] also reported that he had achieved the enzymatic synthesis of cystathionine from homocysteine and serine by a liver preparation freed from cystathionine cleavage enzyme, serine dehydratase, and homoserine deaminase. Since homoserine deamination and cystathionine cleavage are properties of the same enzyme, this is a discrepancy.

Snell [65a] has reported that the enzyme develops rapidly in fetal rat liver starting on the twentieth day of gestation and reaching its adult level on the second day after birth.

A. Isolation of Crystalline Enzyme

γ-Cystathionase (EC 4.4.1.1) was crystallized from liver by Matsuo and Greenberg [66–69] and its properties studied. Rat liver was selected because it showed the highest enzyme activity among the mammalian livers tested. In addition to liver, considerable activity of this enzyme was observed in kidney tissue.

The assay methods used by Binkley and co-workers not being altogether satisfactory, Matsuo and Greenberg chose for assay the determination of the α-ketobutyrate liberated in the reaction by the direct method of Friedmann and Haugen [70]. The α-ketobutyrate can also be determined enzymatically in purified enzyme preparations with lactate dehydrogenase by spectrophotometrically measuring the oxidation of NADH. More recently, an elegant method of assay has been introduced by Flavin [71], based on the reaction between alkyl mercaptans and an aromatic disulfide (ArSSAr) that liberates a strongly colored aromatic mercaptan according to Eq. (11):

$$2 \text{ Alkyl—SH} + \text{ArSSAr} \longrightarrow \text{alkyl—S—S—alkyl} + 2 \text{ ArSH} \qquad (11)$$

The aromatic disulfide commonly employed is 5,5′-dithiobis(2 nitrobenzoic acid). The liberated mercaptonitrobenzoate has a strong absorption at 412 nm.

Crystallization of rat liver cystathionase was achieved by the observation that this protein has a negative temperature coefficient of solubility in ammonium sulfate solutions. After a preliminary fractionation by heating, ammonium sulfate, ethanol, and protamine (the latter to remove nucleic acid components), the enzyme was crystallized by lowering the temperature of a solution of the protein in 50% ammonium sulfate to −8°, adding more ammonium sulfate to a faint turbidity and then allowing the temperature to increase slowly to about 5° in a refrigerator [66]. Needlelike crystals of the enzyme precipitated. Kato et al. [72] have improved the case of obtaining enzyme crystals by the addition of 1 mM pyridoxal-P to the ammonium sulfate solutions. The absorption maximum at 427 nm is characteristic of pyridoxal-P enzymes.

Roisin and Chatagner [73] have obtained a highly purified noncrystalline enzyme by the addition of chromatography on CM-cellulose to the above procedure.

B. Substrate Specificity

L-Homoserine is the sole hydroxyamino acid that is reported to be decomposed by the enzyme at a significant rate [69]. On the contrary, a considerable number of sulfur amino acids, particularly those having

TABLE IV

RELATIVE ACTIVITIES OF SUBSTRATES OF γ-CYSTATHIONASE
(L-HOMOSERINE DEHYDRATASE)

Substrate	Comp. rate	Reference
L-Homoserine	100	[69]
L-Cystathionine	80	[69]
L-Djenkolic Acid	30	[69]
L-(and *meso*)Lanthionine	11	[69]
L-Cysteine	5.4	[69]
L-Cystine	3.0	[74]
L-1,2-Diaminopropionate	4.0	[74]

thioether groups, are substrates for γ-cystathionase. A list of the comparative rates of decomposition of the sulfur amino acids with respect to L-homoserine is shown in Table IV. An unusual substrate recently discovered is 1,2-L-diaminopropionate [74]. This compound occurs in 2-N-oxalyl-L-1,2-diaminopropionate in *Lathyrus* peas.

The rate given for cysteine is undoubtedly spurious, since it has been shown by Cavallini *et al.* [75,76] and others [71,73] that the true substrate for the desulfuration of cysteine is the disulfide cystine, and that the initial reaction product is thiocysteine [HOOC—CH(NH$_2$)—CH$_2$S—SH].

Similarly, the enzymatic decomposition of djenkolic acid yields S-thiolmethylcysteine, which then spontaneously decomposes to cysteine, formaldehyde, and H$_2$S [76].

Various investigators have established that cystathionase and cysteine desulfhydrase activity is associated with a single protein [73,75–79].

Roisin and Chatagner [73] reported homocysteine desulfhydrase (EC 4.4.1.2) to be identical with γ-cystathionase.

C. Physical, Chemical, and Kinetic Properties of γ-Cystathionase (L-Homoserine Dehydratase)

A summary of the physical, chemical, and kinetic characteristics of γ-cystathionase (L-homoserine dehydratase) is given in Table V [66,74,80–84]. The purified enzyme is only moderately active, exhibiting a specific activity of 6 IU for DL-homoserine and 4.8 IU for L-cystathionine [66]. Optimum activity is at pH 8.0 [66]. Reports of others are of the same magnitude. The affinity of its major substrates for the enzyme is not high as shown by Michaelis constants of 20–30 mM for L-homoserine and 1–3.7 mM for L-cystathionine (Table V).

TABLE V

PHYSICAL, CHEMICAL, AND KINETIC PROPERTIES OF RAT LIVER
γ-CYSTATHIONASE (L-HOMOSERINE DEHYDRATASE)

Property	Substrate	Units	Value	Ref.
Specific activity	DL-Homoserine	μmole, min^{-1}, mg^{-1} 37°, pH 7.5	6	[66]
	L-Cystathionine	μmole, min^{-1}, mg^{-1} 37°, pH 7.5	4.8	[66]
Optimum activity	DL-Homoserine	pH	8.0	[66]
	L-Cystathionine	pH	8.0	[66]
Absorption peaks	—	nm	280, 427	[66]
Michaelis constants	L-Homoserine	mM	20	[66]
	L-Homoserine	mM	30	[80]
	L-Cystathionine	mM	1	[66]
	L-Cystathionine	mM	3.7	[80]
	L-Djenkolic acid	mM	7.6	[66]
Turnover number	L-Homoserine	moles, min^{-1}, mole^{-1}	6400	[66]
	L-Cystathionine	moles, min^{-1}, mole^{-1}	2340	[66]
Molecular weight	(by ultracentrifugation)	daltons	190,000	[66]
	(by ultracentrifugation)	daltons	185,000	[81]
	(by amino acid analysis)	daltons	170,500	[82]
	(by light scattering)	daltons	160,000	[81]
	(by chromatography)	daltons	160,000	[74,83]
Pyridoxal phosphate		moles per mole	4	[66]
Subunits (number and weight)		daltons	4, 44,000	[84]
		daltons	8, 22,000	[81]

1. MOLECULAR WEIGHT

The molecular weight of the holoenzyme is generally agreed to be between 160,000 and 190,000 daltons. The value of 160,000 now appears more probable since it has been obtained in a number of laboratories by chromatographic analysis and by light scattering. In experiments employing equilibrium centrifugation, A. Nagabhushanam and D. M. Greenberg (unpublished results) obtained a value of 170,000 daltons. However, the exact value for the molecular weight can only be obtained when the amino acid sequence of the enzyme is determined. The figure of 4 moles of pyridoxal-P per mole of holoenzyme remains unchallenged.

Studies on the subunit structure of the enzyme are contradictory. Deme et al. [84] dissociated the enzyme with sodium dodecyl sulfate and subjected the products to polyacrylamide-gel electrophoresis and ultracentrifugation. Their results lead to the interpretation of a tetrameric structure for the homoserine dehydratase, with a molecular weight of

44,000 daltons for the monomer. On the contrary, by a similar procedure, Churchich and Dupourque [81] reported the dissociation of the enzyme into subunits of 22,000 daltons molecular weight.

2. BINDING OF PYRIDOXAL 5-PHOSPHATE

In a study of the binding of pyridoxal-P by the apoenzyme, Oh and Churchich [83] concluded that there were two classes of binding sites, as determined from the increase in absorbance at 425 nm and by fluorescence spectroscopy. The affinity constant of the tighter binding site was estimated to be 7×10^5 M^{-1} and that of the weaker binding site 1.4×10^4 M^{-1}.

3. AMINO ACID COMPOSITION

Only one analysis has been published, made on the enzyme from rats treated with ethionine [82]. The results are reproduced in Table VI. The figures, as stated by the authors, are not all necessarily definitive because no estimate was made of the influence of the time of acidic hydrolysis, nor of the amide groups possibly present in the enzyme. The authors observed that cystathionase has a predominantly apolar amino acid composition.

4. EFFECT OF SH GROUPS

Matsuo and Greenberg [67] reported that cystathionase is a thiol enzyme, as determined by inhibition of enzyme activity by a variety of reagents that react with SH groups. The thiol groups of the enzyme have been more thoroughly studied in Chatagner's laboratory [85].* These investigators reported that cystathionase contains approximately 12 titratable SH groups per mole of native enzyme and 20 SH groups per mole of urea-denatured enzyme. Titration of the 12 SH groups of the native enzyme resulted in complete loss of enzyme activity. The SH groups were classified by these authors into four classes on the basis of their reactivity toward SH reagents. The presence of pyridoxal-P had no effect on the inhibition by PCMB. Homoserine was found to mask 4 SH groups and L-alanine, 8 SH groups.

Cystathionase is also strongly inhibited by heavy metal ions known to react with thiol groups, namely, Hg^{2+}, Cu^{2+}, and Cd^{2+}.

---

* More recently, Brown and De Foor [85a], found that four of the twelve −SH groups of this enzyme form stable bonds with DTNB, leading to inactivated enzyme that can be isolated. Cystathionine at high concentrations partially relieves the inhibition. Dithiothreitol decomposes the inactivated enzyme with restoration of 65–85% of the original enzyme activity.

TABLE VI

AMINO ACID COMPOSITION OF γ-CYSTATHIONASE[a]

Amino acid residue	Mean[b]	Standard deviation	Number of assays	Samples[c]	Number of residues/mole	
					Monomer	Dimer
Cys (as cysteic acid)	0.2036	0.0013	2	a	41	82
Asp	0.3004	0.0139	4	b + a	60	120
Thr	0.2279	0.0038	4	b + a	46	92
Ser	0.2624	0.0101	2	b	52	104
Glu	0.3923	0.0079	4	b + a	78	156
Pro	0.1882	0.0093	6	b + c + a	38	76
Gly	0.2605	0.0061	4	b + a	52	104
Ala	0.3609	0.0058	4	b + a	72	144
Val	0.2088	0.0031	4	b + a	42	84
Met	0.0777	0.0051	4	b + c	16	32
Ile	0.1509	0.0023	4	b + a	30	60
Leu	0.4297	0.0036	6	b + c + a	86	172
Tyr	0.0712	0.0039	3	b + c	14	28
Phe	0.1747	0.0011	3	b + c	35	70
Lys	0.2453	0.0026	4	b + a	49	98
His	0.1335	0.0136	2	b	27	54
Trp	0.0340	0.0014	2	c	7	14
Arg	0.1662	0.0053	4	b + a	33	66
Molecular weight					85,227	170,450

[a] From Loiselet and Chatagner [81].
[b] Values are expressed as moles of amino acids.
[c] a, Oxidized acid-hydrolyzed sample; b, acid-hydrolyzed sample; c, alkaline-hydrolyzed sample.

5. INHIBITORS OF γ-CYSTATHIONASE

The inhibition of this enzyme by certain cyclic amino compounds has important implications for the enzyme structure. The inhibitors include homocysteine, thiol acetone, α-aminobutyrolactone, cycloserine, and α-aminobutyro-γ-selenolacetone [86]. The L forms of these cyclic amines competitively inhibit the enzyme with respect to the substrates, homoserine and cystathionine. It is presumed that the amino groups of the inhibitors react with the formyl group of the pyridoxal-P, forming complexes through transaldimination, via a nucleophilic attack of the cyclic compounds on a protonated aldimine situated on the enzyme. The reaction is accompanied by a shift in the absorption maximum at 420 nm to 495–530 nm, depending on the cyclic compound. A similar shift in absorption is observed on reaction with pyridoxal phosphate alone.

From a study of the effect of pH on the dissociation constant of the homocysteine complex, Brown *et al.* [86] deduced that the enzyme contains an acidic group of pH 8.6 that participates in complex formation, and a more acidic group of pH 7.3 that does not, but is ionized in the complex.

Cystathionase is inhibited by reagents that react with the formyl group of pyridoxal-P and with SH groups. Among the former is hydroxylamine and the latter, *p*-hydroxymercuribenzoate. The site of action of the highly potent inhibitor, β-cyano-L-alanine, is not known.

6. Reversibility of the Cystathionine Cleavage Reaction

It would not be anticipated that cystathionine could be synthesized from the cleavage products, cysteine, α-ketobutyrate, and ammonia. However, formation from cysteine and homoserine would not be expected to have an insuperable free energy barrier. That this reaction is possible was indicated by an experiment of Matsuo and Greenberg [68] in which DL-[2-^{14}C]homoserine was incubated with L-cysteine and L-cystathionine in the presence of enzyme. The cystathionine isolated by chromatography was found to have become labeled with ^{14}C. This reaction has been confirmed by Chatagner *et al.* [87] and Wong *et al.* [87a], who observed the formation of cystathionine from L-homoserine and L-cysteine on incubation with dialyzed liver supernatants. Purified preparations of cystathionase, however, were much less effective than the crude supernatants.

D. Cystathionine γ-Cleavage Enzyme in Mold

Flavin and co-workers [88–92] have demonstrated the presence of and purified an enzyme from *Neurospora crassa* that predominantly catalyzes the γ-cleavage of cystathionine. A similar enzyme may be present in yeast, but not in bacteria [93].

The *Neurospora* γ-cleavage enzyme has been purified about 400-fold [89]. In many respects it is similar to the liver enzyme. Optimum activity is at about pH 7.4. It decomposes the disulfides, cysteine, lanthionine, and *meso*-cysteine. In contrast to the liver enzyme, the purified *Neurospora* enzyme has only slight activity on homoserine. Michaelis constants found were 0.52 mM for L-cystathionine-D-allocystathionine, 0.39 mM for lanthionine, and 0.032 mM for L-cysteine.

In the study of the mechanism of the γ-elimination reaction by cystathionase, Flavin and Slaughter [93] discovered that N-ethylmaleimide reacts with a transient intermediary precursor of α-ketobutyrate. The reaction product is formed from either L-homoserine or L-cystathionine when these compounds are decomposed enzymatically. Carbon derivatives that yield pyruvate did not react with N-ethylmalemide under the experimental conditions employed. Evidence has been obtained that the compound formed is α-keto-3-[3'-(N'-ethyl-2',5'-dioxopyrrolidyl)] butyric acid (see Chapter 11, Section II,A,1,c for formula).

The compound is postulated to be formed through a nucleophilic attack on the N-ethylmaleimide double bond by an intermediate with a carbanion character on the third carbon of the C_4 intermediate.

VII.   OXIDATION OF CYSTEINE TO CYSTINE

Very little is known about the enzymatic oxidation of cysteine to cystine. The reaction occurs readily nonenzymatically in the presence of oxygen and is catalyzed by a variety of heavy metal ions. For example, cupic ion in ammoniacal solution oxidizes cysteine according to Eqs. (12) and (13) where C represents cysteine [94]

$$4 \ CS^- + 2 \ Cu^{2+} \longrightarrow CS\!-\!SC + 2 \ CSCu + 4 \ H^+ \tag{12}$$

$$2 \ CSCu + 2 \ Cu^{2+} \rightleftharpoons CS\!-\!SC + 4 \ Cu^+ \tag{13}$$

The real intermediate in the oxidation of cysteine is postulated to be a copper chelate of cysteine represented by the following structural formula [95–98].

Enzymatic oxidation was observed to be catalyzed by cytochrome c and cytochrome oxidase [99], and more recently by enzyme systems involving NAD [100,101] in extracts of pea seeds and yeast. The enzyme appears to be different from the NAD glutathione reductase of these same materials. The nature of a cysteine-cystine oxidoreductase system may further be inferred to be analogous to the extensively studied oxidation of glutathionine by a variety of glutathione oxidoreductases (see Chapter 5).

The oxidation of free cysteine to cystine may be an uncommon event in intact cells. In the biosynthesis of proteins, cysteine is incorporated into peptide chains via cysteine-tRNA. Cystine disulfide bonds are formed in the subsequent process of forming the particular conformation of the protein. Free cystine can then result from the enzymatic hydroylsis of protein. The formation of disulfide bonds in protein can occur in numerous ways, including a variety of transsulfuration reactions.

REFERENCES

1. K. Schlossmann and F. Lynen, *Biochem. Z.* **328,** 591 (1957).
2. K. Schlossmann, J. Bruggemann, and F. Lynen, *Biochem. Z.* **336,** 258 (1962).
3. J. Bruggemann and M. Waldschmidt, *Biochem. Z.* **335,** 408 (1962).
4. J. Bruggemann, K. Schlossmann, M. Merkenscharger, and M. Waldschmidt, *Biochem. Z.* **335,** 392 (962).
5. M. Waldschmidt, *Biochem.Z.* **335,** 400 (1962.
6. J. A. Huovinen and B. E. Gustafsson, *Biochim. Biophys. Acta* **136,** 44 (1967).
7. I. S. M. De Issaly, *Biochim. Biophys. Acta* **151,** 473 (1968).
8. J. Giovanelli and S. H. Mudd, *Biochim. Biophys. Res. Commun.* **27,** 150 (1967).
9. J. F. Thompson and D. P. Moore, *Biochim. Biophys. Res. Commun.* **31,** 281 (1968).
10. J. L. Wiebers and H. R. Garner, *J. Biol. Chem.* **242,** 5644 (1967).
11. J. Giovanelli and S. H. Mudd, *Biochim. Biophys. Res. Commun.* **31,** 275 (1968).
12. E. C. Wolff, S. Black, and P. O. Downey, *J. Amer. Chem. Soc.* **78,** 5958 (1956).
13. N. M. Kredich and G. M. Tomkins, *J. Biol. Chem.* **241,** 4955 (1966).
14. N. M. Kredich and G. M. Tomkins, *in* "Organizational Biosynthesis" (H. J. Vogel, J. O. Lampen, and V. Bryson, eds.), p. 189. Academic Press, New York, 1967.
15. R. C. Clowes, *J. Gen. Microbiol.* **18,** 154 (1958).
16. J. Drefus and K. J. Monty, *J. Biol. Chem.* **238,** 1019 (1963).
17. N. M. Kredich, M. A. Becker, and G. M. Tomkins, *J. Biol. Chem.* **244,** 2428 (1969).
18. M. A. Becker, N. M. Kredich, and G. M. Tomkins, *J. Biol. Chem.* **244,** 2418 (1969).
19. N. M. Kredich, *J. Biol. Chem.* **246,** 3474 (1971).
19a. M. D. Hulanicka, N. M. Kredich, and D. M. Treiman, *J. Biol. Chem.* **249,** 867 (1974).
20. M. C. Jones-Mortimer, *Biochem. J.* **110,** 589 (1968).
21. A. E. Braunstein, E. V. Goryachinkova, and N. D. Lac, *Biochim. Biophys. Acta* **171,** 366 (1969).
22. A. E. Braunstein, E. V. Goryachinkova, E. A. Tolosa, I. H. Willhardt, and L. L. Yefromova, *Biochim. Biophys. Acta* **242,** 247 (1971).
23. F. Chapeville and P. Fromageot, *Biochim. Biophys. Acta* **49,** 328 (1961).
24. F. Chapeville and P. Fromageot, *Biochim. Biophys. Acta* **67,** 672 (1963).
25. A. Sentenac and P. Fromageot, *Biochim. Biophys. Acta* **81,** 289 (1964).
26. E. A. Tolosa, N. K. Chepurnova, R. M. Khomutov, and E. S. Severin, *Biochim. Biophys. Acta* **171,** 369 (1969).
27. V. du Vigneaud, *Harvey Lect.* **38,** (1942–1943).
28. V. du Vigneaud, "Trail of Research in Sulfur Chemistry and Metabolism and Related fields." Cornell Univ. Press, Ithaca, New York, 1952.
29. S. K. Shapiro and F. Schlenk, "Transmethylation and Methionine Biosynthesis." Univ. of Chicago Press, Chicago, Illinois, 1965.
30. A. Meister, "Biochemistry of the Amino Acids," Vol. 2, p. 757. Academic Press, New York, 1965.
31. D. M. Greenberg, *in* "Chemical Pathways of Metabolism" (D. M. Greenberg, ed), 1st ed. Vol. 2, p. 151. Academic Press, New York, 1954.
32. H. Tarver and C. L. A. Schmidt, *J. Biol. Chem.* **130,** 67 (1939).
33. V. du Vigneaud, G. B. Brown, and J. P. Chandler, *J. Biol. Chem.* **143,** 59 (1942).
34. W. P. Anslow, Jr., S. Simonds, and V. du Vigneaud, *J. Biol. Chem.* **166,** 35 (1946).
35. G. de la Haba and G. L. Cantoni, *J. Biol. Chem.* **234,** 603 (1959).
36. J. A. Duerre and F. Schlenk, *Arch Biochem. Biophys.* **124,** 422 (1962).
37. J. A. Duerre, *Arch. Biochem. Biophys.* **124,** 411 (1968).

38. W. A. Dodd and E. A. Cossins, *Arch. Biochem. Biophys.* **133**, 216 (1969).
39. S. K. Shapiro and D. J. Ehringer, *Biochim. Biophys. Acta* **177**, 67, (1969).
40. F. Binkley and V. du Vigneaud, *J. Biol. Chem.* **144**, 1507 (1942).
41. F. Binkley, W. P. Anslow, Jr., and V. du Vigneaud, *J. Biol. Chem.* **143**, 559 (1942).
42. F. Binkley, *J. Biol. Chem.* **155**, 39 (1944).
43. W. P. Anslow, Jr. and V. du Vigneaud, *J. Biol. Chem.* **234**, 603 (1959).
44. F. Binkley, *J. Biol. Chem.* **191**, 531 (1951).
45. S. M. Selim and D. M. Greenberg, *Biochim. Biophys. Acta* **42**, 211 (1960).
46. A. Nagabhushanam and D. M. Greenberg, *J. Biol. Chem.* **240**, 3002 (1965).
46a. P. N. Porter, M. S. Grishaver, and O. N. Jones, *Biochim. Biophys. Acta* **364**, 128 (1974).
47. R. A. Friedland and E. H. Avery, *J. Biol. Chem.* **239**, 3357 (1964).
48. E. Ishikawa, T. Ninagawa, M. Suda, H. Nakagawa, and I. Ishizuka, *J. Biochem. (Tokyo)* **55**, 401 (1964).
49. F. C. Brown, J. Mallady, and J. A. Roszell, *J. Biol. Chem.* **241**, 5220 (1966).
50. H. Nakagawa, H. Kimura, and S. Miura, *Biochem. Biophys. Res. Commun.* **28**, 359 (1967).
51. H. Nakagawa and H. Kimura, *J. Biochem. (Toyko)* **66**, 669 (1969).
52. H. Nakagawa and H. Kimura, *Biochem. Biophys. Res. Commun.* **32**, 208 (1968).
53. H. Kimura and H. Nakagawa, *J. Biochem. (Toyko)* **69**, 711 (1971).
54. S. Kashiwamata and D. M. Greenberg, *Biochim. Biophys. Acta* **212**, 488 (1970).
55. S. Kashiwamata, Y. Kotake, and D. M. Greenberg, *Biochim. Biophys. Acta* **212**, 501 (1970).
56. F. C. Brown and P. H. Gordon, *Can. J. Biochem.* **49**, 484 (1971).
57. S. H. Mudd, J. D. F. Finkelstein, F. Irreverre, and L. Laster, *J. Biol. Chem.* **240**, 4382 (1965).
58. S. Kashiwamata, *Anal. Biochem.* **42**, 293 (1971).
58a. S. Kashiwamata, *Brain Res.* **30**, 185 (1971).
58b. J. J. Volpe and L. Laster, *J. Neurochem.* **17**, 425 (1970).
58c. A. H. Datko, J. Giovanelli, and S. H. Mudd, *J. Biol. Chem.* **249**, 1139 (1974).
59. W. R. Carroll, G. W. Stacy, and V. du Vigneaud, *J. Biol. Chem.* **180**, 375 (1949).
60. F. Binkley, *J. Biol. Chem.* **155**, 39 (1944).
61. F. Binkley, G. M. Christensen, and W N. Jensen, *J. Biol. Chem.* **194**, 109 (1952).
62. F. Binkley and D. Okeson, *J. Biol. Chem.* **182**, 273 (1950).
63. F. Binkley and C. K. Olson, *J. Biol. Chem.* **182**, 273 (1950).
64. F. Binkley *in* "Methods in Enzymology" (C. P. Colowick and N. O. Kaplan, eds.), Vol. 2, p. 311. Academic Press, New York, 1955.
65. F. Binkley, *J. Biol. Chem.* **191**, 531 (1951).
65a. K. Snell, *Enzyme* **14**, 193 (1972–1973).
66. Y. Matsuo and D. M. Greenberg, *J. Biol. Chem.* **230**, 545 (1958).
67. Y. Matsuo and D. M. Greenberg, *J. Biol. Chem.* **230**, 561 (1958).
68. Y. Matsuo and D. M. Greenberg, *J. Biol. Chem.* **234**, 507 (1959).
69. Y. Matsuo and D. M. Greenberg, *J. Biol. Chem.* **234**, 516 (1959).
69a. D. M. Greenberg, P. Mastalerz, and A. Nagabhushanam, *Biochim. Biophys. Acta* **81**, 158 (1964).
70. T. E. Friedemann and G. E. Haugen, *J. Biol. Chem.* **147**, 516 (1943).
71. M. Flavin, *J. Biol. Chem.* **237**, 768 (1962).
72. A. Kato, M. Ogura, H. Kimura, T. Kauai, and M. Suda, *J. Biochem. (Toyko)* **59**, 34 (1966).
73. M.-P. Roisin and F. Chatagner, *Bull. Soc. Chim. Biol.* **51**, 481 (1969).

74. I. K. Mushahwar and R. E. Koeppe, *J. Biol. Chem.* **248**, 7407 (1973).
75. D. Cavallini, C. De Marco, and B. Mondovi, *Arch. Biochem. Biophys.* **87**, 281 (1960).
76. D. Cavallini, B. Mondovi, C. De Marco, and B. G. Mori, *Enzymologia* **21**, 11; **22**, 161 (1960).
77. D. Cavallini, B. Mondovi, and C. De Marco, *in* "Chemical and Biological Aspects of Pyridoxal Catalysis," p. 361. Pergamon, Oxford, 1963.
78. B. Mondovi, A. Scioscia-Santoro, and D. Cavallini, *Arch. Biochem. Biophys.* **101**, 363 (1963).
79. M. T. Costa, A. M. Wolf, and D. Giarnieri, *Enzymologia* **43**, 271 (1972).
80. A. Kato, M. Ogura, H. Kimura, T. Kawai, and M. Suda, *J. Biochem. (Toyko)* **59**, 34 (1966).
81. J. E. Churchich and D. Dupourque, *Biochem. Biophys. Res. Commun.* **46**, 524 (1972).
82. J. Loiselet and F. Chatagner, *Biochim. Biophys. Acta* **130**, 180 (1966).
83. K-J. Oh and J. E. Churchich, *J. Biol. Chem.* **248**, 7370 (1973).
84. D. Deme, G. De Billy, and F. Chatagner, *Biochem.* **46**, 1089 (1972).
85. D. Deme, O. Durieu-Trautmann, and F. Chatagner, *Eur. J. Biochem.* **20**, 269 (1971).
85a. F. C. Brown and M. C. De Foor, *Eur. J. Biochem.* **46**, 1317 (1974).
86. F. C. Brown, W. R. Hudgins, and J. A. Roszell, *J. Biol. Chem.* **244**, 2809 (1969).
87. F. Chatagner, M. Tixier, and C. Portemer, *FEBS (Fed. Eur. Biochem. Soc.) Lett.* **4**, 231 (1969).
87a. P. W. K. Wong, V. Schwarz, and G. M. Komrower, *Pediat. Res.* **2**, 149 (1968).
88. M. Flavin, *J. Biol. Chem.* **237**, 768 (1962).
89. M. Flavin and A. Segal, *J. Biol. Chem.* **239**, 2220 (1964).
90. M. Flavin and C. Slaughter, *Biochemistry* **3**, 885 (1964).
91. M. Flavin, *J. Biol. Chem.* **240**, PC2759 (1965).
92. D. Dulavier-Klutchko and M. Flavin, *J. Biol. Chem.* **240**, 2537 (1965).
93. M. Flavin and C. Slaughter, *Biochemistry* **5**, 1340 (1966).
94. M. Flavin, *Biochem. Biophys. Res. Commun.* **20**, 652 (1965).
95. M. Friedman, *in* "The Chemistry and Biochemistry of the Sulfhydryl Group," p. 38. Pergamon, Oxford, 1973.
96. I. M. Kolthoff and W. Stricks, *J. Amer. Chem. Soc.* **73**, 1728 (1951).
97. W. Stricks and I. M. Kolthoff, *J. Amer. Chem. Soc.* **73**, 1723 (1951).
98. D. Cavallini, D. De Marco, S. Dupré, and G. Rolilio. *Arch. Biochem. Biophys.* **130**, 354 (1969).
99. D. Keilin, *Proc. Roy. Soc., Ser. B* **106**, 418 (1930).
100. W. J. Nickerson and A. H. Romano, *Science* **115**, 676 (1952).
101. A. H. Romano and W. J. Nickerson, *J. Biol. Chem.* **208**, 409 (1954).

CHAPTER 13

Utilization and Dissimilation of Methionine

David M. Greenberg

I. INTRODUCTION

Many varied biological functions have been discovered for methionine. In the form of S-adenosylmethionine it is the methyl donor for a bewildering variety of acceptors. The subject has been reviewed by the author [1], the tRNA methyltransferases have been reviewed more recently by Kerr and Borek [2], and the methylation of specific classes of biologically important compounds are discussed in other volumes of this treatise (e.g., Volume III). Thus this topic will not be discussed in detail in this volume.

Abundant evidence has accumulated that protein synthesis is initiated by the formylation of methionine to N-formylmethionine. This subject has been reviewed in this treatise by Bagliono and Columbo (Volume IV, Chapter 21). An interesting utilization of methionine is in the biosynthesis of spermidine and spermine. A recent detailed review of this subject has been published by Tabor and Tabor. [3].

II. ROLE OF METHIONINE IN THE BIOSYNTHESIS OF SPERMIDINE AND SPERMINE

The precursor roles of putrescine (1,4-diaminobutane) and methionine in the formation of spermidine was demonstrated in *Escherichia coli* by Tabor *et al.* [4]. In the biosynthetic process. *S*-adenosylmethionine is decarboxylated and the aminopropyl group formed from methionine is transferred and coupled to 1,4-diaminobutane, yielding spermidine.

The reactions are catalyzed by two enzymes, *S*-adenosylmethionine decarboxylase and spermidine synthase (propylamine transferase). The same process of spermidine biosynthesis was subsequently shown to occur in bakers' yeast [5], liver [6], and prostatic fluid [7,8].

A. Formation of 1,4-Diaminobutane (Putrescine)

This diamine is formed by the decarboxylation of either L-ornithine or L-arginine followed by the hydrolysis of agmatine as shown in Fig. 1.

In animals diaminobutane apparently is formed only by the decarboxylation of ornithine, since arginine decarboxylase has not been found in animal tissue.

Decarboxylases for ornithine and arginine have been known for many years to be formed in bacteria when the organisms were grown in an acid medium (pH 5) containing the particular amino acid [9]. Although it was assumed initially that these decarboxylases were responsible for the formation of the diamines, it became apparent for many reasons that they did not function under normal physiological conditions.

The decarboxylases responsible for the formation of 1,4-diaminobutane utilized in spermidine and spermine synthesis are constitutive enzymes in *E. coli* with optimum activity at about pH 8.0 [10,11].

Only the biosynthetic orithine decarboxylase is formed when *E. coli* is grown in a minimal salts-glucose medium at neutral pH. The enzyme has been purified about 30-fold and has been shown to have an absolute requirement for pyridoxal-P [11].

The synthetic arginine decarboxylase is formed under similar incubation conditions. This decarboxylase has been purified 200-fold. It has an absolute requirement for pyridoxal-P and also for Mg^{2+}. Optimal activity is at pH 8.4 [12].

Escherichia coli extracts also contain agmatinase (EC 3.5.3.11), which catalyzes cleavage of agmatine to urea and diaminobutane.

Ornithine decarboxylase has also been partially purified from regenerating rat liver [12a].

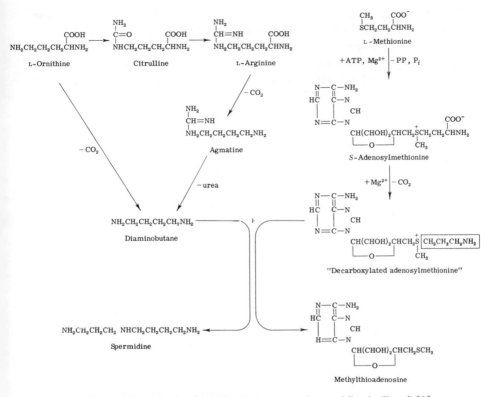

FIG. 1 Biosynthesis of 1,4-diaminobutane and spermidine in *E. coli* [3].

B. Spermidine Biosynthesis

The synthetic steps involve formation of S-adenosylmethionine, decarboxylation of the latter, and condensation of 1,4-diaminobutane with the propylamine moiety of the "decarboxylated" adenosylmethionine (Fig. 1).

Synthesis of S-adenosylmethionine is a well-studied reaction and is described in detail in other chapters of this volume.

1. S-ADENOSYLMETHIONINE DECARBOXYLASE (EC 4.1.1.50)

Decarboxylation of S-adenosylmethionine is catalyzed by the enzyme S-adenosylmethionine decarboxylase. This enzyme from *E. coli* has been purified to homogeneity [13,14]. The purified enzyme has a molecular weight of 113,000 daltons and consists of eight identical subunits of 15,000 daltons. The enzyme requires Mg^{2+} for activity. An interesting property of this enzyme is that the prosthetic group is pyruvate, not the usual pyridoxal-P. The K_m for adenosylmethionine is 0.09 mM and the V_{max} is 1.5 μmoles/minute/mg. Evidence for its SH nature is that it is inhibited by 8-hydroxymercuribenzoate and N-ethylmaleimide. Activity can be restored by addition of such SH compounds as 2-mercaptoethanol and dithiothreitol.

The rat ventral prostate decarboxylase [7,8] has been purified about 500-fold and the yeast decarboxylase about 100-fold [6]. The properties of the animal enzymes differ from the bacterial enzyme by being markedly stimulated by 1,4-diaminobutane and do not show a requirement for Mg^{2+}.

2. PROPYLAMINE TRANSFERASE

The final step in the formation of spermidine is catalyzed by the enzyme propylamine transferase. The propylamine transfer is linked with the reduction of the sulfonium ion of decarboxylated adenosylmethionine to the divalent sulfur of methylthioadenosine [3,15]. The enzyme has been highly purified from *E. coli*. It has a molecular weight of 73,000 daltons and is composed of two identical subunits of 36,000 molecular weight [3]. The purified enzyme was inhibited by SH reagents and by the reaction products. Optimal activity is at about pH 10.3.

Propylamine transferase has also been partially purified from rat prostate [16] and rat brain [17,18]. In rat brain the transferase was separated into two fractions, one active in the synthesis of spermidine and the other active in the synthesis of spermine.

C. Spermine Synthesis

Spermine is formed by the condensation of two molecules of 1,3-diaminopropane with putrescine [$NH_2(CH_2)_3NH(CH_2)_4NH(CH_2)_3NH_2$]. Propylamine transferases from most sources can catalyze the formation of spermine as well as spermidine. A transferase specific for spermine synthesis has been reported to occur in rat brain [17,18].

The enzymatic reactions for the degradation of spermidine have been reviewed by Tabor and Tabor [3].

III.   DISSIMILATION OF METHIONINE

Dissimilation of methionine can occur by a number of different pathways. An important pathway is through formation and cleavage of cysthathionine after demethylation of S-adenosylmethionine. Cleavage of cystathionine, catalyzed by γ-cystathionase, converts the C_4 chain of methionine to α-ketobutyrate and ammonia as described in Chapter 12. The α-ketobutyrate can be oxidatively decarboxylated to propionate and CO_2 as shown by Eq. (1). The pathway for the further dissimilation of propionate is well known.

$$CH_3CH_2COCOO^- + \tfrac{1}{2} O_2 \rightarrow CH_3CH_2COO^- + CO_2 \qquad (1)$$

Cleavage of S-adenosylmethionine to 5'-methylthioadenosine and α-amino-γ-butyrolactone has been shown to occur, catalyzed by enzyme preparations from *Aerobacter aerogenes* [19] and bakers' yeast [20]. Hydrolysis of α-amino-γ-butyrolactone yields homoserine, which also is decomposed to α-ketobutyrate and ammonia by cystathionase. The formation of homoserine has been reported upon injection of DL-[2-^{14}C]-methionine in mice [21]. The yield was very small.

Methionine may also be degraded directly to α-ketobutyrate, ammonia, and methylmercaptan. This type of reaction occurs in both bacterial and liver preparations. The findings of Canellakis and Tarver [22] suggest that the methionine is first converted to α-keto-γ-methiolbutyric acid, catalyzed by L-amino acid oxidase or transaminase. The α-keto-γ-methiolbutyrate is then further decomposed to methyl mercaptan and α-ketobutyrate.

A. Decomposition of Methionine via α-Keto-γ-methiolbutyrate

Methionine may be deaminated either by oxidation or transamination. This amino acid is a good substrate for various L-amino acid oxidases. A well-defined L-amino oxidase has been prepared from rat kidney [23]. This enzyme is a flavoprotein and the prosthetic group has been identified to be flavin mononucleotide.

L-Amino acid oxidases that could be involved in the deamination of methionine occur in various microorganisms, the best-defined ones being from *Neurospora crassa* [24] and *Proteus vulgaris* [25]. A general L-amino acid oxidase has also been highly purified from silk worm eggs [26].

The richest sources of L-amino acid oxidases have been found to be snake venoms. Several of these oxidases have been crystallized. These, however, are not likely to be involved in the physiological disposition of methionine.

Methionine also is moderately active substrate for crude transaminase preparations from heart, liver, and kidney with α-ketoglutarate serving as the amino group acceptor.

Formation of α-ketobutyrate and methyl mercaptan from methionine has been observed in pseudomonads [27] and *E. coli* [28], as well as liver [22]. As mentioned above, this conversion is probably preceded by the deamination of the methionine.

REFERENCES

1. D. M. Greenberg, *Advan. Enzymol.* **25**, 395 (1963).
2. S. J. Kerr and E. Borek, *Advan. Enzymol.* **36**, 1 (1972).
3. H. Tabor and C. W. Tabor, *Advan. Enzymol.* **36**, 203 (1972).
4. H. Tabor, S. M. Rosenthal, and C. W. Tabor, *J. Biol. Chem.* **233**, 907 (1958).
5. J. Jänne, H. G. Williams-Ashman, and A. Schenone, *Biochem. Biophys. Res. Commun.* **43**, 1362 (1971).
6. J. Jänne, *Acta Physiol. Scand., Suppl.* **300**, 1 (1967).
7. A. E. Pegg and H. G. Williams-Ashman, *J. Biol. Chem.* **244**, 682 (1969).
8. J. Jänne and H. G. Williams-Ashman, *Biochem. Biophys. Res. Commun.* **42**, 222 (1971).
9. E. F. Gale, *Advan. Enzymol.* **6**, 1 (1946).
10. D. R. Morris and A. B. Pardee, *J. Biol. Chen.* **241**, 3129 (1966).
11. D. R. Morris, W. H. Wu, D. A. Applebaum, and K. L. Koffron, *Ann. N.Y. Acad. Sci.* **171**, 968 (1970).
12. W. H. Wu and D. R. Morris, *Bacteriol. Proc.* p. 133 (1970).
12a. S. J. Friedman, K. V. Halpern, and E. S. Canellakis, *Biochim. Biophys. Acta* **261**, 181 (1972).
13. R. B. Wickner, C. W. Tabor, and H. Tabor, *J. Biol. Chem.* **245**, 2132 (1970).
14. R. B. Wickner, C. W. Tabor, and H. Tabor, *in* "Methods in Enzymology" (H. Tabor and C. W. Tabor, eds.). Vol. 17B, p. 647. Academic Press, New York, 1971.
15. C. W. Tabor, *in* "Methods in Enzymology" (C. P. Colowick and N. O. Kaplan, eds.), Vol. 5, p. 761. Academic Press, New York, 1962.
16. J. Jänne, A. Schenone, and H. G. Williams-Ashman, *Biochem. Biophys. Res. Commun.* **42**, 758 (1971).
17. P. Hannonen, A. Raina, and J. A. Khawaja, *Scand. J. Clin. Lab. Invest.* **27**, Suppl. 116, 4 (1971).
18. A. Raina and P. Hannonen, *Acta Chem. Scand.* **24**, 3061 (1970).
19. S. K. Shapiro and A. N. Mather, *J. Biol. Chem.* **233**, 631 (1958).
20. S. H. Mudd, *J. Biol. Chem.* **234**, 87 and 1784 (1959).
21. Y. Matsuo and D. M. Greenberg, *J. Biol. Chem.* **215**, 547 (1955).
22. E. S. Canellakis and H. Tarver, *Arch. Biochem. Biophys.* **42**, 387 and 446 (1953).
23. M. Blanchard, D. E. Green, V. Nocito, and S. Ratner, *J. Biol. Chem.* **155**, 421 (1944); **161**, 583 (1945).
24. P. S. Thayer and N. H. Horowitz, *J. Biol. Chem.* **192**, 755 (1951).
25. P. K. Stumpf and D. E. Green, *J. Biol. Chem.* **153**, 387 (1944).
26. S. Kotaka, *J. Gen. Physiol.* **46**, 1087 (1962–63).
27. R. E. Kallio and A. D. Larson, *in* "Amino Acid Metabolism" (W. D. McElroy and B. Glass, eds.), p. 616. Johns Hopkins Press, Baltimore, Maryland, 1955.
28. K. Ohigashi, A. Tsunetoshi, and K. Ichihara, *Med. J. Osaka Univ.* **2**, 111 (1951).

CHAPTER 14

Oxidative Metabolism of Cysteine and Cystine in Animal Tissues

Thomas P. Singer

I.   INTRODUCTION

The intermediary metabolism of sulfur containing amino acids is exceedingly complex and several aspects of it remain obscure. To compound the difficulties, there appears to be more divergence in the catabolic reactions of cysteine and cystine in various forms of life than is the case with most other amino acids. There appears to be little point in presenting the panorama of diverse metabolic reactions that these two amino acids have been shown, or are believed, to enter in different forms of life, all the more since a survey of their metabolism was published a few years ago [1]. Other aspects of the metabolism of cysteine and cystine, including accounts of the earlier literature not covered in the reference quoted, will be found in earlier reviews and monographs [2–5] and in other chapters in this volume.

The present chapter, therefore, concentrates on the oxidative metabolism of cystine and cysteine in animal tissues, with emphasis on the initial events in the conversion of cysteine to inorganic sulfate and pyruvate.

II. CYTOPLASMIC OXYGENATION OF CYSTEINE TO CYSTEINESULFINiC ACID

By 1954 it seemed clear that the main oxidative pathway of cysteine in bacteria and animal tissues involves cysteinesulfinic acid as an intermediate [4,6–10], but the reactions which lead from cysteine to the sulfinic acid were a matter for speculation. The various hypotheses advanced were superseded by the demonstration [11] that in rat liver cytoplasm cysteine is relatively rapidly converted to cysteinsulfinic acid by a reaction that had all the earmarks of an oxygenase, since it had an absolute requirement for O_2 and was stimulated by NADPH and Fe^{2+}, as is true of many reactions catalyzed by monooxygenases ("mixed function oxidases"). Moreover, in an analogous reaction catalyzed by a different enzyme, Cavallini et al. [12] demonstrated that the conversion of cysteamine to hypotaurine (the deaminated form of cysteinesulfinic acid) involves the incorporation of at least 1.5 atoms of oxygen, originating from molecular O_2, rather than water. Hence, by 1966 it was clear that, contrary to earlier speculation, the conversion of cysteine to cysteinesulfinic acid in liver does not involve a series of dehydrogenations with cysteinesulfinic acid as the first product, nor a dehydrogenation followed by a dismutation, but oxygenase action. The postulate of two successive monooxygenase reactions in the conversion of cysteine to cysteinesulfinic acid [11] proved to be incorrect, however, when it was demonstrated [13] by the use of $^{18}O_2$ that all of the oxygen atoms in the sulfinate group of cysteinesulfinic acid arise from molecular O_2, none from water, and that a single dioxygenase is responsible for the overall reaction.

The first evidence for the existence of an oxygenase that converts L-cysteine to L-cysteinesulfinic acid was presented by Sörbo and Ewetz [11]. They reported the net accumulation of cysteinesulfinic acid on aerobic incubation of cysteine with rat liver cytoplasm, in the presence of NADPH, Fe^{2+}, and hydroxylamine. The latter compound served to inhibit decarboxylative and transaminative removal of the cysteinesulfinate formed (both rapid reactions), as well as alternate enzymatic pathways for cysteine metabolism, catalyzed by pyridoxal phosphate enzymes. They also found that NADH could replace NADPH. From the requirement for both O_2 and a reduced pyridine nucleotide, Sörbo and Ewetz [11] speculated that a monooxygenase (mixed function oxidase) might be involved, but, as noted above, this conclusion proved to be incorrect.

About the same time Wainer [14], using [^{35}S]cysteine and carrier cysteinesulfinic acid, documented the net conversion of the former to

[^{35}S]cysteinesulfinic acid in preparations from rat liver cytoplasm in the absence of hydroxylamine. No cofactor requirements were noted, however, in this study.

In extending their studies, Ewetz and Sörbo reported [15] that the cytoplasmic oxygenase appeared to be specific for L-cysteine, that Co^{2+} cannot replace Fe^{2+} as a cofactor, and that a heat labile factor present in mitochondria and microsomes stimulated the activity of the cytoplasmic enzyme. These workers, as well as Wainer [14], were unsuccessful in purifying the oxygenase, presumably because of its instability.

Prior to the appearance of the Ewetz and Sörbo's full paper [15], a systematic study of the conversion of L-cysteine to L-cysteinesulfinic acid was undertaken in the author's laboratory. Stimulation of the reaction by reduced pyridine nucleotides and Fe^{2+} was confirmed, but the detailed findings [16] were at variance with earlier results. Thus, NADH and NADPH were found to be equally effective and maximal stimulation was noted at much lower concentration than in the studies of Ewetz and Sörbo [15]. More importantly, NADH and NADPH did not appear to function as hydrogen donors, as in the action of monooxygenases; in fact, the requirement for reduced pyridine nucleotides was not absolute. In preparations rigorously freed from NADH and NADPH the maximal stimulation by these compounds was only of the order of twofold [16]. Although added NADH disappeared from the reaction mixture during the assay, this oxidation was unrelated to the oxygenation of L-cysteine. Moreover, cysteinesulfinate formation continued long after all added NADH had been destroyed [13]. In accord with this, Ewetz and Sörbo [15] also noted that the oxidation of NADH to NAD is not stoichiometric with the amount of cysteine oxygenated. It is clear, therefore, that reduced pyridine nucleotide is not a cosubstrate but, perhaps, some type of modifier of the enzyme.

It should be noted that there is a discrepancy in the literature as to whether the oxidized forms of the nucleotides stimulate the activity. Sörbo and Ewetz [11] stated that NAD and NADP stimulate only in undialyzed liver extracts, in which systems for reducing these compounds are present, but the oxidized nucleotides are inactive after dialysis. Lombardini et al. [16] also stated that NAD and NADP are inactive. In contrast, Yamaguchi et al. [17], working with straight liver extracts, stated that both NAD and NADP stimulate and yield even greater activity than the reduced forms. Unfortunately, it is difficult to accept this report because of the inappropriate assay conditions used by the latter workers [aged enzyme preparations, omission of the cofactor (cf. below), and inhibitory concentrations of Fe^{2+} and of cysteine].

The role of Fe^{2+} in the action of the enzyme also remains unexplained.

Both Lombardini *et al.* [16] and Ewetz and Sörbo [15] found that the stimulation is only about twofold, but while the former group noted extensive inhibition at concentrations of added Fe^{2+} in excess of the optimum, the latter workers did not observe this. Although the enzyme functions at about half maximal rate in the absence of added iron, even after gel exclusion, this may be due to tightly bound iron in the enzyme, for *o*-phenanthroline inhibits the activity completely. A number of other cations and Fe^{3+} did not stimulate the activity. Thus Fe^{2+} may be a genuine requirement for the action of this enzyme, as it is for several other dioxygenases [18]. Concentrations of L-cysteine and of O_2 beyond those required for maximal activity (~ 2.5 mM and 20 vol%, respectively) strongly inhibited the enzyme [16].

The report that a factor in liver mitochondria and microsomes stimulates the activity [15] could not be confirmed. Lombardini *et al.* [16] discovered, however, that an additional heat stable cofactor present in the cytoplasm is required for activity. Extensive stimulation by the cofactor was noted only in gel-excluded enzyme preparations and had thus been overlooked in earlier studies. The identity of this compound has not been established, although considerable effort was expended in attempts at isolation and characterization [16,19]. The cofactor is not replaced by any of a large number of coenzymes and cofactors of oxygenases and oxidizing enzymes, tested alone or in combination of several of these, including Na_2S. The latter finding is of particular interest since S^{2-} is required for the activity of the closely related enzyme cysteamine oxygenase [20]. The cofactor does not seem to be a common cellular constituent, as judged by its unusual distribution pattern [16,19]. It has been found only in boiled extracts of rat liver and, to a much less extent, beef liver cytoplasm, but appears to be absent from pig liver extracts.

Purification of the compound has been hampered by its instability on storage even at very low temperatures and in the lyophilized state [19]. The evidence accrued suggests that it has a molecular weight of 750, that it is probably an acidic nucleotide, and that it is rather stable to acid but not to alkaline conditions.

The role of this compound in the action of the oxygenase remains as obscure as that of reduced pyridine nucleotides. In contrast to the behavior of NADH (cf. above), however, utilization of the cofactor during the oxygenase assay is dependent on the presence of cysteine. There is suggestive evidence [16] that L-cysteine may form an adduct with the cofactor.

Little progress has been made in purifying the enzyme because of its pronounced lability on storage and to many of the procedures commonly used in enzyme purification [14–16,19]. Despite the absence of a satisfactory purification procedure, it appears quite likely from isoelectric focusing and gel chromatography studies [16,19] that a single enzyme is involved. By means of $^{18}O_2$ and $H_2^{18}O$ it has been shown [13] that both oxygen atoms in the sulfinic acid originate from molecular oxygen, none from water. Thus the enzyme is a dioxygenase and its reaction may be represented as follows:

$$\text{L-Cysteine} + O_2 \xrightarrow[\text{cofactor}]{Fe^{2+},NAD(P)H} \text{L-cysteinesulfinic acid}$$

In accord with this, no intermediate was detected when the enzymatic reaction was carried out with [3-^{14}C]cysteine as substrate [16].

As discussed in Section IV, besides the cytoplasmic oxygenation of cysteine to cysteinesulfinate, there appears to be a mitochondrial pathway for the oxidation of cysteine not involving cysteinesulfinate as an intermediate. In addition, cysteine may also be oxidized to cystine, as well as entering anaerobic reactions. With cysteine as a branching point for its own catabolism, it is not unreasonable to expect that some of the pathways are subject to metabolic regulation. There is, in fact, a report [17] that cytoplasmic cysteine oxygenase activity may be induced in rats by injection of hydrocortisone, nicotinamide, or cysteine. The effects noted appear to be complex, not subject to any simple interpretation, and the findings further complicated by the unsatisfactory assay conditions used for the measurement of cysteine oxygenase, as already discussed above. It appears that all three agents induce a major rise in hepatic cysteine oxygenase activity in 1–3 hours. The effect is transient with hydrocortisone and cysteine, but not with nicotinamide. The effect of cysteine does not appear to be related to the fact that it is a substrate of the enzyme, for its produces a similar rise in tyrosine transaminase activity. For the same reason the effect of nicotinamide seems to be unrelated to the action of reduced pyridine nucleotides on the enzyme. Yamaguchi et al. [17] believed that the mechanism by which nicotinamide and cysteine increase hepatic cysteine oxygenase activity is due, at least in part, to activation of the preexisting enzyme or inhibition of its degradation, rather than to increased synthesis, a conclusion that appears to be at variance with their observation that cycloheximide prevents the effect of cysteine partly and that of nicotinamide entirely. In adrenalectomized rats cysteine oxygenase activity is not significantly lowered; yet hydrocortisone injection causes a much more marked rise in activity than in normal rats, but the effects of nicotinamide and cysteine injection are minimal or nil.

In the writer's opinion these observations merit continuation and extension, using more unambiguous conditions for assaying the level of oxy-genase activity.

III.   METABOLIC FATE OF CYSTEINESULFINIC ACID

Cysteinesulfinate arising from the action of cytoplasmic cysteine ox-ygenase may be, in part, decarboxylated in the cytoplasm to hypotaurine [21], but the major part of the degradation appears to proceed in the mi-tochondria. Once cysteinesulfinate enters the mitochondria it is rapidly transaminated with α-ketoglutarate or oxalacetate to yield β-sulfinyl-pyruvate [8,10]:

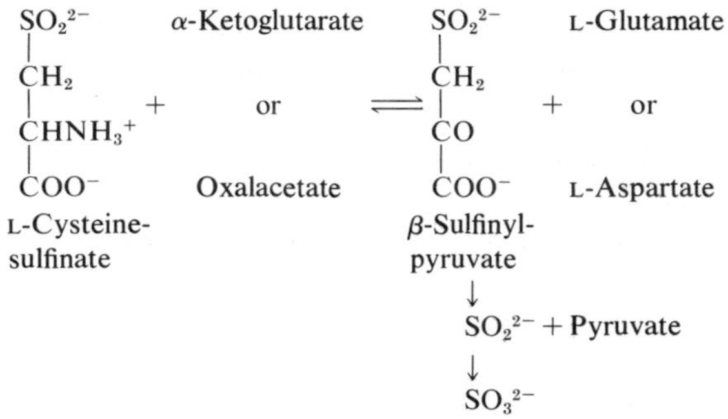

$$
\begin{array}{llll}
SO_2^{2-} & \alpha\text{-Ketoglutarate} & SO_2^{2-} & \text{L-Glutamate} \\
| & & | & \\
CH_2 & & CH_2 & \\
| \quad + \quad \text{or} & \rightleftharpoons & | \quad + \quad \text{or} \\
CHNH_3^+ & & CO & \\
| & & | & \\
COO^- & \text{Oxalacetate} & COO^- & \text{L-Aspartate} \\
\text{L-Cysteine-} & & \beta\text{-Sulfinyl-} & \\
\text{sulfinate} & & \text{pyruvate} & \\
\end{array}
$$

$$
\downarrow
$$
$$
SO_2^{2-} + \text{Pyruvate}
$$
$$
\downarrow
$$
$$
SO_3^{2-}
$$

β-Sulfinylpyruvate is spontaneously desulfinated to pyruvate and sulfite; the latter is then oxidized to sulfate either enzymatically or by reaction with superoxide.

The conversion of cysteinesulfinate to pyruvate and sulfate was first decribed for *Proteus vulgaris* [6,7] and was later demonstrated to be the predominant catabolic pathway of this amino acid in a variety of animal tissues [8–10]. A more detailed discussion of these reactions may be found in earlier reviews [4,5].

IV.   MITOCHONDRIAL OXIDATION OF CYSTEINE TO SULFATE

In addition to the oxidative pathway of cysteine metabolism initiated by the cytoplasmic oxygenase, the existence of a second pathway for the oxidation of cysteine to SO_4^{2-} in rat liver has been reported by Wainer [22,23]. This pathway is thought to occur entirely within the mi-tochondria and not to involve cysteinesulfinic acid as an intermediate.

The mechanism of the mitochondrial oxidation of cysteine is poorly understood and although nearly ten years have elapsed since the initial report of its existence [22], nothing is known about the enzymes involved or about the intermediates in the conversion. Wainer's observations and conclusions may be summarized as follows.

The mitochondrial system requires only cysteine and oxygen for activity in fresh preparations but in frozen-thawed mitochondria, in which SO_4^{2-} production from cysteine is impaired, either oxidized or reduced glutathione stimulate [23]. In experiments using [^{35}S]cysteine, sulfate and cystine were the only products containing significant radioactivity detected, but cystine was not an intermediate, since its oxidation to sulfate occurred only after reduction to cysteine. When ^{14}C-labeled cysteine was used as substrate, pyruvate appeared to be the initial product arising from the carbon skeleton. H_2S, β-mercaptopyruvate, and cysteinesulfinate were ruled out as intermediates in the oxidation. The latter conclusion was based on the observations [22] that the presence of unlabeled cysteinesulfinate during the oxidation of [^{35}S]cysteine did not materially lower the yield of [^{35}S]sulfate and that the specific activity of the reisolated cysteinesulfinate was much lower than that of the sulfate produced. A complication in this experiment is that no keto acid (needed for transamination) was present. When α-ketoglutarate was added along with cysteinesulfinate during the oxidation of [^{35}S]cysteine, the yield of [^{35}S]sulfate was halved, however. This observation is difficult to reconcile with the conclusion that cysteinesulfinate is not an intermediate in the mitochondrial oxidation of cysteine to sulfate, unless one assumes that free sulfite is an intermediate in the reaction, as also in the conversion of cysteinesulfinate to sulfate, and, further, that the oxidation of sulfite is rate limiting in both processes.

Wainer also compared the rates of oxidation of [^{35}S]cysteine to sulfate in various subcellular fractions of rat liver and concluded that the mitochondrial pathway predominates [23]. The data do not justify this conclusion, in the reviewer's opinion, for several reasons. First, since none of the cofactors required for cysteine oxygenase was added, only a fraction of the potential activity of this enzyme could have been measured. Second, it is well known that the metabolic conversion of the cysteinesulfinate formed by the cytoplasmic enzyme to inorganic sulfate occurs in the mitochondria. Thus, if separation of the subcellular fractions had been complete, little or no conversion of cysteine to sulfate should have been observed in the cytoplasmic fraction. Third, conversion of cysteinesulfinate to sulfate requires the presence of α-keto acids but none was added in the experiments discussed. Finally, no provision was made to assure that the activity of sulfite oxidase was not rate limiting. All these considerations tend to obscure the contribution of the cysteinesulfinate pathway to sulfate production in Wainer's experiments

[23], while overemphasizing that of the mitochondrial system, which is reported not to require added cofactors or other cell fractions.

A study initiated some years ago in the author's laboratory* attempted to clarify some of these questions concerning the mitochondrial system. It was found that the assay system used in Wainer's study (precipitation of the [^{35}S]sulfate produced with Ba^{2+} in the presence of carrier sulfate) is subject to considerable error because of coprecipitation of radioactivity other than inorganic sulfate. Among several alternate assays elaborated, separation of the [^{35}S]sulfate by high voltage electrophoresis seemed most satisfactory. With the aid of this method the fact that cysteinesulfinate is not an intermediate in the conversion of [^{35}S]cysteine to [^{35}S]sulfate in rat liver mitochondria was readily confirmed. The conclusion that reduced glutathione stimulates the activity only in damaged mitochondria was not confirmed, since major stimulation was noted in fresh mitochondria as well.

Since o-phenanthroline and an atmosphere of 100% O_2 each inhibit the activity of the cytoplasmic oxygenase almost entirely [16], whereas they inhibit the mitochondrial pathway only slightly,* it was thought that the respective contribution of the two pathways of sulfate formation in intact liver cells could be assessed from the effect of these inhibitors on the oxidation of [^{35}S]cysteine in liver minces. Surprisingly, it was found that the rate of sulfate formation from cysteine in intact cells is far greater than the sum of the known activities of the cytoplasmic and mitochondrial systems measured in subcellular fractions. Moreover, in preparations from both normal and starved rats much more sulfate arose from endogenous substrate than from added cysteine. These findings preclude a decision as to the relative quantitative importance of the two known pathways of sulfate production at this time and suggest either that in isolated cell fractions under the conditions hitherto used the full activity of one of the two systems is not measured or that, possibly, another pathway of sulfate production predominates in intact cells.

V.   CATABOLISM OF CYSTINE

Although cystine may be directly converted to S-(carboxy-methylthio)cysteine and to S-(2-hydroxy-2-carboxyethylthio)cysteine in animal tissues [24], the main route of the catabolism of cystine in animals appears to be initiated by reduction to cysteine, followed by ox-

---

* D. R. Biggs and J. I. Salach, unpublished data (1968).

idative degradation. As already noted, cystine is not a substrate for either the cytoplasmic dioxygenase or for the mitochondrial cysteine oxidase system. There is ample evidence that dietary cystine is rapidly reduced during and after absorption [25] and that the intracellular ratio of cysteine:cystine is very high. In considering the oxidative metabolism of cystine in animals the initial question is, then, the mechanism of its reduction to cysteine.

Although the existence of pyridine nucleotide linked cystine reductases in yeast and higher plants has been well documented [26–28], with the exception of the clothes moth [29], no such enzyme has been found in animal tissues. The reduction is thought to occur, instead, by a thiol–disulfide exchange with reduced glutathione (GSH), followed by regeneration of the GSH by NADH-glutathione reductase:

$$\text{Cystine} + 2\ \text{GSH} \rightleftharpoons 2\ \text{cysteine} + \text{GSSG}$$

$$\text{GSSG} + \text{NADPH} + \text{H}^+ \rightarrow 2\ \text{GSH} + \text{NADP}^+$$

The thiol–disulfide exchange can, of course, occur nonenzymatically, but in extracts of a variety of rat tissues an enzyme catalyzing this reaction has been described [30,31]. The enzyme is localized in lysosomal and soluble fractions of the cell, at least in the case of intestinal mucosa [31]. Similar enzymes have been shown to be present in bovine kidney [32] and human liver [33].

Eriksson and Mannervik [34] have purified the enzyme from rat liver and suggested that its main catalytic activity is the reduction of the mixed disulfide of cysteine and glutathione by GSH:

$$\text{CySSG} + \text{GSH} \rightleftharpoons \text{CySH} + \text{GSSG}$$

According to these authors the reduction of cystine by glutathione probably occurs in two steps:

$$\text{CySSCy} + \text{GSH} \rightarrow \text{CySH} + \text{CySSG}$$

$$\text{CySSG} + \text{GSH} \rightarrow \text{CySH} + \text{GSSG}$$

VI. METABOLIC ORIGIN OF TAURINE

Although the individual metabolic steps in the conversion of cysteine and cystine to taurine in higher animals remain poorly defined, the possible pathways should be briefly mentioned, since taurine is an established product of the catabolism of these amino acids and a constituent of bile acids. The possible reactions in adult mammals are summarized in Scheme 1 below.

Although massive doses of injected cysteic acid give rise to taurine in the rat, the reaction is apparently slow and labeled cysteic acid has not been detected in the livers of rats injected with [35S]cysteine, under conditions where significant formation of [35S]taurine and [35S]hypotaurine was observed [35]. Moreover, cysteic acid is not a major product of oxidative cysteine metabolism, since the transamination and decarboxylation of cysteinesulfinate are reactions far too rapid to permit extensive oxidation of this amino acid to cysteic acid to occur. While the formation of taurine from cysteic acid is thus probably not a significant pathway in adult animals, in embryos it appears to occur readily. Chapeville and Fromageot [36] reported that in chick embryos cysteic acid is rapidly decarboxylated to taurine and Tolosa et al. [37] characterized a pyridoxal enzyme, cysteine lyase, present in the yolk sac, which catalyzes the formation of cysteic acid from cysteine:

$$Cysteine + SO_3^{2-} \rightarrow cysteic\ acid + H_2S$$

On the other hand, in contrast to adult mammals, in chick embryos hypotaurine is not a precursor of taurine [36].

In adult rodents [35S]hypotaurine is rapidly converted to [35S]taurine and to $^{35}SO_4^{2-}$ [38], but the enzyme responsible for the oxidation of hypotaurine has not been isolated. There are two well-defined enzymes for the formation of hypotaurine, however. One is cysteinesulfinate decarboxylase [21,39], a pyridoxal enzyme. The importance of this en-

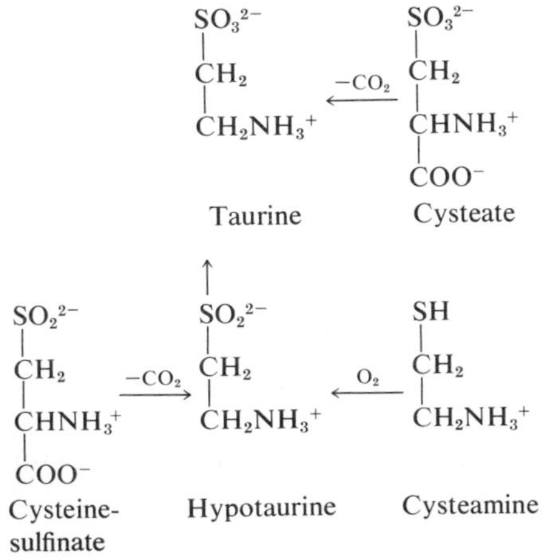

SCHEME 1. Possible sources of taurine in animal tissues.

zyme in the formation of taurine *in vivo* is supported by the finding that vitamin B_6-deficient rats do not excrete taurine, whereas rats on a normal diet do [40]. The alternate route of hypotaurine formation is by oxygenation of L-cysteamine.

Although the metabolic origin of cysteamine remains ill defined, an enzyme of wide distribution is present in animals for the conversion of cysteamine to hypotaurine [41]. Cysteamine oxygenase has been obtained in homogeneous form and its mechanism of action investigated by Cavallini and co-workers [12,42,43]. As with cysteine oxygenase [13], cysteamine oxygenase is also a dioxygenase [12] but neither enzyme acts on the substrate of the other. The molecular weight is 83,000 and the enzyme contains 1 gm-atom of nonheme iron [42]. The action of cysteamine oxygenase requires a cofactor; Na_2S, colloidal S, hydroxylamine, and methylene blue seem to fulfill this requirement [42].

REFERENCES

1. E. Kun, *in* "Metabolic Pathways" (D. M. Greenberg, ed.), 3rd ed., Vol. 3, p. 375. Academic Press, New York, 1969.
2. S. Black, *Annu. Rev. Biochem.* **32,** 399 (1963).
3. G. A. Maw, "The Biochemistry of Sulfur Containing Amino Acids." *Chem. Soc., Annu. Rep.* **63,** 639 (1966).
4. T. P. Singer and E. B. Kearney, *in* "Amino Acid Metabolism" (W. D. McElroy and B. Glass, eds.), p. 558. Johns Hopkins Press, Baltimore, Maryland, 1955.
5. E. Kun, *in* "Metabolic Pathways" (D. M. Greenberg, ed.), 2nd ed., Vol. 2, p. 237. Academic Press, New York, 1961.
6. E. B. Kearney and T. P. Singer, *Biochim. Biophys. Acta* **8,** 698 (1952).
7. E. B. Kearney and T. P. Singer, *Biochim. Biophys. Acta* **11,** 236 (1953).
8. T. P. Singer and E. B. Kearney, *Biochim. Biophys. Acta* **11,** 570 (1954).
9. F. Chatagner, B. Bergeret, T. Séjourné, and C. Fromageot, *Biochim. Biophys. Acta* **9,** 340 (1952).
10. T. P. Singer and E. B. Kearney, *Arch. Biochem. Biophys.* **61,** 397 (1956).
11. B. Sörbo and L. Ewetz, *Biochem. Biophys. Res. Commun.* **18,** 359 (1965).
12. D. Cavallini, R. Scandurra, and F. Monacelli, *Biochem. Biophys. Res. Commun.* **24,** 185 (1966).
13. J. B. Lombardini, T. P. Singer, and P. Boyer, *J. Biol. Chem.* **244,** 1172 (1969).
14. A. Wainer, *Biochim. Biophys. Acta* **104,** 405 (1965).
15. L. Ewetz and B. Sörbo, *Biochim. Biophys. Acta* **128,** 296 (1966).
16. J. B. Lombardini, P. Turini, D. R. Biggs, and T. P. Singer, *Physiol. Chem. Phys.* **1,** 1 (1969).
17. K. Yamaguchi, S. Sakakibara, K. Koga, and I. Ueda, *Biochim. Biophys. Acta* **237,** 502 (1971).
18. O. Hayaishi, *in* "Biological Oxidations" (T. P. Singer, ed.), p. 581. Wiley (Interscience), New York, 1968.
19. J. B. Lombardini, Ph.D. thesis, University of California, San Francisco (1968).
20. D. Cavallini, R. Scandurra, and C. DeMarco, *J. Biol. Chem.* **238,** 2999 (1963).
21. B. Bergeret and F. Chatagner, *Biochim. Biophys. Acta* **9,** 141 (1952).

22. A. Wainer, *Biochem. Biophys. Res. Commun.* **16,** 141 (1964).
23. A. Wainer, *Biochim. Biophys. Acta* **141,** 466 (1967).
24. K. Kobayashi, *Physiol. Chem. Phys.* **2,** 455 (1970).
25. P. C. Jocelyn, "Biochemistry of the SH Group." Academic Press, New York, 1972.
26. W. J. Nickerson and A. H. Romano, *Science* **115,** 676 (1952).
27. A. H. Romano and W. J. Nickerson, *J. Biol. Chem.* **208,** 409 (1954).
28. L. W. Mapson, *Biochem. J.* **55,** 714 (1953).
29. R. F. Powning and H. Irzykiewicz, *Nature (London)* **184,** 1230 (1959).
30. P. L. Wendell, *Biochim. Biophys. Acta* **159,** 179 (1968).
31. B. States and S. Segal, *Biochem. J.* **113,** 443 (1969).
32. S. H. Chang and D. R. Wilken, *J. Biol. Chem.* **241,** 4251 (1966).
33. A. D. Patrick, *Biochem. J.* **83,** 248 (1962).
34. A. Eriksson and B. Mannervik, *FEBS (Fed. Eur. Biochem. Soc.) Lett.* **7,** 26 (1970).
35. J. Awapara and W. J. Wingo, *J. Biol. Chem.* **203,** 189 (1953).
36. F. Chapeville and P. Fromageot, *Biochim. Biophys. Acta* **26,** 538 (1957).
37. E. A. Tolosa, E. V. Goryachenkova, R. M. Khomutov, and E. S. Severin, *in* "Pyridoxal Catalysis" (E. E. Snell *et al.,* eds.), p. 525. Wiley (Interscience), New York, 1968.
38. L. Eldjarn, A. Phil, and A. Sverdrup, *J. Biol. Chem.* **223,** 353 (1957).
39. B. Bergeret and F. Chatagner, *Biochim. Biophys. Acta* **14,** 543 (1954).
40. H. Blaschko, S. P. Datta, and H. Harris, *Brit. J. Nutr.* **7,** 364 (1953).
41. S. Dupré and C. DeMarco, *Ital. J. Biochem.* **13,** 386 (1964).
42. D. Cavallini, C. DeMarco, R. Scandurra, S. Dupré, and M. T. Graziani, *J. Biol. Chem.* **241,** 3189 (1966).
43. D. Cavallini, R. Scandurra, and C. DeMarco, *Biochem. J.* **96,** 781 (1965).

CHAPTER 15

Enzyme Defects in Sulfur Amino Acid Metabolism in Man

James D. Finkelstein

I. INTRODUCTION

Several common clinical syndromes may be characterized by acquired or genetically determined lesions in the pathway for methionine metabolism. Numerous studies have linked nutritional liver disease with deficient dietary intake of methionine and choline. Methionine metabolism may also be impaired in patients with chronic hepatic disease [1–3]. Indeed, there is evidence to suggest that the pathogenesis of hepatic encephalopathy is related to defective methionine assimilation [4]. Several reports indicate that the symptoms of chronic schizophrenia are exacerbated by methionine feeding [5]. Recently, Cohn et al. [6] demonstrated defective transmethylation of catecholamines in erythrocytes of women with affective disorders. When fed to rabbits, excessive homocystine may result in lesions that simulate those of arteriosclerosis. McCully has emphasized the need to define the relationship between methionine intake and metabolism and the development of human cardiovascular disease [7–9]. Finally, abnormalities in the methylation of nucleic acids may be significant in oncogenesis [10].

Despite the clinical significance of the above disorders, the major stimulus to the study of methionine metabolism in man has been the recognition of ten inborn errors of sulfur amino acid metabolism (Table I). In three of these disorders, hypermethioninemia, cystine storage disease (cystinosis), and β-mercaptolactate-cysteinuria, the pathophysiology is undefined. Defects in normal membrane transport mechanisms characterize cystinuria and the methionine malabsorption syndrome. Impairment of specific enzyme reactions are the bases for the remainder of the disorders. In this discussion I will focus on the known enzymatic defects, since studies of these patients have illuminated the normal patterns of methionine metabolism. In light of the considerable number of excellent reviews dealing with the clinical consequences of these de-

TABLE I

DISEASES OF SULFUR AMINO ACID METABOLISM

Membrane transport defects
  Methionine malabsorption
  Cystinuria
Enzyme defects
  Cystathioninuria (cystathionase deficiency)
  Sulfocysteinuria (sulfite oxidase deficiency)
  Homocystinuria (cystathionine synthase deficiency)
  Homocystinuria due to decreased homocysteine remethylation
    (a) Deranged vitamin B_{12} metabolism
    (b) Decreased methylene tetrahydrofolate reductase
Defects unknown
  Hypermethioninemia
  Cystine storage disease (cystinosis)
  β-Mercaptolactate-cysteinuria

fects, I will emphasize studies that define the enzymatic defects, studies that emphasize the biochemical distortions that result from the defects, and studies that relate the biochemical and clinical abnormalities.

II.   CYSTATHIONINE SYNTHASE DEFICIENCY

In 1962 clinical investigators in Northern Ireland [11], Wisconsin [12], and Philadelphia [13] identified patients who excreted excessive amounts of homocystine in the urine. In the exciting decade that followed these observations, a clinically significant syndrome was defined, the enzymatic basis for the defect was elucidated, the biochemical consequences were evaluated, the pattern of inheritance was established, and several potential therapeutic modalities were advanced. Studies of patients with this disorder have provoked significant questions concerning more common clinical problems including arteriosclerotic cardiovascular disease, and schizophrenia. At the same time, we have gained considerable insight into the normal pattern of methionine metabolism from observations made in patients with this discrete defect.

A. Clinical Abnormalities

Carson and Neill identified their first two patients when they surveyed the urinary excretion of amino acids by 2081 mentally retarded individuals in Northern Ireland [11,14]. Retardation also characterized the two patients reported by Gerritsen and Waisman [15]. Similarly, Barber and Spaeth [13] described homocystinuria in a child with mental retardation and dislocation of the ocular lenses. Indeed, the latter ophthalmologic abnormality was present in both cases in Northern Ireland, as well as one of the two children in Wisconsin. For this reason, investigators in search of additional cases studied children with mental retardation and/or ocular defects. In 1965, McKusick and his co-workers broadened the means of ascertainment. Stimulated by an observation of Professor Charles Dent that "homocystinuria" could simulate the clinical picture of Marfan syndrome, McKusick screened urine from patients with dislocation of the lenses or with presumed Marfan syndrome. They found 38 cases in 20 families [16]. This has enlarged to 83 cases in 47 sibships and 45 kindreds [17,18]. Other groups have contributed large series. In general, they too have derived their patients from selected clinical populations. There is no doubt that the means employed in ascertainment bias the individual reports. But the studies of families of readily identified patients have provided a more complete appreciation of the varied clinical manifestations of cystathionine synthase deficiency.

TABLE II

CLINICAL FINDINGS IN CYSTATHIONINE SYNTHASE DEFICIENCY

Ectopia lentis
Skeletal abnormalities
Osteoporosis
Dolichostenomelia
Scoliosis
Pectus excavatum or carinatum
Arterial and venous thromboses
Mental retardation
Abnormal hair pigmentation

I have listed the more common clinical findings in Table II. Dislocation of the ocular lenses is a frequent finding in these patients. Often the abnormality is absent at birth and appears during the subsequent course. In two patients, the subluxation was detected during the third decade of life [18,19]. We do not know the true incidence of ectopia lentis in cystathionine synthase deficiency. Cross and Jensen report this finding in 90% of their patients [20]. But subluxation of the lenses was one criterion for ascertainment. Presley and co-workers report a lower frequency [21]. Both groups emphasize the frequency of other ophthalmic pathology including retinal degeneration and detachment, cataract, and optic atrophy. Henkind and Ashton [22] studied the histopathology of the lens in cystathionine synthase deficiency. They found abnormalities in the zonular fibers and in their attachments. The number of zonules decreased and they appeared disorganized and granular. Normal zonules demonstrated fibrillar morphology.

Patients with cystathionine synthase deficiency may resemble those with Marfan syndrome. Long, thin extremities, scoliosis, and chest deformities are common to both conditions. Presumably osteoporosis results from quantitative or qualitative disturbances in the collagen matrix of bone. McKusick's observation that there was an increased frequency of osteoporosis in cystathionine synthase deficiency was one of the earliest suggestions that the enzyme defect might lead to abnormal collagen.

Vascular occlusion is a major complication of cystathionine synthase deficiency. Thromboses may involve major arteries and veins and may result in death or severe disability. Both myocardial infarction and cerebrovascular accidents have been reported in patients under the age of ten years. It is possible that the mental retardation results from thromboses of cerebral vessels. Trauma, surgery, or vascular catheterization may precipitate thromboses. For this reason, surgery and invasive diagnostic techniques are avoided, if possible. The basis for the diathesis to thromboses remains obscure. McDonald *et al.* [23] suggested that homocysteine increased platelet "stickiness." Ratnoff [24] demonstrated that homocystine activated the Hageman factor and he proposed that abnormal accumulations of the amino acid could initiate clotting. However, the thromboses may be due to pathological changes within the vessel walls rather than to intravascular coagulation. In the first study of the pathological changes in homocystinuria, Gibson *et al.* [25] described areas of intimal fibrosis often encroaching on the vessel lumen, disruption of the elastic fibrils, and the deposition of metachromatic material in the media. Carson *et al.* [26] confirmed these findings in their review of ten cases. McCully [7] has suggested the concept that the vascular lesions represent an accelerated arteriosclerosis. He produced similar lesions experimentally by feeding *dl*-homocysteine thiolactone to rabbits [8]. Recently, McCully has proposed a possible molecular basis for these abnormalities [9].

Mental retardation is a frequent, but not invariable, finding in patients with cystathionine synthase deficiency. In Schimke's study 22 of 38 patients were judged to be retarded [16]. Only three of the 20 families had some affected members with normal intelligence whereas others were mentally retarded. In the other 17 families either all or none of the homocystinuric relatives were retarded. There are several possible explanations. Different abnormal alleles may cause cystathionine synthase deficiency. Not all may result in mental retardation. Other evidences for genetic heterogeneity will be discussed subsequently. It is also possible that expression of the abnormal gene is modified by other genetic or environmental factors.

Abnormalities of the hair and skin are common in patients with cystathionine synthase deficiency. Their hair is coarse, sparse, and light in color. Although the facial skin may be coarse with wide pores, the skin of the trunk and extremities is often thin, displaying a prominent venous pattern.

B. The Enzyme Defect

The accumulation of homocyst(e)ine in tissue, plasma, and urine was a consistent finding in the first reported cases of homocystinuria. The tissue content of methionine was also increased. The presence of homocystinemia made it unlikely that the basic defect was a failure of renal tubular transport similar to that observed in cystinuria. The data suggested impairment of the intermediary metabolism of homocysteine (and methionine). As illustrated in Fig. 1, there are at least four enzymes that utilize L-homocysteine as a substrate. Deficiency of S-adenosyl-L-

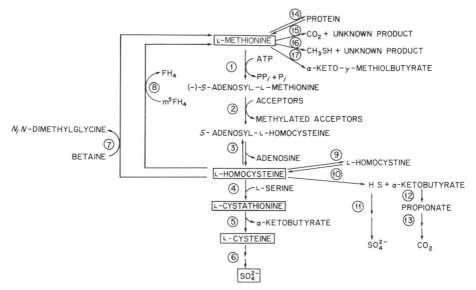

Fɪɢ. 1. Known pathways of mammalian metabolism of methionine and homocysteine. Abbreviations: ATP, adenosine triphosphate; PP_i, inorganic pyrophosphate; P_i, inorganic phosphate; FH_4, tetrahydrofolic acid; m^5FH_4, N^5-methyltetrahydrofolic acid.

homocysteine hydrolase (EC 3.3.1.1, Fig. 1, Reaction 3) would not result in an accumulation of both homocysteine and methionine. The thermodynamic equilibrium favors the synthesis of S-adenosylhomocysteine *in vitro* [27]. In the physiological setting, the equilibrium is shifted toward hydrolysis by the removal of the products, homocysteine and adenosine. In the absence of the enzyme, S-adenosylhomocysteine might accumulate but homocysteine would not. Increased tissue concentrations of S-adenosylhomocysteine might inhibit several transmethylation reactions (Fig. 1, Reaction 2) [28–30] and might increase the synthesis of N^5-methyltetrahydrofolate [31]. However, the synthesis of S-adenosylmethionine (Fig. 1, Reaction 1) is irreversible. Thus, we could not predict that the accumulation of S-adenosylhomocysteine would result in methionine excess.

Mammalian tissues contain several homocysteine methylases. Only two, betaine–homocysteine methyltransferase (EC 2.1.1.5, Fig. 1, Reaction 7) and N^5-methyltetrahydrofolate–homocysteine methyltransferase (EC 2.1.1.13, Fig. 1, Reaction 8), appear to be significant in normal homeostasis. Failure of homocysteine remethylation could result in the accumulation of homocysteine but we would expect a concomitant decrease in tissue methionine.

A defect in cystathionine synthase could explain the accumulation of both homocysteine and methionine. In the absence of this enzyme, homocysteine would be remethylated. Although cystathionine synthase deficiency seemed most likely, there was another possibility. A partial defect in synthesis of S-adenosylmethionine might be associated with an increase in tissue methionine. Methionine inhibits betaine–homocysteine methyltransferase [32] and represses the synthesis of N^5-methyl-tetrahydrofolate–homocysteine methyltransferase [33–35]. If, despite the limitation in the synthesis of S-adenosylmethionine, the quantity of homocysteine derived from methionine exceeded the combined capacities of cystathionine synthase and the homocysteine methylases, then homocysteine might accumulate.

Mudd *et al.* [36,37] designed sensitive and specific microassays by which they measured the activities of S-adenosylmethionine synthetase (EC 2.5.1.6; ATP:L-methionine S-adenosyltransferase), cystathionine synthase [(EC 4.2.1.22; L-serine hydrolyase (adding homocysteine)], and cystathionase (EC 4.4.1.1; L-cystathionine cysteine-lyase) in tissues obtained by biopsy from children with homocystinuria. In preliminary studies only liver proved to be a suitable source of enzyme. Skin, intestinal mucosa, erythrocytes, and leukocytes lacked detectable amounts of cytathionine synthase. As shown in Table III [36–39], the specific activity of cystathionine synthase in the livers of patients was markedly reduced. There was no detectable activity in one patient (C. T.). At the time of that study the assay could detect a specific activity of 13 nmoles/mg protein/135 minutes. Subsequently, with a more sensitive system, minimal cystathionine synthase activity was demonstrated in three patients (J. H., K. B., and L. B.). Appropriate studies confirmed the fact that the residual activity was due to cystathionine synthase [39–41]. Enzyme activity remained low despite dialysis or gel filtration. Coincubation studies also precluded the possibility that there was an inhibitor of cystathionine synthase in the extracts derived from the livers of the patients. Patient M. A. G. differed significantly from the other patients. She was asymptomatic and lacked any stigmata of the syndrome. Homocystinuria was detected only when Spaeth and Barber [13] screened the urines of relatives of C. T. The specific activity of cystathionine synthase in M. A. G.'s liver was 5–10 times greater than that observed in the livers of other patients. Yet the specific activity was only 12% of the mean control value (23% of the lowest control value) and was lower than the values observed in the parents of the patients. It is possible that M. A. G. possesses a different abnormal allele. Alternatively, she may be heterozygous for the abnormal gene that is being expressed because of some other concordant genetic or environmental factor.

Since these original observations, three other research groups have

TABLE III

Specific Activities of Enzymes of Methionine Metabolism in Liver from Patients with Cystathionine Synthase Deficiency[a]

Patient	Clinical features	S-AdoMet synthetase (nmoles/ mg prot/60 min)	Cystathionine synthase (nmoles/ mg prot/135 min)	Cystathionase (nmoles/ mg prot/30 min)
Controls				
Number		17	15	12
Mean		7.1	258	14.0
Range		(4.3–14.5)	(133–610)	(6.0–24.0)
Homocystinuric patients				
C.T.	Full syndrome	9.9	0 (<13)	—
M.A.G.	Cousin of C.T., no abnormalities	4.9	31	—
J.H.	Full syndrome	—	5	17.5
K.B.	Full syndrome	9.9	3	24.0
L.B.	Syndrome without mental retardation	—	6	22.0
Relatives of homocystinuric patients				
D.T.	Mother of C.T.	7.1	86	—
L.T.	Father of C.T.	5.5	88	—
S.M.	Maternal grandmother of C.T.	6.0	257	—
Ky.B.	Father of K.B. and L.B.	5.5	80	—
Le.B.	Mother of K.B. and L.B.	5.6	112	—

[a] The table summarizes data previously published [36–39].

reported enzyme data for 14 additional cases [42–44]. Although the investigators used a variety of assay techniques, the findings of a marked deficiency in hepatic cystathionine synthase was constant.

As illustrated by Table III, the defect in cystathionine synthase is specific. Hepatic content of S-adenosylmethionine synthetase and cystathionase are normal. In addition, Brenton *et al.* found normal activities for betaine–homocysteine methyltransferase and thetin–homocysteine methyltransferase in livers that were obtained postmortem from two children with homocystinuria [45].

Cystathionine synthase was deficient in brain tissue obtained from two patients with demonstrated deficiency of the hepatic enzyme [39,40].

The requirement for liver or brain tissue for enzyme studies was a significant problem. Both specific clinical diagnoses and further investigative activities required a biopsy, not devoid of risk, or had to await postmortem examination. In 1968 Uhlendorf and Mudd [46] demonstrated the presence of cystathionine synthase activity in fibroblasts cultured from normal skin and from cells in amniotic fluid. Fibroblast lines from homocystinuric patients contained very low or undetectable levels of cystathionine synthase (Table IV) [46–48]. Cells from the control and patient populations did not differ as regards S-adenosylmethionine synthetase [46] or N^5-methyltetrahydrofolate–homocysteine methyltransferase [47]. Recently these same investigators summarized their studies of fibroblast cultures from 47 patients with homocystinuria [49]. In 40 of the patients collateral evidence supported a diagnosis of cystathionine synthase deficiency. In 38 of these patients cystathionine synthase activity in the extracts of the fibroblasts was significantly lower than in extracts derived from 39 control fibroblast lines. Bittles and Carson [50] as well as Fleisher *et al.* [51] obtained similar data. The

TABLE IV

ENZYME ACTIVITIES IN CELL-FREE EXTRACTS OF FIBROBLASTS[a]

	S-AdoMet synthetase (nmoles/ mg prot/30 min)	Cystathionine synthase (nmoles/ mg prot/135 min)	N^5-CH$_3$-FH$_4$- homocysteine methyltransferase (nmoles/ mg prot/60 min)
Controls	1.7–4.5 (14)[b]	3.7–65 (35)	3.2–4.8 (4)
Cystathionine synthase deficiency	1.6–2.8 (6)	0–0.5 (29)	4.4–7.3 (3)

[a] Data derived from studies previously published [46–48].
[b] Number of samples tested is indicated in parentheses.

mean specific activity in cultures derived from the patients was 3.4–3.7% of that obtained in normal controls.

Goldstein *et al.* [52] reported the appearance of cystathionine synthase in human lymphocytes following stimulation with phytohemagglutinin. Stimulated lymphocytes from six patients lacked the enzyme. In a single case low levels of cystathionine synthase were detected [53].

Thus, deficiency of cystathionine synthase has been demonstrated in liver, brain, cultured skin fibroblasts, cells from anniotic fluid, and phytohemagglutinin stimulated lymphocytes. For this reason the report of normal activity in the ocular lens from one patient merits considerable attention [54]. Gaull and Gaitonde incubated the intact lens from a seven-year-old patient in media containing L-[^{35}S]methionine. After a four hour incubation, they determined the distribution of radioactivity in several fractions that they prepared from the lens. They found that ^{35}S had been incorporated into both cystathionine and cysteine. The ratio of radioactive cystathionine to radioactive homocysteine in the patient's lens was similar to the ratio obtained with a single control lens from a 50-year-old man with a retinal neoplasm. Although limited to one patient (and one control subject) this study indicates that the transsulfuration pathway is intact in the lens. Gaull and Gaitonde suggest the possibility that the lens contains a different isoenzyme form of cystathionine synthase. Certainly, we require additional studies which test directly for the presence of cystathionine synthase before we can measure the significance of their finding.

C. Genetics

Analyses of multiple pedigrees [17,18], as well as direct determinations of hepatic cystathionine synthase in patients and relatives, indicate an autosomal recessive pattern of inheritance. In reviewing 45 kindreds McKusick noted: (1) that both parents of affected persons were clinically normal; (2) a high incidence of proven or suspected parental consanguinity; (3) a frequency of affected sibs that approximated the 25% predicted by segregation analyses corrected for random ascertainment; and (4) a male:female patient ratio that did not differ significantly from 1.0 [18]. The enzyme studies are in agreement. Two families were evaluated at the National Institutes of Health [38,39]. The data are presented in Table III. The mean value for hepatic cystathionine synthase activity in the parents was 35% of the mean control value but was significantly higher than the low activities characteristic of the patients.

1. HETEROZYGOTES

To date no test, other than direct enzyme assays, distinguishes the heterozygote from the normal population. Obligatory heterozygotes are normal clinically. The results of methionine tolerance tests have been inconstant and normal metabolic patterns have been observed frequently. As noted in the preceding section, cystathionine synthase is present in skin fibroblasts grown in tissue culture [46,49–51] and in lymphocytes transformed with phytohemagglutin [52,53]. Studies utilizing both methodologies indicate a potential use of these techniques for the detection of heterozygotes. Uhlendorf et al. [49] compared cystathionine synthase activities in skin fibroblasts derived from parents with the activities in the affected individuals and in a control population. Most of the enzyme specific activities in all lines from the parents fell near the low end of the control range. While there is a statistical distinction between the heterozygote and control populations, it is difficult to define whether a given individual is normal or heterozygotic. Fleisher et al. also studied enzyme activity in cultured fibroblasts [51]. However, this group employed a different method for enzyme assay. They found that the specific activity of cystathionine synthase was 20.97 ± 1.81 nmoles/mg protein/hour in cells from 17 control subjects; 4.40 ± 0.92 in cells from six obligate heterozygotes and 0.77 ± 0.42 in cells from five patients. There was no overlap between the groups which were readily distinguished. Since patients with cystathionine synthase deficiency may be heterogeneous genetically, the ability of this method to define the heterozygote must be tested with larger populations. Unfortunately, the assay method requires an automatic amino acid analyzer. Both the expense as well as the implicit limitation in the number of analyses may restrict the utility of this technique. Certainly large scale screening studies would not be feasible.

In their study of phytohemagglutin stimulated lymphocytes, Goldstein et al. [53] compared 48 control subjects, 17 obligate heterozygotes, and seven affected patients. The control specific activity was 666.9 ± 70.2 pmoles/mg protein/4 hours. The values for the heterozygote and patient populations were 114.4 ± 27.3 and 2.0 ± 1.6, respectively. But whereas the populations differed significantly from one another, there were individual examples of overlap.

2. GENETIC HETEROGENEITY

There is increasing evidence to suggest that more than one genetic defect results in cystathionine synthase deficiency. Clinically, the patients can be divided into two groups determined by whether large dosages of vitamin B_6 result in an improvement in the amino acid abnormalities in serum and urine (see Section II,E,2). Seashore *et al.* [55] have suggested that the pyridoxine responsive patients may be subdivided further by the characteristics of their clinical response as well as by *in vitro* studies.

Uhlendorf *et al.* [49] were able to divide their 39 patients into two groups by examining enzyme activity in the cultured skin fibroblasts. In 29 of the cell lines levels of cystathionine synthase were measurable although markedly reduced. In ten cell lines there was no detectable enzyme activity. An interesting correlation emerged. The efficacy of pyridoxine therapy was judged in 35 of these patients. Residual enzyme activity was found in cell lines derived from 24 of the 25 pyridoxine responders. Cells from nine of ten nonresponders lacked cystathionine synthase.

There was additional evidence for genetic heterogeneity. Using the analysis of variance test, these investigators showed that differences in enzyme activity between sibships were significantly greater than differences in enzyme activities between siblings. The presence of different abnormal genes in the several families could explain this observation. However, environmental factors or interaction with other genes also unique to a family cannot be excluded.

Cystathionine synthase, purified from rat liver, contains two subunit pairs [56]. In all likelihood the human enzyme also possesses an $\alpha_2\beta_2$ structure determined by at least two structural genes, *A* and *B*. Mutations at either locus could result in a variety of abnormal alleles (*A'*, *A''*,. . . and *B'*, *B''*,. . .) in addition to the normal alleles, *A* and *B*. As indicated by Uhlendorf, this hypothesis allows a large number of genotypes in the population [49]. The normal homozygote would be *AABB*. *A'A'BB* and *AAB'B'* would represent persons who are homozygous for the same abnormal allele. Other persons may be double heterozygotes at the same locus (*A'A''BB*, *AAB'B''*) or at each locus (*AA'BB'*). The simple heterozygotes would be *AA'BB* or *AABB'*. At this time we cannot assign any of the phenotypes to a specific genotype. However, these concepts might help to explain the observation that obligate heterozygotes possess less than 50% of the normal activity of cystathionine synthase. We must assume that an abnormal allele alters the function of the polypeptide subunits. A simple heterozygote, *AA'BB*, would synthesize α, α', and β subunits. Enzyme molecules of the types $\alpha_2\beta_2$, $\alpha'_2\beta_2$, and $\alpha\alpha'\beta_2$ would be present in the proportions $1:1:2$. The $\alpha_2\beta_2$ molecule has normal enzyme characteristics whereas the $\alpha'_2\beta_2$ molecule may be functionally inactive. The heterozygous individual will have 25% of the normal enzyme activity plus whatever contribution is made by the $\alpha\alpha'\beta_2$ molecular species. If $\alpha\alpha'\beta_2$ has one-half of the normal $\alpha_2\beta_2$ activity, the total activity would be 50% of the normal. If $\alpha\alpha'\beta_2$ is one-fourth as active as $\alpha_2\beta_2$, the total activity will approximate 37%.

The double heterozygote $(AA'BB')$ would contain four different subunits that would generate nine molecular species for the tetramaeric enzyme. Presumably 44% of the tetramers ($\alpha_2\beta'_2$, $\alpha'_2\beta_2$, $\alpha'_2\beta'_2$, $\alpha'_2\beta\beta'$ and $\alpha\alpha'\beta'_2$) would be inactive. The normal tetramer ($\alpha_2\beta_2$) would represent only 6% of the enzyme molecules. The total enzyme activity would be this 6% plus the contributions of $\alpha_2\beta\beta'$, $\alpha\alpha'\beta_2$ and $\alpha\alpha'\beta\beta'$. If the latter were inactive and if each of the species with a single abnormal polypeptide unit possessed 25% of the normal activity, the total activity would be 12.5%. Perhaps patient M.A.G. (Table III) is a double heterozygote.

We are far from the time when we can assign genotypes. However, a consideration of genetic heterogeneity must be included in any interpretation of the clinical and biochemical studies in cystathionine synthase deficiency.

3. GENE FREQUENCY

The epidemic of case reports following the initial descriptions suggested that the incidence of cystathionine synthase deficiency might be second only to phenylketonuria as a specific, genetically induced metabolic disorder attended by mental retardation. Subsequently mass screening studies of an unselected population of newborns yielded an incidence of two cases in 400,000 births. The survey employed a test for increased levels of methionine in blood. Since this abnormality may be inconstant in the infant with cystathionine synthase deficiency, the estimate might be low. McKusick's educated guess for the frequency is one in 45,000–50,000 births [18].

The wide geographic and ethnic distribution of cases is notable. The clinical syndrome, as well as the biochemical abnormalities, have affected patients in all the European ethnic groups, in Japan and India, as well as in Negro and Spanish-American families in the United States.

D. Biochemical Consequences

The biochemical abnormalities that occur in cystathionine synthase deficient patients demonstrate clearly that this reaction is an essential step in the transsulfuration sequence. Metabolites proximal to the block accumulate whereas those distal to the lesion are present in abnormally low concentrations. The fact that the abnormal concentrations of metabolites can be predicted suggests that methionine metabolism in man is represented adequately by the pathway (Fig. 1) derived from studies in other mammals.

1. ACCUMULATION OF METABOLITES

Hypermethioninemia is present in most of the patients with cystathionine synthase deficiency [13–15,57–59]. Methionine accumulates in other tissues and in physiological fluids. Increased concentrations occur in erythrocytes [13], liver [60,61], and cerebrospinal fluid [57]. It seems clear that methioninuria results from an "overflow" from the blood rather than from a renal tubular cell transport defect [13,15,57,59].

Methionine accumulation can be explained either by decreased catabolism or increased synthesis. Impairment in the synthesis of S-adenosylmethionine is, however, unlikely. As measured by direct assay, the level of the enzyme S-adenosylmethionine synthetase, is normal [36,39]. Furthermore, this enzyme is not inhibited by homocysteine, S-adenosylhomocysteine, S-adenosylmethionine, or any of the other metabolites likely to present in excess concentrations. Increased synthesis of methionine due to the remethylation of homocysteine is the probable explanation for the accumulation of methionine. Two enzymes can catalyze this reaction (Fig. 1, Reactions 7 and 8). Both enzymes are intact in tissues of patients with cystathionine synthase deficiency [45,62], but methionine inhibits betaine–homocysteine methyltransferase [32]. Furthermore, this enzyme may have a limited distribution in mammalian tissues [35]. N^5-Methyltetrahydrofolate homocysteine methyltransferase is present in most tissues [35] and is not inhibited by methionine. Thus, this enzyme may be most significant in the synthesis of methionine from homocysteine. The work of Carey et al. supports this hypothesis [63]. They found that folate treatment of patients with cystathionine synthase deficiency caused a decrease in plasma homocystine and an increase in plasma methionine. They also observed that despite low blood levels of folate (measured with Lactobacillus casei) the patients failed to excrete formiminoglutamate following a load of histidine. It is likely that these patients lack N^5-methyltetrahydrofolate, but maintain the other biologically active forms of folate.

The hydrolysis of S-adenosylhomocysteine (Fig. 1, Reaction 3) is reversible. The thermodynamic equilibrium favors the synthesis [27]. In vivo, hydrolysis occurs only with the removal of the products, adenosine and homocysteine. Since the enzyme is present in most mammalian organs [64], we would expect S-adenosylhomocysteine to accumulate in the tissues of cystathionine synthase deficient patients. Urinary excretion of S-adenosylhomocysteine by these patients indicates indirectly the increased tissue stores [65,66]. When S-adenosylhomocysteine is injected intravenously into rats, the keto derivative appears in the urine [67]. This compound has not been found in the urine of patients who excrete S-adenosylhomocysteine [66], but Perry et al. identified 5-amino-4-imidazole carboxamide-5'-S-homocysteinylriboside in the urine of two patients [65,68] and suggested that it may arise from S-adenosylhomocysteine.

S-Adenosylhomocysteine inhibits several classes of transmethylation reactions [28–30]. If inhibition occurred in patients, the tissue content of S-adenosylmethionine might increase. This might be magnified due to the increased synthesis of S-adenosylmethionine that would be expected as a consequence of the increased availability of methionine. Applegarth et al. found S-adenosylmethionine in urine from two cystathionine synthase deficient children [66].

2. DEFICIENCY OF METABOLITES

The defect in cystathionine synthase limits the capacity for endogenous synthesis of cystathionine, cysteine, and further derivatives of cysteine. Cysteine becomes an essential amino acid.

Cystathionine occurs in high concentrations in the brains of mammals [69], but no biological function has been assigned to this compound. Cystathionine synthase deficient patients have decreased concentrations of cystathionine in their brains [45,70]. This observation, together with the frequency of mental retardation, has stimulated speculation concerning a possible role for cystathionine in higher cortical functions. There is an alternate pathway for the synthesis of cystathionine. Wong et al. [71] found that human and rat liver homogenates catalyzed the condensation of L-cysteine and L-homoserine with the formation of cystathionine. Liver from a cystathionine synthase deficient patient had the same capacity. Gaull et al. [72] confirmed these findings and demonstrated that the enzyme that catalyzed this reaction was indistinguishable from cystathionase. Both groups observed the appearance of cystathionine in the urine of cystathionine synthase deficient subjects when they added homoserine and cysteine to the diet [71,72]. Similarly, when administered with homoserine, L-cysteine is incorporated into the cystathionine of rat brain [73,74]. In all likelihood the synthesis of cystathionine occurs within the brain, since cystathionine in blood does not penetrate into the central nervous system [75].

In cystathionine synthase deficiency the concentrations of cyst(e)ine may be subnormal in plasma [13–15], urine [13,15,57], erythrocytes

[13], and liver [60]. This decrease may be caused by the formation of homocysteine-cysteine mixed disulfide in the presence of the high concentrations of homocysteine. However, the disruption in the transsulfuration pathway appears more significant. Dietary methionine cannot be converted to cysteine in adequate amounts. Laster *et al.* [76] studied the ability of two patients to convert dietary cysteine and methionine to urinary inorganic sulfate. The catabolism of cysteine was normal, but the quantity of sulfate derived from methionine was markedly reduced in the patients. Taken together, these studies define a functional impairment in the conversion of methionine to cysteine. Brenton and Cusworth [77] extended these findings by means of studies in which they employed [35S]methionine. Their data confirmed a marked impairment of cysteine formation from methionine. In other experiments Brenton *et al.* [78] demonstrated a negative nitrogen balance when methionine was the sole source of dietary sulfur for patients with cystathionine synthase deficiency. Cystine supplementation of the diet resulted in a positive balance confirming the essential role of this amino acid.

In all likelihood, deficiencies of the metabolic derivatives of cysteine occur in patients with cystathionine synthase deficiency when the dietary content of cyst(e)ine is inadequate. The urinary excretion of taurine, for example, appears related directly to dietary cyst(e)ine [78].

3. ABNORMAL METABOLITES

In the preceding sections I have discussed the effects of cystathionine synthase deficiency on the concentrations of normal metabolites. In studies of patients with cystathionine synthase deficiency several investigators have identified unusual metabolites. The presence of these compounds suggests the presence of minor or alternate pathways for the metabolism of methionine and homocysteine.

Perry *et al.* found homolanthionine in the urine of their patients [79]. This compound is a higher homologue of cystathionine. Both Mudd and Gaull demonstrated the synthesis of this compound from homoserine and homocysteine [41,72] in the presence of the enzyme, cystathionase.

Perry also reported the isolation of α-hydroxy-γ-mercaptobutyrate-homocysteine disulfide [68]. He suggested that this compound was derived by the transamination and subsequent reduction of one-half of the homocystine molecule. Alternatively, the transamination and reduction may involve one molecule of homocysteine that subsequently forms a disulfide bond to an unreacted homocysteine molecule.

Lemonnier *et al.* believe they have isolated homocystine sulfone from the urine of two affected children [80]. Finally, Kodama and his associates have identified several unusual derivatives. These include: S-(3-

hydroxy-3-carboxy-n-propylthio)cysteine; S-(3-hydroxy-3-carboxy-n-propylthio)homocysteine; S - (3 - hydroxy - 2 - carboxyethylthio)homocysteine; S-(-carboxyethyl)homocysteine; S-(β-carboxyethylthio)-homocysteine; S-(β-carboxyethylthio)cysteine; S-(carboxymethylthio)homocysteine; S-(3-hydroxy-3-carboxy-n-propyl)-homocysteine; homocysteic acid; and homocysteine sulfinic acid [81–83]. The mode of synthesis, as well as the degree of significance, of these compounds must be defined.

E. Management of Cystathionine Synthase Deficiency

Optimal therapy for patients with cystathionine synthase deficiency must await a better understanding of the pathophysiological basis for the syndrome. We know that cystathionine synthase reactivity is impaired but we do not know the molecular nature of the defect. Do some patients have an adequate concentration of an inactive enzyme whereas other patients have an inadequate concentration of an active protein? We have also identified the metabolic consequences of the defect in cystathionine synthase but we do not know whether the clinical manifestations reflect the accumulation of normal metabolites, the presence of abnormal, toxic metabolites, or a deficiency in essential compounds resulting from faulty synthesis or excessive consumption.

1. THE PATHOPHYSIOLOGY

As indicated earlier, a basic defect in collagen formation might explain subluxation of the ocular lenses, skeletal deformities, and the pathology of the intima of the vasculature. In 1966 Harris and Sjoerdsma [84] reported that the solubility of dermal collagen was increased in two of four patients with cystathionine synthase deficiency. Subsequently, Kang and Trelstad studied the effect of sulfur amino acids on collagen formation *in vitro* [85]. They found that 10 mM homocysteine inhibited the formation of insoluble fibrils. They related the inhibition to an interference with the formation of cross-links. It is possible that homocysteine, like penicillamine, binds to the aldehyde groups that are precursors of the cross-links. Kang and Trelstad then studied the collagen cross-links in dermal tissue from three cystathionine synthase deficient patients. The number of cross-links was reduced and the solubility of the collagen was increased [85].

McCully argues that the basis for homocysteine induced abnormalities in connective tissue lies in the synthesis of an abnormal proteoglycan. He noted that fibroblasts that were cultured from skin biopsies of two patients synthesized proteoglycans that were granular, aggregated, and flocculent. Similar abnormalities occurred when normal fibroblasts were cultured in media containing homocysteine [7]. McCully produced a vascular lesion resembling the arteriosclerotic lesion of cystathionine synthase deficiency when he induced sustained homocysteinemia in rabbits [8]. Pyridoxine deficiency in rabbits was associated with similar fibrous vascular plaques. The transition from normal fibrillar proteoglycan to the abnormal granular molecule is accompanied by an increase in binding of inorganic sulfate [9]. McCully emphasized the role of homocysteic acid. This compound, derived from homocysteine, can react with ATP to form 3'phosphoadenosine 5'-phosphosulfate (PAPS), which is the precursor of the sulfate. In cystathionine synthase deficiency, the accumulation of homocysteine may lead to increased synthesis of homocysteic acid and PAPS. Abnormalities in growth and connective tissue might be explicable by this mechanism [9].

Excess serum concentration of homocysteine may predispose to vascular thromboses. McDonald found increased platelet stickiness in blood from patients with homocysteinemia as well as in normal blood to which he added homocysteine [23]. Alternatively, homocysteine induced activation of the Hageman factor may cause the intravascular coagulation [24].

It is possible that the accumulation of other metabolites is significant in the pathophysiology of cystathionine synthase deficiency. Methionine toxicity occurs in experimental animals. In acute experiments methionine excess may cause ATP deficiency [86]. Perhaps a similar mechanism underlies the hepatic steatosis described in several patients [25,26]. Laster *et al.* [76] found that methionine sulfoxide appeared in the plasma and urine of patients with cystathionine synthase deficiency when they ingested methionine supplements. Is it possible that trace amounts of methionine sulfoximine, a known neurotoxin, are also synthesized?

We have already noted that the accumulation of S-adenosylhomocysteine may inhibit significant transmethylation reactions [28–30]. It is possible that the disturbances of the function of the autonomic nervous system reflect an interference with the metabolism of catecholamines.

Cystathionine synthase deficiency impairs the synthesis of cystathionine and cysteine. The concentration of N^5-methyltetrahydrofolate may be subnormal as a result of increased utilization of this substrate in the remethylation of homocysteine. All, or any, of these relative deficiencies may participate in the pathogenesis of the clinical features.

2. ATTEMPTS AT THERAPY

Each of the proposed therapeutic programs attempts to correct one or more of the biochemical abnormalities. Obviously the proponents of each treatment have made a value judgment based on incomplete information. It is particularly unfortunate, although understandable, that no therapy has been tested in an appropriately controlled study. Cystathionine synthase deficiency has varied clinical manifestations. The absence of symptoms in a treated patient may be a tribute either to the efficacy of the therapy or the good fortune of the patient.

   a. Low Methionine, High Cysteine Diet. Restriction of the dietary intake of methionine will reduce the tissue levels of methionine and homocyst(e)ine in patients with cystathionine synthase deficiency. The addition of cyst(e)ine satisfies the need for this amino acid and its metabolites. Indeed, the deficiency of cystathionine is the only biochemical abnormality which is not addressed by this dietary treatment. Carson *et al.* emphasized the importance for the institution of dietary therapy during the neonatal period, particularly in children who are not breast fed [14]. In human milk 60% of the protein sulfur is in the form of cystine whereas in cows' milk 70% of the protein sulfur occurs as methionine. Several groups have reported their experience with this treatment. Perry and his colleagues followed four patients [87]. Three were older children who had suffered major complications of the disorder prior to the beginning of treatment. During the three years of therapy there have been no additional thrombotic episodes and no spontaneous fractures, but there has been no improvement in established clinical defects. In the fourth case Perry *et al.* established the diagnosis of cystathionine synthase deficiency by age 16 days. They restricted methionine to 40 mg/kg and reduced the level gradually to 19 mg/kg by age 4–5 years. Supplemental L-cystine, 100–300 mg/kg/day, was fed throughout the five years. Biochemical improvement was prompt and has been sustained. However, the plasma concentration of methionine and homocysteine remained above normal and the plasma level of cystine was decreased despite increased supplementation. Nevertheless, this child, who had two older siblings with severe clinical manifestations, remains normal.

   Komrower's group has treated four children with methionine restriction (10–30 mg/kg/day) and cystine supplementation (1.0–1.5 gm/day). These patients were unresponsive to pyridoxine [88]. One child has been treated from age nine days to age 76 months. She is asymptomatic but her intellectual development is slow. Environmental factors could account for this.

   Despite these "successes" with diet therapy, the uncontrolled and anecdotal nature of the data precludes a definitive conclusion. The difficulty in creating a palatable, methionine restricted diet is real and the enforcement of dietary prohibitions becomes difficult, particularly with older, asymptomatic children. Longer periods of observation are necessary before we can suggest whether the older child may follow a more liberal diet.

b. Betaine, Choline, and Folic Acid. One means of reducing the tissue content of homocysteine is to facilitate its use as a substrate in methionine biosynthesis. Investigators have added methyl donors to the diet, either alone or in combination with other modalities of treatment. Perry administered choline dihydrogen citrate (10 gm/day) together with a low methionine, cystine supplemented diet. The concentration of homocysteine in the plasma fell but the level of methionine increased concurrently [89]. Komrower and Sardharwala administered 0.75–2.0 gm/day of betaine hydrochloride to a $1\frac{1}{2}$-year-old child and 6 gm/day to a 6-year-old patient. In the younger patient homocysteine-cysteine mixed disulfide disappeared from the plasma and urine and plasma cystine appeared to increase. The older child was ingesting a diet that contained more methionine at the time betaine was added to the regimen. Concurrent with the administration of the betaine, plasma levels of homocystine and homocysteine-cysteine disulfide fell but the level of plasma methionine was dramatically increased [88].

Large dosages of folate may have similar effects. Carey *et al.* [63] reported two patients in whom administration of folate (20 mg/day) caused a significant increase in urinary methionine and the anticipated decrease in urinary homocystine.

Methyl donors can achieve a redistribution of the sulfur amino acids, decreasing homocysteine levels while increasing methionine content. In all likelihood, decreased tissue homocysteine reduces the formation of homocysteine-cysteine disulfide. The consequent increase in free cysteine may be of some value. But whether the net result is advantageous to the patient must be defined; we cannot assume that the plasma levels mirror the situation in all tissues. Since betaine–homocysteine methyltransferase is present in few mammalian tissues, the effects of choline and betaine may be restricted to the liver and possibly the kidney [35,90].

c. Methods of Increasing Urinary Excretion of Homocystine. Homocystine, which is filtered through the glomerulus, can be reabsorbed by the renal tubular cells. It is likely that the reabsorption mechanism is identical to the transport system for cystine and the dibasic amino acids lysine, arginine, and ornithine. A sustained increase in the urinary excretion of homocystine might significantly reduce the tissue concentration of this amino acid in cystathionine synthase deficient patients.

Cusworth and Gattereau studied the effect of an intravenous infusion of arginine glutamate [91]. Plasma arginine increased 30-fold and was associated with the anticipated increase in the urinary excretion of cystine, lysine, and ornithine. Simultaneously, the urinary excretion of both homocystine and homocysteine-cysteine disulfide increased 20–30-fold. When administered by mouth, arginine glutamate had little effect on the urinary excretion of amino acids. Nevertheless, a small increase in homocystine excretion might be of considerable therapeutic value if sustained over several months.

Falchuk et al. [92] found a two- to threefold increase in the renal clearance of homocystine by two cystathionine synthase deficient patients who were fed α-aminoisobutyric acid. The response was related to the plasma concentration of α-aminoisobutyric acid and a greater excretion of homocystine might be achieved with higher dosages. However, we must establish that the chronic administration of α-aminoisobutyric acid is safe before we can consider a role for this amino acid analogue in the treatment of cystathionine synthase deficiency.

d. *Methods of Increasing Tissue Cystathionine.* The methionine restricted, cystine supplemented diet may reduce further the residual capacity for cystathionine synthesis [93]. Because there appears to be no physiological function for cystathionine, this diet has been used by clinicians. Whether there is a need to augment tissue cystathionine remains academic but the problem has stimulated several interesting studies. Cystathionine is not transported across the blood-brain barrier into the central nervous system [75]. An increase in synthesis by the brain is necessary to increase cerebral cystathionine. As noted above (Section II,D,2), two groups have demonstrated the ability of the enzyme cystathionase to catalyze the reaction between homoserine and cysteine [71,72]. In addition, Wong and Tresco found that the content of cystathionine in mouse brain was increased following an injection that combined these two amino acids [74]. Although they confirmed this observation, Sturman et al. [73] raised several cogent questions that should be answered before this therapy is applied in a clinical situation.

e. *Pyridoxine Therapy.* In 1967 Barber and Spaeth [94] demonstrated that massive dosages of pyridoxine hydrochloride (250–500 mg/day) caused a reduction in plasma methionine and eliminated homocystine from plasma and urine in three cystathionine synthase deficient patients. Other groups have confirmed and extended this important observation. Brenton and Cusworth treated 18 patients with pyridoxine (300–500 mg/day) [95]. They classified the results of therapy in this way: Group I, normal fasting plasma methionine concentration with disappearance of homocystine from plasma and urine (eight patients); Group II, decrease in fasting plasma methionine to less than twice normal associated with a reduction, but not cessation, of homocystinemia and homocystinuria (five patients); and Group III, significant reduction in fasting plasma methionine but to levels above two times normal. Plasma and urinary concentrations of homocystine were unchanged (five patients). Carson and her associates treated 16 patients. Their criteria for pyridoxine responsiveness included decreased plasma levels of methionine, homocystine and homocysteine-cysteine disulfide, together with a rise in plasma cystine. They considered the latter to be the most important criterion. By these standards seven patients were regarded as responsive while nine were classified as resistant [96].

In these series and in others, only the response to pyridoxine distinguished the two groups. Another observation, common to both studies, was that the biochemical response to pyridoxine was the same in all affected members of a single family. Recently Uhlendorf et al. [49] demonstrated that cultured skin fibroblasts from pyridoxine responsive patients possess residual cystathionine synthase activity. They found no enzyme activity in fibroblasts derived from the pyridoxine resistant subjects (Section II,C,2).

There is no agreement concerning the mechanism of action of pyridoxine. Cystathionine synthase requires pyridoxal phosphate as a cofactor [75,97–101]. It is possible that excess pyridoxine could enhance cystathionine synthase activity by: (1) saturating available apoenzyme; (2) saturating an abnormal apoenzyme that possesses a lowered affinity for coenzyme; (3) inducing apoenzyme synthesis; or (4) stabilizing the holoenzyme with retardation of degradation. It is also possible that the pyridoxine effect may not involve the cystathionine synthase reaction. The vitamin could activate alternative pathways for methionine and homocysteine catabolism. Pyridoxal and pyridoxal-P can react chemically with homocysteine. The pyridoxine effect could reflect a similar reaction occurring in vivo.

Thus the question as to whether the cystathionine synthase reaction is essential for the response to pyridoxine is of primary importance. The most direct affirmative evidence comes from the fibroblast studies. In the main, residual enzyme is present in fibroblasts from responders and is absent from cells derived from resistant patients [49]. The biochemical characteristics of the response do not define the mechanism. Any reaction that removes any of the metabolites from the cyclic portion of the pathway (methionine → S-adenosylmethionine → S-adenosylhomocysteine → homocysteine → methionine, Fig. 1) could effect these changes. Tissue levels of all intermediates would fall. Homolanthionine synthesis and homocysteine-cysteine disulfide formation would be limited by the reduced availability of homocysteine. Increased cysteine concentrations could be a consequence of the failure to form the mixed disulfide. Several known reactions provide possible alternate pathways. These include (1) methionine transamination; (2) S-adenosyl-methionine decarboxylation [102]; (3) S-adenosylhomocysteine trans-amination [67]; and (4) homocyst(e)ine desulfurase [103]. All are pyridoxal-P-dependent. Despite numerous searches, no investigator has identified the products anticipated from reactions (1), (2), and (3). Homocyst(e)ine desulfurase yields H_2S and α-ketobutyrate. Both products would be diluted in a normal pool and might escape detection. Specific studies employing radioisotopically labeled methionine administered before and after pyridoxine treatment would be necessary to exclude a significant role for homocysteine desulfurase. Nevertheless, the information currently available indicates that it is unlikely that the response to pyridoxine is mediated by the homocysteine desulfurase reaction. Mudd et al. have pointed out that this enzymatic activity is a property of cystathionase. But cystathionase activity did not increase in association with the pyridoxine response in three patients [41,48].

Neither pyridoxine therapy nor the addition of excessive pyridoxal-P to the assay media results in a marked increase in cystathionine synthase in livers of vitamin B_6 responsive patients. Mudd et al. [41] measured the hepatic enzyme activity in two patients before and during treatment with pyridoxine. In both cases there was a three year interval between the biopsies. In one patient (J.H.) the specific activity had increased from 5.0 to 7.6, whereas in the other (K.B.) there was also an increase from 3.0 to 10.5 (Table III). The value of 10.5 nmoles/mg protein/135 minutes is 4.2% of the mean control value. Yet at the time of biopsy, there was a marked decrease in plasma methionine and homocystine in both patients. In earlier studies both Gaull et al. [44] and Hollowell et al. [42] failed to detect cystathionine synthase in livers of patients who were being treated effectively with pyridoxine.

In contrast to patients with various other pyridoxine-dependent enzyme defects [104–106], the addition of pyridoxal-P in vitro fails to augment cystathionine synthase activity in enzyme preparations derived from pyridoxine responsive patients. The tissues studied included liver [36,39,41,42], brain [39], and skin fibroblasts [49]. The literature contains two exceptional cases. Yoshida et al. [43] report that in one liver preparation, cystathionine synthase activity increased from 0 to 450 cpm/mg protein with the addition of $5 \times 10^{-4}\ M$ pyridoxal-P. The stimulated activity was 31% of the control value. Seashore et al. presented the second atypical case in the context of their studies of the mechanism of response to pyridoxine [55]. They studied two patients, both of whom responded to vitamin B_6. One patient required a significantly lower dosage. In addition, his response to an oral methionine load returned toward normal. Cystathionine synthase activity in skin fibroblasts derived from this patient, increased from 0.3 to 1.4 nmoles/mg protein/135 minutes when pyridoxal-P (1.0 mM) was added to the reaction system. Enzyme activity in fibroblasts from the second patient was unaffected by 1.0 mM pyridoxal-P. When the culture medium was supplemented, no change in enzyme activity occurred. Seashore *et al.* have emphasized that the response to pyridoxine therapy may be mediated by different mechanisms in different patients.

Failure to demonstrate the normaliziation of cystathionine synthase by pyridoxine *in vivo*, and pyridoxal-P *in vitro*, does not provide sufficient reason for rejecting the hypothesis that the effect of vitamin B$_6$ is mediated by an increase in this enzyme. It is possible that the normal level of cystathionine synthase greatly exceeds the level required for metabolism of the usual dietary load of methionine. In patients with cystathionine synthase deficiency the reduced level of enzyme impairs metabolism. The degree of impairment, however, is defined by the ratio of the residual enzyme activity to the minimal activity required not by residual enzyme activity/mean control activity. Balance studies performed by Mudd, Laster and their associates [41,76] provide information relevant to this type of analysis. Normal subjects can convert dietary methionine to inorganic sulfate at a rate greater than 0.85 mmoles/kg body weight/day. This is the suggested intake for children (twice the minimum daily requirement) but exceeds the requirement of 5 mmoles/day for adults [107]. For a 70 kg individual, a rate of methionine metabolism of 0.07 mmoles/kg/day would be compatible with adequate nutrition. This is only 8.4% of the minimal capacity defined by the balance studies [76]. Since the true metabolic capacity of normal subjects is likely to be greater and since the daily allotment of 5 mmoles is generous, a patient with residual enzyme activity of less than 8% of the mean control value may be competent nutritionally. For this reason, in cystathionine synthase deficient patients, apparently small changes in cystathionine synthase activity from 4% to 6% of the mean control value may be of considerable physiological importance. Excessive dietary intake of methionine would still unmask the deficit.

Mudd *et al.* [41] provide a different analysis. Based on methionine loading tests in two cystathionine synthase deficient subjects, before and during pyridoxine therapy, they calculated mean capacities for methionine metabolism of 118 μmoles/kg/day (untreated) and 230 μmoles/kg/day (during therapy). The calculated total capacities for a 70 kg individual are 8.3 mmoles/day and 16.1 mmoles/day. In turn, they relate these metabolic capacities to the measured specific activities of hepatic cystathionine synthase. The resulting values ranged from 22 to 42

μmoles of methionine converted to sulfate/kg/day/unit specific activity. By means of several crude approximations, they obtained a similar figure starting with the enzyme activity in liver. One specific activity unit = 1 nmole/mg protein/135 minutes = 10.7 nmoles/mg protein/day. Assuming a 1500 gm liver with 10% soluble protein the total hepatic cystathionine synthase activity = 1600 μmoles/day/unit or 23 μmoles/kg/day/unit. Mudd *et al.* used a factor of 0.4 to correct for the use of an optimal pH of 8.3 in the assay and used a factor of 1.7 to correct to saturating concentrations of serine. Finally, they multiply by 1.5 to allow for synthesis of cystathionine in extrahepatic tissues. The final value is 23.5 μmoles/kg/day/unit specific activity. This value agrees to a surprising extent with the values that were obtained directly from the methionine metabolizing capacity of the patients and the specific activity of cystathionine synthase in liver.

We can explore the significance of these figures by means of a specific example. The mean control specific activity for cystathionine synthase is 252 units. If 1 unit indicates a methionine metabolizing capacity of 25 μmoles/kg/day then an increase in specific activity of 1% of the mean control value (2.52 units) is equivalent to an increase in methionine metabolizing capacity of 63 μmoles/kg/day. This figure is used in deriving Table V. From the table it is clear that in a 40 kg child, an increase in cystathionine synthase from 1% to 3% of the mean control value might allow the child to ingest 5 mmoles/day without difficulty. An increase from 3% to 5% might allow the child to consume the estimated normal dietary load of 10 mmoles/day without accumulating either methionine or homocysteine. The calculations for a 70 kg individual are more dramatic. An increase to 3% of the mean control specific activity could eliminate homocystinemia and homocystinuria. Unless an adult consumes more methionine than the older child, homocystinuria might

TABLE V

METHIONINE METABOLIZING CAPACITY[a]

Cystathionine synthase (% mean control value)	Body weight (kg)		
	10	40	70
1	0.63	2.5	4.4
3	1.9	7.6	13.2
5	3.2	12.6	22.0
7	4.4	17.6	30.9
9	5.7	22.7	39.7

[a] Measured in mmoles/day.

diminish with maturity, providing that the specific activity of cystathionine synthase remains constant. In the small (10 kg) infant a different situation pertains. The estimated minimal requirement for methionine is 65 mg/kg although twice that amount is recommended. The
child requires at least 7% of the control enzyme activity in order to
tolerate the lower methionine intake. For this reason it is unlikely that
an infant with cystathionine synthase deficiency would respond biochemically to pyridoxine therapy alone. Dietary restriction might also be
necessary.

III. DEFECTS IN HOMOCYSTEINE METHYLATION

N^5-Methyltetrahydrofolate–homocysteine methyltransferase is present in most mammalian tissues [35]. Early studies had shown that
in rats who were fed adequate diets "the neogenesis of methyl groups is
comparatively small . . . utilization of preformed methyl groups for
transmethylation is preponderant" [108,109]. Although folate, vitamin
B_{12}, and homocysteine could replace methionine in the diet, transmethylation from betaine appeared to be the major reaction in the normally fed animal. Based on these conclusions we would not anticipate
that impairment of N^5-methyltetrahydrofolate–homocysteine methyltransferase would affect tissue concentrations of methionine and homocysteine. In fact, defects in this enzyme result in homocystinemia and
hypomethioninemia, thus modifying our concepts of methionine metabolism.

Several genetic disorders impair the N^5-methyltetrahydrofolate–
homocysteine methyltransferase reaction. These include (1) a possible
defect in the synthesis of the apoenzyme [110]; (2) deficient synthesis
of the substrate, methyltetrahydrofolate; and (3) inadequate concentrations of methyl B_{12}, the presumed coenzyme. Either malabsorption of vitamin B_{12} or a derangement in the metabolism of vitamin B_{12} can cause a
deficiency of methyl B_{12}.

A. Vitamin B_{12} Deficiency

Deficiency of vitamin B_{12} due to inadequate dietary intake of this vitamin is rare in Europe and the United States. The current enthusiasm
for vegetarianism in certain segments of society may increase the incidence. In most instances vitamin B_{12} deficiency results from either
generalized or highly specific malabsorption syndromes. These include
(1) defects in the synthesis or function of intrinsic factor; (2) absence of
intestinal "releasing factor"; and (3) absence of specific ileal receptors
for B_{12} [111].

We know of only two reactions in mammalian tissues that require derivatives of vitamin B_{12} as cofactors. Vitamin B_{12} deficiency may impair these reactions. $5'$-Deoxyadenosyl-B_{12} is the coenzyme for methyl-malonyl-CoA mutase and methylmalonic aciduria is a common finding in B_{12} deficiency of any etiology. There are fewer clinical studies of the effect of B_{12} deficiency on the N^5-methyltetrahydrofolate–homocysteine methyltransferase reaction. Recently, Parry reported the frequency of decreased levels of methionine in the plasma of patients with pernicious anemia [112]. Prior to his study, most of the investigations involved studies of children with rarer genetic forms of vitamin B_{12} malabsorption. Hollowell et al. [113] did observe methylmalonic aciduria and homocystinuria in a 10-year-old boy with Immerslund's syndrome. In this patient plasma homocystine was increased and cysteine-homocysteine disulfide was present in plasma and urine. Following treatment with parenteral vitamin B_{12} all of the above biochemical abnormalities reverted to normal. In a brief report Shipman et al. [114] noted the occurrence of homocystinuria in a child with infantile pernicious anemia due to intrinsic factor deficiency.

B. Homocystinuria with Deranged B_{12} Metabolism

In 1969, Mudd, Levy, and Abeles [115,116] described an infant who died at $7\frac{1}{2}$ weeks of age after an illness characterized by poor appetite, failure to thrive, gastrointestinal bleeding, and normocytic anemia. He was found to have increased concentrations of homocystine, cystathionine, and homocysteine-cysteine mixed disulfide in plasma and urine. Plasma levels of methionine and cystine were below normal. Biochemical analyses of postmortem specimens of brain and liver revealed an increase in cystathionine in both tissues and a decrease in the cerebral content of methionine. Homocysteine-cysteine disulfide was present in brain but absent from liver. Decreased levels of methionine together with increased levels of homocyst(e)ine and cystathionine suggested the possibility of a defect in the remethylation of homocysteine. The enzyme content of liver and kidney was studied [90]. Activities of S-adenosylmethionine synthetase, cystathionine synthase, and cystathionase were normal. The activity of betaine–homocysteine methyltransferase was reduced slightly to 75% of the mean control activity in liver and to 39% of the mean control activity in kidney. The decrease in levels of N^5-methyltetrahydrofolate–homocysteine methyltransferase was more marked. Specific activities were 4% and 23% of the control values for liver and kidney, respectively. Coincubation experiments failed to demonstrate an enzyme inhibitor in the patient's tissues.

The patient had received large doses of folic acid and of vitamin B_{12}. Serum folate levels were high when they were measured with L. casei but were low when they were measured with *Pediococcus cerevisiae* or *Streptococcus faecalis*. This indicated the availability of N^5-methyltetrahydrofolate, the substrate for the impaired reaction. S-Adenosylmethionine participates in the N^5-methyltetrahydrofolate–homocysteine methyltransferase reaction. Hepatic levels of S-adenosylmethionine were normal despite low tissue levels of methionine [90]. Deficiency of this reactant could not explain deficient remethylation.

The patient also excreted methylmalonic acid in his urine [115]. The presence of this abnormality, in addition to the deficiency in N^5-methyltetrahydrofolate–homocysteine methyltransferase, suggested that the basic defect was an inadequate supply of active coenzyme forms of vitamin B_{12}. Serum B_{12} content and total hepatic B_{12} were normal but renal and hepatic concentrations of deoxyadenosyl-B_{12} were reduced markedly [115].

Mudd and his associates [47] proceeded to extend their observations and to support their hypothesis by employing sophisticated studies of cultured skin fibroblasts. Unlike control cells, fibroblasts from these patients failed to grow in a culture medium in which homocystine replaced methionine. Some growth occurred when large amounts of hydroxy-B_{12} were added with the homocystine. N^5-Methyltetrahydrofolate–homocysteine methyltransferase activity was markedly reduced in cell free extracts of the fibroblasts when routine assay conditions were used (Table VI, patient 1). With the addition of methyl-B_{12} to the assay system, enzyme activity increased to levels approximating the control values. Subsequently, Goodman *et al.* [117] reported two brothers with homocystinuria in combination with methylmalonic aciduria. Both brothers are relatively healthy and have developed normally. Despite the marked difference in clinical status between these boys and the initial patient, there is a similarity in the results of the studies of N^5-methyltetrahydrofolate–homocysteine methyltransferase in cultured skin fibroblasts (Table VI). When assayed under routine conditions, cells from the boys (patients 2 and 3) were deficient in the enzyme. The addition of methyl B_{12} to the assay media resulted in a marked increase in enzyme activity similar to the effect observed with cells from patient 1. It seems likely that extracts of fibroblasts derived from these patients contained the apoenzyme but lacked the required coenzyme.

The defect in methylmalonyl-coenzyme A isomerase was similarly defined [90]. The addition of deoxyadenosyl-B_{12} to the assay system caused a marked stimulation of the enzyme activity in an extract of the patient's liver. Specific activity increased from 5% to 31% of the lowest control value.

Failure to synthesize the coenzyme forms of vitamin B_{12} could result

TABLE VI

ENZYME ACTIVITIES IN CELL-FREE EXTRACTS OF FIBROBLASTS[a]

	Cystathionine synthase (nmoles/ mg protein/ 135 min)	N^5-CH_3-FH_4-homocysteine methyltransferase[b] (nmoles/mg protein/60 min)	
		No B_{12}	+ Methyl-B_{12}
Controls	3.7–65 (35)[c]	3.2–4.8 (4)	7.0–9.2 (4)
Cystathionine synthase deficiency	0.0–0.5 (29)	4.4–7.3 (3)	7.8–11.3 (3)
Deranged B_{12} metabolism			
Patient 1	23.6	0.03	4.5
Patient 2	23.1	0.6	3.0
Patient 3	10.4	0.4	2.7
Methylene-FH_4 reductase deficiency			
Patient 1	25.9, 37.6	3.5	5.5
Patient 2	36.7	–	–
Patient 3	23.2	4.1	7.0

[a] Data from previously published studies [48,62,117].
[b] Assays performed in the absence of vitamin B_{12} or in the presence of 5 μM methyl-B_{12}.
[c] Numbers in parenthesis indicate number of subjects studied.

either from a failure of cellular uptake or from a defect in the cellular metabolism of vitamin B_{12}. The fact that the total content of vitamin B_{12} in the liver of the first patient was 56 ng/gm wet weight (control 52–150) is not compelling evidence against the failure of uptake. Therapeutic doses of vitamin B_{12} resulted in extremely high plasma levels. Vitamin B_{12} may have entered the cell by a passive process despite a defect in a specific transport mechanism. The ratio of deoxyadenosyl B_{12}/total B_{12} was 0.23 compared to a control value of 0.65. However, studies of additional patients will be necessary before we can assign significance to this difference. More recent studies with fibroblasts cultured from this patient demonstrate that total B_{12} accumulation was one third of that observed in control lines, whereas accumulations of deoxyadenosyl-B_{12} and methyl-B_{12} are significantly lower [62]. As yet we cannot define whether the primary defect is failure of uptake or failure of metabolism. The preliminary data appear to support the latter hypothesis.

C. Homocystinuria with Decreased Methylenetetrahydrofolate Reductase

Another form of homocystinuria has been demonstrated in three teenage patients [48]. None of the patients had the typical symptoms of cystathionine synthase deficiency. Patient 1 was a 17-year-old girl who was evaluated at Johns Hopkins Hospital for schizophrenia and mental deterioration; patient 2 was her asymptomatic 15-year-old sister; and patient 3 was a 16-year-old boy who was studied at the Massachusetts General Hospital where he presented with muscle weakness, seizures, and an abnormal EEG. Homocystinuria, homocystinemia, and decreased levels of plasma methionine were found in all three patients.

In culture fibroblasts from these patients behaved like those obtained from patients with deranged metabolism of vitamin B_{12}, except that they contained normal levels of N^5-methyltetrahydrofolate–homocysteine methyltransferase (Table VI). The failure of fibroblast growth could not be attributed to a deficiency of N^5-methyltetrahydrofolate–homocysteine methyltransferase. Theoretically, deficiency of betaine–homocysteine methyltransferase could explain it. This seemed unlikely since cultured fibroblasts normally lack detectable amounts of this enzyme. More significantly, the hepatic level of betaine–homocysteine methyltransferase was normal in one patient. The failure of homocystine to support fibroblast growth despite adequate concentrations of N^5-methyltetrahydrofolate–homocysteine methyltransferase could be explained by a deficiency of the second substrate rather than of the enzyme. Subsequent studies showed that methylene–tetrahydrofolate reductase (EC 1.1.1.68) activity was deficient in extracts prepared from the fibroblasts (Table VII).

TABLE VII

METHYLENETETRAHYDROFOLATE REDUCTASE ACTIVITY IN CELL-FREE EXTRACTS OF FIBROBLASTS[a]

	Specific activity[b] (nmoles/ mg protein/60 min)		$\frac{-FAD}{+FAD} \times 100$
	−FAD	+FAD	
Controls (10)[c]	1.04–4.64	2.87–7.09	36.8–74.6
Cystathionine synthase deficiency (2)	3.34, 5.25	4.22, 8.95	58.8, 78.8
Deranged B_{12} metabolism (3)	2.02–5.07	3.03–7.71	65.5–66.9
Methylene-FH₄ reductase deficiency			
Patient 1 (a)	0.34	0.62	54.9
(b)	0.54	0.82	66.3
Patient 2	0.61	0.77	78.4
Patient 3	0.27	1.39	19.4

[a] Data from Mudd et al. [48].
[b] Assays were performed in the absence of flavin adenine dinucleotide (−FAD) or in the presence of 60 μM FAD (+FAD).
[c] The number of individuals in each group is indicated parentheses.

The assay data suggest that patient 3 may differ from the other two subjects with deficiency of methylene–tetrahydrofolate reductase. The addition of FAD (60 μM) caused a fivefold increase in the enzyme activity in fibroblasts from patient 3. The FAD induced increase was significantly lower in the other two patients as well as in the other subjects studied. It is possible that the genetic defect in the third patient results in decreased binding of cofactor by the apoenzyme. Formation of adequate amounts of holoenzyme can be achieved only in the presence of high concentrations of FAD.

Since methylene–tetrahydrofolate reductase is necessary for the synthesis of methyltetrahydrofolate, we assume that the concentration of this compound is reduced in the tissues of these three patients. Secondarily, as a consequence of deficient substrate, the N^5-methyltetrahydrofolate–homocysteine reaction is impaired.

IV. CYSTATHIONINURIA

A. Classification

Cystathioninemia with cystathioninuria results from an imbalance between the cystathionine synthase and the cystathionase reactions. As shown in Table VIII, the known causes of cystathioninuria can be divided into two categories, those that are related to augmented synthesis and those that are based on decreased degradation. It is likely that the rate of cystathionine synthesis depends on the availability of the substrate, homocysteine, as well as on the tissue content of the enzyme. For these reasons we would anticipate cystathionine accumulation when

TABLE VIII

CAUSES OF CYSTATHIONINURIA

Increased synthesis of cystathionine
  Decreased homocysteine methylation
  Neoplasms
    Neuroblastoma
    Hepatoblastoma
Decreased degradation of cystathionine
  Primary cystathionase deficiency
  Secondary cystathionase deficiency
    Vitamin B_6 deficiency
    Thyroid excess
    Hepatic diseases

there is more available homocysteine and normal levels of cystathionine synthase. This combination occurs in patients deficient in N^5-methyltetrahydrofolate–homocysteine methyltransferase and in whom cystathioninuria accompanies the homocystinuria. Future studies may document a similar amino aciduria in patients who are deficient in vitamin B_{12} or folic acid, as well as those who receive folate antagonists. An increase in cystathionine synthesis in excess of cystathionine degradation is the probable explanation for the cystathioninuria associated with several neoplasms [118–122]. It is likely that neuroblastoma, like normal neural tissue, possesses cystathionine synthase, but lacks cystathionase [35,64,123,124]. A similar imbalance in enzyme content may characterize hepatoblastomas [125–127]. Grossman found a marked increase in cystathionine synthase relative to cystathionase in two of six lines of Morris rat hepatomas [128].

A reduction in cystathionase activity may result in cystathioninuria. In experimental animals, the administration of thyroid hormone causes a significant reduction in hepatic cystathionase [129–132]. Similar effects may occur in clinical thyrotoxicosis [133]. Cystathioninuria occurs with vitamin B_6 deficiency [134,135]. Both cystathionine synthase [75,97–101] and cystathionase [136] require pyridoxal-P as a cofactor. In the experimental animal, there seems to be a marked difference in the affinities of the two apoenzymes for the coenzyme. Binding to apocystathionase is less avid and functional impairment of this enzyme occurs earlier in the course of vitamin depletion [98,99,137]. Finally, cystathioninuria may be due to specific, inherited defects in the cystathionase apoenzyme.

B. Clinical Manifestations

Since the initial report by Harris *et al.* in 1959 [138], approximately 20 additional cases of primary cystathioninuria have been described. In his review article Frimpter stresses the absence of well-defined and consistent clinical findings in patients with this abnormality [139]. Many of the patients are asymptomatic and it is likely that cystathioninuria itself reflects a benign defect. Of interest is the frequent association with other genetic defects including phenylketonuria [140], nephrogenic diabetes insipidus [141], cystinuria [142], and renal iminoglycinuria [143]. Whether this is a true association or whether it reflects the methods of case finding remains to be established.

C. The Enzyme Defect

Harris and his co-workers suggested that "there was a metabolic block at the point where normally cystathionine is cleaved . . . "

[138]. This conclusion, based on their finding of an increased level of cystathionine in a patient's liver and kidney, has been substantiated by *in vitro* enzyme assays in four subsequent patients [144–146]. Unfortunately, the *in vitro* studies were performed by three different groups, using different assay systems. Direct comparisons between the studies are difficult. The apparent heterogeneity of results may reflect the variations in methodology rather than true differences in the nature of the enzyme defect.

Frimpter studied two patients [144]. His assay system measured the conversion of L-[^{3}H]cystathionine to L-[^{3}H]cystine. In the absence of additional pyridoxal-P, the activity of control extracts ranged from 5200–14,500 dpm cystine/mg protein with a mean value of 10,200. Cystathionase activities in extracts prepared from liver biopsies from the patients had activities of 600 and 200 (5.9% and 2.0% of the mean control value). The addition of 0.2 mg pyridoxal-P to the reaction media (total volume not specified) resulted in a modest increase in the specific activity of the control extracts. The values were 96–244% of those obtained without added cofactor. In contrast, when 0.1 mg pyridoxal-P was added to the patients' extracts, the specific activities increased from 600 to 2400 and from 200 to 9900 cpm/mg protein. In the second patient (patient D) the value fell in the normal range.

Finkelstein *et al.* studied one patient [145]. Their assay measures the amount of α-[^{14}C]ketobutyrate derived enzymatically from [^{14}C]cystathionine. In the presence of 0.05 mM pyridoxal-P the specific activity of cystathionase in the patient's liver was 5% of the mean obtained in 12 control studies and was 13% of the lowest control value. Without pyridoxal-P, specific activities were reduced by 25%. In this respect, the patient's extract did not differ from the control specimens. The effect of greater increases in the concentration of pyridoxal-P was not studied.

Most recently, Tada and Yoshida [146] studied an additional patient. Their assay is based on the conversion of L-[^{14}C]serine and DL-homocysteine first to [^{14}C]cystathionine and then [^{14}C]cystine. Thus, the assay depends on the presence of both cystathionine synthase and cystathionase in the crude extract. The results are expressed as cystathionine formed (cpm/mg protein), cystine formed (cpm/mg protein), and the ratio of cystine/cystathionine. Tada and Yoshida found that the conversion of serine to cystine was reduced markedly in the patient's liver. The ratio of cystine/cystathionine was 16% of the mean value obtained from three control preparations. In this assay system, the concentration of pyridoxal-P was 0.45 mM.

Several reviews have stressed the differences in the results in these four patients. The similarities have not been emphasized. Cystathionase activity in each patient was significantly lower than the activity observed in the control population. In three specimens enzyme activity increased when pyridoxal-P was added *in vitro,* but approximated the control range in only one of these extracts. Similarly, three extracts were assayed in the presence of high concentrations of cofactor, and only one extract possessed a specific activity in the control range.

There is no evidence to suggest either factors that inhibit cystathionase or alternate reaction sequences competing for cystathionine in extracts prepared from the liver of a patient with cystathioninuria [145]. It is most likely that the deficient enzyme activity reflects an abnormality in the synthesis of the apoenzyme. Frimpter's observation that the *in vitro* addition of coenzyme enhanced enzyme activity to a greater degree in cystathioninuric livers suggests that, in some patients, the apoenzyme defect may impair binding of pyridoxal-P. This is minimized by increasing concentrations of coenzyme.

Cystathionase, crystallized from rat liver, catalyzes other reactions including deamination of homoserine [147] and desulfuration of cystine [148], cysteine [149–151], and homocysteine [103]. Homoserine dehydratase activity was decreased in the liver of a patient with cystathioninuria [145]. In control extracts, values for homoserine dehydratase were proportional to values for cystathionase, suggesting that a single protein also catalyzes both reactions in human liver [145]. As yet there have been no similar studies with human liver of other enzymatic activities which are properties of rat liver cystathionase.

D. Biochemical Consequences of the Defect

In considering the consequences of the defect in cystathionase, we must consider several significant points.

1. The impairment of cystathionase is incomplete and some residual enzyme activity is present [144–146]. Unlike the normal condition, the cystathionase reaction may be rate limiting when the patient is synthesizing cystathionine at a normal or at an augmented rate. Thus, the amount of dietary methionine as well as the tissue content of cystathionine synthase are important determinants. Similarly, the rate of homocysteine remethylation is significant. Deficiencies in folate, vitamin B_{12}, choline, etc., that reduce homocysteine remethylation may enhance the synthesis of cystathionine, thereby magnifying the defect in cystathionase.

2. Cystathionase deficiency may affect primarily those tissues in which this enzyme is normally active. In the rat the activity of cystathionase relative to cystathionine synthase is low in brain, adipose

tissue, and spleen [35,64,124]. If similar patterns pertain in man, further reduction of cystathionase may be insignificant in these tissues. An indirect effect on these tissues is possible, although unlikely. Increased plasma cystathionine, resulting from the failure of catabolism in other organs, may lead to tissue deposition through out the body. In the rat, however, the uptake, metabolism, and excretion of plasma L-cystathionine appears to occur largely in the kidney [75,152].

3. The biological functions as well as the toxic effects are undefined. Excess cystathionine does not affect the enzymes of methionine metabolism, either directly or indirectly. The cystathionine synthase reaction appears irreversible *in vivo* and is not inhibited by its product [37]. Cystathionine, added *in vitro,* has no effect on S-adenosylmethionine synthetase [153], S-adenosylhomocysteine synthase [64], betaine-homocysteine methyltransferase [32], N^5-methyltetrahydrofolate-homocysteine methyltransferase [153]. Finkelstein *et al.* [145] also demonstrated normal hepatic activities of S-adenosylmethionine synthetase and cystathionine synthase in a patient with cystathionase deficiency. The rate of synthesis of these enzymes seems to be unaffected by the excessive tissue content of cystathionine.

4. Cystathionase deficiency is accompanied by a deficiency of homoserine dehydratase [145]. In all likelihood cyst(e)ine and homocysteine desulfurase activities are equally depressed. But none of these reactions appears to be essential to normal metabolism. Alternate pathways are available for cyst(e)ine and homocysteine. Homoserine is a minor constituent of the normal diet but tissue levels of this amino acid increase following methionine ingestion [154,155]. The homoserine dehydratase reaction is the major catabolic route for homoserine [156]. However, another enzymatic reaction that yields α-keto-γ-hydroxybutyric acid is present in rabbit liver, normal human liver, and in liver from a patient with cystathionase deficiency [145,157].

In conclusion, there appears to be only one potential hazard to the patient with a defect in cystathionase — the inadequate synthesis of cyst(e)ine. The cystathionase reaction is an integral part of the transsulfuration sequence (Fig. 1). Absence of this enzyme transforms cysteine into an essential amino acid. In this regard the patient with cystathionase deficiency resembles the normal fetus [158]. However, the presence of some residual enzyme activity together with the content of cyst(e)ine in the normal diet reduce the likelihood of a clinically significant deficiency of this amino acid.

Few studies are available that test these predictions. There is an excess of cystathionine in cerebrospinal fluid [159–161], liver, kidney, and brain [45,138], as well as in plasma and urine. Concentrations of methionine and homocysteine are normal [159]. The administration of methionine results in a further increase in plasma cystathionine while the concentrations of methionine and homocyst(e)ine remain, as predicted, normal [140,143,160].

It is particularly unfortunate that there are few data evaluating the dietary cyst(e)ine requirement in patients with cystathionase deficiency. Plasma cystine levels have been normal in patients consuming a normal diet. Tada and Yoshida [146] measured the effect of methionine loading on urinary excretion of inorganic sulfate. Unlike the control subjects, the patient with cystathionase deficiency converted little, if any, methionine to sulfate.

E. Mode of Inheritance

In at least six of the reported studies of cystathioninuria, the authors noted additional instances of abnormal excretion of cystathionine in relatives of the index patient [122,143,159–162]. A familial pattern is clear and three groups can be identified. The patients and some relatives (usually siblings) have plasma cystathionine concentrations of 10–38 μM and a ratio of urinary cystathionine to creatinine (mg/mg) of 0.13–1.8. Frequently parents, other siblings, and other relatives have lower urinary cystathionine/creatinine ratios that still exceed the highest normal ratio of 0.01. Spontaneous cystathioninemia is absent in this group. Finally, other family members lack detectable abnormalities. For this reason, it appears likely that the patient with "typical" cystathioninuria is homozygous for an abnormal gene. However, we cannot apply the terms "dominant" and "recessive" since we can demonstrate lesser abnormalities in presumed heterozygotes.

The several reports reveal that there is considerable variation in the absolute values for plasma cystathionine, urinary cystathionine/creatinine, or daily urinary excretion of cystathionine. Under some circumstances, presumed heterozygotes may excrete as much cystathionine as do the homozygotes. In all likelihood this reflects changes in the rate of synthesis of cystathionine induced by differences in diet. For example, supplements of methionine may repress the synthesis of N^5-methyltetrahydrofolate–homocysteine methyltransferase [35] and increase the tissue content of cystathionine synthase [93]. Simultaneously, the methionine provides additional homocysteine. The net result would be increased cystathionine synthesis with a consequent increase in cystathioninemia and cystathioninuria. Indeed, methionine supplements are useful in the detection of the heterozygotes.

F. Treatment

There is no evidence to suggest that cystathionine excess is detrimental to the patient. But the benign nature of the defect remains to be established with certainty. For this reason, most patients are treated.

Frimpter et al. [159] observed a marked decrease in cystathionine excretion accompanied by a significant increase in urinary sulfate when they treated their patient with large doses of pyridoxine. Berlow [160] demonstrated that pyridoxine therapy caused a reduction in plasma cystathionine. Subsequently, several groups have confirmed the effectiveness of large doses of pyridoxine. Only Tada [146] reported a patient in whom large dosages of vitamin B_6 failed to lessen the biochemical abnormalities.

Thus, most cases of primary or familial cystathioninuria illustrate a dependency on an excessive intake of vitamin B_6. Since other pyridoxine requiring enzyme systems are intact, it is unlikely that there is a defect in the conversion of the vitamin to active, coenzyme forms. Several possible explanations have been advanced. Pyridoxine therapy could stimulate an alternate pathway for the metabolism of cystathionine or for one of its metabolic precursors. Until we have detailed balance studies, this unlikely possibility cannot be excluded. Frimpter's suggestion that the genetic defect resulted in decreased affinity between apoenzyme and coenzyme is attractive. As indicated in an earlier section, this hypothesis is supported by in vitro enzyme assays in two patients [144]. The only contradictory data come from the enzyme studies by Finkelstein et al. [145] performed on tissue obtained from Berlow's patient who responded to pyridoxine therapy. The potential reasons for the differences between these studies have been discussed [103,163].

Not all cases of primary cystathioninuria are pyridoxine responsive. Tada's patient showed no biochemical improvement with vitamin B_6 treatment and this was paralleled by the failure of pyridoxal-P to stimulate cystathionase activity in vitro.

In summary, we must still resolve the question of therapy in familial cystathioninuria. As yet, we have not established the clinical need to reduce tissue cystathionine concentrations nor have we demonstrated that massive doses of vitamin B_6 have no untoward effects [164].

V. SULFITE OXIDASE DEFICIENCY

Only one case of sulfocysteinuria caused by a deficiency of sulfite oxidase has been reported to date [165]. Yet the studies of this one child have yielded important information concerning the terminal step in sulfur amino acid metabolism in man.

The patient died at age 31 months. He had neurological abnormalities from birth and was almost decorticate by age nine months. Subluxation of his ocular lenses occured during the first year. Three of seven older siblings had died with marked central nervous system dysfunction during infancy. Postmortem examinations failed to reveal specific or consistent abnormalities. The parents were not consanguineous and the parents and the unaffected siblings were clinically normal.

A. Biochemical Abnormalities

Detailed analyses of the patient's urine revealed several notable abnormalities [165]. Irreverre *et al.* isolated S-sulfo-L-cysteine by crystallization. Subsequently they established the identity of the compound by comparing it with authentic S-sulfo-L-cysteine. They also found a marked decrease in urinary inorganic sulfate. In normal subjects inorganic sulfate represents 60% or more of the urinary sulfur. Inorganic sulfate in the patient's urine accounted for less than 5% of the total sulfur. Other sulfur containing compounds were present in increased amounts. Thiosulfate represented 16–24% of the total urinary sulfur, taurine represented 15–20%, and inorganic sulfite, 11%. Following dietary supplementation with L-cysteine, total urinary sulfur increased appropriately, but the increase was distributed proportionally among all the urinary forms. The patient's parents and four normal siblings showed normal patterns for sulfur excretion in the urine.

S-Sulfo-L-cysteine appeared to be present in the patient's plasma at a concentration of 47–70 μM. Normal plasma contains no more than 11 μM. However, neither S-sulfo-L-cysteine nor any other sulfo derivatives were present in extracts of brain, kidney, or liver.

B. The Enzyme Defect

Mudd *et al.* [166] proposed that a defect in sulfite oxidase (EC 1.8.3.1; sulfite: oxygen oxidoreductase) could explain the biochemical abnormalities. They confirmed this hypothesis by the direct determination of the enzyme activity in tissues obtained postmortem from the patient (Table IX). Sulfite oxidase was present in control specimens of liver, brain, and kidney but was undetectable in the patient's tissues. The fact that normal activities for S-adenosylmethionine synthetase, cystathionine synthase, and cystathionase were detected in the patient's liver suggests that the deficiency in sulfite oxidase did not result from nonspecific postmortem changes. Using coincubation studies, Mudd *et al.* excluded the possibility of an inhibitor of sulfite oxidase in the patient's tissues. They also tested the possibility that enzyme activity

TABLE IX

SULFITE OXIDASE IN HUMAN TISSUES[a]

	Liver	Brain	Kidney
Control subjects			
Number	10	5	4
Range	22–87	10–13	22–103
Mean	51	11.5	63
Patient	0	0	0
Sensitivity	$\leqslant 2$	$\leqslant 1.5$	$\leqslant 1.0$

[a] Assays were performed on acetone powders prepared from tissues obtained postmortem. Specific activity is expressed as units/mg protein [166].

would be detected if they employed oxygen, rather than cytochrome c, as the electron acceptor in their assay system. Even under aerobic conditions, extracts of the patient's liver showed less than normal ability to convert sulfite to sulfate.

C. Metabolic Consequences of the Defect

As stated by the authors [166], the most significant aspect of their study is the demonstration that "enzymatic catalysis by sulfite oxidase is essential *in vivo*." Autooxidation of sulfite may be limited in animal tissues by low concentrations of free metal ions that are required for catalysis as well as by the presence of numerous compounds that are capable of breaking the free radical chain reaction.

The defect in sulfite oxidase can be correlated directly with the increased urinary sulfite and decreased urinary sulfate but abnormalities in the excretion of thiosulfate, S-sulfo-L-cysteine, and taurine require other factors. Presumably the thiosulfate is formed by the enzyme catalyzed reaction of sulfite with β-mercaptopyruvate as described by Sörbo [167] [Eq. (1)]:

$$SO_3^{2-} + HSCH_2COCO_2H \rightarrow S_2O_3^{2-} + CH_3COCO_2H \tag{1}$$

It is interesting that the patient did not excrete excessive quantities of thiocyanate since an excess of thiocyanate could be anticipated from the rhodanese reaction [Eq. (2)]. The availability of cyanide may limit this reaction [165].

$$S_2O_3^{2-} + CN^- \rightarrow SO_3^{2-} + SCN^- \tag{2}$$

Perhaps a high concentration of sulfite also inhibits thiocyanate formation. In mammals the metabolism of thiocyanate is limited [168].

Therefore, the presence of normal thiocyanate levels in the urine of this patient makes it unlikely that the excess of sulfite is generated from the rhodanese reaction.

S-Sulfo-L-cysteine is the product of the chemical reaction between sulfite and cystine. A similar, nonenzymatic reaction can produce S-sulfo-L-homocysteine and S-sulfo-L-glutathione. Neither of these compounds appeared in the patient's tissues or urine, suggesting the possibility of an enzymatic mechanism for the synthesis of S-sulfo-L-cysteine.

Finally, the conversion of sulfite to cysteine sulfinic acid may explain the excessive urinary excretion of taurine [169,170].

D. Pathophysiology

Inadequate tissue concentrations of inorganic sulfate might result from the failure to oxidize sulfite. In turn, the decrease in inorganic sulfate might impair the synthesis of the organic sulfate esters which are constituents of myelin, cartilage, and acid mucopolysaccharides. However, Percy et al. [171] found that the sulfatides of brain and kidney were normal qualitatively and were present in normal concentrations. Since the brain weight was reduced, the total content of sulfatides was reduced, rendering their observation inconclusive. Decreased content of sulfatides may be either the cause or the consequence of the generalized cerebral atrophy. Since the patient excreted normal quantities of tyrosine O-sulfate and indoxyl 3-sulfate, it is likely that tissue concentrations of sulfate were adequate [171]. Undetected, residual sulfite oxidase activity is one possible explanation. An alternative source of organic sulfate is dietary sulfate. Finally, activated sulfate may be derived from the oxidation of homocysteine [9].

The patient with sulfite oxidase deficiency developed osteoporosis and subluxation of the ocular lenses. Both abnormalities characterize most patients with cystathionine synthase deficiency. Perhaps the common abnormality is a defective collagen due to the formation of the fibril in the presence of excessive concentrations of either sulfite or homocysteine. We must study additional patients in order to relate the specific abnormalities to the deficiency in sulfite oxidase.

VI. HYPERMETHIONINEMIA

Transient neonatal hypermethioninemia is not uncommon [172,173]. In most cases the abnormality disappears when the amount of protein in the diet is reduced. This benign increase in plasma methionine results from a delay in the maturation of the capacity for transsulfuration. Gaull *et al.* [158] have shown that the specific activities of S-adenosyl-methionine synthetase and cystathionine synthase in fetal and neonatal human liver are 25% of the values found in mature control subjects. Fetal liver lacks cystathionase but livers of full term babies contain the enzyme in concentrations approximating those of mature liver. The specific activity of betaine–homocysteine methyltransferase is comparable in mature and neonatal liver, but activity of N^5-methyltetrahydrofolate–homocysteine methyltransferase is greater than that observed in the control specimens [174]. These enzyme patterns suggest that methionine conservation, rather than transsulfuration, is the predominent metabolic pathway in the infant. Indeed, Gaull has emphasized the likelihood of the essential role of cyst(e)ine in the diet of infants [158].

Hypermethioninemia may occur in older children. In this age group, the clinical picture is rarely benign. Increases in the level of plasma methionine may result from cystathionine synthase deficiency (see Section II,D,1). Hypermethioninemia may occur in the course of tyrosinosis, although the relationship between the defects in the metabolism of tyrosine and methionine remains unclear [175]. Hypermethioninemia may appear either concurrently with tyrosinemia or may appear after the levels of plasma tyrosine return to normal. Scriver postulated that the defect in tyrosine metabolism (deficiency of *p*-hydroxyphenylpyruvic acid oxidase) is primary, causing the child to suffer severe liver damage; one manifestation being impaired metabolism of methionine [176]. Even when dietary management controls the tyrosine accumulation, the hypermethioninemia may persist because of previous irreversible liver damage or perhaps as a self-sustaining manifestation of methionine toxicity. It is difficult to reconcile several of the reported cases with Scriver's hypothesis. We are limited by the paucity of studies that measure the tissue content of the relevant metabolites and/or enzymes.

In 1965 Perry *et al.* described a family in which three children died in infancy with hepatic cirrhosis [177]. The third child was studied in detail. He was 11 weeks old when he was admitted to the hospital with ascites and hepatomegaly. He died one week later. The pathological findings were identical to those observed in his two siblings. The liver showed a diffuse intralobular fibrosis with occasional regenerative nodules. The hepatocytes demonstrated a "reversion to ductular structures" and contained iron and fat. The kidneys were grossly enlarged and tubular dilation was marked. Hyperplasia of the islets of Langerhans characterized the pancreas. The biochemical abnormalities included (1) a generalized aminoaciduria including homocystine and cystathionine with a disproportionate increase in the excretion of tyrosine and methionine; (2) a 30-fold increase in serum methionine; (3) increased cerebral methionine with normal content of cystathionine; and (4) urinary excretion of the keto acid derivatives of methionine, tyrosine, and phenylalanine. The mother was tested with a methionine load. Her response was normal as measured by serum levels of methionine.

Perry *et al.* believed that the magnitude of the abnormalities in methionine metabolism was incompatible with nonspecific, severe hepatic failure. In particular the presence of α-keto-γ-methiolbutyric acid was unique. Subsequently, Mudd [178] studied the enzyme activities in liver tissue obtained at autopsy. His findings are summarized in Table X. Mudd used postmortem specimens from children less than four years of age as controls. While the conditions of storage prior to analysis may introduce considerable variation, the activities of the four enzymes studied did not deviate markedly from the control range. The value for cystathionine synthase was 20% of the mean control activity. The presence of cystathionine in the urine and the brain of this patient suggests that the level of cystathionine synthase was not grossly deficient. The level of S-adenosylmethionine synthetase was also in the control range. The tissue content of S-adenosylmethionine was increased markedly, precluding the presence of a defect in the cellular uptake of methionine. It seemed possible that the studies *in vitro* failed to reflect the effect of several abnormal metabolities *in vivo*. *In vitro* studies demonstrated no effect of derivatives of tyrosine on S-adenosylmethionine synthetase prepared from rat liver [153]. Tyrosine, phenylalanine, hydroxyphenyllactate, hydroxphenylpyruvate, and hydroxyphenylacetate did not inhibit the reaction when present in concentrations equal to 20 times that of the substrate methionine.

TABLE X

ENZYME STUDIES IN HYPERMETHIONINEMIA[a]

Investigator	S-AdoMet synthetase	Cystathionine synthase	BH enz[b]	mTHF enz[b]	Cystathionase	S-AdoHcys[b] synthase
Mudd	193	20	17	220	—	—
Gaull I	27	29	—	—	83	—
II	21	6	—	—	86	—
Finkelstein and Dieterlen	32	42	113*	54*	91	100*

[a] Values are presented as percentage of mean control specific activity except that values designated * are percentage of highest control specific activity.

[b] Abbreviations: BH enz, betaine-homocysteine methyltransferase; mTHF enz, N^5-methyltetrahydrofolate-homocysteine methyltransferase; S-AdoHcys synthase, S-adenosylhomocysteine synthase.

Gaull and his co-workers studied two patients [179]. Patient I was five months old when she was evaluated for hepatosplenomegaly, ascites, and a bleeding diathesis. Analysis of her urine demonstrated a generalized aminoaciduria characterized by a disproportionate increase in the excretion of tyrosine and methionine. Following restriction of the dietary intake of tyrosine and phenylalanine, the patient improved dramatically and clinical evidence of hepatocellular dysfunction became minimal. The plasma concentration of methionine persisted despite a reduction in the concentration of tyrosine. Subsequently, the level of plasma methionine declined when there was a reduction in the intake of this amino acid. Ingestion of tyrosine, however, resulted in renewed tyrosinemia. At age 28 months the patient was eating a normal diet. Plasma amino acids were normal. The only residual clinical abnormality was persistent hepatomegaly.

Gaull et al. studied a liver biopsy obtained during a period characterized by hypermethioninemia without tyrosinemia. As shown in Table X, hepatic tissue from patient I contained cystathionase at a normal specific activity. There was a reduction in the specific activities of both S-adenosylmethionine synthetase and cystathionine synthase.

Gaull's second patient died at age 11 weeks. Her illness included hepatosplenomegaly and hematemesis. The concentrations of tyrosine and methionine were increased in plasma and cerebrospinal fluid. The patient died despite a reduction in plasma methionine following dietary restriction. Acute purulent peritonitis was found at autopsy. Hepatic enzyme activities in patient II resembled those found in the first patient (Table X).

In 1970 Dr. M. Dieterlen (in Prof. Beaudoing's department in Grenoble, France) detected methionine in the urine of an asymptomatic infant. Hypermethioninemia was present and persisted for 27 months (2.1–8.0 mg%). There were no other biochemical abnormalities. The plasma concentration of tyrosine remained normal. The urine contained neither homocystine nor α-keto-γ-methiolbutyric acid. Growth and development were unremarkable. Renal and hepatic function were normal. Finkelstein measured the activities of six enzymes in a liver biopsy specimen [180]. As shown in Table X, he found normal activities of betaine–homocysteine methyltransferase, cystathionase, and S-adenosylhomocysteine synthase. The activities of S-adenosylmethionine synthetase and cystathionine synthase were below the range of control values. The specific activity of N^5-methyltetrahydrofolate–homocysteine methyltransferase was within the control range. (Indeed, considering the presence of increased levels of methionine, the activity of this enzyme may be disproportionately high.) The decreased activity of S-adenosylmethionine synthetase was not due to the presence of any soluble inhibitory factor.

Limited studies comparing the kinetics of S-adenosylmethionine synthetase in the patient's liver with that in a control specimen, sent simultaneously from France, were of considerable interest [180]. With the concentration of ATP fixed at 15 mM and the concentration of methionine varied from 20 μM to 200 μM, the apparent Michaelis constants for methionine were 24 μM (patient's enzyme), 360 μM (control human liver), and 250 μM (rat liver enzyme). In addition, the specific activity of S-adenosylmethionine synthetase in the patient's liver approached 7.6% of the control value when derived from the calculated maximum velocities. This is in contrast to the value of 32% (Table X) derived from studies with 100 μM methionine.

From these four cases it is apparent that there are marked variations in the clinical findings, as well as in the biochemical abnormalities, in the cases of hypermethioninemia. In all likelihood this is not a single disease entity but a heterogenous population sharing one common finding, an increase in plasma methionine. Gaull's second patient and the patient studied by Perry et al. and Mudd had similar clinical findings. Yet the biochemical characteristics differed markedly in the two cases. Conversely Gaull's two patients and the patient of Finkelstein and Dieterlen shared common enzyme abnormalities. In all three children hepatic levels of S-adenosylmethionine synthetase and cystathionine synthase were decreased but cystathionase activity was normal. Gaull et al. suggest that a normal value for cystathionase precludes the possibility that acquired liver disease causes the enzyme changes. But in the same paper, these investigators described exactly this pattern of enzymes in five patients with various types of cirrhoses. For this reason, the studies of Finkelstein and Dieterlen are significant. Their patient lacked both tyrosinemia and evidence of hepatic dysfunction.

The data from these four cases suggest two mechanisms that alone, or in combination, could result in the accumulation of methionine. The first is a disproportionate decrease in transsulfuration relative to homocysteine remethylation as reflected by the decrease in cystathionine synthase relative to N^5-methyltetrahydrofolate–homocysteine methyltransferase. The second possibility is a primary defect in S-adenosylmethionine synthetase. As noted above, this may be an unusual mutation that increases the affinity of the enzyme for its substrate. As a result, the mutant enzyme may be saturated at concentrations of methionine within the physiological range. Further increases in substrate concentration would not "overcome" the metabolic block—in contrast to the usual situation in which the abnormal enzyme has a decreased affinity for substrate.

Obviously this hypothesis, provoked by limited studies in a single case, must be supported by additional observations in other patients. Furthermore, it must be reconciled with Mudd's finding of an increase in hepatic S-adenosylmethionine.

VII. CONCLUSIONS

Victor McKusick concludes the preface to his monograph with the following quotation from William Harvey [18].

> Nature is nowhere accustomed more openly to display her secret mysteries than in cases where she shows traces of her workings apart from the beaten path; nor is there any better way to advance the proper practice of medicine than to give our minds to the discovery of the usual law of Nature by careful investigation of cases of rare forms of disease. For it has been found, in almost all things, that what they contain of useful or applicable nature is hardly perceived unless we are deprived of them, or they become deranged in some way.

The appropriateness of this citation to the value of the studies of the genetic disorders of sulfur amino acid metabolism is obvious. Certainly, investigations of these "rarer forms of disease" have illuminated much of the usual "law of Nature."

We can derive several major conclusions from the studies of these patients including the following:

1. The metabolic pathway for methionine catabolism in man is the transsulfuration sequence requiring both the cystathionine synthase and the cystathionase reactions. Impairment of either enzyme converts cyst(e)ine into an essential nutrient.

2. The intracellular concentration of methionine is regulated, at least in part, by the distribution of homocysteine among several competing reactions. Two enzymes are of major importance. Cystathionine synthase removes homocysteine thereby lowering tissue methionine. N^5-Methyltetrahydrofolate–homocysteine methyltransferase conserves homocysteine and maintains tissue methionine. The unopposed functioning of cystathionine synthase, in patients with impairment of N^5-methyltetrahydrofolate–homocysteine methyltransferase, leads to decreased tissue methionine. Conversely, in patients with cystathionine synthase deficiency, augmented remethylation of homocysteine results in increased concentrations of methionine.

3. In the absence of N^5-methyltetrahydrofolate–homocysteine methyltransferase, betaine–homocysteine methyltransferase alone cannot prevent either depletion of methionine or accumulation of homocysteine in the tissues. Previous concepts of the preeminent role of betaine–homocysteine methyltransferase in methionine homeostasis must be reconsidered.

4. Enzymatic conversion of inorganic sulfite to sulfate is an essential reaction in sulfur amino acid metabolism.

ACKNOWLEDGMENTS

I wish to thank Dr. Leonard Laster who initiated my interest in sulfur amino acid metabolism. From the beginning, I have benefitted from my continued formal and informal collaborations with Dr. S. H. Mudd. I am also grateful to Walter E. Kyle and Barbara Harris for the sustained excellence of their technical assistance. In particular, I am indebted to Ann-Marie Pick for her help in all aspects of the preparation of this article.

Our research activities have been supported by the Veterans Administration and by Grant AM-13048 from the National Institutes of Health.

REFERENCES

1. L. W. Kinsell, H. A. Harper, H. C. Barton, M. E. Hutchin, and J. R. Hess, *J. Clin. Invest.* **27,** 677 (1948).
2. L. W. Kinsell, H. A. Harper, G. K. Giese, S. Margen, D. P. McCallie, and J. R. Hess, *J. Clin. Invest.* **28,** 1439 (1949).
3. F. L. Iber, H. Rosen, S. M. Levenson, and T. C. Chalmers, *J. Lab. Clin. Med.* **50,** 417 (1957).
4. S. Chen, L. Zieve, and V. Mahadevan, *J. Lab. Clin. Med.* **75,** 628 (1970).
5. D. M. Israelstam, T. Sargent, N. N. Finley, H. S. Winchell, M. B. Fish, J. Motto, M. Pollycove, and A. Johnson, *J. Psychiat. Res.* **7,** 185 (1970).
6. C. K. Cohn, D. L. Dunner, and J. Axelrod, *Science* **170,** 1323 (1970).
7. K. S. McCully, *Amer. J. Pathol.* **59,** 181 (1970).
8. K. S. McCully and B. D. Ragsdale, *Amer. J. Pathol.* **61,** 1 (1970).
9. K. S. McCully, *Amer. J. Pathol.* **66,** 83 (1972).
10. R. K. Datta and B. Dattam *Exp. Mol. Pathol.* **10,** 129 (1969).
11. N. A. J. Carson and D. W. Neill, *Arch. Dis. Childhood* **37,** 505 (1962).
12. T. Gerritsen, J. G. Vaughn, and H. A. Waisman, *Biochem. Biophys. Res. Commun.* **9,** 493 (1962).
13. G. L. Spaeth and G. W. Barber, *Trans. Amer. Acad. Ophthalmol. Otolaryngol.* **69,** 912 (1965).
14. N. A. J. Carson, D. C. Cusworth, C. E. Dent, C. M. B. Field, D. W. Neill, and R. G. Westall, *Arch. Dis. Childhood* **38,** 425 (1963).
15. T. Gerritsen and H. A. Waisman, *Pediatrics* **33,** 413 (1964).
16. R. N. Schimke, V. A. McKusick, T. Huang, and A. D. Pollack, *J. Amer. Med. Ass.* **193,** 711 (1965).
17. V. A. McKusick, J. G. Hall, and F. Char, *in* "Inherited Disorders of Sulphur Metabolism" (N. A. J. Carson and D. N. Raine, eds.), p. 179. Churchill, London, 1971.
18. V. A. McKusick, "Heritable Disorders of Connective Tissue." Mosby, St. Louis, Missouri, 1972.
19. J. W. Cline, R. A. Goyer, J. Lipton, and R. G. Mason, *S. Med. J.* **64,** 613 (1971).
20. A. D. Jensen and H. E. Cross, *Amer. J. Ophthalmol.* **75,** 405 (1973).
21. G. D. Presley, I. N. Stinson, and J. G. Sidbury, Jr., *S. Med. J.* **62,** 944 (1969).
22. P. Henkind and N. Ashton, *Trans. Ophthalmol. Soc. U.K.* **85,** 21 (1965).
23. L. McDonald, C. Bray, C. Field, F. Love, and B. Davies, *Lancet* **1,** 744 (1964).
24. O. D. Ratnoff, *Science* **162,** 1007 (1968).

15. DEFECTS IN SULFUR AMINO ACID METABOLISM 593

25. J. B. Gibson, N. A. J. Carson, and D. W. Neill, *J. Clin. Pathol.* **17**, 427 (1964).
26. N. A. J. Carson, C. E. Dent, C. M. B. Field, and G. E. Gaull, *J. Pediat.* **66**, 565 (1965).
27. G. De La Haba and G. L. Cantoni, *J. Biol. Chem.* **234**, 603 (1959).
28. V. Zappia, C. R. Zydek-Cwick, and F. Schlenk, *J. Biol. Chem.* **244**, 4499 (1969).
29. T. Deguchi and J. Barchas, *J. Biol. Chem.* **246**, 3175 (1971).
30. J. K. Coward, M. D'Urso-Scott, and W. D. Sweet, *Biochem. Pharmacol.* **21**, 1200 (1972).
31. C. Kutzbach and E. L. R. Stokstad, *Biochim. Biophys. Acta* **139**, 217 (1967).
32. J. D. Finkelstein, B. J. Harris, and W. E. Kyle, *Arch. Biochem. Biophys.* **153**, 320 (1972).
33. H. Dickerman, B. G. Redfield, J. G. Bieri, and H. Weissbach, *J. Biol. Chem.* **239**, 2545 (1964).
34. C. Kutzbach, E. Galloway, and E. L. R. Stokstad, *Proc. Soc. Exp. Biol. Med.* **124**, 801 (1967).
35. J. D. Finkelstein, W. E. Kyle, and B. J. Harris, *Arch. Biochem. Biophys.* **146**, 84 (1971).
36. S. H. Mudd, J. D. Finkelstein, F. Irreverre, and L. Laster, *Science* **142**, 1443 (1964).
37. S. H. Mudd, J. D. Finkelstein, F. Irreverre, and L. Laster, *J. Biol. Chem.* **240**, 4382 (1965).
38. J. D. Finkelstein, S. H. Mudd, F. Irreverre, and L. Laster, *Science* **146**, 785 (1964).
39. L. Laster, G. L. Spaeth, S. H. Mudd, and J. D. Finkelstein, *Ann. Intern. Med.* **63**, 1117 (1965).
40. S. H. Mudd, L. Laster, J. D. Finkelstein, and F. Irreverre, *in* "Amines and Schizophrenia" (H. N. Himwich, J. R. Smytheis, and S. S. Kety, eds.), p. 247. Pergamon, Oxford, 1966.
41. S. H. Mudd, W. A. Edwards, P. M. Loeb, M. S. Brown, and L. Laster, *J. Clin. Invest.* **49**, 1762 (1970).
42. J. G. Hollowell, Jr., M. E. Corvell, W. K. Hall, J. K. Findley, and T. G. Thevaos, *Proc. Soc. Exp. Biol. Med.* **129**, 327 (1968).
43. T. Yoshida, K. Tada, Y. Yokoyama, and T. Arakawa, *Tohoku J. Exp. Med.* **96**, 235 (1968).
44. G. E. Gaull, D. K. Raooin, and J. A. Sturman, *Neuropaediatrie* **1**, 199 (1969).
45. D. P. Brenton, D. C. Cusworth, and G. E. Gaull, *Pediatrics* **35**, 50 (1965).
46. B. W. Uhlendorf and S. H. Mudd, *Science* **160**, 1007 (1968).
47. S. H. Mudd, B. W. Uhlendorf, and K. R. Hinds, *Biochem. Med.* **4**, 215 (1970).
48. S. H. Mudd, B. W. Uhlendorf, J. M. Freeman, J. D. Finkelstein, and V. E. Shih, *Biochem. Biophys. Res. Commun.* **46**, 905 (1972).
49. B. W. Uhlendorf, E. B. Conerly, and S. H. Mudd, *Pediat. Res.* **7**, 645 (1973).
50. A. H. Bittles and N. A. J. Carson, *J. Med. Genet.* **10**, 120 (1973).
51. L. D. Fleisher, H. H. Tallan, N. G. Beratis, K. Hirschhorn, and G. E. Gaull, *Biochem. Biophys. Res. Commun.* **55**, 38 (1973).
52. J. L. Goldstein, B. K. Campbell, and S. M. Gartler, *J. Clin. Invest.* **51**, 1034 (1972).
53. J. L. Goldstein, B. K. Campbell, and S. M. Gartler, *J. Clin. Invest.* **52**, 218 (1973).
54. G. E. Gaull and M. K. Gaitonde, *J. Med. Genet.* **3**, 194 (1966).
55. M. R. Seashore, J. L. Durant, and L. E. Rosenberg, *Pediat. Res.* **6**, 187 (1972).
56. S. Kashiwamata, Y. Kotake, and D. M. Greenberg, *Biochim. Biophys. Acta* **212**, 501 (1970).
57. C. Kennedy, V. E. Shih, and L. R. Rowland, *Pediatrics* **36**, 736 (1965).
58. D. P. Brenton, D. C. Cusworth, and G. E. Gaull, *J. Pediat.* **67**, 58 (1965).
59. H. G. Dunn, T. L. Perry, and C. L. Dolman, *Neurology* **16**, 407 (1966).

60. K. Tada, T. Yoshida, H. Hirono, and T. Arakawa, *Tohoku J. Exp. Med.* **92,** 325 (1967).
61. K. Tada, T. Yoshida, and T. Arakawa, *Tohoku J. Exp. Med.* **101,** 232 (1970).
62. M. J. Mahoney, L. E. Rosenberg, S. H. Mudd, and B. W. Uhlendorf, *Biochem. Biophys. Res. Commun.* **44,** 375 (1971).
63. M. C. Carey, J. Fennelly, and O. Fitzgerald, *Amer. J. Med.* **45,** 26 (1968).
64. J. D. Finkelstein and B. J. Harris, *Arch. Biochem. Biophys.* **159,** 160 (1973).
65. T. L. Perry, S. Hansen, D. MacDougall, and P. D. Warrington, *Clin. Chim. Acta* **15,** 409 (1967).
66. D. A. Applegarth, D. F. Hardwick, F. Ingram, N. L. Auckland, and G. Bozoian, *N. Engl. J. Med.* **285,** 1265 (1971).
67. J. A. Duerre, C. H. Miller, and G. G. Reams, *J. Biol. Chem.* **244,** 107 (1969).
68. T. L. Perry, *in* "Inherited Disorders of Sulphur Metabolism" (N. A. J. Carson and D. N. Raine, eds.), p. 224. Churchill, London, 1971.
69. H. H. Tallan, S. Moore, and W. H. Stein, *J. Biol. Chem.* **230,** 707 (1958).
70. T. Gerritsen and H. A. Waisman, *Science* **145,** 588 (1964).
71. P. W. K. Wong, V. Schwarz, and G. M. Komrower, *Pediat. Res.* **2,** 149 (1968).
72. G. E. Gaull, Y. Wada, K. Schneidman, D. R. Rassin, H. H. Tallan, and J. A. Sturman, *Pediat. Res.* **5,** 265 (1971).
73. J. A. Sturman, K. Schneidman, and G. E. Gaull, *Biochem. Med.* **5,** 404 (1971).
74. P. W. K. Wong and R. Fresco, *Pediat. Res.* **6,** 172 (1972).
75. F. C. Brown and P. H. Gordon, *Biochim. Biophys. Acta* **230,** 434 (1971).
76. L. Laster, S. H. Mudd, J. D. Finkelstein, and F. Irreverre, *J. Clin. Invest.* **44,** 1708 (1965).
77. D. P. Brenton and D. C. Cusworth, *Clin. Sci.* **31,** 197 (1966).
78. D. P. Brenton, D. C. Cusworth, C. E. Dent, and E. E. Jones, *Quart. J. Med.* **35,** 325 (1966).
79. T. L. Perry, S. Hansen, and L. MacDougall, *Science* **152,** 1748 (1966).
80. A. Lemonnier, J. L. Pousset, C. Charpentier, and N. Moatti, *Clin. Chim. Acta* **33,** 359 (1971).
81. H. Kodama, S. Ohmori, M. Suzuki, S. Mizuhara, T. Oura, G. Isshiki, and I. Uemura, *Physiol. Chem. Phys.* **3,** 81 (1971).
82. H. Kodama, *Physiol. Chem Phys.* **3,** 159 (1971).
83. S. Ohmori, H. Kodama, and S. Mizuhara, *Physiol. Chem. Phys.* **4,** 286 (1972).
84. E. D. Harris and A. Sjoerdsma, *Lancet* **2,** 707 (1966).
85. A. H. Kang and R. L. Trelstad, *J. Clin. Invest.* **52,** 2571 (1973).
86. D. F. Hardwick, D. A. Applegarth, D. M. Cockcroft, P. M. Ross, and R. J. Calder, *Metab., Clin. Exp.* **19,** 381 (1970).
87. T. L. Perry, *in* "Inherited Disorders of Sulphur Metabolism" (N. A. J. Carson and D. N. Raine, eds.), p. 245. Churchill, London, 1971.
88. G. M. Komrower and I. B. Sardharwalia, *in* "Inherited Disorders of Sulphur Metabolism" (N. A. J. Carson and D. N. Raine, eds.), p. 254. Churchill, London, 1971.
89. T. L. Perry, S. Hansen, D. L. Love, L. E. Crawford, and B. Tischler, *Lancet* **2,** 474 (1968).
90. S. H. Mudd, H. L. Levy, and G. Morrow, III, *Biochem. Med.* **4,** 193 (1970).
91. D. C. Cusworth and A. Gattereau, *Lancet* **2,** 916 (1968).
92. Z. M. Falchuk, W. A. Edwards, and L. Laster, *Metab., Clin. Exp.* 22, 605 (1973).
93. J. D. Finkelstein and S. H. Mudd, *J. Biol. Chem.* **242,** 873 (1967).
94. G. W. Barber and G. L. Spaeth, *Lancet* **1,** 337 (1967).
95. D. P. Brenton and D. C. Cusworth, *in* "Inherited Disorders of Sulphur Metabolism" (N. A. J. Carson and D. N. Raine, eds.), p. 264. Churchill, London, 1971.

96. N. A. J. Carson *in* "Inherited Disorders of Sulphur Metabolism" (N. A. J. Carson and D. N. Raine, eds.), p. 284. Churchill, London, 1971.
97. H. Nakagawa and H. Kimura, *Biochem. Biophys. Res. Commun.* **32**, 208 (1968).
98. J. A. Sturman, P. A. Cohen, and G. E. Guall, *Biochem. Med.* **3**, 244 (1969).
99. J. D. Finkelstein and F. T. Chalmers, *J. Nutr.* **100**, 467 (1970).
100. S. Kashiwamata and D. M. Greenberg, *Biochim. Biophys. Acta* **212**, 488 (1970).
101. F. C. Brown and P. H. Gordon, *Can. J. Biochem.* **49**, 484 (1971).
102. A. E. Pegg and H. G. Williams-Ashman, *J. Biol. Chem.* **244**, 682 (1969).
103. M. P. Roisin and F. Chatagner, *Bull. Soc. Chim. Biol.* **51**, 481 (1969).
104. L. E. Rosenberg, *N. Engl. J. Med.* **281**, 145 (1969).
105. S. H. Mudd, *Fed. Proc., Fed. Amer. Soc. Exp. Biol.* **30**, 970 (1971).
106 C. R. Scriver, *Metab., Clin. Exp.* **22**, 1319 (1973).
107. P. L. Altman and D. S. Dittmer, "Metabolism," Fed. Amer. Soc. Exp. Biol., Bethesda, Maryland, 1968.
108. V. DuVigneaud, "A Trail of Research." Cornell Univ. Press, Ithaca, New York, 1952.
109. V. DuVigneaud and J. R. Rachele *in* "Transmethylation and Methionine Biosynthesis" (S. K. Shapiro and F. S. Schlenk, eds.), p. 1. Univ. of Chicago Press, Chicago, Illinois, 1965.
110. T. Arakawa, K. Narisawa, K. Tanno, K. Ohara, O. Higashi, Y. Hnoda, T. Tamura, Y. Wada, T. Mizuno, T. Hayashi, Y. Hirook, T. Ohno, and M. Ikeda, *Tohoku J. Exp. Med.* **93**, 1 (1967).
111. J. J. Corcino, S. Waxman, and V. Herbert, *Amer. J. Med.* **48**, 562 (1970).
112. T. E. Parry, *Brit. J. Haematol.* **16**, 221 (1969).
113. J. G. Hollowell, W. K. Hall, M. E. Coryell, J. McPherson, Jr., and D. A. Hahn, *Lancet* **2**, 1428 (1969).
114. R. T. Shipmen, R. R. W. Townley, and D. M. Danks, *Lancet* **2**, 693 (1969).
115. S. H. Mudd, H. L. Levy, and R. H. Abeles, *Biochem. Biophys. Res. Commun.* **35**, 121 (1969).
116. H. L. Levy, S. H. Mudd, J. D. Schulman, P. M. Dreyfus, and R. H. Abeles, *Amer. J. Med.* **48**, 338 (1970).
117. S. I. Goodman, P. G. Moe, K. B. Hammond, S. H. Mudd, and B. W. Uhlendorf, *Biochem. Med.* **4**, 500 (1970).
118. L. R. Gjessing, *Scand. J. Clin. Lab. Invest.* **15**, 474 (1963).
119. L. R. Gjessing, *Scand. J. Clin. Lab. Invest.* **15**, 479 (1973).
120. C. F. Geiser and M. L. Efron, *Cancer* **22**, 856 (1968).
121. W. V. Studintz, *Acta Paediat. Scand.* **59**, 80 (1970).
122. S. K. Wadman, D. V. D. Heiden, F. J. Van Sprang, and P. A. Voute, *in* "Inherited Disorders of Sulphur Metabolism" (N. A. J. Carson and D. N. Raine, eds.), p. 56. Churchill, London, 1971.
123. J. D. Finkelstein, *in* "Sulfur in Nutrition" (J. E. Oldfield and O. H. Muth, eds.), p. 46. Avi Press, Westport, Connecticut, 1970.
124. J. D. Finkelstein, *in* "Inherited Disorders of Sulphur Metabolism" (N. A. J. Carson and D. N. Raine, eds.). p. 1. Churchill, London, 1971.
125. L. R. Gjessing and K. Mauritzen, *Scand. J. Clin. Lab. Med.* **17**, 513 (1965).
126. E. Lieberman, K. N. F. Shaw, and G. N. Donnell, *Pediatrics* **40**, 828 (1967).
127. P. A. Voute, Jr. and S. K. Wadman, *Clin. Chim. Acta* **22**, 373 (1968).
128. M. Grossman, J. D. Finkelstein, W. W. Kyle, and H. P. Morris, *Cancer Res.* **34**, 794 (1974).
129. M. N. D. Goswami, A. R. Robblee, and L. W. McElroy, *J. Nutr.* **68**, 671 (1959).

130. F. Chatagner, B. Jolles-Bergeret, and O. Trautmann, *Biochim. Biophys. Acta* **59**, 744 (1962).
131. J. Fernandez and A. Horvath, *Enzymologia* **26**, 113 (1963).
132. J. D. Finkelstein, *Arch. Biochem. Biophys.* **122**, 583 (1967).
133. L. R. Gjessing, *Scand. J. Clin. Lab. Invest.* **16**, 680 (1964).
134. C. R. Scriver and J. H. Hutchison, *Pediatrics* **31**, 240 (1963).
135. P. Fourman, J. W. Summerscales, and D. M. Morgan, *Arch. Dis. Childhood* **41**, 273 (1966).
136. Y. Matsuo and D. M. Greenberg, *J. Biol. Chem.* 230, 561 (1958).
137. D. B. Hope, *Biochem. J.* **66**, 486 (1957).
138. H. Harris, L. S. Penrose, and D. H. H. Thomas, *Ann. Hum. Genet.* **23**, 442 (1958).
139. G. W. Frimpter, *in* "The Metabolic Basis of Inherited Disease" (J. B. Stanbury, J. B. Wyngaarden, and D. S. Fredrickson, eds.), 3rd ed., p. 413. McGraw-Hill, New York, 1972.
140. K. N. F. Shaw, E. Lieberman, R. Koch, and G. N. Donnell, *Amer. J. Dis. Child.* **113**, 119 (1967).
141. T. Perry, G. C. Robinson, J. M. Teasdale, and S. Hansen, *N. Engl. J. Med.* **276**, 721 (1967).
142. G. W. Frimpter, *Amer. J. Med.* **46**, 832 (1969).
143. D. T. Whelan and C. R. Scriver, *N. Engl. J. Med.* **278**, 924 (1968).
144. G. W. Frimpter, *Science* **149**, 1095 (1965).
145. J. D. Finkelstein, S. H. Mudd, F. Irreverre, and L. Laster, *Proc. Nat. Acad. Sci. U.S.* **55**, 865 (1966).
146. K. Tada and T. Yoshida, *Tohoku J. Exp. Med.* **95**, 235 (1968).
147. Y. Matsuo and D. M. Greenberg, *J. Biol. Chem* 230, 545 (1958).
148. D. Cavallini, B. Mondovi, C. DeMarco, and A. Scioscia-Santora, *Enzymologia* **24**, 253 (1962).
149. Y. Matsuo and D. M. Greenberg, *J. Biol. Chem.* **234**, 516 (1959).
150. J. Loiselet and F. Chatagner, *Biochim. Biophys. Acta* **89**, 330 (1964).
151. B. Jolles-Bergeret and F. Chatagner, *Arch. Biochem. Biophys.* **105**, 640 (1964).
152. J. A. Sturman, P. A. Cohen, and G. E. Gaull, *Biochem. Med.* **3**, 510 (1970).
153. J. D. Finkelstein, unpublished observations.
154. Y. Matsuo and D. M. Greenberg, *J. Biol. Chem.* **215**, 547 (1955).
155. F. Chatagner, *Nature (London)* **203**, 1177 (1964).
156. Y. Matsuo, M. Rothstein, and D. M. Greenberg, *J. Biol. Chem.* **221**, 679 (1956).
157. J. D. Finkelstein and W. E. Kyle, *Clin. Res.* **17**, 43 (1969).
158. G. Gaull, J. A. Sturman, and N. C. R. Raiha, *Pediat. Res.* **6**, 538 (1972).
159. G. W. Frimpter, A. Haymovitz, and M. Horwith, *N. Engl. J. Med.* **268**, 333 (1963).
160. S. Berlow, *Amer. J. Dis. Child.* **112**, 135 (1966).
161. T. L. Perry, D. F. Hardwick, S. Hansen, D. L. Love, and S. Israels, *N. Eng. J. Med.* **278**, 590 (1968).
162. C. R. Scott, S. W. Dassell, S. H. Clark, C. Chiang-Teng, and K. R. Swedberg, *J. Pediat.* **76**, 571 (1970).
163. J. D. Finkelstein, S. H. Mudd, F. Irreverre, and L. Laster, *Amer. J. Dis. Child.* **115**, 388 (1968).
164. P. A. Cohen, K. Schneidman, F. Ginsberg-Fellner, J. A. Sturman, J. Knittle, and G. E. Guall, *J. Nutr.* **103**, 143 (1973).
165. F. Irreverre, S. H. Mudd, W. D. Heizer, and L. Laster, *Biochem. Med.* **1**, 187 (1967).
166. S. H. Mudd, F. Irreverre, and L. Laster, *Science* **156**, 1599 (1967).

167. B. Sörbo, *Biochim. Biophys. Acta* **24,** 324 (1957).
168. J. L. Wood, E. F. Williams, Jr., and N. Kingsland, *J. Biol. Chem.* **170,** 251 (1947).
169. F. Chapeville and P. Fromageot, *Biochim. Biophys. Acta* **14,** 415 (1954).
170. F. Chapeville, P. Fromageot, A. Brigelhuber, and N. Henry, *Biochim. Biophys. Acta* **20,** 351 (1956).
171. A. K. Percy, S. H.Mudd, F. Irreverre, and L. Laster, *Biochem. Med.* **2,** 198 (1968).
172. H. L. Levy, V. E. Shih, P. M. Madigan, V. Karolkewicz, J. R. Carr, A. Lum, A. A. Richards, J. D. Crawford, and R. A. MacCready, *Amer. J. Dis. Child.* **117,** 96 (1969).
173. G. M. Komrower and A. J. Robins, *Arch. Dis. Childhood* **44,** 418 (1969).
174. G. E. Gaull, W. von Berg, N. C. R. Raiha, and J. A. Sturman, *Pediat. Res.* **7,** 527 (1973).
175. M. Partington, C. R. Scriver, and A. Sass-Kortsak, *Can. Med. Ass. J.* **97,** 1045 (1967).
176. C. R. Scriver, *Can. Med. Ass. J.* **97,** 1073 (1967).
177. T. L. Perry, D. F. Hardwick, G. H. Dixon, C. L. Dolman, and S. Hansen, *Pediatrics* **36,** 236 (1965).
178. S. H. Mudd, *in* "Symposium: Sulfur in Nutrition" (O. H. Muth and J. E. Oldfield, eds.), p. 222. Avi Press, Westport, Connecticut, 1970.
179. G. E. Gaull, D. K. Rassin, G. E. Solomon, R. C. Harris, and J. A. Sturman, *Pediat. Res.* **4,** 337 (1970).
180. J. D. Finkelstein and M. Dieterlen, in preparation.

Subject Index

A

Actithiazic acid, 45–46
Acetobacter suboxydans, 1, 4
Acetoin synthesis, 70
Acetylcholine, 73
Acetylcholinesterase, 446
Acetyl CoA carboxylase, 28, 34–35
(+)-6-*S*-Acetyldihydrolipoic acid, 96
N-Acetyl-*d*-galactosamine, 400
N-Acetyl-D-glucosamine, 402, 403
N-Acetylhexosamine, 345
O-Acetylhomoserine, 470, 473–474
O-Acetylhomoserine sulfhydrylase, 24,
   477–479, 482–484
p-Acetylphenylsulfate, 369
O-Acetylserine, 238, 240, 505, 507, 509
O-Acetyl-L-serine acetate-lyase, 507
O-Acetylserine sulfhydrylase (OASS), 238
   OASS-A, 238
   OASS-B, 238
N-Acetylsphingosine sulfate, 340
Achromobacter, 43
Acidemia, 77
Acidomycin, 45
Acyl carrier protein (ACP), 8
O-Acylserine, *see O*-Acetylserine
Acyl-transfer reaction, 95
Adenine, 68
   mutant, 63
Adenosine, 68
   of *Salmonella typhimurium,* 63
Adenosine diphosphate sulfurylase, 269,
   301–302
   from bacteria, 301
   from yeast, 301
Adenosine monosulfate, 305
Adenosine 5'-phosphosulfate (APS), 221,
   289–290
   biosynthesis, 292–301
   degradation by enzyme, 303–304
      by venom, 303
   formation, 292–301
   pathway in algae, 225–228

reduction, 225–230, 246–249
   structure, 410
Adenosine 5'-phosphosulfate kinase, 224,
   300–301
   from yeast, 300
Adenosine 5'-phosphosulfate reductase,
   246–248, 261–262
Adenosine 5'-phosphosulfate sulfohydrolase, 304, 413–415
   analog, synthetic, 305
Adenosine 5'-phosphosulfate:sulfate aden-
   yltransferase, 301
Adenosine 5'-sulfatopyrophosphate, 305
Adenylylsulfate, *see* Adenosine 5'-
   phosphosulfate
Adenosine triphosphate, 224
Adenosine triphosphate:adenylylsulfate
   3'-phosphotransferase, 224, 300–301
Adenosine triphosphate:sulfate adenylyl-
   transferase, *see* Adenosine triphos-
   phate:sulfurylase
Adenosine triphosphate: sulfurylase,
   221–223, 244, 294–299, 415–416
   bacterial, 296, 298
   fungal, 294
   mechanism of action, 299
   substrates, 297
   from yeast, 294, 298
S-Adenosyl-L-homocysteine hydrolase,
   514, 551, 560
S-Adenosylmethionine, 65, 529
   decomposition, 514, 533
   formation, 531
S-Adenosylmethionine decarboxylase,
   530–531
S-Adenosylmethionine synthetase, 590
Aerobacter aerogenes, 374, 376, 448, 533
Agaritine, 104
Agmatine, hydrolysis of, 530
Agrobacterium, 43
Alanine, 66
Alcaligenes, 437
   A. metalkaligenes, 375–378
Alcohol sulfate ester, 379

599

A 5
B 6
C 7
D 8
E 9
F 0
G 1
H 2
I 3
J 4